D1477099

LIVERPOOL JMU LIBRARY
3 1111 00947 9252

Applications of Pharmacokinetic Principles in Drug Development

Applications of Pharmacokinetic Principles in Drug Development

Edited by

Rajesh Krishna, Ph.D., F.C.P.
Aventis Pharmaceuticals, Inc.
Bridgewater, New Jersey

Kluwer Academic / Plenum Publishers
New York, Boston, Dordrecht, London, Moscow

Library of Congress Cataloging-in-Publication Data

Applications of pharmacokinetic principles in drug development/edited by Rajesh Krishna.
p. ; cm.
Includes bibliographical references and index.
ISBN 0-306-47766-1
1. Drug development. 2. Drugs—Testing. 3. Pharmacokinetics. I. Krishna, Rajesh.
[DNLM: 1. Chemistry, Pharmaceutical. 2. Drug Evaluation. 3. Models, Chemical. 4. Pharmacokinetics. QV 744 A6526 2004]
RM301.25.A67 2004
615'.19—dc22

2003061170

ISBN 0-306-47766-1

233 Spring Street, New York, New York 10013

http://www.wkap.nl/

10 9 8 7 6 5 4 3 2 1

A C.I.P. record for this book is available from the Library of Congress

Printed in the United States of America

DEDICATION

On February 2, 2003, the drug development community lost one of the best-respected and valued colleagues in Dr. Wayne Colburn. Wayne authored the first chapter of this book, on efficient and effective drug development.

At his death, Wayne was Vice President and Senior Research Fellow at MDS Pharma Services. He was also the President and Fellow of the American College of Clinical Pharmacology. With over 260 publications, memberships in journal editorial boards, and rich experience in drug development, Wayne was a key leader in the area of drug development.

This book is dedicated to the memory of Wayne Colburn.

CONTRIBUTORS

Xavier Barbaut
Hospice de Beaune
France

Jeffrey Barrett
Head, Global Biopharmaceutics
Aventis Pharmaceuticals, Inc.
1041 Rt. 202-206
Bridgewater, New Jersey

Nathalie Bleyzac
Hospices Civils de Lyon, France

Irina Bondareva
Institute of Physical and Chemical Medicine
Moscow, Russia

Andreas Botnen
Center for Bioinformatics
University of Oslo, Norway

Aida Bustad
Laboratory of Applied Pharmacokinetics
USC School of Medicine
CSC 134-B
2250 Alcazar Street
Los Angeles, California

Wayne A. Colburn (deceased)
Vice President & Senior Research Fellow
MDS Pharma Services
4747 East Beautiful Lane
Phoenix, Arizona

Alyson L. Connor
Pharmaceutical Profiles
Mere Way
Ruddington Fields
Ruddington, Nottingham G11 6JS
United Kingdom

Christopher D. Ellison
Center for Drug Evaluation and Research
Office of Generic Drugs
MPN II Room 285, 7500 Standish Place
Rockville, Maryland

David Fleisher
3058 College of Pharmacy
The University of Michigan
Ann Arbor, Michigan

Ashutosh Gandhi
Laboratory of Applied
Pharmacokinetics
USC School of Medicine
CSC 134-B
2250 Alcazar Street
Los Angeles, California

Barry Gertz
Vice President
Clinical Pharmacology
Merck Research Laboratories
West Point, Pennsylvania

Elora Gupta
Regulatory Affairs
Bristol-Myers Squibb
Hopewell, New Jersey

Gene Heath
Regulatory Affairs Science
1640 South 58th Street
Lincoln, Nebraska

Shiew-Mei Huang
Deputy Office Director for Science
Office of Clinical Pharmacology and Biopharmaceutics
Center for Drug Evaluation and Research
Food and Drug Administration
5600 Fishers Lane, PKLN, HFD-850, Rm 6A/19
Rockville, Maryland

Ajaz S. Hussain
Center for Drug Evaluation & Research
Food and Drug Administration
6615 Hunter Trail Way
Frederick, Maryland

Roger W. Jelliffe
Professor of Medicine
USC Laboratory of Applied Pharmacokinetics
2250 Alcazar St
Los Angeles, California

John Kovarik
Novartis Pharma AG
Clinical Pharmacology Department
Building WSJ 27.4093
4002 Basel, Switzerland

Thierry Lavé
Preclinical Pharmacokinetics
F. Hoffmann-La Roche AG
Pharmaceuticals Division
68-329B
Grenzacherstrasse 124
CH-4070 Basel, Switzerland

Robert Leary
San Diego Supercomputer Center
University of California, San Diego
San Diego, California

Olivier Luttringer
F. Hoffmann-La Roche AG
Pharmaceuticals Division
68-329B
Grenzacherstrasse 124
CH-4070 Basel, Switzerland

Pascal Maire
Hospices Civils de Lyon
France

Nancy E. Martin
Global Biopharmaceutics
Aventis Pharmaceuticals, Inc.
1041 Rt. 202-206
Bridgewater, New Jersey

Timothy J. Maziasz
COX-2 Technology
Pharmacia R&D
4901 Searle Pkwy
Skokie, Illinois

Neil Parrott
F. Hoffmann-La Roche AG
Pharmaceuticals Division
68-329B
Grenzacherstrasse 124
CH-4070 Basel, Switzerland

Susan K. Paulson
Senior Director
Clinical Pharmacokinetics
NeoPharm, Inc.
150 Field Drive, Suite 195
Lake Forest, Illinois

Arturo G. Porras
WP26-372
Merck Research Laboratories
West Point, Pennsylvania

Patrick Poulin
F. Hoffmann-La Roche Ltd.
Pharmaceuticals Division
Non-Clinical Development-Drug Safety
PRNS Bau: 69/101
CH-4070 Basel, Switzerland.

Mark L. Powell
Vice President
Analytical Research and Development
Bristol-Myers Squibb
New Brunswick, New Jersey

David V. Prior
Pharmaceutical Profiles
Mere Way, Ruddington Fields
Ruddington, Nottingham NG11 6JS
United Kingdom

Laurie Reynolds
Globomax LLC
7250 Parkway Drive, Suite 430
Hanover, Maryland

Shashank Rohatagi
Global Biopharmaceutics
Aventis Pharmaceuticals, Inc.
1041 Rt. 202-206
Bridgewater, New Jersey

Alan Schumitzky
Laboratory of Applied Pharmacokinetics
USC School of Medicine
CSC 134-B
2250 Alcazar Street
Los Angeles, California

Michael Sinz
Group Leader
Metabolism and Pharmacokinetics
Bristol-Myers Squibb
5 Research Parkway
Wallingford, Connecticut

Steve E. Unger
Clinical Discovery Analytical Sciences
Bristol-Myers Squibb
PO Box 191, One Squibb Drive
New Brunswick, New Jersey

Michael Van Guilder
Laboratory of Applied Pharmacokinetics
USC School of Medicine
CSC 134-B
2250 Alcazar Street
Los Angeles, California

Xin Wang
Laboratory of Applied Pharmacokinetics
USC School of Medicine
CSC 134-B
2250 Alcazar Street
Los Angeles, California

Ian R. Wilding
Pharmaceutical Profiles
Mere Way
Ruddington Fields
Ruddington, Nottingham
NG11 6JS, United Kingdom

Elizabeth Yamashita
CMC Regulatory Affairs
Bristol-Myers Squibb
Hopewell, New Jersey

Lawrence X. Yu
Food and Drug Administration
Center for Drug Evaluation and Research
Office of Generic Drugs
MPN II Room 285, 7500 Standish Place
Rockville, Maryland

FOREWORD

This volume is an important advancement in the application of pharmacokinetic (PK) and pharmacodynamic (PD) principles to drug development. The series of topics presented deal with the application of these tools to everyday decisions that a pharmaceutical scientist encounters. The ability to integrate these topics using PK and PD methods has optimized drug development pathways in the clinic. New technologies in the areas of *in vitro* assays that are more predictive of human absorption and metabolism and advancement in bioanalytical assays are leading the way to minimize drug failures in later, more expensive clinical development programs.

Pharmacokinetics and pharmacodynamics have become an important component of understanding the drug action on the body and is becoming increasingly important in drug labeling due to it's potential for predicting drug behavior in populations that may be difficult to study in adequate numbers during drug development. The ability to correlate drug exposure to effect and model it during the drug development value chain provides valuable insight into optimizing the next steps to derive maximum information from each study. These principles and modeling techniques have resulted in an expanded and integrated view of PK and PD and have led to the expectations that we may be able to optimally design clinical trials and eventually lead us to identifying the optimal therapy for the patient, while minimizing cost and speeding up drug development.

There is wide utility for the book both as a text and as a reference. For the novice scientist, it will provide their first access to application of fundamental principles of PK/PD used in drug development. For the more experienced scientist, it provides useful examples of some complex development programs where these principles have been used and provides approaches to solving many of the common problems encountered during the course of drug development. This book is a good example of integration of *in vitro*, preclinical, and clinical PK/PD, which is required to fully utilize the capabilities of these principles.

Vijay Bhargava, Ph.D.
Global Head and Vice President, Drug Metabolism and Pharmacokinetics
Aventis Pharmaceuticals, Inc.
Bridgewater, NJ

PREFACE

A key deliverable for a new drug development program is a successful new drug application that has gained regulatory approval for marketing. Numerous scientists from various functional disciplines collaborate effectively and productively in an issue-driven approach to optimally derive knowledge on the preclinical and clinical characterization of the new drug candidate. More importantly then, it becomes critical for the team of drug development scientists to be able to collate, manage, and integrate this knowledge so as to prepare a comprehensive and coherent package for regulatory filing and approval. The net result of this campaign is the delivery of safe and effective medicines to enable optimal patient care. One such functional area within new drug development, which is the theme of this book, is the discipline of pharmacokinetics and drug metabolism science.

The science of pharmacokinetics and drug metabolism as it pertains to drug development has consistently evolved over the past several years. Focus has shifted from merely generating the information to integrating the information generated. Using preclinical data advantageously to enable dose selection in First-In-Human trial, utilizing experimental biomarker and pharmacogenomic strategies in Phase I studies to optimally design Phase II/III trials, identifying the sources of pharmacokinetic variability in Phase I studies to devise inclusion/exclusion criteria in Phase II/III, and applying advanced modeling and simulation tools to integrate pharmacokinetic and pharmacodynamic data to derive meaningful PK/PD relationships that support dose and regimen optimization in Phase III, are just a few of the strategies that apply pharmacokinetic principles. Furthermore, effective knowledge management is founded on effective synchrony of creative thoughts and ideas from all major fronts, namely, industry, academia, and regulatory agencies. To this end, the book has clearly benefited from the diverse but important perspectives originating from experts in industry, academia, and regulatory agencies. The collection of articles presented in this book represent individual and expert perspectives in both preclinical and clinical development, including case studies on real-life examples of successful drugs that add value to the pharmacokinetic principles learned and applied.

The objectives of the book are three-fold: 1) provide the reader with a collection of expert articles that present a perspective on applying pharmacokinetic principles to new drug development, 2) provide the reader with the diversity of perspectives representing academia, industry, and regulatory agencies, and 3) stress the importance of effective knowledge derivation, integration, and management for effective drug development solutions through case studies. It is assumed that the reader has a fair and basic understanding of pharmacokinetic theory. It is anticipated that the subject matter will appeal to established scientists in industry, academia, and regulatory sciences as well as to advanced undergraduate and graduate students in medicine, pharmacy, pharmacology, biochemistry, biology, and allied sciences. Individuals that directly work in pharmacokinetic sciences as well individuals that interact with pharmacokinetic scientists may also find the book helpful. While it will no doubt take volumes to fully capture the essence of pharmacokinetic sciences within the realm of new drug development, the current book attempts to provide the intrigued reader with a snapshot of drug development where pharmacokinetic concepts and principles are integrated and applied.

Successful drug development is both an art and a science and is founded on the collective wisdom of scientific diversity and constructive debate. Not all new chemical entities discovered become successful drugs. The failure rate in new drug development underscores the importance of creative thinking and innovation in our relentless pursuit of developing new drugs that add value in our armamentarium of treatment modalities for unmet medical needs.

Finally, I remain thankful to the team of expert scientists that have contributed immensely to this volume, to my wife Bhuvana for help with proofreading, and to the publishing team at Kluwer Academic Publishers with specific acknowledgement to Kathleen Lyons, Marissa Kraft, and Brian Halm.

Rajesh Krishna, Ph.D., F.C.P.
Aventis Pharmaceuticals, Inc.
Bridgewater, NJ

CONTENTS

Efficient and Effective Drug Development 1
Wayne A. Colburn and Gene Heath

Bioanalytical Methods: Challenges and Opportunities In Drug Development 21
Mark L. Powell and Steve E. Unger

Predicting Human Oral Bioavailability using *In Silico* Models 53
Lawrence X. Yu, Christopher D. Ellison, and Ajaz S. Hussain

Drug Metabolism in Preclinical Development 75
Michael W. Sinz

Interspecies Scaling 133
Thierry Lavé, Olivier Luttringer, Patrick Poulin, and Neil Parrott

Human Drug Absorption Studies in Early Development 177
David V. Prior, Alyson L. Connor, and Ian R. Wilding

Food-Drug Interactions: Drug Development Considerations 195
David Fleisher and Laurie Reynolds

Global Regulatory and Biopharmaceutics Strategies in New Drug Development: Biowaivers 225
Elora Gupta and Elizabeth Yamashita

Special Population Studies in Clinical Development: Pharmacokinetic Considerations 245
John M. Kovarik

Drug-Drug Interactions 307
Shiew-Mei Huang

Pharmacokinetic/Pharmacodynamic Modeling in Drug Development 333
Shashank Rohatagi, Nancy E. Martin, and Jeffrey S. Barrett

Population Pharmacokinetic and Pharmacodynamic Modeling 373
Roger Jelliffe, Alan Schumitzky, Aida Bustad, Michael Van Guilder, Xin Wang, and Robert Leary

Role of Preclinical Metabolism and Pharmacokinetics in the Development of Celecoxib 405
Susan K. Paulson and Timothy J. Maziasz

The Role of Clinical Pharmacology and of Pharmacokinetics in the Development of Alendronate – A Bone Resorption Inhibitor 427
Arturo G. Porras and Barry J. Gertz

Optimizing Individualized Dosage Regimens of Potentially Toxic Drugs 477
Roger W. Jelliffe, Alan Schumitzky, Robert Leary, Andreas Botnen, Ashutosh Gandhi, Pascal Maire, Xavier Barbaut, Nathalie Bleyzac, and Irina Bondareva

Appendix 529

Index 545

EFFICIENT AND EFFECTIVE DRUG DEVELOPMENT

Wayne A. Colburn[1] and Gene Heath[2]*

1. INTRODUCTION

Development of new drugs and other treatments for patients around the world requires an enduring commitment to conduct scientific research in a world that is changing every day. Research commitments are complicated by a limited but growing understanding of disease processes as well as the spread of new afflictions such as AIDS, West Nile Fever, etc. to larger and larger populations. In addition, improved medical treatments of disease have resulted in an increasing incidence of diseases associated with aging such as Alzheimer's disease, Parkinson's disease and various forms of cancer.

Drug development encompasses many disciplines and requires information about the impact of drugs as well as other treatments on a particular disease. Looking at the end product of the development process; i.e., packaging and product information (package insert), it is easy to appreciate the complexity of the process. Every statement contained in the package insert must be supported by information captured during the non-clinical and/or clinical development processes. New technologies provide an opportunity to develop drugs more effectively and efficiently.

The process starts with discovery and some basic research on the new molecular entity (NME) followed by screening for pharmacological activity, characterization and validation of the lead compound. The NME is also characterized with respect to its physical and chemical properties to begin planning potential approaches for developing oral and/or parenteral formulations that may be needed during the developmental process. After selecting the potential lead compound, a full pharmacology as well as acute and sub-acute toxicology profile is developed in animal models of human disease and animal models of toxicity, respectively. At the same time, a bio-analytical method is developed to evaluate the pharmacokinetics and exposure-response in animals. In addition, the lead candidate is characterized to evaluate its absorption, distribution, metabolism and excretion (ADME) in animals. The NME is then evaluated with respect to more definitive toxicology in animals, usually a rodent and a non-rodent species.

Information gathered from pre-clinical studies must be sufficient to support progressing into human studies. Before a NME can be administered to human, an investigator's brochure must be developed which includes all relevant information

---

[1] Deceased
[2] Regulatory Affairs Science, 1640 S. 56th St., Lincoln, NE 68506

Applications of Pharmacokinetic Principles in Drug Development
Edited by Krishna, Kluwer Academic/Plenum Publishers, 2004

available for the lead compound and its intended human formulation. The information must be enough to convince an institutional review board/ethics committee that the NME in its formulation should be safe enough for evaluation in first time in human studies at the planned doses and duration of exposure. The investigator's Brochure is a living document that is updated several times as the NME progresses through the clinical development process and as more information on the NME, its toxicity, and experience from clinical trials becomes available.

The clinical development of the NME can be divided into exploratory studies, usually small numbers of healthy volunteers or patients[1] as well as larger confirmatory studies. In exploratory studies, data are captured to design protocols for later studies. Confirmatory studies are then conducted in an effort to prove statistically that the compound is safe and efficacious for the treatment of individuals with a particular disease.

2. PHASES OF DRUG DEVELOPMENT

Drug development can be divided into phases for descriptive purposes and better understanding (Figure 1). The phase descriptors are 0-IV.

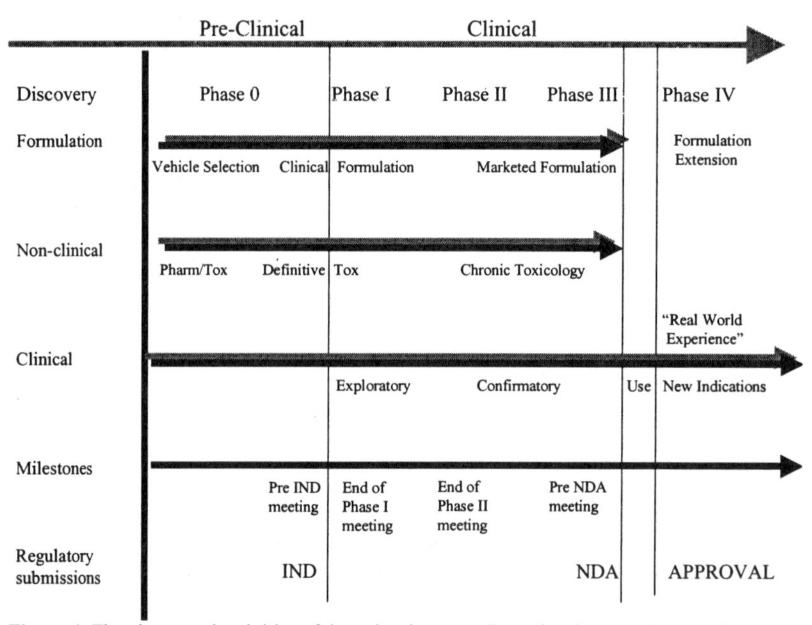

Figure 1. The phases and activities of drug development. Drug development is a continuous process with no real beginning and no real end. Functional activities such as toxicology start in phase 0 and continue through Phase III. Phase I activities start with first in human studies but continue throughout the NME/product life cycle.

Following the discovery, characterization and selection of a NME, Phase 0 pre-clinical development begins. Studies include expanded pharmacological evaluation, in vitro and in vivo toxicology evaluations including genotoxicology and cytotoxicology as well as formulation development and evaluation of animal models for efficacy[2, 3] (Table 1). The non-clinical evaluation of the NME includes the effects of the compound on the whole animal, the selection of possible biomarkers of disease and pharmacological effects[4, 5].

Table 1. Phase 0: Non-clinical drug development studies

Vehicle selection	Cytotoxicology
Pharmacology screening	Genotoxicology
Cell-based models	Acute toxicology
Tissue-based models	ADME
Bio-analytical development	Toxicokinetics /Definitive toxicology
Whole animal pharmacology	Chronic toxicology
Disease models	Clinical formulation development
Biomarker evaluation	Reproductive toxicology
Exploratory toxicology	Carcinogenicity

Studies are conducted to evaluate the ADME and initial bio-analytical methods development. Definitive toxicology and toxicokinetics generates data for further progression of the development in the clinic. Chronic toxicology, reproductive and carcinogenicity data are collected. These studies are designed to meet the non-clinical requirements for authorization to sell the drug throughout the world.

Clinical development of a drug or biologic is usually divided into four or five phases. Phase I studies usually encompass studies conducted in healthy volunteers to evaluate preliminary safety, tolerability, and pharmacokinetics of the drug (Table 2).

Table 2. Phase I: Safety and tolerability studies

First time in human, single-dose	Mass Balance/metabolism
First time in human, multiple-dose	Effect of food
Pharmacokinetics	Effect of age
BA/BE studies	Effect of gender
Dose proportionality	Renal insufficiency
Drug interactions	Hepatic insufficiency
Pharmacodynamics	Proof of concept (if possible)
Biomarker evaluation	

In some cases, it may even be possible to evaluate efficacy or pharmacological action (pharmacodynamics) in healthy subjects [6, 7, 8, 9]. In some cases, Phase I studies may be conducted in patients when it is unethical to use healthy subjects or when the patient is generally healthy, such as mild asthmatics using only PRN beta-agonists or Type 2 diabetics controlled with diet and exercise. Other phase I studies evaluate ADME, and interactions with food, other drugs or treatments while others are conducted to evaluate the NME in the elderly, effects due to gender, and treatment of patients with other diseases such as renal and hepatic insufficiency.

Phase II studies are conducted in the patient population of interest (Table 3).

Table 3. Phase II: Early patient studies

Clinical endpoints	Experimental designs
First time in patients	Mechanism of action
Dose ranging	Proof of concept
Dose regimen	Surrogate endpoints
Biomarker evaluation	Confirmatory designs

Phase II studies are generally classified as either Phase IIa or Phase IIb. Phase IIa studies are conducted first in small numbers of patients to explore the initial dose and dosing regimen to produce the desired clinical effects. Evaluation of the mechanism of actions, effects on the chosen biomarkers, as well as proof that the drug are effective in treating the disease are also conducted [10, 11]. Later phase IIb studies are conducted in larger numbers of patients to confirm or refine the choice of dose and dosing regimen, verify the efficacy endpoints, validate a surrogate or clinical endpoint [12], and to provide variability information to support the number of patients needed for statistical and modeling procedures that are used in later confirmatory phase III studies.

Phase III studies are primarily focused on collecting information that confirms the safety and efficacy of the treatment being evaluated (Table 4).

Table 4. Phase III: Confirmatory patient studies

Phase III: Confirmatory patient studies
Confirmatory safety and efficacy for each indication
Population pharmacokinetics and pharmacodynamics
Treatment use studies in large populations (Phase IIIb)
Pharmacoeconomics and Pharmacoepidemiology

The studies are generally conducted in 400-5000 patients or even more to evaluate clinical safety and efficacy endpoints or surrogate endpoints that predict outcomes. Data collected during these studies must demonstrate that the new treatment is as good as or better than current therapy with respect to safety, efficacy or both. Studies conducted after a New Drug Application is submitted are usually called treatment use studies and are conducted to collect information on the use of the therapy under conditions that are more likely to represent the treatment of patients in the general population. These studies are generally described as Phase IIIb.

Phase IV studies are also called post-marketing studies (Table 5).

Table 5. Phase IV: Post-marketing/First Approval

Phase IV: Post-marketing/First Approval
First "real world" clinical experience
Surveillance for safety and efficacy databases
New indications
Product extension with new formulations

These studies are conducted to fulfill commitments to regulatory authorities, gather more information on safety and efficacy of the therapy in actual practice using post-marketing surveillance tools. The regulatory authorities have withdrawn approval during this phase). In addition, clinical studies are done after marketing to investigate additional indications as well as to investigate alternate delivery systems such as transdermal, modified release, inhalation, etc. Developing alternative delivery systems has also been called Phase V to differentiate it from Phase IV commitment or new indication studies.

3. DRUG APPROVAL PROCESS

The International Conference on Harmonization (ICH) was established in 1990 as a three-region, joint regulatory/industry project to improve the efficiency of processes for developing and registering new medicinal treatments in Europe, Japan, and the United States without compromising regulatory requirements for safety and effectiveness [13]. In recent years, regulatory authorities and industry associations have undertaken many important initiatives to promote global harmonization of regulatory requirements. One of the goals of harmonization is to identify and reduce differences in the technical requirements for NME development among regulatory agencies. ICH was organized to provide an opportunity for harmonization initiatives to be developed with input from both regulatory and industry representatives. ICH efforts are ongoing and have produced agreements, processes and formats that are used for the global registration and approval of new treatments. Although much progress has been made and harmonization will continue, many differences still exist between the three regions and some differences such as regional preferences for formulations, difference in toxicology requirement, regional submission processes, etc. may never be resolved.

United States, European, and Japanese regulatory authorities are addressing the regulatory processes within each of their constituencies by adopting a more open discussion opportunities with pharmaceutical companies during the drug development process. An example is the initiatives outlined by the United States Food and Drug Administration that request meetings at critical times. These meetings are opportunities to review data and to plan and agree on requirements for the development process. All major regulatory bodies have developed/adopted guidances, points to consider, and review standards that improve the development process. However, for most efficient and effective drug developers, points to consider are more useful than guidances that can be applied as directives. Implementation of the common technical document for technical and scientific applications can simplify global registration. With the current modular structure, regional and national requirements can be addressed in the first module with other modules containing technical and scientific data presented in a standardized format[14, 15, 16, 17].

In the United States, legislation that authorized the Prescription Drug Users Fee Act (PDUFA) in 1992 and was re-authorized with the FDA Modernization Act of 1997 (FDAMA) redirected the FDA to focus on the management of risk during the drug development process. It also establishes timelines for review of the applications. The FDA has become more proactive in the process by suggesting that certain interactions be formalized to improve government and industry discussions during pre-clinical and clinical phases of drug development [18]. A pre-IND meeting should be held to discuss the clinical development plans of the company to review all pre-clinical data, from the NME, toxicology information, exposure to animals, decisions on the safety of the first dose in humans to the overall clinical development. An end of Phase I meeting may be required if the NME is being developed for life threatening or a severely debilitating disease. It may be possible to provide sufficient information on safety and efficacy during early studies in patients to support approval, with more extensive data gathered during post-marketing studies. End of Phase II meetings are held to review data from Phase I/II clinical studies to develop consensus on Phase III requirements with emphasis on the number of successful studies, number of patients and the relevant clinical or surrogate endpoints needed for approval. The regulations clearly indicate that only one Phase III clinical trial is needed if sufficient supporting documentation such as mechanism of disease, mechanism of drug action and PK/PD exposure response information is provided. Discussions also focus on the clinical endpoints to be evaluated and how the new therapy will impact the practice of medicine. Pre-NDA meetings allow the drug company to discuss the results of their Phase III program and assure that data convince FDA reviewers that the new therapy is safe, efficacious and is approvable for marketing in the United States.

Another United States initiative, the fast track process for drugs for serious and life threatening with no adequate therapy has been implemented and may include priority review, accelerated approval and other special procedures during drug development. This also establishes timelines for evaluating an NDA submission[19, 20].

4. NON-CLINICAL AND CLINICAL DEVELOPMENT

The timeline from discovery of a NME to marketing and acceptance as a new effective treatment requires approximately 800 million US dollars and 7-12 years of research. For efficient and effective drug development, studies may be conducted in parallel rather than in sequence if the information gained is not needed to design the next studies in the series. For example, toxicology and clinical studies may overlap as long as the toxicology information supports the duration of human dosing. Formulation development is a never-ending process with new vehicles being developed just in time for their use in pre-clinical protocols and dosage forms being developed just in time for their clinical application. Information needed to move through the development process depends on the requirements for the next study in the series. However, all pertinent information to properly use a new therapy must be submitted to the FDA in the New Drug Application or the Biological License Application. Every statement in the Package Insert must by supported by studies. A team approach is required to coordinate the planning and interaction of many disciplines. Input from discovery, pharmacology, toxicology, formulation development, medical, regulatory as well as others can lead to a continuous evaluation and risk control during the development process. Stopping development is just as important as continuing [21, 22].

4.1. New Molecular Entity and Formulation

Production of the NME and its formulation is a continuing process; balancing the costs of production versus the need for an acceptable formulation to conducts the needed studies. Ultimately, the goal is to produce an elegant formulation for marketing. Only small amounts of the drug are needed for early pharmacological screening and initial exploratory pharmacology and toxicology studies. However, as more of the compound is needed for whole animal pharmacology and toxicology, a suspension or solution formulation can be developed that is absorbed from the GI tract. Major scale-up is required later to complete the toxicology studies and prepare a formulation for early clinical trials. This is followed with the development of a commercial formulation (See Figure 1).

4.2. Pharmacology/Toxicology

The initial focus of the pre-clinical development of a NME is to provide sufficient information for its safe use during clinical development. A variety of in vitro studies are conducted to evaluate the biological activity of the NME. These are followed with in vivo experiments that will confirm the pharmacological activity of the NME and evaluate exploratory toxicology. The development of bio-analytical methods for the measurement of the NME and possible metabolite and biomarkers are also ongoing. These studies will help define the dose, dosing schedule, animal exposure (pharmacokinetics, drug metabolism and excretion [23, 24]), biochemical and pharmacological endpoints used for definitive acute toxicology studies. These studies are usually followed with formal single /multiple dose toxicology studies that will support limited exposure in humans. Longer toxicology studies are required to extend the length of exposure in clinical trials increases. Other studies are required for registration including carcinogenicity studies and reproductive and developmental toxicology studies.

4.2. Clinical Development

Once the information from pre-clinical studies is sufficient to safely administer the NME to humans, the first clinical trials are started (Figure 2).

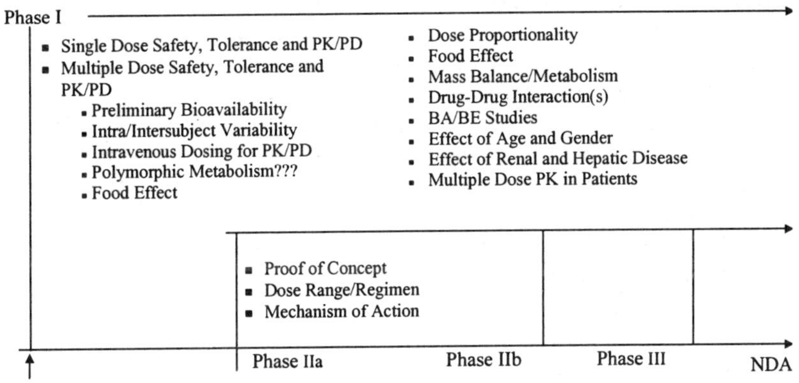

Figure 2. Efficient and effective early drug development. Pre-clinical development sets the foundation for efficient and effective clinical development. All aspects should be considered in series as well as in parallel.

First, a single dose study is conducted followed by a multiple dose study to gather information on the safety, side effects and pharmacokinetics in humans. If the safety profile is acceptable, further trials gather information on the possible pharmacological evaluation in healthy volunteers, metabolism of the NME, effect of food and other drugs to guide in the design of the following studies in patients. The first studies in patients are usually small, with evaluations of pharmacokinetics, effects of the NME on disease progression, confirmation of mechanism of action and proof of concept. These exploratory studies help define the dose, dose regimen, time course of pharmacological effects, effects on biomarkers or clinical endpoint for the conduct of the confirmatory studies in larger numbers of patients. During clinical development, many other studies such as drug-drug interactions, special populations, population PK/PD and use studies for treatment the broader patient populations are needed to support registration [25, 26, 27].

Large clinical studies are conducted to confirm the information gathered during early exploratory clinical development. Studies are designed to prove that the NME is safe and effective for the treatment of patients in a broader population. These studies generally use a larger number of patients and are statistically powered to demonstrate that the new treatment is efficacious while gathering information of the safety profile of the treatment, i.e. types of side effects, incidence of these adverse reactions and interactions with other treatment. The FDA requires only one confirmatory study for a biologic and may only require a single study if sufficient exploratory information can demonstrate scientific and medical rational for the new treatment [28]. Following the filing of the BLA or NDA, further studies are conducted to gather more information on the practical use of the new treatment in larger, less restrictive populations of patients. After market approval of a new treatment, studies are conducted to fulfill any requests by the FDA and to continue to gather more information about adverse events.

The drug development process can be radically altered as new technologies and approaches to drug development come to bear on discovery through pre-clinical and clinical development to marketing of new disease treatments.

5. ROLE OF NEW TECHNOLOGIES IN DRUG DEVELOPMENT

New technologies are made available every day and their roles in drug development are increasing with every passing moment. In fact, the technologies and their specific applications that are described in this chapter will have changed even before the book is published. Therefore, this is not a comprehensive discussion of the current technology, but will highlight some general concepts that are of interest today and should have expanded application in the coming years. Drug discovery technologies including combinatorial chemistry and high throughput screening (HTS) plus pharmacogenetics, pharmacogenomics, functional proteomics and improved animals models of human disease are contributing to improve drug development. In addition, better information technology (IT), electronic data capture (EDC) and e-clinical trials including patient recruitment as well as electronic CRFs, data warehousing contribute to improved clinical development. Nanotechnology for diagnosis as well as drug and gene therapy delivery, bio-analytical methods including biosensors for *in vivo* biomarker and drug assays, imaging technologies such as PET, SPECT, functional MRI, etc., novel drug and gene delivery systems also contribute. The advent of technologies is only a small part of the overall process. New ways of looking at drug development and innovative approaches to combine various technologies also leads to more effective and efficient drug development, and has the potential to contribute much more. Some technologies and combinations are discussed below.

- Combinatorial chemistry and high throughput screening (HTS) based on biotechnology and computer assisted drug design provided a foundation to improve lead identification and optimization[29-32]. The promise still remains. Technology has not yet resulted in more and better lead-compounds entering development or reaching the market. This lack of progress may be the result of the long development cycles, but it may also be that drug developers have not yet been able to assimilate the information into knowledge that can be applied to make drug development more efficient and effective.

- NME selection/rejection based on the computer data sorting, biotechnology derived procedures and pharmacogenetics including such things as *in silico*

ADME screening[33, 34], *in vitro* pre-clinical lead optimization technologies[35] and *in vitro* transporter/absorption screening[36]. This has resulted in some good, some bad choices. It is not always in the best interest of medicine or the pharmaceutical industry to discontinue the development of an efficacious compound because it is metabolized through CYP 2D6[37] or because a compound model indicated that the compound may be less than optimally absorbed or have less than perfect pharmacokinetic profile[38-40]. Metabolic drug-drug interactions can be managed to minimize risk and can even be used to advantage under certain circumstances or formulators may be able to improve delivery via alternate dosage forms or delivery sites.

- Selecting appropriate patients for treatment using gene-based diagnostics can alter drug development and the marketing of the resulting product. For example, identifying an appropriate population makes Herceptin® an acceptable therapy for breast cancer because the patients are diagnosed with elevations in HER2 before prescribing Herceptin®. This approach leads to appropriate prescribing during development as well as co-marketing of the diagnostic and the drug after approval. At the end of 2001, the sponsor received FDA approval to include the HER2 gene detection test in the Herceptin® product labeling [41, 42].

- Pharmacogenetics, pharmacogenomics and functional proteomics during drug discovery can lead to better therapeutic target identification that can result in improved biomarkers and surrogate endpoint selection for drug development [43-48]. Improved biomarkers and surrogate endpoints can lead to more efficient and effective drug development. Personalized drugs may be an opportunity, but not today. Today, using gene therapy to convert the entire world population to fast or to slow drug metabolizers is as cost effective as designer drugs[49]. Tomorrow, who knows what will happen, but I would put my money on big pharma going the gene therapy route. Who, other than niche companies, need a bunch of micro markets?

- Improved animal models of human disease are needed. Transgenic or gene knock- out animal models of human disease can be used to study human disease and therapeutic intervention prior to entering human studies [50-53]. In the early days of creating and using transgenic and gene knock out animals, many of the animal models diverted biochemical processes to circumvent genetic changes and function normally[54,55]. New technological intervention has lead to more human-like diseases in these altered animals, but the human-like nature of the disease model needs to be documented and monitored over time.

- Improved information technology (IT), electronic data capture (EDC) and e-clinical trials including patient recruitment and electronic case report forms (CRFs), data warehousing are critical to be more efficient and effective drug development[56, 57]. EDC needs to start as early in drug discovery or development as possible. EDC eliminates transcription errors throughout drug discover and development and the need for anything except electronic source documents.

Data warehousing can help make information available to everyone in the development process and convert information to knowledge to make drug development more efficient and effective. A primary application for data warehousing is to institutionalize PK/PD information from pre-clinical, early clinical and late clinical information so it's available to drug developers as soon as it is captured and to create a complete population PK/PD data based that can be used to optimize future drug development.

- New technologies such as nanotechnology and aptamer technology with all of their varied applications can revolutionize drug development and the practice of medicine[58-61]. Nanotechnology should be self explanatory, but because of the specialized nature of aptamer technology, a definition is in order. Aptamers are small nucleic acid ligands that bind to target molecules by proving a limited number of specific contact points embedded in a larger 3-D structure. Aptamers are isolated from combinatorial libraries using systematic evolution of ligands by exponential enrichment (SELEX) procedures The new technologies that are coming to the fore today as well as those that will come into play in the near future represent practical applications that have resulted from the convergence of biotechnology with information technology[62]. Nanotechnology and aptamer technology can be used to develop novel delivery systems for drugs and gene therapy[63, 64] as well as to provide a platform for biotechnology–derived diagnostics and therapeutics[65-68], artificial organ systems[69], and ***in vitro*** as well as ***in vivo*** drug or biomarker assay devices/systems[70-72] including ***in vivo*** biosensors for diagnosis and therapy[65, 73]. In addition, other novel approaches such as combining liposomal and aptameric delivery[74] or flow-based microimmunoassays with aptamers[75] can be used to improve drug development and, ultimately, the practice of medicine.

- Imaging technologies such as positron emission tomography (PET), single positron emission computed tomography (SPECT), morphological and functional magnetic resonance imaging (MRI), etc. are being used for more effective and efficient drug development. Imaging technologies are used to investigate drug distribution[76] and as biomarkers of disease and therapeutic intervention[77-80]. In addition, miniaturization has made some of these techniques applicable to both pre-clinical and clinical development work[81, 82].

Even with these technological advances, it still comes down to having the right people; those people who know what to do as well as why, when, where and how to do it. Although many drug development cookbooks have been published, there is no good cookbook for this process. Successful application of who, what, when, where, why and how that works in one development program may or may not work for another program. One size does not fit all. Effective and efficient drug development is an interactive process that requires a great deal of innovation, determination and communication to operate effectively and efficiently, as well as the fortitude and foresight to know when to kill the compound and move on. Although there are common, general processes for effective and efficient drug development that can be applied across programs, actual program specifics are a function of pathologic disease processes, mechanism of the therapeutic intervention and a good understanding of the regulators who will evaluate investigational and new drug market approval submissions.

The complexity of the technologies and processes that are needed for effective and efficient drug development necessitate cross-fertilization for effective communication. Efforts have been put forth to teach physicians to communicate with molecular pharmacologists, chemists, biotechnologists, etc. with some success. Other efforts have been put forth to approach this obstacle to efficient and effective drug development from another perspective wherein PharmDs and PhDs are trained to communicate more effectively with clinicians so that early therapeutic approaches can be communicated among bench scientists and those who are responsible for delivering the therapy in the clinical setting. The ultimate goal is to ensure that the basic scientists, applied scientists and clinical scientists are striving to accomplish the same objectives; more effective and efficient drug development programs that result in better therapeutics.

Effective and efficient drug development requires a multitude of expertise to optimize the process. Therefore, drug development needs to be orchestrated by a team of specialists that not only have their own expertise, but also are knowledgeable about drug development processes that are in current use as well as those that may be useful in the future. The team must be empowered to find better ways to ensure that the best technologies, medical practice as well as drug development science and regulation are integrated to achieve their full potential and communication is continuous so that unanticipated synergies can be realized. This allows the team to optimize the development process and make it even better in the future. The overall development process can be no better than the least effective functional input or member of the team. Cohesive multifaceted teams are needed to optimize development processes and institutionalize their good points within the team and the organization while remembering why certain approaches were not effective.

Innovation is critical to effective and efficient drug development. Creativity leads to totally new ways to do things, but innovation leads to ways to combine existing technologies and processes to do things that are even more efficient and effective. For example, combining increased IT capacity with PK/PD modeling and virtual clinical trials leads to synergies that could not have been anticipated only a few years ago. In most cases, this is the result of teamwork rather than a single individual burning the midnight oil until a revolutionary new idea is discovered. How many times has a team taken a new concept that was brought to the team by one individual and worked it until the idea was molded into an even better conceptual approach to the issue at hand? It is not just the technology; it is technology, together with innovative teams, that leads to improved drug development using novel applications of new technologies.

6. INTEGRATING PK AND PD INTO DRUG DEVELOPMENT

Several recent articles discuss the potential for PK/PD modeling and simulation in drug discovery and development[83-96]. Therefore, this chapter will not provide an in depth review, but will focus more on concepts that can be applied to biomarker selection and use in PK/PD modeling and simulation to improve drug development. PK/PD concepts, modeling and simulation can lead to better communication and help to institutionalize processes and conceptual approaches that result in better drug development paradigms and better peer acceptance/marketing.

Biomarkers can create tremendous value throughout the drug discovery and development process by providing early decision making information and input for PK/PD modeling and simulations for the later phases. Moving into the future, models are fitted to data so the models can then be used to predict future events. Predictive capability will require that the models are mechanism-based and that they provide unique solutions so predictions are meaningful, relevant and broadly applicable to future drug development paradigms [96, 97].

Model development and application should move through a series of steps. PK/PD data needs to be captured starting in pre-clinical development and models need to be built during pre-clinical development and then added to and modified as human data comes available. Before attempting to model PK/PD effects of therapeutic intervention, models should be created, tested and applied to model the mechanism-based disease process without intervention. The underlying model needs to accurately reflect the natural progression of disease biomarkers and clinical endpoints. Without this disease-only starting point, changes seen during therapeutic intervention could reflect natural variability in the markers and endpoints, circadian rhythms in the markers and endpoints, analytical or measurement variability, or a combination thereof [98].

Models of disease and, subsequently, models of disease with therapeutic intervention require input from a variety of expertise. Multifunctional teams with members from all aspects of drug development including synthetic chemists, biotechnologists, molecular biologists, pharmacologists, toxicologists, clinical pharmacologists, PK/PDists, therapeutic area specialists, biostatisticians, bio-analytical chemists, clinical chemists and multi-center clinical experts provide their perspectives for PK/PD model creation and evaluation that lead to optimized tools for PK/PD modeling and simulation. A model built by a team of pharmacologists and toxicologists will most likely be different than one built by a team of clinical chemists and clinical pharmacologists, but the most functional and useful models integrate best ideas of all drug development functions. Model building and validation is an ongoing process that will continue through drug development and into post-approval marketing.

Understanding the temporal differences between drug and biomarker concentrations in the PK/PD-disease model are critical and a mechanistic understanding of disease processes as well as efficacy and toxicity resulting from therapeutic intervention will drive effective and efficient PK/PD modeling and simulation. Poor or inappropriate understanding of temporal relationships can limit the potential of PK/PD modeling and simulation, thereby resulting less efficient and effective drug development. Progress is being made with respect to modeling disease processes and therapeutic intervention, but the there is still a long way to go. Overstating the current power of PK/PD modeling and its application to clinical trial simulation can spell the end of this approach because the medical community and drug developers will discount any future claims.

Plasma drug concentrations are markers that serve as surrogates for drug concentrations in tissues including the sites of action for beneficial and adverse effects. In contrast, markers that are not grounded in a sound theoretical foundation, disease process-based rationale and therapeutic mechanism-based intervention can limit or destroy the potential usefulness of other science-based PK/PD modeling and simulation approaches to drug discovery and development. Ultimately, all of the best intentions can be destroyed if all components of the modeling process do not come together to accomplish the intended objective. For example, if all components are in place except the GLP-like assay method does not measure the correct biomarker, does not adequately measure changes in the biomarker concentrations or does not have adequate sensitivity to characterize the observed biomarker profile, the results of the modeling are doomed. Alternatively, if all aspects of the data capture and analytical results were done meticulously, but the PK/PD-disease model development and validation was haphazard, the PK/PD output can be meaningless or misleading.

Preparing to provide PK input for PK/PD modeling and simulation starts with the answer to the question "What needs to be measured?"

- Are trough samples, sparse sampling or complete profiles required to accomplish the PK/PD modeling objective? First, if PK/PD modeling and simulation is needed, trough sampling will not suffice. Sparse sampling with population PK/PD analysis or complete profiles with conventional PK/PD modeling and simulation are needed to achieve study analysis objectives. Biomarker responses often do not parallel drug and/or active metabolite concentrations and that is why models are used to better understand the relationship and how it might change as a function of drug input and other variables. Any decision on which approach to take is based on logistics as well as experience and receptivity of the study sites and investigators to each approach.

- Do drug or drug plus active metabolites need to be measured? Having only one active moiety or one active moiety that dominates the beneficial and toxic effects of therapeutic intervention makes PK/PD work easier. PK/PD models can be complex even when only one drug **or** one active metabolite is used as a single model input function, but it becomes even more complex when there is more than one input function such as drug plus one or more active metabolites with varying activity and affinity for the receptor.
- Will free or total concentrations be assayed? Is plasma protein binding linear or nonlinear across the therapeutic/toxic drug concentration range? In addition, determining whether to measure unbound or total drug concentrations can be another variable. If drug-plasma binding is linear in the therapeutic and toxic range, unbound and total concentrations are simple ratios. In contrast, nonlinear plasma protein binding requires that unbound drug concentrations be measured because unbound drug concentrations increase disproportionately with increasing total drug concentrations.

- Does the drug compete with endogenous ligands for receptor binding? Endogenous ligands compete with drugs for receptor binding[98]. Competitive endogenous ligands complicate the PK and PD inputs[99] for PK/PD models. Since environmental, dietary and endogenous substances compete with drugs for receptor occupancy ***in vivo***, should drug developers identify these competitive substances and use them as biomarkers to track the effects of their drug development programs? Or, are these competing substances just obstacles that complicate assay development and validation as well as modeling efforts? Since they compete with the drug for receptor binding, they must be acting to stimulate or inhibit physiologic activity that could be a predecessor to pharmacologic activity at greater concentrations. It can be anticipated that changes in endogenous ligand concentrations following drug dosing may be similar to changes observed in angiotensin-I, angiotensin-II and plasma renin following doses of an angiotensin-II antagonist[100]. With what is known about competing environmental, dietary and endogenous substances, it is likely that there is baseline activity even in the absence of drug substance. For comprehensive PK/PD modeling, it is important to determine concentrations of the competing ligand(s) after both active and placebo treatment so that the competition can be built into the models[99].

Specific types of models have been reviewed and will not be discussed here. It will suffice to say that model selection and validation should ensure that the model is able of provide insight and guidance for drug developers[83-96]. PK/PD models should not be more complicated than is required to fit existing data and to simulate anticipated outcomes for future studies. PK/PD models should only include rate-limiting and mechanism-based components of the system that are required to describe the PK and the PD data. For example, if the rate-limiting PD step is receptor mediated, then the model should reflect receptor binding. In contrast, if the rate-limiting step is hormone secretion or protein production, then the model should reflect the secretion process so the modeling process reflects biologic, physiologic and pharmacologic reality. PK/PD models need to be simple enough to provide unique solutions based on the PK and PD data that are used for the modeling process[101, 102]. If the PK/PD models are not simple and unique, use of the models to predict endpoints under different conditions may provide errant outputs and thereby lead drug development teams astray. The combined impact of selecting 1) mechanism-based biomarkers of disease and drug action, 2) appropriate bio-analytical methods for biomarker assays, 3) appropriate clinical endpoints for comparison, and 4) unique mechanism-based PK/PD models will lead to more efficient and effective drug development as follows:

- If biomarkers are selected and validated with an understanding of disease processes and mechanisms of drug action for drug efficacy and toxicity and used in PK/PD modeling and simulation during drug development, biomarkers will predict clinical proof of concept and surrogate endpoints will predict clinical outcomes, thereby making drug development more efficient and effective.

- Efficient and effective drug development uses biomarkers for discovery, pre-clinical and early clinical development to learn more about the NME.

- Efficient and effective drug development uses biomarkers for discovery, pre-clinical and early clinical development to eliminate unwarranted new molecular entities as quickly as possible, thereby freeing up resources to be applied to other programs.

- Efficient and effective drug development uses discovery, pre-clinical and early clinical development knowledge to optimize the design of confirmatory proof of safety and efficacy trials.

Although our current PK/PD models are not sufficiently mechanism-based for effective prospective predictions[103, 104], biomarkers, together with PK/PD modeling and simulation, will provide a continuous process to link what has been learned during the current drug development program to the next generation of biomarkers, assays and models to improve mechanism-based models with each development cycle. Biomarkers together with PK/PD modeling and simulation can provide a foundation to bridge effective, multidirectional communication among marketing, drug discovery, pre-clinical research and development and clinical research and development

7. FUTURE DIRECTIONS

Most-critical variables for drug success must be evaluated earlier in development. Discontinuing development as early as possible for compounds that have little potential for success will focus limited resources on only those compounds that have high probability of success. Much of the information needed for these decisions is determined from PK/PD and ADME studies. The old school approach of waiting until marketing is confident that the drug will make it to the market before investing in ADME or PK/PD studies is obsolete. Early use of biomarkers and PK/PD analyses to guide and optimize drug selection and development is a requisite for company success in the coming years.

As drug development moves from discovery to pre-clinical research and development, mechanism-based PK/PD modeling should be initiated. Although pre-clinical models may not be directly applicable to the clinical setting, they can establish a foundation for future clinical models[105]. For example, transgenic animals or other animal models that closely reflect human disease can be used to start capturing pharmacokinetic data for PK/PD modeling that will eventually be moved into human testing. This foundation can then be expanded and modified to incorporate PK/PD model components that reflect human pharmacokinetics, biology, physiology and pharmacology as drug development moves from Phase I through Phase IIa. By the end of Phase IIa, the model should be sufficiently established to implement through the rest of the development and marketing phases. These same modeling tools are also being built into regulatory review and approval processes [95, 106].

As disease, pharmacologic, and adverse event mechanisms become better understood, biomarkers become more mechanism-based. Underlying disease processes and models of disease need to be understood before starting to create PK/PD models. Incorporating biomarker and modeling concepts into drug development plans needs to occur prior to entering the discovery phase. HTS targets can serve as sources of potential biomarkers. For example, if the discovery process is going to target a specific pathogen, enzyme or receptor, these targets should serve as the source of potential biomarkers that can be used for pre-clinical and early clinical assessments. As a result, as biomarker assay methods become even more robust, biomarkers will provide increasing value to the drug development process. As mechanism-based PK/PD models are implemented, the models will provide more unique solutions, resulting in better simulation quality and, therefore, prediction will improve. Simulations will be used to anticipate and predict proof of concept or principle during Phase I/IIa. As biomarkers are linked to clinical outcomes to create surrogate endpoints, simulations will be used to anticipate and predict clinical endpoints for Phase IIb/III. Clinical trials will truly become confirmatory.

Processes to design, conduct, model, interpret and report biomarker studies are not as well defined as processes for pharmacokinetic studies within most pharmaceutical companies. This situation may reflect the relative newness of biomarkers in drug development. But, more likely it is a function of the varied inputs that are needed from the information producers and knowledge users in this area of research and development. Multidisciplinary teams that provide input, direction and oversight as well as an interface for producer-user communication is one way to better integrate the biomarker concept as a central focus of drug development within a company. The team is needed because there are too many sophisticated and complex inputs into too many diverse processes to allow one person or a small group of like-minded individuals make this complex process work effectively. This true today, but will become even more important in the future.

Pre-clinical and clinical biomarker study protocols should be written using a rationale that parallels protocols development for pharmacokinetic studies. Questions that should be asked include what is known, what must be confirmed and what do we still need to learn? Proposed bio-analytical methods as well as detailed clinical sample collection and handling instructions must be included in the protocol. Clinical protocols should be written to help provide a link between the biomarker and the clinical endpoints. Even a perfect biomarker or surrogate endpoint has no value if it is not widely accepted by all stakeholders including the producer, users, management and the medical community. This leads to a need for constant, effective communication. Communication must be effective within the company, between the company and academia, between the company and regulators, and between the company and other stakeholders such as physicians, pharmacists, hospitals, healthcare management companies, insurance companies and patients.

If this all come to fruition, biomarkers and PK/PD models will guide compound selection and retention. Even if the biomarker cannot be linked to clinical endpoint, mechanism-based biomarkers will dramatically improve the early development of drugs by simplifying decision-making processes. Only compounds with a high probability of medical and commercial success will enter Phases IIb and III. The cost of drug development will decrease in response to this more efficient and effective process.

References

1. Sheiner LB. Commentary: Learning versus confirming in clinical drug development. *Clin. Pharmacol Ther.*61: 275-291 (1997). Placke ME. Development of New Candidate Drugs, Part 1: In Vitor Studies. *Appl Clin .Trials* 6(10): 36: 38 (1997).
3. Placke ME. Development of New Candidate Drugs, Part 2: In Vivo Studies. *Appl Clin .Trials* 6(11): 42- 46 (1997).
4. Lee JW, Hulse JD and Colburn WA. Biochemical Markers: Maximizing the Use in New Drug Development. *Appl Clin Trials* 5(10): 24-32 (1995).
5. Colburn WA: Selecting and Validating Biologic Markers for Drug Development. *J Clin Pharmacol* 37: 355-362 (1997).
6. Posvar EL and Sedman AJ. First-in-Human Studies of synthetic Molecules. *Appl Clin Trials* 5(10): 70-74 (1996)
7. Froehlich J. First-in-Human Studies of Biologicals. *Appl Clin Trials* 5(10): 65-68 (1996).
8. Heath EC, Pierce CH. Inducing "disease" in healthy volunteers for early evaluation. *Appl. Clin Trials* 8(5): 42-48 (1999).
9. Eldon MA. Clinical Pharmacokinetics during Drug Development. *Appl Clin Trials* 5(10): 56-64 (1996).
10. Colburn WA: Optimizing the Use of Biomarkers, Surrogate Endpoints and Clinical Endpoints for More Efficient Drug Development. *J Clin Pharmacol* 40: 1419-1427 (2000).
11. Colburn WA. Decision Making During New Molecular Entity Development. *Appl Clin Trials* 5(10): 44-55 (1996).
12. Biomarkers Definitions Working Group. Biomarkers and surrogate endpoints in clinical trials: Proposed definitions and conceptual framework. *Clin Pharmacol Ther* 69: 89-95 (2001).
13. S'Arcy P, Harron D (1992) Proceedings of the first International Conference on Harmonization. Belfast: Queens University Press. pp 189-191.
14. ICH Consensus Guideline: Organization of the Common Technical Documents for Registration of Pharmaceuticals for Human Use (Released for consultation, July, 2000).
15. ICH Consensus Guideline: The Common Technical Document for the Registration of Pharmaceuticals for Human Use - Safety (Released for consultation, July, 2000).

16. ICH Consensus Guideline: The Common Technical Document for the Registration of Pharmaceuticals for Human Use - Efficacy (Released for consultation, July, 2000).
17. ICH Consensus Guideline: The Common Technical Document for the Registration of Pharmaceuticals for Human Use - Quality (Released for consultation, July, 2000).
18. FDA Guidance for Industry: Formal Meetings With Sponsors and Applicants for PDUFA Products (Feb., 2000).
19. FDA Guidance for Industry: Fast Track Drug Development Programs- Designation, Development, and Application Review (Sept., 1998).
20. FDA Guidance for Industry: Information Program on Clinical Trials for Serious or Life-Threatening Diseases and Conditions (March, 2002).
21. Lesko JL, Rowland M, Peck CC, and Blaschke TF. Optimizing the science of Drug Development:Opportunities for Better Candidate Selection and Accelerated Evaluation in Humans. *Pharm. Research* 17: 1335-1344 (2000).
22. Colburn WA. Early Clinical Development Moves into the 21st Century. *Appl Clin Trials* 8(10): 54-56 (1998)
23. FDA Guidance for Industry: Drug Metabolism/Drug Interaction Studies in the Drug Development Process: Studies In Vitro (April, 1997).
24. FDA Guidance for Industry: In Vivo Drug Metabolism/Drug Interaction Studies-Study Design, Data Analysis, and Recommendations for Dosing and Labeling (Nov., 1999).
25. FDA Guidance for Industry: Guideline for Studying Drugs Likely to be Used in the Elderly (Nov., 1989).
26. FDA Guidance for Industry: Guideline for the Study and Evaluation of Gender Differences in the Clinical Evaluation of Drugs (July, 1993).
27. FDA Guidance for Industry: General Considerations for Pediatric Pharmacokinetic Studies for Drugs and Biological Products (Draft: Nov., 1998).
28. Peck CC and Wechsler MA. Report of a Workshop on Confirmatory Evidence to Support a Single Clinical Trial as a Basis for New Drug Approval. *Drug Inf J* 36: 517-544 (2002).
29. Labute P, Nilar S, Williams C. A probabilistic approach to high throughput drug discovery, *Comb Chem High Throughput Screen*, 5: 135-145 (2002).
30. Gedeck P, Willett P., Visual and computational analysis of structure – activity relationships in high-throughput screening data, *Curr Opin Chem Biol*, 5: 389-395 (2001).
31. Guillouzo A., Applications of biotechnology to pharmacology and toxicology, *Cell Mol Biol (Noisy-le-grand)*, 47: 1301-1308 (2001).
32. Seneci P, Miertus S., Combinatorial chemistry and high-throughput screening in drug discovery: different strategies and formats, *Mol Divers*, 5: 75-89 (2000).
33. Keseruu GM, Molnar L.,METPRINT: a metabolic fingerprint. Application to cassette design for the high-throughput ADME screening, *J Chem Inf Comput Sci*, 42: 437-444 (2002).
34. van de Waterbeend H., High-throughput and in silico techniques in drug metabolism and pharmacokinetics, *Curr Opin Drug Discov Develop*, 5: 33-43 (2002).
35. Atterwill CK, Wing MG., In vitro preclinical lead optimization technologies (PLOTs) in pharmaceutical development, *Toxicol Lett*, 127: 143-151 (2002).
36. Bohets H, Annaert P, Van Beijsterveldt L, Anciaux K, Verboven P, Meuldermans W, Lavrijsen K., Strategies for absorption screening in drug discovery and development, *Curr Top Med Chem*, 1: 367-383 (2001).
37. Preskorn SH, Reducing the risk of drug-drug interactions: a goal of rational drug development, *J Clin Psychiatry*, 57(Suppl 1): 3-6 (1996).
38. Norris DA, Leesman GD, Sinko PJ, Grass GM., Development of predictive pharmacokinetic simulation models for drug discovery, *J Control Release*, 65: 55-62 (2000).
39. Langowski J, Long A., Computer systems for the prediction of xenobiotic metabolism, *Adv Drug Deliv Rev*, 57: 407-415 (2002).
40. Poulin P, Theil FP., Prediction of pharmacokinetics prior to In Vivo studies. II. Generic physiologically based pharmacokinetic models of drug distribution, *J Pharm Sci*, 91: 1358-1370 (2002).
41. Lewis R, Bagnall A, Forbes C, Shirran E, Duffy S, Kleijnen J, Riemsma R, Ter Riet G., The clinical effectiveness sof trastuzumab for breast cancer: a systematic review, *Health Technol Assess*, 6: 1-71 (2002).
42. Genentech receives FDA approval to include HER2 gene detection test in Herceptin product labeling. Business Wire via Newsedge Corp. August 30, 2002
43. Rothberg BE., The use of animal models in expression pharmacogenomic analyses, *Pharmacogenomics J*, 1: 48-58 (2001).

44. Juang JX, Mehrens D, Wiese R, Lee S, Tam SW, Daniel S, Gilmore J, Shi M, Lashkari D., High-throughput genomic and proteomic analysis using microarray technology, *Clin Chem*, 47: 1912-1916 (2001).
45. van Ommen GJ, The Human Genome Project and the future of diagnostics, treatment and prevention, *J. Inherit Metab Dis*, 25: 183-188 (2002).
46. Jain KK. Applications of biochip and microarray systems in pharmocogenomics, *Pharmacogenomics*, 1: 289-307 (2000).
47. Figeys D., Functional proteomics: mapping protein-protein interactions and pathways, *Curr Opin Mol Ther, 4:* 210-215 (2002).
48. Brookes PS, Pinner A., Ramachandran A., Coward L., Barnes S., Kim H., Darley-Usmar VM, High throughput two- dimensional blue –native electrophoresis: A tool for functional proteomics of mitochrondria and signaling complexes,*Proteomics*, 2: 969-977 (2002).
49. Sengupta LK, Sengupta A., Sarker M., Pharmacogenic applications of the post genomic era, *Curr Pharm Biotechnol*, 3: 141-150 (2002).
50. Harris S., Trangenic knockout as part of high-throughput, evidence-based target selection and validation strategies, *Drug Discov Today*, 6: 628-636 (2001).
51. Petters RM., Sommer JR., Transgenic animals as models for human disease, *Trangenic Res*, 9: 347-351 (2000).
52. Harris S., Foord SM., Transgenic gene knock-outs: functional genomics and therapeutic target selection, *PhArmacogenomics*, 1: 433-443 (2000).
53. Ge R., genetically manipulated animals and their use in experimental research, *Ann Acad Med Singapore*, 9: 560-564 (1999).
54. Christen U., Von Herrath MG., Apoptosis of autoreactive CD8 lymphocytes as a potential mechanism for the abrogation of type 1 diabetes by isletspecific TNF-alpha expression at a time when the autoimmune process is already ongoing., *Ann N Y Acad Sci*, 958: 166-169 (2002).
55. Mikkola I., Heavey B., Horcher M., Busslinger M., Reversion of B cell commitment upon loss of Pax5 expression, *Science*, 297: 110-113 (2002).
56. Marks RG, conion M., Ruberg SJ, Paradigm shifts in clinical trials enabled by information technology, *Stat Med*, 20: 2683-2696 (2001).
57. Hardison CD, Schnetzer T., Using information technology to improve the quality and efficiency of clinical trial research in academic medical centers, *Qual Manag Health Care*, 7: 37-44 (1999).
58. Davis SS., Biomedical application of nonotechnology- -implications for drug targeting and gene therapy, *Trends Biotechnol*, 15: 217-224 (1997).
59. Lockman PR, Mumper RJ, Khan MA, Allen DD., Nanoparticle technology for drug delivery across the blood-brain barrier, *Drug Dev Ind Pharm*, 28: 1-13 (2002).
60. Hoppe-seyler F., Butz K., Peptide aptamers: powerful new tools for the molecular medicine, *J. Mol Med*, 78: 426-430 (2000).
61. Robert AF Jr., The future of nanofabrication and molecular scale devices in the nanomedicine, *Stud Health Technol Inform*, 80: 45-59 (2002).
62. Zajtchuk R., New technologies in medicine: biotechnology and nanotechnology, *Dis Mon*, 45: 449-495 (1999).
63. Scherer F., Anton M., Schillinger U., Henke J., Bergemann C., Kruger A., Gansbacher B., Plank C., Magnetofection:enhancing and targeting gene delivery by magnetic force and in vivo, *Gene Ther*, 9: 102-109 (2002).
64. Martell RE, Nevins JR, Sullenger BA, Optimizing Aptamer Activity for Gene Therapy Applications Using Expression Cassette SELEX, *Mol Ther*, 6: 30-34 (2002).
65. Freitas RA Jr., The future of nanofabrication abd molecular scale devices in nanomedicine, *Stud Health Technol Inform*, 80: 45-59 (2002).
66. Bogunia-Kubik K. Sugisaka M., From molecular biology to nanotechnology and nanomedicine, *Biosystems*, 65: 123-38 (2002).
67. Rusconi CP, Sacrdino E., Layzer J., Pitoc GA, Ortel TL, Monroe D., Sullenger BA, RNA aptamers as reversible antagonists of coagulation factor IXa, *Nature*, 419: 90-94 (2002).
68. Li JJ, Fang X.,Tan W., Molecular aaptamer beacons for real-time protein recognition, *Biochem Biophys Res Commun*, 292: 31-40 (2002).
69. Prokop A., Bioartificial organs in the twenty-first century: Nanobiological devices, *Ann N Y Acad Sci*, 944: 472-490 (2001).
70. Laval JM, Mazeran PE, Thomas D., Nanobiotechnology and its role in the development of new analytical devices, *Analyst*, 125: 29-33 (2000).
71. Clark Sl, remcho VT, Aptamers as analytical reagents, *Electrophoresis*, 23: 1335-1340 (2002).

72. Green LS, Bell C., Janjic N., Aptamers as reagents for high throughput screening, *Biotechniques,* 30: 1094-4, 1098, 1100 (2001).
73. O'Sullivan CK, Aptasensors- - the future of biosensing, *Anal Bioanal Chem,* 372: 44-48 (2002).
74. Willis M., Forssen E., Ligand-targeted liposomes, *Adv Drug Deliv Rev,* 29: 249-271 (1998).
75. Hayes MA, Polson TN, Phayre AN, Garcia AA, Flow-based microimmununoassay, *Anal Chem,* 15: 5896-5902 (2001).
76. Fischman AJ, Alpert NM, Rubin RH, Pharmacokinetic imaging: a noninvasive method for determining drug distribution and action, *Clin Pharmacokinet,* 41: 581-602 (2002).
77. Carroll TJ, Tenneggi V.,Jobin M., Squassante L., Treyer V., Hany TF, Burger C., Wang L., Bye A., Von Schulthess GK, Buck A., Absolute Quantification of Cerebral Blood Flow With Magnetic Resonance, Reproducibility of the Method, and Comparison With H215O Positron Emission Tompography, *J Cereb Blood Flow Metab,* 22: 1149-1156 (, 2002).
78. Muramoto S., Uematsu H., Sadato N. Tsuchida T., Matsuda T., Hatabu H., Yonekura Y., Itoh H., H215O Positron Emission Tomography Validation of Semiquantitative Prostate Blood Flow Determined by Double-Echo Dynamic MRI: A Preliminary Study, *J Comput Assist Tomogr,* 26: 510-514 (2002).
79. Nadeau SE, McCoy KJ, Crucian GP, Greer RA, Rossi F., Bowers D., Goodman WK, Heilman KM, Triggs WJ, Cerebral blood flow changes in depressed patients after treatment with repetitive transcranial magnetic stimulation: Evidence of individual variability, *Neuropsychiatry Neuropsychol Behav Neurol,* 15: 159-175 (2002).
80. Bartolini M., candela M., Brugni M. Catena L., Mari F., Pomponio G., Provinciali L., Danieli G., Are behaviour and motor performances of rheumatoid arthritis patients influenced by subclinical cognitive impairments? A clinical and neuroimaging study, *Clin Exp Rheumatol,* 20: 491-497 (2002).
81. Weissleder R., Scaling down imaging: molecular mapping of cancer in mice, *Nat Rev Cancer,* 2: 11-18 (2002).
82. Mitchell P., Turning the spotlight on cellular imaging, *Nature Biotech,* 19: 1013-1017 (2001).
83. Derendorf H, Lesko LJ, Chaikin P, Colburn WA, Lee P, Miller R, Powell R, Rhodes, G, Stanski D, Venitz J. Pharmacokinetic/pharmacodynamic modeling in drug research and development. *J Clin Pharmacol* 40:1399-1418 (2000).
84. Meibohm B, Derendorf H. Basic concepts of pharmacokinetic/ pharmacodynamic (PK/PD) modeling. *Int J Clin Pharmacol Ther,* 35: 401-413 (1997).
85. Colburn WA, Lee JW. Biomarkers in PK/PD Modeling and Simulation. *Clin Pharmacokinet,* (Requested article in print).
86. Danhof M. Applications of pharmacokinetic/pharmacodynamic research in rational drug development. *Meth Find Exp Clin Pharmacol,* 18:53-54 (1996).
87. Balant LP, Balant-Gorgia AE. Advantages and synergism of combined pharmacokinetic and pharmacodynamic studies during drug development. *Boll Chim Farm,* 132: 212-213 (1993).
88. Sheiner LB, Steimer J-L. Pharmacokinetic/pharmacodynamic modeling in drug development. *Annu Rev Pharmacol Toxicol,* 40: 67-95 (2000).
89. Gieschke R, Steimer JL. Pharmacometrics: Modelling and simulation tools to improve decision making in clinical drug development. *Eur J Drug Metab Pharmacokinet* 25: 49-58 (2000).
90. Colburn WA and Eldon MA. Simultaneous Pharmacokinetic/Pharmaco-dynamic modeling *in* Pharmacodynamics and drug development: Perspectives in clinical pharmacology (Cutler NR, Sramek JJ and Narang PK, eds.). London, John Wiley, 1994, pp 19-44.
91. Park K, Verotta D, Gupta SK, Sheiner LB, Use of a pharmacokinetic/ Pharmacodynamic model to design an optimal dose input profile. *J Pharmacokinet Biopharm,* 26: 471-493 (1998).
92. Verotta D, Beal SL, Sheiner LB. Semiparametric approach to pharmacokinetic-pharmacodynamic data. *Amer Physiol Soc,* R1005-R1010 (1989).
93. Jelliffe R, Schumitzky A, Van Guilder M. Population pharmacokinetics/ pharmacodynamics modeling: Parametric and nonparametric methods. *Ther Drug Monit,* 22: 354-365 (2000).
94. Derendorf H, Meibohm B. Modeling of pharmacokinetic/ pharmacodynamic (PK/PD) relationships: Concepts and perspectives. *Pharm Res,* 16: 176-185 (1999).
95. Machado SG, Miller R, Hu C. A regulatory perspective on pharmacokinetic/pharmacodynamic modelling. *Stat Meth Med Res,* 8: 217-245 (1999).
96. Colburn WA. Optimizing the use of biomarkers, surrogate endpoints and clinical endpoints for more efficient drug development. *J Clin Pharmacol,* 40: 1419-1427 (2000).
97. Levy G. Mechanism-based pharmacodynamic modeling. *Clin Pharmacol Ther,* 56:356-358 (1994).
98. Colburn WA. Drugs and endogenous ligands compete for receptor occupancy. *J Clin Pharmacol,* 34: 1148-1152 (1994).

99. Colburn WA and Gibson DM. Endogenous agonists and pharmacokinetic/ pharmacodynamic modeling of baseline effects *in* Pharmacodynamic research: Current problems and potential solutions (Kroboth PD, Smith RB and Juhl RP, eds.) Cincinnati, OH: Harvey Whitney Books, 1988, pp 167-184.
100. Colburn WA. Selecting and validating biologic markers for drug development. *J Clin Pharmacol*, 37: 355-362 (1997).
101. Gibson DM, Taylor ME, Colburn WA. Curve fitting and unique parameter identification. *J Pharm Sci*, 76: 658-659 (1987).Nestorov IA. Sensitivity analysis of pharmacokinetic and pharmacodynamic systems: I. A structural approach to sensitivity analysis of physiologically based pharmacokinetic models. *J Pharmacokinet Biopharm*, 27: 577-597 (1999).
103. Kimko HC, Reele SSB, Holford NHG, Peck CC, Prediction of the outcome of a Phase 3 Clinical Trial of an Antischizophrenic Agent (quetiapine fumurate) by Simulation with a Population Pharmacokinetic and Pharmacodynamic Model. *Clin Pharmacol Ther*, 68: 568-577 (2000).
104. Veyat-Follet C, Bruno R, Olivares R, Rhodes GR, Chaikin P. Clinical Trial Simulation of Docetaxel in Patients with Cancer as a Tool for Dosage Optimization. *Clin Pharmacol Ther*, 68: 677-687 (2000).
105. Obach RS , Baxter JC, Liston TC, Silber BM,Jones BC,MacIntyre F, et.al. The Prediction of Human Pharmacokinetic Data from Preclinical and In Vitro Metabolism Data. *J Pharmacol Exp Ther*, 283:46-58 (1997).
106. Gobburu JV, Marroum PJ. Utilisation of pharmacokinetic-pharmacodynamic modeling and simulation in regulatory decision-making. *Clin Pharmacokinet*, 40: 883-892 (2001).

BIOANALYTICAL METHODS: CHALLENGES AND OPPORTUNITIES IN DRUG DEVELOPMENT

Mark L. Powell and Steve E. Unger[1]

1. INTRODUCTION

The development and validation of quantitative bioanalytical methods to measure drug and biotransformation products (metabolites) in biological matrices has been evolving for decades. The uninterrupted introduction of new technologies and the increasing attention being paid by global regulatory authorities to validation issues has continued to shape this scientific field. However, in the last five years, there has been an unprecedented increase both in the pace and breadth, of research activities related to drug discovery and development.[1] The high-throughput screening of compound libraries has dramatically increased the number of potential drug candidates in discovery programs. The resulting need to evaluate these candidates during lead optimization has fueled extraordinary growth in the number of samples being analyzed in bioanalytical laboratories for the quantitative determination of both drugs and putative metabolites in biological matrices such as blood, plasma, serum and urine. Additionally, significantly greater attention is being focused on the metabolic liabilities of these compounds at earlier stages in drug development than ever before. In a continuing effort to keep up with the increasing demands for higher sample throughput, greater sensitivity, and increased metabolic information, bioanalytical scientists are continually accelerating their efforts to search for technological advances. This pattern of change began three decades ago and continues today.

During the 1970's, high-performance liquid chromatography (HPLC) was first introduced and quickly gained initial acceptance in bioanalytical laboratories. The ability to analyze a wide variety of structurally dissimilar compounds without the need for chemical derivatization offered significant advantages over gas chromatography (GC)

[1] Analytical Research and Development and Clinical Discovery Analytical Sciences
Pharmaceutical Research Institute, Bristol-Myers Squibb Inc., PO Box 191, One Squibb Drive, New Brunswick, NJ

and even GC coupled with mass spectrometry (GC/MS), both of which had been in widespread use for some time.[2] Like the common glass GC columns which were widely available at the time, initial HPLC columns were generally hand packed and primarily available in one standard length. This resulted in relatively long chromatographic run times and fairly broad peak shapes.

During the 1980's, HPLC continued to "mature" with the introduction of commercially available columns of different lengths and packing materials with broader applications. Chromatographic cycle times began to shorten as equipment improved, but not as much as might have been predicted. It was the successful coupling of HPLC with atmospheric pressure ionization (API) MS (LC/MS) that provided bioanalytical scientists with the alternative to GC/MS that they had been eagerly anticipating. The MS detector, based on either electrospray ionization (ESI) or atmospheric pressure chemical ionization (APCI), dramatically increased sensitivity and specificity (selectivity) of the bioanalytical methods as the need for baseline resolution of all components disappeared.

In the 1990's, advances continued to be made in HPLC. A broader variety of column configurations and materials, rugged narrow and microbore columns, precision LC pumps, and fast and reliable autosamplers became available to bioanalytical scientists. As a result of the widespread and enthusiastic acceptance of the API-based LC/MS techniques by bioanalytical and other analytical scientists, the sophistication and reliability of the technique dramatically improved as more analytical instrument companies joined in the design and manufacture of a variety of LC/MS instruments. Tandem mass spectrometric systems (MS-MS) that incorporated triple-quadrupole technology (consisting of two mass analyzers and a dissociation cell) became increasingly more reliable and cost-effective for routine use in biological sample analyses. By the end of the decade, bioanalytical scientists had replaced a majority of their "conventional" HPLC-UV systems with hybrid LC/MS/MS systems. LC/MS had moved from the research category to routine laboratory use. All of which brings us to today where modern bioanalytical laboratories are all equipped with LC/MS-MS systems, GC/MS systems, as well as HPLC systems.

Despite the fact that bioanalytical scientists are constantly striving for increased sensitivity in their drug and metabolite assays, the driving force behind this dramatic shift in mainstream laboratory instrumentation, was *not* increased sensitivity. Instead, it was an emerging emphasis on speed of analysis and overall cycle times. This is best illustrated by the fact that despite the dramatic advances in HPLC over the last twenty years, a survey of approximately 150 published HPLC bioanalytical methods revealed that the median retention time for a primary analyte was identical from 1983-1989 compared to 1990-1999 (15 min). In just the last six years within the Clinical Discovery Analytical Sciences Department at Bristol-Myers Squibb, the use of HPLC for the quantitative determination of small-molecule drugs and metabolites decreased from over 75% in 1994 to less than 10% in 2000.

This chapter will focus on two areas of significant ongoing challenges in current bioanalytical laboratories: continuing changes in chromatographic approaches designed to increase the speed of bioanalysis as well as the dramatically increasing emphasis on obtaining significantly more metabolic information earlier in drug development programs. Because the metabolite quantitation area is much newer, more attention will be devoted to it in this chapter. These two areas of interest are providing rich scientific opportunities for bioanalytical scientists and both are being increasingly influenced by regulatory authorities.

2. REGULATORY INFLUENCES

Recognizing the rapid pace of technological advances in bioanalysis and the critical role that it plays in the pharmacokinetic evaluation of drugs in development, The U.S. Food and Drug Administration (FDA) along with the American Association of Pharmaceutical Sciences (AAPS) sponsored two scientific conferences designed to foster harmonization in scientific approaches to quantitative bioanalyses, and form the basis for an eventual Guideline on Bioanalysis. The first conference took place in 1990, with an eventual conference report appearing in 1992.[3] That initial conference was the first attempt to gather bioanalytical scientists and pharmacokineticists from all over the world with the purpose of beginning the process of agreeing on standardizing approaches to bioanalysis. Although some of the topics debated at the conference proved to be controversial, the eventual report achieved a level of consensus for the first time and served as a type of regulatory "guide" in the absence of published policy. Ten years later a follow-up conference was held to revisit what progress had been made over that time period. In the ensuing years, an FDA DRAFT Guidance, *Bioanalytical Methods Validation for Human Studies*, had been published for comment in 1997, but had not been finalized. Thus, whereas the initial conference was a first look at standardization and building consensus, this follow-up conference was designed to resolve the remaining scientific issues relating to bioanalytical method development and validation and to discuss comments on the existing DRAFT FDA Guidance. The conference report appeared later in 2000.[4] The FDA Final Guidance (Bioanalytical Method Validation) eventually was published in 2001,[5] and remains in effect today. Although similar discussions and meetings have taken place outside of the US, it appears that most, if not all, other countries are utilizing the FDA Bioanalytical Method Validation Guidance to evaluate the bioanalytical sections of their regulatory submissions.

This guidance, and the scientific conferences that supported it, focused almost exclusively on small molecules. Macromolecules, on the other hand, present unique bioanalytical challenges, which are partially, but not completely, addressed in the small molecule guidance. There has been one FDA/AAPS sponsored conference focusing on macromolecules and there will likely be follow-up meeting(s) aimed at a probable eventual separate FDA Guidance for the bioanalysis of macromolecules.

Although not as much regulatory attention has been focused on metabolite characterization and determinations, to date, a recent guidance has appeared which will influence bioanalytical scientists. This will be covered later in the quantitation of metabolites section of the chapter.

3. HIGH-THROUGHPUT BIOANALYSIS

Because of the inherent selectivity/specificity of LC/MS/MS, overall run times of bioanalytical methods can now be decreased to less than 2.0 min. This has allowed for the rapid turn-around of sample analysis that is required in today's environment of fast-paced drug discovery and development. Consequently, the bottleneck in LC/MS/MS bioanalysis had now shifted to sample preparation.[6] One technique that is becoming routine in an increasing number of laboratories to alleviate the sample preparation bottleneck is the use of a high-flow on-line extraction systems which allow the direct injection of a biological sample into an LC/MS/MS system without any sample extraction.

Fast chromatography LC/MS/MS coupled with a direct injection system is a powerful technique. This is essentially a "dilute and shoot" approach with no off-line sample preparation except for the addition of an internal standard. Following the injection of a biological sample, extraction occurs on-line, prior to analyses. In the last few years, direct-injection approaches based on the use of ultra fast flow-through extraction columns packed with relatively large stationary phase particle sizes ($\geq 30\ \mu m$) have successfully been implemented and a variety of systems with different configurations have been developed.[7-11] Such systems are normally obtained commercially. However, at Bristol-Myers Squibb, we have demonstrated that a system can be assembled "in-house" using existing laboratory equipment without significant capital outlay. This simple, versatile system assembled in our laboratories is shown in Figure 1. Two HPLC columns are used in this system. The first column serves as a sample extraction column (typically an Oasis® HLB column, 1.0 x 50 mm, 30 μm) and

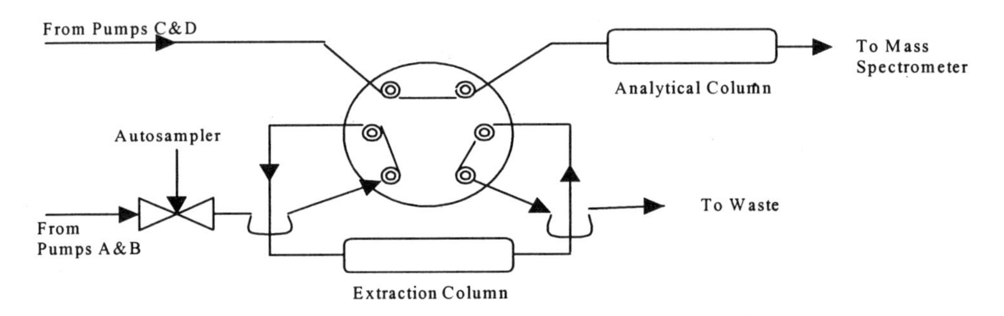

Configuration (a): Loading sample, extraction and equilibration

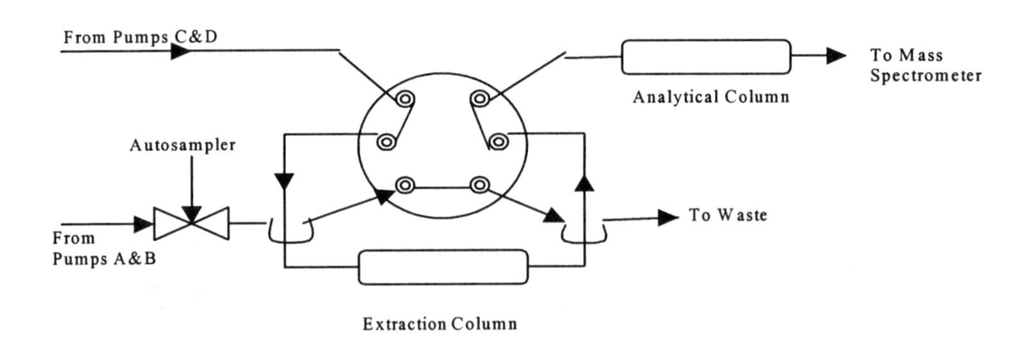

Configuration (b): Elution

Figure 1. A schematic representation of on-line extraction LC-MS-MS system

the second column serves as an analytical column (typically C18, 2.0 x 50 mm, 5 μm). A 10-μL portion of the biological plasma sample is transferred by the autosampler onto the Oasis® column, using a largely aqueous mobile phase (typically ≤ 10% acetonitrile and ≥ 90% aqueous 3.0-mM formic acid) at a flow rate of 4.0 mL/min, with the effluent directed to waste (Figure 1, configuration (a)). This is the sample extraction and cleanup stage and lasts for a fraction of a minute (typically ≤ 0.3 min). The valve is then switched (Figure 1, configuration (b)) so that the Oasis® column is in line with the analytical column and the mass spectrometer, and the mobile phase is switched to one with a higher percentage of the organic component (e.g., 60% acetonitrile and 25% aqueous 3.0-mM formic acid) at a lower flow rate (typically 1.0 mL/min). This is the elution stage. The analytes are eluted from the Oasis® column to the analytical column for detection by the mass spectrometer. The elution stage is accompanied by the equilibration stage to first wash the extraction column, autosampler and other parts of the system with a reverse gradient mobile phase (from 100% organic to ≤ 10% organic) and then recondition the extraction column with the mobile phase composition used in the extraction stage (Figure 1, configuration (a)). With this system, total run times can be very short (typically ≤ 2.0 min). Our experience with this type of simple, but versatile, system has been very fulfilling. Single extraction columns can be routinely used for the analysis of several thousand samples, with no clogging or chromatographic peak shape problems. The direct injection approach has been further simplified by using robotic liquid handlers for aliquotting samples into 96-well plates for direct injection from the plates. The simple system depicted in Figure 1 can be further modified and improved by using two extraction columns in parallel. In this modified system, the equilibration of one extraction column occurs while analysis is ongoing on the other extraction column. Thus, equilibration time does not add to the overall chromatographic run time. Consequently, cycle times are faster and the working life of the system is effectively doubled.

It would seem that this approach could be utilized to explore run times considerably shorter than 2 min. However, there is a practical limit to how far the chromatographic run times in LC/MS/MS bioanalytical methods can be shortened. In spite of the tremendous specificity/selectivity provided by the selected-reaction-monitoring (SRM) used in LC/MS/MS, experience in the last 3-5 years has shown that there is often a need for *some* chromatographic separation in LC/MS/MS bioanalysis. Among the many factors which govern this is one representative scenario in which a biotransformation product (or a prodrug) may give a signal in the SRM channel used for the quantitation of the drug.[12-15] This occurs as a direct result of in-source collision-induced-dissociation (CID) of the biotransformation product (or a prodrug). If the drug is not chromatographically separated from such interfering drug-related compounds, the measured concentration of the drug will be inflated. This type of interference is illustrated with the mass spectrometric behavior of fosinopril (prodrug) and fosinoprilat (drug). The chemical structures of the two compounds are shown in Figure 2 and their full-scan electrospray mass spectra (Q1) are presented in Figure 3. The Q1 spectra shows significant responses from the protonated and sodiated adduct species of the compounds. From the MS/MS product ion spectra of the $[M+H]^+$ precursor ions for the two compounds (MS/MS spectra data not shown), the SRM channels chosen for the quantitation of fosinoprilat and fosinopril were *m/z* 436 → *m/z* 390 and *m/z* 564 → *m/z* 436, respectively.

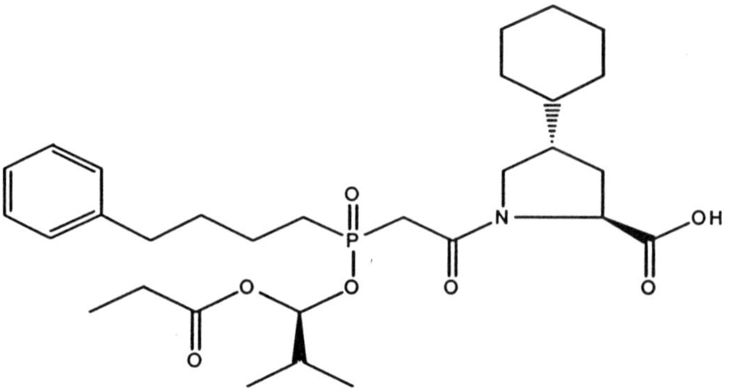

Mol. Wt.: 563.66

Fosinopril (prodrug)

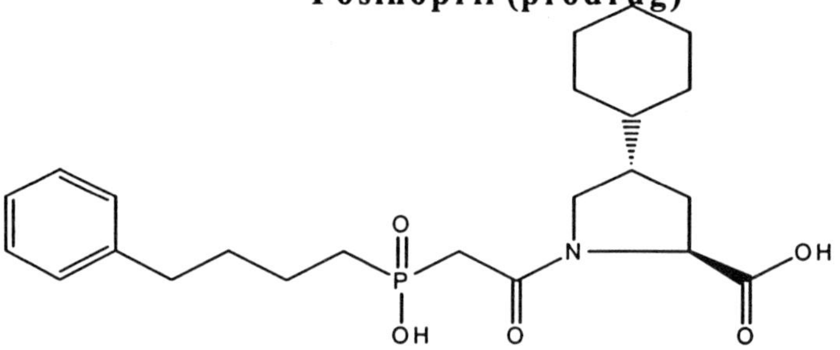

Mol. Wt.: 435.49

Fosinoprilat (drug)

Figure 2. Chemical structures of fosinopril (prodrug) and fosinoprilat (drug)

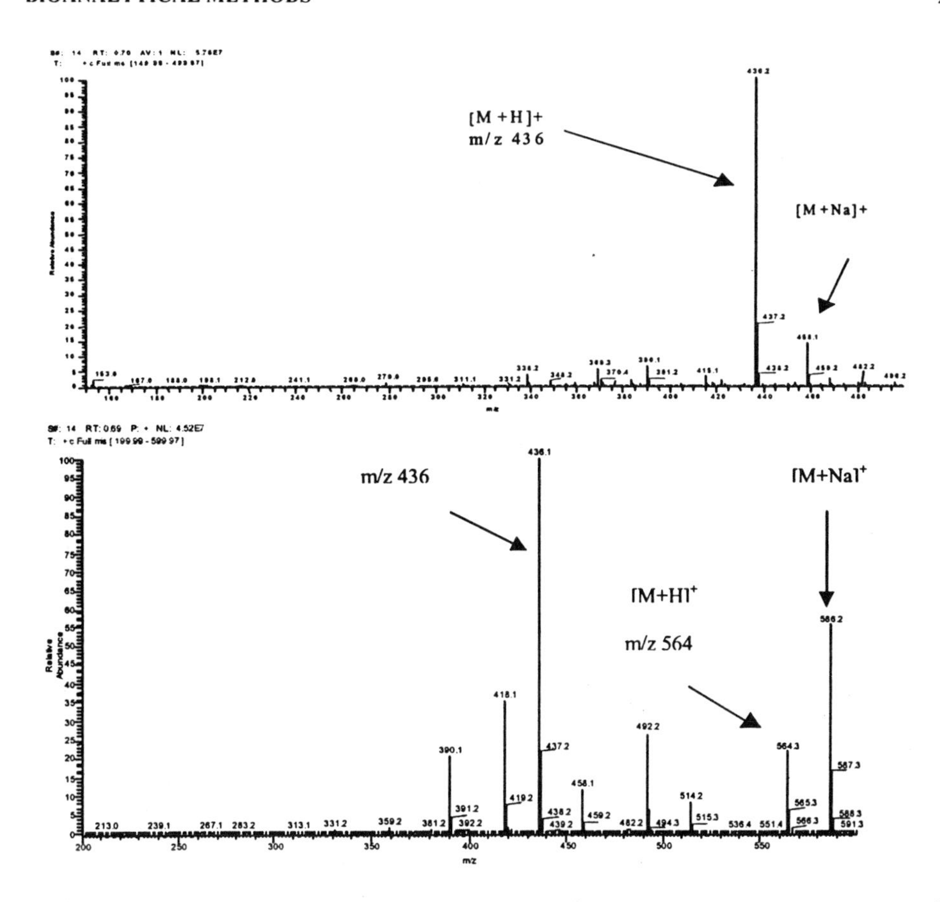

Figure 3. Full scan mass spectra: top panel: fosinoprilat ($[M+H]^+ = 436$); bottom panel: fosinopril ($[M+H]^+ = 564$)

It should be noted, however, that the *m/z* 436 ion is present not only in the Q1 spectrum of fosinoprilat but also in that of fosinopril. Consequently, the two compounds will have a common product ion of *m/z* 390, both arising from the same *m/z* 436 precursor ion. Since fosinopril will produce a signal in the *m/z* 436 → *m/z* 390 SRM channel used for the quantitation of fosinoprilat, it is necessary to achieve adequate chromatographic separation between the two compounds when analyzing samples that contain both compounds even when the objective is to quantitate only the fosinoprilat concentration. This phenomenon is illustrated in Figure 4, where The *m/z* 436 → *m/z* 390 and *m/z* 564 → *m/z* 436 SRM channel chromatograms obtained from a 500 ng/mL fosinopril QC sample (spiked serum sample contains no fosinoprilat – only fosinopril) are shown.

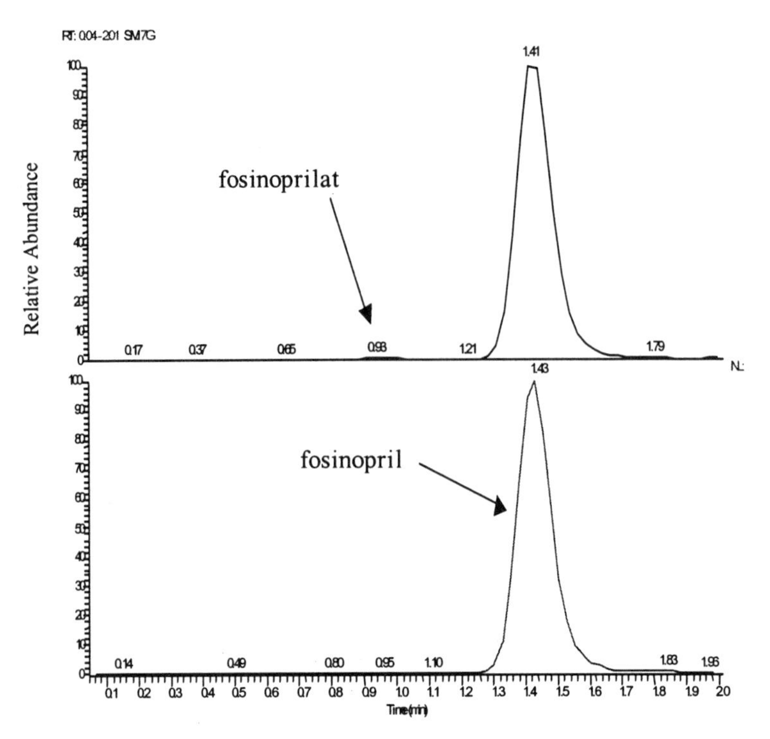

Figure 4. Selected reaction monitoring chromatograms of a fosinopril QC 500 ng/mL serum sample, with the two SRM channels shown for fosinoprilat (*m/z* 436 → *m/z* 390, top) and fosinopril (*m/z* 564 → *m/z* 436, bottom)

Under the HPLC conditions used, there is baseline chromatographic resolution between fosinopril (retention time 1.4 min) and fosinoprilat (retention time 0.9 min). The peak that is shown at the retention time of 1.4 min in the fosinoprilat channel (*m/z* 436 → *m/z* 390) is due to in-source hydrolysis of fosinopril. This in-source hydrolysis of fosinopril gave the same SRM transition as that of fosinoprilat. On the other hand, the fosinoprilat channel does not show a peak at the retention time of fosinoprilat (0.9 min) indicating the absence of fosinoprilat in the sample injected. This would be expected for the fosinopril-only sample used.

A second example supporting the need for chromatographic separation is seen when some components in a biological matrix that do not give a signal in the SRM channel used for the quantitation of the drug may, however, affect the response of the drug.[16-21] This phenomenon, commonly referred to as a matrix effect, is usually caused by endogenous components in the biological matrix, but can also be caused by components related to, or administered with, the drug. The presence of matrix effects can dramatically decrease the response of the analyte,[21] thus affecting sensitivity, or it can adversely affect the accuracy/precision of a bioanalytical method.[18, 19] The presence/absence of matrix effects must be ascertained during the development of a LC/MS/MS bioanalytical method. A simple procedure for accomplishing this has previously been published.[20] Matrix effects will be discussed in more detail in the following section on metabolite quantitation.

4. QUANTITATION OF METABOLITES

As previously alluded to, worldwide regulatory agencies are requiring a better understanding of drug safety, which is increasingly including exposure to both parent drug and relevant metabolites. Regulatory agencies and the pharmaceutical industry have recently issued a guidance that is directed at metabolite exposure.[22, 23] The challenge to drug metabolism and bioanalytical scientists is to provide exposure data which is sufficiently precise and accurate to satisfy these emerging requirements.

A decision regarding what metabolites should be measured and when and how they should be measured in support of drug development is essential. Good metabolism data must be available to make good analytical decisions. In vitro and in vivo profiling in the discovery animal models and anticipated safety species, along with in vitro human profiling, provides much of the needed information. For metabolites deemed worthy of quantification in studies submitted to regulatory agencies, exact structures are needed for synthesis. The input of other departments, including chemistry, is needed for analytical reference standards and/or internal standards.

Prior to the availability of synthetic standards, method development may begin using crude isolates, often from microsomes, urine or bile. When diluted into plasma, these are effective sources for early bioanalytical work. Extraction, separation and detection conditions may be established. Many important analytical figures of merit such as stability, recovery, as well as an estimate of dynamic range and sensitivity can be obtained from crude metabolite standards.

An important initial question that must be answered is "should there be separate assays or a single multi-component assay?" Since the introduction of LC/MS/MS, there is a common expectation that multi-component assays can be developed quickly. Their use in N-in-1 or cassette dosing PK screening[24, 25] for discovery optimization has furthered these expectations. However, for discovery studies, acceptance criteria are normally relaxed relative to what is ultimately acceptable for regulatory filings.

Recognition of the fact that failure rates will statistically increase when larger numbers of metabolites are quantified is important. Analytical qualifiers (QCs) assure assay performance within a given confidence interval (C.I.). Under ideal conditions, the failure rate among analytes is mutually exclusive.[26] As shown below, the study's overall statistical failure rate among all n analytes can be approximated as ($1- C.I.^n$).

Table 1. Relationship between major metabolite definition and assay feasibility

Metabolite (minimum %)	# of Analytes (Parent & metabolites)	Relative Detection Limit	Study failure rate (95% C.I.)
N/A	1	1.0	5 %
30	3	1.7	14 %
20	5	2.2	23 %
10	10	3.2	40 %
5	20	4.5	65 %

As the number of metabolites approaches 20% of the total circulating drug-equivalents, the likelihood of assay failure for any one analyte is 23%. Almost one in four analytes would need to be re-determined to adhere to conventional acceptance criteria. The table also illustrates that unless individual segmented scanning is possible, the relative detection sensitivity will be impacted. Achieving the same LLOQ or low failure rate is not simply defined by the analytical procedure but by the number of metabolites being analyzed.

Concurrent analyses are more efficient because they allow simultaneous assays of all analytes. However, the physicochemical properties of metabolites are frequently quite diverse. Consequently, final assay conditions are usually a compromise between the ideal conditions of each individual analyte. Assay ruggedness can be affected and the single analyte failure rate may also increase. If separate assays are run for each analyte, optimization is made easier but additional risks of non-statistical assay failure are attendant. Additionally, more assays increase the likelihood of unexpected but inevitable instrument malfunctions, analyst error or other problems.

Method development should quickly establish the possible detection sensitivity and dynamic range. Profiling data from metabolism or animal studies is needed to establish assay requirements. Bioanalytical decisions cannot be made without detailed in vivo biotransformation data. In vitro metabolism data can serve many purposes, but it should not guide assay requirements. The process of establishing assay requirements must be based upon a thorough understanding of what is circulating in plasma or present in excreta.

Differences in instrument sensitivity or procedures can complicate the transfer of profiling to bioanalytical methods. Crude calibrators, mixtures of metabolites isolated from in vitro preparations or excreta, can serve as system suitability standards to help transition a profiling method to a bioanalytical method.

The first step towards transitioning a profiling method to a quantitative method is to establish what can be measured and to what standards. If exploratory studies are undertaken, the quantitation of numerous metabolites to GLP standards is inappropriate. There are considerable scientific challenges associated with generating and qualifying metabolite standards for use as calibrators. For regulatory studies, metabolite pharmacokinetic data needs to be obtained with the same precision and accuracy as parent drug. Since metabolite reference standards are generally produced in smaller quantities, they are rarely characterized in the same detail as parent drug.

The preparation and characterization of metabolic bioanalytical reference standards is a time-consuming process. Organic synthesis of the metabolite serves first to confirm its identification. The use of microbial[27] biological systems or fractionation from excreta[28] can serve to generate standards for preliminary studies but they lack independent confirmation of structure. Enzymes are unpredictable and spectra can be mis-interpreted. Structural assignments of metabolites should always be considered preliminary until proven by synthesis. To assess activity, define metabolic pathways and preserve intellectual property, synthetic metabolites may be prepared in milligram quantities. If desired for dosing in safety studies, gram quantities of properly characterized material will be needed.

Qualitative analysis of metabolite reference standards is often not as problematic as accurate quantitation. Nuclear magnetic resonance (NMR) and elemental microanalysis indicate gross impurities and should agree with HPLC determinations. Calculated theoretical levels of impurities may be factored into the elemental analysis. However, levels must be high before significant deviations are noted. Ion chromatography can address counterion composition while ^{19}F and ^{31}P-NMR may serve to quantify drug or impurity levels that contain these atoms. Besides HPLC impurities, water is usually the most significant impurity affecting a use-as value and can be determined using a Karl Fischer coulometric titrator. Next are residual solvents, especially from preparative isolations. Thermal gravimetric analysis (TGA) may give an indication of all volatiles and most, but not all, compounds release moisture sufficiently to be used for quantitation. The goal is to obtain the best quantitative measure of mass balance with as little material as possible. An illustrative example follows.

Roxifiban, a glycoprotein (GP) IIb/IIIa antagonist, and its metabolites were determined in human plasma. To accurately measure the metabolites, reference standards were synthesized and isolated using ion-pairing HPLC. The metabolite standards showed various amounts of TFA counterion, as determined in Table 2 (shown below) using ^{19}F-NMR.

Table 2. Analytical reference standards of roxifiban metabolites

Analyte	Molecular Formula	Formula Weight	**HPLC Purity**
XV459	$C_{20}H_{27}N_5O_6 \bullet 1.00\ C_6H_6O_3S$	591.65	99.2
M1a	$C_{20}H_{27}N_5O_7 \bullet 1.15\ C_2HF_3O_2$	580.58	>99%
M1b	$C_{20}H_{27}N_5O_7 \bullet 1.20\ C_2HF_3O_2 \bullet 1.5\ H_2O$	613.32	>99%
M2	$C_{20}H_{27}N_5O_7 \bullet 1.00\ C_2HF_3O_2$	563.48	>95%
M3	$C_{20}H_{27}N_5O_7 \bullet 0.14\ C_2HF_3O_2$	465.42	>98%
M8a	$C_{20}H_{27}N_5O_7 \bullet 1.00\ C_2HF_3O_2$	563.48	>99%
M8b	$C_{20}H_{27}N_5O_7 \bullet 1.00\ C_2HF_3O_2$	563.48	>99%
D4-XV459	$C_{20}H_{23}D_4N_5O_6 \bullet 1.00\ C_2HF_3O_2$	551.52	>99%
DMP 728 Displacer	$C_{25}H_{36}N_8O_7 \bullet 1.00\ CH_4O_3S$	656.72	>99%

As noted, the exact composition (isolated salt or zwitterion) can be in question. A major problem with analytical reference standards prepared by HPLC fractionation is the presence of counterions and solvents. As the counterion or solvent approaches the molecular weight of the analyte, even a small uncertainty can have great impact. For a compound of MW 500, uncertainty in the monohydrate will add at most 3.5 % while uncertainty in its identity as a free base or mono-TFA salt will add 18.6%.

To avoid delays, method development should not be limited by the availability of analytical reference standards. By using crude isolates as calibrators, method development can proceed in parallel with metabolite synthesis. Samples from exploratory or discovery animal studies are particularly useful in setting assay requirements. Effects of induction in subsequent multiple-dose studies or dose-limited exposure due to dissolution rate-limited absorption, first-pass metabolism or transporter saturation are unpredictable. Projections made from single-dose studies using low doses may under or over-estimate metabolite concentrations. The bioanalyst must be prepared to adjust the dynamic range requirements appropriately.

The need to prepare stable isotope internal standards should also be factored into the synthetic efforts. Knowing that a metabolite is unstable, absorptive or significantly different from others will guide decisions on whether a stable isotope is needed. For instance, the d_5-labeled isotopomers of atorvastatin, atorvastatin lactone and its hydroxy metabolites were synthesized as internal standards to quantify atorvastatin and its hydroxy metabolites in human serum using LC/MS/MS. Clinical drug development justified preparing both the d_5-atorvastatin and its lactone.[29]

The impact of using synthetic resources to prepare stable isotopes rather than analytical reference standards is routinely evaluated. The analyst's goal should be to develop an assay that measures as many significant metabolites as is practical without needing a new internal standard. Discovery chemistry often provides a wealth of good structural analogues and more than one may be needed to accurately measure parent and all significant metabolites.

As previously mentioned, the metabolites of roxifiban, a glycoprotein (GP) IIb/IIIa antagonist, were characterized using NMR and MS.[30] Synthesis provided milligram quantities of both referenced materials. Because a human radiolabeled mass balance study was not possible due to drug retention in specific tissues, quantitation of the metabolites using LC/MS/MS provided mass balance in clinical studies.

Roxifiban is metabolized to its primary active zwitterionic form, XV459, and several minor active, hydrolyzed and hydroxylated metabolites (Figure 5). These included two pairs of epimers (M1a/M1b and M8a/M8b) and one pair of geometric isomers (M2/M3). Due to lack of specificity of MS/MS detection, three critical chromatographic pairs needed to be resolved. Use of large quantities of DMP 728, another GP IIb/IIIa antagonist to liberate XV459 and metabolites from the GP IIb/IIIa receptor caused severe ion suppression. An automated ion exchange solid phase extraction (IX-SPE) procedure was developed to selectively extract the seven metabolites of roxifiban and its deuterated internal standard and to specifically exclude DMP 728. Low doses were administered, requiring simultaneous determination of all metabolites in human plasma over the concentration range of 0.5 to 80 nM.

Since DMP 728 eluted together with several metabolites on reverse or normal phase systems, attempts to develop an extraction based on a difference in polarity were unsuccessful. The guanidine group (pKa ca. 12) of DMP 728 was much more basic than the benzamidine group (pKa ca. 10) of roxifiban or its metabolites, suggesting that the former molecule could be readily differentiated by ion exchange chromatography. At pH $\approx$ 11, metabolites should be negatively charged but DMP 728 will remain as a neutral zwitterionic species. Strong Anion Exchange (SAX) SPE provided a selective extraction, removing more than 99.9% of DMP 728. As expected, the SPE was sensitive to pH of the loading solution and required close attention.

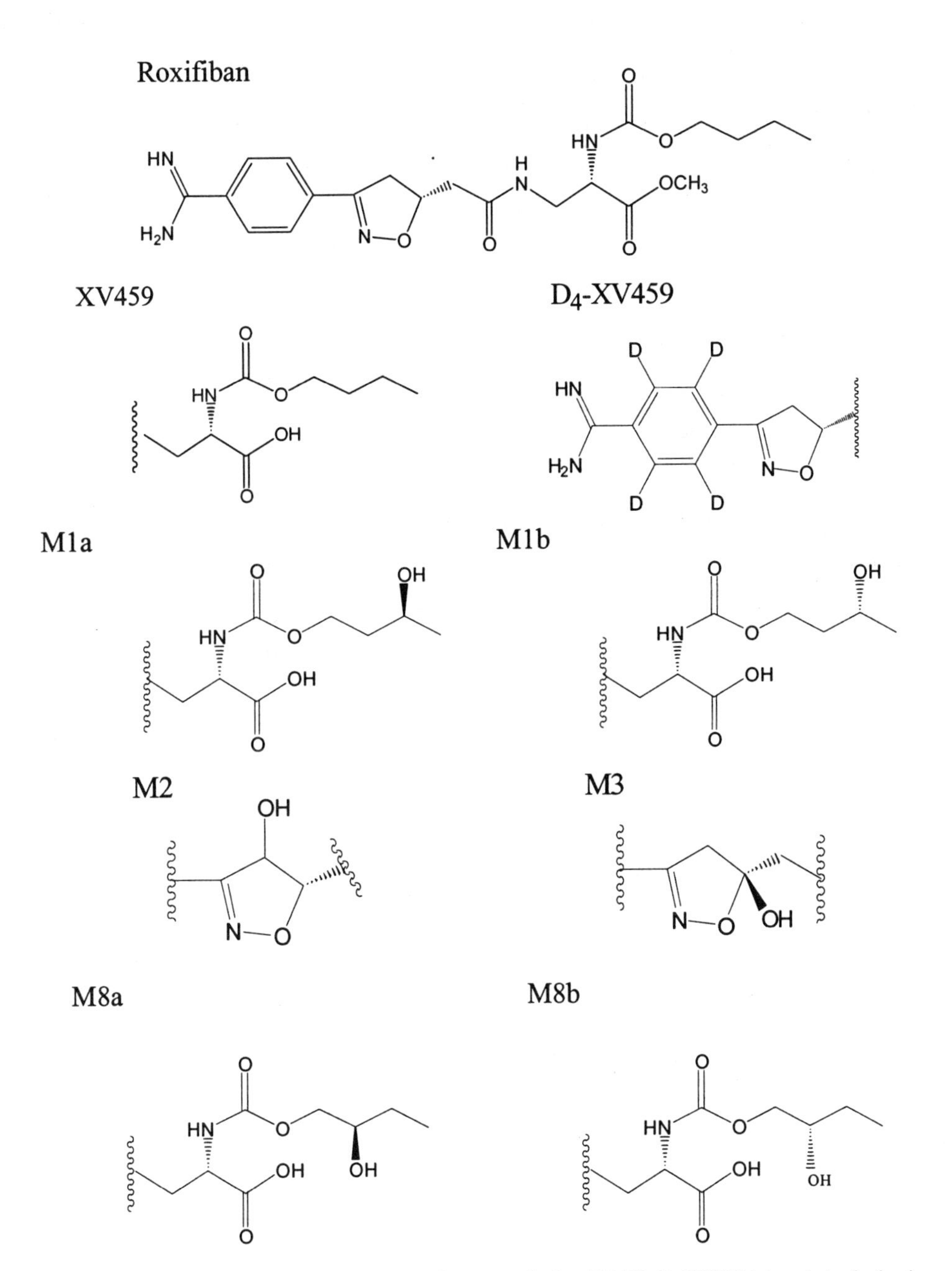

Figure 5. Chemical structures of roxifiban, its primary metabolite, XV459, D_4-XV459 internal standard, minor metabolites, M1a, M1b, M2, M3, M8a, and M8b.

The extraction procedure was also less efficient than that normally seen for routine SPE procedures. However, the lack of a matrix effect indicated a clean extraction for MS detection. For roxifiban and all of its metabolites, recoveries ranged from 41 to 59% and the matrix effect[31] varied from 0.94 to 1.10. Because all hydroxylated metabolites shared common MS/MS product ions, no SRM transition could be found that was totally specific and adequately sensitive. An HPLC separation of all metabolites prior to MS/MS detection became necessary. Numerous combinations of different types of stationary and mobile phases were tested before an optimal chromatographic separation was achieved (Figures 6 and 7).

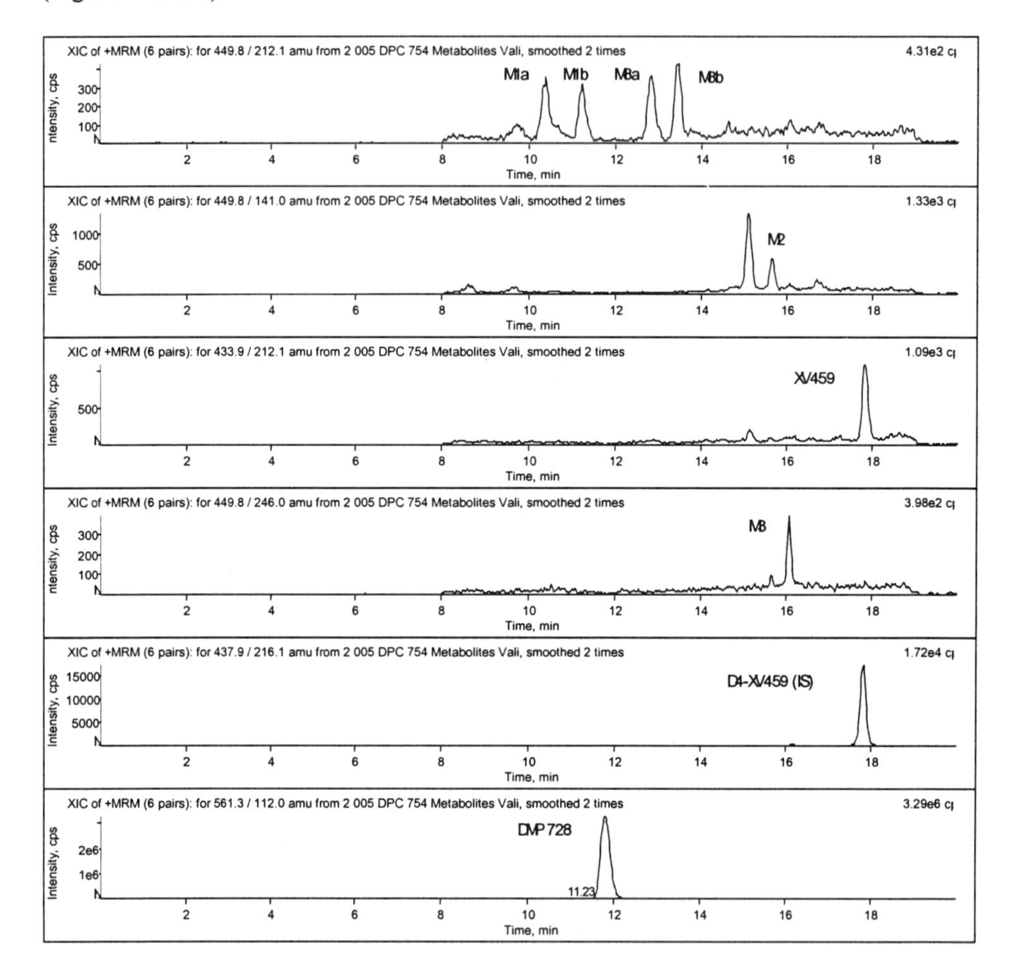

Figure 6. Representative LC/MS/MS chromatograms of seven roxifiban metabolites in human plasma at a concentration of 0.5 nM (LLOQ).

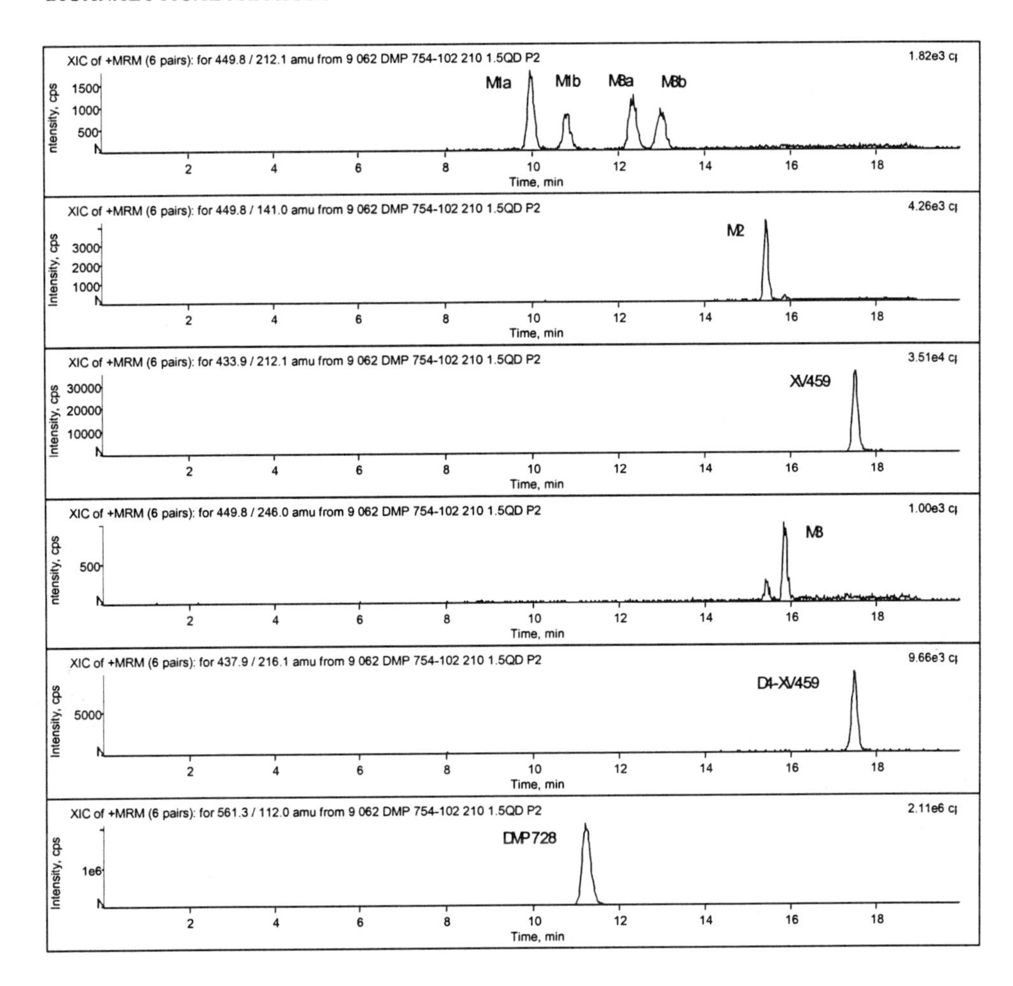

Figure 7. Representative LC/MS/MS chromatograms of seven roxifiban metabolites in human plasma 3 hours following dosing 1.5 mg once daily.

Before undertaking pre-clinical assay development, specifications should be established using information from discovery metabolism and pharmacology studies, as well as considering what, if any, toxicological liabilities may be associated with metabolite exposure during safety studies. LC/MS profiling of in vitro (liver microsomes, S9 and hepatocytes) or in vivo (plasma, urine or bile) metabolites provides much of the analytical information. Quantitation of metabolites prior to clinical trials is not usually done unless:

- The metabolite comprises a major portion of the overall exposure
- It is pharmacologically active
- It has a known toxicity
- There is a structural alert (reactive or formed via a reactive intermediate)

- There is insufficient exposure information from radiolabeled metabolism studies in Safety Assessment species

LC/MS profiling of human plasma and excreta from single and multiple dose clinical studies provides a first determination of whether there are unique metabolites not seen in animal studies. With animal profiling data, this work establishes an estimate of relative metabolite exposures across species. A radioactive metabolite profiling arm added to the multiple dose escalation/tolerability study or a separate ^{14}C mass balance study provides more definitive early human metabolite exposure data. The use of radiolabeled studies to determine metabolite concentrations avoids many analytical and synthetic problems but does not provide individual exposure data from toxicology or clinical studies. Alternatively, plasma and urine samples collected during safety studies can be preserved for later profiling or quantitation; however, metabolite stability will eventually need to be confirmed.

Considering that metabolites can have quite different polarities affecting their extraction efficiency, chromatographic behavior and MS response, establishing a common procedure routinely requires compromise during method optimization. Accepting non-optimum performance can result in further failures for specific analytes. Non-linear calibrations due to dimer formation when a co-eluting internal standard was used, interference from unsuspected metabolites and ion suppression adversely affect assays.[32] Ion suppression due to matrix effects can be particularly problematic, since MS response is impacted differently over the course of the separation. Seldom does one have the luxury of using multiple stable isotope internal standards to correct for these variations. Co-elution of analytes and internal standard, while correcting for temporal differences, creates new ionization and fragmentation problems that can further complicate the analysis.[15, 12] An extraction that provides adequate enrichment to achieve the desired detection sensitivity and is class-specific for drug and metabolites should provide high recovery. When coupled with an effective separation, there should be minimum matrix effects[17, 18] and a clean chemical background.

Many HPLC/UV assays, particularly for analytes with high extinction coefficients at selective wavelengths, are sufficiently sensitive to accurately determine metabolite concentrations. As apparent in an HPLC/UV assay to determine concentrations of paracetamol and six glucuronide, sulfate, cysteine or mercapturate metabolites in urine,[33] these assays are limited by chemical background. Bupropion and its three major basic metabolites were measured in plasma using liquid-liquid extraction and HPLC/UV. Dual-wavelength UV detection was used to optimize sensitivity, obtaining detection limits of 5 ng/mL for bupropion and 100 ng/mL for the major metabolites.[34]

To measure greater numbers of metabolites at lower concentrations often requires more selective detection. Fluorescence (FL) and electrochemical (EC) detection are more selective. While an advantage in eliminating endogenous chemical noise, both FL and EC detection can be limited by their inability to give responses to all metabolites. An HPLC/FL method to measure concentrations of hydroxychloroquine and three major metabolites used a cyano column for both analytical and preparative separations. Low ng/mL concentrations could be determined directly while isolated fractions were prepared for subsequent chiral analysis using an AGP column.[35] Tamoxifen and its metabolites were determined directly using UV detection. However, sensitivity was improved considerably by post-column photolysis to form the fluorescent phenanthrene derivatives.[36] Coupling FL derivatization with an on-line semi-permeable surface extraction column and cyano analytical column allowed low ng/mL concentrations of tamoxifen and four of its metabolites to be measured.[37]

Detection using mass spectrometry has become preferred for its selectivity, sensitivity and speed in method development. Provided the separation does not require complex additives in the mobile phase, detection sensitivity and selectivity is not compromised by chromatographic conditions. Assays are limited by electronic noise, and predicted limits of quantitation are routinely achieved. Assay development is straightforward and seldom limited by the number of metabolites or chemical background.

The choice of the extraction procedure is sometimes difficult. In an attempt to achieve high enrichment, more selective extractions can exclude important metabolites. Procedures that use extremes of pH or temperature, while suitable for parent drug analysis, may be unsatisfactory for metabolites. Often, a high enrichment extraction of parent drug excludes important metabolites.

Protein precipitation is often a first choice for multiple component assays since it generally ensures high recovery of all analytes, including unstable, absorptive and structurally-dissimilar metabolites. Matrix effects can complicate individual determinations, and the lack of an enrichment step can limit detection sensitivity. Assay development and refinement is aided by accurately determining recovery and matrix effects. Recovery should be calculated to a blank extract that, following extraction, has been spiked with an amount equivalent to 100% recovery of the analyte. This post-extraction spiked blank also serves to determine the matrix effect by measuring the ratio of its response to that from a mobile phase standard. For all analytes, extraction and chromatographic method development should optimize recovery while minimizing matrix effects. What is sacrificed in using a non-selective extraction can be regained using a more efficient chromatographic method. With the introduction of monolithic columns, analytical speed is no longer compromised to achieve a good separation.[38]

Liquid-liquid and solid phase extraction can be used to recover structurally-similar metabolites, providing cleaner extracts, fewer matrix effects and lower limits of detection. Loperamide and its N-demethyl metabolite were extracted from human plasma using a liquid-liquid extraction with methyl t-butyl ether (recoveries of 72% and 79%, respectively). Analysis using a reversed-phase separation and positive ESI MRM quantitation provided sub-nM limits of quantitation.[39] Solid-phase extraction can enhance detection sensitivity by enriching analytes, however, individual metabolites will have very different recoveries. Cocaine and its 12 metabolites were measured in rat blood, amniotic fluid, placental and fetal tissue homogenates using LC/MS/MS.[40] Use of a silica solid-phase extraction gave a range of recoveries for the 12 metabolites from rat blood (47-100%), amniotic fluid (61-100%), placental homogenate (31-83%), and fetal homogenate (39-87%).

The development of generic SPE materials provides a good start to enrich metabolite mixtures. Pibutidine, an H_2-receptor antagonist, and its four metabolites were measured in human plasma. Plasma samples were prepared using an Oasis HLB extraction prior to LC/ESI-MS/MS analysis over the concentration range of 0.1 to 25 ng/mL.[41] In another study,[42] diazepam and its three major active metabolites oxazepam, temazepam (3-hydroxydiazepam) and desmethyldiazepam were determined in dog plasma. Using an Empore C-18 high performance extraction disk plate and a Packard MultiProbe II liquid handler, dog plasma was extracted for LC/MS/MS analysis. D_5-Diazepam was used as the internal standard. Various bonded phases including ODS (C18), octyl (C8), phenyl,

CN and silica were evaluated. Recoveries were 76.6% for oxazepam, 59.8% for temazepam, 86.6% for desmethyldiazepam and 43.7% for diazepam. The matrix effect varied from 0.71 for temazepam to 1.01 for diazepam. An inverse correlation between recovery and matrix effect suggested that any changes to the extraction procedure would affect either diazepam detection or temazepam accuracy. Diazepam and its three metabolites were determined in dog plasma over the range of 1 to 500 nM.

Two-step combinations of liquid-liquid and solid-phase extraction have been used to provide assays of greater selectivity and sensitivity. An LC/MS/MS method to measure cilostazol, a quinolinone derivative, and its three active metabolites in human plasma used a combination of liquid-liquid and solid-phase extraction prior to a gradient C18 separation and positive ESI MS/MS quantitation.[43]

Drugs whose metabolites maintain unique chemical functionality such as amines or acids present an opportunity for more extensive separation using anion or cation exchange. Provided the chromatographic separation is efficient, low levels of metabolites may be measured with minimum sample preparation. Lysates of human peripheral blood mononuclear cells, extracted by protein precipitation, were separated by weak anion exchange prior to MRM detection. Weak anion exchange with step gradient elution of mono, di and tri-phosphates provided a fast (2 min) MS-compatible separation. A lower limit of quantitation of 5 fmol/10^6 cells (6 fmol on-column) was achieved.[44]

More selective extraction can be achieved using materials such as molecular imprinted polymers[45, 46] or antibodies[47, 48] that recognize common antigenic determinants in the drug and its metabolites. Molecular imprinted polymers have the advantage that they are more easily generated while non-specific antibodies, produced for other assays, may serve for immunoprecipitation. While it remains to be seen what role molecular imprinted polymers will play in multiple component quantitation, immunoprecipitation has been used to enrich LSD and its metabolites in urine prior to MRM detection.[47] The analytical system used three columns, a protein G immunoaffinity column with noncovalently immobilized antibody, packed capillary trapping column and a packed capillary analytical column. The on-line system increased detection limits 20-fold over an existing SPE method. More generally, immunoprecipitation is reserved for highly potent drugs, biologics or protein biomarkers. A quantitative method for detection of amyloid β peptides used immunoprecipitation prior to LC/ESI-MRM detection to measure the peptides Aβ 1-40 and Aβ 1-42 at pM concentrations. Comparison of the immunoprecipitation LC/MS with ELISA revealed comparable results[48]. Immunoprecipitation has also been widely used with matrix assisted laser desorption (MALDI) of biomolecules, but mostly for qualitative analysis.

Unstable metabolites, often important to understanding toxicity, present unique analytical challenges. Procedures to stabilize drug or metabolites used within a research laboratory are often unavailable to clinical sites. Studies must be undertaken to demonstrate that all analytes, especially metabolites, are stable during the sample collection process. If the assay is capable of measuring whole blood concentrations, stability determinations are relatively straightforward. Otherwise, plasma prepared at different times following spiking of analyte into fresh blood can be analyzed. Extrapolation to t=0 min affords an estimate of the initial concentration. Analysis at several temperatures helps to determine what sampling conditions are acceptable.

Many in vitro experiments only approximate real samples. Potentially labile conjugates such as sulfates or glucuronides, labile esters or thiols, and equilibration effects can affect analyte concentrations. Some examples include interconverting isomers, lactone hydrolysis/acid dehydration or RBC/plasma partitioning. Metabolic or chemical interconversion may be sensitive to relative concentrations, formulation, anticoagulant or other factors not controlled in an artificial QC sample. Stability samples must adequately mimic real samples; therefore, collection process stability should be evaluated using fresh incurred samples. Normal QC samples containing interconverting analytes should also contain varying ratios of analytes such that their relative concentrations approximate those seen in study samples.

Determining analyte stability in tissue, bone and other matrices is often difficult, requiring that stability be determined only in homogenate. For rare matrices, dilution into other matrices or the same matrix of another species is common but must not destabilize analytes. For biomarkers, stability should be demonstrated in normal plasma using incurred samples as well as stripped plasma using normal QC samples.

Drugs and metabolites that are light, temperature or pH-sensitive must be stabilized prior to analysis. Stabilizing procedures may be as simple as excluding UV light, heat or a pH known to catalyze decomposition. Addition of an antioxidant such as ascorbic acid may be needed. Absorption onto glass or plastics during sample collection can be incorrectly interpreted as instability. Container types should be evaluated during method development to avoid unnecessary analyte loss.

When these procedures are infective, in situ derivatization is another means of stabilizing analytes. Salicylic acyl glucuronide was converted to salicylic hydroxamic acid and quantified, along with salicylic acid, gentisic acid and salicyluric acid in human urine. Samples were treated with hydroxylamine prior to injection onto an HPLC. Salicylic phenolic glucuronide was quantified by difference after hydrochloric acid hydrolysis at 65°C with no loss of salicylic acid.[49] Cyclophosphamide and its oxime metabolite hydroxycyclophosphamide were determined in mouse plasma and tissue using LC/MS/MS. To avoid conversion to phosphoramide mustard or other reaction products, hydroxycyclophosphamide was derivatized during sample collection.[50]

Omapatrilat, a vasopeptidase inhibitor, and four of its metabolites were determined in human plasma. Omapatrilat and two of the metabolites are sulfhydryl-containing compounds that rapidly oxidize to disulfides. Methyl acrylate was used to form a stable derivative with the reactive thiols. The reaction was demonstrated to be quantitative and, unlike previous N-ethylmaleimide derivatives, did not introduce another chiral center. Omapatrilat and one metabolite could be measured to 0.5 ng/mL and the other three metabolites to 2 ng/mL. For sample collection, ethyl acrylate was added to Vacutainer tubes.[51]

Some derivatives afford both stability and enhanced detection sensitivity. Pentafluorobenzyl (PFB) derivatives have long been used to enhance negative chemical ionization detection in GC/MS. Recent work has demonstrated its use to increase negative atmospheric pressure chemical ionization (APCI) sensitivity by two orders of magnitude.[52] The derivatives undergo dissociative electron capture, yielding the molecular anion. Using MRM detection, attomole limits of detection were obtained. A normal-phase separation of the PFB derivatives of estradiol and its metabolites was accomplished within 7 min. The use of PFB bromide or similar electron capturing derivatives may provide enhanced detection sensitivity for acid-containing drugs and their metabolites.

Because of the inherent sensitivity of LC/MS, for many applications derivatization is unnecessary. However, as noted for omapatrilat, it can be a highly effective approach to stabilize a drug and its metabolites. Introduction of the stabilizing agent into the sampling tube affords the best result. Drugs or metabolites unstable to enzymes may be stabilized using inhibitors. If the enzyme responsible for the degradation is known, highly selective inhibitors are available. Otherwise, use of a more general inhibitor may be needed. The rapid introduction of a stable isotope internal standard can correct for losses that may occur before sample analysis.

Remifentanil, a potent analgesic, contains a labile methyl ester, which is highly susceptible to chemical and enzymatic hydrolysis resulting in a short half-life. A GC/HR-MS method for the determination of remifentanil in blood used immediate introduction of a $^{2}H_{4}$-stable isotope internal standard in acetonitrile to precipitate blood proteins, followed by extraction with methylene chloride.[53] The sampling procedures required extensive chemical manipulations. An alternative assay added 50% citric acid to blood to stop the hydrolysis. Extraction of chilled blood at pH 7.4 with butylchloride afforded a sufficiently clean sample for HPLC/UV analysis.[54]

One of the more complex issues in measuring drugs and their metabolites is their possible inter-conversion. Every effort should be made to minimize inter-conversion at all steps of the analytical process. In the quantitative analysis of two analytes that can potentially undergo inter-conversion, it is essential to minimize the conversion during every step of the bioanalytical procedure. An important example is the study of HMG-CoA reductase inhibitors such as Pravastatin that readily undergo conversion between their δ-hydroxy acid and lactone forms. The adjustment of pH is critical to accurately measure the active drug. A narrow pH range of 4-5 stabilizes both acid and its lactone. Increasing pH results in base-catalyzed hydrolysis of the lactone while decreasing the pH facilitates dehydration of the acid. The ratio of acid and lactone affects the equilibrium. QC samples should include varying proportions of drug and inter-converting metabolite. A small amount of conversion of a major component can easily be detected from its impact on a minor component. Other examples included acyl glucuronides and geometric (E/Z) isomers.[55]

To be assured of specificity, all possible isomeric standards should be prepared and resolved on the basis of either their chromatographic or spectral properties. ^{1}H and ^{19}F-NMR, as well as LC/NMR have provided great insight into complex isomeric metabolite mixtures, including the stability and rearrangement of acyl glucuronides. Ester glucuronides may undergo internal acyl migration reactions, resulting in the formation of new positional isomers with both α- and β-anomers. Inter-converting isomers were resolved using HPLC prior to introduction into a flow probe NMR spectrometer. The reaction kinetics were determined and compared to a simulated model.[56] Having an adequate understanding of the liability of new chemical entities to produce a reactive acyl glucuronide is aided by LC/NMR. In a separate study, S-naproxen glucuronide isomers formed in a biosystem were studied and showed that the initial intramolecular rearrangement of the 1- to 2-isomer can be reversible and that, by acyl migration of the corresponding β-1-O-acyl glucuronide, α-1-O-acyl isomers can occur under physiological conditions.[57]

For less concentrated samples in complex biological matrices, LC/MS is needed. The decomposition of α- and β-anomers of ifetroban 1-O-acyl glucuronide in an aqueous solution was studied using LC/MS. Similar to lactones, best stability was observed at a pH of 4 while hydrolysis occurred at both acidic and basic pH. First-order rate constants, as a function of pH, were established according to a model that included both hydrolysis and acyl migration processes. Concern regarding exposure to methanol prompted an investigation of trans-esterification in this solvent. Methanolysis was rapid and, at a given pH and temperature, increased with methanol concentration.[58]

On-line restricted access media extraction and column switching LC/MS/MS was used to measure entacapone glucuronide concentrations in rat plasma. An extraction column was used to purify the E- and Z-isomers of entacapone glucuronides prior to reversed-phase HPLC analysis. The aglycone fragment from the deprotonated molecule was used for MRM quantitation, permitting detection of 1 ng/mL of each isomer.[59]

R- and S-enantiomers of the β-agonist albuterol and its 4-O-sulfate metabolite were measured in human plasma and urine. Samples were prepared using a 96-well SPE and analyzed using LC/MS/MS. The assay could measure as little as 100 pg/mL of albuterol in plasma and 5 ng/mL of the enantiomers and its 4-O-sulfate metabolite. By eliminating the urine and sulfate metabolite analysis, the enantiomers could be measured in plasma to 25 pg/mL. By further elimination of the chiral separation, an LC/MS/MS method was developed for racemic drug with a three min per sample throughput and an LLOQ of 5 pg/mL.[60] The three assays demonstrate necessary compromises to achieve a multiple component assay. Resolution, speed, sensitivity and applications were sacrificed to provide a single assay.

Once the metabolite(s) are stabilized and available for analysis, a decision regarding how they will be determined must be reached. One of the most unpredictable problems in method development is dealing with phase II conjugates. While it is preferred to measure metabolites directly, using two assays to determine free and total plasma concentrations may be necessary. Sometimes, the lack of an analytical reference standard may require the use of an indirect method or biological systems as generators. For many studies, these procedures are effective means to measure metabolites for which there are no analytical reference standards.

If procedures are unavailable to rapidly generate conjugate reference standards, selective cleavage procedures are a first alternative. A general procedure to cleave drug conjugates using immobilized β-glucuronidase and arylsulfatase was developed. β-Glucuronidase and arylsulfatase were purified and co-immobilized on an agarose gel matrix and packed into columns. Morphine conjugates could be completely hydrolyzed within 60 min in urine.[61] In a separate study, HPLC/FL was used to measure concentrations of moxisylyte metabolites in human plasma and urine. Deacetylmoxisylyte glucuronide was hydrolyzed enzymatically using β-glucuronidase and quantified by difference.[62]

Two HPLC assays were used to determine salicylamide and its five metabolites in serum, urine and saliva. One method measured salicylamide, its glucuronide and sulfate metabolites while a second method measured the hydroxylated metabolite (gentisamide), and its glucuronide and sulfate metabolites. Due to the absence of analytical reference standards for the sulfate and glucuronide conjugates, calibration curves from partial enzymatic hydrolysis were used. The decrease of the conjugate peaks and the concomitant increase of free salicylamide or gentisamide concentrations was used to determine peak area ratio vs. concentration relations.[63]

The effect of probenecid on the pharmacokinetics of indomethacin was studied using an LC/MS/MS assay that measures indomethacin, its O-demethylated, acyl and 1-O-glucuronide metabolites in human plasma and urine. Using HPLC, all could be measured directly without enzymatic deglucuronidation. Glucuronide conjugates were isolated by preparative HPLC from urine and calibration curves constructed by enzymatic deconjugation of samples containing different concentrations of the isolated indomethacin acyl glucuronide, O-demethylated acyl glucuronide and ether glucuronide.[64]

A HPLC method was developed to quantify propranolol and its metabolites in urine, bile and serum. To quantify free and total (conjugates), analysis was performed before and after incubation with β-glucuronidase-arylsulfatase. Fractionation of basic, neutral and acidic metabolites was performed prior to HPLC/FL analysis to low ng/mL levels.[65]

Enzymatic hydrolysis with a mixture of β-glucuronidase and arylsulfatase was used to determine N-hydroxymexiletine glucuronide enantiomers in plasma. The glucuronide was determined as the quantity of mexiletine released by hydrolysis. Liquid-liquid extraction was used prior to o-phthalaldehyde derivatization and HPLC/FL analysis.[66]

Less selective procedures are sometimes required, particularly if the metabolite is a poor substrate or has already undergone rearrangement. The reactivity of acyl glucuronide conjugates was investigated using LC/MS/MS and a combination of radiolabeled material and synthetic standards. In the absence of standards, conjugate concentrations were measured indirectly by hydrolysis.[67]

An HPLC/FL assay was developed to simultaneously measure diflunisal, the phenolic and acyl glucuronides and sulfate conjugate in human urine. Chromatographically purified standards of the conjugates were isolated from human urine. Urine samples contained rearrangement products of the 1-O-acyl glucuronide, making quantitation of the acyl glucuronide inaccurate. The acyl glucuronide was therefore indirectly determined using alkaline hydrolysis.[68] An HPLC method for measuring gemfibrozil and four metabolites in human plasma and urine and rat tissue homogenates was developed. The acyl glucuronides and covalently bound protein adducts were determined following alkaline hydrolysis.[69]

R- and S-enantiomers of ketoprofen and their acyl glucuronide conjugates were determined in plasma and dialysate using chiral HPLC. In plasma, the glucuronides could not be directly quantified due to matrix interference. Therefore, the glucuronides were isolated using reversed-phase HPLC and quantified following alkaline hydrolysis.[70] These examples demonstrate the utility of selective enzymatic or chemical processes for indirect determinations.

Even having limited quantities of an impure conjugate can help. High purity standards of phase II conjugates are needed for GLP studies and are usually generated by synthesis.[71] When unavailable, other approaches have been tried. In an effort to prepare milligram quantities of analytical reference standards, β-D-glucuronide synthesis using UDP-glucuronyl transferase was optimized. The right amount of enzyme and co-factor were important, with too much enzyme affecting yield. UDPGA concentration was optimized at twice that of substrate. Choice of co-solvent was also important.[72] Diastereomeric β-D-glucuronides of the bronchodilator procaterol were prepared using immobilized rabbit liver microsomes. Microsomes from a phenobarbital-induced rabbit were immobilized using cyanogen bromide-activated Sepharose beads.[73] Recombinant sulfotransferases, UDP-glycosyltransferases and glutathione transferases are available today. These baculovirus expressed enzymes have quite different substrate specificity.

As generating systems, pooled isozymes or some knowledge of substrate-isozyme preference is needed.[74]

Radiolabeled metabolites from a biosynthesizer can serve as calibrators or a means to validate other methods. Quantitation of intracellular triphosphate of emtricitabine (FTC) in peripheral blood mononuclear cells from HIV+ patients used anion-exchange SPE to separate mono, di and tri-phosphate (TP) nucleotides. The triphosphate was digested using alkaline phosphatase and analyzed by HPLC/UV. Peripheral blood mononuclear cells from donors exposed to ^{3}H-labeled drug of known specific activity lead to the formation of intracellular FTC-TP that was quantified using anion-exchange HPLC/radiochemical detection. These levels of FTC-TP determined using radiochemical response served to calibrate the HPLC/UV assay.[75]

To directly determine metabolite concentrations without analytical reference standards requires an instrument that provides equivalent molar response for drug and all of its metabolites. Like radiochromatography, this instrument also needs to be sensitive and selective. Much effort has been made at developing an element selective detector that meets these criteria. Not surprisingly, most success has been made with compounds that possess a unique element or isotope not present in biological fluids.

Many drug candidates contain one or more fluorine atoms. Total fluoride concentration in plasma is approximately 5 uM with inorganic fluoride accounting for the majority.[76] Slightly higher concentrations appear in urine.[77] Some have taken advantage of the quantitative nature of NMR and the relatively low abundance of fluorine in biological fluids and tissues.[78] ^{1}H and ^{19}F-NMR was used to measure the rat urinary metabolites of 15 substituted phenols. Glucuronide or sulfate conjugates were noted with ortho-substituted phenols showing more glucuronidation than other isomers. NMR spectroscopy provided a rapid quantitative assessment of urinary metabolites from their relative NMR response, without needing reference standards. In another study, the metabolism of a fluorinated nucleoside reverse transcriptase inhibitor was studied by examining the ^{19}F-NMR spectrum of human urine.[79] The number and relative proportions of the drug metabolites were obtained from ^{19}F NMR spectra of whole human urine. Similar procedures have been used to study to elimination and metabolism of fluorinated nucleosides to 5-F-cytidine, 5-F-uracil and related degradants.

While capable of resolving minor structural differences, NMR sensitivity is limited and requires long acquisition times to enhance low level signals. Spectral resolution of individual metabolites is required. Chromatographic separation is possible using LC/NMR.[80] However, problems estimating peak volumes within the flow probe and differences in acquisition conditions preclude its use for quantitation. Cryogenic probes[81] afford a 3 to 4-fold improvement in detection sensitivity. Alternatively, microbore LC with smaller volume probes[82] has shown some promise to extend sensitivity. However, the complexity and expense of NMR makes it a methodology that will likely see limited use for multiple component or metabolite quantititation. Its best application may be to serve as a calibrator of more concentrated samples of metabolites that can subsequently be used with more sensitive instrumentation.

Detectors that provide universal response include evaporative light scattering and nitrogen chemilumiscence. Both have been used extensively to determine drug purity. Because of its more selective nature, chemiluminescent nitrogen detection (CLND) has seen more application to biological systems. For CLND, linear response from 25 to 6400 pmol of nitrogen for a set of structurally diverse N-containing compounds was noted. The response was independent of mobile-phase composition and required only a single external standard.[83] In a metabolism study, a ^{14}C-labeled N-containing compound was administered orally to rats. Plasma, urine and bile were profiled using a combination of mass spectrometry, radiochromatography and CLND.[84] Radiochromatographic and CLND profiles showed good quantitative correlation. Total mass balance also showed excellent agreement between the two approaches. Metabolite concentrations were significantly greater than background nitrogen levels. The wavelength range for chemiluminescence is very broad from 600 to 2800 nm, centering at 1200 nm. While element specific and very sensitive, unlike mass spectrometry, it is not isotope specific.

Mass spectrometers that employ quantitative combustion prior to isotopic analysis have also been developed. Inductively-coupled plasma mass spectrometry (ICP-MS) has been adapted to LC. For fluorine, detection sensitivity suffers from its high ionization potential. Due to its lower ionization potential, Cl sensitivity is better but is limited by chemical background. Single digit ppb levels can be achieved in water. However, in the presence of organic mobile phases, background increases significantly.

Urinary metabolites of diclofenac were monitored using LC/TOF/ICP-MS. Isotope selective detection of ^{35}Cl, ^{37}Cl and ^{32}S was achieved using ICP-MS while metabolites were identified using TOF-MS.[85] On-line chorine detection was achieved from rat urine at mid-ng (on-column) levels. Useful metabolic profiles were obtained from the chlorine chromatogram. Glucuronide acid and sulfate conjugates, mono- and dihydroxylated diclofenac and an N-acetylcysteinyl conjugate were identified. High level of endogenous sulfur-containing compounds made it difficult to detect drug-related metabolites from the sulfur chromatogram. The detection of ^{32}S is particularly helpful in detecting and measuring phase II conjugates such as sulfates, glutathione or related conjugates. For ^{32}S, not only endogenous sulfur-containing molecules complicate the S-chromatogram but interference from components that produce O_2^+. Using a dynamic reaction cell to convert S^+ into SO^+, which is interfered less then S^+, may help to reduce background levels.

For ICP-MS, detection limits for the halogens are related to their ionization potential. Because of its low ionization potential, metabolic studies of brominated drugs are easily achieved using LC/ICP-MS. The rat urinary metabolites of 4-bromoaniline were studied using a combined LC/ICP-MS and TOF-MS system. The majority of the LC effluent was directed to the ICP-MS where bromine and sulfur-containing metabolites were selectively detected, including a sulfate, several glucuronides, N-oxanilic acid and an N-acetylcysteinyl conjugate.[86] In another study, human and rat plasma were incubated with a Br-labeled form of bradykinin and metabolites monitored using LC/TOF/ICP-MS. Quantification of the bromine-containing metabolites was performed using ICP-MS.[87] In a third study, LC/TOF/ICP-MS was used to profile, identify and quantify rat urinary metabolites following administration of 2-bromo-4-trifluoromethylacetanilide. Bromine-containing metabolites were detected and quantified using ICP-MS. A ring hydroxyl-substituted metabolite, N-sulfate, N-hydroxylamine glucuronide, and N- and N-hydroxyglucuronides were also identified.[88]

Besides sensitivity issues related to ionization potential and background of the biological matrix, other practical considerations may limit the use of ICP-MS for metabolite quantitation. The composition of mobile phases is limited. Methanol is preferred to reduce plasma graphite contamination observed with acetonitrile. Organic mobile phases may produce significant background on some ion channels. Post-column make-up to the ICP may be needed to maintain response during gradient LC. Interface dead volume and absorption can degrade the separation. While useful for heavier elements, ICP is less sensitive to lighter elements that comprise most drug candidates.

An isotope-selective detector for LC based upon solvent drying and analyte transport in the particle-beam-interface coupled to a helium microwave chemical reaction interface mass spectrometer (CRIMS) was used to detect deuterium-labeled analytes. HD could be measured from 20 to 490 ng using $^{2}H_{3}$-labeled cortisol. The relative abundance of three labeled cortisol metabolites was determined in human urine.[89] In another study, hamster urine and bile was profiled after dosing with $^{2}H_{4}$-acetaminophen. Seventeen metabolites, of which 14 were conjugates, were quantified from a single chromatogram of urine. Sulfur-selective CRIMS detection assisted the identification of glutathione and related conjugates in bile.[90] In a third study, profiles of hydrolyzed urinary metabolites of $^{2}H_{3}$-cortisol in six human subjects were obtained using LC/CRIMS. The relative amounts of tetrahydrocortisol, tetrahydrocortisone, and cortolones were determined.[91] In a validation study, metabolites from incubations of ^{15}N, ^{14}C-buspirone with rat liver slices were analyzed by gradient LC/CRIMS and radioactivity counting. Metabolite response from CRIMS was comparable to that obtained by radioactivity counting.[92] All four studies illustrated the potential of LC/CRIMS to measure isotopically-labeled drug metabolites.

As with ICP-MS, drug candidates containing a unique element or isotope can be profiled and measured using CRIMS. The anti-psychotic drug clozapine forms adducts with glutathione. Clozapine contains chlorine and glutathione sulfur atoms that can be monitored using CRIMS. In vitro incubations generated conjugates as noted by the simultaneous appearance in both the Cl and S-specific chromatograms.[93]

In mass balance and metabolism studies, drug metabolites enriched with small amounts of ^{14}C are monitored. Tritium can be used but is normally avoided due to possible metabolic loss. Detection sensitivity is excellent since the natural abundance of this isotope is low (100 amol/mg carbon). Dosimetry calculations derived from animal tissue distribution studies define the maximum allowable human organ exposures to radiolabeled drug. MicroCurie doses are used to detect metabolites by liquid scintillation counting to low dpm/mL levels. Exposure to sensitive organs, low doses of radioactivity due to drug potency, radiolytic instability or inability to produce a compound of high specific activity and pharmacokinetic dilution complicate human mass balance studies.

Accelerator mass spectrometry (AMS) has been used for early drug metabolism studies in humans. AMS has several orders of magnitude better limits of detection than conventional scintillation counting.[94] As evident from its use to determine ^{14}C in depleted samples for radiocarbon dating, it can easily be used to measure ^{14}C in enriched drug and metabolite samples. Dosing drug candidates containing low nanoCuries ($< 1\ \mu Sv$) of radiolabeled drug reduces radiation hazards and, in some countries, are excluded from regulatory approval for radioactive substances. Analysis using AMS affords mass balance and metabolism information without the limitations of scintillation counting. However, AMS requires extensive sample preparation to afford attomole limits of detection. For metabolism studies, individual LC fractions are needed and contamination is a problem. Planning, time, expertise and money are needed to successfully complete a study.

A collaboration between Novartis and the Center Biological Accelerator Mass Spectrometry (CBAMS) illustrated that AMS could be used to measure plasma, urine and fecal concentrations from rats given a very low dose of a radiolabeled drug candidate. Mean peak plasma concentrations (total radioactivity) of 2.21 dpm/mL and levels of 0.18

dpm/mL at 120 h were determined. Fecal elimination was 87% while urinary elimination was only 1.1%, similar to a previous study in rats. Due to its lack of urinary clearance, individual metabolites were not quantified.[95] In a collaboration between Glaxo and CBAMS, the metabolism and mass balance of ^{14}C-fluticasone propionate was studied in male Han Wistar rats. Drug was administered intravenously to male rats (11.5 μCi/kg) and samples of blood, urine and feces were collected up to 96 h. A second group was administered a lower dose of radiolabeled FP (18.1 nCi/kg). Samples from the high dose group were diluted 900-fold prior to AMS analysis. LSC was performed without dilution. A good correlation was observed between LSC and AMS results.[96] The paper also reported collaboration with Pfizer comparing LSC and AMS results for the analysis of ^{14}C-fluconazole in human plasma. Following subtraction of response from the blank, AMS results showed excellent agreement with LSC. The authors noted several possible reasons for AMS response from blanks, including memory effects due to saturation of the gas ionization source or the ingestion of food containing ^{14}C.

Of even greater value may be the use of AMS to study DNA adducts using AMS. DNA adducts following tamoxifen dosing to rats could be detected in the liver at levels of 0.18 pmol/mg DNA. Differentiation of responsible metabolites could provide great insight into the relative safety across species.[97]

AMS instrumentation is large, complex and expensive. Collaboration with physicists at laboratories such as the Center for Accelerator Mass Spectrometry at Lawrence Livermore National Laboratory and CBMAS has provided most of the biological examples of AMS to date. Other AMS facilities exist to provide service for geological, archeological, environmental and other life science studies that need carbon dating or the analysis of other isotopes.[98]

A low energy AMS for detecting ^{3}H and ^{14}C that accepts chromatographic effluent is being developed by Newton Scientific and MIT. The compact and relatively inexpensive system includes a gas-fed ion source and a 1 MV tandem accelerator that incorporates a foil stripper.[99] If successful, more routine applications in drug metabolism studies would be possible. Presently, the cost and complexity of AMS studies prohibits broad application.

5. CONCLUDING REMARKS

Current practices in today's bioanalytical laboratories are dramatically different than what was commonly employed only 5 years ago. The introduction and widespread implementation of API-based LC/MS/MS has revolutionized quantitative bioanalysis. Bioanalytical scientists have been able to drastically shorten chromatographic run times as the result of the specificity/selectivity provided by LC/MS/MS techniques. This has given impetus to the subsequent development and implementation of high-throughput sample preparation techniques, including high-flow on-line extraction followed by direct injection of biological samples. The combination of rapid chromatography and fast extraction has reduced typical total sample analysis times to less than two minutes per sample, resulting in a significant increase in sample throughput in most modern laboratories. In spite of the great specificity/selectivity of LC/MS/MS, there is still a need for some chromatographic separation in order to eliminate potential interferences from components that are related to the drug or are found naturally in a biological matrix.

Without the option of returning to longer chromatographic run times in the face of ever increasing sample volumes, bioanalysts must continue to be judicious in their choice of chromatographic and mass spectrometric parameters in order to maximize assay performance.

Along with these dramatic advances in speed of drug analysis has come significant interest in generating quantitative metabolic information much earlier than ever before. The scarcity of appropriate metabolic reference standards is presenting bioanalytical scientists with growing challenges as they balance supporting internal drug development programs with compliance with emerging regulatory guidelines. Questions regarding metabolite exposure often occur in parallel to their identification. Routinely, analytical methods are expected to determine the concentrations of drug and all *significant* metabolites. New problems arise that require compromise, non-standard approaches or the use of new technology. Complex or unstable metabolites may require unique conditions for sample collection or analysis, while structurally diverse mixtures of metabolites often require class-specific approaches to analysis. Enrichment is needed during sample preparation; however, too selective an extraction can result in low recovery of some metabolites. As discussed, the availability of internal standards greatly impacts an assay's performance and, for some assays, more than one is needed. New instrumentation such as ICP-MS, CRIMS or AMS has the ability to provide information without the use of reference standards. Established technologies such as NMR and LC/MS, when combined, continue to provide powerful knowledge of both metabolite structure and quantity. New sources for calibration standards allow analysts to quickly make preliminary determinations and establish assay conditions. Internal drug development decision can go forward using these approaches. However, more established procedures are often required for regulatory submissions.

References

1. Powell ML, Jemal M. Rapid chromatography coupled with direct injection LC/MS/MS. American Pharmaceutical Review. 2001; 4: 63-69.
2. Garland WA, Powell ML. Quantitative selected ion monitoring (QSIM) of drugs and/or drug metabolites in biological matrices. J. Chromatogr. Sci. 1981; 19: 392-434.
3. Shah VP, Midha KK, Dighe S, McGilveray IJ, Skelly JP, Yacobi A, et al. Analytical methods validation: bioavailability, bioequivalence and pharamcokinetic studies. Pharmaceutical Research. 1992; 9(4): 588-92.
4. Shah VP, Midha KK, Findlay JWA, Hill HM, Hulse JD, McGilveray IJ et al. Bioanalytical method validation - revisit with a decade of progress: Pharmaceutical Research, 2000; 17(12): 1551-57.
5. Guidance for Industry. Bioanalytical Method Validation, U.S. Food and Drug Administration, Center for Drug Evaluation and Research; May 2001.
6. Jemal M. High-throughput quantitative bioanalysis by LC/MS/MS. Biomed. Chromatogr. 2000; 14: 422-29.
7. Ayrton J, Dear GJ, Leavens WJ, Mallett DN, Plumb RS. The use of turbulent flow chromatography/mass spectrometry for the rapid, direct analysis of a novel pharmaceutical compound in plasma. Rapid Commun. Mass Spectrom. 1997; 11: 1953-58.
8. Ayrton J, Dear GJ, Leavens WJ, Mallett DN, Plumb RS. Optimisation and routine use of generic ultra-high flow-rate liquid chromatography with mass spectrometric detection for the direct on-line analysis of pharmaceuticals in plasma. J. Chromatogr. 1998; 828: 199-207.
9. Jemal M, Xia Y-Q, Whigan DB. The use of high-flow high performance liquid chromatography coupled with positive and negative ion electrospray tandem mass spectrometry for quantitative bioanalysis via direct injection of the plasma/serum samples. Rapid Commun. Mass Spectrom. 1998; 12: 1389-99.

10. Jemal M, Ouyang Z, Xia Y-Q, Powell ML. A versatile system of high-flow high performance liquid chromatography with tandem mass spectrometry for rapid direct-injection analysis of plasma samples for quantitation of a β-lactam drug candidate and its open-ring biotransformation product. Rapid Commun. Mass Spectrom. 1999; 13: 1462-71.
11. Xia Y-Q, Whigan DB, Powell ML, Jemal M. Ternary-column system for high-throughput direct-injection bioanalysis by liquid chromatography/tandem mass spectrometry. Rapid Commun. Mass Spectrom. 2000; 14: 105-11.
12. Jemal M, Xia Y-Q. The need for adequate chromatographic separation in the quantitative determination of drugs in biological samples by high performance liquid chromatography with tandem mass spectrometry. Rapid Commun. Mass Spectrom. 1999; 13(2): 97-106.
13. Naidong W, Lee JW, Jiang X, Wehling M, Hulse JD, Lin PP. Simultaneous assay of morphine, morphine-3-glucuronide and morphine-6-glucuronide in human plasma using normal-phase liquid chromatography-tandem mass spectrometry with a silica column and an aqueous organic mobile phase. J. Chromatogr. B 1999; 735: 255-69.
14. Romanyshyn L, Tiller PR, Hop CECA. Bioanalytical applications of "fast chromatography" to high-throughput liquid chromatography/tandem mass spectrometric quantitation. Rapid Commun. Mass Spectrom. 2000; 14: 1662-68.
15. Jemal M, Ouyang Z. The need for chromatographic and mass resolution in liquid chromatography/tandem mass spectrometric methods used for quantitation of lactones and corresponding hydroxy acids in biolgical samples. Rapid Commun. Mass Spectrom. 2000; 14: 1757-65.
16. Clarke SD, Hill HM, Noctor TAG, Thomas D. Matrix-related modification of ionization in bioanalytical liquid chromatography-atmospheric pressure ionization tandem mass spectrometry. Pharmaceutical Sciences 1996; 2: 203-07.
17. Buhrman DL, Price PI, Rudewicz PJ. Quantitation of SR 27417 in human plasma using electrospray liquid chromatography-tandem mass spectrometry: a study of ion suppression. J. Am. Soc. Mass Spectrom. 1996; 7(11): 1099-1105.
18. Fu I, Woolf EJ, Matuszewski BK. Effect of the sample matrix on the determination of indinavir in human urine by HPLC with turbo ion spray tandem mass spectrometric detection. J. Pharm. Biomed. Anal. 1998; 18: 347-57.
19. Matuszewski BK, Constanzer ML, Chavez-Eng CM. Matrix effect in quantitative LC/MS/MS analyses of biological fluids: a method for determination of finasteride in human plasma at picogram per milliliter concentrations. Anal. Chem. 1998; 70: 882-89.
20. Bonfiglio R, King RC, Olah TV, Merkle K. The effects of sample preparation methods on the variability of the electrospray ionization response for model drug compounds. Rapid Commun. Mass Spectrom. 1999; 13: 1175-85.
21. Choi BK, Hercules DM, Gusev AI. Effect of liquid chromatography separation of complex matrices on liquid chromatography-tandem mass spectrometry signal suppression. J. Chromatogr. A 2001; 907: 337-42.
22. Guidance for industry. Carcinogenicity study protocol submissions, U.S. Food and Drug Administration, Center for Drug Evaluation and Research; May 2002.
23. Baille TA, Cayen MN, Fouda H, Gerson R, Green JD, Grossman SJ, et al. Drug metabolites in safety testing. Toxicol. Appl. Pharmacol. 2002; 182(3): 188-96.
24. Berman J, Halm KA, Adkison K, Shaffer J. Simultaneous pharmacokinetic screening of a mixture of compounds in the dog using API LC/MS/MS analysis for increased throughput. J. Med. Chem. 1997; 40(6): 827-29.
25. Olah TV, Mcloughlin DA, Gilbert JD. The simultaneous determination of mixtures of drug candidates by liquid chromatography/atmospheric pressure chemical ionization mass spectrometry as an in vivo drug screening procedure. Rapid Commun. Mass Spectrom. 1997; 11(1): 17-23.
26. Anderson RL. Practical Statistics for Analytical Chemists, Van Nostrand Reinhold: New York: 1987.
27. Yang W, Jiang T, Acosta D, Davis PJ. Microbial models of mammalian metabolism: involvement of cytochrome P450 in the N-demethylation of N-methylcarbazole by Cunninghamella echinulata. Xenobiotica 1993; 23(9): 973-82.
28. Mutlib AE, Chen H, Nemeth GA, Markwalder JA, Seitz SP, Gan LS et al. Identification and characterization of efavirenz metabolites by liquid chromatography/mass spectrometry and high field NMR: species differences in the metabolism of efavirenz. Drug Metab. Disp. 1999; 27(11): 1319-33.
29. Chen B-C, Sundeen JE, Guo P, Bednarz MS, Hangeland JJ, Ahmed SZ, Jemal M. Synthesis of deuterium-labeled atorvastatin and its metabolites for use as internal standards in a LC/MS/MS method

developed for quantitation of the drug and its metabolites in human serum. J. Label. Comp. Radiopharm. 2000; 43(3): 261-70.

30. Mutlib AE, Diamond S, Shockcor J, Way R, Nemeth G, Gan L, Christ, DD. Mass spectrometric and NMR characterization of metabolites of roxifiban, a potent and selective antagonist of the platelet glycoprotein IIb/IIIa receptor. Xenobiotica 2000; 30(11): 1091-1110.
31. Zheng JJ, Lynch ED, Unger SE. Comparison of SPE and fast LC to eliminate mass spectrometric matrix effects from microsomal incubation products. J. Pharm. Biomed. Anal. 2002; 28(2): 279-285.
32. Foltz RL, Edom RW. Problems and solutions in quantitative determination of drugs and metabolites in physiological specimens by LC-MS/MS. J. Mass Spectrom. Soc. Japan 1998; 46(3): 235-39.
33. Ladds G, Wilson K, Burnett D. Automated liquid chromatographic method for the determination of paracetamol and six metabolites in human urine. J. Chromatogr. 1987; 414(2): 355-64.
34. Cooper TB, Suckow RF, Glassman A. Determination of bupropion and its major basic metabolites in plasma by liquid chromatography with dual-wavelength ultraviolet detection. J. Pharm. Sci. 1984; 73(8): 1104-07.
35. Wei Y, Nygard GA, Khalil SKW. HPLC method for the separation and quantification of the enantiomers of hydroxychloroquine and its three major metabolites. J. Liq. Chromatogr. 1994; 17(16): 3479-90.
36. Brown R, Bain R, Jordan VC. Determination of tamoxifen and metabolites in human serum by high-performance liquid chromatography with post-column fluorescence activation. J. Chromatogr. 1983; 272: 351-58.
37. Fried KM, Wainer IW. Direct determination of tamoxifen and its four major metabolites in plasma using coupled column high-performance liquid chromatography. Journal of Chromatogr. B: Biomed. Appl. 1994; 655: 261-68.
38. Wu J-T, Zeng H, Deng Y, Unger SE. High-speed liquid chromatography/tandem mass spectrometry using a monolithic column for high-throughput bioanalysis. Rapid Commun. Mass Spectrom. 2001; 15(13): 1113-19.
39. He H, Sadeque A, Erve JCL, Wood AJJ, Hachey DL. Quantitation of loperamide and N-demethyl-loperamide in human plasma using electrospray ionization with selected reaction ion monitoring liquid chromatography-mass spectrometry. J. Chromatogr. B: Biomed. Appl. 2000; 744(2): 323-31.
40. Srinivasan K, Wang P, Eley AT, White CA, Bartlett, MG. Liquid chromatography-tandem mass spectrometry analysis of cocaine and its metabolites from blood, amniotic fluid, placental and fetal tissues: study of the metabolism and distribution of cocaine in pregnant rats. J. Chromatogr. B: Biomed. Appl. 2000; 745(2): 287-303.
41. Kato K, Jingu S, Ogawa N, Higuchi S. Determination of pibutidine metabolites in human plasma by LC-MS/MS. J. Pharm. Biomed. Anal. 2000; 24(2): 237-49.
42. Liu Z, Short J, Rose A, Ren S, Contel N, Grossman S, Unger S. The simultaneous determination of diazepam and its three metabolites in dog plasma by high-performance liquid chromatography with mass spectroscopy detection. J. Pharm. Biomed. Anal. 2001; 26(2): 321-30.
43. Bramer SL, Tata PNV, Vengurlekar SS, Brisson JH. Method for the quantitative analysis of cilostazol and its metabolites in human plasma using LC/MS/MS. J. Pharm. Biomed. Anal. 2001; 26(4): 637-50.
44. Shi G, Wu J-T, Li Y, Geleziunas R, Gallagher K, Emm T, Unger S. Novel direct detection method for quantitative determination of intracellular nucleoside triphosphates using weak anion exchange liquid chromatography/tandem mass spectrometry. Rapid Commun. Mass Spectrom. 2002; 16(11): 1092-99.
45. Andersson LI. Molecular imprinting for drug bioanalysis. A review on the application of imprinted polymers to solid-phase extraction and binding assay. J. Chromatogr., B: Biomed. Appl. 2000; 739: 163-73.
46. Venn RF, Goody RJ. Synthesis and properties of molecular imprints of darifenacin - does molecular imprinting have a future in ultra-trace bioanalysis? Methodol. Surv. Bioanal. Drugs 1998; 25: 13-20.
47. Cai J, Henion J. Online immunoaffinity extraction-coupled column capillary liquid chromatography/tandem mass spectrometry: Trace analysis of LSD analogs and metabolites in human urine. Anal. Chem. 1996; 68(1): 72-78.
48. Clarke NJ, Tomlinson AJ, Ohyagi Y, Younkin S, Naylor S. Detection and quantitation of cellularly derived amyloid β peptides by immunoprecipitation-HPLC-MS. FEBS Letters 1998; 430(3): 419-23.
49. Mallikaarjun S, Wood JH, Karnes, HT. High-performance liquid chromatographic method for the determination of salicylic acid and its metabolites in urine by direct injection. J. Chromatogr. 1989; 493(1): 93-104.
50. Sadagopan N, Cohen L, Roberts B, Collard W, Omer C. Liquid chromatography-tandem mass spectrometric quantitation of cyclophosphamide and its hydroxy metabolite in plasma and tissue for determination of tissue distribution. J. Chromatogr. B: Biomed. Appl. 2001; 759(2): 277-84.

51. Jemal M, Khan S, Teitz DS, McCafferty, JA, Hawthorne DJ. LC/MS/MS determination of omapratrilat, a sulfydryl-containing vasopeptidase inhibitor, and its sulfhydryl- and thioether-containing metabolites in human plasma. Anal. Chem. 2001; 73(22): 5450-456.
52. Singh G, Gutierrez A, Xu K, Blair IA. Liquid chromatography/electron capture atmospheric pressure chemical ionization mass spectrometry: analysis of pentafluorobenzyl derivatives of biomolecules and drugs in the attomole range. Anal. Chem. 2000; 72(14): 3007-013.
53. Grosse C.M, Davis I M, Arrendale RF, Jersey J, Amin J. Determination of remifentanil in human blood by liquid-liquid extraction and capillary GC-HRMS-SIM using a deuterated internal standard. J. Pharm. Biomed. Anal. 1994; 12(2): 195-203.
54. Selinger K, Lanzo C, Sekut A. Determination of remifentanil in human and dog blood by HPLC with UV detection. J. Pharm. Biomed. Anal. 1994; 12(2): 243-8.
55. Jemal M, Xia Y-Q. Bioanalytical method validation design for the simultaneous quantitation of analytes that may undergo interconversion during analysis. J. Pharm. Biomed. Anal. 2000; 22(5): 813-27.
56. Sidelmann UG, Hansen SH, Gavaghan C, Carless HAJ, Farrant RD, Lindon JC, Wilson ID, Nicholson JK. Measurement of internal acyl migration reaction kinetics using directly coupled HPLC-NMR: application for the positional isomers of synthetic (2-fluorobenzoyl)-D-glucopyranuronic acid. Anal. Chem. 1996; 68(15): 2564-72.
57. Corcoran O, Mortensen RW, Hansen SH, Troke J, Nicholson JK. HPLC/1H NMR spectroscopic studies of the reactive α-1-O-acyl isomer formed during acyl migration of S-naproxen β-1-O-acyl glucuronide. Chem. Res. Toxicol. 2001; 14(10): 1363-70.
58. Khan, Sanaullah, Teitz, Deborah S, Jemal, Mohammed. Kinetic analysis by HPLC-electrospray mass spectrometry of the pH-dependent acyl migration and solvolysis as the decomposition pathways of ifetroban 1-O-acyl glucuronide. Anal. Chem. 1998; 70(8): 1622-28.
59. Keski-Hynnila H, Raanaa K, Forsberg M, Mannisto P, Taskinen J, Kostiainen R. Quantitation of entacapone glucuronide in rat plasma by on-line coupled restricted access media column and liquid chromatography-tandem mass spectrometry. J. Chromatogr. B: Biomed. Appl. 2001; 759(2): 227-36.
60. Joyce KB, Jones AE, Scott RJ, Biddlecombe RA, Pleasance S. Determination the enantiomers of salbutamol and its 4-O-sulfate metabolites in biological matrixes by chiral liquid chromatography tandem mass spectrometry. Rapid Commun. Mass Spectrom. 1998; 12(23): 1899-1910.
61. Toennes SW, Maurer HH. Efficient cleavage of conjugates of drugs or poisons by immobilized β-glucuronidase and arylsulfatase in columns. Clinical Chemistry 1999; 45(12): 2173-82.
62. Marquer C, Bressolle F. High-performance liquid chromatographic determination of the conjugate metabolites of moxisylyte in human plasma and urine. J. Chromatogr. B: Biomed. Appl. 1997; 691(2): 389-96.
63. Morris ME, Levy G. Determination of salicylamide and five metabolites in biological fluids by high-performance liquid chromatography. J. Pharm. Sci. 1983; 72(6): 612-17.
64. Vree TB, van den Biggelaar-Martea M, Verwey-van Wissen CPWGM. Determination of indomethacin, its metabolites and their glucuronides in human plasma and urine by means of direct gradient high-performance liquid chromatographic analysis. Preliminary pharmacokinetics and effect of probenecid. J. Chromatogr., Biomed. Appl. 1993; 616(2): 271-82.
65. Kwong EC, Shen DD. Versatile isocratic high-performance liquid chromatographic assay for propranolol and its basic, neutral and acidic metabolites in biological fluids. J. Chromatogr. 1987; 414(2): 365-79.
66. Lanchote VL, Santos VJ, Cesarino EJ, Dreossi SAC, Mere Y Jr, Santos SRCJ. Enantioselective analysis of N-Hydroxymexiletine glucuronide in human plasma for pharmacokinetic studies.. Chirality 1999; 11(2): 85-90.
67. Rindgen D, Grotz D, Clarke NJ, Cox KA. The application of HPLC/tandem mass spectrometry for the assessment of acyl glucuronide metabolite formation in in vitro and in vivo systems in a drug discovery environment. American Pharmaceutical Review 2001; 4(4): 52-58.
68. Loewen GR, Macdonald JI, Verbeeck RK. High-performance liquid chromatographic method for the simultaneous quantitation of diflunisal and its glucuronide and sulfate conjugates in human urine. J. Pharm. Sci. 1989; 78(3): 250-55.
69. Hermening A, Grafe AK, Baktir G, Mutschler E, Spahn-Langguth H. Gemfibrozil and its oxidative metabolites: quantification of aglycones, acyl glucuronides, and covalent adducts in samples from preclinical and clinical kinetic studies. J. Chromatogr. B: Biomed. Appl. 2000; 741(2): 129-44.
70. Grubb NG, Rudy DW, Hall SD Stereoselective high-performance liquid chromatographic analysis of ketoprofen and its acyl glucuronides in chronic renal insufficiency. J. Chromatogr. B: Biomed. Appl. 1996; 678(2): 237-44.

71. Kaspersen FM, van Boeckel CAA. A review of the methods of chemical synthesis of sulphate and glucuronide conjugates. Xenobiotica 1987; 12: 1451-71.
72. Stevenson DE, Hubl U. Optimization of β-D-glucuronide synthesis using UDP-glucuronyl transferase. Enzyme and Microbial Technology 1999; 24(7): 388-96.
73. Woolf T F, Chang T. Preparation of diastereomeric β-D-glucuronides of the bronchodilator procaterol using immobilized rabbit liver microsomal enzymes. Eur. J. Drug Metab. Pharmacokinet. 1989; 14(2): 111-16.
74. Burchell B, Brierley CH, Rance D. Specificity of human UDP-glucuronosyltransferases and xenobiotic glucuronidation. Life Sciences 1995; 57(20): 1819-95.
75. Darque A, Valette G, Rousseau F, Wang LH, Sommadossi JP, Zhou XJ. Quantitation of intracellular triphosphate of emtricitabine in peripheral blood mononuclear cells from human immunodeficiency virus-infected patients. Antimicrobial Agents and Chemotherapy 1999; 43(9): 2245-50.
76. Lentner C, Ed. Geigy Scientific Tables Vol. 3; Ciba-Geigy: 1984.
77. Lentner C, Ed, Geigy Scientific Tables Vol. 1; Ciba-Geigy: 1984.
78. Bollard ME, Holmes E, Blackledge CA, Lindon JC, Wilson ID, Nicholson JK. ^{1}H and ^{19}F-NMR spectroscopic studies on the metabolism and urinary excretion of mono- and disubstituted phenols in the rat. Xenobiotica 1996; 26(3): 255-73.
79. Shockcor JP, Unger SE, Savina P, Nicholson JK, Lindon JC. Application of directly coupled LC-NMR-MS to the structural elucidation of metabolites of the HIV-1 reverse-transcriptase inhibitor BW935U83. J. Chromatogr. B: Biomed. Appl. 2000; 748(1): 269-79.
80. Shockcor JP, Unger SE, Wilson ID, Foxall PJD, Nicholson JK, Lindon JC. Combined HPLC, NMR spectroscopy, and ion-trap mass spectrometry with application to the detection and characterization of xenobiotic and endogenous metabolites in human urine. Anal. Chem. 1996; 68(24): 4431-35.
81. Russell DJ, Hadden CE, Martin GE, Gibson AA, Zens AP, Carolan JLA. Comparison of inverse-detected heteronuclear NMR performance: conventional vs. cryogenic microprobe performance. J. Nat. Prod. 2000; 63(8): 1047-49.
82. Lacey ME, Tan ZJ, Webb AG, Sweedler JV. Union of capillary high-performance liquid chromatography and microcoil nuclear magnetic resonance spectroscopy applied to the separation and identification of terpenoids. J. Chromatogr. A. 2001; 922(1-2): 139-49.
83. Taylor EW, Qian MG, Dollinger GD. Simultaneous online characterization of small organic molecules derived from combinatorial libraries for identity, quantity, and purity by reversed-phase HPLC with chemiluminescent nitrogen, UV, and mass spectrometric detection. Anal. Chem. 1998; 70(16): 3339-47.
84. Taylor EW, Jia W, Bush M, Dollinger GD. Accelerating the drug optimization process: identification, structure elucidation, and quantification of in vivo metabolites using stable isotopes with LC/MS^{n} and the chemiluminescent nitrogen detector. Anal. Chem. 2002; 74(13): 3232-38.
85. Corcoran O, Nicholson JK, Lenz EM, Abou-Shakra F, Castro-Perez J, Sage AB, Wilson ID. Directly coupled liquid chromatography with inductively coupled plasma mass spectrometry and orthogonal acceleration time-of-flight mass spectrometry for the identification of drug metabolites in urine: application to diclofenac using chlorine and sulfur detection. Rapid Commun. Mass Spectrom. 2000; 14(24): 2377-84.
86. Abou-Shakra FR, Sage AB, Castro-Perez J, Nicholson JK, Lindon JC, Scarfe GB, Wilson ID. High-performance liquid chromatography-UV diode array, inductively coupled plasma mass spectrometry (ICPMS) and orthogonal acceleration time-of-flight mass spectrometry (oa-TOFMS) applied to the simultaneous detection and identification of metabolites of 4-bromoaniline in rat urine. Chromatographia 2002; 55(Suppl.): S9-S13.
87. Marshall P, Heudi O, McKeown S, Amour A, Abou-Shakra F. Study of bradykinin metabolism in human and rat plasma by liquid chromatography with inductively coupled plasma mass spectrometry and orthogonal acceleration time-of-flight mass spectrometry. Rapid Commun. Mass Spectrom. 2002; 16(3): 220-28.
88. Nicholson JK, Lindon JC, Scarfe GB, Wilson ID, Abou-Shakra F, Sage AB, Castro-Perez J. High-performance liquid chromatography linked to inductively coupled plasma mass spectrometry and orthogonal acceleration time-of-flight mass spectrometry for the simultaneous detection and identification of metabolites of 2-bromo-4-trifluoromethyl-[13C]-acetanilide in rat rrine. Anal. Chem. 2001; 73(7): 1491-94.
89. Teffera Y, Abramson FP, McLean M, Vestal M. Development of an isotope-elective high-performance liquid chromatography detector using chemical-reaction-interface mass spectrometry: Application to deuterated cortisol metabolites in urine. J. Chromatogr. B Biomed. Appl. 1993; 620(1): 89-96.

90. Teffera Y, Abramson F. Application of high-performance liquid chromatography/chemical reaction interface mass spectrometry for the analysis of conjugated metabolites: a demonstration using deuterated acetaminophen. Biol. Mass Spectrom. 1994; 23(12): 776-83.
91. Yergey AL, Teffera Y, Esteban NV, Abramson FP. Direct determination of human urinary cortisol metabolites by HPLC/CRIMS. Steroids 1995; 60(3): 295-8.
92. Goldthwaite CA Jr, Hsieh F-Y, Womble SW, Nobes BJ, Blair IA, Klunk LJ, Mayol RF. Liquid chromatography/chemical reaction interface mass spectrometry as an alternative to radioisotopes for quantitative drug metabolism studies. Anal. Chem. 1996; 68(17): 2996-3001.
93. Teng J, Teffera Y, Mclean M, Abramson FP. Studying the reaction between clozapine and glutathione with element-selective detection. Res. Commun. Mol. Path. Pharmacol. 1998; 99(2): 131-42.
94. Barker J, Garner RC. Biomedical applications of accelerator mass spectrometry-isotope measurements at the level of the atom. Rapid Commun. Mass Spectrom. 1999; 13: 285-93.
95. Dain J, Warsheski J, Garner RC, Fischer V. Accelerator mass spectrometry: Application to an early phase drug metabolism study in humans. Synthesis and Applications of Isotopically Labeled Compounds, Proceedings of the Seventh International Symposium. June 18-22, 2000; John Wiley & Sons.
96. Garner RC, Barker J, Flavell C, Garner JV, Whattam M, Young GC, et al. A validation study comparing accelerator MS and liquid scintillation counting for analysis of ^{14}C-labelled drugs in plasma, urine and fecal extracts. J. Pharm. Biomed. Anal. 2000; 24(2): 197-209.
97. White INH, Martin EA, Mauthe RJ, Vogel JS, Turteltaub KW, Smith LL. Comparisons of the binding of [^{14}C]radiolabeled tamoxifen or toremifene to rat DNA using accelerator mass spectrometry. Chem. Biol. Interact. 1997; 106: 149-60.
98. Tuniz C, Bird JR, Fink D, Herzog GF. Accelerator Mass Spectrometry: Ultrasensitive Analysis for Global Science. CRC Press; 1998.
99. Skipper PL, Wishnok JS, Tannenbaum SR, Hughey BJ, Klinkowstein RE, Shefer RE. Development of a chromatograph-interfaced accelerator mass spectrometer for detection of ^{3}H and ^{14}C. 216th ACS National Meeting. August 23-27, 1998; American Chemical Society.

PREDICTING HUMAN ORAL BIOAVAILABILITY USING *IN SILICO* MODELS

Lawrence X. Yu*, Christopher D. Ellison, and Ajaz S. Hussain

1. INTRODUCTION

Finding a safe and effective compound amongst the hoards of available chemical moieties is challenging and costly, and involves the alliance of chemistry, biology, pharmacology, clinical sciences, and other disciplines (Drews, 2000). To bring a single drug to market may take years, cost hundreds of millions of dollars, and generally require testing in thousands of human subjects. Most drug candidates never make it as far as human testing and even many that do are rejected for various reasons. In fact, recent reports suggest that the success rate in developing drugs from the lead candidate stage to market is only 10-20% (Prentis et al., 1988; Kennedy, 1997). While many factors contribute to a high failure rate, it is somewhat surprising that poor pharmacokinetic properties, such as poor oral bioavailability, account for almost half of the failures in clinical development.

In an effort to minimize time and costs of drug discovery and development, pharmaceutical scientists have developed numerous screening techniques, including *in silico* models, to identify drug candidates most likely to make it to market. *In silico* models are one of the fastest and most effective means for screening large numbers of compounds for bioavailability. For drugs that have survived the screening process, *in silico* models can continue to play an important role. In particular, *in silico* models can be used to predict the effects of formulation changes on bioavailability, to suggest an appropriate dose and dosing schedule for multiple doses, and to analyze aberrations discovered during *in vivo* testing.

In silico models for predicting oral drug absorption models have been reviewed in a recent book chapter (Yu et al. (2000)) and review articles (Agoram et al. 2001; Grass and Sinko, 2002; Stenberg et al., 2002; Ekins et al., 2000). This chapter will examine two types of *in silico* models and how they are applied to predicting the bioavailability of orally administered immediate release (IR) drug products. The first is mechanistic compartment modeling, in which physiological processes are described using differential

---

* Lawrence X. Yu, Christopher D. Ellison, and Ajaz S. Hussain, Food and Drug Administration, Center for Drug Evaluation and Research, Office of Generic Drugs, 7500 Standish Place, Rockville, MD

Applications of Pharmacokinetic Principles in Drug Development
Edited by Krishna, Kluwer Academic/Plenum Publishers, 2004

equations, and parameters are determined from *in vitro* and *in vivo* tests. The other is quantitative structure bioavailability relationships (QSBR) model, a subclass of quantitative structure activity relationship (QSAR). The QSAR approach learns and establishes the relationship between aspects of the structure and the desired biological property from structurally diverse compounds.

Before continuing, a distinction needs to be made between what a model is, and what it is not. For a given system, a mechanistic model is a representation with as many features of the primary system built into it as observations or data will allow (Balant and Gex-Fabry, 2000). Such a model should be consistent with the observed behavior of the system – retroaction; it should further be predictive of the system's future behavior under perturbation – prediction. To build the mechanistic model requires some knowledge of the primary system in terms of structural connectivity and functional mechanisms. It may be created to focus on particular system aspects while ignoring the complexities of the full system. On the other hand, a QSAR model sets mechanism aside, and does not require any knowledge of the system of interest. Numerically, such statistical models are generally easier to handle than many mechanistic models. A statistical model is retroactive, and may be locally predictive. It should be noted that any model, whether it is *in vitro* or *in silico*, is usually a simplification of the original system, and may miss important complexities of the modeled system, and therefore, caution must be exercised when the prediction results are interpreted.

2. MECHANISTIC COMPARTMENT MODELING

The Food and Drug Administration defines bioavailability as "the rate and extent to which the active ingredient or active moiety is absorbed from a drug product and becomes available at the site of action." (FDA/CDER Guidance for Industry, 2000). This definition focuses on the processes by which the active ingredients or moieties are released from an oral dosage form and move to the site of action. It is often difficult, if not impossible, to truly assess bioavailability as defined, namely, as it affects the site of action. Orally administered drugs typically reach the site of action via the circulatory system, and we can gauge bioavailability by monitoring levels of drug in the blood. Thus, oral bioavailability of drugs can be documented by developing a systemic exposure profile obtained by measuring the concentration of active ingredients and/or active moieties over time in samples collected from the systemic circulation. Bioavailability studies focus on determining the process by which a drug is released from the oral dosage form and moves to the systemic circulation. So to predict bioavailability of an oral drug mechanistically, we want to model how it gets into the blood, how much gets into the blood, how fast it gets into the blood, and how long it stays there, i.e., absorption, distribution, metabolism, and excretion (ADME).

A traditional type of model, and one that is used quite frequently for biological processes such as drug absorption, is the compartmental model (Jacquez, 1984; Holz and Fahr, 2001). The model tries to mimic the mechanics of a system by portraying each major component as a single compartment. By segregating the processes associated with bioavailability, compartmental models make it possible to study the effects of changing controllable parameters. For example, the solubility and dissolution of drugs may be pH dependent. The intestinal absorption of drugs may be region dependent. Changes in product formulation may alter a drug's dissolution profile, leading to faster or slower absorption of the drug. Some formulation excipients may affect intestinal transit, solubility, and permeability. Co-administration of other drugs may affect transporters or otherwise alter drug bioavailability.

The course of oral bioavailability can be broken down into compartments that reflect delivery to the intestine (gastric emptying and intestinal transit), absorption from the lumen (dissolution, permeability, particle size, intestinal efflux, and carrier-mediated transport), first-pass metabolism, and elimination (Yu et al., 1996)). In this section, we tour the basic paths of oral drugs, from administration to appearance in the systemic circulation, and describe pharmacokinetic processes affecting bioavailability, to the extent that they are typically modeled. We then describe models to describe these paths/processes.

2.1. Gastric Emptying

Though the stomach is not a major site of absorption for most drugs, activities in the stomach can greatly influence the rate at which drug becomes available for absorption in the small intestine. Models of stomach activities related to bioavailability primarily consist of drug product disintegration and dissolution, gastric emptying, and in some cases, acid-based drug degradation. For a highly acid soluble drug dosed in a rapidly dissolving dosage form, drug product disintegration and dissolution likely to be complete in the stomach, and dissolved drug will be emptied from the stomach. While for a poorly acid soluble drug dosed in a rapidly dissolving dosage form, drug product disintegration and limited drug particle dissolution will occur in the stomach. Dissolved and undissolved drug will be emptying from stomach.

After administration, an oral drug product quickly reaches the stomach. During fasting conditions, the stomach is generally quiescent, and maintains a small volume of acidic juice. The gastric pH during fasting conditions varies from 1.4 to 2.1. In the absence of food, gastric emptying generally follows first order dynamics, with half the contents transferred to the intestine in approximately 15 minutes (Oberle et al., 1990). Drug particles of 1000 μm or less are small enough to pass through the closed pylorus and can behave more as a solution than a solid when administered (Kelly, 1981). Thus, for both liquids and solid drug particles of 1000 μm or less under fasting conditions, gastric emptying can be modeled by first order kinetics.

2.2. Small Intestinal Transit

As indicated in the introduction to this chapter, a model may greatly simplify the system being studied. This readily becomes apparent when modeling the small intestine, which is the primary site of absorption for most orally administered drugs. It is typical to think of the intestine as a long cylindrical tube with uniform diameter, and thus constant volume and surface area (Ho et al., 1977). Such portrayals of the bowel could not be further from the truth.

The small intestine is both collapsible and elastic so that the volume of a particular portion of the intestine changes dramatically depending on its contents. The flow of material is controlled primarily by peristaltic waves, i.e. rhythmic muscular contractions along the length of the intestines. The frequency and intensity of these contractions varies with location and intestinal content. Despite all of the complexity of the small intestine, the cylindrical tube image and the compartmental approach seems remarkably sufficient for modeling purposes (Yu et al., 2000). After passing the stomach through the pyloric sphincter, the drug enters the duodenum, which measures 20 to 30 cm in length and 3 to 5 cm in diameter (Shiner, 1994). Contents are transported rapidly through this short section to the jejunum. The jejunum is the primary site of absorption for most nutrients, and most drugs. The remaining small intestine makes up the ileum. The pH of the intestine rises gradually throughout its length. Under the fasting conditions, the pH varies from 4.9 to 6.4 in the duodenum, 4.4 to 6.6 in the jejunum, and 6.5 to 7.4 in the ileum (Dressman, 1997). Drugs passing from the ileum through the ileocecal valve enters the large bowel, where most remaining water and some salts are removed.

Unlike gastric emptying, a simple linear equation is not sufficient for describing the transit process through the small intestine. Material entering at one time will undergo some dispersion during its trip through the intestine, and there will be a lag before it starts exiting the intestine. One approach to modeling such non-linear behavior is to divide the intestine into multiple compartments, each of which is small enough to act as a well-mixed chamber. Then linear processes can connect the compartments. Based on published data of small intestinal transit time, Yu determined that the intestine should optimally be broken down into seven compartments (Figure 1, Yu et al., 1996b; Yu and Amidon, 1998a). When considering absorption of drug into the bloodstream, each of the seven compartments may be considered separately. This can greatly complicate the model, but can also provide vast flexibility – different regions of the intestine may have different drug absorption properties and dissolution media. For this approach, each segment of the intestine has the same *residence time*, rather than the same length or volume. This avoids the issues of variable rate of progress through the intestine, and variable volumes of the segments. The residence time for each of the seven segments is approximately 28.4 minutes so that the total residence time passes through the intestine in 199 minutes.

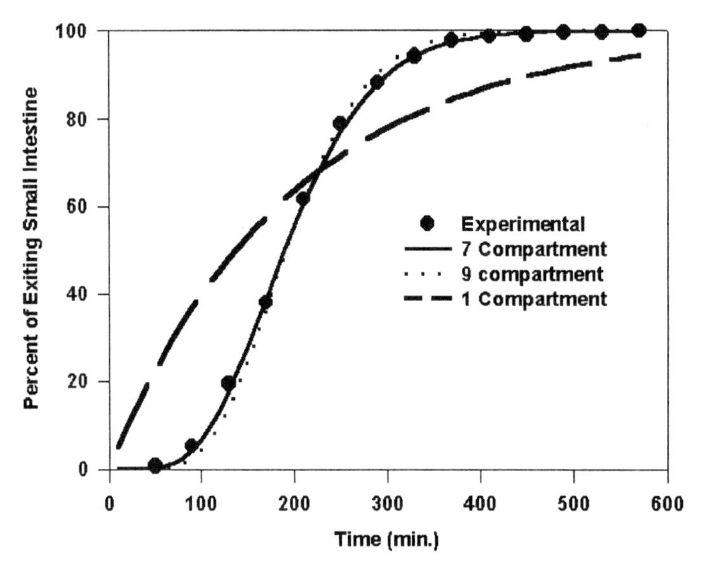

Figure 1. Predicting human small intestinal transit flow (After Yu et al., 1996b)

2.3. Intestinal Absorption

There are two pathways a drug might take to get through the intestinal epithelial membrane: the paracellular pathway, in which diffusion occurs throughout water-filled pores of the tight junctions between the individual cells; and the transcellular pathway, which requires the drug to penetrate the intestinal cell membranes. Both passive and active processes may contribute to the permeability of drugs transported by the transcellular pathways. These two pathways are distinctly different, and the molecular properties that influence drug transport by these pathways are also different. Paracellular absorption is generally limited to relatively small hydrophilic molecules. Other drug substances, especially lipophilic drugs, may take a transcellular route of absorption. Active transporters may become saturated in the presence of high drug concentrations. Influx transporters, which aid absorption, place an upper limit on the active absorption rate. Efflux transporters, which hinder absorption, may fail to work on excess drug amounts. A major area of important drug-drug interaction centers on these transporters—drugs that compete for the same influx transporter will each suffer lower bioavailability. Similarly, a drug affected by an efflux transporter may enjoy improved bioavailability when coadministered with another compound competing for the same transporter.

Yu et al. (1996a) developed a compartmental absorption and transit (CAT) model. For passively transported absorption, Yu and Amidon (1999a) derived an equation that can be used to predict the extent of intestinal absorption (fraction of dose absorbed):

$$F_a = 1 - (1 + 0.54 P_{eff}(cm/hr))^{-7} \qquad (1)$$

where P_{eff} is the effective human intestinal permeability. Equation (1) assumes that dissolution is instantaneous, and absorption from the stomach and colon is minor compared to the small intestine. Equation (1), along with the single compartmental model, was used to predict the extent of intestinal absorption for ten drugs covering a wide range of absorption characteristics. Overall, the single compartment model underestimates the extent of intestinal absorption, whereas the CAT models give a much better fit to the data.

The rate constant of a linear absorption process is independent of the amount of drug involved. Some processes, however, are concentration dependent. Processes involving carrier-mediated transporters with limited capacity are called "saturable" and require a variable rate "constant" when modeled (Yu and Amidon, 1998b). To model a saturable process, we typically replace the rate constant with a Michaelis-Menton factor, $V_{max}/(K_m+C)$, where V_{max}, K_m, and C are the maximum rate of absorption, Michaelis constant of saturable absorption, and the concentration of the drug. Coupled with the Michaelis-Menton equation, the CAT model was used to predict the observed plasma concentration-time profiles of cefatrizine at the doses of 250, 500, and 1000 mg (Figure 2). The reported absolute bioavailability was 75% and 50% at 250 and 1000 oral doses, respectively, compared favorably to the theoretical extent of intestinal absorption or fraction of dose absorbed (74% and 48%).

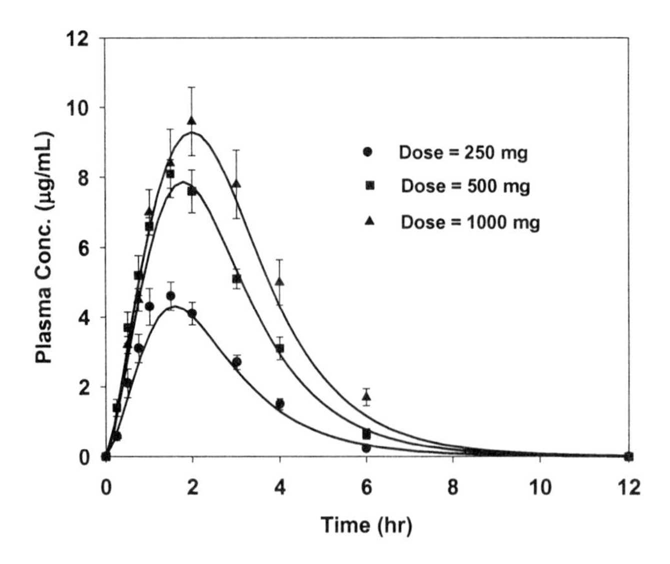

Figure 2. Estimating the plasma concentration profiles of cefatrizine at oral doses of 250, 500, and 100 mg, where (—) represents the prediction from the CAT model. (After Yu and Amidon, 1998b)

The CAT model was also used to characterize the intestinal absorption of ranitidine without a *prima facie* assumption of non-absorption in the jejunum. The simulation indicates that ranitidine is absorbed in the duodenum/proximal jejunum and the terminal ileum, with first-order rates 0.45 hr^{-1} and 0.51 hr^{-1}, respectively. Ranitidine absorption is negligible in the central portion of the small intestine. The model also indicates that the administration of ranitidine to the stomach or duodenum leads to double peaks while the administration to the jejunum and ileum does not lead to double peaks (Figure 3), in agreement with the literature observations (Pithavala, 1998). This model provides evidence that the two-peak phenomenon for ranitidine is caused by site-specific absorption.

An important feature of this ranitidine model is the fact that there is neither a *prima facie* assertion that absorption of ranitidine is discontinuous nor an attempt to force absorption at only the most proximal and terminal regions of the small intestine. In fact, the model runs were each initialized with uniform absorption across the entire small intestine, i.e., the initial "guess" for each compartment's absorption was identical. The absorption rate constant of each compartment was determined by minimizing the difference between the calculated and experimental plasma concentration profiles. The computer determined values of absorption rate indicate that the primary sites of absorption are the proximal and terminal regions of the small intestine. This outcome is due to the fitting to the experimental data rather than the explicit design of the model. This gives extra credence to previous models that *assumed* a mid-intestinal region of no absorption

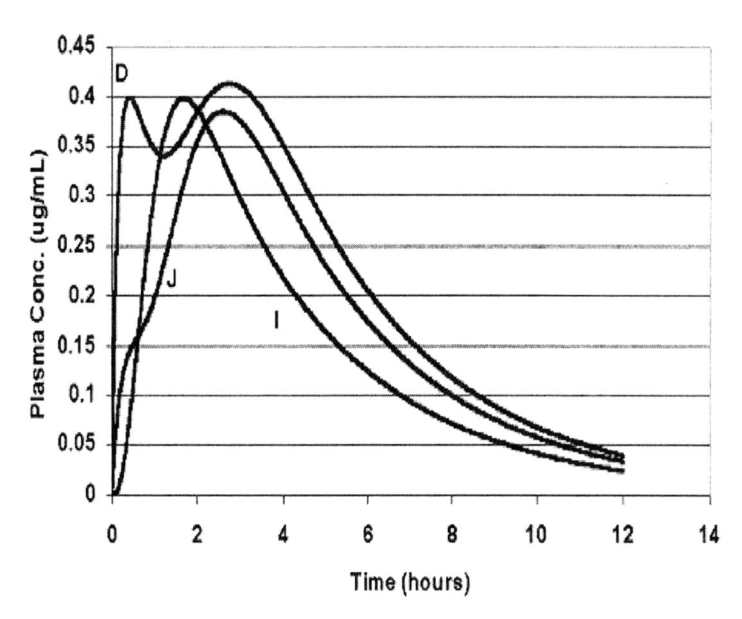

Figure 3. Simulated plasma concentration profiles of 150 mg ranitidine when administered into different locations within the small intestine. D: drug administered in intestinal compartment 1 (approximately duodenum). J: drug administered in compartment 2 (jejunum). I: drug administered in compartment 4 (ileum). Note that only D displays a double peak

2.4. Drug Particle Dissolution

For a poorly soluble drug dosed in an immediate release product, dissolution of drug particles can greatly affect its bioavailability. Dissolution is based on a number of factors, including solubility (and thus pH), volume of the dissolution media, the presence of fats and/or surfactants, and particle size of the drug being dissolved. Other potential sources are changes in the volume of media, or changes in available surfactants and/or fats, but for most systemic models these issues are not specifically portrayed. Many drugs are administered as tablets or other solid forms, and dissolution can be a major part of an oral absorption model. A drug product often consists of drug particles embedded within a tablet or capsule that must disintegrate to expose the drug particles to the dissolution media. After liberated, the drug particles can begin to dissolve. The basic differential equation (Hintz and Johnson , 1989; Yu, 1999b) for dissolution of drug particles is derived from Fick's Law:

$$\frac{dM_s}{dt} = \frac{3DM_s}{\rho hr}\left(C_s - \frac{M_l}{V}\right) \tag{2}$$

where M_s is the amount of solid drug at time t, M_l is the amount of dissolved drug at time t, D is diffusion coefficient, ρ is the solid drug density, h is the aqueous boundary thickness, r is the particle radius, C_s is the solubility of the drug, and V is the volume of dissolution medium. This equation assumes that all of the particles are spherical, with the same initial radius, and begin dissolving at the same time. Clearly, in reality, none of these assumptions are met. However, in cases where the dissolution is controlling absorption, the dissolution model in conjunction with the CAT model seems to predict the absorption reasonably well (Figure 4).

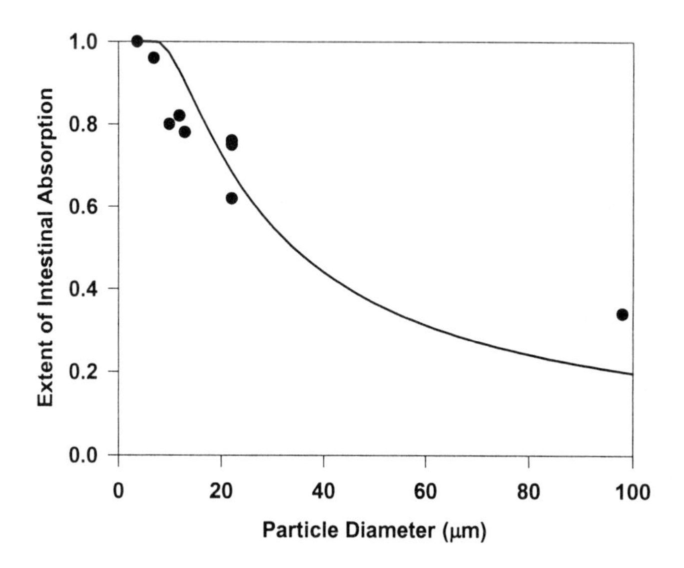

Figure 4. Experimental and predicted extent of digoxin intestinal absorption as a function of particle size, where (—) predicted value and (•) experimental data (After Yu 1999b)

2.5. Permeability, Dissolution, and Solubility Limited Absorption

Without considering the transit flow, the absorption can be limited by dissolution rate and permeation rate (permeability), where permeation rate refers to the flux of drug across the intestinal membrane. The supply rate of dissolution and the uptake rate of permeation determine the concentration of drug in the gastrointestinal (GI) tract. However, the concentration in the GI tract is also limited by the solubility of drug. When the supply rate is far more than the uptake rate, the drug concentration in the gastrointestinal fluid approaches its solubility limit. The diffusion coefficient (*D*), solid drug density (ρ), and aqueous boundary thickness (h) in Eq. (2) are generally kept very much constant, the dissolution rate is, thus, mainly determined by the drug particle size and concentration difference (C_s-M_l/V). Therefore, poor dissolution can be caused either by particle size (r) or solubility (C_s). To emphasize the importance of solubility, Yu (1999b) referred to the dissolution/solubility-limited case as solubility-limited absorption. The dissolution/particle size-limited case is still called dissolution-limited absorption. As a result, permeability, solubility, and/or dissolution can limit the absorption of poorly absorbable drugs.

To determine dissolution-, solubility-, and permeability-limited absorption, Yu (1990b) defined three characteristic parameters, dissolution time (T_{diss}), effective permeability (P_{eff}), and absorbable dose (D_{abs}). Dissolution time is an estimate of the minimum time required to dissolve a single particle under sink conditions. Absorbable dose is the amount of drug that can be absorbed during the period of transit time when the solution contacting the effective intestinal surface area for absorption is saturated with the drug. Based on these three characteristic parameters, Yu (1999b) proposed a framework for determining the corresponding absorption limiting processes.

The section on intestinal absorption discussed the permeability-limited absorption while the section on drug particle dissolution discussed the case of dissolution limited absorption. In this section, we will discuss a case of solubility limited absorption using ganciclovir as a model drug. Ganciclovir is approved by the FDA for treatment of cytomegalovirus in immunocompromised patients. Ganciclovir has human bioavailability less than 10% (Anderson et al., 1995). Figure 5 shows the simulated mass absorbed as a function of dose in humans (Norris et al., 2000). Figure 5 suggests that the absorption of ganciclovir is limited by its permeability and solubility. At the dose less than 0.2 mg/kg, the absorption of ganciclovir is mainly limited by its permeability, solubility has little effect. Between 0.2 mg/kg to 1 mg/kg, solubility begins to play a more significant role. When the dose is above 1 mg/kg, the absorption of ganciclovir is mainly limited by its solubility, where the absolute mass absorbed does not increase with increasing dose. But for either permeability or dissolution limited absorption, the absolute mass absorbed will increase with the increased dose (Yu, 1999b).

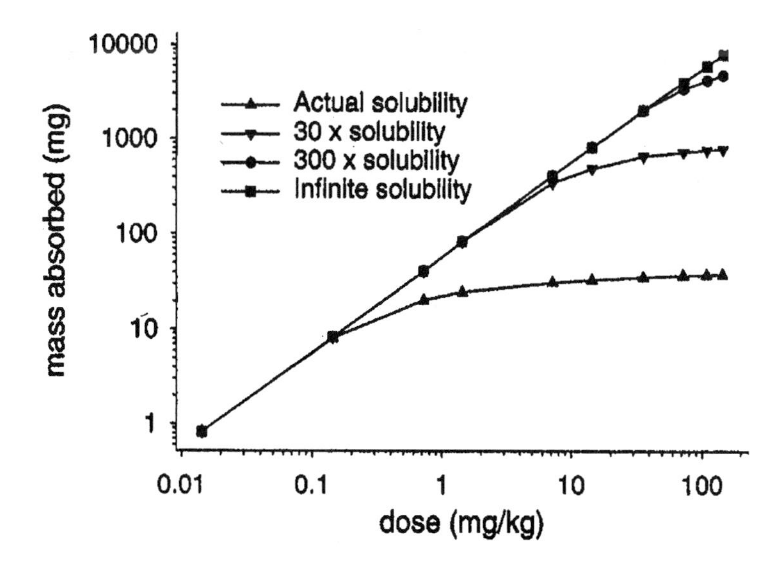

Figure 5. Simulated mass absorbed of ganciclovir as a function of dose to show the solubility limited absorption (From Norris et al., 2000, with kind permission from Elsevier Science-NL, Sara Burgerhartstraat 25, 1055 KV Amsterdam, The Netherlands)

Solubility-limited absorption may occur more frequently than we expect, especially during the preclinical toxicity evaluation. Ganciclovir is an example used to illustrate that the absolute amount of drug absorbed will not increase with higher doses. Other potential examples include some HIV protease inhibitors. Certainly, formulation techniques could change these phenomena. For example, solubility-limited absorption will not likely occur for such HIV protease inhibitor drugs if they can be formulated into self-emulsified delivery systems.

2.6. Hepatic Metabolism

Models of drug metabolism have centered on hepatic clearance because the liver is a major site of metabolism. Because the liver functionally resides between the gastrointestinal tract and systemic circulation, drugs absorbed from the intestine are subject to metabolism by the liver during the forced "first pass" before reaching the systemic circulation. Since some of the same enzyme systems that are found in the liver also exist in the intestinal tissue, drug metabolism can occur before reaching the liver. In the analysis of clinical data of metabolized drugs, this difference may be undetectable and both intestinal and hepatic first pass metabolisms may be accounted for within the same mathematical compartments.

Physiologically based models of metabolism that use *in vitro* data as input to predict *in vivo* results have generally used microsomal fractions or hepatocytes as assays of *in vitro* metabolism (Houston and Carlile, 1997). The microsomal fraction method has the advantage of being somewhat easier to carry out, while the hepatocyte method has the advantage of including Phase 1 and Phase 2 metabolism.

There exist several models to predict hepatic metabolism: namely, the well stirred model, the parallel-tube model, and the dispersion model (Pang and Rowland, 1977; Roberts and Rowland, 1986). The well-stirred model often has been used because of its simplicity:

$$F_H = \frac{fu_b \cdot Q_H}{Q_H + fu_b \cdot Cl_{int}} \tag{3}$$

where F_H is the fraction of drug escaping elimination by the liver, Q_H is hepatic blood flow , Cl_{int} is intrinsic clearance relating the rate of metabolism to unbound concentrations at the enzyme site, and fu_b is the fraction unbound in plasma.

2.7. Integrated Model to Predict Bioavailability

If a drug is not in solution when administered, its absorption from the GI tract can be described by a four step process: first, delivering the drug into its absorption site (gastric emptying and small intestinal transit flow); second, getting the drug into solution (dissolution); third, permeating the dissolved drug through the intestinal membrane (permeation/absorption); and finally, moving the drug through the liver and to the systemic circulation. All four of these processes have been modeled in previous sections. Thus, combining these four processes gives an integrated model to predict the extent of oral drug absorption or bioavailability (Figure 6).

One of the utilities of such a model is to process biopharmaceutical profiles data. With the advent of combinatorial chemistry and high throughput screening, the drug discovery and development process has undergone a major change. Thus, a large number of compounds with good potency and specificity have been identified. While the ideal would be to screen a large number of drug candidates by *in vivo* pharmacokinetic studies, the *in vivo* models are impractical because these studies are inherently slow, labor intensive, and not amenable to automation. As a result, the pharmaceutical industry has recently adopted an approach to acquire biopharmaceutical profiles data, such as solubility, permeability, and metabolism. Although good models exist to correlate these *in vitro* assay with *in vivo* (Mandagere et al., 2002; Arthursson and Karlsson, 1991; Obach et al., 1997), few published models exist that can tie solubility, permeability, and metabolism all together (Argoram, 2001; Grass and Sinko, 2002).

Figure 7 portrays the relationship between solubility, permeability, and intrinsic hepatic clearance., and potency (effective dose). This surface was obtained by assuming: (1) Bioavailability is at least 30% from solid dosage forms; (2) Particle size is 20 μm; (3) Solubility is measured in a pH 6.8 buffer; (4) Human permeability is measured, and (5) *In vitro* intrinsic hepatic clearance. If a drug candidate is above this surface, this candidate likely has oral bioavailability below 30%. Otherwise, it will probably have bioavailability above 30%. If a candidate has bioavailability below 30%, we may able to change its structure to obtain a highly bioavailable candidate. For example, if a candidate has high permeability, low solubility, and high clearance, we will need to focus on improving its solubility and reducing intrinsic hepatic clearance. On the other hand, if a candidate has reasonable solubility and intrinsic hepatic clearance, we may need to focus on improving its permeability.

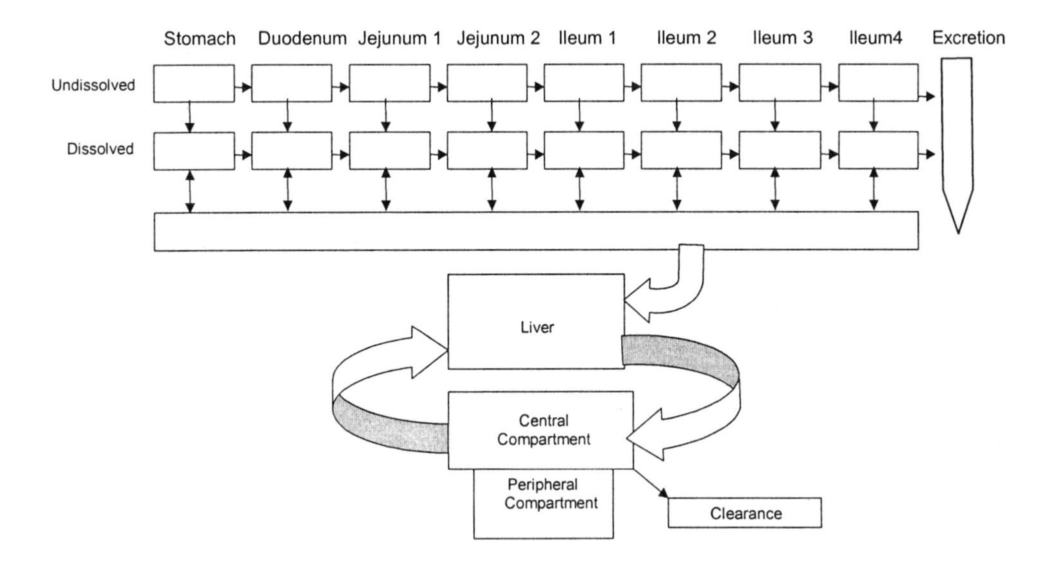

Figure 6. Schematic of the CAT based integration model for predict human oral bioavailability

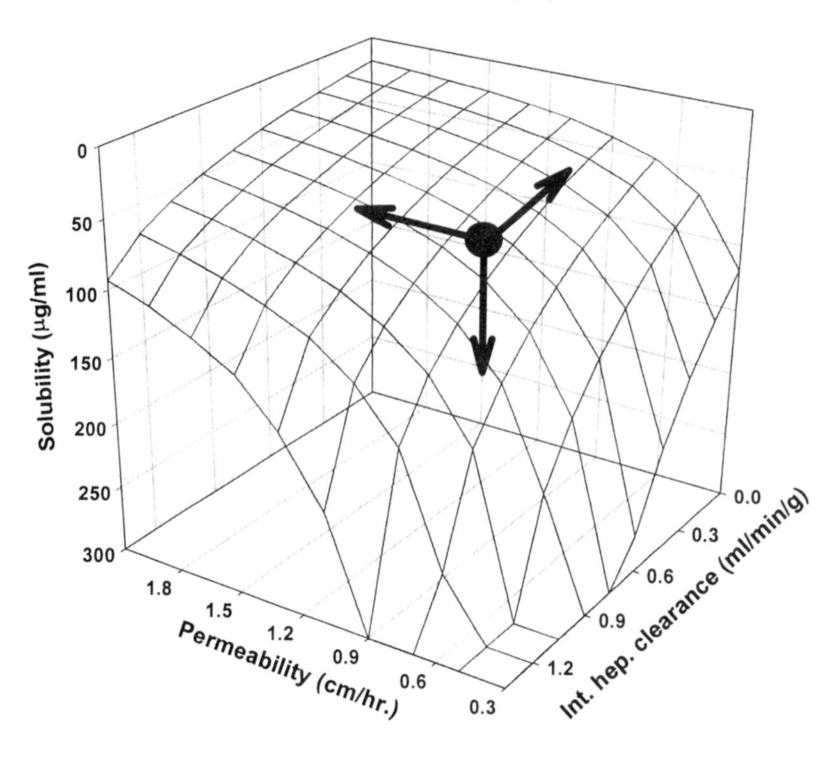

Figure 7. Simulated surface of preferable properties as a function of solubility, permeability, intrinsic hepatic clearance, and effective dose.

2.8. Limitations

One limitation to mechanistic compartment modeling is its heavy reliance on data. For many linear processes, there may be good *in vivo/in vitro* correlation (IVIVC), and the data might be readily obtained. For example, the Caco-2 in vitro model works well for estimating the permeability of passively absorbed drugs (Lennernas, 1998). However, for many non-linear processes (such as active drug transport) no good IVIVC correlation exists, necessitating more expensive, more time consuming *in vivo* tests. Such a need for data also makes modeling new drug moieties difficult. For such drugs, QSAR methods may be more appropriate (see section 3, below).

Finally, the integrated absorption model is still a simplification of a complex system. Sufficient models do not yet exist for some complex processes, such as *in vivo* precipitation. Thus, when these processes play an important role in the oral drug absorption process, the prediction may not be as good as what we would like. In such cases, investigating the causes of poor prediction may be more important than obtaining a good prediction.

3. QUANTITATIVE STRUCTURE BIOAVAILABILITY RELATIONSHIPS

The other type of *in silico* model that we will examine is quantitative structure bioavailability relationships (QSBR) model, a subclass of quantitative structure activity relationship (QSAR) models. Although various QSAR studies of congeneric compounds have been reported concerning different processes affecting oral bioavailability (Stenberg et al., 2002), an overall quantitative relationship between the oral bioavailability of structurally diverse compounds and their physicochemical/structural properties has proved to be an elusive goal (Yoshida and Topliss, 2000). However, the ability to predict the human oral bioavailability of drugs from their structure prior to synthesis would be of invaluable benefit in the discovery of new drugs. Recently, Yoshida and Topliss (2000) assigned the human oral bioavailability of 232 compounds into one of four ratings and analyzed their bioavailability in relation to physicochemical and structural properties. Meanwhile, Andrews et al. (2000) developed a quantitative structure bioavailability relationship (QSBR) model using fragment descriptors based on 591 structurally diverse compounds.

The primary goal of the QSBR modeling is to develop mathematical expressions correlating bioavailability to structure descriptors and physicochemical properties. Structural descriptors used in the literature include fragment descriptors, hydrophphobicity, hydrogen bonding descriptors, topological indices, polar surface area, and quantum chemical parameters (Yamashita et al., 2002). The general form of QSAR models is: F = f(Y), Where F is the activity and f(Y) is a mathematical expression correlating F with a matrix of structural parameters (Y). Consider stepwise linear regression analysis:

$$f(Y) = \sum_{i=1}^{n} a_i y_i + b \qquad (4)$$

where y_i are the individual fragment descriptors, a_i are coefficients, and b is a constant. Both a_i and b are determined from multiple regression analysis by fitting the equation above to the structure activity data, provided the number of drugs exceeds the number of fragment structures.

To develop the relationship between these molecular descriptors and bioavailability, multiple linear regression, partial least squares, artificial neural network, and genetic algorithm can be used (Yamashita et al., 2002). The number of variables involved in such calculations can be daunting. Stepwise multiple linear regression (MLR) is one method of reducing the number of variables. MLR works by incorporating one variable at a time into the model, but only if that variable improves the quality of the model. Moreover, as additional variables are included in the model, previously included variables may be dropped if they become redundant. In this way, only the most significant variables are retained. This identification of important variables is a major advantage of MLR, especially when the system variables can be individually manipulated to achieve a desired result. Such variable exclusion can, however, cause some sources of variability to be overlooked.

3.1. General QSAR Procedures

A typical QSAR approach is to clearly define a property of interest (e.g., bioavailability) and then create a list of contributing variables likely to affect that property (e.g., fragment structures and hydrophobicity). The values of these variables should be relatively easy to establish by computational method or by direct measurement.

Next a "training set" of compounds must be selected. This is a set of compounds for which the desired property as well as the contributing variables are well quantified. The training set serves two important purposes. First, it establishes the "chemical space" in which we can work. For example, if we use molecular weight as one of our contributing variables, then the molecular weights in our training set provide an allowable range for compounds being tested. Compounds falling outside our chemical space may fail to be adequately modeled. Second, the training set provides the data upon which the model is built.

Before a robust QSAR model is developed and used, it should be validated. Many QSAR models in the literature have little practical utility because they have not gone through rigorous validation. One method of validating such a model is to reserve a randomly chosen portion of the training set from model development, and determine how well the model functions on the omitted samples. A popular method is to omit only one test sample, using the rest to create the model, and test the one remaining sample. Then repeat for each sample separately. A model that frequently yields erroneous output may be the result of insufficient input. Then we must consider whether we have overlooked factors that contribute to the property being studied. We may need to consider variables that seem only marginally related to the property of interest. The magic of QSAR techniques is that they frequently discover hidden relationships between contributing factors.

After a model has been validated, it may be used to examine the set of unknowns. It is assumed that each unknown falls within the chemical space defined by the test set, and that each contributing variable can readily be measured for the unknowns.

3.2. Bioavailability and Lipinski's Rule of Five

Andrews et al. (2000) considered 591 compounds with known human bioavailability values as well as experimental variability (if available). Any compounds whose bioavailability is strongly affected by the dose and formulation was not collected. The mean human bioavailability of the chosen compounds was 57%. Experimental errors (standard deviations) were 12% (from 282 compounds, Figure 8) In agreement with previous work (Hellriegel et al., 1996), Figure 8 shows that experimental coefficient of variance generally decreases with increasing bioavailability.

Lipinski's Rule of Five predicts that oral activity is likely to be poor when there are more than 5 H-bond donors or more than 10 H-bond acceptors, the molecular weight is greater than 500, or the calculated Log P is greater than 5 (Lipinski et al., 1997). When the Rule of Five was applied to the data set, little correlation was found between human oral bioavailability and molecular weight while reasonable correlations were obtained between the human oral bioavailability and calculated Log P, H-bond donors, or H-bond acceptors, respectively. Figure 9 shows the human oral bioavailability as a function of calculated Log P. Figure 9 suggests that a lower limit of –2 should also be imposed.

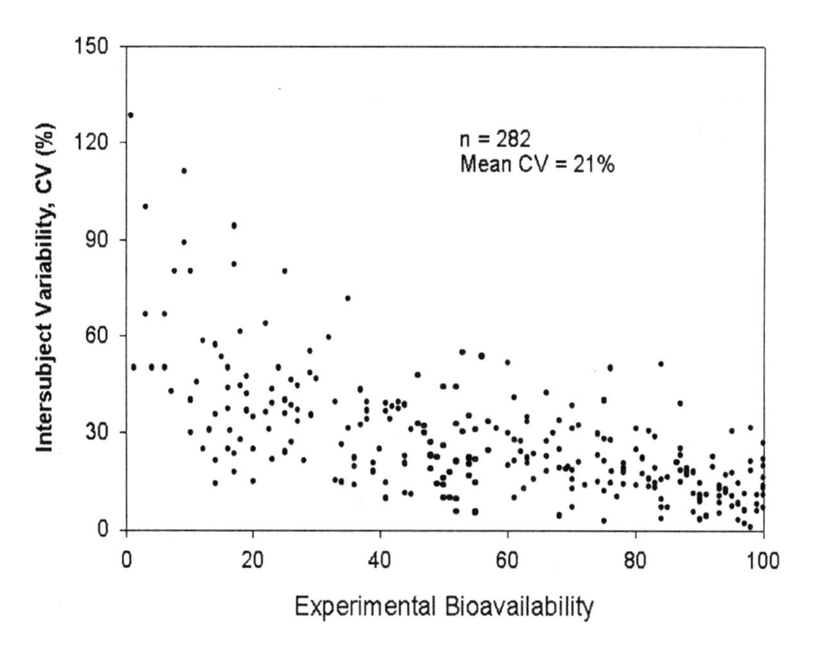

Figure 8. Intersubject variability as a function experimental bioavailability

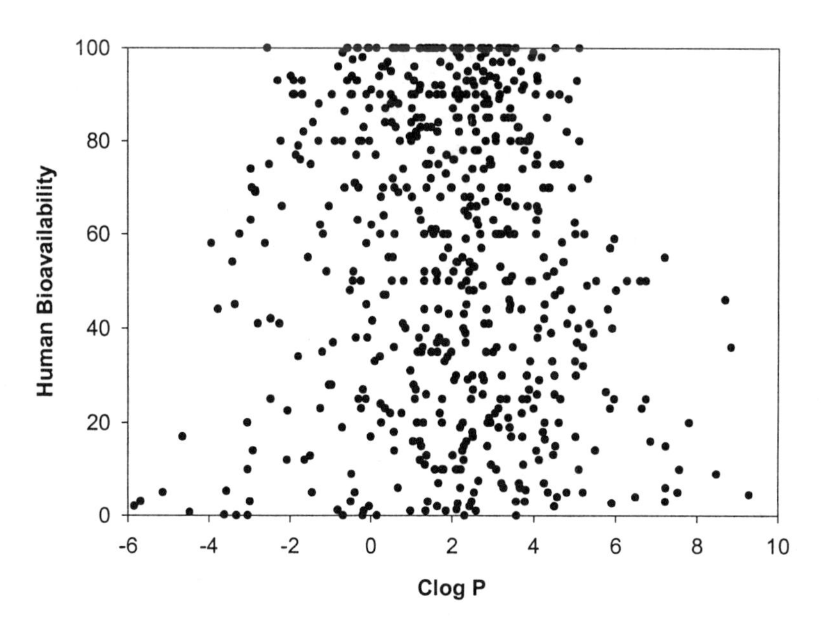

Figure 9. Human bioavailability as a function of calculated log P

3.3. QSBR Models

Yoshida and Topliss (2000) used the ordered multicategorical classification method to classify drugs into four bioavailability classes (class 1: $\leq$ 20%; class 2: 20-49%; class 3: 50-79%; and class 4: $\geq$ 80%). They then selected from a list of structural descriptors that described functional groups liable to undergo hydrolysis, oxidation, conjugation, and other metabolic fates, together with conventional descriptor, Δ log D (log $D_{6.5}$ – log $D_{7.4}$). Yoshida and Topliss developed a model that used 18 descriptors and the model correctly classified 165 of 232 compounds (71%). Classification accuracy is highly dependent on the bioavailability of compounds. Approximately 88% of class 4 compounds were correct, and a similar result was obtained for class 3 compounds. Accuracy dropped off for class 2 and class 1 compounds, to 52% and 43% correct, respectively. This is not surprising since model descriptors, except lipophilicity, are primarily related to well-known metabolic processes. As a result, the effect of solubility on bioavailability has not been well characterized by this model.

Yoshida and Topliss' method suffers from two theoretical drawbacks (Bains et al., 2002). The descriptors used to characterize the molecules to the system have to be preselected by the user, as there are many thousands of ways that a molecule can be described. Although the body of knowledge about molecular features affecting bioavailability is very extensive, it is probable that a computer model will be improved by including descriptors that human experts would not consider because a computer can consider more complex statistics than the human experts can embrace. Further, the method mentioned above is not efficient at partitioning the data: it excludes the possibility of artificial intelligence statements such as "IF...THEN." Since biological processes generally occur in more than one mechanism, and different molecular types may impinge on different mechanisms, the artificial statements are found to be critical.

One way to circumvent both these problems is to use genetic algorithm. A genetic algorithm is an optimization algorithm based on the mechanisms of Darwinian evolution that uses random mutation, crossover and selection procedures to breed better models or solutions from an originally random starting population or sample (Rogers and Hopfinger, 1994). It begins with a population of randomly constructed QSAR models; these models are rated using error measure that estimates each model's relative productiveness. The population is evolved by repeatedly selecting two better-rated models to serves as parents, then creating a new child model by using terms from each of the parental models. As evolution proceeds, the population becomes enriched with higher and higher quality modes. Bains et al. (2002) used the genetic algorithm to predict oral bioavailability, and found the genetic algorithm generates better results than Yoshida and Topliss' method.

Another way to circumvent these problem is to use a technique called recursive partitioning, where a very large number of descriptors are searched for the ability to partition the data into two classes. Then, within these classes, all the remaining descriptors are searched again for the ability to partition, and so on. Any number of descriptors can be used. This recursive partitioning approach has recently been applied to bioavailability on a set of 591 compounds (Andrews et al., 2000). The method screened 608 molecular fragment descriptors. There are a range of such descriptors, but all characterize a molecule by whether it does or does not contain a fragment that is either chemically defined (acid, alcohol, etc) or topologically defined as a group of atoms connected together.

The QSBR model developed by Andrews et al. (2000) had an R^2 of 0.71 and contained 85 substructural descriptors. The RMSE, an estimate of the error in the model, is 18. Given the mean experimental error of 12, this model error seems reasonable. The cross-validated leave-one-out R^2 is 0.63, indicating that unique compounds are not a particularly bad problem in the data set. A predicted versus actual plot is shown in Figure 10 using leave-one-out predictions. Recently, this QSBR model was used to predict the human oral bioavailability of recently approved drugs (Table 1, Yu and Andrews, 2001). Figure 11 shows that predicted R^2 is 0.62, which is very close to the cross-validated R^2 (0.63). This suggests that the QSBR model was reasonably predictive and stood up to cross-validation.

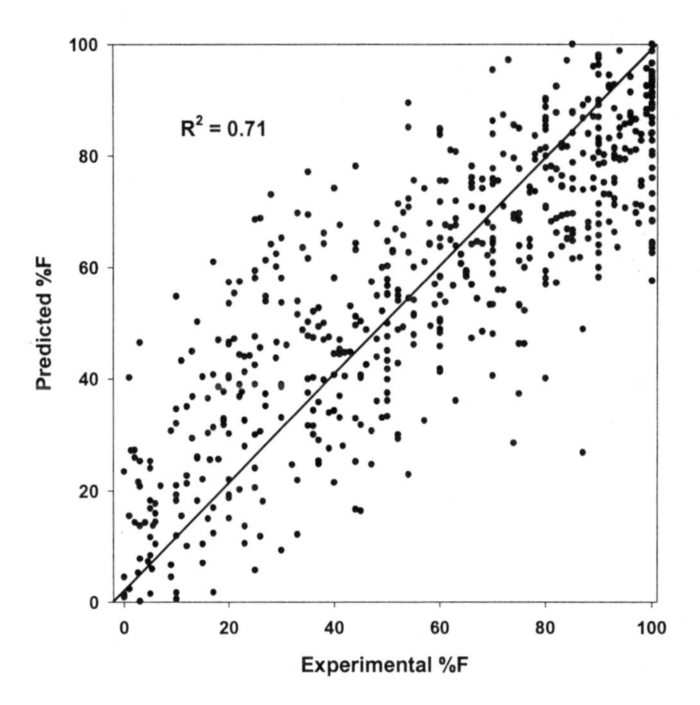

Figure 10. Predicted versus experimental bioavailability for model development (After Andrews et al., 2000)

Table 1. List of recently approved drugs (post model development) for evaluation of the QSBR model

No.	Drug Name	Experimental Bioavailability	Approval Date	Reference*
1	Alosetron	60	02/00	PDR
2	Dofetilide	90	10/99	PDR
3	Entacapone	35	10/99	PDR
4	Esomeprazole	50	02/01	Hassan-Alin et al., 2000
5	Gatifloxacin	96	12/99	PDR
6	Letrozole	99.9	01/01	Sioufi et al., 1997
7	Levetiracetam	100	12/99	PDR
8	Linezolid	100	4/00	PDR
9	Meloxicam	89	4/00	PDR
10	Moxifloxacin	90	12/9	PDR
11	Pantoprazole	77	02/00	Huber et al., 1996
12	Pravastatin	17	06/00	PDR
13	Rivastigmine	54	4/1/00	PDR
14	Rizatriptan	45	06/00	PDR
15	Rofecoxib	93	5/99	PDR
16	Rosiglitazone	99	4/00	PDR
17	Terbinafine HCl	40	03/00	PDR
18	Tolterodine	40	12/00	Brynne et al., 1997
19	Zaleplon	30	9/00	PDR
20	Zolmitriptan	40	2/01	PDR

* PDR – Physicians' Desk Reference

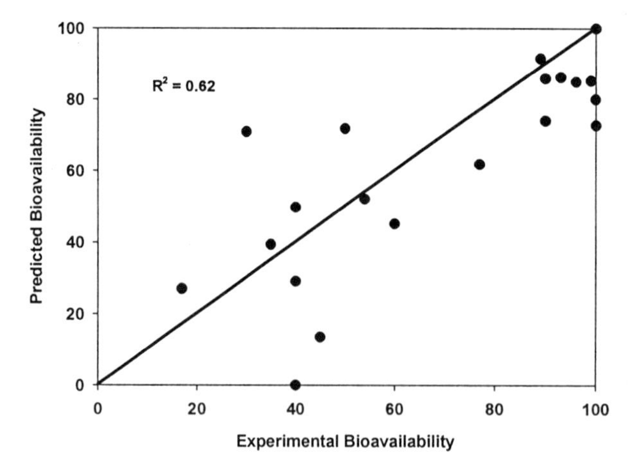

Figure 11. Predicted and experimental human bioavailability for 20 recently approved drugs (post QSAR model development)

3.4. Limitations

The QSAR model would be expected to perform well for predicting the bioavailability of compounds within the chemical space represented in the data set from which the model was derived. However, for compounds outside of the chemical space, this would usually not be the case. The QSAR approach is the dependency on a sufficient test set of chemicals with well-known characteristics. An accurate model depends on accurate data. Another issue is that experimental samples may fall outside of the chemical space defined by the test compounds, i.e., outside the range of validity for the model. Despite these potential drawbacks, QSAR models are valuable, generally accurate tools for estimating activity parameters for drugs. This provides drug developers an invaluable tool for screening large numbers of new potential drugs in rapid fashion, and gives conventional modelers additional means for establishing parameters related to their compartmental models.

4. CONCLUSION

The mechanistic and QSAR approaches to predicting human oral bioavailability presented in this chapter can be used to assess human oral bioavailability during allstages of drug discovery and development. These models are not yet perfect, but they can be further develped using experimental data obtained in various projects. There is no doubt that *in silico* models will play a more significant role in the future of drug discovery, development, and regulatory approval. The QSBR model, which requires only structural features of the compound and not experimental data, is best used for rational drug design and high throughput screening. The mechanistic approach, which requires experimental data, is best utilized for finding causes of poor bioavailability and analyzing high throughput screening data.

5. FUTURE DIRECTIONS

Among the areas requiring further effort in developing predictive bioavailability models are:

- Establishing databases that include accurate pharmaceutical and pharmacokinetic information for commercialized and uncommercialized drugs so that the e-ADME concept (Yu, 2001) can truly be implemented. The difficult element of model development may not be the creating of models, but rather the gathering and acquiring of sufficient data. Many models, which are based on data sets with limited number of compounds, need further development and validation. QSAR techniques in particular generally rely on having a sufficiently large collection of data for training and validation of the model. If the pharmacokinetic database is sufficiently large enough to cover diverse chemical spaces, predictive QSAR models may no longer be elusive.
- Further developing mechanistic approaches to include influx and efflux transporters, gut wall metabolism, and *in vivo* precipitation. In addition, mechanistic pharmacokinetic models should be connected to pharmacodynamic models. Model development should emphasize evaluation and validation, especially the determination of appropriate model parameters. Currently available models should

be reevaluated based on the new data and new information so that the existing models can be modified and improved. In this age of computational biology wherein we are striving to develop models of cellular traffic and organ function (Noble, 2001), we have not yet developed a perfect model that can fully predict *in vivo* dissolution.

- Developing user-friendly interfaces for input and output of prediction software. Ultimately, a good rapport between the laboratory and the modeler will continue to yield improvements in efficiency of drug discovery, development, and regulation.

ACKNOWLEDGMENTS

The assistance of P. Sathe, L. Gross, and P. Downs in preparing this book chapter is gratefully acknowledged.

REFERENCE

Agoram B, Woltosz WS, Bolger MB. Predicting the impact of physiological and biochemical processes on oral drug bioavailability. Adv. Drug Del. Rev. 2000;50:S41-S67.

Anderson RD, Griffy KG, Jung D, Dorr A, Hulse JD, Smith RB. Ganciclovir absolute bioavailability and steady-state pharmacokinetics after oral administration of 3000-mg/d dosing regiments in human immunodeficiency virus and cytomegalorvirus-seropositive patients. Clin. Ther. 1995;17:425-432.

Andrews W, Bennett L, Yu LX. Predicting human oral bioavailability of a compound: Development of a novel quantitative structure bioavailability relationship. Pharm. Res. 2000;17:639-644.

Arthursson P, Karlsson J. Correlation between oral drug absorption in humans and apparent drug permeability coefficients in human intestinal epithelial (Caco-2) cells. Biochem Biophys. Res. Commun. 1991;175:880-885.

Bains W, Gilbert R, Sviridenko L, Gascon J, Scoffin R, Birchall K, Harvey I, and Caldwell J. Evolutionary computational methods to predict oral bioavailability QSPRs. Cur. Opin Drug Dis Dev. 2002,5:44-51.

Balant LP, Gex-Fabry M. Modeling during drug development. Euro. J. Pharm. Biopharm. 2000,50:13-26.

Brynne N, Stahl MM, Hallęn B, Edluud PO, Palmer L, Hoglund P, Gabrielsson J. Pharmacokinetics and pharmacodynamics of tolterodine in man: a new drug for the treatment of urinary bladder overactivity. Int. J. Clin. Pharmacol. Ther. 1997;35(7):287-295.

Dressman JB. Physiological aspects of the design of dissolution tests. In Amidon GL, Robinson RL , Williams RL, editors. Scientific Foundations for Regulating Drug Product Quality. AAPS Press; 1997. p. 155-168.

Drews J. Drug discovery: A historical perspective. Science; 2000;287:1960-1964.

Skins S, Waller CL, Swaan PW, Cruciani G, Wrighton SA, Wikel JH. Progress in predicting human ADME parameters in silico. J. Pharmacol. Tox. Meth. 2000;44:251-272.

Grass GM, Sinko PJ. Physiologically-based pharmacokinetic simulation modeling. Adv. Drug Del. Rev. 2002;54:433-451..;

Guidance for industry, Bioavailability and Bioequivalence Studies for Orally Administered Drug Products – General Considerations. October 2000, CDER/FDA.

Hassan-Alin M, Andersson T, Bredberg E, Rohaa K, Pharmacokinetics of esomeprazole after oral and intravenous administration of single and repeated dose to healthy subjects. Eur. J. Clin. Pharmacol. 2000;56(9-10):665-670.

Hellriegel ET, Bjornsson TD, Hauck WW. Interpatient variability in bioavailability is related to the extent of absorption: Implications for bioavailability and bioequivalence studies. Clin. Pharm. Ther. 1996;60:601-607.

Hintz RJ, Johnson KC. The effect of particle size distribution on the dissolution rate and oral absorption. Int. J. Pharm. 1989;51:9-17.

Ho NFH, Park JY, Morozowich W, Higuchi WI. Physical model approach to the design of drugs with improved intestinal absorption. In: Roche EB, editor. Design of Biopharmaceutical Properties Through Prodrugs and Analogs. American Pharmaceutical Association, Academy of Pharm. Sci.: Washington, D. C., 1977. p. 136-227.

Holz M; Fahr A. Compartment modeling. Adv. Drug Del. Rev. 2001;48:249-264.

Houston JB, Carlile DJ. Prediction of hepatic clearance from microsomes, hepatocytes, and liver slices. Drug Metab. Rev. 1997;29:891-922.

Huber R, Hartmann M, Bliesath H, Luhmann R, Steinijans VW, Zech K. Pharmacokinetics of pantoprazole in man. Int. J. Clin. Pharmacol. Ther. 1996;34(1 Suppl):S7-16.
Jacquez JA. Compartmental Analysis in Biology and Medicine. 2nd, The University of Michigan Press, Ann Arbor, 1984.
Kelly KA. In L. R. Johnson (Editor-in-Chief), Physiology of the Gastrointestinal Tract, Vol 1, Raven Press, New York; 1981.p. 393-410.
Kennedy T. Managing the drug discovery and development interface. Drug Discovery Today. 1997, 2(10):436-444.
Lennernas H. Human intestinal permeability. J. Pharm. Sci. 87:403-410 (1998).
Lipinski CA, Lombardo F, Dominy BW, Feeney PJ. Experimental and computational approaches to estimate solubility and permeability in drug discovery and development settings. Adv. Drug Del. Rev. 1997;23:3-25.
Noble D. The rise of computational biology. Nature Rev. Mol. Cell. Bio. 2002;3:460-463.
Norris DA, Leesman GD, Sinko PJ, Grass GM. Development of predictive pharmacokinetic simulation models for drug discovery. J. Control. Res. 2000;65:55-62.
Mandagere AK, Thompson TN, Hwang KK. Graphical model for estimating oral bioavailability of drugs in humans and other species from their Caco-2 permeability and in vitro liver enzyme metabolic stability rates. J. Med. Chem. 45:304-311 (2002).
Obach RS, Baxter JG, Liston TE, Silber MR, Jones BC, MacIntyre F, Rance DJ, Wastall P. The prediction of human pharmacokinetic parameters from preclinical and in vitro model metabolism data. J. Pharmacol. Exp. Ther. 1997;283:46-58
Oberle RL, Chen TS, Lloyd C, Barnett JL, Owyang C, Meyer J, Amidon GL. The influence of the interdigestive migrating myoelectric complex on the gastric emptying of liquids. Gastroenterology. 1990;99:1275-82.
Pang KS, Rowland M. Hepatic clearance of drugs. I. Theoretical considerations of a "well-stirred' model and a "parallel tube" model. Influence of hepatic blood flow, plasma and blood cell binding, and hepatocellular enzymatic activity on hepatic drug clearance. J. Pharmacokinet. Biopharm. 1977; 5:625-653.
Pithavala YK, Heizer WD, Parr AF, O'Connor-Semmes RL, Brouwer KLR. Use of the InteliSite® capsule to study ranitidine absorption from various sites within the human intestinal tract. Pharm. Res. 1998;15:1869-1875.
Prentis RA, Lis Y, Walker SR. Pharmaceutical innovation by the seven UK-owned pharmaceutical companies (1964-1985). Br. J. Clin. Pharmacol. 1998, 25:387-396.
Roberts MS and Rowland. Correlation between in vitro microsomal enzyme activity ad whole organ hepatic elimination kinetics; analysis with a dispersion model. J. Pharm. Pharmacol. 1986;38:177-181.
Shiner M. Anatomy of the small intestine. In: Haubrich WS, Schaffner F, Berk JE, editors, Gastroenterology, W. B. Sanders Co., 1994. p. 885-898.
Sioufi A, Gauducheau N, Pineau V, Marfil F, Jaouen A, Cardot JM, Godbillon J, Czendlik C, Howald H, Pfister C, Vreeland F. Absolute bioavailability of letrozole in healthy postmenopausal women. Biopharm. Drug Dispos. 1997;18(9):779-789.
Stenberg P, Bergstrom CAS, Luthman K, Artursson P. Theoretical prediction of drug absorption in drug discovery and development. Clin. Pharmacokinet. 2002;41(11):877-899.
Yamashita F, Wanchana S, Hashida M. Quantitative structure/property relationship analysis of Caco-2 permeability using a genetic algorithm-based partial least squares method. J. Pharm. Sci., 91:2230-2239 (2002).
Yoshida F, Topliss J. QSAR model for drug human oral bioavailability. J. Med. Chem. 2000;43:2575-2585.
Yu LX, Crison JR, Lipka E, Amidon GL. Transport approaches to the biopharmaceutical design of oral drug delivery systems: prediction of intestinal absorption. Adv. Drug Del. Rev. 1996a;19:359-376.
Yu LX, Crison JR, Amidon GL. Compartmental transit and dispersion model analysis of small intestinal transit flow in humans. Int. J. Pharm. 1996b;140:111-118.
Yu LX, Amidon GL. Characterization of small intestinal transit time distribution in humans. Int. J. Pharm. 1998a;171:157-163.
Yu LX, Amidon GL. Saturable small intestinal drug absorption in humans: modeling and explanation of the cefatrizine data. Eur. J. Pharm. Biopharm. 1998b;45:199-203.
Yu LX, Amidon GL. A compartmental absorption and transit model for estimating oral drug absorption. Int. J. Pharm. 1999a;186:119-125.
Yu LX. An integrated absorption model for determining causes of poor oral drug absorption. Pharm. Res. 1999b;16:1883-1887.
Yu LX, Gatlin LA, Amidon GL. Predicting gastrointestinal drug absorption in humans. In: Amidon GL, Lee PL, Topp EM, editors. Transport Processes in Pharmaceutical Systems. Marcel Dekker, Inc., 2000; p. 377-409.

Yu LX, Andrews CW. Predicting human oral bioavailability: Applications to 20 recently approved drugs. AAPS PharmSci 3 (2001) http://www.aapspharmaceutica.com/scientificjournals/pharmsci/am_abstract/2001/ 177.html
Yu LX, e-ADME: Predicting bioavailability and permeability. Gordon Research Conference Drug Metabolism, July 31-13, 2001.

DRUG METABOLISM IN PRECLINICAL DEVELOPMENT

Michael W. Sinz*

1. INTRODUCTION

The multi-faceted discipline of drug metabolism involves the identification of metabolites, the identification of enzymes involved in the metabolism of drugs, determination of the rates of these enzyme reactions, and the induction or inhibition of drug metabolizing enzymes. The concepts of drug metabolism are incorporated into most aspects of ADME (absorption, distribution, metabolism, and elimination), toxicology, and pharmacology. The importance of drug metabolism can be realized with an appreciation of how drug metabolism effects are integrated into the disposition and efficacy of pharmaceutical agents. Extensive metabolism by either the gut wall or liver can effect the extent of absorption and bioavailability of a drug by reducing the systemic exposure. In addition to being metabolized by drug metabolizing enzymes, drugs can also effect the enzymatic activity of drug metabolizing enzymes by either decreasing their intrinsic activity or increasing the amount of available enzyme which can lead to significant drug-drug interactions resulting in loss of efficacy or toxicity. The metabolic elimination or clearance of a drug is defined by the rate and extent of metabolism which often determines the time frame of pharmacological activity. In some instances, the rate and extent of drug metabolism is directly related to the efficacy of the drug, such as in the case of prodrugs or pharmacologically active metabolites. The presence or more importantly the absence of certain drug metabolizing enzymes in certain individuals (polymorphic enzyme expression) can have profound effects on the elimination or efficacy of drugs that are predominately metabolized by these enzymes. The biotransformation of drugs can also lead to toxic events in animals and humans by formation of reactive intermediates or metabolites that interact with endogenous macromolecules, such as proteins and nucleic acids. All of these brief highlights illustrate the importance of drug metabolism in the discovery, development, and eventual registration of a new drug.

---

* Michael W. Sinz, 5 Research Parkway, Bristol-Myers Squibb, Pharmaceutical Research Institute, Wallingford, CT.

Applications of Pharmacokinetic Principles in Drug Development
Edited by Krishna, Kluwer Academic/Plenum Publishers, 2004

As of 2002, the estimated cost of bringing a new drug to the market was between $500-800 million dollars and in total time takes 10-15 years from first synthesis to patient[1,2]. In today's pharmaceutical environment, a new chemical entity (NCE) will spend about 4 years in drug discovery, 2 years in preclinical development, 6-7 years in clinical development, and 19 months in regulatory review[1,3]. It has been projected that of 30,000 newly synthesized drug candidates only 8 would be predicted to eventually be approved by the FDA and only 1 of 30,000 would result in a positive return on investment for the pharmaceutical company[2,4]. Given this incredible loss of drug candidates during a long and expensive development process, pharmaceutical companies are exploring new and better ways to bring forward drug candidates with increased efficacy and increased probabilities of success. The aforementioned aspects of drug metabolism comprise a significant proportion of the reasons why many of the new drug candidates drop out along the development process or in rare cases result in removal of the drug from the market.

Although there is a clear line of demarcation between preclinical development and clinical development of a new drug candidate (progressing from *in vivo* animal testing to human testing), there is less clarity between preclinical development and drug discovery. Any discussion of drug metabolism-related experiments conducted during preclinical development should be prefaced with a brief description of similar assays conducted in drug discovery. In general, any experiment conducted in drug development (for the purposes of regulatory filings, such as an IND or NDA) is worthy of evaluation during the drug discovery phase either to gain a greater knowledge about a candidate or remove it from further progression due to unwanted liabilities. For example, the metabolic clearance of a drug can have profound effects on the plasma levels of the drug (relating to efficacy) and the daily dose. Characterization of the *in vivo* clearance of drug candidates are standard preclinical studies in order to predict the eventual human plasma concentrations, doses, and dosing schedule. These same studies can be conducted during drug discovery either through conventional animal dosing and pharmacokinetic analysis or by *in vitro* methods which afford greater throughput, such as the ability to study 100's of compounds per week.

Table 1 illustrates the typical timeframe for a variety of drug metabolism-related studies conducted from discovery through clinical development. These same studies will be discussed in greater detail in the upcoming portions of this chapter. In general, drug discovery is dominated by *in vitro* studies employing animal and human systems due to the increased throughput of these assays, whereas clinical development is conducted with normal healthy volunteers and patients. In between, during preclinical development, the studies are a mix of animal and human *in vitro* and *in vivo* studies all meant to predict the actual outcomes of the drug candidate in human patients. Depending on the size and philosophy of the pharmaceutical company, the studies conducted during discovery vs. preclinical development are blurred and quite varied.

2. DRUG METABOLIZING ENZYMES AND REACTIONS

The human body metabolizes or biotransforms drugs in an attempt to deactivate the pharmacological effects of a drug (xenobiotic) by converting these non-polar

Table 1. Comparison of *in vitro* and *in vivo* metabolism related studies typically performed during different stages of drug discovery and development.

	Discovery	Preclinical Development	Clinical Development
In Vitro	Metabolic Stability	Enzyme Inhibition	
	Enzyme Inhibition	Enzyme Induction	
	Enzyme Induction	Reaction Phenotyping	
	Reaction Phenotyping	Cross Species Metabolic Profiles	
	Formation of Reactive Metabolites	Formation of Reactive Metabolites	
		Prediction of Human Metabolic Clearance	
In Vivo	Animal Pharmacokinetics	Animal Pharmacokinetics	Human Pharmacokinetics
		Animal Metabolic Profiling	Human Metabolic Profiling
			Drug-Drug Interactions

chemicals into more polar and readily excretable metabolites, that are eliminated either via urine or bile. These reactions typically result in non-toxic and pharmacologically inactive metabolites. However there are numerous examples where drugs are metabolized to toxic and/or active metabolites. For example, the antineoplastic agent cyclophosphamide (actually a prodrug) is metabolized to both an active metabolite (the methylating agent, a phosphoramide mustard) as well as a toxic metabolite (acrolein) which is responsible for producing hemorrhagic cystitis of the bladder[5], Figure 1. Also, the first thiazolidinedione antidiebetic drug, Rezulin® (troglitazone) was removed from the market in 1999 due to severe hepatotoxicity in some patients where the toxicity was proposed to be mediated by one or more of the metabolites of troglitazone[6,7]. The cholesterol lowering agent Lipitor® (atorvastatin) is the major active component for inhibiting HMGCoA reductase, however the hydroxylated metabolites, ortho and para-hydroxy atorvastatin also contribute to the overall efficacy of the drug[8,9], Figure 1.

The panoply of drug metabolizing enzymes can be broadly categorized into two groups: Phase I and Phase II enzymes. The Phase I enzymes are considered functionalization enzymes because they incorporate or uncover a polar functional groups on the drug molecule. These functionalization reactions can be oxidative, reductive, or hydrolytic in nature. The most common type of Phase I reactions are the oxidative reactions performed by the cytochromes P450 (CYP450), flavin monooxygenases, and alcohol/aldehyde dehydrogenases. Hydrolysis or cleavage reactions are also considered Phase I type reactions, of which epoxide hydrolase, esterase, and amidase are the predominant enzymes found during xenobiotic metabolism. Enzymatic reactions in which a part of the molecule undergoes reduction is a third and minor biotransformation pathway found in Phase I reactions. Typical reduction reactions are catalyzed by CYP450 enzymes or by enzymes found

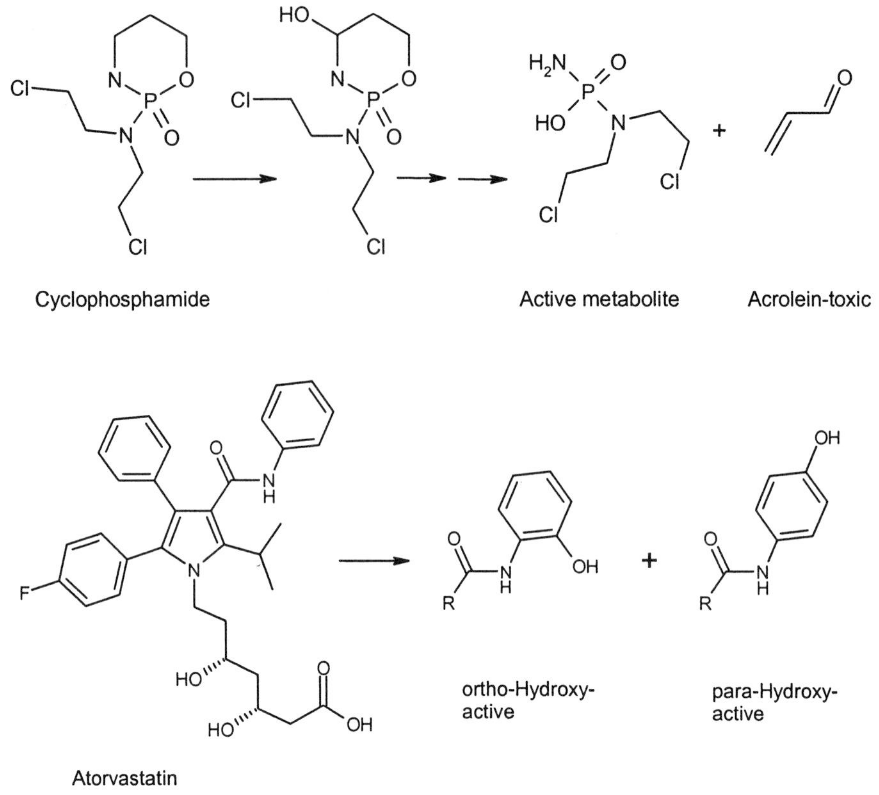

Figure 1. Metabolism of cyclophosphamide to active and toxic metabolites, as well as atorvastatin to two active metabolites.

Table 2. Phase I drug metabolizing enzymes.

Phase I – Functionalization Enzymes	
Oxidation	Cytochrome P450
	Flavin Monooxygenase
	Monoamine Oxidase
	Alcohol & Aldehdye Dehydrogenase
	Aldehyde Oxidase
	Xanthine Oxidase
Hydrolysis	Epoxide Hydrolase
	Esterase
	Amidase
	Phosphatase
	Peptidase
	Beta-Oxidation
Reduction	Cytochrome P450
	DT-Diaphorase, (NADPH quinone oxidoreductase)
	Aldehyde & Ketone Reductase
	Intestinal Microflora

in bacteria located in the host organism's gastrointestinal tract. Table 2 lists the majority of enzymes that catalyze Phase I (functionalization) reactions. Greater details into the mechanisms of Phase I enzymes and examples of reactions catalyzed by these enzymes can be found in several excellent reviews and texts[10-18].

Phase II enzymes perform conjugative reactions, where a polar endogenous cofactor, such as a sugar or sulfate is covalently bound to a metabolite or sometimes the parent drug. Of the Phase II reactions, glucuronidation, sulfation, glutathione conjugation, and acetylation are found most often in drug metabolism. Table 3 lists the drug metabolizing enzymes that participate in Phase II-conjugative reactions typically found in mammals. In general, both Phase I and II reactions will result in metabolites that are more polar, except for the Phase II reaction of methylation which can actually conceal a polar functional group, such as a thiol or alcohol and make it less polar. Additional information and examples of conjugation reactions can be found in several reviews[10-14, 18, 19, 20].

Table 3. Phase II conjugative enzymes.

Phase II – Conjugative Enzymes
Glucuronidation (glucuronic acid conjugation)
Sulfation
Glutathione Conjugation
Acetylation
Methylation
Amino Acid Conjugation
Glucose Conjugation

Nearly all of the aforementioned drug metabolizing enzymes are comprised of families of enzymes from as few as 2 enzymes in the case of monoamine oxidase A and B and some with greater than 10 members, such as the CYP450's. Table 4 illustrates the diversity and extent to which the CYP450 are involved in the metabolism of known drugs. The table represents the relative amounts of CYP450 enzymes found in the human liver and the proportion of known drugs metabolized by the CYP450 enzymes[21,22]. It is clear to see that CYP3A is the predominant isoform present in the liver, however in regards to the proportion of drugs metabolized by the CYP450's both CYP3A4 and the minor form CYP2D6 are the major enzymes involved in metabolism. Overall, the most prevalent enzyme systems involved in drug biotransformation reactions are the cytochromes P450 along with the Phase II reaction of glucuronidation. These Phase I and Phase II systems often work in concert by first oxidizing the drug (e.g., hydroxylation) followed by conjugation of the newly formed functional group (e.g., glucuronidation). Representative examples of the both Phase I and Phase II reactions are illustrated in Figure 2. Carbamazepine is hydroxylated by CYP450 (oxidation) followed by ring opening by epoxide hydrolase (hydrolysis). Phenytoin is hydroxylated by CYP450 (oxidation) followed by glucuronidation (conjugation). Acetaminophen is directly conjugated forming both glucuronide and sulfate conjugates as the major metabolites.

Have we found all of the drug metabolizing enzymes? Although we have found the majority of mammalian drug metabolizing enzymes, new enzymes are periodically elucidated. For example, nitroglycerin (glyceryl trinitrate) used to treat angina and heart failure for the past 130 years, is bioactivated to its 1,2-glyceryl dinitrate metabolite. The enzyme involved in this bioactivation was recently found to be an unidentified mitochondiral aldehyde dehydrogenase enzyme. Although several enzyme systems have been shown to metabolize nitroglycerin, such as CYP450 and glutathione transferase, this new dehydrogenase is thought to play a central role in nitroglycerin bioactivation[23].

Table 4. Relative amounts of CYP450 enzymes found in human liver and the proportion of drugs that are metabolized by various CYP450 enzymes.

CYP450	**Relative Amount**	**CYP450**	**Proportion of Drugs Metabolized by CYP Enzyme**
1A2	13%	1A2	6%
2A6	4%	2A6	2%
2B6	<1%	2B6	----[a]
2C	18%	2C9	10%
	(2C9>2C19>2C8)	2C19	4%
2D6	2%	2D6	30%
2E1	7%	2E1	5%
3A	30%	3A	40-50%
Other	26%	Other	3-13%

[a] not well characterized

3. MODEL SYSTEMS EMPLOYED IN DRUG METABOLISM STUDIES

There are multiple systems or models used in the evaluation of drug metabolism related events and reactions, such as *in vitro*, *in vivo*, in situ, and in silico methods. By far the *in vitro* methods are used early in the drug development process and constitute the majority of data generated in this area. As a drug progresses through preclinical and clinical development, *in vivo* studies become more relevant. The overwhelming majority of *in vitro*, in situ, and in silico methods employed in drug metabolism studies are derived from a single tissue (liver), therefore confirmation of *in vitro* results are prudent when considering all of the ADME properties functioning in a whole animal or human system. Ultimately, nearly all of the *in vitro* conclusions, especially all of the positive results, are addressed in some form using *in vivo* studies for confirmation of results. As stated in the FDA Guidance for Industry: *In vivo* drug metabolism/drug interaction studies[24]. "A complete understanding of the relationship between *in vitro* findings and *in vivo* results of metabolism/drug-drug interaction studies is still emerging. Nonetheless, *in vitro* studies can frequently serve as an adequate screening mechanism to rule out the importance of a metabolic pathway and drug-drug interaction that occur via this pathway so that subsequent *in vivo* testing is unnecessary........In contrast, when positive findings arise in *in vitro* metabolic and/or drug-drug interaction studies, clinical studies are recommended because of the limited ability at present of *in vitro* findings to give a good quantitative estimate of the clinical importance of the metabolic pathway or interaction.". The following sections will provide a description of each *in vitro* and *in vivo* model used to study drug metabolism along with a discussion of application during preclinical development.

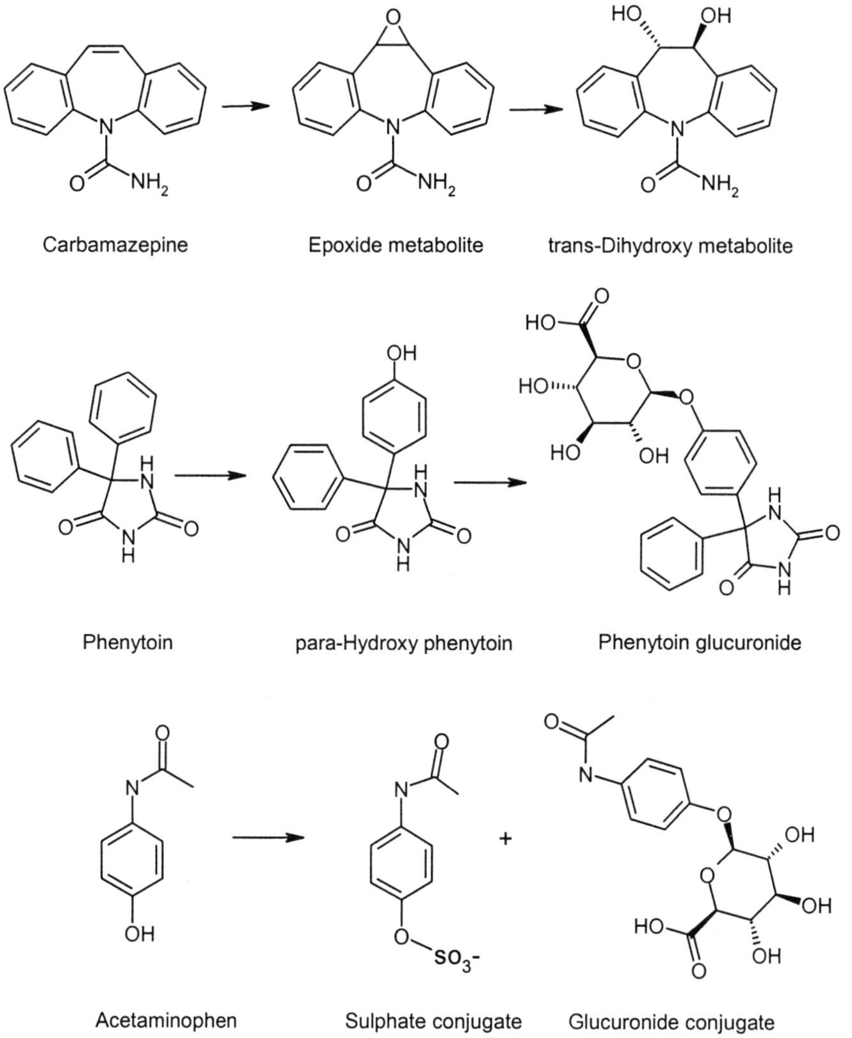

Figure 2. Metabolism of carbamazepine, phenytoin, and acetaminophen.

3.1. *In Vitro* Methods

In vitro methods or model systems used to study drug metabolism come in many different forms, each having their own advantages and disadvantages[25]. As with home improvement, a hammer can't be used for every job, the same logic applies to *in vitro* drug metabolism studies, a single system can't be used for all purposes. As mentioned, the majority of *in vitro* systems are derived from liver tissue and each system contains a subset of drug metabolizing enzymes, therefore each constitutes a different level of complexity and capability[18, 25, 26]. The variety of *in vitro* methods, the ease at which they can be manipulated (concentration, pH, and temperature), and their utility in higher throughput screening methods make these models more popular in early pharmaceutical research or for more detailed mechanistic studies. From the simplest *in vitro* system to the most complex are: individual drug metabolizing enzymes, subcellular fractions of liver tissue, hepatocytes, liver slices, and liver perfusion.

3.1.1. Single Enzyme Systems

Whether derived from cDNA expression systems or isolated from various tissues, single enzyme systems are powerful tools to study individual enzyme reactions and the metabolites of these reactions. These isolated enzymes can be stored at -80°C for extended periods of time (years) making them readily available whenever necessary and are easy to manipulate either by hand or through the use of high throughput liquid handling systems commonly employed in drug discovery. The most common expressed enzymes used in drug metabolism are available through commercial vendors. For example, the following Phase I and Phase II human drug metabolizing enzymes are readily available: CYP450's 1A1, 1A2, 2A6, 2B6, 2C8, 2C9, 2C19, 2D6, 2E1, 3A4, 3A5, 4A11, and CYP19; MAO A and B; FMO1, 3, 5; UGT1A1, 1A3, 1A4, 1A6, 1A7, 1A8, 1A9, 1A10, 2B7, and 2B15; SULT1A1, 1A2, 1A3, 1E, and 2A1; and GSTA1, M1, and P1. Single enzyme systems are often employed in specific situations during discovery, such as the determination of drug inhibition potential of particular enzymes (e.g., CYP3A4) or in evaluating polymorphic drug metabolism (e.g., CYP2D6). The single enzyme system is also used extensively during drug development to perform more complete reaction phenotyping studies and in the biosynthesis of metabolites that can be used in further pharmacological testing or as analytical standards.

The power of this system is fueled by its simplicity of components and lack of confounding factors, such as other enzymes or membranes allowing for a direct assessment of the interaction between drug and enzyme, however this simplicity also has some disadvantages. Single enzyme systems are good for exploring and delving into the mechanisms of an enzymatic reactions, however they are not amenable to studying processes that involve additional enzymes or cofactors. By removing all of the surrounding enzymes, cofactors, organelles, and cells from the system, only a small portion of what actually occurs *in vivo* is represented and observed. Furthermore, enzymes isolated from organs or tissues may be degraded or modified during preparation, such as the internalization of the UDP-glucuronyl transferase enzymes during microsomal isolation[20]. Such degradations or alterations that can effect the enzyme activity thereby skew the final outcome or prediction of the experiment.

Most reports of single enzyme systems demonstrate that expression systems exhibit catalytic activities and properties comparable to those attributed to liver microsomes. For example, Lee, et al., reported that CYP3A4 coexpressed with

CYP450 reductase in baculovirus infected insect cells gave comparable results to microsomes in regard to kinetic properties, regioselectivity of metabolism, and modulation by inhibitors and activators when using testosterone as a substrate[27]. However, these individual enzymes are expressed in vectors that do not always resemble the endoplasmic reticulum of mammals, such as membranes formed in E. coli or insect cell lines. Kisselev, et al., have shown that variations in membrane properties (type and charge of bilayer lipids) altered the enzyme kinetics and stereoselectivity of the (+/-) isomers of benzo[a]-7,8-dihydrodiol epoxidation by expressed human CYP1A1 by potentially altering the substrate binding or changing the interaction between CYP450 and CYP450 reductase[28]. Furthermore, with the CYP450 expressed enzymes, the ratios of expression levels or presence of CYP450, CYP450 reductase, and cytochrome b_5 can influence the activity and stability of CYP450. For example, coexpression of CYP450 and reductase along with cytochrome b_5 was shown to stabilize CYP3A4 and increase its activity[29]. Also, the expression level of cytochrome P450 reductase can regulate CYP450 activity by inactivation of CYP450 via radical and heme oxygenase mediated pathways as was shown with CYP2D6[30]. Whether these membrane or coenzyme changes are important in making *in vivo* predictions of metabolism, clearance, or inhibition potential are not fully understood at this time. Additional research into these subtle modifications may be necessary to understand the complex interactions between CYP450, CYP450 reductase, b_5, and membrane components, to the eventual level of expression that occurs in these systems compared to more intact systems derived directly from liver tissue or *in vivo*. In general, single enzyme systems should be used for the qualitative estimation of kinetic parameters, which can then be confirmed in more complex *in vitro* or *in vivo* systems[31]. This makes them excellent tools in drug discovery and early preclinical development where initial estimates of metabolic turnover or inhibition potential are most relevant.

3.1.2. Subcellular Fractions

Subcellular fractions are derived from liver tissue that has been homogenized thereby disrupting the cellular integrity of the tissue and repartitions several cellular components that can then be separated by differential centrifugation[32, 33]. Subcelluar fractions include (from most complex to simplest forms): liver homogenate, S9 fraction, cytosol, microsomes, peroxisomes, and mitochondria. These were some of the first *in vitro* systems employed to study drug metabolism and toxicology with microsomes still being the predominant *in vitro* system employed today in drug development. Microsomes contain a variety of drug metabolizing enzymes the most common of which are the Phase I oxidative enzymes, such as the CYP450 and flavin monooxygenases. In addition to these enzymes, microsomes contain glucuronosyltransferases, microsomal epoxide hydrolase, esterases, amidases, microsomal glutathione S-transferase, and methyltransferases. Microsomes are the principle *in vitro* system employed to evaluate metabolic stability (predict metabolic clearance), drug inhibition potential, cross-species comparisons of metabolic profiles, and reaction phenotyping in preclinical development. General reviews of subcellular systems, their isolation, and uses can be found in the following references[18, 34].

Each subcellular fraction contains a subset of liver enzymes, except for the liver homogenate which is a diluted whole liver mixture. Therefore, when deciding which subcellular fraction to employ, of primary importance are the enzymes found in each fraction. As mentioned, the microsomes contain the majority of oxidative enzymes, however the cytosol contains the majority of Phase II conjugating enzymes and oxidoreductase enzymes, such as sulfotransferases, acylating and methylating enzymes, glutathione conjugation, alcohol and aldehyde dehydrogenases, as well as xanthine oxidase, aldehyde oxidase, esterases, amidases, and cytosolic epoxide hydrolase. Mitochondria contain such enzymes as MAO and amino acid conjugation enzymes, whereas peroxisomes contain β-oxidation (hydrolytic) enzymes[17, 35]. Fractions such as cytosol, mitochondria, and peroxisomes, due to their lack of oxidative enzymes, are not commonly used in drug development. Typically these more specialized enzymes become important if they are found to be part of the biotransformation pathway during drug development. Some biotransformation processes, such as β-oxidation of alkyl chains, may actually involve enzymes from different parts of the cell. In the case of β-oxidation of alkyl side chains, the initial hydroxylation reaction is carried out by CYP450 enzymes in the endoplasmic reticulum, followed by oxidation of the alcohol to a carboxylic acid presumably by cytosolic dehydrogenase enzymes, which is then followed by the β-oxidation enzymes found in peroxisomes that eventually lead to the successive cleavage of two-carbon units from the parent molecule[17]. This is a situation where the ultimate metabolite would not be detected in anything short of a multi-enzyme system, such as hepatocytes, liver slices, or a liver perfusion model and exemplifies the need to carefully consider the subcellular fraction in regard to the metabolic process under.

3.1.3. Cellular Systems

Hepatocytes or liver cells (fresh isolates or cryopreserved) are probably the second most utilized *in vitro* system to study drug metabolism-related reactions or interactions in preclinical development, however their use in drug discovery is more limited due to the cost and availability of human hepatocytes. In comparison to the aforementioned *in vitro* systems, hepatocytes have an intact cell membrane, the enzymes within the cell are not subjected to harsh isolation conditions, liver cells contain the majority of liver drug metabolizing enzymes, and have a normal complement and concentration of enzymes and cofactors. This is in contrast to the single enzyme and subcellular systems that often use abnormally high concentrations of endogenous cofactors, such as 1.0 mM NAPDH or 3.0 mM UDPGA (uridine diphospho-glucuronic acid) for CYP450 and glucuronidation reactions, respectively. Also, the earlier systems lacked a membrane barrier that drugs needed to cross to interact with enzymes. Cell surface or intracellular transporters may be involved in increasing or decreasing the concentration of drug and/or metabolites inside of the cell, thereby potentially altering the concentration of drug presented to the enzymes for metabolism or to different cellular locations, such as the nucleus when considering such changes as enzyme induction. Overall, the hepatocyte system is more reflective of the actual *in vivo* situation than the preceding *in vitro* systems. There are several good references describing general uses of hepatocytes as well as the isolation and characterization of hepatocytes from multiple species[36, 37, 38].

Hepatocytes can be employed in two different formats: isolated suspension or primary culture. The major differences between the two formats are: 1) isolated suspension are generally used for shorter term incubations (1-3 hours) whereas primary cultures can be used for several days, and 2) isolated suspensions contain similar levels of drug metabolizing enzyme activity as found in fresh liver tissue, however primary cultures of hepatocytes lose significant enzyme activity after 1-2 days in culture. Isolated suspensions are best suited for experiments that require optimal concentrations of enzyme, such as comparisons of metabolic stability[39], estimations of metabolic clearance[40], cross-species comparisons of metabolic profiles[41-44], reaction phenotyping, or drug inhibition studies[45]. Cultured primary hepatocytes, are commonly used to study changes that take longer time to evolve, such as CYP450 induction[43, 46].

Hepatocytes can be employed from two different starting points: freshly isolated or cryopreserved cells. Freshly isolated cells can be employed in all types of drug metabolism studies, either in suspension or culture. The ability to store cryopreserved hepatocytes in liquid nitrogen from multiple different species and utilize them for drug metabolism studies at any time has an advantage over the laborious nature of isolating cells when needed and the spurious access to human liver donors. The major oxidative enzymes (CYP450) and conjugation enzymes (glucuronidation and sulfation) have been shown to be similar between fresh and cryopreserved cells[47, 48] making them suitable for metabolite identification, species comparisons, and drug inhibition studies[39, 49]. More recent data suggests that certain lots (individual donors) of cryopreserved cells may be used to study enzyme induction when cells are properly prepared although this is still controversial[47, 50.]

Alternate cell types of liver origin that have shown promise in the areas of metabolic stability and CYP450 induction are human hepatoblastoma cells, such as the HepG2/C3A[51, 52]. Although these cell lines can mimick certain aspects of liver cells, they often do not maintain all hepatocyte functions. These 'liver' cells offer a unique advantage similar to cryopreserved cells in that they can be used when ever necessary. They offer an advantage over human hepatocytes in that the variability often encountered with donor hepatocytes (due to quality of liver tissue, cells, and interindividual differences in enzyme levels) is diminished by employing a well characterized cell line that does not change. However, this also eliminates the normal differences that each individual naturally has with the level and function of drug metabolizing enzymes which may be important in judging the human variability of a response[53]. Needless to say, these new cell types will find a place in the armamentarium of *in vitro* systems, albeit not as a direct replacement for fresh tissue at the present time.

3.1.4. Liver Slices

Liver slices are one of the oldest *in vitro* systems to study drug metabolism, however their utilitiy decreased for a period of time due to the irreproducible nature of preparing the slices[25, 54]. In the 1980's and 90's, the use of slices resurfaced as an important tool to study metabolism and other biochemical functions with the release of new and reproducible slicing instruments. They have been employed in the study of metabolite identification, cross species comparisons, and prediction of metabolic clearance[44, 55, 56]. Liver slices represent an *in vitro* system with a higher level of structural (cellular) design, where all of the liver cell types are present and cellular junctions remain intact. Similar to the hepatocytes, liver slices contain all of the phase I and phase II enzymes necessary to obtain a complete metabolic profile. An issue with liver slices is the need for a drug substance to penetrate multiple layers of cells by diffusion, as opposed to the *in vivo* situation where the circulatory system 'pumps' the drug into the tissue. Therefore, cells on the outer surface of the slice are exposed to higher concentrations of drug and the concentration tends to decrease while moving toward the center of the slice. This concentration gradient across the slice may help explain why the predicted clearances from liver slices tend to underestimate the clearance predicted by microsomes, hepatocytes, and *in vivo* data[56].

3.1.5. Organ Perfusion – Isolated Perfused Liver

Although debatable, I have place the isolated perfused liver in the section on *in vitro* models, however this could also be considered an *in situ* technique. The focus here is on the liver as it constitutes the major organ for biotransformation reactions although other organs can be employed in this capacity, albeit to a lesser extent[57]. The isolated perfused liver can be used to study the uptake of drugs into or out of the liver[58], identification of metabolites formed by the liver[59], and estimates of pharmacokinetic parameters, such as hepatic clearance, hepatic extraction ratio, and in assessing non-linear pharmacokinetics[57, 60]. There are several advantages of the liver perfusion system over the previously mentioned *in vitro* techniques, such as: 1) more representative, anatomically and physiologically, to the *in vivo* situation, ie., drug input is through the normal vasculature and the drug metabolizing enzymes are appropriately distributed along the vasculature (periportal or pericentral)[57, 60]; 2) the ability to vary the flow rate and media components entering the liver[61]; and 3) sampling of drug components from both the vena cava and bile duct for metabolite identification and quantitation[59].

Although organ perfusion is an excellent model to study the whole organ effects of a drug (metabolism, route of elimination, pharmacokinetics) it is typically a single experiment with a single drug, therefore it is best utilized to answer specific questions concerning a drug candidate during preclinical development. In addition, the experimental set-up and surgical skills necessary to properly perform this experiment are much more rigorous than the *in vitro* models previously described[57]. Once mastered, this technique is extremely powerful in assessing the ADME properties of a new drug by the liver.

3.2. *In Vivo* Methods

In vivo animal or human studies are used during all phases of drug development for the determination of pharmacokinetic properties and identification of metabolites. They act as the confirmation of predictions made by *in vitro* experiments and serve as the final results or interpretation on how a drug is metabolized or will interact with other drugs. Clinical studies related to metabolism are essential in identifying the variability in patient populations due to genetic changes, age, or race, as well as the confirmation or absence of potential drug interactions observed from *in vitro* studies conducted earlier in preclincal development. Clinical studies in humans are paramount to the regulatory acceptance of any new drug candidate, whereas these variables can not be addressed in preclinical animal models. Not only the confirmation of a drug interaction in humans is important, the magnitude of the interaction with other commonly prescribed drugs is essential to determine if dose adjustments are warranted or certain coadministered drugs should not be given in combination.

In vivo studies can take multiple forms and many considerations must be taken into account when designing *in vivo* studies for drug metabolism, such as: 1) the dose of drug; 2) route of administration (typically oral or IV); 3) single drug administration or cassette dosing (N-in-one dosing)[62, 63], where single drug administration is the standard for preclinical studies and cassette dosing is common during drug discovery; 4) radiolabeled (hot) or unlabeled (cold) drug administration; 5) intact animals or bile-duct cannulated (BDC) animals; 6) multiple vs single dose administation to elucidate time dependent changes; and 7) animal species. Generally the dose, animal species, and route of administration are determined by the drug or stage of development, whereas the use of radiolabled drug or BDC animals are typically determined by the question being addressed. For example, to fully characterize the metabolic profile from an *in vivo* study, a radiolabled drug is necessary. The radiolabel, typically a C-14 or H-3 label, allow for the detection and quantitation of all metabolites present. Any other technique, such as UV, fluorescence, or even mass spectrometry have limitations on what can be observed in the chromatographic profile of metabolites, whereas radioactivity profiling is considered a universal technique. Also, radioactivity does not require an analytical standard for each metabolite to perform quantitative analysis because the radioactivity allows for a direct comparison of the amount of each metabolite formed along with the amount of parent. Bile duct cannulated animals are useful in further characterizing the route of metabolite and parent drug elimination without having to extract feces, the final elimination matrix for many biliary metabolites. Bile samples are cleaner and easier to extract than fecal samples making them more amenable to chromatographic separation and identification. Bile duct cannulated studies are common in rats and can be performed in dogs and monkeys if necessary, however this type of study tends to be rare in the case of humans. This is unfortunate because a great deal of metabolite identification occurs from animal bile and these metabolites are more difficult to identify from human feces. In particular, metabolites generated by intestinal microflora that contaminate fecal samples and represent metabolites not generated by the host animal. A recent example of a human bile duct cannulated study (T-tube study) was performed with the cholesterol lowering agent, Lipitor® [64]. The majority of Lipitor® metabolites were originally identified from animal bile and attempts were made to identify these same metabolites from human feces without much success. A T-tube study in two patients where Lipitor® was administered and bile was collected, allowed for the identification of multiple metabolites that would have been difficult to identify by standard means.

Clearly, the use of BDC animals and a radiolabeled drug (in conjunction with mass spectrometry for metabolite identification) yield a wealth of information on the metabolites formed, amount of metabolite present, and route of elimination. This is typically known as a mass balance (radiolabeled) study where a great deal of metabolism information can be assimilated through a single *in vivo* study. The term mass balance comes from the other purpose of the study which measures the total amount of radioactivity eliminated in all collected routes of elimination (typically urine and feces) compared to the amount of radioactivty administered. Mass balance is achieved if the majority (~90%) of the radioactivity is recovered, otherwise low recovery may indicate some form of retention (covalent binding of drug or metabolites to macromolecules) within the body, a volatile metabolite, or an extremely slow elimination process.

Most *in vivo* preclinical studies are conducted with an eye as to how the drug will perform in humans where the animals are actually surrogates for predicting what will occur in humans. Similar to the aforementioned BDC study, *in vivo* studies can also be used to compare the metabolic profile of different animal species used in pharmacology or toxicology studies with those of human *in vitro* samples or patient samples. This is to show that from a pharmacological and toxicological standpoint, the animals metabolize the drug in a similar fashion as humans (more about this in the proceeding section on cross-species comparisons). In contrast, drug interaction studies, either due to inhibition or induction of drug metabolizing enzymes are a common *in vivo* study performed during human clinical trials, however these same studies are not as useful in animals due to their low predictability to humans[65]. Although inhibition and induction experiments in animals can help explain unusual pharmacokinetic-pharmacodynamic situations found in animals during preclinical testing their use in predicting human interactions is limited.

4. DRUG METABOLISM STUDIES

The amount of drug metabolism related information utilized in the development and regulatory submission of new drug candidates has steadily increased over the past two decades. This increase in drug metabolism information has been fueled by our ability to more accurately predict human outcomes, such as metabolic clearance, drug interactions due to inhibition or induction, patient variability due to polymorphic drug metabolizing enzymes, or toxicological events, from *in vitro* methods. Our improved understanding of *in vitro* methods, such as those described in the preceding section, revolutionized our ability to predict and understand the human *in vivo* situation. In general, preclinical studies are designed to characterize the metabolism of drug candidates in the animal species employed in pharmacokinetic, pharmacologic, or toxicologic studies, as well as predict human metabolism and interactions. From a preclinical standpoint, there are numerous studies that contribute to our understanding of drug biotransformation and prediction of human metabolism or interactions. A central theme in most of the preclinical studies deals with the identification of metabolites from *in vitro* animal and human tissues and *in vivo* animals studies. Surrounding this theme are: the rate of metabolism, the enzymes involved in forming the metabolites, and the prediction of human drug-drug interactions. These experiments generally take on the following descriptions: metabolite identification, cross species comparisons of metabolic profiles, prediction of human metabolic clearance from *in vitro* methods, reaction phenotyping, and drug-drug interaction studies (inhibition and induction).

Although the focus here is on preclinical drug metabolism studies, many if not all of the following experimental descriptions can be used during different stages of drug discovery[18]. At the other extreme, clinical drug metabolism studies are used to characterize the human metabolism of a drug candidate and quantitate the extent of any predicted drug interactions. However, when unexpected results occur during clinical development these same preclinical studies (*in vitro* and *in vivo*) can be extremely useful in elucidating the mechanism behind unusual or unpredicted clinical findings. For example, when pharmacological efficacy extends well beyond measurable parent drug concentrations, loss of efficacy occurs after multiple dosing of drug, or unpredicted toxicological effects are found in patients, these effects could be due to active metabolites, induction of drug metabolizing enzymes, or formation of reactive metabolites, respectively. Each of these events, uncovered during clinical trials, could be explored mechanistically with preclinical *in vitro* and *in vivo* models.

4.1. Formation of Metabolites: Routes and Rates of Metabolism

4.1.1. Metabolite Identification

As previously mentioned, metabolite identification forms the cornerstone to many of the current preclinical drug metabolism. Preclinical studies that utilize metabolite identification are: identification of potential active or toxic metabolites; comparison of metabolic profiles from different species used in preclinical development; and most importantly, the first look at human metabolites from *in vitro* models several years prior to clinical studies. In nearly every instance, the use of an appropriately radiolabeled parent drug will aid in the identification of metabolites. Although unlabeled drug can be used for this purpose, a radiolabeled drug (and therefore radiolabeled metabolites) can be readily detected and quantitated with a radiochemical detector. An unlabeled parent compound may not allow for all of the metabolites to be observed in a chromatogram, eg., metabolites in the solvent front. Therefore, for the purpose of complete metabolite identification from either *in vitro* or *in vivo* experiments, a radiolabeled parent drug is the preferred method of metabolite identification and quantitation in preclinical studies.

The metabolites and parent drug are generally separated by HPLC after which there are a variety of detection and identification instruments available for the characterization and identification of metabolites. These detectors most often include UV detectors, mass spectrometers, and radiochemical detectors. Stand alone detectors that utilize direct injections, such as mass spectrometry, NMR, and to a lesser extent UV and IR spectral characterization are also used to identify metabolites once they have been sufficiently isolated and purified. The most common tool used to identify metabolites is the LC-MS[18]. The molecular weight of a metabolite can often yield clues as to the type of metabolism that has occurred. For example, the addition of 16 amu is indicative of a hydroxylation reaction or a decrease in molecular weight may indicate that a hydrolysis or dealkylation reaction has taken place. Fragmentation of the metabolite gives further clues to the type of metabolic reaction, as well as information pertaining to the site of the metabolic event[18]. Sometimes this fragmentation information is sufficient to characterize the structure of the metabolite, other times several characterization techniques are necessary to fully confirm the structure of the metabolite. A typical secondary tool after LC-MS analysis is NMR which is extremely useful in determining which exact position on the parent molecule has been modified. Two powerful techniques developed in recent years have been the LC-NMR and the coupling of several detectors in tandem LC-NMR-MS, both providing separation and identification in a more efficient manner. These techniques have been recently reported in the identification of metabolites of tacrine and DPC-963 [66, 67]. A good review of this technique is given by Wilson, Nicholson, and Lindon citing several additional examples [68].

Metabolites can be generated and obtained from a variety of different sources. Typical *in vitro* sources are liver microsomes, hepatocytes, *in vivo* matrices such as plasma, bile, and urine, as well as bacterial systems. Sometimes it becomes necessary to measure or detect metabolites in specific organs, such as brain or tumors, if these are target organs or it is hypothesized that metabolism may be occurring in these tissues. Often each of these matrices need to be extracted with organic solvents to remove macromolecules and eventually concentrated to increase the amount of minor components. Sometimes even larger amounts (0.1-10 mg) of material are necessary to clearly identify metabolites and larger scale *in vitro* reactions are needed. These *in vitro* bioreactors, which use cDNA expressed enzymes, liver microsomes, hepatocytes, or microorganisms can generate larger quantities of metabolites for the purposes of metabolite identification, testing in pharmacological models (as possible active metabolites), or for use in analytical assays where metabolite concentrations need to measured [69]. Sometimes due to the novel nature of the metabolite or the stereochemistry inherent in the metabolite, chemical synthesis of the metabolite is very difficult and biosynthetic methods are a useful alternative. Rushmore, et al. were able to show that suspension cultures of Sf21 insect cells expressing either CYP3A4 or CYP2C9 were able to metabolize testosterone, diazepam, and diclofenac to milligram quantities of their respective metabolites, 6β-hydroxytestosterone, temazepam, and 4-hydroxydiclofenac, respectively [69]. Also, Trejtnar was able to demonstrate that rat hepatocytes could produce similar amounts of 3-hydroxyoracin from the cytostatic agent oracin as could be isolated from *in vivo* rat experiments, however the *in vitro* approach was found to be much more efficient and less time consuming than the *in vivo* approach [70].

4.1.2. *Active and Toxic Metabolites*

Prodrugs are often created by design, whereby an inactive form of the drug is given that requires some form of enzymatic or chemical transformation to release the active product which elicits a pharmacological effect[71]. However many prodrugs are discovered serendipitously through the process of metabolite identification and subsequent pharmacological testing, ie., active metabolites. Although metabolites are designed to be rapidly eliminated from the body and generally do not contribute to the efficacy of a drug, they can sometimes have equal or greater efficacy than the parent drug. Formation of active metabolites contribute to the total pharmacological effect and the duration of the effect by often circulating in plasma after the parent drug has been eliminated. There are some classic examples of active metabolites, such as the glucuronide conjugate of morphine and the metabolism of terfenadine

(Seldane®) to fexofenadine (Allegra®)[72, 73], Figure 3. A more recent example is the identification of para- and ortho-hydroxy metabolites of the cholesterol lowering agent, atoravastatin (Lipitor®). These metabolites were also HMG-CoA reductase inhibitors which contriubuted ~70% to the overall circulating activity[8], Figure 1. Also, the marketing of Clarinex® (desloratidine), an active metabolite of the commonly prescribed antihistamine Claritin® (loratidine) by Schering in 2002[74], Figure 3.

In contrast to increased pharmacological activity, some metabolites can be harmful or toxic to living systems. Certain metabolic reactions are indicative of potential toxicity, such functional group reactions include α,β-unsaturated ketones, terminal alkynes, and a variety of quinone-type metabolites which can be reactive enough to covalently bind to essential macromolecules and cause toxicity either directly or indirectly through an autoimmune response[75, 76]. For example, glutathione adducts are typically formed by the addition of the nucleophilic glutathione sulfhydryl group to an electrophilic drug or metabolite. This metabolic reaction acts as a protective mechanism to scavange reactive metabolites that could potentially bind to important macromolecules and cause toxicity. An example of this reaction is the hepatotoxicity of the antidiabetic agent troglitazone (Rezulin®) which may be mediated through reactive intermediates that have been trapped as a series of glutathione adducts in human liver microsomes and isolated from rats after dosing with troglitazone[7], Figure 4. An example of immune mediated toxicity due to metabolic activation is halothane. Halothane is metabolized by CYP2E1 and to a lesser extent by other CYP450 isoforms to a reactive species that becomes covalently bound to cellular proteins creating neoantigens which then elicit an immune reaction causing severe liver necrosis[77], Figure 4.

In vitro or *in vivo* metabolism studies can reveal possible reactive metabolic pathways which can be further explored through the use of carbon-14 labeled drugs and covalent binding studies. *In vitro* studies can determine the extent of binding through exhaustive extraction or dialysis experiments and *in vivo* studies can elucidate covalent binding through the use of mass balance studies. The mass balance study (recovery of a radioactive dose) is conducted during preclinical testing whereby a radiolabeled parent drug is dosed to animals and bile, urine, feces, and sometimes expired air are collected. The total amount of radioactivity eliminated from the animal should be between 90-100% of the total amount of radioactivity given in the original single dose. When less than total recovery occurs, this is indicative of the drug binding to or being stored in tissues which may be the result of covalent binding. Drugs with extremely long half-lives *in vivo* may not result in total recovery of radiolabeled material unless urine and feces are collected for extended periods of time. Covalent binding of drug or metabolites from either *in vivo* or *in vitro* studies merely indicate an increased risk for toxicity to occur. These studies in of themselves are not toxicology studies and extrapolations from covalent binding data to toxicity should not be made directly without supporting *in vivo* data. Ultimately, toxicity is demonstrated through the use of *in vivo* studies and *in vitro* experiments merely indicate a potential for such reactions to occur.

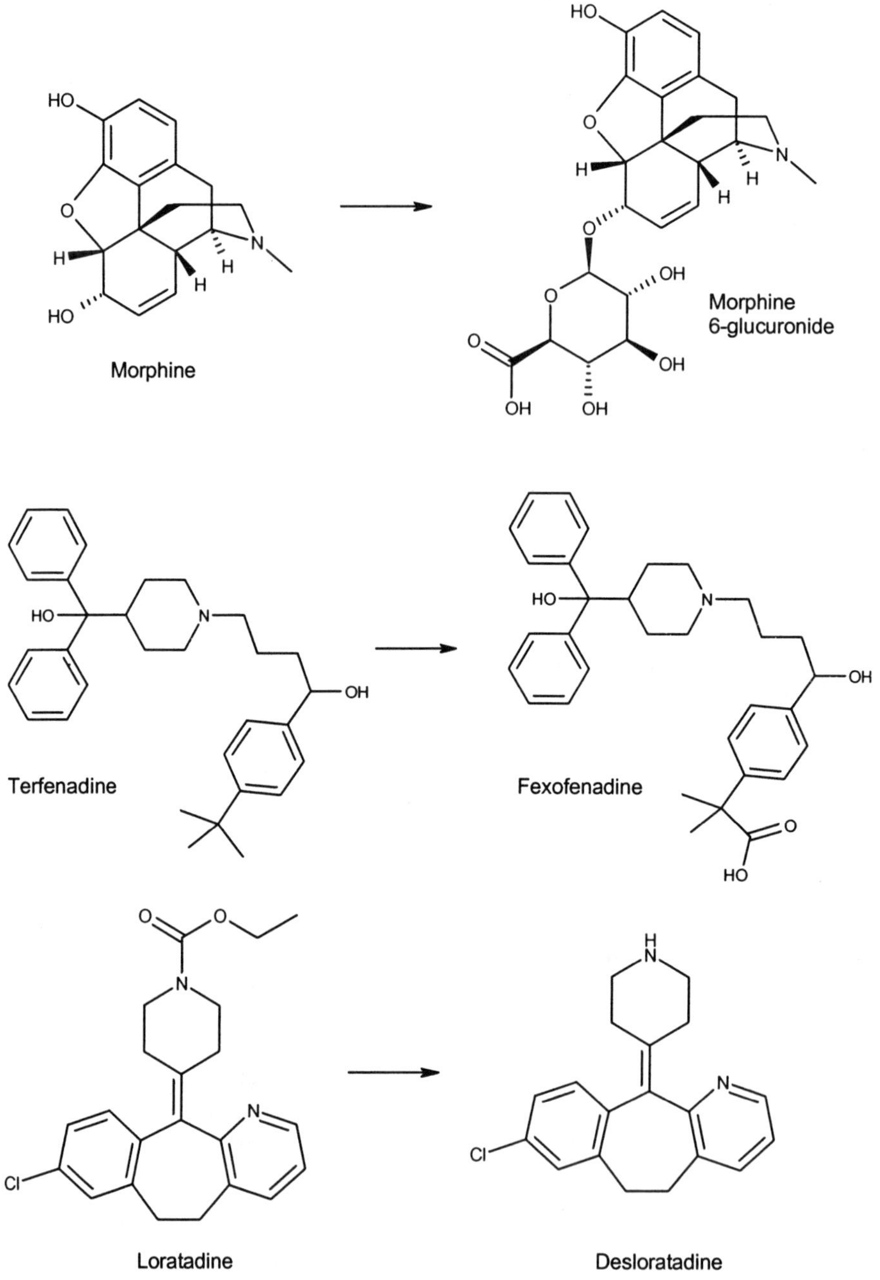

Figure 3. Metabolism of morphine, terfenadine, and loratadine to active metabolites.

4.1.3. Cross-Species Comparison of Metabolic Profiles

During preclinical development, animal models are employed to determine or predict the pharmacology, pharmacokinetics, and toxicology that will ultimately be demonstrated by human patients. Animal models used during the preclinical evaluation of toxicology are deemed appropriate if the metabolites observed in humans (by *in vitro* methods or human clinical studies) are also formed in the animals models. The animal models may actually produce more types of metabolites, however qualitatively the human metabolites must be represented in the animal *in vitro* or *in vivo* metabolic profiles to ensure that the animals were exposed to similar metabolites as human patients. In addition to choosing the appropriate animal species for preclinical toxicology testing, this information can be useful in explaining pharmacokinetic or efficacy differences between preclinical animal models used in pharmacokinetic and pharmacology studies, respectively.

The characterization of metabolites and pathways of metabolism across several species is known as cross-species comparisons of metabolic profiles. These studies generally use such preclinical species as mice, rats, dogs, monkeys (preclinical), and patients (clinical). These comparisons aid in the identification and interpretation of animal data that most resemble the human ADME properties, in this case biotransformation[78]. Toxicology studies are conducted in rodent and non-rodent species, typically the rat is the rodent species, however there is usually a choice between monkey and dog as the non-rodent species. Data obtained from cross species metabolic profiling studies contribute to the selection of dog or monkey as the second toxicological species[79]. These studies are conducted early in preclinical development with *in vitro* preparations (microsomes and hepatocytes) and subsequently in animals. Although a radiolabeled drug is not required for these studies, a radiolabel certainly expedites the analysis and identification of metabolites, as well as ensures a complete metabolic profile.

As an example, the HIV protease inhibitor CI-1029, was found through a series of *in vitro* and *in vivo* cross-species studies to be metabolized primarily to an N-acetylated metabolite in mice, rats, monkeys, and human, however this metabolite was absent from dogs, Figure 5. In this situation, the non-rodent toxicology species (dog or monkey) should be the monkey given dogs would not produce the N-acetylated metabolite and therefore would not be exposed to the major human metabolite of CI-1029[80]. Although this result was in theory predicted, given that dogs are known to be deficient in cytosolic N-acetyltransferase and would not be expected to produce such a metabolite, the *in vitro* and *in vivo* studies were necessary to confirm this fact, as well as confirm that monkeys would be an appropriate animal model[81].

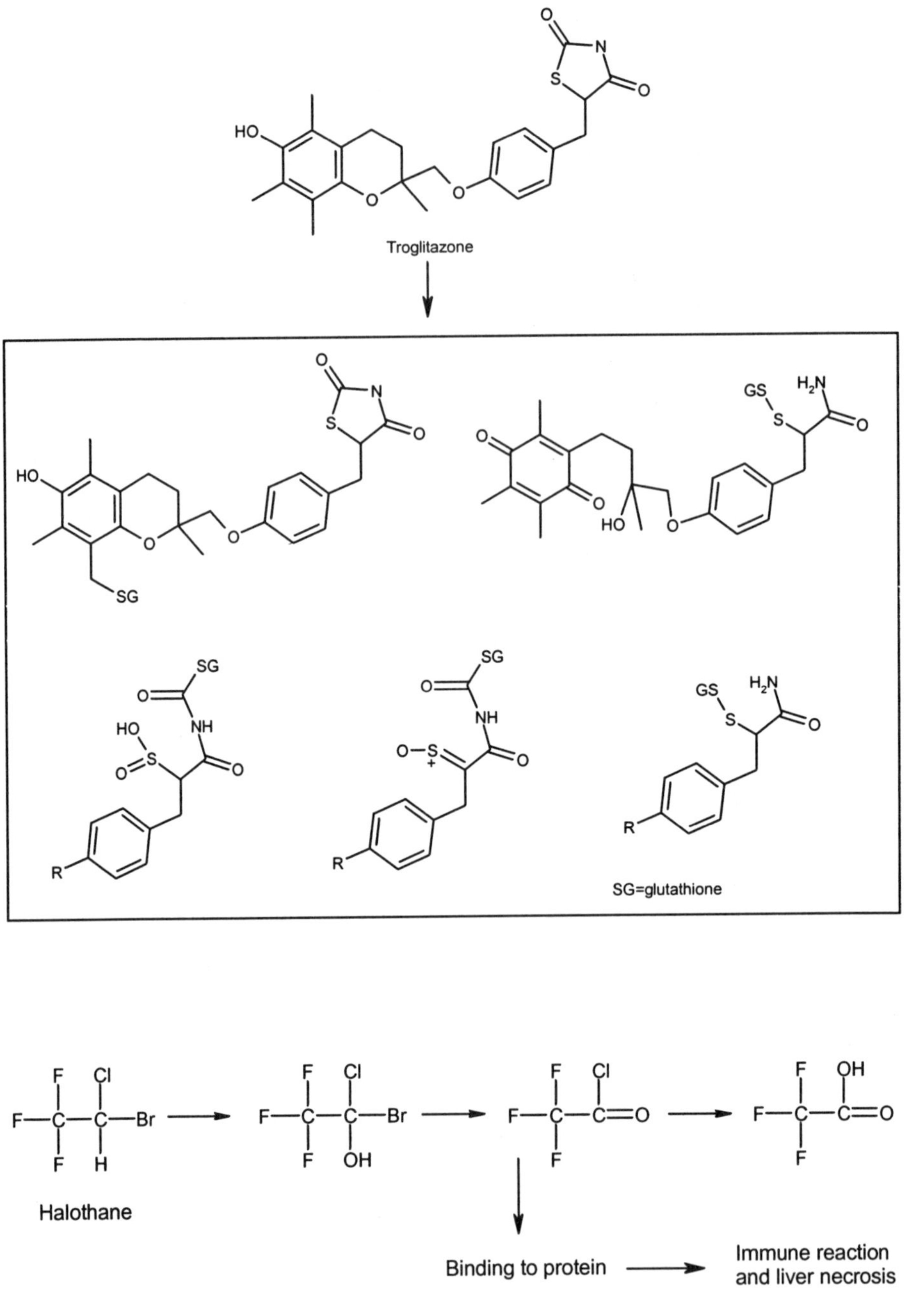

Figure 4. Metabolism of troglitazone to glutathione conjugates and halothane to a reactive - toxic metabolite.

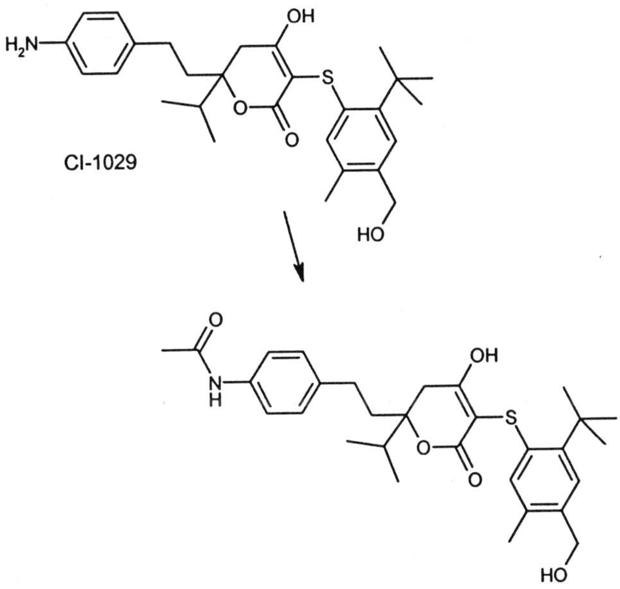

Figure 5. N-Acetylation of CI-1029

4.1.4. Rates of Metabolism – Prediction of Human Metabolic Clearance

The accurate estimation of human hepatic clearance can provide useful information during the drug discovery process to exclude rapidly metabolized compounds and throughout the preclinical process where this information can be useful in selecting the starting dose in the first clinical trial, in describing potential drugs that may exhibit non-linear or saturable clearance (ie., low K_m), or compounds that may be susceptible to high first pass elimination by the liver[82-86]. In general, low clearance drugs are more desirable because they require less frequent dosing, have greater bioavailability, and longer half-lives (assuming reasonable absorption and volume of distribution). The estimation of hepatic clearance should not be confused with the total body clearance which is commonly referred to when all clearance pathways are combined. Hepatic clearance does not predict total body clearance unless 100% of the clearance is through liver metabolism and all of the metabolic pathways are accounted for in the calculations (this assumes no renal or biliary elimination of parent drug). However, liver metabolism is a predominant elimination pathway for most drugs, therefore often these hepatic clearance determinations are reflective or good approximations of the total clearance. It is generally accepted that these models can yield predictions that fall within a two-fold range of the actual *in vivo* clearance value when properly executed[87, 88]. Literature references, such as diazepam, ethoxycoumarin, and almokalant have described the rationale behind the prediction of clearance and the utility of these estimates in drug development[89-91].

Basically, there are four distinct decision points that need to be addressed in order to obtain a prediction of hepatic clearance, with multiple other decisions to be made at each point along the process, Figure 6. A general description of the calculations necessary to determine hepatic clearance based on intrinsic clearance from enzyme kinetic data will be described in the following subsections.

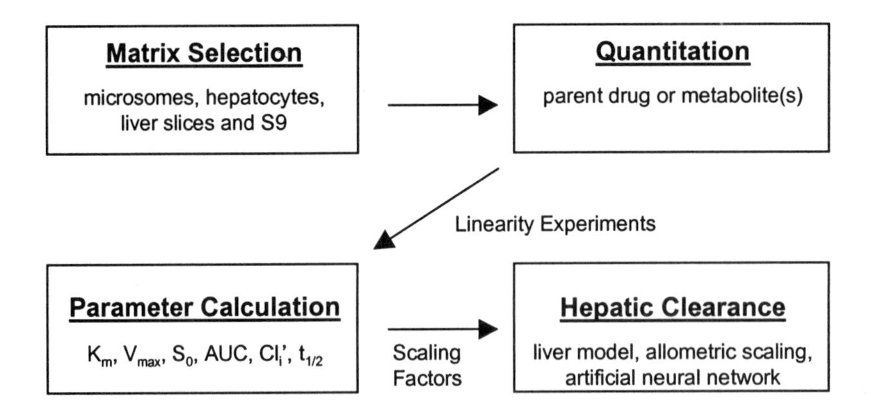

Figure 6. Decision points in estimation of hepatic clearance.

4.1.4a. Choice of In Vitro Matrix. As described earlier, there are different matrices, with different levels of metabolic capability, that can be used in the determination of hepatic clearance: microsomes, S9, hepatocytes, or liver slices. The most common systems today are microsomes with hepatocytes as a close second. Selection of matrix should be based on a knowledge of the primary metabolic pathways known to act on the drug candidate. If oxidative metabolism is the predominant primary pathway, then microsomes are an appropriate choice, however if other non-microsomal routes of metabolism significantly contribute to the metabolism then hepatocytes should yield a more accurate prediction. Keep in mind that we are only considering primary metabolic events that occur with the parent drug, not secondary reactions that occur with metabolites. When using hepatocytes or slices from individual liver donors, the variability between individual replicates is typically greater due to the interindividual differences between donors in drug metabolizing enzyme levels. Although this is an accurate representation of the differences between patients, it can make an estimate of the average clearance more difficult. However, Lave et al, was able to successfully place a series of 19 test compounds into the appropriate categories of low, intermediate, and high hepatic extraction when using human hepatocytes as the *in vitro* model to determine *in vitro* intrinsic clearance and ultimately hepatic extraction[92]. Pooled human liver microsomes provide an estimate of the average clearance across the population. There are many examples over the past decade that have employed rat and human liver microsomes in similar predictions which has been succinctly summarized by Obach[93].

4.1.4b. What to Quantitate? The next question to address is whether to measure parent drug disappearance or formation of metabolite(s). This actually depends on what is available at the time of the experiment. Early in drug discovery, unlabeled (cold) parent drug is available and no metabolite standards are available, in this situation measuring the disappearance of parent is appropriate. Later in discovery or development, when major metabolites are available, directly measuring the formation of metabolites is a more accurate approach (e.g., it is more accurate to measure the formation of a metabolite than measure a small change in a large parent peak). Both the parent drug disappearance and metabolite formation are typically quantitated by LC-MS which provides fast and accurate quantitative data. Radiolabeled parent drug provides the opportunity to measure both the disappearance of parent drug, as well as formation of all metabolites, however this also incurs the need for HPLC and a radiometric detector.

4.1.4c. Determination of In Vitro Parameter. Next to consider is the *in vitro* parameter to be calculated. The most common parameter in drug development is determination of intrinsic clearance CL'_i derived from an enzyme kinetic experiment where V_{max} and K_m are determined, Eq 1.

$$CL'_i = V_{max} / K_m \qquad (1)$$

When using this approach, linearity experiments regarding time and protein concentration are necessary in order to insure linear metabolism conditions which can then be used during incubations with multiple substrate concentrations. Another common method which measures the disappearance of parent is the substrate depletion method which determines CL'_i by determining the area under the curve (AUC), Eq 2, where S_0 is the starting amount of drug.

$$CL'_i = S_0 / AUC \quad (2)$$

A third approach first brought forth by Obach involves determination of an *in vitro* half-life ($t_{1/2}$) to determine intrinsic clearance through substrate depletion[93]. In this approach, the *in vitro* half-life of the compound is measured at a low substrate concentration ($S \ll K_m$) and related to intrinsic clearance by the following equation, Eq 3.

$$CL'_i = 0.693 / (in\ vitro\ t_{1/2} \times \text{microsomal protein concentration}) \quad (3)$$

4.1.4c. Determination of Hepatic Clearance. The determination of intrinsic clearance, *in vitro* half-life, or merely the rates of metabolism can be used to rank order compounds directly as a way to screen discovery compounds. In addition, parameters such as intrinsic clearance can be used in conjunction with scaling factors to determine hepatic clearance, which is the more common approach in drug development. Scaling factors are used to transform the *in vitro* intrinsic clearance to the whole liver (*in vivo*) intrinsic clearance. Some of the most common scaling factors are listed in Table 5 for human and rat liver microsomes and hepatocytes[94-97].

Table 5. Human and rat liver scaling factors.

Human		Rat	
Microsomes (mg/g liver)	Hepatocytes (cells/g liver)	Microsomes (mg/g liver)	Hepatocytes (cells/g liver)
45-52.5	1.20×10^8	25-60	$1.0\text{-}1.3 \times 10^8$
20 g liver/kg body weight		45 g liver/kg body weight	

At this juncture, *in vivo* intrinsic clearance, there are several options to determine hepatic clearance: 1) liver model, 2) allometric scaling, or 3) artificial neural network. The liver models mathematically bring together the terms of liver blood flow and free concentration of drug in plasma (with the assumption that the drug in the hepatocyte is in equilibrium with the unbound drug in plasma). In order to understand the models and how they are interrelated, we need to introduce two primary pharmacokinetic parameters, hepatic extraction ratio and liver blood flow. The hepatic extraction ratio (E_h) is the extent to which the drug is extracted by the liver and is represented by the ratio of drug concentration entering the liver and exiting the liver, C_{in} and C_{out}, respectively, Eq 4.

$$\text{Hepatic Extraction Ratio: } E_h = 1 - (C_{out} / C_{in}) \quad (4)$$

The hepatic extraction ratio is related to the hepatic clearance by incorporation of liver blood flow, Eq 5[60, 98].

$$\text{Hepatic Clearance: } Cl_H = Q_h \times E_h \quad (5)$$

The most common and simplest liver model is the well stirred (venous equilibrium) model shown in Eq 6. Where Cl_H is the hepatic clearance, Q_h is the human liver blood flow (20 ml/min/kg body weight) and f_{ub} is the fraction of drug unbound in human blood.

$$Cl_H = \frac{Q_h \times CL'_i \times f_{ub}}{Q_h + CL'_i \times f_{ub}} \quad (6)$$

Other models used for the determination of hepatic clearance are sinusoidal (parallel tube), distributed sinusoidal perfusion, and dispersion models each with increasing complexities[87, 95, 98]. Each model differs in the representation of drug concentration through the liver (E_h) therefore the greater the difference between C_{in} and C_{out} (ie., high extraction drugs) the greater the difference between models. In general, the hepatic clearances determined by the well stirred model are lower than those predicted from either the distributed or dispersion models with the parallel tube model giving the highest value prediction. Overall, no one model has been proven to be more accurate for general purposes[87].

Alternative approaches to the determination of hepatic clearance involve allometric scaling or artificial neural networks. Allomeric scaling involves collecting both *in vitro* and *in vivo* clearance data from animals, as well as human *in vitro* data. Here, the *in vitro* parameters are used as a correction or normalization factor with the *in vivo* clearance data[99]. These correction factors are similar to the standard allometric scaling factors of brain weight (BW) or maximum life span (MLS), however both BW and MLS have been found to be less predictive for hepatic clearance than the *in vitro* clearance factors[100]. Lave et al, using animal and human hepatocytes was able to demonstrate a reasonable prediction of human hepatic clearance using this type of allometric approach on 10 different drugs[85]. A more recent approach to hepatic clearance predictions involves artificial neural networks. Here a model is developed with input nodes which are *in vitro* and/or *in vivo* clearance parameters and an output node which is the predicted human hepatic clearance. Subsequently the model is trained with a set of known compounds and then employed to evaluate unknown compounds[99, 101]. In comparing the aforementioned methods for predicting hepatic clearance, the liver model approach is practiced to a greater extent, followed by allometric scaling, and finally artificial neural networks. Direct comparisons of these methods can be found in several general references[88, 95, 99, 102].

One point that has been missing from the following discussion is that of protein binding, whether it be binding to plasma proteins or binding to endogenous proteins within an *in vitro* matrix. This is a controversial topic and we most likely do not fully understand the full impact of binding or have the complete means to interpret protein binding into our *in vitro* models. Nonetheless, there have been several interesting papers in this area that are beginning to shed more light on this complex aspect of human predictions, eg., hepatic clearance, drug inhibition or induction. When considering liver microsomes, lipophilic drugs can bind significantly to microsomal membranes, thereby decreasing the free concentration available to interact with the enzyme. The decrease in free (unbound) drug has been shown to result in the determination of a higher K_m (indicating less binding affinity) with no apparent effect on the V_{max} of the reaction. The effect of higher K_m and unchanged V_{max} will result in an underestimation of intrinsic clearance (Eq 1)[93, 94]. Obach et al, have shown in a series of papers that using plasma protein binding by itself in the prediction of hepatic clearance gave poorer results compared to ignoring protein binding all together. Furthermore, the inclusion of microsomal protein binding along with plasma protein binding resulted in more accurate predictions[102, 103]. In this study, hepatic clearance of twenty nine drugs of differing physicochemical properties were determined. In the calculation of hepatic clearance, the microsomal protein binding or blood binding were considered alone or in conjunction with each other. Overall the predictions of basic, neutral, and acidic drugs were accurately projected when all binding (*in vitro* and *in vivo*) was considered in the calculations[104]. Some researchers have advocated the addition of serum albumin to microsomal incubations as a means to impart this *in vivo* physiological situation on the *in vitro* system. This idea stems predominately from the work of Ludden et al., who demonstrated that the in vitro K_m for para-hydroxylation of phenytoin better predicted the *in vivo* K_m by the addition of bovine serum albumin (2-4%) to human liver microsomes. In fact, the *in vitro* K_m value decreased from 30.8 μM to 1.5 μM with the addition of albumin which now accurately reflects the *in vivo* steady state K_m value of 2-3 μM[105]. This indicated that an accurate estimate of clearance could be predicted when albumin was added to the reaction mixture, but that a significant underprediction would occur if the reaction was run without albumin. Subsequently, Carlile et al confirmed these results with phenytoin, however they were unable to demonstrate an improvement in clearance prediction with a second test compound, tolbutamide[106]. To date, the effects of adding albumin to microsomal incubations are unclear as we do not yet understand which drugs or how certain drugs will be affected[107]. Furthermore, albumin does not directly interact with CYP450 enzymes *in vivo* as they are separated by the hepatocyte membrane thereby creating a situation *in vitro* which does not exist *in vivo*. Therefore it is not recommended to add albumin to microsomal incubations to predict clearance or drug interactions[31, 106]. In contrast to microsomes, the addition of serum albumin to hepatocyte incubations has been shown to be a more convenient and equally predictive method to assess intrinsic clearance. Shibata et al., have dubbed this new method the "serum incubation method" which incorporates the incubation of rat or human hepatocytes with rat or human serum, respectively[96, 108]. The major advantage of this technique as opposed to a standard method of hepatocyte incubation without 100% serum is the practical incorporation of plasma protein binding directly into the incubation, therefore eliminating the need to determine plasma protein binding in a separate experiment. On the whole, intrinsic clearance or hepatic clearance predictions using human hepatocytes can be conducted with or without the addition of serum as long as the appropriate variables are accounted for in the final determination.

I would be remise if I did not reveal several situations which can lead to the erroneous determination of hepatic clearance. Oscar Wilde once said "the pure and simple truth is rarely pure and never simple" and the same can be said for the estimation of human hepatic clearance which in the final result is based on a myriad of assumptions and sometimes unknowns. Several situations can lead to deviations from the actual clearance value, such as: 1) significant extrahepatic metabolism or metabolism not accounted for in the liver matrix used to estimate clearance, 2) an incorrect assumption concerning the equilibrium between unbound drug in plasma and the concentration within the hepatocyte (liver model assumption) which can be drastically modified by transporters or by not accounting for binding to liver microsomes or plasma, 3) inappropriate incubation condition, such as high substrate concentrations or non-linear reaction conditions, 4) inaccurate scaling factors, or 5) inaccurate determination of enzyme kinetic parameter due to atypical or complex enzyme kinetics[109]. Although not straight forward, accurate estimates of clearance can be achieved with a proper understanding of the principles and science behind the derivations.

4.2. Reaction Phenotyping – Identification of Enzymes Involved in Metabolism

Reaction phenotyping is the determination of the drug metabolizing enzymes involved in the biotransformation of a preclinical drug candidate. In the pharmaceutical industry this centers predominately around the major oxidative pathways mediated by the cytochromes P450, however other drug metabolizing enzymes can and are often characterized to a lesser extent. The identification of specific isoforms of the cytochromes P450 are typically conducted and in a broader sense other pathways are identified, such as glucuronidation or hydrolysis reactions, however specific isoforms of non-P450 reactions are typically not performed unless they represent a major pathway of elimination.

Why is reaction phenotyping necessary during preclinical development? Reaction phenotyping answers several questions pertaining to the clinical development of drug candidates[24]. For example, in the prediction and planning of drug interaction studies it is useful to know which coadministered drugs may interact with the drug candidate. Furthermore, by knowing which enzymes metabolize a drug candidate, we can identify certain patient populations that may be susceptible to increased toxicity, decreased efficacy, or exaggerated efficacy due to the presence of genetic polymorphisms of drug metabolizing enzymes.

One important question pertaining to reaction phenotyping that relates to drug-drug interactions is the number of enzymes involved in the metabolism of a drug candidate: Is the drug metabolized by a single metabolic pathway or multiple pathways? When a compound is metabolized by a single pathway, any effect (inhibition or induction) of that pathway will have a significant effect on the metabolic elimination of the compound. For example, a compound only metabolized by CYP3A4 will be greatly effected by co-administration of the CYP3A4 inhibitor ketoconazole or the CYP3A4 inducer rifampicin. Where as, if the same compound were metabolized equally by CYP3A4, CYP2C9 and CYP1A2, the inhibition or induction effect on metabolic elimination will be minimal due to metabolic switching to alternate enzymes. Anytime two coadministered drugs are substrates for the same enzyme, there exists the potential for a drug-drug interaction. In this situation, it is useful to know what enzymes are involved in the metabolism of the drug candidate in order to predict which co-administered drugs (using the same enzyme) may interact with the drug candidate or visa versa. For example, a candidate drug that is solely metabolized by CYP2C19 when co-administered with S-mephenytoin (another CYP2C19 substrate) may create an interaction between the two compounds.

Another important question to ask regarding reaction phenotyping: Is the drug candidate metabolized by an enzyme that is known to have a genetic polymorphism that distributes throughout the population? Reduced or complete loss of enzyme activity contributes greatly to the variability of drug response and drug toxicity. The most characterized example of this is CYP2D6, which is deficient in 5-10% of the

Caucasian population and segregates the population into what are characterized as extensive (normal CYP2D6 activity) and poor (CYP2D6 deficient) metabolizers[110, 111]. A third minor group, identified as ultrarapid metabolizer, have greater than normal CYP2D6 activity beyond that of extensive metabolizers due to gene amplification[111]. Other significant polymorphic enzymes include CYP1A1, CYP2E1, CYP2C9, CYP2C19, N-acetyltransferase, UGT1A1, glutathione S-transferase, aldehyde dehydrogenase, and DT-diaphorase[14, 112-117]. The distribution of these polymorphic drug metabolizing enzymes is varied across several racial groups as can be illustrated in Table 6 for Caucasian and Oriental populations, a fact that should be considered when evaluating the drug in different populations or markets[118-121]. Lack of a drug metabolizing enzyme in a patient can result in reduced clearance of a drug that requires the enzyme for elimination. This can result in elevated concentrations of parent drug leading to exaggerated pharmacology or toxicity. For example, the anticoagulant warfarin is administered as a racemic mixture and the more potent isomer, S-warfarin is primarily metabolized by CYP2C9. Patients with allelic variants of CYP2C9 have been shown to have increased plasma levels of warfarin, reduced clearance of warfarin, and an increased frequency of bleeding[122]. In contrast, when a prodrug requires a polymorphic enzyme for conversion to the active drug, patients deficient in enzyme activity will not properly convert the prodrug resulting in decreased efficacy. This situation is illustrated by the conversion of codeine to morphine by CYP2D6. Poor metabolizers of CYP2D6 are unable to convert codeine to morphine and therefore do not benefit from the analgesic effects of morphine[111].

The significance of adverse events due to genetic polymorphisms are again related to the number of enzymes involved in metabolism, the specific polymorphic enzyme, and the contribution that this enzyme plays in the overall elimination of the drug. For example, a drug metabolized only by CYP2D6 will have significant consequences in poor metabolizers of CYP2D6, however a compound that is metabolized by CYP2D6 and CYP3A4 will be of less significance in CYP2D6 poor metabolizers. In addition, the effect is greatest when metabolic elimination is the major elimination pathway, ie., renal or biliary elimination are minor. The greatest adverse events due to genetic polymorphisms are encountered when metabolic elimination is the major pathway for drug elimination and a single polymorphic enzyme is the major enzyme involved in the biotransformation of the drug, in this situation significantly higher than normal plasma levels of drug can be anticipated in patients with the allelic variant.

Table 6. Racial differences in polymorphic enzymes.

	Ethnic Group	
Enzyme	Caucasian	Oriental
CYP2D6	5-10%	~1%
CYP2C19	3-5%	12-23%
N-Acetyltransferase (NAT2)	40-60%	5-25%

4.2.1. Reaction Phenotyping Experiments

There are several types of experiments and *in vitro* matrices that can be used to determine the enzymes involved in the metabolism of a preclinical candidate. Given the predominance and clinical significance of the CYP2D6 polymorphisms, many pharmaceutical companies have incorporated discovery screening strategies to determine if this enzyme plays a significant role in CYP450-mediated metabolism of discovery candidates[123]. However, due to the lack of other information at this early stage (as described above), compounds are generally not eliminated from the discovery process, nevertheless they are flagged and their status reevaluated as more information is generated.

There are three typical forms that a reaction phenotyping experiment can take: metabolism in cDNA expressed enzymes, chemical or antibody inhibition of metabolism, and correlation analysis. Each of these experiments are described in greater detail below, however a common characteristic of each method is specificity or individuality, ie., they each have the ability to examine individual enzymes, one at a time either through individual expression, specific chemicals, antibodies, or probe reactions. Typically, the results from reaction phenotyping are presented as a summary of at least two of the methods described above where the results between experiments agree with one another. There are multiple examples in the literature of reaction phenotyping experiments using a combination of these techniques[124-126]. For example, Hyland et al., determined through a combination of correlation analysis, chemical inhibition, and expressed recombinant CYP450 enzymes, that the N-demethylation of Viagra® (Sildenafil) was mediated by CYP3A4 and CYP2C9[124]. Also, Ring et al., determined through the use of correlation analysis, expressed CYP450 isoenzymes, and antibody inhibition that both the R- and S- enantiomers of norfluoxetine were formed by CYP2D6, but that only R-norfluxetine could also be formed by CYP2C9[125].

Reaction phenotyping as determined through biotransformation by cDNA expressed enzymes is a very simple system that really determines which of the enzymes has the capability to biotransform the drug, not necessarily the enzyme that will predominate *in vivo*. The fact that each enzyme is expressed at very different levels in the liver and that other enzymes could simultaneously metabolize the compound need to be taken into consideration. The CYP450 isoforms are present in human liver at different concentrations with the greatest activity found with CYP1A2, CYP2C family, and CYP3A, Table 4. Equal turnover by one of these enzymes and an enzyme of lower concentration (eg., CYP2D6 at 1.5%) may not yield similar results *in vivo*. The prediction of *in vivo* contribution of various CYP-isoforms to the overall biotransformation of a preclinical candidate typically involves some form of normalization of recombinately expressed enzymes[127] or application of a relative activity factor[128] both of which take into consideration the enzyme kinetics of the reaction and the abundance of each CYP-isoform in human liver.

The chemical or antibody inhibition method uses human liver microsomes (or hepatocytes) and specific inhibitors (chemical or antibody) that inhibit a single enzyme thereby shutting down a single enzyme pathway[12, 127, 129]. Incubations of the drug candidate with and without the specific inhibitors are performed and the changes in the rate of parent disappearance or metabolite formation are determined. If metabolism is inhibited, then this would indicate that the inhibited enzyme participates in the biotransformation of the drug. If only partial inhibition of metabolism is observed, then this would be an indication that additional enzymes are involved in the overall metabolism of the compound. The antibody inhibition experiment is similar to chemical inhibition, with the exception of using specific antibodies that inhibit a single enzyme in human liver microsomes[130]. In effect, antibody inhibition is in some ways better than chemical inhibition due to the specific nature of antibodies and the fact that chemical inhibitors are themselves substrates for metabolism and are also metabolized, thereby changing their concentration and possibly inhibitory potential during the incubations. Unfortunately, the availability of antibodies against some of the human CYP-enzymes is limited. Therefore, microsomal inhibition studies will typically involve both chemical and antibody inhibition in order to cover a broader range of CYP450 isoforms.

The correlation analysis compares the rate of drug candidate metabolism with the rate of metabolism of a probe substrates specific for an individual CYP450 isoform in human liver microsomes. For example correlating the formation of a metabolite with formation of 6β-hydroxy testosterone from testosterone (a reaction that is mediated by CYP3A4). This experiment involves measuring the rate of metabolism for a drug candidate in $\sim 10^{+}$ individual preparations of human liver microsomes and correlating the rate of reaction with the activity of individual CYP-enzymes in the same microsomal preparations. Typically a good range of activities are found with each CYP-isoform given there is considerable variability in levels of CYP-enzymes found in human liver microsomes, 5-200 fold[53]. This experiment becomes more difficult to interpret when multiple enzymes are involved in the metabolism of a compound, thereby resulting in mixed correlations which may be resolved by multivariate analysis[124, 125, 128].

4.2.2. *Reaction Phenotyping Case Study*

CI-1007 is a dopamine agonist that had undergone clinical trials as an antipsychotic agent[131, 132]. The metabolism of CI-1007 in animals and humans was primarily to two monohydroxylated metabolites which were further metabolized to a dihydroxy and sulfate conjugates, Figure 7. Incubations of ^{14}C-CI-1007 with cDNA expressed CYP450 isoforms and chemical inhibition experiments in human liver microsomes indicated that CYP2D6 was the predominate enzyme involved in the formation of the primary metabolites with some minor participation by CYP1A2[132, 133], Figure 8. Similar results were obtained by measuring parent drug disappearance or metabolite formation by HPLC-RAM. Also, metabolism was inhibited in human hepatocytes by the CYP2D6 inhibitor quinidine (50% decrease in metabolite formation)[80, 134]. All of these data, expressed enzymes, chemical inhibition in microsomes, and hepatocytes indicated that CYP2D6 was the predominate enzyme involved in the metabolic turnover of CI-1007. Animal studies had previously indicated that CI-1007 was extensively metabolized and that urinary excretion of parent drug was negligible[132]. Given this information, it would be reasonable to expect that CYP2D6 deficient patients (poor metabolizers) would not be able to efficiently eliminate CI-1007 and that this patient population would have elevated plasma concentrations of CI-1007. An early clinical study in poor and extensive metabolizers of CYP2D6 confirmed this hypothesis[80, 134]. While the plasma concentrations in normal patients (extensive metabolizers) fell between the minimum effective concentration and the maximum no effect concentrations of 8 and 90 ng/ml,

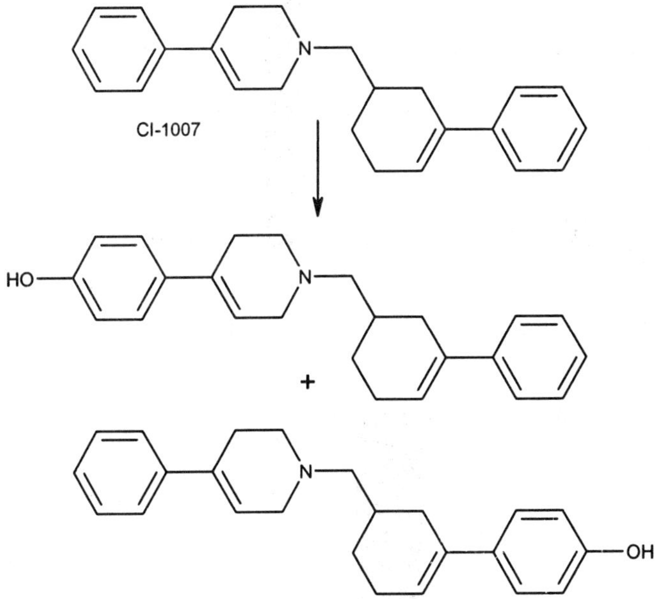

CI-1007

HO

N

+

OH

Figure 7. Metabolism of CI-1007 to hydroxylated metabolites mediated by CYP2D6.

the plasma concentration in poor metabolizers ranged from 60-250 ng/ml in most cases exceeding the toxicology limit of 90 ng/ml. This is a practical example of how *in vitro* reaction phenotyping can direct early clinical trials and highlights the potential for serious drug reactions due to polymorphisms in drug metabolizing enzymes.

4.3. Drug-Drug Interactions: Enzyme Inhibition or Induction

Metabolic drug interactions can lead to significant changes in human pharmacokinetics, often resulting in changes in efficacy, safety, or tolerability of a new drug. The most common metabolic drug interactions occur when the enzyme activity of drug metabolizing enzymes are altered, through inhibition or induction. For example, increased exposure (greater AUC) of a drug due to inhibition of the principle enzyme involved in the elimination of the drug can lead to toxicity or an exaggerated pharmacological response. In contrast, decreased drug exposure (reduced AUC) due to an increase in activity of an enzyme principally involved in

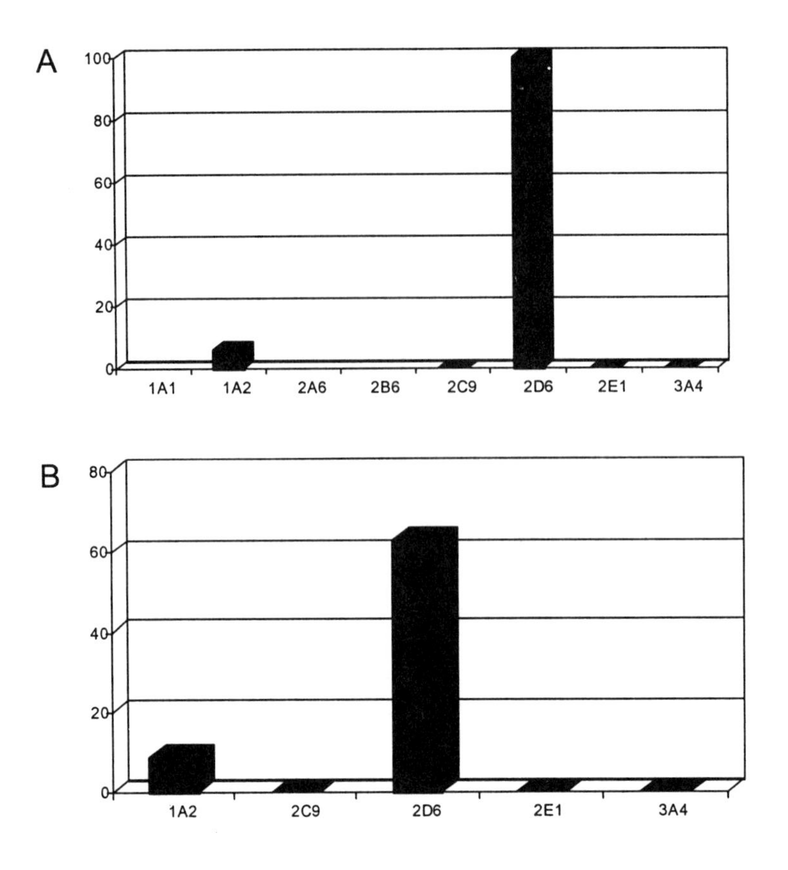

Figure 8. Percent metabolism of CI-1007 in cDNA expressed human CYP450 enzymes (A) and percent inhibition of metabolism of CI-1007 by various chemical inhibitors in human liver microsomes (B).

the metabolic elimination of a drug can lead to a decrease in efficacy and in some situations toxicity. Due to these the seriousness of these situations, drug-drug interactions are a primary focus for the pharmaceutical industry, regulatory authorities, and clinicians. As a result, the U.S. Food and Drug Administration has issued guidances for the conduct of *in vitro* and *in vivo* drug interaction studies[24, 135].

Furthermore, these interactions are a major cause of candidate attrition in drug discovery and a serious liability in drug development as certain drugs have been rejected or removed from the market due to life threatening drug interactions[136]. The early, reliable prediction of drug-drug interactions using human *in vitro* methodologies play an integral part in the drug development processes. When used in conjunction with the anticipated therapeutic concentration of the drug, these predictions can be used to assess the likelihood of an interaction in patients. If a significant interaction is predicted, then clinical studies can be planned to assess the actual clinical significance of the interaction as early as possible during clinical trials.

Most of the following discussion will focus on the cytochromes P450 enzymes whereas we have already noted they are the major drug metabolizing enzyme system responsible for metabolic elimination of drugs. Even amongst the CYP450 enzymes, there is a predilection to focus on CYP3A4 in regards to drug-drug interactions. CYP3A4 represents the most abundant CYP450 enzyme in the human liver and is responsible for the metabolism (all or in part) of 50-60% of marketed drugs. Hence, due to this preponderance for CYP3A4 in the elimination of so many drugs, our focus on drug-drug interactions with this particular enzyme is warranted. Therefore, when studying either enzyme inhibition or induction greater focus will be on CYP3A4, as well as additional emphasis on the significance of CYP3A4 interactions compared to other CYP's or drug metabolizing enzymes.

4.3.1. Enzyme Inhibition

The evaluation of inhibitory potential of drug candidates is a long process beginning as early as drug discovery, progressing throughout preclinical development, clinical development, and often times post-marketing through continued surveillance of adverse events. The most common preclinical enzyme systems employed to evaluate inhibition are cDNA expressed CYP450 enzymes and human liver microsomes during drug discovery and preclinical development, respectively[137, 138]. More complex *in vitro* systems, such as human hepatocytes have been used to study enzyme inhibition, however liver slices have not found much utility in this regard[139-140]. Although hepatocytes are clearly more similar to the *in vivo* situation than subcellular fractions, their use as a routine predictive tool of inhibition requires further investigation. To date, most of the studies involving hepatocytes deal with rat hepatocytes with limited data available on human hepatocytes. Zomorodi and Houston performed one of the most comprehensive *in vitro-in vivo* comparisons of inhibition using hepatocytes, microsomes, and *in vivo* rat studies[142]. In this study, the *in vivo* K_i of omeprazole inhibition of diazepam metabolism was determined to be 57 uM which was more similar to the results obtained with hepatocytes (28 and 76 μM) as compared to microsomes (108 and 226 μM) for the 3-hydroxylation and N-demethylation of diazepam. For additional background information on *in vitro* drug inhibition studies, some exemplary references describing both *in vitro* and *in vivo* drug-drug interactions can be found in the following references on the subject[10, 143-149].

Although the following discussion focuses on the parent drug as the inhibitory agent, we must always remember that circulating metabolites can also be potent inhibitors. Whenever metabolite concentrations begin to reach or exceed parent drug levels, they should be evaluated for their potential to inhibit drug metabolizing enzymes. Furthermore, when circulating metabolites are present for extended periods of time (long half-life), they should also be evaluated. For example, it was suggested that the drug interactions caused by amiodarone may be caused by the major N-deethylated metabolite of amiodarone instead of the parent drug. The metabolite, desethylamioradone, has lower K_i values for several CYP450 isoforms and is found in higher concentrations than the parent in liver tissue. Moreover, both metabolite and parent drug have extremely long half-lifes[150]. The evaluation of metabolites does impose some difficulties in that the metabolites often need to be isolated from *in vitro* or *in vivo* matrices or chemically synthesized in order to properly perform the interaction experiments. Therefore, the testing of metabolites for inhibition potential is often limited to the aforementioned scenarios during preclinical development.

4.3.1a. Inhibition Kinetics. Before delving deeper into the specifics of how inhibition studies are conducted or interpreted, we need to set the stage with some basics on inhibition kinetics and the various forms of enzyme inhibition. There are two basic forms of inhibition, reversible and irreversible. Reversible inhibition can occur in a variety of different forms depending on how the inhibitor interacts with the enzyme and these are described kinetically based on derivations of the Michaelis-Menten equation. The most common form of reversible inhibition is competitive inhibition where the inhibitor is typically an analog of the substrate and will compete with the substrate for the active site of the enzyme. The inhibitor can only bind to free enzyme and the binding of substrate and inhibitor are considered mutually exclusive. This competition results in a raising of the K_m for the substrate, but has no effect on the maximum velocity of the reaction (V_{max}). Uncompetitive inhibition occurs when the inhibitor binds to the enzyme-substrate complex, but not the free enzyme. This is not a common form of inhibition found with drug metabolizing enzymes (eg., CYP450) and has the effect of decreasing both the K_m and V_{max} of the substrate reaction. Non-competitive inhibition is a combination of both competitive and uncompetitive because the inhibitor can bind to both the free enzyme and the enzyme-substrate complex. The possibility for two separate binding situations yields two inhibition constants (K_i, αK_i) for the overall reaction. If the two K_i values are equal (α=1), then the inhibition is characterized as simple non-competitive inhibition, if they are different then the inhibition is described as mixed inhibition. In most instances the inhibition of CYP450 enzymes is a result of either competitive or mixed inhibition. The forms of reversible inhibition along with the corresponding changes in K_m and V_{max} are illustrated in Table 7[147, 151-153]. In addition, the fractional inhibition formulas and relationship between IC_{50} and K_i for each form of inhibition are presented.

Irreversible inhibition occurs when the enzyme-substrate complex produces a product that is chemically reactive enough to interact with the enzyme (either through covalent binding or some form of tight binding) as to inactive the enzyme and eliminate it from the pool of active enzyme. Similar to reversible, irreversible inhibition is concentration dependent, that is inhibition increases with increasing inhibitor concentration, however irreversible inhibition is also time dependent in that the degree of inhibition increases with time of exposure as more enzyme becomes inactivated. Often times, irreversible inhibition can be confused with reversible inhibition if the appropriate studies are not conducted to distinguish between these two forms of inhibition. Note that the changes to K_m and V_{max} for irreversible inhibition are the same as those for non-competitive inhibition (decreased V_{max} and unchanged K_m). Irreversible inhibition can often be detected by a simple preincubation experiment where the inhibitor is preincubated with the enzyme prior to addition of a probe substrate. This modification of the typical inhibition incubation illustrates the time dependent nature of irreversible inhibition. Often, potent irreversible inhibitors exhibit a combination of reversible and irreversible inhibition, making the separation between these distinctly different forms of inhibition difficult. A more detailed description of metabolism dependent inhibition is given toward the end of this section.

4.3.1b. Derived Inhibition Parameters: IC_{50} and K_i. Inhibition experiments are typically conducted in expressed CYP450 enzymes or human liver microsomes and the most commonly derived parameters are IC_{50} and K_i. Ultimately, the choice of kinetic parameter and *in vitro* matrix are dependent on the phase of development and the desired outcome of the experiment. In general, expressed enzymes and IC_{50} determinations occur early in preclinical development in an attempt to generate preliminary inhibition data on a large set of compounds across a broad set of enzymes[154, 155]. As a compound progresses in development, more detailed information is gathered on potential cytochrome P450 inhibitors, such as determination of the mechanism of inhibition along with a K_i and an assessment as to whether the inhibition is reversible or irreversible. The most common form of inhibition experiments conducted in preclinical drug metabolism laboratories are those involving human liver microsomes with a determination of mechanism of inhibition and K_i determination[31, 145]. Drug candidates are typically evaluated for their ability to inhibit the major forms of CYP450, such as CYP1A2, CYP2C8, CYP2C9, CYP2C19, CYP2D6, or CYP3A4, however other forms can also be tested (CYP2A6, CYP2B6, CYP2E1). Each CYP450 is tested against a single probe substrate because enzyme kinetics assumes that the K_i value should be independent of the substrate used to measure inhibition (except for uncompetitive inhibition). More recently researchers have determined that this does not apply when assessing the inhibition potential of CYP3A4 substrates[156-159] and possibly CYP2C9 substrates[160], although other CYP450's appear to follow this substrate independence. For this reason, multiple CYP3A4 substrates are employed in assessing the potential of preclinical candidates to inhibit this enzyme, such as testosterone, midazolam, and nifedipine.

The simplest form of measuring inhibition is determination of the IC_{50} which is the concentration of inhibitor that causes 50% inhibition of the enzyme activity under the given experimental condition. The determination of IC_{50} is appropriate when large numbers of compounds require testing which allows for the identification of potential inhibitors, a rank order of compounds as to their inhibitory potency, and the CYP450 enzymes inhibited. An advantage of the IC_{50} determination is that it is independent of the mechanism of inhibition and requires fewer samples to generate a meaningful result. However there are limitations to the use of IC_{50} as a means to predict the magnitude of an *in vivo* drug interaction. The IC_{50} determination is dependent on the experimental and incubation conditions under which the value is determined[151]. Therefore, the IC_{50} value is only meaningful at the substrate concentration for which the IC_{50} was determined for all forms of inhibition (except non-competitive inhibition, see Fractional Inhibition in Table 7). Therefore, the relationship between IC_{50} and K_i will vary for most forms of inhibition as the probe substrate concentrations changes. Depending on the concentration of substrate used in the preliminary IC_{50} experiment, there can be a correlation between the IC_{50} and the K_i which can be used as an early approximation of K_i, Table 7. The K_i value is an inhibition constant that defines the affinity of an inhibitor to the enzyme, much like a K_m value. In contrast to the IC_{50} value, K_i determinations are more reproducible because they are less dependent on experimental conditions as they are determined over a range of substrate and inhibitor concentrations. K_i determinations are dependent on the mechanism of inhibition which must first be determined by graphical or statistical methods. K_i determinations are the accepted method of predicting the interaction potential by pharmaceutical companies and regulatory agencies. Moreover the majority of *in vitro-in vivo* correlation reported in the literature are with human liver microsomes and K_i determinations. This valuable database of information provides a foundation for the prediction of drug interactions based on these values.

4.3.1c. Interpretation of Enzyme Inhibition and Drug Interactions. When forecasting the degree of *in vivo* drug interaction there are several factors that need to be considered: 1) relationship between inhibition potency and plasma concentration, 2) extent to which the inhibited CYP enzyme is involved in the metabolism of a compound, and 3) the form of inhibition (reversible or irreversible). The interrelationship between inhibition potency and plasma concentration proposed by many researchers involves the K_i value in relation to the anticipated or known human plasma concentration of the drug (C_{ss} or C_{max}), this is known as the $[I]/K_i$ ratio. Equation 7 illustrates how the ratio of inhibited AUC_i to the control AUC is proportional to this ratio (based on competitive inhibition)[31].

$$AUC_i / AUC = 1 + [I] / K_i \qquad (7)$$

This ratio is interpreted as follows: when the ratio of $[I]/K_i$ is less than 0.1 (inhibitor concentration is much less than the K_i) then an interaction is unlikely to occur. When the ratio is between 0.1 and 1.0 then the inhibitor concentration is similar to circulating plasma concentrations and an interaction is possible. When plasma concentrations exceed the K_i then the likelihood of an interaction increases with increasing values greater than 1.0. A similar approach involving the same parameters is determination of the fractional inhibition or percent inhibition predicted from the enzyme kinetic equations with certain assumptions on the relationship between substrate concentration and K_m, Table 7.

In regards to protein binding and *in vitro* inhibition experiments, it is generally accepted that the unbound K_i value should be determined taking into account the binding of inhibitor to matrix proteins and non-specific binding to glass or plastic[31, 161]. Although in practice and is often found in the literature, the total inhibitor

Table 7. Types of enzyme inhibition and inhibition kinetics.

Type of Inhibition	Inhibition Equation	Changes in K_m and V_{max} with ↑ [I]	Fractional Inhibition (FI)	Relationship Between IC_{50} and K_i ($[S] = K_m$)
Competitive	$v = \frac{V_{max} \times [S]}{K_m \times (1+[I]/K_i) + [S]}$	$V_{max} \leftrightarrow$, $K_m \uparrow$	$FI = \frac{[I]}{K_i (1+ [S]/K_m) + [I]}$	$IC_{50} = 2K_i$
Non-Competitive	$v = \frac{V_{max} \times [S]}{K_m \times (1+[I]/K_i) + (1+[I]/K_i) \times [S]}$	$V_{max} \downarrow$, $K_m \leftrightarrow$	$FI = \frac{[I]}{K_i + [I]}$	$IC_{50} = K_i$
Uncompetitive	$v = \frac{V_{max} \times [S]}{K_m + (1+[I]/K_i) \times [S]}$	$V_{max} \downarrow$, $K_m \downarrow$	$FI = \frac{[I]}{K_i (1+ K_m/[S]) + [I]}$	$IC_{50} = 2K_i$
Mixed	$v = \frac{V_{max} \times [S]}{K_m \times (1+[I]/K_i) + (1+[I]/\alpha K_i) \times [S]}$	$V_{max} \downarrow$ $K_m \uparrow$, $\alpha K_i > K_i$ $K_m \downarrow$, $\alpha K_i < K_i$	$FI = \frac{[I]}{K_i \left(\frac{K_m + [S]}{K_m + [S]/\alpha}\right) + [I]}$	$IC_{50} =$ $K_i (2\alpha/1+\alpha)$

concentration has been used in the determination of K_i. The use of total or unbound inhibitor concentrations (I or I_u, respectively) in the ratio of $[I]/K_i$ is more controversial in that total inhibitor concentrations tend to give better overall predictions, however this goes against the hypothesis that the inhibitor concentrations inside of the hepatocyte is in equilibrium with the unbound concentration of drug in the plasma[145, 162-164]. This controversy is most likely based on our inability to accurately determine the actual concentration of inhibitor at the enzyme site (liver). The participation of drug transporters on the hepatocyte surface may significantly increase or decrease the intracellular inhibitor concentration from the unbound plasma concentration. Another significant issue with accurately predicting a drug interaction is that we use post-liver plasma concentrations (C_{max} or C_{ss}) to make predictions of drug interactions realizing that the actual liver concentration is most likely greater than the plasma concentration. Using the unbound plasma concentration is still an appropriate approach assuming an accurate liver (hepatocyte) concentration can be determined. A more conservative approach commonly employed uses the total plasma concentration in the $[I]/K_i$ prediction.

The above mentioned criteria work well for compounds that are predominately metabolized by the inhibited CYP450 enzyme. For example when an inhibitor results in a ratio of 10 for CYP2C9 and a coadministered drug is exclusively eliminated by CYP2C9. In many instances the elimination of coadministered drugs are through multiple P450 enzymes or other non-metabolic routes of elimination. In these cases, the degree of inhibition predicted will overestimate the overall inhibitory effect. This relationship between inhibitor concentration, K_i and fraction metabolized (f_m) can be addressed by application of the Rowland-Matin equation which is similar to Equation 7 but also incorporates the fraction of drug metabolized by the inhibited pathway, Eq 8[145, 165]. Thus allowing for the optimal design of clinical drug-drug interaction studies.

$$AUC_i / AUC = \frac{1}{[f_m / (1+[I]/K_i)] + (1- f_m)} \qquad (8)$$

The importance of fraction metabolized can be appreciated visually when plotted against the change in AUC as shown in Figure 9.

In summary, a likelihood of drug interaction can be gauged by the use of the $[I]/K_i$ ratio using an unbound K_i value and a good approximation of liver concentration, [I]. This, in conjunction with a knowledge of the fraction metabolized and other non-metabolic elimination pathways, will generally yield a good approximation of the drug interaction.

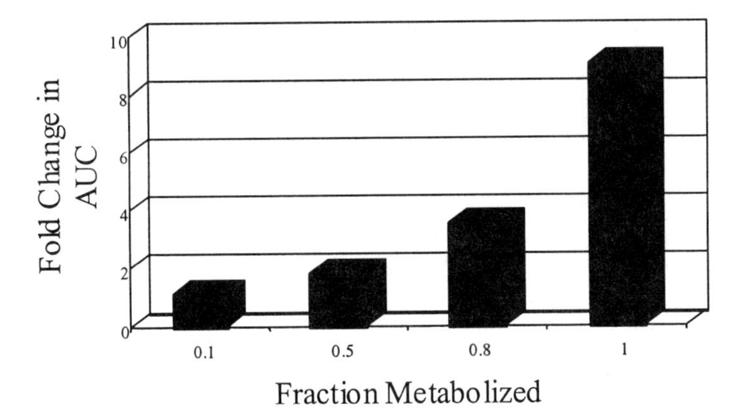

Figure 9. Relationship between fraction metabolized and fold change in AUC due to enzyme inhibition ($K_i = 0.5\ \mu M$, $I = 4\ \mu M$).

4.3.2. Metabolism Dependent Inhibition

As was noted previously, distinguishing between reversible and irreversible inhibition is essential in predicting the eventual degree of *in vivo* inhibition. There are many terms associated with irreversible inhibition of drug metabolizing enzymes, such as time dependent inhibition, mechanism-based inactivation, suicide inhibition, or metabolism dependent inhibition[166-169]. True irreversible inhibition which results in covalent binding and enzyme inactivation is assessed by meeting several criteria characteristic of this form of inhibition as described by Silverman[170, 171]. There is another term used when describing irreversible inhibition of CYP450 enzymes, that being quasi-irreversible inhibition. This situation arises when the reactive intermediate forms a stable complex with the prosthetic heme of the CYP450 known as a metabolite-intermediate complex (MI complex). This complex is not covalent in nature and therefore is not definitively an irreversible reaction. The complex can be dissociated in *in vitro* preparations, however this same disassociation is extremely slow *in vivo* and for all intensive purposes is considered irreversible. The term metabolism dependent inhibition is most appropriate for describing irreversible inhibition of drug metabolizing enzymes because in order to create this form of inhibition the substrate must be metabolized to the reactive species (inhibitor) as a prerequisite to the inhibition. This is in contrast to the reversible situation where only inhibitor binding is required for inhibition to occur. Some distinguishing characteristics of metabolism dependent inhibition compared to typical reversible inhibition include: 1) greater potential for non-linear pharmacokinetics, 2) inhibition effect that extend beyond the elimination of the parent drug (due to enzyme inactivation), 3) inhibition effects tend to accumulate after multiple (frequent) dosing, 4) inhibition effect is generally greater than that predicted by the reversible K_i or IC_{50} values, 5) rare instances of hepatotoxicity associated with covalently bound adducts, and 6) metabolism dependent inhibition is more difficult to predict than reversible inhibition due to the confounding factors of enzyme inactivation and resynthesis rates.

The concern over metabolism dependent inhibition has been exemplified by the recent experience with mibefradil. In June of 1997, Roche Laboratories received FDA approval to market a new calcium channel blocker for the treatment of hypertension, mibefradil (Posicor®). In June of 1998, the developers of mibefradil withdrew the drug from the market as a result of its potential to cause serious drug interactions[136]. Interactions were originally identified with simvastatin, cyclosporin, and terfenadine in which the plasma concentrations of these drugs increased significantly resulting in serious toxicity[172]. Additional studies indicated that the most serious interactions appeared to with substrates of CYP3A4. Prueksaritanont and colleagues were able to demonstrate that at therapeutic concentrations, mibefradil was a potent mechanism based inhibitor of CYP3A4 using a series of cholesterol lowing agents known as statins[173]. Furthermore, the long half-life of mibefradil (17-25 hrs) had lead to drug-drug interactions up to 5 days after the last dose of drug was taken[172, 174]. In addition to mibefradil, many compounds prior to and after this incident have been identified as metabolism dependent inhibitors, such as the hormone drugs ethynylestradiol[175, 176], gestodene[177], and raloxifene[178], macrolide antibiotics (erythromycin, clarithromycin, troleandomycin)[179], HIV protease and reverse transcriptase inhibitors (ritonavir and delavirdine)[180-182], calcium channel blocker (diltiazem, nicardipin, and verapamil)[183-184], and several other compounds, such as a grapefruit juice component, dihydroxybergamottin[185], mifepristone (RU486)[186], furafylline[187], and tienilic acid[188]. Due to the prevalence of this phenomenon and the seriousness of the interaction it has been recommended that all preclinical candidates be tested for time dependent or metabolism dependent inhibition[31].

We previously discussed the rationale and interpretations for reversible inhibition, however these same equations and principles do not apply in the situation of metabolism dependent inhibition due to the time dependent nature of the reaction. Recall that reversible enzyme kinetics make the assumption that the enzyme concentration throughout the reaction is a constant which is not applicable in the case of irreversible inhibition. Metabolism dependent inhibition is assessed by alternate set of inhibition experiments, typically performed in human liver microsomes where a K_I and K_{inact} are determined. K_I is the concentration of inhibitor at half maximal rate of inhibition and K_{inact} is the maximal rate of inactivation[170]. Also note the distinction between the reversible and irreversible inhibition constants (K_i and K_I, respectively) for clarity and to distinguish that they are derived from separate kinetic equations. These kinetic parameters by themselves provide information on the intrinsic potency of the metabolism dependent inhibitor, however a more accurate assessment of actual interaction potential can be made when these values are combined with the known or predicted plasma concentration of the drug (similar to the reversible inhibition interpretation). This new parameter, the rate of inactivation (λ) is determined as shown in Eq 9[169, 178].

$$\lambda = (k_{inact} \text{ x } I) / (K_I + I) \qquad (9)$$

The λ value of a tested drug candidate can be compared directly to the same value calculated for known metabolism dependent inhibitors and their degree of clinical interaction to predict the likelihood of an interaction. Other researchers have also taken the concept of the λ (rate of inactivation) one step further to develop models that incorporate the rate of enzyme inactivation with the rate of enzyme turnover, as well as plasma protein binding to predict the extent of liver enzyme inactivation and ultimately increases in AUC[169, 170, 189]. This model has worked well in predicting the magnitude of *in vivo* drug interactions for the known inhibitors clarithromycin, fluoxetine, and diltiazem (N-desmethyl diltiazem), however it does involve generation of additional data and assumptions on the turnover of enzyme.

Another characteristic parameter common to this form of inhibition is the partition ratio (PR) which is defined as the ratio of total metabolism of the drug to the inactivation process, Eq 10. This parameter places into perspective the amount of inactivation to the overall metabolism of the compound and can also be depicted as the ratio of k_{cat} to k_{inact}. Additional references into the kinetics and experiments behind metabolism dependent inhibition can be found in the following references[151, 170, 178, 187, 190, 191].

$$PR = k_{cat}/k_{inact} \quad (10)$$

4.3.3. Enzyme Induction

In comparison to enzyme inhibition, enzyme induction occurs less frequently in humans and it has been estimated that only 2% of drugs fail in development due to clinical induction[192]. Although induction occurs with less frequency and generally does not result in toxicity, it still remains as an important liability to assess preclinically and clinically as to the potential and magnitude of possible drug-drug interactions. Presently our interpretation of enzyme induction data from preclinical *in vitro* model systems is more qualitative than quantitative. Therefore, our *in vitro* systems, when they demonstrate increased enzyme activity, can be used to guide multiple dose clinical studies at therapeutic concentrations to determine the actual effect on either the parent drug or other coadministered drugs. Enzyme induction is a common occurrence in preclinical animal studies when new drug candidates are tested at high doses generally associated with toxicology studies. However, over the years we have learned that *in vivo* animal models are inappropriate to predict human enzyme induction and therefore the observed animal induction typically does not preclude a candidate from moving forward to clinical development. For example, pregnenolone 16α-carbonitrile is an effective CYP3A inducer in the rat but not humans, and rifampin induces CYP3A in humans but not in rats[193]. These findings are not unexpected as it has been established that there are marked species differences in the effects of different xenobiotics on CYP3A expression due to the nuclear hormone receptor PXR (SXR)[193, 194]. These differences highlight the importance of using human *in vitro* systems to evaluate human induction, especially where marked species differences have been noted in the induction profiles of drug-induced increases in CYP450 gene expression[194, 195].

For our purposes, enzyme induction will be defined as an increase in enzyme activity due to an increase in the amount of enzyme present. The increase in enzyme concentration can occur by several different mechanisms, the most common is transcriptional activation which is found for most CYP450 enzymes. A known caveat to this mode of enzyme induction is found with CYP2E1 where enzyme stabilization is known to occur and results in increased activity due to a decrease in enzyme degradation[196]. Other minor mechanisms that may play a part in enzyme induction can include mRNA stabilization and increased translation of protein.

Overall, the inducing agent (drug) elicits some form of cellular process that with increasing time and dose results in elevated enzymatic activity. The elevated enzyme activity results in an increased rate of metabolism for compounds that are biotransformed by the induced enzyme(s). For greater detail on the pharmacokinetic changes and consequences of enzyme induction the reader is referred to a review by Worboys and Carlile[197]. The drug affected is most often a co-administered drug that uses the induced enzyme as an elimination pathway (hetero-induction), however the affected drug can also be the parent drug that caused the induction, this is known as autoinduction (homo-induction). This increased biotransformation can lead to dramatic decreases in plasma levels of drugs that go through the induced enzymatic pathway, thereby creating a situation where the drug is no longer at efficacious concentrations. Overall this results in a loss of efficacy and the drug is no longer effective as a treatment. A recent example is that of troglitazone, an antihyperglycemic agent that causes significant CYP3A4 induction. Troglitazone has been shown to decrease plasma exposure of oral contraceptives, cyclosporine, terfenadine, and atorvastatin[198-201]. In almost all cases, doses were adjusted to maintain efficacy or loss of efficacy and noted in the package insert. Another example is that of rifampin which can cause a 20-fold increase in total body clearance of the CYP3A4 substrate midazolam[202]. In rare instances, a drug may be metabolized to a minor toxic metabolite that under normal circumstances is innocuous, however if that particular pathway is induced, greater amounts of the toxic metabolite can be formed, thereby creating a situation that is now detrimental. A classic example is that of CYP2E1 induction and coadministration of acetaminophen. A minor metabolite of acetaminophen, N-acetyl-p-benzoquinone imine (NAPQI), is readily detoxified through conjugation reactions, however when CYP2E1 becomes induced and greater quantities of NAPQI can no longer be detoxified liver necrosis occurs[203, 204]. Another example of both loss of activity and adverse event involves a reduction in immunosuppressant activity with organ transplant patients taking CYP3A4 enzyme inducers. Subtherapeutic levels of cyclosporine (immunosuppressant) have been noted in several instances when coadminstered to transplant patients receiving phenobarbital, a known CYP3A4 inducer[205]. Although our discussion here surrounds drugs as inducing agents, we should always keep in mind that social activities and diet can also lead to induction. For example, smoking, alcohol abuse, and the herbal remedy St. John's Wort can lead to CYP1A, CYP2E1, and CYP3A4 induction and drug interactions, respectively[203, 206].

4.3.3a Induction Model: Human Hepatocytes. Cultures of primary human hepatocytes are considered the most relevant *in vitro* model to evaluate the potential for enzyme induction of drug metabolizing enzymes[31, 145]. This *in vitro* system, along with its animal counterparts have been used to study enzyme induction for many years with a multitude of drug examples, such as rifampin, ethanol, phenobarbital, phenytoin, metyrapone, and cyclophosphamide[36, 207-211]. For recent reviews on the use and methodology of human hepatocytes as an induction model see LeClyuse and Silva[212, 213]. Enzyme induction is a cellular response and therefore the hepatocyte model is most appropriate to study this phenomenon. Also, hepatocytes offer an advantage in that many other drug metabolizing enzymes can be studied in addition to the cytochromes P450. Other enzymes, such as epoxide hydrolase and glutathione transferase, as well as those involved in glucuronidation can be elevated by various inducing agents[214-217]. The major drawback to this system is the availability of good quality fresh human liver tissue or cells. Tissue discarded from surgical procedures and transplant rejected livers are the most common sources of human liver tissue. Access to these tissues is random and the competition between academic and industrial laboratories to obtain tissue is growing. Needless to say, using human hepatocytes as a model for induction is rarely something that can be adopted as high throughput or placed on a defined schedule. The varied phenotype of donor cells can be a complicating factor as the genotype and medical/social background of each individual will determine the existence of or level of enzyme activity which has been shown to vary greatly amongst individuals[53]. Patients taking known enzyme inducers as part of their medical treatment or individuals with naturally high levels of enzyme at the time of cell isolation may have elevated enzyme levels that can not be (or minimally) induced. There appears to be an upper limit to the extent of induction, ie., limit of maximal inducability. Kostrubsky, et al were able to show that the maximal inducible level of CYP3A4 enzyme activity in human hepatocytes was similar between 6 individuals, however the fold induction varied between individuals[46]. Those individuals with the lowest basal CYP3A4 levels had the greatest fold induction of testosterone hydroxylation when treated with the CYP3A4 inducer taxol. These interindividual differences in response to inducing agents complicates interpretation of the final results in that a range of values is typically generated, eg., 2-20 fold increase in enzyme activity.

In addition to interindividual differences in enzyme activity and quality of cells, culture and study conditions between labs are not consistent which contributes to the variability of results. The major differences in culture conditions revolve around media components and extracellular matrices. For short term cultures (~1 week) many media types are adequate, however addition of media supplements, such as hormones (dexamethasone) at various concentrations can lead to significant changes in cell dynamics and even induce some enzymes[212]. Another media component often disregarded is the vehicle in which the drug is delivered. The best situation is when the drug can be dissolved in culture media, although this is rarely the case and organic solvents are necessary to dissolve non-polar drugs. The most common of which is DMSO that has been shown to increase the levels of CYP3A4 in human hepatocytes at concentrations as low as 0.2%[212] underscoring the need to include vehicle controls in all hepatocyte experiments that use anything other than culture media to deliver the drug. In order for hepatocytes to survive over the longer periods of time needed to study enzyme induction (several days) they require attachment to some form of extracellular matrix. The type (collagen / Matrigel®) and architecture (single layer substratum or overlay- sandwich culture) of extracellular matrix can be species dependent and lead to changes in cell morphology or extent of induction, however all variations will typically lead to the correct qualitative response[212]. In general, the sandwich or overlay technique is preferred due to the 3-diminsional environment of cells which is more *in vivo*-like than the 2-dimensional single substratum situation.

Study conditions comprise such variables as drug concentrations and time of incubation, as well as how the induction will be measured. As with inhibition studies, the test compound should be incubated at concentrations encompassing the anticipated or known plasma concentrations for efficacy. Typically ±10 times the therapeutic concentration are attempted at 3-6 different concentrations recognizing that drug solubility or cell toxicity may limit the upper concentration range. This number of concentrations and range should provide a reasonable dose response curve of induction effect. After cell isolation, a typical hepatocyte induction experiment will begin with placing cell in culture for 2-3 days without drug, followed by daily treatment with drug for another 2-3 days. After a total of about 5 days in culture, induction can be assessed by several different methods. Measurements of induction are generally a combination of one or more of the following changes: mRNA, protein, or enzyme activity. The activation of mRNA is a common mechanism of CYP450 induction and one of the first events to occur. The mRNA is typically measured by real time reverse transcription polymerase chain reaction (real time RT-PCR)[213, 218]. An alternative to mRNA measurements is quantitating the amount of protein expressed by Western blotting. Ultimately, the preferred method of measurement is enzyme activity (using specific enzyme probes) whereas it is this change that results in drug interactions. Although the earlier events of mRNA activation and increased protein expression generally correlate to increased activity, the correlation is not 100%, nor is it a direct 1:1 correlation. Therefore, when reporting potential drug interaction information related to enzyme induction, the increases in enzyme activity are the most pertinent. Combining, these various techniques can be useful in determining the mechanism of induction, such as mRNA activation or protein stabilization should this level of detail become necessary.

Once measurements of enzyme activity are measured they need to be normalized and put into a context that rank orders compounds or gives some indication of potency. A common derived parameter is the 'fold increase' in activity compared to control incubations. This is a simple means of rank ordering compounds from the same liver preparation, however as mentioned previously, different livers have different control levels and therefore the fold induction can change between different liver preparations. Similar to fold increase is the 'potency index' which normalizes the increase in activity to the increase in activity of a positive control inducer. For example the increase in enzyme activity of a test compound compared to the increase in activity found with the CYP3A4 inducer rifampicin. This method helps to normalize the differences between liver preparations and will be subject to less variability between preparations. A third method is determination of EC_{50} (concentration at which 50% maximal induction is observed) which gives a reasonable measure of potency, however can suffer from differences between hepatocyte preparations. Typically, when induction is observed in hepatoctye preparations, a range of induction values from three or more preparations derived from individual donors are employed to determine the variability and extent of potential induction.

After obtaining some estimate of increase in enzyme activity, this *in vitro* parameter must be rolled into an overall assessment of the induction potential and possible liability of drug-drug interaction. Other factors to consider in relation to the *in vitro* data include the following: 1) Is this a single dose drug or is it intended for chronic use? Whereas induction occurs after multiple dosing, a single dose drug would not be susceptible to induction. Similarly each drug has a different time course to reach maximal induction, some quickly others longer, short term dosing

(<2 weeks) is unlikely to lead to significant induction. 2) Is this anticipated to be a high or low dose drug? Induction typically occurs with high dose drugs and drug doses will be a major determining factor as to whether clinically significant induction will occur[192]. Similar and related to dose are the plasma concentrations necessary for efficacy and the relationship between *in vitro* hepatocyte data and plasma concentrations. 3) Is this drug intended for a life threatening disease or a disease that is already treated with a class of enzyme inducers? Fatal disease areas where efficacious drugs are not available are more tolerable to the issues surrounding drug-drug interactions due to enzyme induction (or inhibition). Anticonvulsants and many drugs to treat HIV are well known CYP450 inducers and physicians are accustomed to managing the drug interactions in these areas. Albeit a new drug in these therapeutic areas that is not an enzyme inducer will have a greater safety profile and marketability. 4) What are the commonly prescribed coadministered drugs given in this therapeutic area? Perhaps the majority of coadministered drugs are eliminated through other non-induced P450's and the interaction potential is minimized. The opposite scenario is of greater concern when the majority of coadministered drugs are known to be eliminated through the induced pathway. 5) What is the enzyme being induced? As mentioned previously, there is a special emphasis placed on the significance of CYP3A4 induction in comparison to other drug metabolizing enzymes due to the preponderance of CYP3A4 substrates on the market. All of these factors can be considered in the overall fate of a new drug candidate, however, eventually the actual significance or occurrence of drug interactions due to enzyme induction has to be determined in the clinic.

4.3.3b. Other Cell Based Induction Models. Our understanding of the intricate mechanisms of CYP450 induction have increased dramatically over the past 5 years with the discovery and understanding of orphan nuclear hormone receptors and their role in CYP450 induction. We recognized early that mRNA activation was a means to increase enzyme activity, however the mechanisms that lead to elevated gene expression were not well understood. We now know that increased CYP450 gene expression is through a variety of receptor dependent mechanisms involving the Ah receptor (aryl hydrocarbon receptor), CAR (constitutive androstane receptor), PXR (pregnane X receptor or SXR, steroid / xenobiotic receptor), and PPAR (peroxisome proliferator activated receptor) for induction of the CYP1A, CYP2B, CYP3A, and CYP4A families[219-224]. With a better knowledge of the initial mechanistic steps involved in CYP450 induction, cell based models incorporating these receptors have been developed to evaluate the potential for enzyme induction[195, 225-227]. In addition, well characterized subclones of HepG2 cells have demonstrated the ability to respond to prototypical inducers, such as the C3A subclone for assessing CYP1A2 induction[228]. Although these are useful models for discovery screening, their application in drug development is limited to the elucidation of mechanisms of induction while primary human hepatocytes remain the most relevant *in vitro* induction model to predict human induction *in vivo*[31]. A retrospective analysis of 14 drugs to elicit PXR transactivation and CYP3A4 induction in human primary hepatocytes (transcriptional activation mRNA, Western blotting of microsomal protein, and enzyme activity) was able to demonstrate a good correlation with known CYP3A4 inducers between the induction in hepatocytes and PXR transactivation[229]. However, we would be justified in assuming that in a prospective study, a significant number of drug candidates that indicated binding to PXR or that caused PXR transactivation would not lead to induction of CYP3A4. This could be due to binding and/or activation that did not ultimately lead to significant increases in enzyme activity. Moreover, the use of other cell types (non-hepatocyte), such as HepG2, which can be different in regards to cell membrane permeability or cellular contents compared to primary human hepatocytes. Therefore we would expect that known inducers would indicate positive in most binding or cell-based (non-hepatocyte) assays, however there will be many compounds that are positive in these cell-based models that do not demonstrate induction of CYP3A4 in hepatocytes or *in vivo*, hence the lower predictive power.

4.3.3c. Induction Case Study – Rezulin. Rezulin® (troglitazone) was approved in 1997 by the FDA as a drug having antidiabetic activity[201]. The drug was more recently removed from the market in 2000 due to rare instances of hepatotoxicity and the approval of two other drugs in the same class with lower risks of hepatotoxicity. Troglitazone has been associated with significant clinical drug interactions originally proposed to be due to CYP3A4 induction. Troglitazone was shown to decrease the plasma concentrations of oral contraceptives (ethinyl estradiol and norethindrone) and cyclosporin by 30% and 32%, respectively and reduce the plasma AUC's of terfenadine and atorvastatin by 60-70% and 33%, respectively[198-201]. Interestingly, while the effects of troglitazone on oral contraceptives, cyclosporin, and terfenadine were due to a loss of efficacy, the cholesterol lowering pharmacodynamic effect on atorvastatin adminsitration was not greatly changed. Atorvastatin is metabolized to two primary hydroxylated metabolites via CYP3A4 and both metabolites have cholesterol lowering activity. CYP3A4 induction caused a decrease in atorvastatin plasma levels, however at the same time increased the formation of active metabolites, therefore causing little change in efficacy.

Many studies were conducted with the previously described induction models to elucidate the mechanism of drug interactions with troglitazone. In similar studies using primary human hepatocyte cultures, troglitazone resulted in a 5-15 fold increase in CYP3A4 enzyme activity (EC_{50} 5-10 uM) with corresponding increases in CYP3A4 mRNA and protein levels[230, 231]. Troglitazone has also been shown to bind to and activate PXR at similar concentrations as those found in the hepatocyte studies[231]. This compilation of results clearly indicated that the interactions of troglitazone could have been predicted with other CYP3A4 substrates and was due to induction of CYP3A4 mediated through PXR.

4.3.4. Why Do Interaction Prediction Fail?

As a general rule with any *in vitro* drug metabolism assay (reaction phenotyping, inhibition, or induction), positive *in vitro* results should be followed-up during clinical trials to confirm or deny the predicted outcome and assess the magnitude of pharmacokinetic changes. Negative *in vitro* results may or may not be followed-up in clinical trials to confirm the negative result. The literature is full of retrospective predictions of drug interactions that illustrate good correlations between *in vitro* and *in vivo* results. However, there are several issues that must be considered whenever a prospective analysis (the actual situation during preclinical development of a candidate drug) is performed that may lead to inaccurate or erroneous predictions. Some of these issues have already been mentioned, such as distinguishing between the mechanisms of inhibition, plasma or matrix protein binding, accurate estimates of actual liver concentrations, and the myriad of assumptions and inaccuracies in determining enzyme kinetic parameters[109, 232]. Other characteristics that contribute to inaccurate predictions include the static nature of systems and direct access to enzymes. Most of the *in vitro* models are static systems which means that metabolites will accumulate and parent drug may be rapidly metabolized leaving only metabolites. This situation with parent drug and metabolite concentrations are not relevant to the *in vivo* situation where concentrations of both constantly vary with time. Whenever using cell free systems, the drug has direct assess to receptors and enzymes which they may or may not have when an appropriate cell membrane is present to enhance or limit the drug concentration within the cell. Therefore, many cell free systems can give false positive results due to direct assess to receptors or enzymes. Drug candidates can also give erroneous results when tested at very high concentrations in cell free systems that may actually be toxic if conducted in a cellular model or *in vivo*.

A situation that is extremely difficult to predict occurs when a test compound causes significant enzyme inhibition and induction simultaneously. Such is the case with ritonavir where it has been shown to both increase the metabolism of dapsone and ethynyl estradiol and decrease the metabolism of clarithromycin and indinavir in patients[229, 233]. *In vitro*, ritonavir is a potent reversible and irreversible inhibitor of CYP3A4 and has been shown to transactivate PXR, increase CYP3A4 mRNA and protein levels[229]. Depending on the concentration of ritonavir *in vitro* or the dose (liver exposure) *in vivo*, one effect will predominate. Furthermore, the type of interaction changes with time *in vivo*. Whereas reversible inhibition is a direct action of the drug on the enzyme, this event occurs as early as the first dose of drug, followed by irreversible inhibition, and eventually enzyme induction. In this situation, one may expect such a drug, like ritonavir, to be a potent inhibitor early on and have the inhibitory drug interaction replaced by interactions due to enzyme induction after repeated dosing. In some situations you may anticipate that the combined effects of inhibition and induction may equally offset one another as was hypothesized for the clinical interaction between ritonavir and alprazolam[233]. Situations of simultaneous drug inhibition and induction are nearly impossible to predict from *in vitro* studies alone and will tend to vary considerable between patients depending on the dose, time on drug, and interindividual differences in normal enzyme concentrations. Well controlled clinical trials are the best means to assess the predominant type and magnitude of interaction, as well as the interindividual variability of such interactions.

5. SUMMARY AND FUTURE DIRECTIONS

The preceding information and examples on preclinical drug metabolism illustrate how diverse the field can truly be. In today's environment, one must have a background in multiple disciplines, such as biochemistry, molecular biology, organic chemistry, and analytical chemistry to understand the interplay amongst the various model systems and tools employed in modern day drug metabolism. The experiments and results surrounding biotransformation, reaction phenotyping, and drug interactions are major areas of focus for all preclinical drug candidates. The importance and impact of drug metabolism data is realized by all in the pharmaceutical industry and regulatory agencies. This is best demonstrated by a quick walk through the most recent Physicians' Desk Reference[234] which contains more drug metabolism related information, both *in vitro* and *in vivo*, than previous editions.

Certainly the future holds additional breakthroughs and enhancements in our armamentarium of *in vitro* and *in vivo* model systems to predict human drug metabolism and drug interactions. Recent advancements include the use of cryopreserved hepatocytes for CYP450 induction studies and the elucidation of receptor mediated CYP450 induction mechanisms[47], such as PXR[220]. Even though both still require a great deal of continued research, they have already made an impact on the way in which we study preclinical drug candidates. The development of transgenic animals, such as mice that express human CYP2D6[235] or human PXR[236], hold great promise in our ability to study human enzymes, reactions, and interactions in an *in vivo* setting, albeit animal vs. human. A burgeoning area of research is the development of in silico models for predicting biotransformation reactions[237, 238], intrinsic clearance[239], drug inhibition[240], and enzyme induction[241]. In silico models allow researchers to evaluate many virtual compounds and actually study a smaller subset of compounds at the bench expediting the experimental process. The advent of gene chip technology which allows us to simultaneously measure the expression of 1000's of genes has aided our understanding of the regulation of drug metabolizing enzymes[242]. Perhaps the newer technology of protein chips (similar to gene chips) will also provide us with yet another means to study protein – drug interactions[243].

Ultimately, these improvements in models and technology should yield greater predictability and correlations between *in vitro* and *in vivo* data. Clearly every area (biotransformation, reaction phenotyping, and drug interactions) of drug metabolism requires increased predictive power. In particular the correlation between toxic metabolite formation and progression of toxicological events in patients and the prediction of CYP450 induction in human hepatocytes to the magnitude of induction observed in patients. This increased predictive power may come through newer models or a better understanding of our current systems. Nonetheless, this is a key area for advancement where improved correlations between *in vitro* human preparations (models) to the outcomes in patients will allow safer and more efficacious drugs to enter the marketplace.

6. ACKNOWLEDGEMENTS

The author would like to thank Drs. Kenneth Santone, Susan Hurst, Jasminder Sahi, and Renke Dai for their critical review of the manuscript.

7. REFERENCES

1) K.J. Watkins, Fighting the clock, Chem. Eng. News, Jan. 28, 27-34 (2002).
2) C.A. Shillingford and C.W. Vose, Effective decision making: progressing compounds through clinical development, Drug Disc. Today, 6(18), 941-946 (2001).
3) A. Barrett, Commentary: How drugmakers should handle a cautious FDA, Bus. Week, May 20 (2002).
4) J. Kuhlman, Drug research: From idea to the product, Int. J. Clin. Pharm. Ther. 35, 541-552 (1997).
5) A.V. Boddy and S.M. Yule, Metabolism and pharmacokinetics of oxazaphosphorines, Drug Disp. 38(4), 291-304 (2000).
6) D.C. Klonoff, Rezulin to be withdrawn from the market, Diab. Tech. Ther. 2(2), 289-290 (2000).
7) K. Kassahun, P.G. Pearson, W. Tang, I McIntash, K. Leung, C. Elmore, et al, Studies on the metabolism of troglitazone to reactive intermediates *in vitro* and *in vivo*. Evidence for novel biotransformation pathways involving quinone methide formation and thiazolidinedione ring scission, Chem. Res. Toxicol. 14, 62-71 (2001).
8) A.P. Lea and D. McTavish, Atorvastatin, a review of its pharmacology and therapeutic potential in the management of hyperlipidaemias, Drugs, 53(5), 828-847 (1997).
9) W.W. Bullen, R. A. Miller, and R.N. Hayes, Development and validation fo a high performance liquid chromatography tandem mass spectrometry assay for atorvastatin, ortho-hydroxy atorvastatin and para-hydroxy atorvastatin in human, dog, and rat plasma, J. Am. Soc. Mass Spectrom. 10, 55-66 (1999).
10) T. F. Woolf, Handbook of drug metabolism. 1st ed. Marcel Dekker, New York (1999).
11) C. Ionnides, Enzyme systems that metabolise drugs and other xenobiotics, 1st ed. John Wiley & Sons, New York (2002).
12) A. Parkinson, Biotransformation of xenobiotics. In: C.D. Klaassen, ed. Casarett and Doull's Toxicology, the basic science of poisons, 5th ed. McGraw-Hill, New York, pp. 113-186 (1996).
13) L.K. Low, Metabolic changes of drugs and related organic compounds. In: R.F. Doerge ed. Wilson and Grisvold's textbook of organic medicinal and pharmaceutical chemistry, 8th ed. Lippincott, Philadelphia, 43-121 (1982).
14) W. Kalow, Pharmacogenetics of drug metabolism. 1st ed. Pergamon Press, New York (1992).
15) T. Satoh, P. Taylor, W.F. Bosron, S.P. Sanghani,M. Hosokawa, and B.N. La Du, Current progress on esterases: From molecular structure to function, Drug Metab. Dispos. 30(5) 488-493 (2002).
16) M.S. Benedetti, Biotransformation of xenobiotics by amine oxidase, Fund. Clin. Pharm.15, 75-84 (2001).
17) M.W. Sinz, A.E. Black, S.M. Bjorge, A. Holmes, B.K. Trivedi, and T.F. Woolf, *In vitro* and *in vivo* disposition of 2,2-demethyl-N-(2,4,6-trimethoxyphenyl) dodecanamide (CI-976), Drug Metab. Dispos. 25(1) 123-130 (1997).
18) M.W. Sinz and T. Podoll, The mass spectrometer in drug metabolism. In: D.T. Rossi and M.W. Sinz, eds. Mass spectrometry in drug discovery. Marcel Dekker, New York, pp. 271-335 (2002).
19) G.J. Mulder, Conjugation reactions in drug metabolism: An integrated approach. 1st ed. Taylor & Francis, New York (1990).
20) M.B. Fisher, M.F. Paine, T.J. Strelevitz, and S.A. Wrighton, The role of hepatic and extrahepatic UDP-glucuronyltransferases in human drug metabolism, Drug Met. Rev. 33(3&4), 273-297 (2001).
21) T. Shimada, H. Yamazake, M. Mimura, Y. Inui, F.P. Guengerich, Interindividaul variations in human liver cytochrome P-450 enzymes involved in the oxidation of drugs, carcinogens and toxic chemicals: Studies with liver microsomes of 30 Japanese and 30 Caucasions, J. Pharm. Exp. Ther. 270(1), 414-423 (1994).
22) J.M. Lasker, M.R. Wester, E. Aramsombatdee, and J.L. Raucy, Characterization of CYP2C19 and CYP2C9 from human liver: Respective roles in microsomal tolbutamide, S-mephenytoin, and omeprazole hydroxylations, Arch. Biochem. Biophy. 353(1), 16-28 (1998).
23) Z. Chen, J. Zhang, and J.S. Stamler, Identification of the enzymatic mechanism of nitroglycerin bioactivation, PNAS, 99(12), 8306-8311 (2002).
24) Food and Drug Administration, Guidance for Industry: Drug metabolism/drug interaction studies in the drug development process, studies *in vitro*. www.fda.gov/cder/guidance.htm, April (1997).
25) S. Ekins, B.J. Ring, J. Grace, D.J. McRobie-Belle, and S.A. Wrighton, Present and future *in vitro* approaches for drug metabolism, J. Pharm. Tox. Met. 44, 313-324 (2000).
26) R.N. Hayes, W.F. Pool, M.W. Sinz, and T.F. Woolf, Recent developments in drug metabolism methodology. In: P.G. Welling an F.L.S. Tse, eds. Pharmacokinetics, regulatory, industrial, and academic perspectives. Marcel Dekker, New York, pp. 201-234 (1995).

27) C.A. Lee, S.H. Kadwell, T.A. Kost, and C.J. Serabjit-Singh, CYP3A4 expressed by insect cells infected with a recombinant baculovirus containing both CYP3A4 and human NADPH-cytochrome P450 reductase is catalytically similar to human liver microsomal CYP3A4, Arc. Biochem. Biophy. 219(1), 157-167 (1995).
28) P. Kiselev, D. Schwarz, K.-L. Platt, W.-H. Schunck, and I. Roots, Epoxidation of benzo[a]pyrene-7,8-dihydrodiol by human CYP1A1 in reconstituted membranes, Eur. J. Biochem. 269, 1799-1805 (2002).
29) M.W. Voice, Y. Zhang, C.R. Wolf, B. Burchell, and T. Friedberg, Effects of human cytochrome b_5 on CYP3A4 activity and stability *in vivo*, Arch. Biochem. Biophys. 366(1), 116-124 (1999).
30) S. Ding, D. ao, Y.Y. Deeni, B. Burchell, C.R. Wolf, and T. Friedberg, Human NADPH-P450 oxidoreductase modulates the level of cytochrome P450 CYP2D6 holoprotein via heam oxygenase-dependent and independent pathways, Biochem. J. 356, 613-619 (2001).
31) G.T. Tucker, J.B. Houston, and S.-M. Huang, Optimizing drug development: Strategies to assess drug metabolism/transporter interaction potential-toward a consensus, Pharm. Res. 18(8), 1071-1080 (2001).
32) F.P. Guengerich, Analysis and characterization of enzymes. In: A.W. Hayes, ed. 2^{cd} ed. Raven Press, New York, pp. 777-814 (1989).
33) B.A. Fowler, G.W. Lucier, and A.W. Wallace, Organelles as tools in toxicology. In: A.W. Hayes, ed. 2^{cd} ed. Raven Press, New York, pp. 815-833 (1989).
34) S. Ekins, J. Maenpaa, and S.A. Wrighton, *In vitro* metabolism: subcellular fractions. In: T. F. Woolf, Handbook of drug metabolism. 1st ed. Marcel Dekker, New York, pp. 363-399 (1999).
35) H. Osmundsen, J. Bremer, and J.I. Pedersen, Metabolic aspects of peroxisomal beta-oxidation, Biochem. Biophys. Acta, 1085, 141-158 (1991).
36) M.W. Sinz, *In vitro* metabolism: hepatocytes. In: T. F. Woolf, Handbook of drug metabolism. 1st ed. Marcel Dekker, New York, pp. 401-424 (1999).
37) A.P. Li, Primary hepatocyte cultures as an *in vitro* experimental tool for xenobiotic metabolism and toxicology, Comm. Tox. 6(3), 199-220 (1998).
38) M.N. Berry, A.M. Edwards, and G.J. Barritt, Isolated hepatocytes: Preparation, properties and applications. 1st ed. Elsevier, New York (1991).
39) A.P. Li, C. Lu, J.A. Brent, C. Pham, A. Fackett, C.E. Ruegg, and P.M. Silber, Cryopreserved human hepatocytes: Characterization of drug metabolizing enzymes activities and applications in higher throughput screening assays for hepatotoxicity, metabolic stability, and drug-drug interaction potential, Chem.-Biol. Interact. 121, 17-35 (1999).
40) Y. Shibata, H. Takahashi, and Y. Ishii, A convenient *in vitro* screening method for predicting *in vivo* drug metabolic clearance using isolated hepatocytes suspended in serum, Drug Metab. Dispos. 28(12), 1518-1523 (2000).
41) R.A. Kemper, R.J. Krause, and A.A. Elfarra, Metabolism of butadiene monoxide by freshly isolated hepatocytes from mice and rats: Different partitioning between oxidative, hydrolytic, and conjugation pathways, Drug Metab. Dispos. 29(6), 830-836 (2001).
42) N.J. Hewitt, K.-U. Buhring, J. Dasenbrock, J. Haunschild, B. Ladstetter, and D. Utesch, Studies comparing *in vivo:in vitro* metabolism of three pharmaceutical compounds in rat, dog, monkey, and human using cryopreserved hepaocytes, microsomes, and collagen gel immobilized hepatocyte cultures, Drug Metab. Dispos. 29(7), 1042-1050 (2001).
43) C. Lu and A.P. Li, Species comparisons in P450 induction: Effects of dexamethasone, omeprazole, and rifampicin on P450 isoforms 1A and 3A in primary cultured hepatocytes from man, Sprague-Dawley rat, minipig, and beagle dog, Chem-Biol. Interact. 134, 271-281 (2001).
44) G.N. Kumar, V. Jayanti, R.D. Lee, D.N. Whittern, J. Uchic, S. Thomas, et al, *In vitro* metabolism of the HIV-1 protease inhibitor ABT-378: Species comparison and metabolite formation, Drug Metab. Dispos. 27(1), 86-91 (1999).
45) L. Lacidi, E.C. Scott, D. Eckoff, S. Bynon, and J.-P. Sommadossi, Metabolic drug interactions between angiogenic inhibitor, TNP-470 and anticancer agents in primary cultured hepatocytes and microsomes, Drug Metab. Dispos. 27(5), 623-626 (1999).
46) V.E. Kostrubsky, V. Ramachandran, R. Venkataramanan, K. Dorko, J.E. Esplen, S. Zhang, et al, The use of human hepatocyte cultures to study the induction of cytochrome P450, Drug Metab. Dispos. 27(8), 887-894 (1999).
47) J.M. Silva, S.H. Day, and D.A. Nicoll-Griffith, Induction of cytochrome-P450 in cryopreserved rat and human hepatoctyes, Chem. Biol. Int. 121, 49-63 (1999).
48) A. Guillouzo, L. Rialland, A. Fautrel, and C. Guyomard, Survival and function of isolated hepatocytes after cryopreservation, Chem. Biol. Int. 121, 7-16 (1999).

49) J.G. Hengstler, D. Utesch, P. Steinberg, K.L. Platt, B. Diener, M. Ringel, et al, Cryopreserved primary hepatocytes as a constantly available *in vitro* model for the evaluation of human and animal drug metabolism and enzyme induction, Drug Met. Rev. 32(1), 81-118 (2000).
50) A.P. Li, P.D. Gorycki, J.G. Hengstler, G.L. Kedderis, H.G. Koebe, R. Rahmani, et al, Present status of the application of cryopreserved hepatocytes in the evaluation of xenobiotics: Consensus of an international expert panel, Chem. Biol. Int. 121, 117-123 (1999).
51) M. Walterscheid, A. Nouraldeen, N.L. Sussman, and J.H. Kelly, CYP3A4 induction using a human liver cell line, T3512, AAPS (2001).
52) T. Butler, M. Walterscheid, A. Nouraldeen, N.L. Sussman, and J.H. Kelly, Metabolic stability and clearance measurements using a human liver cell line, T3152, AAPS (2001).
53) K.R. Iyer and M.W. Sinz, Characterization of phase I and Phase II hepatic drug metabolism activities in a panel of human liver preparations, Chem. Biol. Int. 118, 151-169 (1999).
54) A.R. Parrish, A.J. Gandolfi, and K. Brendel, Precision cut tissue slices: Applications in pharmacology and toxicology, Life Sci. 57(21), 1887-1901 (1995).
55) K. Yeleswaram, L.G. McLaughlin, J.O. Knipe, and D. Schabdack, Pharmacokinetics and oral bioavailability of exogenous melatonin in preclinical animal models and clinical implications, J. Pineal Res. 22(1), 45-51 (1997).
56) D.J. Carlile, N. Hakooz, and J.B. Houston, Kinetics of drug metabolism in rat liver slices: IV. Comparison of ethoxycoumarin clearance by liver slices, isolated hepatocytes, and hepatic microsomes from rats pretreated with known modifiers of cytochrome P-450 activity, Drug Metab. Dispos. 27(4), 526-532, (1999).
57) M. Kukan, The isolated perfused liver as a tool in drug metabolism studies. In: T. F. Woolf, Handbook of drug metabolism. 1st ed. Marcel Dekker, New York, pp. 425-442 (1999).
58) M. Yoshida, R.I. Mahato, K. Kawabata, Y. Takakura, and M. Hashida, Disposition characteristics of plasmid DNA in the single pass rat liver perfusion system, Pharm. Res. 13(4), 599-603 (1996).
59) H. Inoue, H. Yokota, T. Makino, A. Yuasa, and S. Kato, Bisphenol A glucuronide, a major metabolite in rat bile after liver perfusion, Drug Metab. Dispos. 29(8), 1084-1087 (2001).
60) T. Iwatsubo, N. Hirota, T. Ooie, H. Suzuki, and Y. Sugiyama, Prediction of *in vivo* drug disposition from *in vitro* data based on physiological pharmacokinetics, Biopharm. Drug Disp. 17, 273-310 (1996).
61) F.S. Chow, W. Piekoszewski, and W.J. Jusko, Effect of hematocrit and albumin concentration on hepatic clearance of tacrolimus (FK506) during rabbit liver perfusion, Drug Metab. Dispos. 25(5), 610-616 (1997).
62) R.E. White and P. Manitpisitkul, Pharmacokinetic theory of cassette dosing in drug discovery screening, Drug Metab. Dispos. 29(7), 957-966 (2001).
63) D.D Christ, Cassette dosing pharmacokinetics: valuable tool or flawed science? Drug Metab. Dispos. 29(7), 935 (2001).
64) W.F. Pool, Clinical drug metabolism studies. In: T. F. Woolf, Handbook of drug metabolism. 1st ed. Marcel Dekker, New York, 583 (1999).
65) A.R. Boobis, D. Sesardic, B.P. Murray, R.J. Edwards, A. M. Singleton, K.J. Rich, et al, Species variation in the response of the cytochrome P-450-dependent monooxygenase system to inducers and inhibitors, Xenobiotica, 20(11), 1139-1161 (1990).
66) H. Chen, J. Shockcor, W. Chen, R. Espina, L.S. Gan, and A.E. Mutlib, Delineating novel metabolic pathways of DPC 963, a non-nucleoside reverse transcriptase inhibitor, in rats. Characterization of glutathione conjugates of postulated oxirene and benzoquinone imine intermediates by LC/MS and LC/NMR, Chem. Res. Toxicol. 15, 388-399 (2002).
67) D. Bao, V. Thanabal, and W.F. Pool, Determination of tacrine metabolites in microsomal incubate by high performance liquid chromatography-nuclear magnetic resonance/mass spectrometry with a column trapping system, J. Pharm. Biom. Anal. 28, 23-30 (2002).
68) I.D. Wilson, J.K. Nicholson, and J.C. Lindon, The role of nuclear magnetic resonance spectroscopy in drug metabolism. In: T. F. Woolf, Handbook of drug metabolism. 1st ed. Marcel Dekker, New York, pp. 523-550 (1999).
69) T. H. Rushmore, P.J. Reider, D. Slaughter, C. Assang, and M. Shou, Bioreactor systems in drug metabolism: Synthesis of cytochrome P450 generated metabolites, Metab. Eng. 2, 115-125 (2000).
70) F. Trejnar, L. Skalova, B. Szotakova, and V. Wsol, Use of rat hepatocytes immobilized in agarose gel threads for biosynthesis of metabolites of potential cytostatics, Exp. Toxicol. Pathol. 51(4-5), 432-435 (1999).
71) V.J. Stella, W.N.A. Charman, and V.H. Naringrekar, Prodrugs, do they have advantages in clinical practice? Drugs, 29, 455-473 (1985).
72) I. Paakkari, Cardiotoxicity of new antihistamines and cisapride, Tox. Lett. 127, 279-284 (2002).

73) J. Lotsch and G. Geisslinger, Morphine-6-glucuronide, an analgesic of the future? Clin. Pharmacol. 40(7), 485-499 (2001).
74) D.K. Agrawal, Pharmacology and clinical efficacy of desloratadine as an anti-allergic and anti-inflammatory drug, Expert. Opin. Investig. Drugs, 10(3), 547-560 (2001).
75) G.N. Kumar and S. Surapaneni, Role of drug metabolism in drug discovery and development, Med. Res. Rev. 21(5), 397-411 (2001).
76) F.P. Guengerich, Common and uncommon cytochrome P450 reaction related to metabolism and chemical toxicity, Chem. Res. Tox. 14(6), 612-650 (2001).
77) A. Madan and A. Parkinson, Characterization of the NADPH-dependent covalent binding of [14C] halothane to human liver microsomes, role for cytochrome P4502E1 at low substrate concentrations, Drug Metab. Dispos. 24(12), 1307-1313 (1996).
78) J.M. Collins, Inter-species differences in drug properties, Chem. Biol. Int. 134, 237-242 (2001).
79) T.A. Baillie, M.N. Cayen, H. Fouda, R.J. Gerson, J.D. Green, S.J. Grossman, et al, Drug metabolites in safety testing, Toxicol. Appl. Pharmac. 182, 188-196 (2002).
80) M.W. Sinz, Metabolic profiling, reaction phenotyping and drug-drug interactions. In: N. Rudolph and M. Tulloch, eds, The rationale for predictive drug metabolism, Advanced Tech Monitor, Wobern, MA (2000).
81) G.N. Levy and W.W. Weber, Arylamine acetyltransferases. In: C. Ionnides, Enzyme systems that metabolise drugs and other xenobiotics, 1st ed. John Wiley & Sons, New York, pp. 441-457 (2002).
82) K.A. Bachmann and R. Ghosh, The use of *in vitro* methods to predict *in vivo* pharmacokinetics and drug interactions, Curr. Drug Metab. 2, 299-314 (2001).
83) Y. Naritomi, S. Terashita, S. Kimura, A. Suzuki, A. Kagayama, and Y. Sugiyama, Prediction of human hepatic clearance from *in vivo* animal experiments and *in vitro* metabolic studies with human liver microsomes from animals and humans, Drug Metab. Dispos. 29(10), 1316-1324 (2001).
84) B.A. Hoener, Predicting the hepatic clearance of xenobiotics in humans from *in vitro* data, Biopharm. Drug Dispos. 15, 295-304 (1994).
85) T. Lave, S. Dupin, C. Schmitt, R.C. Chou, D. Jaeck, and P. Coassolo, Integration of *in vitro* data into allometric scaling to predict hepatic metabolic clearance in man: Application to 10 extensively metabolized drugs, J. Pharm. Sci. 86(5), 584-590 (1997).
86) R.S. Obach and A.E. Reed-Hagen, Measurement of Michaelis constants for cytochrome P450-mediated biotransformation reactions using a substrate depletion approach, Drug Metab. Dispos. 30(7), 831-837 (2002).
87) J.B. Houston and D.J. Carlile, Prediction of hepatic clearance from microsomes, hepatocytes and liver slices, Drug Metab. Rev. 29(4), 891-922 (1997).
88) T. Lave, P. Coassolo, and B. Reigner, Prediction of hepatic clearance based on interspecies allometric scaling techniques and *in vitro-in vivo* correlations, Clin. Pharmacol. 36(3), 311-231 (1999).
89) K. Zomorodi, D.J. Carlile, and J.B. Houston, Kinetics of diazepam metabolism in rat hepatic microsomes and hepatocytes and their use in predicting *in vivo* hepatic clearance, Xenobiotica, 25(9), 907-916 (1995).
90) D.J. Carlile, A.J. Stevens, E.I. L. Ashforth, D. Waghela, and J.B. Houston, *In vivo* clearance of ethoxycoumarin and its prediction from *in vitro* systems, Drug Metab. Dispos. 26(3), 216-221 (1998).
91) T.B. Andersson, H. Sjoberg, K.J. Hoffmann, A.R. Boobis, P. Watts, R.J. Edwards, et al, An assessment of human liver derived *in vitro* systems to predict the *in vivo* metabolism and clearance of almokalnat, Drug Metab. Dispos. 29(5), 712-720 (2001).
92) T. Lave, S. Dupin, C. Schmitt, B. Valles, G. Ubeaud, R.C. Chou, et al, The use of human hepatocytes to select compounds based on their expected hepatic extraction ratios in humans, Pharmacol. Res. 14(2), 152-155 (1997).
93) R.S. Obach, The prediction of human clearance from hepatic microsomal metabolism data, Curr. Opin. Drug Disc. Dev. 4(1), 36-44 (2001).
94) K. Venkatakrishnan, L.L. Von Moltke, R.S. Obach, and D.J. Greenblatt, Microsomal binding of amitriptyline: Effect on estimation of enzyme kinetic parameters *in vitro*, J. Pharmacol. Exp. Ther. 293(2), 343-350 (2000).
95) T. Iwatsubo, N. Hirota, T. Ooie, H. Suzuki, N. Shimada, K. Chiba, et al, Prediction of *in vivo* drug metabolism in the human liver from *in vitro* metabolism data, Pharmacol. Ther. 73(2), 147-171 (1997).
96) Y. Shibata, H. Takhashi, and Y. Ishii, A convenient *in vitro* screening method for prediction *in vivo* drug metabolic clearance using isolated hepatocytes suspended in serum, Drug Metab. Dispos. 28(12), 1518-1523 (2000).

97) D.J. Carlile, K. Zomordori, and J.B. Houston, Scaling factors to relate drug metabolic clearance in hepatic microsomes, isolated hepatocytes, and the intact liver, Drug Metab. Dispos. 25(8), 903-911 (1997).
98) L. Shargel and A.B.C. Yu, Applied biopharmaceutics and pharmacokinetics, 2[cd] ed., Appleton-Century-Crofts, Norwalk, CT, pp. 184-191 (1985).
99) J. Zuegge, G. Schneider, P. Coassolo, and T. Lave, Prediction of hepatic metabolic clearance. Comparison and assessment of prediction models, Clin. Pharm. 40(7), 553-563 (2001).
100) J.H. Lin, Applications and limitations of interspecies scaling and *in vitro* extrapolation in pharmacokinetics, Drug Metab. Dispos. 26(12), 1202-1212 (1998).
101) G. Schneider, P. Coassolo, T. Lave, Combining *in vitro* and *in vivo* pharmacokinetic data for prediction of hepatic drug clearance in humans by artificial neural networks and multivariate statistical techniques, J. Med. Chem. 42, 5072-5076 (1999).
102) R.S. Obach, J.G. Baxter, T.E. Liston, B.M. Silber, B.C. Jones, F. MacIntyre, et al, The prediction of human pharmacokinetic parameters from preclinical and *in vitro* metabolism data, J. Pharmacol. Exp. Ther. 283, 46-58 (1997).
103) R.S. Obach, Nonspecific binding to microsomes: Impact on scaling-up of *in vitro* intrinsic clearance to hepatic clearance as assessed through examination of warfarin, imipramine, and propranolol, Drug Metab. Dispos. 25(12), 1359-1369 (1997).
104) R.S. Obach, Prediction of human clearance of twenty-nine drugs from hepatic microsomal intrinsic clearance data: An examination of *in vitro* half-life approach and nonspecific binding to microsomes, Drug Metab. Dispos. 27(11), 1350-1359 (1999).
105) L.K. Ludden, T.M. Ludden, JM. Collins, H.S. Pentikis, and J.M. Strong, Effect of albumin on the estimation, *in vitro* of phenytoin V_{max} and K_m: Implications for clinical correlation, J. Pharmacol. Exp. Ther. 282(1), 391-396 (1997).
106) D.J. Carlile, N. Hakooz, M.K. Bayliss, and J.B. Houston, Microsomal prediction of *in vivo* clearance of CYP2C9 substrates in humans, Br. J. Clin. Pharmacol. 47, 625-636 (1999).
107) C. Tang, Y. Lin, A.D. Rodrigues, and J.H. Lin, Effect of albumin on phenytoin and tolbutamide metabolism in human liver microsomes: An impact more than protein binding, Drug Metab. Dispos. 30(6), 648-654 (2002).
108) Y. Shibata, H. Takahashi, M. Chiba, and Y. Ishii, Prediction of hepatic clearance and availability by cryopreserved human hepatocyte: An application of serum incubation method, Drug Metab. Dispos. 30(8), 892-896 (2002).
109) J.M. Hutzler and T.S. Tracy, Atypical kinetic profiles in drug metabolism reactions, Drug Metab. Dispos. 30(4), 355-362 (2002).
110) U.A. Meyer, R.C. Skoda, U.M. Zanger, M. Heim, and F. Broly, The genetic polymorphism of debrisoquine/spartein metabolism-molecular mechanisms. In: W. Kalow, Pharmacogenetics of drug metabolism. 1st ed. Pergamon Press, New York, pp. 609-623 (1992).
111) H.K. Kroemer and M. Eichelbaum, “It’s the genes, stupid”, Molecular bases and clinical consequences of genetic cytochrome P450 2D6 polymorphism, Life Sci. 56(26), 2285-2298 (1995).
112) W.E. Evans and M.V. Relling, Pharmacogenoics: Translating functional genomics into rational therapeutics, Science, 286, 487-491 (1999).
113) W.W. Weber, Population and genetic polymorphisms, Mol. Diagn. 4(4), 299-307 (1999).
114) D.M. Grant, G.H. Goodfellow, K.S. Sugamori, and K. Durette, Pharmacogenetics of the human arlyamine N-acetyltransferases, Pharmacol. 61, 204-211 (2000).
115) A. Rannug, A.K. Alexandrie, I. Persson, M. Ingelman-Sundberg, Genetic polymorphism of cytochrome P450 1A1, 2D6, and 2E1: Regulation and toxicological significance, J. Occup. Environ. Med. 37(1), 25-36 (1995).
116) H. Norppa, Genetic polymorphisms and chromosome damage, Int. J. Hyg. Environ. Health, 204(1), 31-38 (2001).
117) T. Nakajima and T. Aoyama, Polymorphism of drug metabolizing enzymes in relation to individual susceptibility to industrial chemicals, Ind. Health, 38(2), 143-152 (2000).
118) M. Ingelman-Sundberg, I. Johansson, I. Persson, M. Oscarson, Y. Hu, L. Bertilsson, et al, Genetic polymorphism of cytochrome P450. Functional consequences and possible relationship to disease and alcohol toxicity, EXS, 71, 197-207 (1994).
119) U.A. Meyers and U.M. Zanger, Molecular mechanisms of genetic polymorphisms of drug metabolism, Annu. Rev. Pharmacol. Toxicl. 37, 269-296 (1997).
120) J.A. Golstein, Clinical relevance of genetic polymorphisms in the human CYP2C subfamily, Br. J. Clin. Pharmacol. 52, 349-355 (2001).

121) J. Green, E. Banks, A. Berrington, S. Darby, H. Deo, and R. Newton, N-Acetyltransferase 2 and bladder cancer: an overview and consideration of the evidence for gene-environment interaction, Br. J. Cancer, 83(3), 412-417 (2000).
122) A.R. Redman, Implications of cytochrome P450 2C9 polymorphism on warfarin metabolism and dosing, Pharmacotherapy. 21(2), 235-242 (2001).
123) D.F. McGinnity, A.J. Parker, M. Soars, and R.J. Riley, Automated definition of the enzymology of drug oxidation by the major human drug metabolizing cytochrome P450s, Drug Metab. Dispos. 28(11), 1327-1334 (2000).
124) R. Hyland, E.B.H. Roe, B.C. Jones, and D.A. Smith, Identification of the cytochrome P450 enzymes involved in the N-demethylation of sildenafil, Br. J. Clin Pharmacol. 51, 239-248 (2001).
125) B.J Ring, J.A. Eckstein, J.S. Gillespie, S.N. Binklyey, M. Vandenbranden, S.A. Wrighton, Identification of the human cytochrome P450 responsible for *in vitro* formation of R- and S-norfluoxetine, J. Pharmacol. Exp. Toxicol. 297(3), 1044-150 (2001).
126) Y. Hu and D. Kupfer, Metabolism of the endocrine disruptor pesticide methoxychlor by human P450s: pathways involving a novel catechol metabolite, Drug Metab. Dispos. 30(9), 1035-1042 (2002).
127) A.D Rodrigues, Integrated cytochrome P450 reaction phenotyping, Biochem. Pharmacol. 57, 465-480 (1999).
128) K. Ohyama, M. Nakajima, S. Nakamura, N. Shimada, H. Yamazaki, and T. Yokoi, A significant role of human cytochrome P450 2C8 in amiodarone N-deethylation: an approach to predict the contribution with relative activity factor, Drug Metab. Dispos. 28(11), 1303-1310 (2000).
129) H. Suzuki, M.B. Kneller, R.L. Haining, W.F. Trager, and A.E. Rettie, (+)-N-3-Benzyl-nirvanol and (-)-N-3-benzyl phenobarbital: new potent and selective inhibitors of CYP2C19, Drug Metab. Dispos. 30(3), 235-239 (2002).
130) Q. Mei, C. Tank, Y. Lin, T.H. Rushmore, and M. Shou, Inhibition kinetics of monoclonal antibodies against cytochromes P450, Drug Metab. Dispos. 30(6), 701-708 (2002).
131) J.J. Sramek, M.A. Eldon, E. Posvar, M.R. Feng, S.S. Jhee, J. Hourani, et al, Initial safety, tolerability pharmacodynamics and pharmacokinetics of CI-1007 in patients with schizophrenia, Psycopharm. Bull. 34(1), 93-99 (1998).
132) M.R. Feng, J. Loo, and J. Wright, Disposition of the antipsychotic agent CI-1007 in rats, monkeys, dogs and human cytochrome P450 2D6 extensive metabolizers, Drug Metab. Dispos. 26(10), 982-988 (1998).
133) M. Sinz, B. Michniewicz, and T. Woolf, Metabolic species comparison of the antipsychotic agent CI-1007 using microsomes, hepatocytes and purified enzymes, Proceedings of the 4th North American ISSX meeting, 8, 118 (1995).
134) R.J. Guttendorf, How nonclinical PK-ADME data are used to streamline human investigations – one company's perspective, AAPS PharmSci Supp. 2(4), (2000).
135) Food and Drug Administration, Guidance for Industry: *In vivo* drug metabolism/drug interaction studies-study design, data analysis, and recommendations for labeling. www.fda.gov/cder/guidance.htm, November (1999).
136) S.J. Billups, B.L.Carter, Mibefradil withdrawn from the market, Ann. Pharmcother. 32, 841 (1998).
137) A. Nomeir, C. Ruegg, M. Shoemaker, L. Favreau, J. Palamanda, P. Silber, and C.C. Lin, Inhibition of CYP3A4 in a rapid microtiter plate assay using recombinant enzyme and in human liver microsomes using conventional substrates, Drug Metab. Dispos. 29(5), 748-753 (2001).
138) K. Komatsu, K. Ito, Y. Nakajima, S.I. Kanamitsu, S. Imaoka, Y. Funae, et al, Prediction of *in vivo* drug-drug interactions between tolbutamide and various sulfonamides in humans based on *in vitro* experiments, Drug Metab. Dispos. 28(4), 475-481 (2000).
139) G. Ubeaud, J. Haenback, S. Vandenschrieck, L. Jung, and J.C. Koffel, *In vitro* inhibition of simvastatin metabolism in rat and human liver by naringenin, Life Sci. 65(13), 1403-1412 (1999).
140) L.H. Cohen, R.E.W. van Leeuwen, G.C.F. van Thiel, J.F. van Pelt, and S.H. Yap, Equally potent inhibitors of cholesterol synthesis in human hepatocytes have distinguishable effects on different cytochrome P450 enzymes, Biopharm. Drug Dispos. 21, 353-364 (2000).
141) K. Maheo, f. Morel, S. Langouet, H. Kramer, E. Le Ferrec, B. Ketterer, and A. Guillouzo, Inhibition of cytochromes P-450 and induction of glutathione S-transferase by sulforaphane in primary human and rat hepatocytes, Cancer Res. 57, 3649-3652 (1997).
142) K. Zomorodi and J.B. Houston, Effect of omeprazole on diazepam disposition in the rat: *in vitro* and *in vivo* studies, Pharm. Res. 12(11), 1642-1646 (1995).
143) R. Levy, K. Thummel, W. Trager, P. Hansten, M. Eichelbaum, eds. Metabolic drug interactions, 1st ed. Lippincott Williams and Wilkins, New York (2000).
144) A.D. Rodrigues, Drug-drug interactions, 1st ed. Marcel Dekker, New York (2002).

145) R.J. Weaver, Assessment of drug-drug interactions: concepts and approaches, Xenobiotica, 31(8-9), 499-538 (2001).
146) R.J. Betz and G.R. Granneman, Use of *in vitro* and *in vivo* data to estimate the likelihood of metabolic pharmacokinetic interactions, Clin. Pharmacokinet. 32(3), 210-258 (1997).
147) A.D. Rodrigues and S.L. Wong, Application of human liver microsomes in metabolism based drug-drug interactions: *in vitro-in vivo* correlations and the Abbott laboratory experience, Adv. Pharmacol. 41, 65-101 (1997).
148) K. Ito, T. Iwatsubo, S. Kanamitsu, K. Ueda, H. Suzuki, and Y. Sugiyama, Prediction of pharmacokinetic alterations caused by drug-drug interactions: metabolic interactions in the liver, Pharmacol. Rev. 50(3), 387-411 (1998).
149) K. Venkatakrishnan, L.L. von Moltke, D.J. Greenblatt, Human drug metabolism and cytochrome P450: application and relevance of *in vitro* models, J. Clin. Pharmacol. 41, 1149-1179 (2001).
150) K. Ohyama, M. Nakajima, M. Suzuki, N. Shimada, H. Yamazaki, and T. Yokoi, Inhibitory effects of amiodarone and its N-deethylated metabolite on human cytochrome P450 activities: prediction of *in vivo* drug interactions, Br. J. Clin. Pharmacol. 49, 244-253 (2000).
151) A. Madan, E. Usuki, L. Burton, B. Ogilvie, and A Parkinson, *In vitro* approaches for studying the inhibition of drug metabolizing enzymes and the identifying the drug metabolizing enzymes responsible for the metabolism of drugs. In: A.D. Rodrigues, Drug-drug interactions, 1st ed. Marcel Dekker, New York, pp. 217-294 (2002).
152) I.H. Segel, Enzyme kinetics, John Wiley, New York (1975).
153) L.L. von Moltke, D.J. Greenblatt, J. Schmider, C. Wright, J. Harmatz, and R. Shader, *In vitro* approaches to predicting drug interactions *in vivo*, Biochem. Pharmacol. 55, 113-122 (1998).
154) Z. Yan and G. Caldwell, Metabolism profiling and cytochrome P450 inhibition and induction in drug discovery, Curr. Top. Med. Chem. 1(5), 403-425 (2001).
155) C. Crespi and D. Stresser, Fluorometric screening for metabolism based drug-drug interactions, J. Pharmacol. Toxicol. Meth. 44, 325-331 (2000).
156) K. Kenworthy, J. Bloomer, S. Clarke, and J.B. Houston, CYP3A4 drug interactions: correlation of 10 *in vitro* probe substrates, Br. J. Clin. Pharmacol. 48, 716-727 (1999).
157) R.W. Wang, D. Newton, N. Liu, W. Atkins, and A. Lu, Humand cytochrome P-450 3A4: *in vitro* drug-drug interaction patterns are substrate dependent, Drug Metab. Dispos. 28(3), 360-366 (2000).
158) K. Kenworthy, S.E. Clarke, J. Andrews, and J.B. Houston, Multisite kinetic models for CYP3A4: simultaneous activation and inhibition of diazepam and testosterone metabolism, Drug Metab. Dispos. 29(12), 1644-1651 (2001).
159) P. Lu, Y. Lin, A.D. Rodrigues, T. Rushmore, T. Baillie, and M. Shou, Testosterone, 7-benzyloxyquinoline, and 7-benzylosy-4-trifluormethyl coumarin bind to different domains within the active site of cytochrome P450 3A4, Drug Metab. Dispos. 29(11), 1473-1479 (2001).
160) K. Korzekwa, N. Krishnanmachary, M. Shou, A. Ogai, R. Parise, A. Rettie, et al, Evaluation of atypical cytochrome P450 kinetics with two substrate models: evidence that multiple substrates can simultaneously bind to cytochrome P450 active sites, Biochemistry. 37, 4137-4147 (1998).
161) J.A McLure, J.O. Miners, and D.J. Birkett, Nonspecific binding of drugs to human liver microsomes, Br. J. Clin. Pharmacol. 49, 453-461 (2000).
162) S. Kanamitsu, K. Ito, and Y. Sugiyama, Quantitative prediction of *in vivo* drug-drug interactions from *in vitro* data based on physiological pharmacokinetics: use of maximum unbound concentration of inhibitor at the inlet to the liver, Pharm. Res. 17(3), 336-343 (2000).
163) M. Ishigam, M. Uchiyama, T. Kondo, H. Iwabuchi, S. Inoue, W. Takasaki, et al, Inhibition of *in vitro* metabolism of simvastatin by itraconazole in humans and prediction of *in vivo* drug-drug interactions, Pharm. Res. 18(5), 622-631 (2001).
164) J.H. Lin, Sense and nonsense in the prediction of drug-drug interactions, Curr. Drug Metab. 1,305-311 (2000).
165) M. Rowland and S.B. Martin, Kinetics of drug-drug interactions, J. Pharmacokinet. Biopharm. 1, 553-567 (1973).
166) P.R. Ortiz de Montellano and M. A. Correia, Suicidal destruction of cytochrome P450 during oxidative drug metabolism, Ann. Rev. Pharmacol. Toxicol. 23, 481-503 (1983).
167) M. Murray, Drug mediated inactivation of cytochrome P450, Clin. Exp. Pharmacol. Physiol. 24, 465-470 (1997).
168) T. Maurer, M. Tabrizi-Fard, and H.L. Fung, Impact of mechanism based enzyme inactivation on inhibitor potency: implications for rational drug discovery, J. Pharm. Sci. 89(11), 1404-1414 (2000).
169) B. Mayhew, D. Jones, and S.D. Hall, An *in vitro* model for predicting *in vivo* inhibition of cytochrome P450 3A4 by metabolic intermediate complex formation, Drug Metab. Dispos. 28(9), 1031-1037 (2000).

170) D.R. Jones and S.D. Hall, Mechanism based inhibition of human cytochromes P450: *in vitro* kinetics and *in vitro-in vivo* correlations. In: A.D. Rodrigues, Drug-drug interactions, 1st ed. Marcel Dekker, New York, pp. 387-413 (2002).
171) R.B. Silverman, Mechanism based enzyme inactivation: chemistry and enzymology, Vol. 1, CRC Press, Boca Raton, FL, pp. 3-30 (1988).
172) S.J. Billups and B.L. Carter, Mibefradil: a new class of calcium channel antagonists, Ann. Pharmacotherapy. 32, 659-670 (1998).
173) T. Prueksaritanont, B. Ma, C. Tang, Y. Meng, C. Assang, P. Lu, et al, Metabolic interactions between mibefradil and HMG-CoA reductase inhibitors: an *in vitro* investigation with human liver microsomes, Br. J. Clin. Pharmacol. 47, 291-298 (1999).
174) J.C. Krayenbul, S. Vozeh, M. Kondo-Oestreicher, and P. Dayer, Drug-drug interactions of new active substances: midbfradil example, Eur. J. Clin. Pharmacol. 55, 559-565 (1999).
175) F.P. Guengerich, Oxidation of 17α-ethynylestradiiol by human liver cytochrome P-450, Mol. Pharmacol. 33, 500-508 (1988).
176) F.P. Guengerich, Metabolism of 17α-ethynylestradiiol in humans, Life Sci. 47, 1981-1988 (1990).
177) F.P. Guengerich, Mechanism based inactivation of human liver microsomal cytochrome P-450 IIIA4 by gestodene, Chem. Res. Toxicol. 3, 363-371 (1990).
178) Q. Chen, J. Ngui, G. Doss, R. Wang, X. Cai, F. DiNinno, et al, Cytochrome P450 3A4 mediated bioactivation of raloxifene: irreversible enzyme inhibition and thiol adduct formation, Chem. Res. Toxicol. 15, 907-914 (2002).
179) N. von Rosenstiel and D. Adam, Macrolide antibiotics, drug interactions of clinical significance, Drug Safety. 13(2), 105-122 (1995).
180) L.L. von Moltke, A. Durol, S. Duan, D. Greenblatt, Potent mechanism based inhibition of human CYP3A *in vitro* by amprenavir and ritonavir: comparison with ketoconazole, Eur. J. Clin. Pharmacol. 56, 259-261 (2000).
181) T. Koudriakova, E. Iatsimrskaia, I. Utkin, E. Gangl, P. Vouros, E. Storozhuk, et al, Metabolism of the human immunodeficiency virus protease inhibitors indinavir and ritonavir by human intestinal microsomes and expressed cytochrome P4503A4/5: mechanism based inactivation of cytochrome P4503A by ritonavir, Drug Metab. Dispos. 26(6), 552-561 (1998).
182) R.L. Voorman, S.M. Maio, N.A. Payne Z. Zhao, K.A. Koeplinger, and X. Wang, Microsomal metabolism of delavirdine: evidence for mechanism based inactivation of human cytochrome P450 3A, J. Pharmacol. Exp. Ther. 287(1), 381-388 (1998).
183) B. Ma, T. Prueksaritanont, and J.H. Lin, Drug interactions with calcium channel blockers: possible involvement of metabolite-intermediate complexation with CYP3A, Drug Metab. Dispos. 28(2), 125-130 (2000).
184) D. Jones, J.C. Gorski, M. Hamman, B. Mayhew, S. Rider, and S.D. Hall, Diltiazem inhibition of cytochrome P450 3A activity is due to metabolite intermediate complex formation, J. Pharmacol. Exp. Toxicol. 290(3), 1116-1125 (1999).
185) P. Schmiedlin-Ren, D. Edwards, M. Fitzsimmons, K. He, K. Lown, P. Woster, et al, Mechanism of enhanced oral availability of CYP3A4 substrates by grapefruit constituents, Drug Metab. Dispos. 25(11), 1228-1233 (1997).
186) K. He, T. Woolf, and P. F. Hollenberg, Mechanism based inactivation of cytochrome P450 3A4 by mifepristone (RU486), J. Pharmacol. Exp. Toxicol. 288(2), 791-797 (1999).
187) K.L. Kunze and W.F. Trager, Isoform selective mechanism based inhibition of human cytochrome P450 1A2 by furafylline, Chem. Res. Toxicol. 6, 649-656 (1993).
188) M.P. Lopez-Garcia, P. M. Dansette, and D. Mansuy, Thiophene derivatives as new mechanism based inhibitors of cytochromes P450: inactivation of yeast expressed human liver cytochromes P450 2C9 by tienilic acid, Biochemistry. 33, 166-175 (1994).
189) S. Kanamitsu, K. Ito, C.E. Green, C.A. Tyson, N. Shimada, and Y. Sugiyama, Prediction of *in vivo* interactions between triazolam and erythromycin based on *in vitro* studies using human liver microsomes and recombinant human CYP3A4, Pharm. Res. 17(4), 419-426 (2000).
190) F.P. Guengerich, Inhibition of drug metabolizing enzymes: molecular and biochemical aspects. In: T. F. Woolf, Handbook of drug metabolism. 1st ed. Marcel Dekker, New York, pp. 203-238 (1999).
191) S.G. Waley, Kinetics of suicide substrates, Biochem. J. 185, 771-773 (1980).
192) D.A. Smith, Induction and drug development, Eur. J. Pharm. Sci. 11, 185-189 (2000).
193) T.A. Kocerek, E.G. Schuetz, S.C. Strom, R.A. Fisher, and P.S. Guzelian, Compartive analysis of cytochrome P4503A induction in primary cultures of rat, rabbit, and human hepatocytes, Drug Metab. Dispos. 23(3), 415-421 (1995).

194) S.A. Jones, L.B. Moore, J.L. Shenk, G.B. Wisely, G.A. Hamilton, D.D. McKee, et al, The pregnane X receptor, a promiscuous xenobiotic receptor that has diverged during evolution, Mol. Endocrinol. 14, 27-39 (2000).
195) L. Moore, D. Parks, S. Jones, R. Bledsoe, T. Consler, J. Stimmel, et al, Orphan nuclear receptors constitutive androstane receptor and pregnane X receptor share xenobiotic and steroid ligands. J. Biol. Chem. 275, 15122-15127 (2000).
196) J.Y. Chein, K.E. Thummel, and J.T. Slattery, Pharmacokinetic consequences of induction of CYP2E1 by ligand stabilizatoin, Drug Metab. Dispos. 25(10), 1165-1175 (1997).
197) P.D. Worboys and D.J. Carlile, Implications and consequences of enzyme induction on preclinical and clinical development, Xenobiotica, 31 (8-9), 539-556 (2001).
198) C.M. Loi, R. Stern, R. Abel, J. Koup, A. Vassos, and A. Sedman, Effect of troglitazone on PK and PD of atorvastatin, Clin. Pharmacol. Therap. PIII-39, 186 (1999).
199) C.M. Loi, J. Koup, A. Vassos, and A. Sedman, Effect of troglitazone on fexofenadine pharmacokinetics, Clin. Pharmacol. Therap. PIII-40, 186 (1999).
200) S. Burgess, G. Singer, and D. Brennan, Effect of troglitazone on cyclosporine whole blood levels, Transplant. 66(2), 272 (1998).
201) C. Chen, Troglitazone, an antidiabetic agent, Am. J. Health-Syst. Pharm, 55, 905-925 (1998).
202) J. Backman, K. Olkkola, and P. Neuovnen, Rifampin drastically reduces plasma concentrations and effects of oral midazolam, Clin. Pharmocol. Ther. 59, 7-13 (1996).
203) O. Pelkonen, J. Maenpaa, P. Taavitsainen, A. Rautio, and H. Raunio, Inhibition and induction of human cytochrome P450 enzymes, Xenobiotica, 28(12), 1203-1253 (1998).
204) E. Tanaka, K. Yamazaki, and S. Misawa, Update: the clinical importance of acetaminophen hepatotoxicity in non-alcoholic and alcoholic subjects, J. Clin. Pharm. Therap. 25, 325-332 (2000).
205) H. Carstensen, N. Jacobsen, and H. Dieperink, Interaction between cyclosporin A and phenobarbitone, Br. J. Clin. Pharmacol. 21, 550-551 (1986).
206) J.M. Wentworth, M. Agostini, J. Love, J.W. Schwabe, and V.K.K. Chatterjee, St John's wort, a herbal antidepressant, activates the steroid X receptor, J. Endo. 166, R11-R16 (2000).
207) A.P. Li, A. Rasussen, L. Xu, and D.L. Kaminski, Rifampicin induction of lidocaine metabolism in cultured human hepatocytes, J. Pharmacol. Exp. Ther. 274(2), 673-677 (1995).
208) V.E. Kostrubsky, S.C. Strom, S.G. Wood, S.A. Wrighton, P. Sinclair, and J.F. Sinclair, Ethanol and isopentanol increase CYP3A and CYP2E in primary cultures of human hepatocytes, Arch. Biochem. Biophys. 322(2), 516-520 (1995).
209) R. Nims, P. Sinclair, J. Sinclair, P. Thomas, C. Jones, D. Mellini, et al, Pharmacodynamics of cytochrome P450 2B induction by phenobarbital, 5-ethyl-5-phenylhydantoin, and 5-ethyl-5-phenyloxazolidinedione in the male rat liver or in cultured rat hepatocytes, Chem. Res. Toxicol. 6, 188-196 (1993).
210) J. Harvey, A. Paine, P. Maurel, and M. Wright, Effect of the adrenal 11-β-hydroxylase inhibitor metyrapone on human hepatic cytochrome P450 expression: induction of cytochrome P450 3A4, Drug Metab. Dispos. 28(1), 96-101 (2000).
211) C. Lindley, G. Hamilton, J. Mccune, S. Faucette, S. Shord, R. Hawke, et al, The effect of cyclophophamide with and without dexamethasone on cytochrome P450 3A4 and 2B6 in human hepatocytes, Drug Metab. Dispos. 30(7), 814-821 (2002).
212) E.L. LeCluyse, Human hepatocyte culture systems for the *in vitro* evaluation of cytochrome P450 expression and regulation, Eur. J. Pharm. Sci. 13, 343-368 (2001).
213) J.M. Silva and D.A. Nicoll-Griffith, *In vitro* models for studying induction of cytochrome P450 enzymes. In: A.D. Rodrigues, Drug-drug interactions, 1st ed. Marcel Dekker, New York, pp. 189-216 (2002).
214) C.J. Omiecinski, Epoxide hydrolase. In: R. Levy, K. Thummel, W. Trager, P. Hansten, M. Eichelbaum, eds. Metabolic drug interactions, 1st ed. Lippincott Williams and Wilkins, New York, pp. 205-214 (2000).
215) D.L. Eaton and T.K. Bammler, Glutathione S-transferase. In: R. Levy, K. Thummel, W. Trager, P. Hansten, M. Eichelbaum, eds. Metabolic drug interactions, 1st ed. Lippincott Williams and Wilkins, New York, pp. 175-189 (2000).
216) A. Kern, A. Bader, R. Pichlmayr, and K.F. Sewing, Drug metabolism in hepatocyte sandwich cultures of rats and humans, Biochem. Pharmacol. 54, 761-772 (1997).
217) M.J.J. Ronis and M. Ingelman-Sundberg, Induction of human drug metabolizing enzymes: mechanisms and implications. In: T. F. Woolf, Handbook of drug metabolism. 1st ed. Marcel Dekker, New York, pp. 239-262 (1999).

218) M. Burcznksi, M. McMillina, J. Parker, S. Bryant, A. Leone, E. Grant, et al, Cytochrome P450 induction in rat hepatocytes assessed by quantitative real time reverse transcription polymerase chain reaction and the RNA invasive cleavage assay, Drug Metab. Dispos. 29(9), 1243-1250 (2001).
219) U. Fuhr, Induction of drug metabolising enzymes, Clin. Pharmacokinet. 38(6), 493-504 (2000).
220) D. Waxman, P450 gene induction by structurally diverse xenochemicals: central role of nuclear receptors CAR, PXR, and PPAR, Arch. Biochem. Biophys. 369(1), 11-23 (1999).
221) S. Safe, Molecular biology of the Ah receptor and its role in carcinogensis, Toxicol. Lett. 120, 1-7 (2001).
222) G. Bertilsson, J. Heidrich, K. Svensson, M. Asman, L. Jendeberg, M. Sydow-Backman, et al, Identification of a human nuclear receptor defines a new signaling pathway for CYP3A induction, Proc. Natl. Acad. Sci. USA, 95, 12208-12213 (1998).
223) J.M. Lehmann, D. McKee, M. Watson, T. Willson, J. Moore, and S. Kliewer, The human orphan nuclear receptor PXR is activated by compounds that regulate CYP3A4 gene expression and cause drug interactions, J. Clin. Invest. 102(5), 1016-1023 (1998).
224) W. Xie, J. Barwick, C. Simon, A. Pierce, S. Safe, B. Blumberg, et al, Reciprocal activation of xenobiotic response genes by nuclear receptors SXR/PXR and CAR, Genes Dev. 14, 3014-3023 (2000).
225) J.T. Moore and S.A. Kliewer, Use of nuclear receptor PXR to predict drug interactions, Toxicol. 153, 1-10 (2000).
226) W. El-Sankary, G.G. Gibson, A. Ayrton, and N. Plant, Use of a reporter gene assay to predict and rank the potency and efficacy of CYP3A4 inducers, Drug Metab. Dispos. 29(11), 1499-1504 (2001).
227) J. Raucy, L. Warfe, M. Yueh, S. Allen, A cell based reporter gene assay for determining induction of CYP3A4 in a high volume system, J. Pharmacol. Exp. Ther. 303(1), 412-423 (2002).
228) J. Kelly and N. Sussman, A fluorescent cell based assay for cytochrome P450 isozyme 1A2 induction and inhibition, J. Biomol. Screen. 5(4), 249-253 (2000).
229) G. Luo, M. Cunningham, S. Kim, T. Burn, J. Lin, M. Sinz, et al, CYP3A4 induction by drugs: correlation between a pregnane X receptor reporter gene assay and CYP3A4 expression in human hepatocytes, Drug Metab. Dispos. 30(7), 795-804 (2002).
230) V. Ramachandran, V. Kostrubsky, B. Komoraski, S. Zhang, K. Dorko, J. Esplen, et al, Troglitazone increases cytochrome P450 3A protein and activity in primary cultures of human hepatocytes, Drug Metab. Dispos. 27(10), 1194-1199 (1999).
231) J. Sahi, G. Hamilton, M. Sinz, S. Barros, S.M. Huang, J. Lesko, and E. LeClyuse, Effect of troglitazone on cytochrome P450 enzymes in primary culture of human and rat hepatocytes, Xenobiotica, 30(3), 273-284 (2000).
232) Y. Lin, P. Lu, C. Tang, Q. Mie, G. Sandig, A.D. Rodrigues, et al, Substrate inhibition kinetics for cytochrome P450 catalyzed reactions, Drug Metab. Dispos. 29(4), 368-374 (2001).
233) A. Hsu, G.R. Granneman, and R. Bertz, Ritonavir, Clin Pharmacokinet. 35(4), 275-291 (1998).
234) Physicians' Desk Reference, 56th ed. Medical Economics, Montvale, NJ (2002).
235) J. Corchero, C. Granvil, T. Akiyama, G. Hayhurst, S. Pimprale, et al, The CYP2D6 humanized mouse: effect of the human CYP2D6 transgene on HNF4α on the disposition of debrisoquine in the mouse, Mol. Pharmacol. 60(6), 1260-1267 (2001).
236) W. Xie, J. Barwick, M. Downes, B. Blumberg, C. Simon, M. Nelson, et al, Humanized xenobiotic response in mice expressing nuclear receptor SXR, Nature, 406, 435-439 (2000).
237) D.F.V. Lewis and M. Dickins, Substrate SARs in human P450s, DDT, 7(17), 918-925 (2002).
238) J. Langowski and A. Long, Computer systems for the prediction of xenobiotic metabolism, Adv. Drug Del. Rev. 54, 407-415 (2002).
239) S. Ekins and R.S. Obach, Three dimensional quantitative structure activity relationship computational approaches for prediction of human *in vitro* intrinsic clearance, J. Pharmacol. Exp. Ther. 295(2), 463-473 (2000).
240) P. Bonnabry, J. Sievering, T. Leemann, and P. Dayer, Quantitative drug interactions prediction system (Q-DIPS), Clin. Pharmacokinet. 40(9), 631-640 (2001).
241) D.F.V. Lewis, M.N. Jacobs, M. Dickins, and B.G. Lake, Quantitative structure activity relationships for inducers of cytochromes P450 and nuclear receptor ligands involved in P450 regulation within the CYP1, CYP2, CYP3, and CYP4 families, Toxicol. 176, 51-57 (2002).
242) T. Rushmore and A. Kong, Pharmacogenomics, regulation and signaling pathways of phase I and II drug metabolizing enzymes, Curr. Drug Metab. 3(5), 481-490 (2002).
243) A. Sinskey, S. Finkelstein, and S. Cooper, The promise of protein microarrays, Pharmagen. July-August, 20-24 (2002).

INTERSPECIES SCALING

Thierry Lavé, Olivier Luttringer, Patrick Poulin, and Neil Parrott*

1. INTRODUCTION

Although technical advances in drug discovery are identifying an increasing number of biologically active compounds, many of these are still eliminated during the selection and development phases. Historically, a high proportion of these failures have been due to poor pharmacokinetic properties. To reduce this failure rate, candidate compounds are now being screened for DMPK properties (absorption, distribution, metabolic stability and excretion) and the derived parameters are then being used to predict their human pharmacokinetic profiles. These predicted profiles not only help to select the best candidates for development, but can also provide a starting dose for the first clinical studies. Such predictions can, therefore, drastically reduce the time and expense of drug research and development[1]. Furthermore, because DMPK issues are considered during the selection process, fewer compounds are now dropping out of development because of pharmacokinetic reasons.

The approaches used to predict human pharmacokinetics tend to fall into two categories: empirical interspecies scaling and physiologically-based pharmacokinetic (PBPK) modeling [2-12]. Both approaches rely to a large extent on the underlying physiological processes. Due to their relative simplicity, during the past 20 years the empirical approaches (such as the allometric scaling of *in vivo* pharmacokinetic parameters from different animal species) have been more widely used. By contrast, although PBPK models were developed over 30 years ago, until very recently their use to support drug discovery and development has been very limited (for examples see [13-15]).

One of the main reasons for this has been the high levels of resources that are needed to characterize the input parameters for the various pharmacokinetic processes. However with the recent developments of *in silico* and *in vitro* tools together with a

* F. Hoffmann - La Roche AG, Pharmaceuticals Division, CH-4070 Basel, Switzerland

marked increase in computing power, PBPK modelling is rapidly becoming a powerful tool for predicting human pharmacokinetics.

The present chapter reviews recent developments in both the empirical and physiologically-based methods for predicting human pharmacokinetics. First some general background information is given on these methods, which are then reviewed in the context of absorption, distribution, metabolism and excretion.

A number of examples illustrating the use of these approaches in drug research are also described, together with some strategic issues that may be considered when applying such methods.

2. METHODS

2.1. Physiologically Based Models

Physiologically based pharmacokinetic models (PBPK) divide the body into compartments [16-19 20], including the eliminating organs for instance kidney and liver and non-eliminating tissue compartments such as fat, muscle, and brain which are connected by the circulatory system (Figure 1).

The models use physiological and species specific parameters such as blood flow rates and tissue volumes to describe the pharmacokinetic processes. These physiological parameters are coupled with physicochemical, biochemical and compound specific parameters (e.g. tissue/blood partition coefficients and metabolic clearance) to predict the plasma and tissue concentration versus time profiles of a compound in an *in vivo* animal or human system.

Once a model has been developed, the concentrations in the various tissues can be determined by using the mass balance equations below: Drug concentration in tissue = rate of drug distribution into tissue - rate of drug distribution out - rate of drug elimination within the tissue.

Depending on the drug and tissue, the distribution can be perfusion-rate or diffusion-rate limited. Perfusion-rate limited kinetics tend to occur with relatively low molecular weight, hydrophobic drugs which have no problem crossing the lipid barrier of the cell wall. In this case the process limiting the penetration of the drug into the cells is the rate at which it is delivered to the tissue, so that blood flow is the limiting process.

By contrast diffusion-rate limited kinetics occur with more polar and/or larger drugs that do not freely dissolve in the lipid of the cell membrane and, therefore, have difficulty in penetrating into the cell. In this case the diffusion of drug across the membrane, which is independent of blood flow, becomes the limiting process.

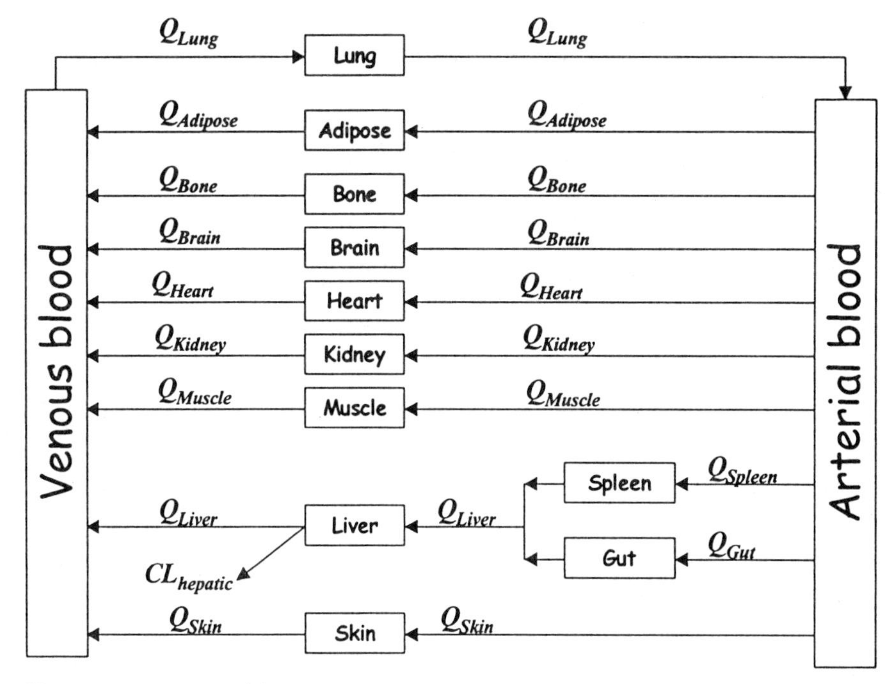

Figure 1. PBPK model

For perfusion limitation, the rate of change of drug concentration in a tissue where no elimination occurs can be described as:

$$V\frac{dC}{dt} = Q\left(Cin - \frac{Cin}{Kp}\right) \qquad [1]$$

where V is the physical volume of the tissue, C is the drug concentration in tissue, Q is the blood flow to the tissue, Cin is the drug concentration entering the tissue and Kp is the partition of the drug between tissue and blood. When diffusion rate limitation occurs, diffusion to and from the extra cellular space must be taken into account:

$$Ve\frac{dCe}{dt} = Q(Cin - Ce) - P(Ce - Ci) \qquad [2]$$

where P is the membrane permeability coefficient, Ce is the free extra cellular drug concentration, Ci is the free intracellular drug concentration and Ve is the anatomical extra cellular volume.

With organs such as liver, where elimination can occur, the rate of elimination must be included in the mass-balance equation. The rate of elimination is described by the equation:

$$\text{Rate of elimination} = CLCt \qquad [3]$$

where CL is the intrinsic clearance of the drug from that organ and Ct is the concentration in tissue. In some instances, Ct might also be replaced by Cvt (concentration in the venous blood leaving the tissue equivalent to Cin/Kp). A comparative validation is still needed to find out which one of these terms (Ct or Cvt) is most relevant for the *in vivo* situation.

This term can then be inserted into the mass balance equation, such as that for perfusion limitation:

$$V\frac{dC}{dt} = Q\left(Cin - \frac{Cin}{Kp}\right) - CLCt \qquad [4]$$

As the above equations indicate, a considerable amount of information is required to construct physiologically based pharmacokinetic models. In the past this has limited their use within the pharmaceutical industry. Thus, estimates of tissue/blood partitioning and intrinsic clearance are required for each drug, and generating this information frequently involved extensive *in vitro* and *in vivo* experimentation. Recently, however, tissue composition models have been developed which allow the tissue/blood partition coefficients to be estimated either *in silico* or to be measured experimentally as physicochemical descriptors. These approaches have dramatically reduced the amount of experimentation that is needed to support the use of flow models, considerably extending their utility [21]. Such approaches to predict tissue distribution are briefly discussed in the "Distribution" section below and elsewhere [22].

The major advantage of a physiologically based pharmacokinetic model is that it allows information that has been obtained in one species to be used to predict the time course of drug in another [23-25]. This is achieved by replacing the biochemical and physiological parameter values from the test species (e.g. a laboratory animal) with the corresponding values for the species of interest (e.g. human). A number of excellent reviews that outline the basis for physiologically based interspecies scaling, and compare this procedure with the more commonly used allometric approach, have been published [5, 6, 26]. In this respect, the major benefit of physiologically based scaling is that it incorporates information related to the species that is being predicted (in contrast to allometry, where all of the information utilized originates from the test species, eg. the rat). The species-specific information is most often in the form of physiological parameters (e.g. human tissue flows and weights when the pharmacokinetics are being scaled to human); however compound-specific information can also be incorporated (e.g. *in vitro*-derived absorption and/or metabolic data, protein binding parameters, etc.).

A number of publications illustrate the potential of this approach, both for predicting human pharmacokinetics and for examining species differences in pharmacokinetics. For example, Sawada predicted the human plasma concentration versus time profiles of seven drugs (phenobarbital, phenytoin, hexobarbital, quinidine, tollbutamide, valproate, and diazepam) [25]. Using a physiologically based approach he incorporated the intrinsic (unbound drug) clearance and unbound tissue-to-plasma concentration ratios that were extrapolated from rat data, plus the human plasma protein binding and blood-to-plasma concentration ratios. The concentration-time curves that were predicted for phenytoin, hexobarbital, quinidine and phenobarbital showed comparatively good agreement with the values observed in human plasma. By contrast, for tolbutamide, valproate and diazepam, because of poor clearance predictions the areas under the concentration-time curve were poorly estimated [25].

More recently, lumped physiologically based models (in which the tissues are subdivided into groups, according to their kinetic properties) have been used to predict human plasma concentration-time curves for the melatonin agonist S20098[27]. By combining *in vitro* metabolism data obtained from human cytochrome P450 enzymes with partition coefficients in fat and liver, the range of plasma concentrations observed in human could be successfully predicted[27]. In another study[9] a lumped physiologically based pharmacokinetic model originally developed by Arundel for the rat [22, 28] was extended and applied to human for ten extensively metabolised compounds. The model consists of two blood compartments (venous and arterial), with the tissues lumped into six groups. With this approach, the volume of distribution at steady state can be combined with a standard set of human kTi x Vdss values to estimate the kinetics of the compound in (lumped) tissues and organs (excluding adipose tissue). This is based on the empirical finding that there is a relationship between Vdss *in vivo* and the disappearance rate constant from the tissues (kTi). This approach allows the tissue-to-blood partition coefficient (Kp) and the rate constant for the exit of drug from tissues (kTi) to be predicted without the need to take any tissue samples. Overall, the results showed that this model might be used to predict concentration versus time profiles in human. Irrespective of the compounds' pharmacokinetic and physicochemical characteristics, the average error for the predicted pharmacokinetic parameters (CL, Vdss, T1/2) of the ten compounds was less than twofold [9].

2.2. Interspecies Scaling

Interspecies scaling has been investigated extensively as a way of predicting human concentration-time profiles and pharmacokinetic parameters from animal data.

The methods used include animal-human correlations, allometric scaling and superposition of concentration time profiles.

The first of these approaches involves correlating a given pharmacokinetic parameter between one test animal species and humans [29, 30, 31]. This has been used to establish relationships between animals and human for a variety of pharmacokinetic parameters; for example: the fraction absorbed in the rat (or the dog or the monkey) and humans [32-34], the renal clearance in monkeys and human [30, 31]. As described in the various sections below, this empirical correlation approach can provide reasonable and useful predictions.

Allometric scaling is an alternative approach for correlating pharmacokinetic parameters in animal and human. This empirical approach, which is still widely used to predict human pharmacokinetic parameters from animal data [4, 35-40] is based on a power function of the form:

$$y = a \cdot W^x \quad [5]$$

where y is the dependent variable (e.g. clearance, volume of distribution, half-life), W (body weight) is the independent variable, and a and x are the allometric coefficient and exponent respectively. The values of a and x can be estimated by linear least squares regression of the log transformed allometric equations:

$$\log(\text{clearance, volume of distribution, half-life, etc.}) = \log(a) + x \log(W) \quad [6]$$

The sign and magnitude of the exponent indicate how the physiological or pharmacokinetic variable is changing as a function of W [5, 41]: for $b < 0$, Y decreases as W increases; for $0 < b < 1$, Y increases as the species get larger, but does not increase as rapidly as W; when $b = 1$, Y increases in direct proportion to the increases in W; and when $b > 1$, Y increases faster than W.

A number of physical and physiological processes have been shown to vary with body weight according to a power function [42]. From the power equation, Adolph [42] compiled 33 equations that can be used to relate variables such as physiological periods and organ weights to body weight. Boxenbaum has also shown that log liver weight and log liver blood flow are linearly related to log body weight in mammals [43] (Figure 2). In general, the exponents for biological frequencies (e.g. heartbeat time) tend to be around -0.25, those for biological time periods (e.g., circulation time, maximum lifespan potential) tend to be about 0.25, while biological rates (e.g., clearance, physiologic-flow rates, and metabolism) tend to have an exponent of 0.75 and volumes tend to have an exponent of about 1.0 [5, 41].

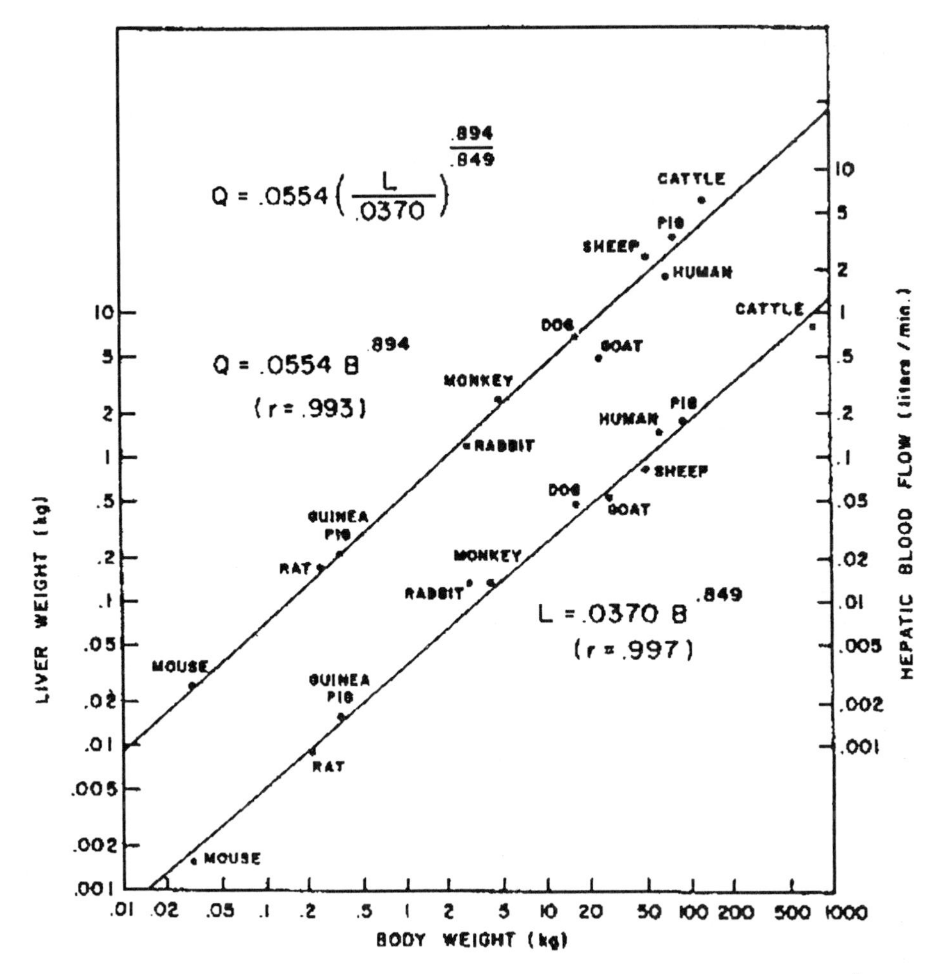

Figure 2. Allometric scaling of liver weight and hepatic blood flow (from [3])

Allometric scaling techniques have been used to extrapolate pharmacokinetic data for small organic molecules from laboratory animals to humans, and the results have appeared in several reviews [5, 11]. The allometric equations for the pharmacokinetic parameters tend to be of the same order of magnitude as those for the corresponding physiologic variables, i.e., half-life (min) ~ $aW^{0.2}$ to $aW^{0.4}$, clearance (ml/min) ~ $aW^{0.6}$ to $aW^{0.8}$, and the volume of distribution (ml) ~ $aW^{0.8}$ to $aW^{1.0}$ [5, 41].

The allometric approach is particularly useful for compounds that are primarily eliminated by physical transport processes and when distribution occurs through passive processes. In such cases, when their values depend upon underlying physical or physiological processes, which scale according to body weight, parameters such as clearance and volume of distribution do lend themselves well to interspecies scaling. However, as discussed in the sections below, it may be less reliable for compounds that exhibit marked species differences in distribution, metabolism or excretion mechanisms.

Although prospective interspecies allometric scaling can yield satisfactory estimates of human PK parameters, a number of authors have cautioned against its use for selecting doses for the first human studies [44], [11], [45]. The mathematical techniques that are used in allometric scaling has been criticised on a number of grounds. Firstly, the use of a log-log plot to represent the pharmacokinetic parameters as a function of body weight tends to minimize the error in the relationship. Thus, a predicted value can differ considerably from an observed value, yet still appear visually from the plots to be providing a satisfactory relationship [11]. Furthermore, as the experimental observations are normally obtained in laboratory animals that have a lower body weight than humans, the extrapolation to human involves the extremely hazardous practice of projecting beyond the observed range [44]. Also, it is generally recognized that the method of calculation employed in regression analysis does not take equal account of all the data placing less emphasis on the intermediate values than the extremities, and giving greater emphasis to 'clustered' values. The frequently used practice of measuring pharmacokinetic parameters in a number of low body-weight animals plus one heavier species (such as the dog or a primate) tends to compound such errors. Consequently, an inaccuracy in the value(s) for the heavier species unduly influences the parameter estimates for human [46]. However, such problems can be overcome by using pharmacokinetic time units (see the section below), where each animal species potentially has a similar impact on the prediction in human [46]. Some of these criticisms of allometric scaling can also be reduced by using a population approach [47], [48]. In this way, the coefficients and exponents that characterize the relationship between pharmacokinetic parameters and the species characteristics (e.g. body weight, brain weight, maximum lifespan, and/or other covariates that describe the status of the animal) are estimated in a single step directly from the concentration vs. time profiles in the various species [47], [48].

2.3. Physiological Time Scales

Physiological time is the unit of time that is required to complete a physiological event. While the physiological event occurs in all species (i.e. it is species-independent) the time taken to complete the event is species-dependent. Good examples of physiological times are breathing time and heart beat. In chronological time smaller animals have higher heart and respiratory rates than larger animals. However, in their life span all mammals have the same number of heartbeats and breaths. Physiological rates are known to follow allometric relationships that are characterized by comparable body weight exponents for the power function:

(physiological rate = constant x $W^{0.25}$) [7]

Examples for Breath time and Heart beat time are as follows[29]:

Breath time = 0.169 x $W^{0.28}$ [8]

Heart beat time = 0.0428 x $W^{0.28}$ [9]

Consequently, the ratios of these internal biological rates are relatively constant across the species. For example the ratio of respiratory rate/heart rate is approximately 4, indicating that all mammals have four heartbeats for each breath. Many biological rates show this type of relationship, and the following general trend has been proposed:

Biological rate = constant X $W^{0.25}$ [10]

Dedrick et al. [49] first introduced the concept of equivalent time (also called pharmacokinetic time) across species. Using data for methotrexate after parenteral administration to several mammalian species, they dose-normalized the plasma concentrations by dividing by milligrams per kilogram body weight. Adjusting the dose-normalized plasma levels for clearance, by dividing the chronological time by $W^{0.25}$, then gave plots that were superimposable (Figure 3).

Boxenbaum and others [29, 46, 50-53] have also used this adjustment in the time scale for different species, to produce superimposable concentration-equivalent time curves. Such conversions of chronological time to pharmacokinetic time are due to an allometric relationship between volume of distribution and clearance, one simply substitutes allometric expressions into the appropriate disposition function. For example, a monoexponential disposition gives the following relationships:

$V = a . W^{y}$ [11]

$CL = b . W^{z}$ [12]

And the plasma concentration (C) after intravenous bolus administration is equal to:

$C = (Dose/V)\ e\text{-}((CL/V)x\ t) = (D/bW^{y})e^{-(a/b)(Wz-y)\ .\ t}$ [13]

where V is volume of distribution (liters), W is body mass (kg) CL is the clearance (ml/min) and C is the plasma concentration at time t. Based on this equation, interspecies superimposability of plasma concentrations will occur when plasma concentrations divided by dose per unit W^{y} are plotted as a function of time divided by W^{y-z}.

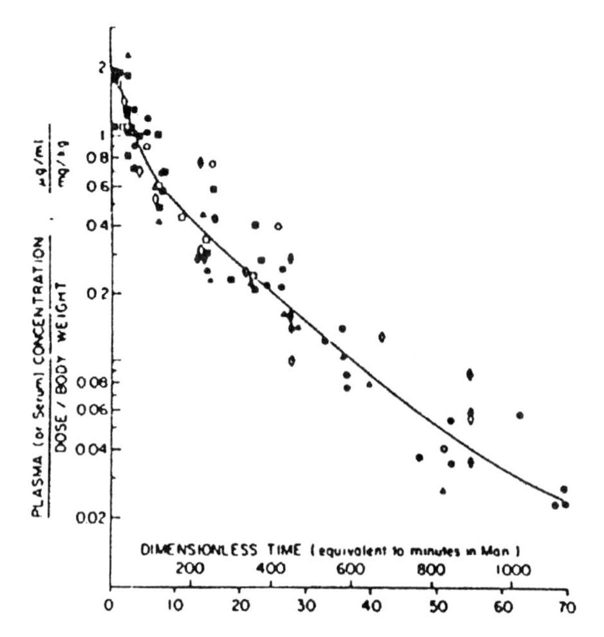

Figure 3. Superimposition of methotrexate plasma concentration data (from [49]).

Depending on the allometric exponents for clearance and volume of distribution, different adjustments to the allometric coordinate can be considered. These include Kallynochrons, leading to the elementary Dedrick plot, or Apolysichrons which give the complex Dedrick plot [3, 29, 50, 51]. Both the elementary and complex Dedrick plots assume that CL and V are related to body weight through a simple allometric equation.

- Kallynochrons and the elementary Dedrick plot

In this pharmacokinetic time unit, it is assumed that y=1. Thus, one kallynochron is equivalent to W^{1-z} units of chronological time. Based on these transformations, the time from each species may be converted into human time using the following equation:

$$t_{human}=t_{animal} \times (W_{human}/W_{animal})^{1-z} \qquad [14]$$

- Apolysichrons and the complex Dedrick plot

This pharmacokinetic time unit considers the case where V is not directly proportional to body weight (ie. $y \neq 1.0$). In this case, superimposability can be achieved by dividing concentration by dose per W^y and plotting this as a function of time divided by W^{y-x}.

Thus, with Apolysichrons the time from each species may be converted into human time using the following equation:

$t_{human}=t_{animal} \times (W_{human}/W_{animal})^{y-z}$ [15]

For hepatically metabolised drugs, it has been suggested that the interspecies superimposability of plasma concentrations can be improved by correcting the allometric equation for clearance by factors such as maximum lifespan potential (MLP) [3, 54-56] or *in vitro* data [3, 54-56]).

- Pharmacokinetic time units for hepatically metabolised drugs

In this pharmacokinetic time unit, correcting the allometric scaling of clearance by empirical factors such as MLP and brain weight (BRW) leads to the following allometric relationships:

$CL = b \cdot W^x/MLP$ [16]

$CL = b \cdot W^x \cdot BRW^z$ (references [3, 54, 55]) [17]

Integrating the latter into the monoexponential equation for plasma concentrations leads to the following expression:

$C = (Dose/V)$ e-((CL/V)x t)= $(D/bW^y)e^{-(a/b).(Wx-y).(BRWz)\,.\,t}$ [18]

In this case, superimposability is achieved by dividing concentration by dose per W^y and plotting this as a function of time divided by W^{y-x}. BRW^z. When the allometric equation for clearance is adjusted by BRW or MLP the pharmacokinetic time units obtained are named syndesichrons and dienetichrons respectively.

Some other "physiological" factors based on *in vitro* metabolism data in animals and human have also been suggested for the allometric scaling of clearance. Such factors can also be incorporated into the calculation of pharmacokinetic time units [56].

A number of examples in which time transformations of preclinical data have been used to predict the full plasma concentration versus time profiles in man have been published [52, 56-62]. Thus, based on the elementary Dedrick plot, satisfactory predictions have been obtained for renally and biliary excreted compounds, such as lamifiban and napsagatran. In both cases, the species that gave the best predictions for man was identified as the cynomolgus monkey [57, 58]. Using the complex and/or elementary Dedrick approaches also gave reasonable predictions of the human situation for a number of other non metabolised compounds, such as: interferon [63], metazocin [60], 2',3'-Dideoxycytidine [61], panipenem and betamipron [64] and for the 50 kDa MW Digoxin-specific Fab [62].

Superimposability of plasma concentration vs time profiles has also been achieved for metabolised compounds such as antypirine, chlordiazepoxide, lorazepam and amphotericin B [3, 55, 59]. In these cases, superimposability was obtained by using syndesichrons or dienetichrons that incorporate empirical correction factors such as BRW and MLP. An attempt to replace the empirical correction factors by physiologically based experimental data, namely animal and human hepatocyte data, also successfully predicted the plasma concentrations versus time profile of tolcapone in man from preclinical data [56].

3. ABSORPTION

3.1. Introduction to Absorption

Drug absorption is influenced by a variety of physiological and physicochemical factors. The physiological factors are species dependent and include: gastric and intestinal transit time, blood flow rate, gastrointestinal pH and first-pass metabolism, while the physicochemical factors correspond to drug specific and species-independent properties, such as pKa, molecular size, solubility, and lipophilicity [65].

The oral bioavailability of a drug is defined as the fraction of an oral dose that reaches the systemic circulation unchanged. The oral bioavailability (F) can be described as:

$$F = fabs \times (1-fg) \times (1-fh) \qquad [19]$$

where fabs is the fraction of dose absorbed from the gastrointestinal lumen and fg and fh are the fractions of drug metabolised by the gut wall and liver, respectively, during the first pass [8]. The fabs is determined mainly by solubility and stability of the drug in the gastro-intestinal tract and its permeability across the intestinal membrane. Where marked species differences in oral bioavailability have been reported [66] (Figure 4) they are usually due to species differences in first-pass metabolism by the gut and liver (fg and fh). This observation suggest that no one species can be relied upon to accurately predict the oral bioavailability in man [66]. As discussed below, this poor predictability is most likely due to physiological differences between the species. Nevertheless attempts have been made to correlate single parameters involved in absorption between animals and man. Thus for fabs, although some similarities across species [32-34, 67] have been reported, differences were also observed as a consequence of the species differences in anatomy and physiology of the GI tract. These observations suggest that caution should be exercised when human oral bioavailability is predicted from animal data. For this purpose PBPK simulation models, in which the various parameters involved in absorption can be modelled independently, represent an attractive approach [68]. Extensive evaluations of these approaches are ongoing and are briefly discussed below.

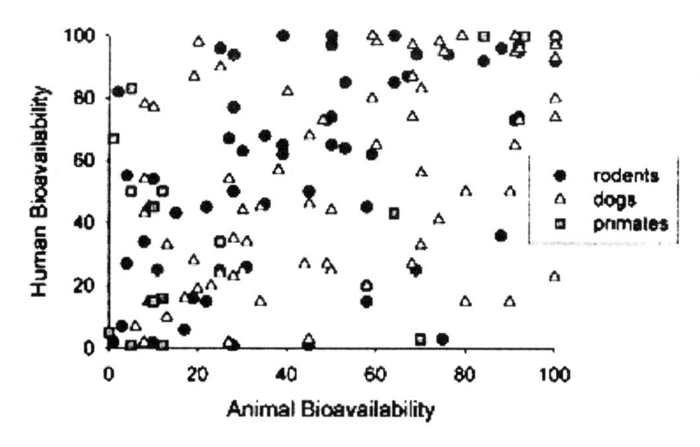

Figure 4. Species differences in oral bioavailability (from [66])

3.2. Species Differences in Anatomy and Physiology of the GI Tract

A recent review article has highlighted a number of similarities and differences in anatomy and physiology of the GI tract in rats and humans [69]. The authors provided some evidence that the human GI tract is capable of absorbing materials faster and to a greater extent than the rat. Such differences are likely to influence the extent to which drugs and toxicants are absorbed, and therefore need to be taken into consideration when developing physiologically-based pharmacokinetic models to predict absorption in different species. Some of the anatomical differences that are likely to influence the rate and extent of absorption in different species are summarised below.

Overall, the rat's gastrointestinal tract is organised in the same way as the human's, but with a few important differences. For example, the relative lengths of the small intestine in the rat differ from those in man, in that the jejunum makes up nearly the entire small intestine in the rat. Another outstanding difference is that the human intestinal tract is only about 5.5 times the length of that in the rat, despite man's much larger body size (70 kg) compared to the rat (0.25 kg). The absolute surface area of the human intestinal tract is about 200 times that of the rat which when normalized on the basis of body surface area amounts to a factor of 4 times. The physiological consequences of these anatomical differences are twofold. First, well-absorbed substances are likely to be absorbed more quickly in humans (i.e. man will exhibit a higher absorption rate). Second, substances that are poorly or incompletely absorbed by both species are likely to be absorbed to a greater extent by man [69].

With regard to interspecies differences in secretions into the intestinal tract, an important anatomical difference between rats and humans is the lack of a gall bladder in rats. This means that, in rat, bile enters the duodenum continuously as it is made. By contrast the bile in humans is released only when chyme is present. These anatomical differences can be responsible for pronounced species differences in the concentration versus time profile of drugs that undergo enterohepatic recycling.

3.3. Species Differences and Similarities for Fraction Absorbed

As just described there are physiological reasons suggesting that species differences should be observed in rat and human however, nearly identical oral absorptions in have been recently reported for 64 drugs, despite markedly varying physicochemical properties and a wide range of intestinal permeabilities (Figure 5) [34]. In addition, similar oral bioavailabilities in rats and humans have been reported for 16 PEG compounds with molecular weights that ranged from 280 to 950 [70]. These similarities in the extent of oral absorption between human and rat were rationalized by comparable first-order absorption rate constants and similar mean small intestinal transit times in the two species [34]. In another report based on a purely empirical approach, Chiou et al. [67] suggested that the rat might be a good animal to predict the extent of human oral absorption for drugs that exhibit non-linear absorption. These authors reported that for 4 drugs (chlorothiazide, acyclovir, miglitol and pafenolol), dose-dependent differences in the fraction absorbed were observed when the dose was expressed relative to body weight (mg/kg), but these differences disappeared if the dose was based on body surface area (mg/m^2). Such an approach certainly warrants further studies with a larger number of compounds.

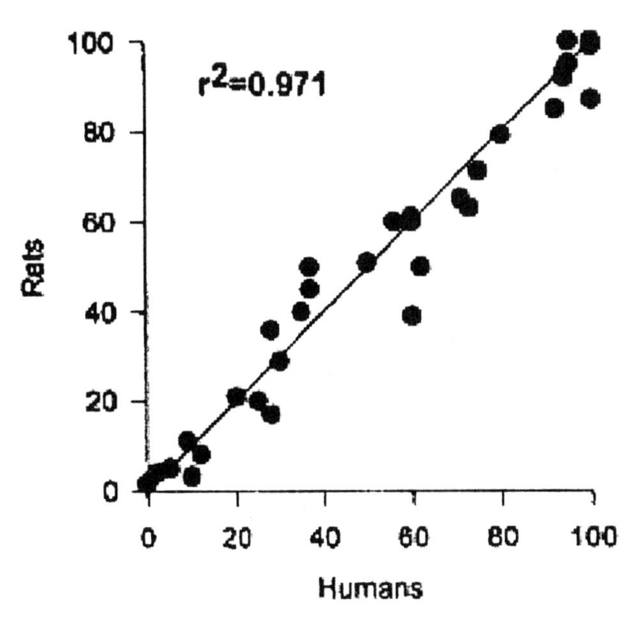

Figure 5. Fraction absorbed in rats vs. Humans (from [34])

In contrast to rat, the dog appears to be less predictive for the situation in man. For the 16 PEG compounds the absorption in dogs was better than in humans [70]. Also for 43 drugs, the correlation coefficient between the fraction absorbed in humans and dogs (r^2=0.5123) was much lower than the one (r^2=0.971) reported between humans and rats [34], [32] (Figure 6). This tendency for dogs to absorb better than man has also been observed with a number of other compounds. For acyclovir and nadolol, the fraction absorbed in dogs was about 100% while humans only absorbed about 20% of the dose. Great differences were also found for atenolol (50% in man vs. 100% in dogs), methyldopa (43% in man vs. 100% in dogs), ranitidine (63% in man vs. 100% in dogs), sumatriptan (60% in man vs. 97% in dogs) and xamoterol (8.6% in man vs 36% in dogs).

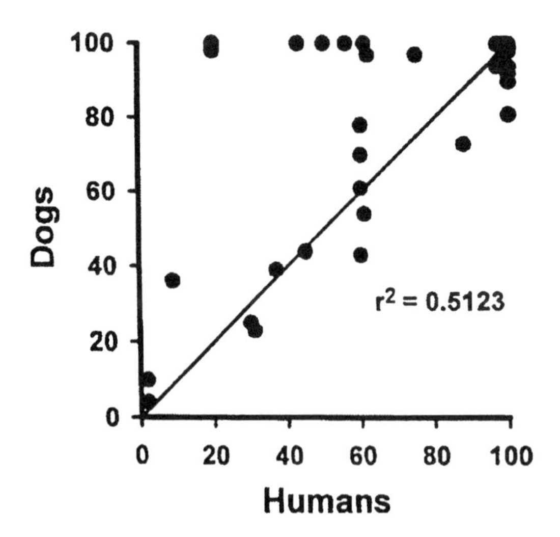

Figure 6. Fraction absorbed in dogs vs. Humans (from [32])

A number of hypotheses have been proposed to explain the higher absorption in dog than man. Compared to man, the dog has longer villi, which may compensate for a shorter intestinal transit time. Furthermore, the higher bile salt secretion rate in the dog could increase the permeability of the intestinal membrane and might, through a solubilizing effect, also facilitate the absorption of poorly water-soluble drugs [34], [32]. In addition the size and frequency of the tight junction for paracellular transport may be greater in dogs than in humans, which might explain the greater extent of absorption of small hydrophilic compounds, such as polyethylene glycol oligomers [70]. Also, the higher (about one unit) intestinal pH in fasted dogs relative to humans may contribute to the more efficient absorption of many weak bases in dogs drugs [34], [32].

3.4. Physiologically Based Modelling Approaches to Predict Oral Absorption

Modelling approaches to predict oral absorption are numerous and date back to the 1950's (see Yu et al [71] for a review). However models for the scaling of absorption across species have not been reported in the literature. Recently a physiologically based model for predicting both the extent and the rate of oral absorption in various species became commercially available, and its application to rat, dog and man is briefly described below.

The ACAT (Advanced Compartmentalized Absorption and Transit) model in GastroPlus™ (Simulations Plus Inc.). [65] is based upon an original CAT (Compartmental and Transit Model) model described by Yu [71]. It is a semi-physiological model, with 9 compartments corresponding to different segments of the digestive tract (stomach, colon and 7 small intestinal compartments). The model takes into account the pH dependency of dissolution and permeability, transport down the gastrointestinal tract, and absorption through the intestinal wall into the portal vein. For different species the following variables are changed:

- pH values - pH in each of the intestinal compartments. Defaults are provided for each species, and also for fed and fasted states.
- transit times - the overall small intestine transit time. A default value is provided for each species.
- geometric parameters - small intestine length and radius, volume of stomach and colon.
- permeability – a built in linear transformation function between man, rat and dog.

To evaluate GastroPlus™ we collected fraction absorbed data for a set of 13 drugs in man, dog and rat. These were taken from the papers published by Chiou, which include extensive literature searches for data from dog and rat [32, 34].

For these 13 drugs we measured the thermodynamic solubilities using our company assay. Where data on human jejunal permeability was available in the literature [72] this was used as the primary permeability input. For 4 drugs where human permeability data were not available we estimated this using rabbit colon in the Ussing chamber. These data were transformed using a correlation that we had established previously, based upon reference drugs measured in our assay:

$$\text{Log(HumanPeff)} = 1.2807 * \text{Log(Rabbit)} + 2.5574 \qquad [20]$$

The following approaches were then used to evaluate GastroPlus:

1. Fraction absorbed in man for the 13 drugs. Compare predicted and observed clinical data (Table 1).
2. Change the physiology model to dog. Compare predicted to observed data (Table 2).
3. Change the physiology model to rat. Compare predicted to observed data (Table 2).

In man 9 of the 13 drugs were correctly categorized as low (<33%), medium or high (>66%) absorption, and the mean error of the prediction was 26%. In dog an overall trend towards higher absorption was correctly predicted, and 10 of the 11 drugs were correctly categorized, with a mean error of 13%. In rat 6 of the 11 drugs are correctly categorized and the mean error was 22%. Although this study was clearly too small to constitute a validation of the GastroPlus model, it does show how the PBPK approach can be applied to absorption, and indicates that physiologically based modelling has the potential to become valuable a tool in this area.

Table 1. Compounds for Gastroplus evaluation

Drug	Mol. Wt	Dose in man	Solubility pH 6.5 mg/ml	Measured pKa	Human Peff (cm/s) x 10E-4	Fabs Man	Fabs Dog	Fabs Rat
Aciclovir	225	350	3.65	2.27B, 9.25A	0.09a	23	100	21
Atenolol	266	50	34800	9.54B	0.2	50	100	49
Bretylium	243	200	8.95		0.01a	23b		20
Cimetidine	252	200	22.4	7.1B	0.33a	64	98	100
Fluvastatin	411	2	0.47	4.32A	2.4	100	100	
Furosemide	331	80	8.4	10.63A, 3.52A	0.05	61	55	60
Hydrochloro-thiazide	297	50	0.843	9.96A, 8.87A	0.04	69		65
Methyldopa	211	200	11.9	2.28A,9.82A,12.39A ,8.98B	0.2	41	100	
Naproxen	230	500	3.51	4.01A	8.3	99		92
Propranolol	259	240	44.3	9.52B	2.9	100	100	99
Ranitidine	314	60	196	8.64B	0.27	64	100	63
Sumatriptan	295	200	21	11.03A, 9.24B	0.31a	57	97	50
Terbutaline	225	10	67.6	8.72A,10.00A,11.10 B	0.3	62	78	60

Table 2. Predicted vs observed fraction absorbed

	Drug	HUMAN		DOG		RAT	
		Pred %	Obs %	Pred %	Obs %	Pred %	Obs %
ac	Acyclovir	23	20	75	100	40	21
at	Atenolol	29	50	81	100	52	49
ci	Cimetidine	46	64	94	98	76	100
fu	Furosemide	15	61	59	55	20	60
me	Methyldopa	26	43	77	100	na	na
ra	Ranitidine	38	61	91	100	67	63
te	Terbutaline	39	60	89	78	65	60
br	Bretylium	8	23			13	20
hy	Hydrochloro-thiazide	10	71	57	71	18	65
su	Sumatriptan	43	62	93	97	72	50
pr	Propranolol	94	100	100	100	99	99
fl	Fluvastatin	99	98	100	98	na	na
na	Naproxen	100	94	na	na	100	92

4. DISTRIBUTION

The PK parameter that is most commonly used to characterize the distribution of a drug is its volume of distribution at steady-state (Vss). This represents the equivalent volume into which a given dose of drug is apparently distributed within the body and, as such, it includes the extent to which the drug is bound to tissue and/or plasma proteins. The numerical value of Vss that is recorded can be as small as the blood volume or, for compounds that are extensively bound to tissues, it can exceed the volume of total body water. The following equation has been proposed to relate the volume of distribution (Vss) to plasma and tissue binding:

$$Vss = V_P + f_{up} V_E + V_P R_{E/I} (1\text{-}fu) + V_R f_{up} / f_{ut} \quad [21]$$

where V_P , V_E and $R_{E/I}$ are the plasma and extracellular volumes and the ratio of extravascular to intravascular proteins. V_R is the volume in which the drug distributes minus the extracellular space. f_{up} and f_{ut} are the fractions of unbound drug in plasma and tissue [73].

Drugs that are more extensively bound to plasma proteins than to tissues ($f_{ut} >> f_{up}$) are characterised by small volumes of distribution, and their Vss approximates extracellular space. Since, based on total body weight the volumes of extracellular space tend to be similar in different species, the Vss values for compounds whose distribution is restricted to the extracellular space are likely to be similar in all species by contrast to compounds more extensively distributed.

Drugs that are more extensively bound to tissues ($f_{ut} << f_{up}$) are more widely distributed. For these drugs:

$$Vss = V_R f_{up} / f_{ut} \quad [22]$$

At first sight predicting the volume of distribution in such cases appears to be more challenging, since the apparent volume will depend upon species differences in V_R, f_{up} and f_{ut}. The percentages of total body water and the relative sizes of major organs are more similar across species, and since the allometric exponent for volume with respect to body size is close to 1.0 [5, 41] such factors are likely to have only a small effect on inter-species differences. While plasma protein binding can be subject to large interspecies differences, its values are easy to determine experimentally, so correction factors for its effects in the various animal species can be readily applied. Vss can also be represented in terms of tissue:plasma partition coefficients (Kp) and the plasma volume (V_P):

$$Vss = V_P + V_{Ery} \times Ery/Plasma + \sum V_t \times Kp_t \quad [23]$$

If elimination occurs in one or more tissues, the 3rd term of the equation becomes:

$$\sum V_t \times Kp_t(1-E) \quad [24]$$

where E is the extraction ratio for the elimination. Kp therefore indicates the extent to which a drug will accumulate in tissues, relative to plasma, under steady-state conditions. The extrapolation methods for distribution parameters (Kp and Vss) are presented below.

4.1. Interspecies Correlations

A number of reports have dealt with interspecies variations in volume of distribution. Boxenbaum reported a good correlation between body weight and volume of distribution across species for compounds (such as methotrexate, cyclophosphamide and antipyrine) which are not extensively bound to plasma proteins [29]. Large species differences have, however, been reported for compounds which are highly protein bound [74], due mainly to species differences in plasma binding. These differences may reflect differences in the plasma protein concentration, its affinity for the test compound(s) and/or the number of binding sites. As plasma protein binding in various animal species is easy to measure, but species variations in fut are more difficult to determine, attempts to predict Vss in man have been based on species comparisons of Vt/fut , which equates to:

$$(Vss-Vp)/fup \quad [25]$$

This ratio measures the extravascular distribution of unbound drug [75, 76]. Sawada et al. [74, 77] have reported that the Vt/fut values for both basic and acidic drugs are quite similar among different species (mouse, rat, dog, monkey, rabbit and human), irrespective of whether they are bound by albumin, alpha1-glycoprotein, and/or to lipoproteins. For basic compounds, a better correlation (r=0.944) between animal and human Vt/fut was achieved compared to the correlation between Vss only (r=0.748) [74]. Similar results were obtained for 9 weakly acidic drugs [77]. Based on these results, the authors suggested that Vss in humans can be predicted from the Vt/fut values for laboratory animals, combined with fup and the blood/plasma ratio determined in human.

Obach et al. (1997) [40] have also emphasized that plasma protein binding has to be considered when predicting volumes of distribution. Of 4 methods that they tested on a dataset of 16 compounds, the one in which protein binding was disregarded was relatively unreliable. For example, allometric scaling of unbound volume of distribution (average error = 1.83 fold) was substantially more predictive than allometric scaling which excluded interspecies differences in plasma protein binding (average error = 2.78 fold). The other 2 approaches, which estimated Vss in humans from Vt/fut values in preclinical species and which incorporate the effects of plasma protein binding were also more successful (average fold error = 1.56).

4.2. Scaling *In Vivo* Kp Values From Animal to Human

Since it is difficult to determine the Kp for human tissues experimentally, such values are usually obtained from the scaling of *in vitro* or *in vivo* Kp data from animals to human. In these procedures, it is usually assumed that the tissue-to-plasma unbound concentration ratio (Kpu) [78], and hence the distribution volume for unbound drug (Vssu) [22] are similar for animals and human as discussed in the previous section. The rat and rabbit Kpu for various tissues (e.g. lung, adipose, muscle, heart, bone, brain, skin, gut, kidney) have been shown to be highly correlated for biperiden (r=0.933), but the lung, adipose and gut seem to be less correlated [78]. It was shown that rat and human Vssu of acidic and basic drugs differ by an average factor of only 0.97 (+/- 0.38, n=24) [22]. This assumption on similar Kpu and Vssu between species is probably more valid for hydrophilic compounds, which distribute mainly into the water compartment, since the tissues and plasma of different mammalian species tend to have similar water contents [22]. However, for more lipophilic compounds that distribute into and bind more extensively to tissues, another study demonstrates some interspecies differences of up to 3-fold in Kpu values of heart muscle and lung determined in rat and rabbit [22, 21]. These differences may be due to differences between rodents and humans in the lipid content and composition especially of certain tissues, such as muscle, heart and lung [22, 21]. For drugs whose distribution is diffusion rate limited, species differences in membrane permeability could also influence the Kpu values. In addition, different degree of active transport processes (e.g. Na/K ATPase) in animal and human can lead to significant interspecies differences in Vssu.

For example, Vssu of digitoxin is 2.1 L/kg in rat but 9.1 L/kg in human [22]. Despite these numerous factors which might influence Kpu values in different species, a series of 40 drugs with widely differing physicochemical characteristics did show close interspecies correlations between the binding to muscle tissue [76] (Figure 7). Overall, therefore, species differences in tissue binding seem to be less pronounced than the corresponding differences in plasma binding.

Physiological approaches to predict Kp and Vdss have also been developed recently. These approaches are based on species differences in tissue composition data or the rate of disappearance of compound from tissues. Empirical and semi-empirical models (e.g. QSAR) have also been reported in the literature. Those methods do not involve conventional species scaling but rather predict directly distribution parameters in various species. These methods have just been reviewed [22].

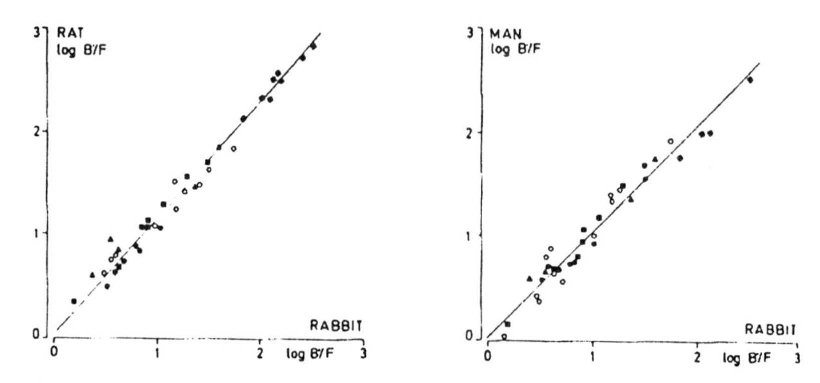

Figure 7. Correlation between binding of drugs to muscle tissue of different species *in vitro*. B/F represent the ratio between bound (B) and free (F) concentrations (from [76]).

5. METABOLIC CLEARANCE PREDICTION

The prediction of metabolic clearance is particularly important, as it affects both drug bioavailability and elimination. Especially important during drug discovery and development are accurate predictions of the hepatic clearance, since the liver is the most important organ for metabolism.

Hepatic clearance can be described by theoretical models which relate the intrinsic clearance (CLint) to the other physiological parameters that influence clearance, namely hepatic blood flow and blood binding. The most frequently used models are: the well-stirred model, the dispersion model and the parallel tube model, all of which represent approximations of liver physiology and share a number of assumptions [79]. The main assumptions are that the distribution of drug into the liver is perfusion rate limited and the diffusion into hepatocytes is assumed to be rapid and not subject to diffusional barriers. Furthermore, it is assumed that only the free drug crosses the hepatocyte membrane and drug-metabolizing enzymes are homogeneously distributed within the liver acinus. Of these assumptions at least the last one is incorrect, as the drug-metabolizing are known to be distributed heterogeneously across the acinus [80, 81].

For simplicity, only the well-stirred model will be considered in this chapter. Hepatic clearance in the well-stirred model is described by:

$$CL = \frac{LBF x fu x CL\mathrm{int}}{LBF + fu x CL\mathrm{int}} \qquad [26]$$

where LBF is the liver blood flow, fu is the fraction unbound in the blood and CLint (the intrinsic clearance) is a measure of the maximal ability of the liver to metabolise a drug in the absence of protein binding or blood flow limitations.

For drugs with high extraction ratios (fu x CLint >> LBF) the value of CL approaches LBF and hepatic clearance is limited by the liver blood flow. As indicated previously, LBF correlates with body weight [43], so for high extraction drugs the hepatic clearance is expected to scale allometrically across the species. For drugs with low extraction ratios (fu x CLint << LBF) the value of CL is directly related to fu and Clint (CL = fu x CLint). Thus, the clearance of such compounds is mainly determined by their plasma protein binding and intrinsic clearance (both of which can be readily measured *in vitro*). Since Clint reflects the amount of hepatic enzymes, and this correlates well with body weight, in theory the clearance of low –extraction ratio drugs will also scale allometrically across the species. However, over the past decade research into the 30 isoforms that constitute the cytochrome P450 superfamily has not only demonstrated marked variation in the total amounts of these enzymes, but also species-specific differences in individual isoforms [82]. Therefore, when metabolism is not limited by liver blood flow, the allometric scaling of hepatic metabolic clearance that is based *in vivo* preclinical data alone [4, 5, 83] often leads to poor predictions. As discussed below, further adjustments are required to improve the accuracy of such scaling of clearance [3, 10, 39, 84].

Both empirical and physiologically based approaches have been developed to predict the *in vivo* hepatic metabolic clearance in animals and humans. Due to the improved availability of human liver samples, the number of predictions that include *in vitro* human data have markedly increased during the past few years. However the body of *in vitro* and/or *in vivo* data that is needed to predict human clearance varies with the particular approach used. Drug clearance can be predicted from human *in vitro* data alone, either by using physiologically based direct scaling or *in vitro*-to-*in vivo* correlations [40, 85-88]. Allometric scaling of *in vivo* preclinical data can also be expanded, by incorporating animal and human *in vitro* data, to adjust for the hepatic metabolic clearance in human [10, 56, 89, 90]. More recently, approaches that use artificial neural networks to predict hepatic metabolic clearance from *in vitro* data have also been proposed [91]. These various methods, together with the accuracy of the predictions, are reviewed below.

5.1. *In Vitro* to *In Vivo* Physiologically Based Direct Scaling

By applying pharmacokinetic principles, it has recently been shown that human hepatic metabolic clearance can be predicted reasonably well from *in vitro* data [79, 92-96]. Excellent reviews that cover all aspects of the *in vitro* to *in vivo* scaling have recently been provided [79, 93-95]. The strategy that allows this extrapolation of *in vitro* clearance to the *in vivo* situation is depicted in the Figure 8 [95]. In summary, the first step is to obtain Clint from *in vitro* data. Various *in vitro* systems (e.g. hepatocytes, liver microsomes) are available for these metabolism studies. The second step consists of scaling the activity measured *in vitro* to the whole liver. Depending on the test system, this is achieved by using microsomal recovery factors and hepatocellularity values. The third stage involves the use of a liver model, which incorporates the effects of liver blood flow and blood binding, to convert the estimated Clint into a predicted CL.

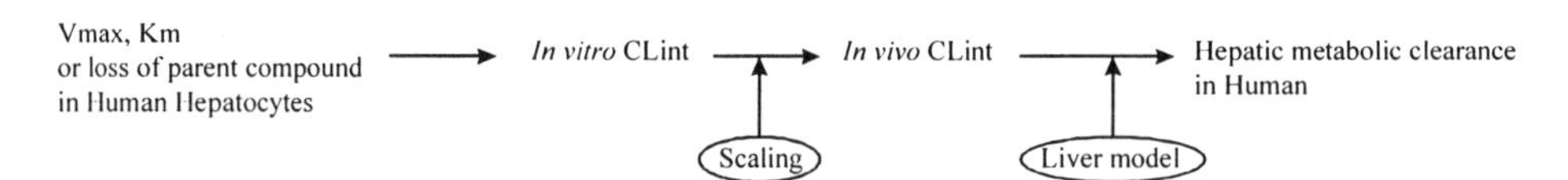

Figure 8. Physiologically based direct scaling approach (from [10]).

This method was originally developed and validated using microsomal data obtained in the rat [79, 93-95]. The *in vivo* hepatic clearance in rats of many drugs that are metabolised by CYP450 has now been successfully predicted from data obtained with both rat liver microsomes and isolated hepatocytes (J.B. Houston, 1994). Due to the increasing availability of human liver, the same methods can now be applied to human. Based on microsomal preparations from about 12-20 different human liver samples, Hoener [96] successfully predicted the *in vivo* CLh for chlorzoxazone 6-hydroxylation, and lidocaine and nifedipine oxidation. Using human literature data from 25 metabolic reactions, Iwatsubo et al. [97] have also compared the Clint values obtained *in vitro* and *in vivo*. According to the report, although Clint *in vitro* generally did correlate with the *in vivo* values, more than a 3-fold difference was observed for about 50% of the 25 metabolic reactions. Based on human hepatocyte data for 22 extensively metabolised compounds, and using approaches that are based on physiologically-based direct scaling, we have also recently obtained reasonable predictions of Clint *in vivo* [12]. This study achieved a success rate close to 65%, with an average error of about 2 fold, although it again tended to underestimate the *in vivo* clearance values. Recently, the *in vivo* metabolic clearance in humans has been predicted for a number of drugs that primarily undergo phase II metabolism [98]. In this evaluation, human hepatocytes did quantitatively predict the hepatic clearances, in contrast to human liver microsomes which consistently underestimated the *in vivo* clearance by approximately 10-fold.

A number of improvements to these approaches have been suggested, including a variation of the physiologically-based direct scaling approach, recently proposed by Naritomi [99]. This approach applies animal *in vitro-in vivo* scaling factors to human *in vitro* data. It is based on the observation that, for 8 test compounds, the animal (rat and dog) scaling factors are within 2-fold of the human values, suggesting that the same scaling factors are applicable to different species. Correcting the human *in vitro* Clint values by the animal scaling factor in this way, yielded predictions that were mostly within 2-fold of the actual values, and which were therefore more accurate than those obtained with the traditional direct scaling approach. Taking account of *in vitro* binding is another approach that can potentially improve the results for direct scaling. The importance of non-specific binding in microsomal incubations has recently received considerable attention [40, 88, 98, 100], and inclusion of microsomal binding has been shown to improve the prediction of clearance from *in vitro* data. Such issues appear to be especially important for acidic compounds (Obach 1999).

5.2. Empirical *In Vitro - In Vivo* Correlations

In our laboratories we have recently used 22 compounds whose *in vivo* hepatic extractions in human were already known, to evaluate the possibility of using human hepatocytes to estimate hepatic extraction ratios (Figure 9) [87]. Despite a variety of metabolic pathways, and wide ranges of protein binding and *in vivo* clearances, *in vitro* clearances did successfully sub-divide the compounds into those with low, intermediate or high hepatic extractions, i.e. the compounds could be classified as having potentially high, intermediate or low oral bioavailabilities. This approach also appeared useful for quantitative predictions, as 73% of the predicted clearance values were within a factor of 2 of the observed clearance in human.

5.3. Predictions Using Allometric Scaling Techniques

Weiss et al [84] first described the use of an allometric equation for the pharmacokinetic scaling of clearance. As discussed above, when the metabolism is not limited by liver blood flow the authors expressed reservations about the use of allometry for scaling hepatic metabolic clearance between species [84]. For several low clearance compounds, such as antipyrine, phenytoin or clonazepam, plotting the intrinsic clearance as a function of body weight dramatically over estimated the human value [3].

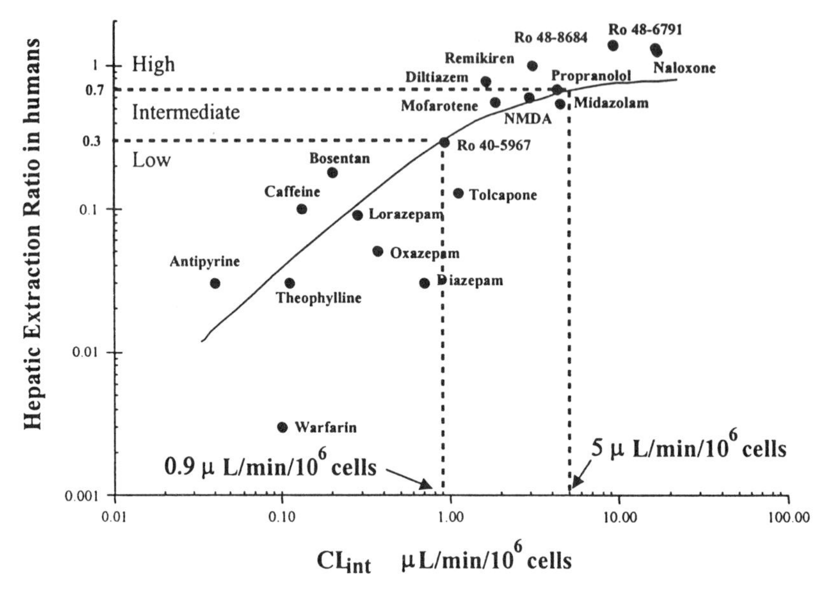

Figure 9. Relationship between *in vivo* hepatic extraction ratios in humans and *in vitro* intrinsic clearances (CLint) in human hepatocytes for a number of 20 reference compounds (from [87])

This discrepancy was ascribed to a lower metabolic activity in human relative to other mammalian species[3]. This shortcoming was initially related to the concept of "neoteny" which, it was suggested, is associated with larger and more complex brain and/or an enhanced longevity in humans compared to other animal species [3, 39, 101]. Such links with oxidative metabolism are however obscure, and it has recently been suggested that neoteny is only a trivial biologic phenomenon which is unrelated to phase I oxidative metabolism [39].

In assessing the impact of neoteny on pharmacokinetic paramaters, Boxenbaum initially related the lower intrinsic clearance of metabolized compounds in humans to two physiological processes which do not scale allometrically according to body weight (W), i.e: brain weight (BRW) and maximum lifespan potential (MLP). It was suggested that the clearance of low extraction drugs which are metabolized by cytochrome P450 could be scaled allometrically, by using an empirical function which contains two power terms. The equations suggested[55] were:

$$CL = a.BW^{x}.B^{y} \qquad [27]$$

$$CL = a \,.\, B^{x} / MLP \qquad [28]$$

where MLP is the maximum documented longevity for a species, and is calculated from the following equation[102]:

$$MLP = 10.839BW^{0.636}.B^{-0.225} \qquad [29]$$

A related approach has also been used, where the clearance in animals is multiplied by the brain weight or maximum lifespan of the species, and then the product is scaled as a function of body weight[37, 103], i.e.:

$$CL \,.\, BW = a \,.\, W^{x} \qquad [30]$$

For example, Mahmood and Balian [37] showed that CL x MLP or CL x BW estimated the clearance of some anti-epileptic drugs more accurately than the simple allometric approach of CL vs body weight. For this purpose, Mahmood and Balian [37] tried to define conditions under which only one of the three methods gave reasonably accurate prediction (a maximum of 30 error) of clearance; i.e:

1. if the exponent of the simple allometry lies between 0.55 and 0.70, they proposed that simple allometry will predict clearance more accurately than CL x MLP or CL x BRW;
2. if the exponent of the simple allometry lies between 0.71 and 1.0, then the CL x MLP approach will predict clearance better than simple allometry or CL x BRW;
3. if the exponent of the simple allometry is $\geq$ 1.0, then CL x BRW is the most suitable approach to predict clearance in humans.

It was further suggested that, if the exponent of the simple allometry is greater than 1.3, then the prediction of clearance from animals to human may not be accurate, and if the exponent of simple allometry is below 0.55, the predicted clearance may be substantially lower than the observed clearance.

This type of approach has been criticized on a number of grounds [104]. With such purely empirical approaches the choice of allometric method (simple scaling of clearance, or correction for MLP or BW) depends purely on the exponent for the allometric equation of CL vs body weight. Considerable uncertainty is, however associated with this exponent, as it is often determined using only three species and the value that is recorded depends upon which species are used. Furthermore, this approach ignores the crucial role of key parameters such as protein binding.

Recently, an empirical allometric approach has been proposed which appears to give better accuracy than those based on brain weight or maximum lifespan. This combines the *in vivo* clearances for preclinical species, scaled allometrically against body weight, with the corresponding *in vitro* (hepatocyte) clearances relative to humans, as follows [10, 56, 90, 105]:

CLanimal (*in vivo*).(CLhuman (hepatocytes) / CLanimal (hepatocytes)) = a x W^x. [31]

The rationale for this *in vitro* normalization is that it applies animal *in vitro-in vivo* scaling factors to the human *in vitro* data.

5.4. Allometric Correlations of Clearance Between Two Species (The "Boxenbaum Approach")

In this procedure, the animal and human data are plotted on a log-log scale, and a linear regression is used to obtain the parameters of the power equation. Initially Boxenbaum [106] used this approach to compare the pharmacokinetic parameters for 12 benzodiazepines between dogs and humans, to predict the values for further compounds in human. Although good predictions could be achieved for some compounds, the differences between the observed and predicted values were relatively large. For example, 3 to 5-fold deviations were observed between the total clearance of triazolam, nordiazepam and diazepam. These might be attributed to species differences in metabolic pathways, as the metabolism of the isolated phenyl ring of benzodiazepines occurs in dogs but not in humans [106]. Sawada [77] subsequently used the same approach for a series of structurally diverse compounds. The main advantage of this approach, which has been used with some success, is that the extrapolation to humans only requires data in one animal species.

5.5. Accuracy of the Various Methods Available to Predict Clearance in Human

An evaluation of the literature indicates that allometric scaling which is based on *in vivo* preclinical data, combined with *in vitro* animal and human data, predicts the clearance in human more accurately than the empirical approaches that use MLP and BW as correction factors. When the former method was applied to a selection of 10 extensively metabolised compounds [105], the predictions for 9 of the compounds were within a factor of 2 of their actual clearance in human, and the average error was +1.6-fold. By contrast the MLP and BRW approaches both resulted in poorer predictions of clearance; for a set of approximately 40 compounds, only 35 to 50 % of the predicted clearances were within a factor of 2 of their actual values [105]. Another approach that did, however, produce acceptable accuracy was the allometric correlation of CLint between rats and humans. In this case, the clearances predicted for 24 out of 30 compounds were within a factor of 2 of their actual values, with an average error of 1.9-fold [77]. Another comparative study compared the accuracy of five different models for predicting human hepatic drug clearance for a diverse set of 22 metabolised drug molecules [107]. The approaches compared included: allometric and physiologically based direct scaling, empirical *in vitro-in vivo* correlation, and supervised artificial neural networks (Table 3). For the multiple allometric scaling approach combined with *in vitro* and *in vivo* data , 68% of the predictions were within a factor of 2 of their actual values, and the average error was close to 2 fold. Similar success rates were achieved for the physiologically based direct scaling approach, the empirical *in vitro-in vivo* correlation and the artificial neural networks, but interestingly those approaches that are based on *in vitro* data alone (e.g. human hepatocyte data) showed average errors that ranged from 1.64 to 2.03 fold, and were therefore more accurate than the allometry.

These findings were consistent with the results of a study by Obach et al. [40], in which they compared the accuracy of several strategies for predicting *in vivo* Vss,

Table 3. Prediction accuracy for clearance using 5 different approaches (from [107])

Approach	PRESS	r2	Average Fold-error	Successful prediction (%)a	Successful prediction (%)b	maximum fold-error
Multiple species allometric scaling	833	0.44	2.03	68.2	81.8	10.2
Rat allometric scaling	1086	0.38	1.99	54.6	72.7	6.4
Physiologically based direct scaling	638	0.77	2.01	63.6	77.3	6.2
Empirical *in vitro-in vivo* correlation	196	0.84	1.64	63.6	95.5	14.6
ANN	267	0.78	1.81	68.2	81.8	21.8

[a] within 2-fold of observed CL
[b] within 3-fold of observed CL

half-life, and CL. Results for the predictions of clearance are shown in Table 4. Interestingly, the average error for both the allometric and direct scaling approaches was <2 -fold; i.e. the two allometric strategies for estimating CL were as accurate as the *in vitro* direct scaling methods. By contrast the allometric method that ignored interspecies differences in plasma binding and life-span was less accurate, but its average error was nevertheless still <3-fold. A subsequent evaluation of the direct scaling approach that is based on microsomal half-life and excludes all binding parameters showed an average error of <2-fold for 26 out of 29 compounds tested, a successful prediction rate (if ±2-fold is set as an acceptable standard) of 90% [88].

Table 4. Accuracy of prediction of *in vivo* human clearance by *in vitro* and allometric methods (from [40])

Method	**Average Fold Error**	**N**
In vitro $t_{1/2}$ excluding protein binding: parallel tube model	1.81	7
In vitro V_{max} and K_m, excluding f_u: well-stirred model	1.63	8
In vitro V_{max} and K_m, excluding f_u: parallel tube model	1.67	8
Allometric scaling, excluding interspecies f_u differences, including MLP factor	1.91	14
Allometric scaling, including interspecies f_u differences, excluding MLP factor	1.79	12
Allometric scaling, excluding interspecies f_u differences, excluding MLP factor	2.67	14

6. EXCRETION-BILIARY CLEARANCE

Renal and biliary excretion are the major pathways that remove xenobiotics and their metabolites from the body. For many endogenous and exogenous compounds (including drugs), carrier-mediated transport contributes to uptake and excretion by the liver and kidney. In the liver, in addition to Phase I (P-450 enzymes) and Phase II metabolism (conjugation), uptake and biliary excretion are now respectively designated as "Phases 0 and III" of the detoxification process (T. Ishikawa, 1992). The role and contribution of transporters to drug disposition has recently been summarized not only in various literature reviews, but also on internet sites such as: http://www.med.rug.nl/mdl/humanabc.htm for human ABC transporters and http://bigfoot.med.unc.edu/watkinsLab/welcome.htm for Human Enterocyte / Hepatocyte Transporters.

These excretion processes are due to specialised cell systems in the hepatocytes and the kidney proximal tubules. Important carrier families for uptake of drugs and endogenous compounds in both liver and kidney are the organic anion transporting polypeptides (OATPs), the organic cation transporters (OCTs) and the organic anion transporters (OATs). These carriers are complemented by the Na+ taurocholate transporting polypeptide for hepatic uptake and by the ATP-dependent multidrug resistance proteins (MDRs) and the multidrug resistance-associated proteins (MRPs) for excretion into bile and urine. The recognition by transporters can affect biliary and urinary excretion and substrate specificities can be subject to large species differences. In addition, the relative contributions of biliary and urinary excretion are dependent on the physicochemical properties of the compound. Thus, for low molecular weight compounds, urinary excretion is high and biliary excretion is low (<10%), while high molecular compounds (>500 in human) are predominantly excreted into bile. Also many of the molecules excreted in bile have amphipatic structures (i.e. they contain both polar and nonpolar groups).

6.1. Species Differences in Biliary Excretion

Most of the information on drug elimination is reported as percent of dose recovered in bile and urine within a given time period [11]. In general, mice, rats and dogs are good biliary excreters, while guinea pigs, monkeys, and humans are relatively poor. The species differences in biliary excretion become less marked when the molecular size of the drug being excreted exceeds 700 Da. Although mammals exhibit certain anatomical (e.g. presence/absence of a gallbladder) and/or physiological differences (e.g. bile flow rates), per se these do not appear to account for the species differences in biliary excretion. Species variation in the efficiency of the transport systems is the more probable cause. However, as discussed earlier, differences in biliary anatomy (i.e. presence or absence of a gallbladder) can give rise to species differences in pharmacokinetics [11].

The role of active transport processes and their impact on species differences in drug disposition has been reviewed recently [108]. Species differences in active transport may result from differences in the nature, levels and/or regulation of the various transport proteins, as well as from their sequence substrate/inhibitor specificities. Where data are available on rat and human orthologues, sequence identity is reasonably high (70-80%) for the OATP [109], OCT [110], and MDR [111] transporter families. There is some evidence to suggest that the rate of transport in rat is greater than in human. For example, some basic compounds were shown to be taken up ~ 10-fold faster by rat than human hepatocytes [112]. Also Km and Ki for a range of bases were significantly lower for rat OCT1 than human OCT1 [110], suggesting a greater substrate affinity in rat. Also, species differences in the transport across the bile canalicular membrane of certain organic anions (temocaprilat, 2,4-dinitrophenyl-S-gluthatione and taurocholate, which are specific substrates of mrp2 and bsep transporters) were shown to be due mainly to differences in Vmax rather than Km [113].

6.2. Prediction of Biliary Clearance from *In Vitro* and *In Vivo* Preclinical Data

In addition to knock-out animals, a battery of *in vitro* models is available for predicting biliary clearance *in vivo*. They include: membrane vesicles, cultured hepatocytes, freshly isolated hepatocyte suspensions, hepatocyte couplets, over expressed cell lines and transfected cell lines. Overall, the ability of such models to predict *in vivo* data is qualitative rather than quantitative. Recently, however, the biliary secretion kinetics of selected compounds have been successfully predicted in some initial studies with 'sandwich' hepatocyte cultures [114], a model which conserves the hepatocytes' integrity and seems to retain the capacity for biliary secretion [115].

Attempts to apply allometric scaling techniques to biliary excretion have resulted in poor predictions of human clearance for compounds that are highly excreted in bile. Thus, the biliary clearances of susalimod [116] and napsagatran [58] were over-estimated by 20- and 7-fold, respectively, and other similar failures to predict the human pharmacokinetics from pre-clinical data have also been reported [117]. These failures are probably due to species differences in the rate of biliary secretion. For example, the mrp2 mediated hepatobiliary transport of temocaprilat displays large differences between human, mouse, rat, guinea pig, rabbit and dog [113]. The accuracy of such predictions may, however be improved by correcting for factors such as the bile flow rate. Thus, a good prediction of human clearance has been obtained for the vitronectin receptor antagonist, SB-265123 [118], a drug that is cleared by a variety of hepatic processes.

7. EXCRETION-RENAL CLEARANCE

Besides biliary excretion, many drugs and drug metabolites are excreted by the kidneys. Renal clearance is the net result of three inter-related processes: glomerular filtration, tubular secretion and tubular reabsorption. Both glomerular filtration and active tubular excretion eliminate a drug from the blood, while tubular reabsorption allows drug that has entered the renal tubules to re-enter the circulation.

A mechanistic relationship, in which the contributions of tubular secretion and glomerular filtration are additive, has been proposed by Levy [119]. This describes renal clearance (Clr) in the following terms:

$$CLr = (1 - Fr)xfuxGFR + (1 - Fr)xQrx\left(\frac{fuxCLus,\text{int}}{Qr + fuxCLus,\text{int}}\right) \qquad [32]$$

where CLr represents renal clearance, Fr is the fraction of drug reabsorbed in the tubules, fu is the free fraction in plasma, GFR the glomerular filtration rate, Qr the renal blood flow and Clus,int is the intrinsic tubular secretion rate. This model assumes that tubular secretion and glomerular filtration are fully independent processes. If tubular secretion is absent then CLr is directly proportional to GFR and the free fraction in plasma. By measuring fu *in vitro*, the unbound renal clearance can be estimated. Then, since GFR has a well-recognized allometric relationship [11], the unbound renal clearance can be correlated across the species. Similarly, if the secretory pathway is very efficient at removing drug then CLr approximates Qr, and since renal blood flow correlates with body weight [11], drugs whose renal clearance is limited by blood flow (e.g. para-aminohippuric acid) will also scale well across the species.

A number of examples support the conclusion that the allometric approach can be used to predict the renal clearance of a drug in humans [30, 31, 120, 121]. In these studies, the best predictions of human renal clearance were achieved when the intrinsic parameters were scaled across species. For example, a successful prediction was achieved for a lactamylvinylcephalosporin, Ro 25-6833, when unbound renal clearance Clur was used for interspecies scaling [30].

Correlation analysis has also been used successfully to predict the human renal clearance from animal data. Sawada obtained good correlation coefficients for six beta-lactam antibiotics, with good predictions of clearance in human based on monkey data [122]. As with allometric scaling, the use of an intrinsic parameter (CLur) gave a better prediction and correlation than the corresponding hybrid parameter (CLr). Such results were confirmed by a study with 11 cephalosporins [30], in which correcting for protein binding again improved the correlations between monkey and human. A limitation of allometric scaling and correlation analysis is an underlying assumption that the excretion mechanisms in animals and human are similar. This is probably true for the passive processes, but is less likely to be the case for active transport. For example, Giacimini et al [123] have shown that the human organic cation transporter (hOCT1) system is functionally different from the OCT1 homolog in other mammals, and it is reasonable to assume that these differences will modify their interactions with drugs. In this context, Dresser at al. [124] determined the kinetics and substrate selectivity of the OCT1 homologs from mouse, rat, rabbit, and human, and found that the human homolog is functionally different from those of the rodents and rabbit. Such differences reinforce the conclusion that it is difficult to extrapolate data for the active transport systems from animal models to humans. However, despite these observations, predictions of renal clearance for a number of drugs that are actively secreted (e.g. lamifiban, atenolol, famotidine, β-lactam antibiotics, ACE inhibitors) were within 2-fold of observed values when the data from animals were extrapolated to humans [58, 117]. Poor predictive capabilities might occur, because of species differences in urinary pHs [125]. Such effects might influence not only the extent of diffusion and reabsorption across the renal tubule, but could also result in artificial differences due to changes in the stability at different urinary pHs [125, 30].

Even when precise rates of renal clearance for humans cannot be predicted from animal data, depending on whether renal clearance is less than, equal to or greater than the GFR, it may be possible to predict the mechanisms for renal elimination in human. Lin [117] provides the following example: famotidine has a renal clearance in rats of 1.68 L/h/kg, which exceeds the GFR (0.522 L/h/kg), indicating that a net tubular secretion is occurring. In humans, the renal clearance of famotidine is 0.266 L/h/kg, which is also greater than GFR 0.108 L/h/kg. The ratio of renal clearance to GFR is 2.5 and 3.2 in humans and rats, respectively. Therefore for famotidine, there is reasonably good agreement between renal clearance in humans and that predicted from animal models. Extending this approach to the data for 6 compounds published by Sawada [31] tends to support this observation. Comparing the unbound renal clearance to GFR in the rat and human gave comparable ratios for cefoperazone (0.5 and 0.9), moxalactam (1.7 and 1.8) and cefazolin (7.7 and 3.6). For cefmetazole, the mechanism of renal excretion was similar in rat and human but the net secretion was much more pronounced in human (ratios of 1.6 in rat and 6 in human). For cefpiramide the ratios in rat (1.1) and human (0.7) were similar but the values indicated different excretion processes, namely a net secretion in rat and a net reabsorption in human.

Given the difficulties in predicting renal clearance when the excretion mechanisms are different, Mahmood has proposed a correction factor. This is calculated as the product of glomerular filtration rate and renal blood flow, divided by the product of glomerular filtration rate and kidney weight [36]. However, although using this factor did appear to improve the accuracy of the prediction, the author did not provide a rationale for its use; therefore caution should be exercised when using this approach.

8. ESTIMATING THE STARTING DOSE FOR ENTRY INTO HUMANS

The different approaches that may be used to calculate the starting dose have recently been reviewed [126]. The US FDA has also issued a draft guidance on dose selection recently. Briefly, the methods can be classified as 4 different approaches: (1) dose by factor methods that utilize the NOAEL (no observable adverse effect level) from preclinical toxicology studies multiplied by a safety factor; (2) the similar drug approach, that may be used when clinical data are available for another compound of the same pharmacologic class as the investigational drug; (3) the pharmacokinetically guided approach that uses systemic exposure rather than dose for the extrapolation from animal to man; and (4) the comparative approach that consists of utilizing two or more methods to estimate a starting dose and then critically comparing the results to arrive at the optimal starting dose [126].

The first method is purely empirical and consists of identifying a dose (usually expressed in mg/kg) associated with a specific effect in preclinical toxicology studies and then multiplying it by one or more sensitivity factors to estimate a safe human starting dose. Different sensitivity or safety factors may be applied. For example, the maximum starting dose for the EIH study can be choosen as the smallest of the following three doses: 1/10 of the highest no effect dose in rodents, 1/6 of the highest no effect dose in dogs or 1/3 of the highest no effect dose in monkeys.

The similar drug approach is also largely empirical and can be used when human safety data are available for a drug similar to the one under investigation. This approach is based on the ratio of the clinically safe dose of the similar drug to its NOAEL and assumes that this ratio is equal to the ratio of the starting dose for the compound under investigation to its NOAEL. The main limitation is that applying a cross-species dosing ratio for one drug to another drug assumes that pharmacokinetic and pharmacodynamic differences between animal and man are the same for both compounds.

The pharmacokinetically guided approach is currently the method of choice as it uses sytemic exposure instead of dose for the extrapolation from animal to man and is currently the most widely used approach at Hoffmann La Roche [126]. With this approach, a desired systemic exposure (usually AUC or Cmax) for humans is targeted. If a NOAEL and its corresponding AUC are available for example from more than one animal species, the animal species with the lowest AUC is used. The clearance (CL) of the drug in humans obtained from allometric or direct scaling[10, 107] can be used to estimate the dose corresponding to that targeted AUC as follows:

$$DOSE = AUCp \times CLh \qquad [33]$$

This approach assumes interspecies similarities in concentration-effect relationships and its success depends on the reliability of the pharmacokinetics predictions in man based on *in vivo* and *in vitro* preclinical data. Calculation of the starting dose is an area that will benefit from the advances made in pharmacokinetic predictions using for example physiologically based approaches. The PK guided approach was shown to yield a starting dose that was higher than a starting dose calculated from a more empirical approach. This may result into increased efficiency in the conduct of the SAD studies with time savings ranging from 2 weeks to 6 months [127].

9. CONCLUSION

The use of both empirical methods and physiologically based models to predict human pharmacokinetic profiles can help to select the best candidates for drug development. They can also help to select doses for the first clinical studies. The present chapter has reviewed the approaches that can be used as well as recent developments in the field.

Allometric scaling explores the mathematical relationships between pharmacokinetic parameters from various animal species, and these can then be used to predict the corresponding values in other species, including human. Such methods are also built to some extent on physiological principles. They are relatively easy to apply but resource demanding for the collection of the *in vivo* data in animals. Nevertheless, their application has lead to useful predictions of individual pharmacokinetic parameters (e.g. clearance, fraction absorbed, volume of distribution). In general, however, such methods can only take account of 'passive' transport differences between species.

Physiologically based models (PBPK) can be used to explore, and help to explain, the mechanisms that lie behind species differences in pharmacokinetics and drug metabolism. Such models can provide a rational basis for interspecies scaling of individual parameters which can then be integrated, to provide quantitative and time dependent estimates of both the plasma and tissue concentrations in humans. Furthermore, being mechanistically based they can be used diagnostically, to generate information on new compounds and to understand the sensitivity of the *in vivo* profile to compound properties. Despite this great potential the use of PBPK models in drug discovery and development has been relatively limited. One reason for this is the large amounts of data (e.g. estimates of tissue/blood partitioning and intrinsic clearance) required for each drug. However, recent developments of in silico and *in vitro* models, which can provide estimates of these input parameters, have dramatically reduced the amount of experimental work required [22]. Another reason for the limited use of PBPK models relates to their mathematical complexity, so that a high level of expertise in needed to develop such models. Again however, recent improvements in the availability of well-validated and user-friendly software packages should remove this barrier. In the very near future, therefore, the use of physiologically based models is likely to increase dramatically both to predict concentration versus time profiles in a variety of species, and also as diagnostic tools to better understand potential development compounds.

9. REFERENCES

1. D.A. Norris, G.D. Leesman, P.J. Sinko, and G.M. Grass, Development of predictive pharmacokinetic simulation models for drug discovery, *J. Control. Release* **65** (1-2), 55-62 (2000).
2. H. Boxenbaum and M. Battle, Effective half-life in clinical pharmacology, *J. Clin. Pharmacol.* **35** (8), 763-766 (1995).
3. B. Boxenbaum and R.W. D'Souza, *Interspecies Pharmacokinetic Scaling, Biological Design and Neoteny*, in *Advances in Drug Research*, B. Testa, Editor. 1990, Academic Press Limited: London. p. 139-196.
4. D.B. Campbell, Can allometric interspecies scaling be used to predict human kinetics, *Drug Inf. J* **28** 235-245 (1994).
5. J. Mordenti, Man versus beast: pharmacokinetic scaling in mammals, *J. Pharm. Sci.* **75** (11), 1028-40 (1986).
6. R.M. Ings, Interspecies scaling and comparisons in drug development and toxicokinetics, *Xenobiotica* **20** (11), 1201-31 (1990).
7. J.H. Lin, Species similarities and differences in pharmacokinetics, *Drug Metab. Dispos.* **23** (10), 1008-21. (1995).
8. J.H. Lin and A.Y. Lu, Role of pharmacokinetics and metabolism in drug discovery and development, *Pharmacol. Rev.* **49** (4), 403-49. (1997).
9. T. Lavé, O. Luttringer, J. Zuegge, G. Schneider, P. Coassolo, and F.P. Theil, Prediction of human pharmacokinetics based on preclinical *in vitro* and *in vivo* data, *Ernst Schering Research Foundation workshop {Ernst Schering Res Found Workshop}* **37** 81-104 (2002).
10. T. Lave, P. Coassolo, and B. Reigner, Prediction of hepatic metabolic clearance based on interspecies allometric scaling techniques and *in vitro-in vivo* correlations, *Clin. Pharmacokinet.* **36** (3), 211-31 (1999).
11. P.J. McNamara, *Interspecies scaling in pharmacokinetics*, in *Pharmaceutical bioequivalence*, P.G. Welling, F.L.S. Tse, and S.V. Dighe, Editors. 1991, Marcel Dekker: New York. p. 267-300.
12. J. Zuegge, G. Schneider, P. Coassolo, and T. Lave, Prediction of Hepatic Metabolic Clearance in Man - Comparison and Assessment of Prediction Models, *Clin. Pharmacokinet* **40** (7) 553-63 (2001).
13. R. Kawai, M. Lemaire, J.L. Steimer, A. Bruelisauer, W. Niederberger, and M. Rowland, Physiologically based pharmacokinetic study on a cyclosporin derivative, SDZ IMM 125, *J. Pharmacol. Exp. Ther.* **22** (5), 327-65 (1994).
14. R. Kawai, D. Mathew, C. Tanaka, and M. Rowland, Physiologically based pharmacokinetics of cyclosporine A: extension to tissue distribution kinetics in rats and scale-up to human, *J. Pharmacol. Exp. Ther.***287** (2), 457-68 (1998).
15. S.B. Charnick, R. Kawai, J.R. Nedelman, M. Lemaire, W. Niederberger, and H. Sato, Physiologically based pharmacokinetic modeling as a tool for drug development, *J. Pharmacokinet. Biopharm.* **23** (2), 217-229 (1995).
16. L.E. Gerlowski and R.K. Jain, Physiologically based pharmacokinetic modeling: principles and applications, *J. Pharm. Sci.* **72** (10), 1103-27 (1983).
17. M.E. Andersen, Development of physiologically based pharmacokinetic and physiologically-based pharmacodynamic models for applications in toxicology and risk assessment, *Toxico. Lett.* **79** (1-3), 35-44 (1995).
18. M.E. Andersen, H.d. Clewell, M.L. Gargas, F.A. Smith, and R.H. Reitz, Physiologically based pharmacokinetics and the risk assessment process for methylene chloride, *Toxicol Appl Pharmacol* **87** (2), 185-205 (1987).

19. J.C. Ramsey and M.E. Andersen, A physiologically based description of the inhalation pharmacokinetics of styrene in rats and humans, *Toxicology* **73** (1), 159-75 (1984).
20. R.C. Ward, C.C. Travis, D.M. Hetrick, M.E. Andersen, and M.L. Gargas, Pharmacokinetics of tetrachloroethylene, *Toxicol. Appl. Pharmacol.* **93** (1), 108-17 (1988).
21. P. Poulin and F.P. Theil, A priori prediction of tissue:plasma partition coefficients of drugs to facilitate the use of physiologically-based pharmacokinetic models in drug discovery, *J. Pharm. Sci* **89** ((1)), 16-35 (2000).
22. F.P. Theil, T.W. Guentert, S. Haddad, and P. Poulin, Utility of Physiologically Based Pharmacokinetic Models to Drug Development and Rational Drug Discovery Candidate Selection, *Toxico. Lett.* (2002), in press.
23. M. Rowland, Physiologic pharmacokinetic models and interanimal species scaling, *Pharmacol. Ther.* **29** (1), 49-68 (1985).
24. H. Harashima, Y. Sawada, Y. Sugiyama, T. Iga, and M. Hanano, Prediction of serum concentration time course of quinidine in human using a physiologically based pharmacokinetic model developed from the rat, *J. Pharm. Sci* **9** (2), 132-8 (1986).
25. Y. Sawada, H. Harashima, M. Hanano, Y. Sugiyama, and T. Iga, Prediction of the plasma concentration time courses of various drugs in humans based on data from rats, *J. Pharmacokinet. Biopharm.* **8** (9), 757-66 (1985).
26. H.W. Ruelius, Extrapolation from animals to man: predictions, pitfalls and perspectives, *Xenobiotica* **17** (3), 255-65 (1987).
27. J.J. Bogaards, E.M. Hissink, M. Briggs, R. Weaver, R. Jochemsen, P. Jackson, M. Bertrand, and P.J. van Bladeren, Prediction of interindividual variation in drug plasma levels *in vivo* from individual enzyme kinetic data and physiologically based pharmacokinetic modeling, *Eur. J. Pharm. Sci.* **12** (2), 117-24 (2000).
28. P.A. Arundel, A multi-compartmental model generally applicable to physiologically-based pharmacokinetics, *The third IFAC symposium. Modeling and control in biomedical systems* University of Warwick 23-26 March (1997).
29. H. Boxenbaum, Interspecies scaling, allometry, physiological time, and the ground plan of pharmacokinetics, *J. Pharm. Sci.* **10** (2), 201-27 (1982).
30. W.F. Richter, P. Heizmann, J. Meyer, V. Starke, and T. Lave, Animal pharmacokinetics and interspecies scaling of Ro 25-6833 and related (lactamylvinyl)cephalosporins, *J. Pharm. Sci.* **87** (4), 496-500 (1998).
31. Y. Sawada, M. Hanano, Y. Sugiyama, and T. Iga, Prediction of the disposition of beta-lactam antibiotics in humans from pharmacokinetic parameters in animals, *J. Pharmacokinet. Biopharm.* **12** (3) 241-61 (1984).
32. W.L. Chiou, H.Y. Jeong, S.M. Chung, and T.C. Wu, Evaluation of using dog as an animal model to study the fraction of oral dose absorbed of 43 drugs in humans, *Pharm. Res.* **17** (2), 135-40 (2000).
33. L. Chiou W. and W. Buehler P., Comparison of oral absorption and bioavailablity of drugs between monkey and human, *Pharm. Res* **19** (6), 868-74 (2002).
34. W.L. Chiou and A. Barve, Linear correlation of the fraction of oral dose absorbed of 64 drugs between humans and rats, *Pharm. Res* **15** (11), 1792-5 (1998).
35. I. Mahmood and J.D. Balian, Interspecies scaling: predicting pharmacokinetic parameters of antiepileptic drugs in humans from animals with special emphasis on clearance, *J. Pharm. Sci.* **85** (4), 411-414 (1996).
36. I. Mahmood, Interspecies scaling of renally secreted drugs, *Life Sci* **63** (26), 2365-71 (1998).
37. I. Mahmood and J.D. Balian, Interspecies scaling: Predicting clearance of drugs in humans. Three different approaches, *Xenobiotica* **26** (9), 887-895 (1996).
38. I. Mahmood and J.D. Balian, The pharmacokinetic principles behind scaling from preclinical results to phase I protocols, *Clin. Pharmacokinet.* **36** (1), 1-11 (1999).
39. H. Boxenbaum and C. Dilea, First-time-in-human dose selection: Allometric thoughts and perspectives, *J. Clin. Pharmacol.* **35** (10), 957-966 (1995).

40. R.S. Obach, J.G. Baxter, T.E. Liston, B.M. Silber, B.C. Jones, F. MacIntyre, D.J. Rance, and P. Wastall, The prediction of human pharmacokinetic parameters from preclinical and *in vitro* metabolism data, *J. Pharmacol. Exp. Ther.* **283** (1), 46-58 (1997).
41. J. Mordenti, S.A. Chen, J.A. Moore, B.L. Ferraiolo, and J.D. Green, Interspecies scaling of clearance and volume of distribution data for five therapeutic proteins, *Pharm. Res.* **8** (11), 1351-9 (1991).
42. E.F. Adolph, Quantitative relations in the physiological contributions of mammals, *Science* **109** 579-85 (1949).
43. H. Boxenbaum, Interspecies variation in liver weight, hepatic blood flow, and antipyrine intrinsic clearance: extrapolation of data to benzodiazepines and phenytoin, *Journal* **8** (2), 165-76 (1980).
44. P.L. Bonate and D. Howard, Prospective allometric scaling: does the emperor have clothes?, *J. Clin. Pharmacol* **40** (6), 665-70 (2000).
45. A.A. Heusner, What does the power function reveal about structure and function in animals of different size?, *Annu. Rev. Physiol* **49** 121-33 (1987).
46. T. Lave, B. Levet Trafit, A.H. Schmitt Hoffmann, B. Morgenroth, W. Richter, and R.C. Chou, Interspecies scaling of interferon disposition and comparison of allometric scaling with concentration-time transformations, *J. Pharm. Sci* **84** (11), 1285-90 (1995).
47. V.F. Cosson, E. Fuseau, C. Efthymiopoulos, and A. Bye, Mixed effect modeling of sumatriptan pharmacokinetics during drug development. I: Interspecies allometric scaling, *J. Pharmacokinet, Biopharm* **25** (2), 149-67 (1997).
48. T. Martín Jiménez and E. Riviere Jim, Mixed-effects modeling of the interspecies pharmacokinetic scaling of oxytetracycline, *Journal* **91** ((2)), 331-41 (2002).
49. R.L. Dedrick, K.B. Bischoff, and D.S. Zaharko, Interspecies correlation of plasma concentration, history of methotrexate (NSC-740), *Cancer Chemother. Rep* **54** 95-101 (1970).
50. H. Boxenbaum and R. Ronfeld, Interspecies pharmacokinetic scaling and the Dedrick plots, *Am. J. Physiol* **245** (6), R768-75 (1983).
51. H. Boxenbaum, Time concepts in physics, biology, and pharmacokinetics, , *J. Pharm. Sci* **75** (11), 1053-62 (1986).
52. C. Efthymiopoulos, R. Battaglia, and M. Strolin-Benedetti, Animal pharmacokinetics and interspecies scaling of FCE 22101, a penem antibiotic *J. Antimicrob. Chemother* **27** (4), 517-26 (1991).
53. M. Chung, E. Radwanski, D. Loebenberg, C.C. Lin, E. Oden, S. Symchowicz, R.P. Gural, and G.H. Miller, Interspecies pharmacokinetic scaling of Sch 34343 *J. Antimicrob. Chemother* **15** 227-33 (1985).
54. D.S.R. W and H. Boxenbaum, Physiological pharmacokinetic models: some aspects of theory, practice and potential, *Toxicology* **4** (2), 151-71 (1988).
55. H. Boxenbaum, Interspecies pharmacokinetic scaling and the evolutionary-comparative paradigm, *Drug Metab. Rev.* **15** (5-6), 1071-121 (1984).
56. T. Lave, S. Dupin, M. Schmitt, M. Kapps, J. Meyer, B. Morgenroth, R.C. Chou, D. Jaeck, and P. Coassolo, Interspecies scaling of tolcapone, a new inhibitor of catechol-O- methyltransferase (COMT). Use of *in vitro* data from hepatocytes to predict metabolic clearance in animals and humans, *Xenobiotica* **26** (8), 839-51 (1996).
57. T. Lave, A. Saner, P. Coassolo, R. Brandt, A.H. Schmitt Hoffmann, and R.C. Chou, Animal pharmacokinetics and interspecies scaling from animals to man of lamifiban, a new platelet aggregation inhibitor, *J. Pharm: Sci.* **48** (6), 573-7 (1996).
58. T. Lave, R. Portmann, G. Schenker, A. Gianni, A. Guenzi, M.A. Girometta, and M. Schmitt, Interspecies pharmacokinetic comparisons and allometric scaling of napsagatran, a low molecular weight thrombin inhibitor, *J. Pharm. Pharmacol* **51** (1), 85-91 (1999).
59. A. Hutchaleelaha, H.H. Chow, and M. Mayersohn, Comparative pharmacokinetics and interspecies scaling of amphotericin B in several mammalian species, *J. Pharm. Pharmacol* **49** 178-83 (1997).
60. R. Lapka, V. Rejholec, T. Sechser, M. Peterkova, and M. Smid, Interspecies pharmacokinetic scaling of metazosin, a novel alpha-adrenergic antagonist, *Biopharm. Drug Dispos* **10** (6), 581-9 (1989).
61. S.S. Ibrahim and F.D. Boudinot, Pharmacokinetics of 2',3'-dideoxycytidine in rats: application to interspecies scale-up, *J. Pharm. Pharmacol.* **41** (12), 829-34 (1989).

62. N.A. Grene Lerouge, M.I. Bazin Redureau, M. Debray, and J.M. Scherrmann, Interspecies scaling of clearance and volume of distribution for digoxin-specific Fab, *Toxicol. Appl. Pharmacol.* **138** (1), 84-9 (1996).
63. T. Lave, B. Levettrafit, A.H. Schmitthoffmann, B. Morgenroth, W. Richter, and R.C. Chou, Interspecies Scaling of Interferon Disposition and Comparison of Allometric Scaling With Concentration-Time Transformations, *J. Pharm. Sci* **84** (11), 1285-1290 (1995).
64. A. Kurihara, H. Naganuma, M. Hisaoka, H. Tokiwa, and Y. Kawahara, Prediction of human pharmacokinetics of panipenem-betamipron, a new carbapenem, from animal data, *Antimicrob. Chemother* **36** (9), 1810-6 (1992).
65. B. Agoram, W.S. Woltosz, and M.B. Bolger, Predicting the impact of physiological and biochemical processes on oral drug bioavailability, *Adv. Drug Deliv. Rev* **50 Suppl 1** S41-67 (2001).
66. M. Grass George and J. Sinko Patrick, Physiologically-based pharmacokinetic simulation modelling, *Adv. Drug Deliv. Rev* **54** (3), 433-51 (2002).
67. W.L. Chiou, C. Ma, S.M. Chung, T.C. Wu, and H.Y. Jeong, Similarity in the linear and non-linear oral absorption of drugs between human and rat, *Int. J. Clin. Pharmacol. Ther. 2000 Nov* **38** (11), 532-9 (2000).
68. N.J. Parrott and T. Lavé, Prediction of Intestinal Absorption: Comparative Assessment of Commercially Available Software, *Eur. J. Pharm: Sci.* **17** 51-61 (2002).
69. J.M. DeSesso and C.F. Jacobson, Anatomical and physiological parameters affecting gastrointestinal absorption in humans and rats, *Food Chem. Toxicol. 2001 Mar* **39** (3)), 209-28 (2001).
70. Y.L. He, S. Murby, G. Warhurst, L. Gifford, D. Walker, J. Ayrton, R. Eastmond, and M. Rowland, Species differences in size discrimination in the paracellular pathway reflected by oral bioavailability of poly(ethylene glycol) and D- peptides, *J. Pharm. Sci. 1998* **87** (5) 626-33 (1998).
71. L.X. Yu, E. Lipka, J.R. Crison, and G.L. Amidon, Transport approaches to the biopharmaceutical design of oral drug delivery systems: prediction of intestinal absorption, *Adv. Drug Deliv. Rev.* **19** (3), 359-76 (1996).
72. S. Winiwarter, N.M. Bonham, F. Ax, A. Hallberg, H. Lennernäs, and A. Karlén, Correlation of human jejunal permeability (*in vivo*) of drugs with experimentally and theoretically derived parameters. A multivariate data analysis approach, *J. Med. Chem.* **41** (25), 4939-49 (1998).
73. M. Rowland and T. Tozer, *Clinical Pharmacokinetics: Concepts and Applications* . 1989, Philadelphia, London: Lea and Febiger.
74. Y. Sawada, M. Hanano, Y. Sugiyama, H. Harashima, and T. Iga, Prediction of the volumes of distribution of basic drugs in humans based on data from animals, , *J. Pharmacokinet, Biopharm* **12** (6), 587-96 (1984).
75. G. Schuhmann, B. Fichtl, and H. Km- z, Prediction of drug distribution *in vivo* on the basis of *in vitro* binding data, *Biopharm. Drug Dispos.* **8** 73- 86 (1987).
76. B. Fichtl, A. Nieciecki, and K. Walter, *Tissue binding versus plasma binding of drugs: general principles and pharmacokinetic consequences*, in *Advances in drug research*, C.a. Mordenti, Editor. 1991.
77. Y. Sawada, M. Hanano, Y. Sugiyama, and T. Iga, Prediction of the disposition of nine weakly acidic and six basic drugs in humans from pharmacokinetic parameters in rats, *J. Pharmacokinet. Biopharm.* **13** (5), 477-492 (1985).
78. K. Yokogawa, J. Ishizaki, S. Ohkuma, and K. Miyamoto, Influence of lipophilicity and lysosomal accumulation on tissue distribution kinetics of basic drugs: a physiologically based pharmacokinetic model, *Methods Find. Exp. Clin. Pharmacol.* **24** (2), 81-93 (2002).
79. J.B. Houston and D.J. Carlile, Prediction of hepatic clearance from microsomes, hepatocytes, and liver slices, *Drug Metab. Rev* **29** (4), 891-922. (1997).
80. K.S. Pang and G.J. Mulder, The effect of hepatic blood flow on formation of metabolites, *Drug Metab. Dispos* **18** (3), 270-5 (1990).
81. K.S. Pang, J.A. Terrell, S.D. Nelson, K.F. Feuer, M.J. Clements, and L. Endrenyi, An enzyme-distributed system for lidocaine metabolism in the perfused rat liver preparation, *J. Pharmacokinet, Biopharm* **14** (2), 107-30 (1986).

82. K. Walton, J.L. Dorne, and A.G. Renwick, Uncertainty factors for chemical risk assessment: interspecies differences in the *in vivo* pharmacokinetics and metabolism of human CYP1A2 substrates, *Food Chem. Toxicol* (7) 667-80 (2001).
83. H. Boxenbaum and J.B. Fertig, Scaling of antipyrine intrinsic clearance of unbound drug in 15 mammalian species, *Eur. J. Drug Metab. Pharmacokinet* **9** (2), 177-83 (1984).
84. M. Weiss, W. Sziegoleit, and W. Forster, Dependence of pharmacokinetic parameters on the body weight, *Int. J. Clin. Pharmacol* **15** 572-5 (1977).
85. Y. Shibata, H. Takahashi, and Y. Ishii, A convenient *in vitro* screening method for predicting *in vivo* drug metabolic clearance using isolated hepatocytes suspended in serum, *Drug Metab. Dispos* **28** (12), 1518-23 (2000).
86. Y. Shibata, H. Takahashi, M. Chiba, and Y. Ishii, Prediction of hepatic clearance and availability by cryopreserved human hepatocytes: an application of serum incubation method, *Drug Metab. Dispos* **30** (8), 892-6 (2002).
87. T. Lave, S. Dupin, C. Schmitt, B. Valles, G. Ubeaud, R.C. Chou, D. Jaeck, and P. Coassolo, The use of human hepatocytes to select compounds based on their expected hepatic extraction ratios in humans, *Pharm. Res* **14** (2), 152-5 (1997).
88. R.S. Obach, Prediction of human clearance of 29 drugs from hepatic microsomal intrinsic clearance data: an examiniation of *in vitro* half-life approach and nonspecific binding to microsomes, *Drug Metab. Dispos* **27** (11), 1350-9 (1999).
89. T. Lave, A.H. Schmitt Hoffmann, P. Coassolo, B. Valles, G. Ubeaud, B. Ba, R. Brandt, and R.C. Chou, A new extrapolation method from animals to man: application to a metabolized compound, mofarotene, *Life Sci.* **56** (26), PL473-8 (1995).
90. T. Lave, P. Coassolo, G. Ubeaud, R. Brandt, C. Schmitt, S. Dupin, D. Jaeck, and R.C. Chou, Interspecies scaling of bosentan, a new endothelin receptor antagonist and integration of *in vitro* data into allometric scaling, *Pharm. Res.* **13** (1), 97-101 (1996).
91. G. Schneider, P. Coassolo, and T. Lavé, Combining *in vitro* and *in vivo* pharmacokinetic data for prediction of hepatic drug clearance in humans by artificial neural networks and multivariate statistical techniques, *J. Med. Chem* **42** (25), 5072-6 (1999).
92. A. Rane, G.R. Wilkinson, and D.G. Shand, Prediction of hepatic extraction ratio from *in vitro* measurement of intrinsic clearance, *J. Pharmacol. Exp. Ther* **200** 420-4 (1977).
93. K. Ito, T. Iwatsubo, S. Kanamitsu, Y. Nakajima, and Y. Sugiyama, Quantitative prediction of *in vivo* drug clearance and drug interactions from *in vitro* data on metabolism, together with binding and transport, *Annu Rev Pharmacol Toxicol* **38** 461-99. 109 Refs. (1998).
94. T. Iwatsubo, N. Hirota, T. Ooie, H. Suzuki, and Y. Sugiyama, Prediction of *in vivo* drug disposition from *in vitro* data based on physiological pharmacokinetics, *Biopharm. Drug Dispos* **17** (4), 273-310 (1996).
95. J.B. Houston, Utility of *in vitro* drug metabolism data in predicting *in vivo* metabolic clearance, *Biochem. Pharmacol* **47** (9), 1469-1479 (1994).
96. B.A. Hoener, Predicting the hepatic clearance of xenobiotics in humans from *in vitro* data, *Biopharm. Drug Dispos* **15** (4), 295-304 (1994).
97. T. Iwatsubo, *et al.*, Prediction of *in vivo* drug metabolism in the human liver from *in vitro* metabolism data, *Pharmacol. Ther* **73** (2), 147-171 (1997).
98. M.G. Soars, B. Burchell, and R.J. Riley, *In vitro* analysis of human drug glucuronidation and prediction of *in vivo* metabolic clearance, , *J. Pharmacol. Exp. Ther* **301** (1), 382-90 (2002).
99. Y. Naritomi, S. Terashita, S. Kimura, A. Suzuki, A. Kagayama, and Y. Sugiyama, Prediction of human hepatic clearance from *in vivo* animal experiments and *in vitro* metabolic studies with liver microsomes from animals and humans, *Drug Metab. Dispos* **29** (10), 1316-24 (2001).
100. J.A. McLure, J.O. Miners, and D.J. Birkett, Nonspecific binding of drugs to human liver microsomes, *Brit. J. Clin. Pharmacol* **49** (5), 453-61 (2000).
101. F.E. Yates and P.N. Kugler, Similarity principles and intrinsic geometries: contrasting approaches to interspecies scaling, *J Pharm Sci* **75** (11), 1019-27 (1986).

102. G.A. Sacher, *Relation of lifespan to brain weight and body weight in mammals.*, in *The lifespan of animals*, G. Wostenholme and M. O'Connor, Editors. 1959, Little Brown & Co: Boston.
103. T. Lave, A.H. Schmitt-Hoffmann, P. Coassolo, G. Ubeaud, B. Vallès, B. Ba, R. Brandt, and R.C. Chou, A new extrapolation method from animal to man ; application to a metabolized compound, mofarotene, *Life Sci.* **56** (26), 473-478 (1995).
104. T. Lave and P. Coassolo, Commentary on "integration of *in vitro* data and brain weight in allometric scaling to predict clearance in humans: Some suggestions", *Journal Of Pharmaceutical Sciences* **87** (4), 530 (1998).
105. T. Lave, S. Dupin, C. Schmitt, R.C. Chou, D. Jaeck, and P. Coassolo, Integration of *in vitro* data into allometric scaling to predict hepatic metabolic clearance in man: Application to 10 extensively metabolized drugs, , *J. Pharm. Sci* **86** (5), 584-590 (1997).
106. M. Boxenbaum, Comparative pharmacokinetics of benzodiazepines in dog and man, *J. Pharmacokinet, Biopharm* **10** 411-26 (1982).
107. J. Zuegge, G. Schneider, P. Coassolo, and T. Lavé, Prediction of hepatic metabolic clearance: comparison and assessment of prediction models, *Clin. Pharmacokinet* **40** (7), 553-63 (2001).
108. A. Ayrton and P. Morgan, Role of transport proteins in drug absorption, distribution and excretion, *Xenobiotica* **31** (8-9), 469-97 (2001).
109. B. Hsiang, Y. Zhu, Z. Wang, Y. Wu, V. Sasseville, W.P. Yang, and T.G. Kirchgessner, A novel human hepatic organic anion transporting polypeptide (OATP2). Identification of a liver-specific human organic anion transporting polypeptide and identification of rat and human hydroxymethylglutaryl- CoA reductase inhibitor transporters, *J Biol Chem* **274** (52), 37161-8. (1999).
110. L. Zhang, M.E. Schaner, and K.M. Giacomini, Functional characterization of an organic cation transporter (hOCT1) in a transiently transfected human cell line (HeLa),), *J. Pharmacol. Exp. Ther* **286** (1), 354-61. (1998).
111. I. Klein, B. Sarkadi, and A. Varadi, An inventory of the human ABC proteins, *Biochim. Biophys. Acta* **1461** (2), 237-62. (1999).
112. G.W. Sandker, B. Weert, P. Olinga, H. Wolters, M.J. Slooff, D.K. Meijer, and G.M. Groothuis, Characterization of transport in isolated human hepatocytes. A study with the bile acid taurocholic acid, the uncharged ouabain and the organic cations vecuronium and rocuronium, *Biochem. Pharmacol* **47** (12), 2193-200. (1994).
113. H. Ishizuka, K. Konno, T. Shiina, H. Naganuma, K. Nishimura, K. Ito, H. Suzuki, and Y. Sugiyama, Species differences in the transport activity for organic anions across the bile canalicular membrane, *J. Pharmacol. Exp. Ther* **290** (3), 1324-30. (1999).
114. X. Liu, J.P. Chism, E.L. LeCluyse, K.R. Brouwer, and K.L. Brouwer, Correlation of biliary excretion in sandwich-cultured rat hepatocytes and *in vivo* in rats, *Drug Metab. Dispos* **27** (6), 637-44. (1999).
115. E.L. LeCluyse, K.L. Audus, and J.H. Hochman, Formation of extensive canalicular networks by rat hepatocytes cultured in collagen-sandwich configuration, *Am. J. Physiol.* **266** (6 Pt 1), C1764-74. (1994).
116. I. Pahlman, S. Andersson, K. Gunnarsson, M.L. Odell, and M. Wilen, Extensive biliary excretion of the sulfasalazine analogue, susalimod, but different concentrations in the bile duct in various animal species correlating to species-specific hepatobiliary toxicity, *Pharmacol. Toxicol.* **85** (3), 123-9 (1999).
117. J.H. Lin, Species similarities and differences in pharmacokinetics, *Drug metabolism and {Drug Metab Dispos} 1995 Oct* **23** (10), 1008-21 (1995).
118. K.W. Ward, *et al.*, Preclinical pharmacokinetics and interspecies scaling of a novel vitronectin receptor antagonist, *Drug Metab. Dispos* **27** (11) 1232-41. (1999).
119. G. Levy, Effect of plasma protein binding on renal clearance of drugs, *J Pharm Sci* **69** (4), 482-3. (1980).
120. H. Matsushita, H. Suzuki, Y. Sugiyama, Y. Sawada, T. Iga, M. Hanano, and Y. Kawaguchi, Prediction of the pharmacokinetics of cefodizime and cefotetan in humans from pharmacokinetic parameters in animals, *J. Pharmacobiodyn* .**13** (10), 602-11. (1990).
121. C. Efthymiopoulos, R. Battaglia, and M. Strolin Benedetti, Animal pharmacokinetics and interspecies scaling of FCE 22101, a penem antibiotic, *J. Antimicrob. Chemother.* **27** (4), 517-26. (1991).

122. Y. Sawada, M. Hanano, Y. Sugiyama, and T. Iga, Prediction of the disposition of beta-lactam antibiotics in humans from pharmacokinetic parameters in animals, *J. Pharmacokinet. Biopharm.* **12** (3), 241-61 (1984).
123. K.M. Giacomini, Membrane transporters in drug disposition, *J. Pharmacokinet. Biopharm.* **25** (6), 731-41. (1997).
124. M.J. Dresser, A.T. Gray, and K.M. Giacomini, Kinetic and selectivity differences between rodent, rabbit, and human organic cation transporters (OCT1), *J. Pharmacol. Exp. Ther.* **292** (3), 1146-52. (2000).
125. P.L. Bonate, K. Reith, and S. Weir, Drug interactions at the renal level. Implications for drug development, *Clin. Pharmacokinet.* **34** (5), 375-404 (1998).
126. G. Reigner Bruno and S. Blesch Karen, Estimating the starting dose for entry into humans: principles and practice, *Eur. J. Clin. Pharmacol.* **57** (12), 835-45 (2002).
127. B.G. Reigner, P.E.O. Williams, J.H. Patel, J.L. Steimer, C. Peck, and P. vanBrummelen, An evaluation of the integration of pharmacokinetic and pharmacodynamic principles in clinical drug development: Experience within Hoffmann La Roche, *Clin. Pharmacokinet.* **33** (2), 142-152 (1997).

HUMAN DRUG ABSORPTION STUDIES IN EARLY DEVELOPMENT

David V. Prior, Alyson L. Connor, Ian R. Wilding[1]

1. INTRODUCTION

Advances in combinatorial chemistry, proteomics, and genomics have led to the potential for an unprecedented number of new molecular entities (NMEs) to exit discovery and enter full-scale development. Although the emphasis remains focused on developing oral products, few drug candidates have ideal biopharmaceutical properties for oral administration. Factors known to limit or inhibit drug absorption via the oral route include poor solubility or inadequate stability in gastrointestinal (GI) fluids, poor permeability across the intestinal epithelium, enzymatic or non-enzymatic degradation/metabolism in certain GI segments, and complexation with chelating ligands or metal cations normally present in the GI tract[1].

Pharmaceutical companies recognise the need to identify those compounds with problematic biopharmaceutical properties long before the first formal, prototype formulations are administered to humans[2]. As a consequence, drug discovery groups now routinely screen potential clinical candidates using a variety of *in silico*, *in vitro* and *ex-vivo* technologies, as well as the more traditional live animal models. Although some of these technologies are suited to high throughput screening (HTS), output predictions are far from definitive and so need to be treated with caution. Taken from a recent publication[3], Figure 1 shows the correlation between bioavailability in humans versus animals (rodent, dog and primate) to be surprisingly poor across a broad range of drugs. Increasingly, pharmaceutical and drug delivery companies are therefore using human drug absorption (HDA) studies to characterize the biopharmaceutical properties of selected drug candidates, including those in need of life cycle management through new line extensions. Results from these studies can give an early indication of unsuitable or problem compounds and so provide a reliable "route-map" for development.

Today, HDA studies are easily conducted using specially designed, engineering-based capsules to provide non-invasive, targeted drug delivery to key sites of the human

[1] Pharmaceutical Profiles Ltd, Mere Way, Ruddington Fields, Nottingham NG11 6JS

intestine. The most advanced technology in this arena is the proprietary Enterion capsule (Phaeton Research, Nottingham UK), which has emerged as the market leader. This device is extremely versatile and capable of delivering a wide variety of formulations, including solutions, viscous suspensions, particulates, pellets and mini-tablets.

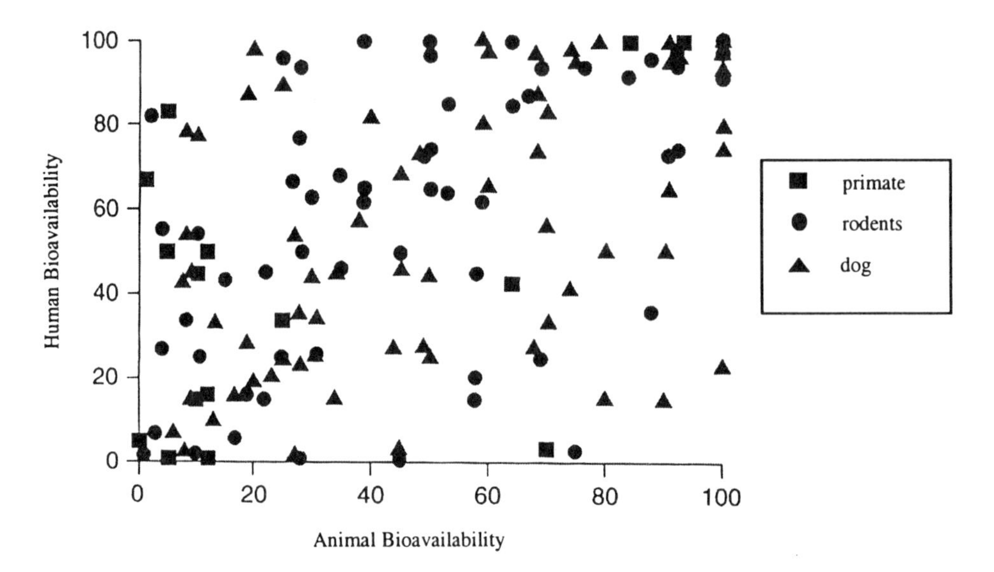

Figure 1. Absolute Bioavailability in Humans Vs. Animals (source: reference 3)

2. BACKGROUND INFORMATION

2.1. Mechanisms of Drug Absorption from the Gastrointestinal Tract

The gastrointestinal (GI) tract is responsible for the extraction and absorption of nutrients from foodstuffs and so provides a natural pathway for the absorption of orally administered drugs[4].

The adult human small intestine is approximately 6 m in length and 3 to 5 cm in diameter. It is arbitrarily divided into three regions; the first 20 to 30 cm is the duodenum, the next 2.4 m the jejunum and the final 3.6 m the ileum. The large intestine (or colon) is 3 to 9 cm in diameter and approximately 1.3 m in length. However, striking differences can be seen in the surface area of the different regions; 1.5 m^2 for the colon compared to approximately 180 m^2 for the jejunum and 280 m^2 for the ileum. It is the effect of this high surface area that primarily accounts for increased absorptive capacity normally associated with the small intestine.

2.1.1. Unmediated Permeation

It is generally accepted[1] that the unmediated permeation of molecules across the intestine occurs by two distinct pathways, paracellular and transcellular. Adjacent enterocytes (epithelial cells) making up the lipoidal epithelial layer are attached at the tight junctional complex near the cell apex. In the case of the paracellular route, molecules penetrate the tight junctions and pass through the epithelial cell and apical membrane to the underlying blood capillary. Paracellular permeation is thought to be maximal in the upper small intestine and limited in the colon[5]. The transcellular pathway considers the epithelium to comprise of a heteroporous barrier, allowing partitioning and diffusion of small molecules through the enterocyte membrane itself.

2.1.2. Mediated Transport Systems

In addition to unmediated permeation, active transport mechanisms intrinsic to the cell membrane, such as carrier and pump systems, are also well recognised[6]. Carriers may be involved in three kinds of transport processes; facilitated diffusion, co-transport (symport) and counter-transport (antiport). Co-transporters and counter-transporters can perform 'secondary active transport' by using energy from the downhill transport of one substrate to drive the uphill transport of another. Pumps use external energy sources, such as hydrolysis of a phosphate bond, to provide 'primary active transport'.

Carrier-mediated processes may be saturated or inhibited, and so may be regulated by a variety of internal or external factors. Saturability of carrier-mediated transport can lead to dose-dependent pharmacokinetics of drugs that are carrier substrates. Similarly, induction or inhibition of carriers may lead to enhanced or reduced absorption of drugs with affinity for these carriers.

2.2. Inhibitory Mechanisms to Drug Absorption

The primary barriers to drug absorption were once considered to be dissolution in the intestinal medium and permeation across the epithelial layer. However, adequate aqueous solubility and permeability across a lipoidal membrane do not guarantee that the drug will be well absorbed. The healthy, properly functioning intestine also serves as a barrier protecting the body from exposure to noxious macromolecules and bacteria.

2.2.1. Intestinal Efflux Systems

Several of the enzyme transport systems known to mediate efflux in the major clearing organs (e.g. liver and kidney) are also expressed in the intestine[7]. The most widely studied of these is P-glycoprotein (P-gp), a secretory transporter with broad specificity located on the mucosal surface of the epithelial cells. Drugs that are substrates for P-gp (or other multidrug resistance associated proteins) can be actively expelled from cells after entering via other routes. Efflux mechanisms therefore act to lower the concentration of drug in the cell and so reduce the rate of diffusion through the intracellular space to the blood capillary. Furthermore, since P-gp is a membrane-bound protein, the efflux system can be up regulated by the cell to increase its pumping rate.

Table 1. Selected Drugs Reported to Interact with p-Glycoprotein[6] and/or Metabolized by Cytochrome P4503A4[10]

Drug Class	**Substrate for p-Glycoprotein**	**Substrate for CYP3A4 and p-Glycoprotein**	**Substrate for CYP3A4**
Antiarrhythmics	Propranolol	Amiodarone Lidocaine Quinidine	Propafenone Disopyramide
Antihistamines			Astemizole Loratadine Terfenadine
Antimicrobials	Cefoperazone Ceftrixone	Erythromycin Itraconazole	Dapsone Troleandomycin
Antiulcer			Omeprazole
Calcium Channel Blockers	Bepradil Tiapamil Nisoldipine Felodipine Nitrenidipine	Diltiazem Nifedipine Verapamil	
Hormones/Steroids	Aldosterone Deoxycorticosterone Clomiphene Dexamethasone Prednisone	Cortisol Progesterone Tamoxifen	Ethinyl estradiol Paclitaxel Testosterone
Immunosuppresents		Cyclosporine	Tacrolimus

Selected drugs reported to interact with P-gp[8] are presented in Table 1. Interestingly, relatively hydrophobic drugs that would be expected to have a high permeability are also those same drugs likely to be most susceptible to the P-gp efflux system. Co-administered drugs that are substrates for P-gp may competitively inhibit the efflux system, which can lead to significant pharmacokinetic interactions. Some pharmaceutical excipients are also P-gp substrates, such as polysorbate 80 and cremaphor-EL[9], and so can be used to modulate P-gp efflux and influence drug absorption.

2.2.2. Intestinal Metabolism

The liver is the major metabolic organ with cytochrome P450 identified as the primary catalyst responsible for oxidative metabolism. However, it is now recognised that there is also substantial P450 activity in the intestinal mucosa[10]. The majority of P450 found in the intestine is the isoform known as cytochrome P4503A4 (or CYP3A4), for which more drugs are known substrates than any other P450 subclass. CYP3A4 levels that have been found in the duodenum, jejunum and ileum are presented in Table 2[11].

Table 2. Levels of Cytochrome P4503A4 in the Human Intestine[8]

Intestinal Region	Length (approx)	Total CYP3A4	Variation in CYP3A4
Duodenum	0.2 m	9.7 nmol	3 to 90 pmol/mg
Jejunum	2.4 m	38.4 nmol	2 to 98 pmol/mg
Ileum	3.6 m	22.4 nmol	2 to 38 pmol/mg

Studies have shown that the extent of so-called "gut-wall" metabolism can be significant and a list of selected drugs metabolised by CYP3A4[12] is included in Table 1. However, interactions are complex and certain drugs (such as rifampin) may induce the activity of CYP3A4 whereas others (such as ketoconazole) inhibit the enzyme system[13, 14].

In addition to P450 enzymes, Phase 1 enzymes, such as alcohol dehydrogenase, monoamine oxidase, esterases, amidases and glucuronidases, are also present in the gastrointestinal mucosa. Most of these are expressed in the duodenum and jejunum with the notable exception of alcohol dehydrogenase, which has a high activity in gastric mucosa.

From a drug development perspective, the overlap of compounds that are subject to both P-gp efflux and CYP3A4 metabolism is striking. If both these systems could be selectively and reversibly inhibited, the bioavailability of many poorly absorbed drugs could be significantly enhanced. Table 1 identifies a partial listing of such "problem drugs" whose bioavailability is limited by both mechanisms.

3. TRADITIONAL Vs CURRENT APPROACHES

Historically, the most popular approach for determining the absorption of drugs from different regions of the intestine has been through the use of perfusion or intubation techniques[15, 16, 17]. These techniques require invasive tubes to be placed at the relevant part of the GI tract via the mouth or rectum. Once located at the correct region, a drug solution or suspension is infused into the gut lumen at a pre-determined rate. Clearly these invasive procedures are associated with significant volunteer discomfort and more importantly, the presence of a tube in the intestines has been demonstrated to alter the function of the GI tract[18]. In particular, intubation has been shown to influence the absorption and secretion balance within the gut, which questions the pharmaceutical relevance of drug absorption data collected via this approach.

Over the last 20 years, a number of engineering-based capsule devices have been developed to allow site-specific measurement of human absorption in a non-invasive manner[19]. The primary emphasis of these different systems has been the ability to control the time and location of drug release. Below is a summary of the operative mechanisms for some of the commercially available technologies.

3.1. High Frequency (HF) Capsule

The HF capsule (Battelle-Institute V, Frankfurt am Main, Germany) was developed in the early 1980s. After swallowing, the passage of the capsule is tracked using x-ray fluoroscopy. On reaching the target location, activation is triggered by a radio-frequency (RF) pulse from a high frequency generator external to the body. Heat generated as a result of the RF-induced current melts a thread and releases a needle, which in turn pierces a latex balloon and allows drug to passively empty from ports in the wall of the capsule.

The capsule has been used successfully to study the regional absorption of a variety of drugs, such as ipsapirone[20] and nifedipine[21]. However, this device is not particularly well suited to delivery of particulate formulations due to its passive release mechanism and difficulty of filling powders into the balloon. Furthermore, the use of x-ray fluoroscopy to track the location of the capsule has further limited its application because of the potentially high radiation dose arising from studies targeting colonic absorption.

3.2. InteliSite Capsule

The InteliSite capsule[22] (Innovative Devices, Raleigh, NC, USA) was commercialised in the late 1990s. At only 10 mm in diameter and 35 mm in length, InteliSite is more compact than its predecessors. It consists of an on-board electronics/actuator assembly, a drug reservoir 0.8 ml in volume, and a radiotracer port. Location in the GI tract is followed using a gamma camera by placing a short half-life, gamma-emitting radionuclide inside the sealed radiotracer port.

The capsule is activated by application of an external oscillating magnetic field, which induces an electric current in a 3-dimensional array of receiving coils. This slowly warms a contact plate transferring heat to a pair of shape memory alloy (SMA) wires. These wires then straighten causing an inner sleeve to rotate and align a series of apertures with corresponding apertures in an outer sleeve. Alignment of the apertures allows passive release of drug from the reservoir.

As with other passive release systems, the lack of free water and agitation does not favour complete and reproducible delivery of particulate formulations to the distal colon, which has compromised the reliable use of InteliSite[23, 24] in commercial studies. Other difficulties encountered with InteliSite include slow or possibly failed activations if the capsule is particularly deep in the body, as well as the potential for pre-activation leakage from the drug reservoir[24].

3.3. Enterion Capsule

The Enterion capsule (Phaeton Research, Nottingham UK) is the latest advance in available HDA technologies and overcomes the primary limitations of the earlier devices[25]. It is a round-ended capsule, 32 mm in length and 11 mm in diameter, with a drug chamber of approximately 1 ml in volume located within the main body (Figure 2). The active delivery mechanism makes this technology extremely versatile and fully effective with a wide variety of formulations, including solutions, viscous suspensions, particulates, pellets and mini-tablets.

The capsule is loaded with the drug (or drug formulation) through an opening 9 mm in diameter, which is then sealed by inserting a push-on cap with a silicone O-ring gasket. The floor of the drug chamber is a piston face, which is held back against a compressed actuation spring by a high tensile strength polymer filament (the spring latch). To track capsule location after administration, a radioactive marker is sealed inside a separate radiotracer port beneath the rounded end cap. This allows capsule position to be followed in real time using a gamma camera.

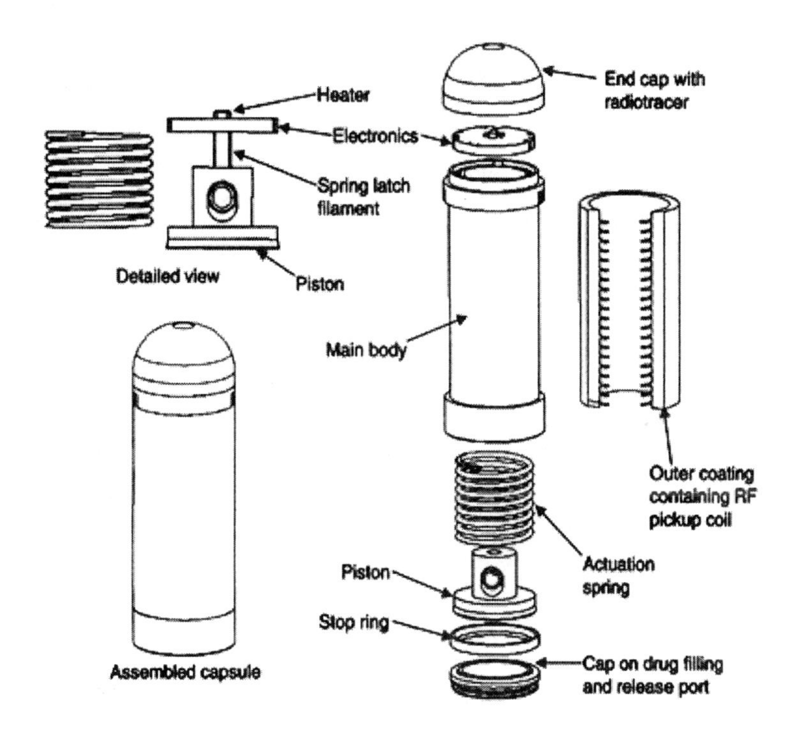

Figure 2. The Enterion Capsule and its Constituent Components

When the capsules arrives at the target site in the GI tract, it is remotely triggered by application of an oscillating electromagnetic field, which is generated over the abdominal cavity by an external radio-frequency (RF) generator. The frequency of the field is 1.8 MHz; which is low enough for negligible absorption of energy by the body tissues, but sufficiently high to induce usable power in a tuned coil receiver (pickup coil) embedded inside the wall of the capsule. The electric current induced by the magnetic field in the receiving coil is fed to a low power (0.0625 W) heater resistor, which is situated within a sealed electronics compartment. The small size of the heater (<1 mm^3) results in a rapid temperature rise in just a few seconds.

The spring latch filament that anchors the piston is in direct contact with the heater. As rapid heat build up occurs, the filament quickly reaches a critical temperature at which point it softens and immediately breaks under the tensile strain of the spring. The energy stored in the spring is only about 0.18 joules; however, the relatively low mass and friction coefficient produces high acceleration of the piston. Once the spring is released, it drives the piston into the drug chamber forcing off the O-ring sealed cap (this force is rapidly dissipated as the cap quickly decelerates in the relatively viscous GI lumenal fluids). Under the continued forward motion of the spring-driven piston, the entire capsule contents are actively expelled into the surrounding GI environment within milliseconds. A restraining (or stop) ring situated near the end of the capsule stops the piston movement. This also maintains the seal and so prevents contact of the electronic components with GI fluids.

As the piston travels the first centimetre immediately after activation, it operates a switch diverting the incoming electrical energy from the heater resistor to a radio-frequency transmitter coil (also embedded inside the capsule wall). This generates a weak radio signal at approximately 500 kHz, which is picked up by an external aerial. Detection of the radio signal confirms that the capsule has opened successfully and is used to initiate the blood sampling protocol.

Table 3. Key Design Features of the Enterion capsule

Essential Attribute	Enterion Design Feature
Biocompatible	Food contact and medical grade plastics used for fabrication of all parts that come into contact with GI-lumenal fluids.
Readily swallowed by volunteers	Round-ended with overall dimensions (32 mm x 11 mm in diameter) comparable to 000-sized gelatin capsule
Easy tracking of capsule location	Short half-life radionuclide sealed inside a radiotracer port allows tracking via a gamma camera.
Suitable for delivering a range of physical forms	Spring-driven piston ensures rapid and complete delivery of particulate, semi-solid, and liquid formulations.
High loading capacity	Drug chamber approximately 1 ml in volume
No drug leakage prior to activation	Compressed silicone O-ring provides a reliable closure system with high seal integrity.
Reliable activation at all intestinal sites	Compressed spring provides an on-board energy source. RF activation frequency selected to avoid absorption by human tissue. Proprietary cap release mechanism based on a unique "rolling" O-ring design.
Feed back signal to confirm drug delivery	RF signal generated on forward motion of the piston.

Phaeton Research developed the Enterion capsule in collaboration with PA Consulting Group (Melbourn, Hertfordshire, UK). It was specifically designed to ensure reliable delivery of both liquid and particulate formulations, especially to the distal colon where there is minimal free water[26] to assist passive drug delivery. The volume ratio of the drug chamber to the overall capsule size was also an important consideration in providing maximum versatility, while ensuring that subjects could swallow the capsule relatively easily without any serious discomfort or gag reflex. A summary of the essential attributes and design features of the Enterion capsule is presented in Table 3.

An in-depth technical description of the Enterion capsule is provided in two published patent applications, which embody both the capsule[27] and the unique features of the cap release mechanism[28].

4. APPLICATIONS OF HDA STUDIES

4.1. Development of New Molecular Entities (NMEs)

Today, product development scientists are regularly faced with complex compounds exiting drug discovery as lead clinical candidates. Many compounds exhibit one or more of the following undesirable biopharmaceutical properties, limiting its intrinsic bioavailability:

- Poor aqueous solubility
- Instability in gastrointestinal fluids
- Complexation with chelating ligands or metal ions present in the GI-tract
- Poor permeability across the intestinal epithelium
- Narrow absorption window in the upper small bowel
- Intestinal efflux
- Gut-wall metabolism

Early characterization and understanding of these properties can point the development scientist towards the right enabling drug delivery technology and thereby accelerate development of the optimal dosage form. Examples where HDA studies can be best applied according to a drug's physiochemical properties and development objectives are presented under the following series of sub-headings:

4.1.1. Poor Aqueous Solubility

Over 40% of the NMEs currently in development are estimated to have poor aqueous solubility[29]. A host of traditional as well as proprietary drug delivery technologies are available to improve the bioavailability of such compounds. Generally, these enabling technologies involve either the classical approach of reducing the drug particle size (e.g. production of stabilised nanoparticles) to enhance dissolution rate, or else utilise non-aqueous solvents and/or surfactants to enhance intrinsic *in vivo* solubility through the creation of a solution or microemulsion[29].

HDA studies are particularly valuable for discriminating between drugs with solubility or permeability-limited absorption. A common study design involves delivery of the drug to one or more intestinal regions in two different forms, a particulate form (such as a powder or granule) and a solution form (or other solubility-enhanced form). If bioavailability is essentially unaffected by the form of the drug regardless of the absorption site, then the drug must be permeability-limited. However, if the solution shows a significantly higher bioavailability at one or more of the absorption sites, then solubility is limiting.

Anecdotal evidence suggests that the number of compounds with sub-optimal aqueous solubility will continue to increase as the industry targets compounds with higher pharmacological activity, potency, selectivity, and specificity[29]. This dilemma may ultimately be overcome through closer cooperation between discovery chemists and development scientists, but in the meantime selection of the enabling technology best suited to the compound in question will remain at the forefront of successful product development. To this end, HDA studies can play an important role. Measurement of drug absorption in key segments of the GI-tract and then comparing these baseline data to the same drug in a solubility-enhanced or dissolution-enhanced formulation can provide valuable data on the maximum threshold of oral bioavailability.

4.1.2. Poor Intestinal Permeability

The biotechnology revolution is providing a growing number of peptides, proteins and genomic drug candidates. These macromolecules are often rapidly degraded in the GI tract and almost always suffer from low intestinal permeability. Achieving acceptable oral dosage forms of such compounds is therefore far from straightforward.

Enabling technologies that may overcome the problem of poor permeability include formulation with custom-synthesised carriers, use of chemical enhancer systems to alter transcellular absorption, and even techniques specifically altering the tight junctions between cells to facilitate paracellular uptake[30]. Although the potential rewards are great, these approaches may be fraught with significant regulatory obstacles since the potential adverse effects of altering the permeability of the intestinal barrier are not yet fully researched.

HDA studies are a valuable tool for determining the oral bioavailability of biotechnology compounds and assessing the performance of potential permeability enhancers. Using a remotely triggered device such as the Enterion capsule, problems caused by gastric instability are overcome by delivering the drug (or drug formulation) directly to the most favourable sites of absorption. This allows rapid and reliable screening of the clinical candidate (with or without any companion delivery technology), which can accelerate key development decisions. For example, a compound suffering a negative human absorption finding could be de-emphasised from development, returned to discovery for re-engineering, or evaluated for alternative routes of delivery such as pulmonary, nasal and parenteral.

4.1.3. Intestinal Metabolism and Efflux

An increasing number of NMEs demonstrate not only complex chemistry, but also low and highly variable pharmacokinetics. CYP3A4 gut-wall metabolism in combination with intestinal transporter systems can provide major development obstacles for NMEs[31].

As more transporter systems are discovered, a growing number of companies are attempting to develop technologies designed to inhibit intestinal metabolism and/or efflux. HDA studies can be used to evaluate the absorption of drugs susceptible to these enzyme systems and directly measure the performance of co-delivered permeation enhancers intended to improve bioavailability.

Targeted drug delivery may also be used to overcome the inhibitory effects of intestinal metabolism and efflux. In a recent HDA study[32], a drug with low oral bioavailability, highly variable pharmacokinetics and a known substrate for CYP3A4 was delivered to the proximal small intestine, distal small intestine and ascending colon. Relative bioavailability (AUC) was measured versus an immediate release reference in a fully randomised, four-way crossover study in eight healthy volunteers. The drug was found to exhibit a higher and less variable bioavailability when delivered to the ascending colon. These results prompted the pharmaceutical company to fast-track development of a modified release formulation and seriously consider parallel development of a colon targeting formulation to maximise drug absorption.

4.1.4. Narrow Absorption Window

The anatomy and physiology of the human intestine means that relatively few drugs are well absorbed throughout the entire GI tract. Indeed, the overwhelming majority show decreasing permeability on descending from the proximal small intestine to the large bowel. Compounds such as L-dopa and ciprofloxacin are classical examples of drugs that exhibit a narrow absorption window in the upper intestine, making development of sustained release formulations extremely challenging.

Several drug delivery companies are developing enabling technologies based on the concept of gastric retention[29], including rapid swelling/slow erosion formulations or multi-particulate, mucoadhesive systems. HDA studies are invaluable for compounds vulnerable to absorption window effects and can provide fundamental data for selecting the optimal enabling technology.

4.2. Life Cycle Management of Established Drugs

Pharmaceutical Companies continue to search for more creative forms of life cycle management (LCM) as they strive to maximise revenue from existing marketed drugs. For oral products, the LCM strategy often involves development of a modified release (MR) dosage form designed to offer clinical, safety or compliance benefits. Examples include extended release (ER) technologies, targeted intestinal approaches, or chronotherapeutic delivery systems. Knowing the rate and extent of drug absorption from different regions of the GI tract is essential to the rational and cost effective development of such products.

The traditional approach for developing an MR delivery system begins with the preparation and *in vitro* dissolution testing of early formulation prototypes to predict the possible drug delivery characteristics *in vivo*. Once acceptable laboratory data are obtained, the pharmaceutical company will typically undertake a straightforward pharmacokinetic study to evaluate actual *in vivo* performance. It is not unusual for the outcome of such studies to be equivocal, necessitating changes to the formulation without a clear indication of how (or even if) drug delivery can be optimised to achieve the development objectives. Modified prototypes are then often evaluated in a series of pharmacokinetic studies, such that the formulation is gradually refined through an iterative process (or until the development is eventually deemed unviable).

A more rational and cost effective approach is to first conduct a straightforward HDA study. Measurement of drug absorption from the key residence regions defines the "art of the achievable" from the outset, thereby enabling informed decision making before valuable development resources are spent in the preparation and testing of complex formulation prototypes. Example applications of HDA studies that have assisted the development of oral MR products are presented under the following series of sub-headings:

4.2.1. Extended (or Sustained) Release

The total gastrointestinal transit time for an extended release (ER) product is about 24 hours. Typical residence times in each of the key regions of the human GI tract are presented in Table 4. This shows that approximately 80% of transit time is normally spent in the colon, with only around 15% in the small intestine[4]. It is therefore essential to have an understanding of drug absorption from the key regions of the human gastrointestinal tract (e.g. proximal jejunum, terminal ileum and colon) before commencing any ER development strategy.

Table 4. Residence Times in Key Regions of the Human Gastrointestinal Tract Following Fasted Dosing of a Tablet Formulation[a]

Region	Residence Time (hrs)		
	Mean	Range	Cumulative Mean
Stomach	0.5	0 – 2	0.5
Jejenum	1.25	0.5 – 2	1.75
Ileum	1.5	0.5 – 2.5	3.25
ICJ	1.25	0 – 12	4.5
Colon	20	0 – 72	24.5

[a] Pooled data from human studies 1985 to present day.

Results from a recent drug absorption study are provided in Figure 3[33]. A proprietary drug was selectively delivered to four target sites (proximal small bowel, mid small bowel, distal small bowel and ascending colon). Bioavailability was demonstrated to be highly dependent on site of delivery in a randomised, five-way crossover study in ten healthy volunteers versus an oral solution reference. Most critically, drug absorption was poor in the distal intestines, which would severely inhibit development of a once-daily product using conventional ER technologies. The availability of human absorption data therefore allowed early definition of possible development strategies.

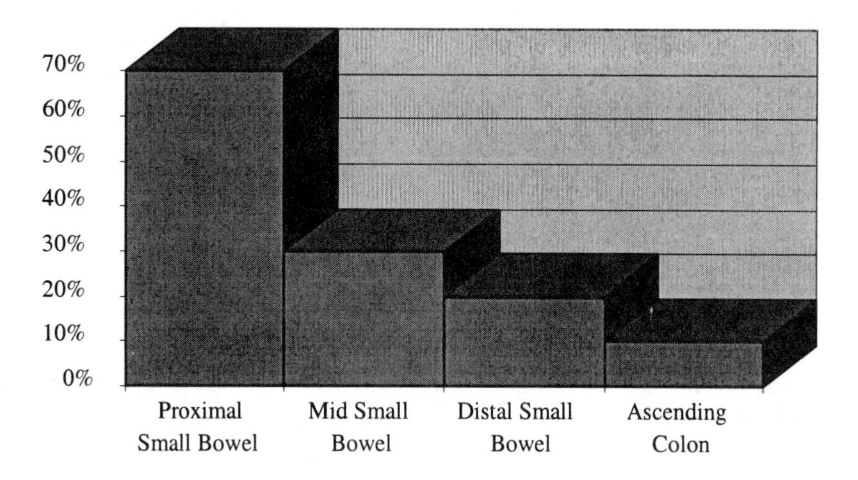

Figure 3. Bioavailability of a proprietary drug in different regions of the human gastrointestinal tract relative to an oral solution (n = 10 healthy volunteers)

4.2.2. Chronotherapeutic Drug Delivery Systems

The recent emergence of chronotherapeutic delivery is demanding ever more sophisticated MR technologies that can control variable drug input rates sympathetic to the body's natural biological rhythms. These technologies often involve a combination of delayed, extended, and pulsed release of the drug.

In a recent HDA study[32,] a drug exhibiting rapid and complete oral absorption was screened as a candidate for pulsed release. The drug (low solubility/high permeability or Class II according to the Biopharmaceutics Classification System) had previously shown good colonic permeability in both the rat and dog pre-clinical models. However, using the Enterion capsule for targeted delivery in humans, only about 30% relative bioavailability was found in both the ascending and descending colon segments. As a result of these findings, the pharmaceutical company abandoned the pulsed-release development and recognised that the earlier pre-clinical data had not been predictive of human absorption.

5. HDA CASE STUDY

HDA studies using the Enterion capsule are performed exclusively by Pharmaceutical Profiles (Nottingham, UK), an early phase development company specialising in the use of imaging technologies to aid decision-making. From launch of the technology (March 2000) to time of writing (August 2002), nearly 800 capsules have been dosed and activated successfully to more than 300 individual subjects. Table 5 presents the use statistics for the capsule, illustrating its excellent versatility and reliability.

Table 5. Use Statistics for the Enterion capsule

Statistic	Cumulative no. of Enterion capsules
Subjects dosed	301
Capsules activated *in vivo*	785
Capsules activated by anatomical region:	
Stomach	90 (11%)
Proximal small bowel	122 (15%)
Distal small bowel	242 (31%)
Ascending colon	214 (27%)
Hepatic flexure	75 (10%)
Transverse colon	32 (4%)
Spleenic flexure	9 (1%)
Sigmoid colon	1 (0.1%)
Capsules activated by formulation type:	
Powder/particulate	330 (42%)
Semi-solid paste/gel	32 (4%)
Solution/suspension	423 (54%)

Most HDA studies are sponsored by major pharmaceutical companies and involve proprietary, early phase compounds; hence opportunities to publish study detailed findings are rare. However, the results of a recent collaboration with Bayer AG (Wuppertal, Germany) to investigate the absorption of faropenem daloxate (FD) have recently been published[34].

Table 6. Pharmacokinetic Parameters of Faropenem (free acid) Following Targeted Delivery of Faropenem Daloxate in Particulate Form Using the Enterion Capsule

PK Parameter (free acid)	IR tablet formulation	Proximal small intestine	Distal small intestine	Ascending colon
AUC (mg.hr/L)[a]	25.80 (1.12)	22.70 (1.17)	20.13 (1.17)	8.64 (1.72)
C_{max}	15.30 (1.44)	11.79 (1.25)	9.96 (1.21)	2.29 (1.65)

[a] Geometric mean (geometric SD)

FD is a novel ester pro-drug of faropenem sodium, a synthetic broad-spectrum oral antibiotic. After oral administration, FD is rapidly absorbed and hydrolysed in serum to the active moiety, faropenem (FAR). The study was designed to compare the bioavailability of FD when delivered to the proximal small bowel (PSB), distal small bowel (DSB) and ascending colon (AC) versus the immediate release (IR) tablet. A single dose (equivalent to 300 mg FAR) was administered in a randomised, four-way crossover study in eight healthy male subjects. The Enterion capsule was loaded with a powder formulation (*i.e.* crushed IR tablet). To further assist with real-time interpretation of each subject's GI anatomy, the water used to administer the Enterion capsule was radiolabelled with ^{99m}Tc-DTPA solution (4 MBq). Following confirmation of capsule activation, blood samples were taken over 24 hours and subsequently analysed using a validated HPLC method with UV detection.

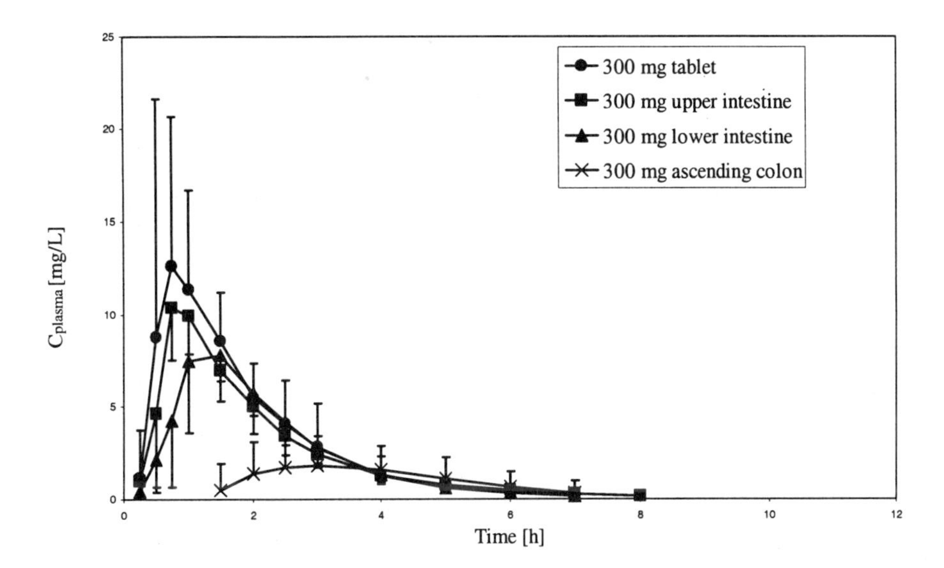

Figure 4. Faropenem (Free Acid) Plasma-Concentration vs Time Profiles [geo.mean/SD] 300mg Tablet vs Capsule Formulations (n=8).

The PK profiles following delivery to the PSB and DSB were similar and comparable to the IR reference tablet (Table 6 and Figure 4). The relative bioavailabilities (AUC) were 87% and 80%, respectively. Significant colonic absorption was also demonstrated for all subjects following delivery to the AC; however, AUC and C_{max} were markedly reduced to 31% and 15%, respectively. The study sponsor considered these results essential in predicting the optimal FD modified release profile and therefore guiding future product development.

6. CONCLUSION

HDA studies are a convenient, reliable and efficient way of gathering fundamental biopharmaceutical data by mapping human intestinal absorption. Delivery of the drug in different physical forms allows assessment of the various barriers to absorption, their contribution to limiting the drug's bioavailability, and whether such barriers might be overcome through selection of an appropriate enabling technology.

Consideration of these data aid key development decisions, such as to fast-track the compound, search for a suitable enabling technology, explore other routes of delivery, or perhaps terminate the development.

7. FUTURE DIRECTIONS

Researchers and regulators continue to search for faster ways to obtain early human ADME (Absorption, Distribution, Metabolism & Excretion) data both safely and ethically. Indeed, advances in plasma analytical techniques have made it possible to conduct pharmacokinetic studies with very low drug doses, well below therapeutic levels. This has led to the concept of "human microdosing" or "cassette dosing", where one or perhaps a cocktail of NMEs are administered to assist in lead optimization. In support of this approach, The European Agency for the Evaluation of Medicinal Products (EMEA) has recently issued a draft position paper outlining the proposed requirements for conducting low dose clinical screening studies[35], where the term "low dose" is defined as less than 1/100 of the (systemic) exposure expected to yield a pharmacological effect in humans. Such studies will allow many more compounds to be screened in humans before candidate selection.

By combining microdosing or cassette dosing with a regional absorption study, it is possible to not only obtain human ADME data, but also gain a very early insight into the biopharmaceutical properties of the compound(s) under study. Studies of this type are likely to prove increasingly popular once the regulatory framework is firmly established.

HDA studies will also advance through developments in commercially available technologies. For example, a likely next generation of the Enterion capsule will be an extended (or perhaps programmable) release version, so that drug can be infused more slowly or pulsed to different intestinal regions. Future versions could even include an on-board pH sensor, miniaturized video camera and/or a system for collecting samples of lumenal fluid.

8. REFERENCES

1. D. Wagner, H. Spahn-Langguth, A. Hanafy, A. Koggel, P. Langguth, Intestinal drug efflux: formulation and food effects. *Advanced Drug Delivery Rev.* 50: S13-31, 2001.
2. E. S. Prosser, The road to drug delivery, *Chem. Ind.* 19:632-634, 2000.
3. G. M. Grass, P. J. Sinko, Effect of diverse datasets on the predictive capability of ADME models in drug discovery, *Drug Discovery Today* 6(12 Suppl): S54-S11, 2001.
4. I. R. Wilding, Site-specific drug delivery in the gastrointestinal tract, *Crit. Rev. Therap. Drug Carrier Syst.* 17: 557-622, 2000.

5. M. Rowland, K. T. Douglas, G. Warhurst, Paracellular permeation, *Eur. J. Pharm. Sci.* 5(Suppl 2): S22-S23, 1997.
6. A. Tsuji, I. Tamai, Carrier-mediated intestinal transport of drugs, *Pharm. Res.* 13 963-977, (1996).
7. K. Amrimori, M. Nakano, Drug exosorption from blood into the gastrointestinal tract, *Pharm. Res.* 15: 371-376, 1998.
8. J. Hunter, B. H. Hirst, Intestinal secretion of drugs: the role of p-glycoprotein and related drug efflux systems in limiting oral drug absorption, *Advanced Drug Delivery Rev.* 25: 129-157, 1997.
9. M. Martin-Facklam, J. Burhenne, R. Ding, R. Fricker, G. Mikus, I. Walter-Sack, W. E. Haefeli, Dose-dependent increase of saquanir bioavailability by the pharmaceutic aid cremophor EL, *Brit. J. Clin. Pharm.* 53(6): 576, 2002.
10. V. J. Wacher, C. Y. Wu, L. Z. Benet, Overlapping substrate specificities and tissue distribution of cytochrome P450 3A and p-glycoprotein: implications for drug delivery and activity in cancer chemotherapy, *Mol. Carcinorg.* 13(3): 129-134, 1995.
11. M. F. Paine, M. Khalighi, J.M. Fisher, D. D. Shen, K. L. Kunze, C. L. Marsh, J. D. Perkins, K. E. Thummel, Characterization of interintestinal and intraintestinal variations in human CYP3A-dependent metabolism, *J. Pharmacol. Exp. Ther.* 283(3): 15562-62, 1997.
12. L. L. von Moltke, D. J. Greenblatt, J. Schmider, J. S. Harmatz, R. I. Shader, Metabolism of drugs by cytochrome P450 3A isoforms; implications for drug interactions in psychopharmacology, *Clin. Pharmacokinet.* 29(suppl): 22-44, 1995.
13. A. P. Li, D. L. Kaminski, A. Rasmussen, Substrates for human hepatic cyctochrome P450, *Toxicology* 104: 1-8, 1995.
14. L. Z. Benet, C. Y. Wu, M. F. Herbert, V. J. Wacher, Intestinal drug metabolism and antitransport processes: A potential paradigm shift in oral drug delivery, *J. Control. Rel.* 39: 139-143, 1996.
15. N. Rouge, P. Buri, E. Doelker, Drug absorption sites in the gastrointestinal tract and dosage forms for site-specific delivery, *Int. J. Pharm.* 136: 117-139, 1996.
16. L. Knutson, B. Odlind, R. Hullgren, A new technique for segmental jejunal perfusion in man, *Am. J. Gastroenterol.* 84(10):1278-1284, 1989.
17. H. Lennernäs, O. Ahrenetet, R. Hallgren, L. Knutson, M. Ryde, L.K. Paalzow, Regional jejunal perfusion, a new in vivo approach to study oral drug absorption in man, *Pharm. Res.* 9(10): 1243-1251, 1992.
18. N. W. Read, M. N. Al Janabi, T. E. Bates, D. C. Barber, Effect of gastrointestinal intubation on the passage of a solid meal through the stomach and small intestine in humans, *Gastroenterology* 84(6):1568-1572, 1983.
19. D. Gardner, R. Casper, F. Leith, I. R. Wilding, The InteliSite capsule: a new easy to use approach for assessing regional drug absorption from the gastrointestinal tract, *Pharm. Technol.* 21: 82-89, 1997.
20. U. Fuhr, A. H. Staib, S. Harder, K. Becker, D. Liermann, G. Schollnahmmer, I. S. Roed, Absorption of ipsapirone along the human gastrointestinal tract, *Br. J. Clin. Pharm.* 38(1): 83-86, 1994.
21. H. Bode, E Brendel, G. Ahr, U. Furh, S. Harder, A. H. Staib, Investigation of nifedipine absorption in different regions of the human gastrointestinal (GI) tract after simultaneous administration of ^{13}C- and ^{12}C-nifedipine, *Eur. J. Clin. Pharmacol.* 50(3): 195-201, 1996.
22. R. A. Casper, L.M. McCartney, W. J. Jochem, A. F. Parr, Medical Capsule Device activated by radiofrequency (RF) signal, US Patent 5,170,801, 1992.
23. V. Mummaneni, W. Doll, E. Sandefer, R. Page, J. Ryo, F. Digenis, S. Kaul, Gamma scintigraphic evaluation of the intestinal absorption of stavudine in healthy male volunteers using InteliSite Capsule, *Pharm. Sci.* 1(4): S-608, 1999.
24. N. J. Clear, A. Milton, M. Humphrey, B. T. Henry, M. Wulff, D. J. Nichols, R. J. Anziano, I. Wilding, Evaluation of the InteliSite capsule to deliver theophylline and frusemide tablets to the small intestine and colon, *Eur. J. Pharm. Sci.* 13: 375-384, 2001.
25. I. R. Wilding, P. H. Hirst, A. L. Connor, Development of a new engineering-based capsule for human drug absorption studies, *Pharm. Sci. Technol. Today.* 3: 385-392, 2000.
26. J. H. Cummings, J. G. Banwell, I. Segal, N. Coleman, H. N. Englyst, G. T. McFarlane, The amount and composition of large bowel contents in man, *Gastroenterology* 98: A408.
27. P. J. Houzego, D. J. Westland, P. N. Morgan, I. R. Wilding, P. H. Hirst. An ingestible device, PCT Patent Application: WO 01/4579 A2, 2001.
28. P. J. Houzego, D. J. Westland, P. N. Morgan, I. R. Wilding, P. H. Hirst, Ingestible device for the release of substances at district locations in alimentary canal, PCT Patent Application: WO 01/45552 A1, 2001.
29. I. R. Wilding, In search of a simple solution for complex molecules, *Scrip Magazine*, May: 9-11, 2001.

30. G. Gwinup, A. N. Elias, E. S. Domurat, Insulin and C-peptide levels following oral administration of insulin in intestinal-enzyme protected capsules, *Gen. Pharmac.* 22:243-246, 1990.
31. I. R. Wilding, Evolution of the biopharmaceutics classification system (BCS) to oral modified release (MR) formulations; what do we need to consider? *Eur. J. Pharm. Sci.* 8: 157-159, 1999.
32. Commercial study undertaken by Pharmaceutical Profiles Ltd using the Enterion capsule (unpublished data).
33. I. R. Wilding, The Enterion capsule: A novel technology for understanding the biopharmaceutical complexity of new molecular entities (NMEs), *Drug Delivery Technology* 1(1): 50.
34. A. L. Connor, H. A. Wray, B. Voith, U. Voigt, J. Nagelschmitz, J. Kuhlmann, I. R. Wilding, Using the Enterion capsule to investigate the absorption of faropenem daloxate delivered in particulate form from different sites in the gastrointestinal tract, Poster Presentation. AAPS Annual Meeting. Denver CO, 2001.
35. Position paper on the non-clinical safety studies to support clinical trials with a single low dose of a compound. CPMP/SWP/2599/02 draft. The European Agency for the Evaluation of Medicinal Products, London, June 2002.

FOOD-DRUG INTERACTIONS: DRUG DEVELOPMENT CONSIDERATIONS

David Fleisher and Laurie Reynolds

1. INTRODUCTION

Regulatory guidance promotes meal effect studies in the industrial development of orally administered drugs. This was prompted by a number of significant cases in which the oral co-administration of marketed drug dosage forms with meals resulted in substantial differences in clinical or toxicological response as compared to administration without a meal (Karim, 1985; Karim, Burns et al., 1985; Karim, Burns et al., 1985; Karim, 1986). One of the more dramatic drug-nutrient interactions with significant clinical repercussions may occur subsequent to administration of a monoamine oxidase inhibitor drug with meal components containing tryptamine. A depressed patient on oral tranylcypromine therapy who ingests cheese as a meal component (Cramer, 1997) illustrates a classic example for such an interaction requiring emergency care for the resultant hypertensive crisis. This extreme example illustrates a direct pharmacological interaction in which a meal component alters clinical response to oral drug administration dictating marketed product insert warnings and special prescription labeling.

Most drug-food interactions are reflected by meal influences on drug pharmacokinetic profiles (Singh, 1999) and are the subject of this chapter. Clinically significant changes in profile correspond to maximum drug plasma levels varying above or below the therapeutic range with meal administration. These changes are most serious for drugs with a narrow therapeutic window for which effective underdose or overdose critically impacts patient health. Such changes dictate prescription labeling to take the drug with or without food and physician and pharmacist counseling on the timing of oral drug administration with respect to meal intake. A more common pharmacokinetic effect is represented by a delay in therapeutic drug concentrations from meal-induced increases

*David Fleisher College of Pharmacy, University of Michigan, Ann Arbor, Michigan 48109-1065.
Laurie Reynolds, Globomax LLC, 7250 Parkway Drive, Suite 430, Hanover, Maryland 21076

Applications of Pharmacokinetic Principles in Drug Development
Edited by Krishna, Kluwer Academic/Plenum Publishers, 2004

in gastric emptying time without a change in systemic availability. While this interaction is not of concern for many drugs, a meal-induced delay in absorption may be a significant clinical event to a patient on an oral analgesic drug hoping to achieve rapid pain relief.

Following initial review of this topic (Welling, 1977), a number of recent review articles on drug-food effects are available in the literature (Schmidt, 2002; Maka, 2000; Jarosz, 2000; Fuhr, 2000; Fleisher, 1999; Brown, 1999). An updating of these reviews is part of the case study component of this chapter. In line with a recent commentary co-authored by this book's editor (Li, 2002), this chapter is focused on drug-food effect considerations for scientists in drug development. From the authors' perspectives, this includes drug developmental scientists in drug candidate screening as well as those involved in regulatory-guided clinical pharmacokinetic studies for oral drug product marketing. With respect to the latter, a number of drug-food effect articles appear in the literature to document that there is not a clinically significant pharmacokinetic food effect on a marketed drug or drug candidate within regulatory guidelines (Zhou, Khalilieh et al., 1999; Cass, Moore et al., 2001; Oscier, Orchard et al., 2001; Delrat, Paraire et al., 2002). In those cases where a significant food effect is observed, mechanistic studies can be helpful. Mechanistically, meal effects on drug pharmacokinetics result from alterations in oral drug delivery component processes including dosage form drug release, intestinal absorption and first-pass metabolism with drug-meal co-administration. For those in drug candidate screening, considerations for assessing the potential for a food effect based on drug candidate physical chemical and pharmacokinetic properties and gastrointestinal (GI) biology constitutes a significant chapter component. At the clinical study end of drug development, the possibility that certain types of meals are less likely to produce a clinically significant food-drug effect (Yeh, Deutsch et al., 1998) may result from mechanistic considerations. Mechanistic understanding of a meal effect on drug pharmacokinetics as a function of drug properties can provide general classification and may permit some projection of the potential for a food-drug interaction in a clinical setting.

2. DRUG PROPERTY CONSIDERATIONS

A starting point for evaluating the potential for a drug-food interaction to alter oral drug candidate absorption is utilization of the Biopharmaceutical Classification System (BCS) (Lobenberg and Amidon, 2000). The system is based on the fact that the maximum rate at which drug is absorbed, the maximum mass flux, is the product of drug permeability across the intestinal barrier and drug aqueous solubility in the intestinal lumen. BCS lends itself to evaluating two primary limitations to oral drug delivery that can be measured in early drug candidate screening; drug solubility in aqueous solution and drug permeability across intestinal cell monolayers. An extended flux definition brings out the potential for food effects on oral drug delivery. Drug flux is the product of the drug concentration gradient driving force from the intestinal lumen to the systemic circulation and the inverse barrier resistance to the flow of drug mass from lumen to systemic blood. The concentration gradient driving force for absorption is, at maximum, equal to the drug solubility, which may be influenced by meal components in the intestinal lumen. The extended definition of permeability (inverse barrier resistance) from *in vitro* transport across a cell monolayer to *in vivo* systemic blood absorption rate contains a first-pass elimination component outside the scope of BCS which can also be influenced by meal intake.

The extent of absorption of a given drug dose is estimated by multiplying the drug flux by the human small intestinal residence time and the intestinal membrane surface area over which the drug can be absorbed. Although small intestinal residence time does not vary appreciably under fed or fasted conditions (Yu, Crison et al., 1996), the rate of gastric emptying and drug delivery to the small intestine is strongly influenced by meals (Dressman, Berardi et al., 1992). While the BCS represents a tremendous simplification of the absorption process, its power, for researchers in drug candidate screening, lies in the partitioning of the absorption process into solubility and permeability rate limits. The system permits a drug absorption rate limit classification on this basis of high versus low permeability and solubility. With these two rate limits, four possible drug categories or classes are defined. In considering the possibility of meal effects on oral drug delivery, the overlaying effects of meals on gastric emptying and the rate and extent of first pass elimination must be brought into play. By incorporating these considerations, meal effect potential on drug candidates on the basis of BCS provides some useful generalizations to initiate this chapter.

In this system, meals might be expected to reduce the rate but not the extent of absorption of class I (high solubility/high permeability) drugs. Given the properties of this class of drugs, by default, gastric emptying should control absorption rate. Since meals slow gastric emptying, drug absorption should be delayed resulting in an increase in t_{max} and a decrease in c_{max}. The analgesic drug, acetaminophen, is a classic example in this category (Nimmo, 1980) and meals delay its absorption without reducing the extent of absorption (Forbes, Sandberg et al., 1998). However, this expectation of a simple absorption delay does not take into account the potential for regional-dependent absorption and/or the influence of rate of absorption on the extent of first-pass metabolism. Both issues will be illustrated in the case study section of this chapter. In addition, class I drugs that degrade in the gastric contents may show decreased bioavailability with food (Schmidt, 2002) since administration with a meal will dictate a greater drug residence time in the stomach. It is also important to note that classification as high permeability and high solubility may depend on administered dose. While regulatory recommendations to drug development are to study meal effects at the highest projected dose, original company dosing recommendations may be expanded in long-term clinical practice. A meal effect may occur when administration greatly exceeds the typical dose range and dose-dependent absorption is observed. This can occur for drugs in class I if normally high carrier-mediated permeability becomes saturated at very high clinical doses (Stewart, Kugler et al., 1993). It may also occur if very high doses produce intestinal concentrations that are in excess of the drug's high solubility and GI residence time is not adequate for complete dissolution of the administered dose (Drusano, Yuen et al., 1992). In the former case a class I drug becomes a class III drug and in the latter case a class I drug becomes a class II drug at high dose.

2.1. Drug Solubility Considerations

Drug discovery activity screening for a particular biological target gauges potency in non-aqueous vehicles resulting in projected lead compounds with low aqueous solubility. BCS class II compounds have low solubility and high permeability. This is typical for lipophilic compounds that are poorly soluble in aqueous media and GI fluids while dissolved drug readily permeates GI membranes. Although there are not many examples of class IV compounds with poor solubility and poor permeability, since limited prospects for oral delivery restrict development and associated literature report, these properties can be observed in candidates with substantial hydrogen bonding potential (Li, Fleisher et al., 2001). These compounds may have high crystallinity resulting in high solid form melting points. For the lipophilic compounds in class II, solubilization by lipid meal components is anticipated. Since lipid membrane permeability of fat-soluble compounds is generally high, drug dissolution rate as a function of low aqueous solubility is rate limiting for intestinal absorption of this class of compounds. Many class II drugs will show higher oral bioavailability when administered with a high-fat meal as compared to administration with water. In fact, for class II drugs with low oral bioavailability, oral delivery can sometimes be improved by formulation with lipid vehicles (Perry and Noble, 1998). A significant increase in absorption will not be observed with a high-fat meal or with formulated solubilization for class IV drugs since permeability will limit absorptive flux even if GI lumenal drug concentration in solution is moderately enhanced (Li, Fleisher et al., 2001).

Class II drugs may be weak acids or bases with pK_a in the range of GI pH variation. Food effects on weak acids in this class are not common (Schmidt, 2002), since ionized drug promotes high solution concentrations in the intestine and permeability of the unionized compound is frequently high enough to shift ionization equilibrium toward favorable absorption. The previously marketed nonsteroidal anti-inflammatory drug (NSAID), bromfenac, may be exceptional in this regard (Forbes, Sandberg et al., 1998). This drug showed a reduced analgesic effect when administered with a meal. This unusual meal effect for a weak acid drug with a pK_a within GI pH variation may be a function of bromfenac's exceptionally low dose as compared to other NSAIDs.

The potential for meal effects on class II weak base drugs with pK_a in the range of GI pH variation is greater than for weak acids. This is a function of their potential to precipitate at intestinal pH or high gastric pH as promoted by some types of meals (Carver, Fleisher et al., 1999). The lower the pK_a and the higher the administered dose, the greater is this possibility. Drug precipitation in the GI tract provides an example of a nonlinear change in a local concentration gradient as a function of administration conditions that is associated with dose-dependent pharmacokinetic behavior. It is around such nonlinearities that the potential for a food-drug interaction is greatest. Factors that influence local GI drug concentrations that may underlie meal effects on drugs showing nonlinear kinetics are discussed in subsequent sections of this chapter.

2.2 Drug Permeability Considerations

Given the absorption pathways of various nutrients and the impact of some meal components on GI physiology, food effects on the oral bioavailability of BCS class III compounds with high solubility and low permeability and class IV compounds with low solubility and low permeability might be expected. Consistent with this expectation, a nutrient effect on drug permeability can often be demonstrated in an isolated intestinal segment (Lu, Sinko et al., 1991; Lu, Thomas et al., 1992). Further, the possibility that a meal would enhance the absorption of a poorly permeable drug candidate might provide dosage formulation insight in drug development. However, clinically significant meal effects on the pharmacokinetics of class III drugs that have low permeability throughout the intestinal tract are not usually observed. Failure to connect a nutrient effect on drug permeability with an effect on clinical pharmacokinetics may be a function of protective gastrointestinal adaptation to limit input of toxic elements with meal administration. Such protective processes may involve coordinated function over the entire GI tract that is not evident when studying permeability in an isolated intestinal segment (Lu, Thomas et al., 1992). However, the potential to salvage poorly absorbed class III drug candidates based on formulated permeability enhancement as related to nutrient absorption warrants further study. The possibility of formulating to increase both permeability and solubility of class IV compounds based on nutrient absorption mechanisms is more remote. The discussion is topically divided on the basis of drug permeability pathways and nutrient classes.

2.2.1. Fat Meals and Lipid Membrane Absorption Pathways

Most drugs are absorbed by membrane-surface partitioning with subsequent diffusion across the lipid membrane barrier to intestinal capillaries. In addition to solubilization of lipophilic drugs, it might be expected that lipid components of a high fat meal would influence the lipid membrane permeability and absorption of some compounds with low permeability. The influence of lipid vehicles on drug permeability has been most often studied in isolated perfusion of animal intestine and some positive influences have been reported (van Hoogdalem, de Boer et al., 1989). Mechanistically, lipid vehicle enhancement of drug permeability has been attributed to changes in intestinal membrane fluidity or viscosity impacting drug partitioning and diffusion across the membrane (Brasitus, Dudeja et al., 1989).

Another potential for lipid meal components to influence the permeability component of absorption would be through enhanced lymphatic absorption (Khoo, Edwards et al., 1599). Since this is a low capacity system compared to the parallel absorption pathway into intestinal blood capillaries, it is projected that only the most lipophilic of compounds with very high partition coefficients would utilize lymphatic absorption to an appreciable extent (Porter and Charman, 2001). Such compounds would fall into BCS class II and not have a permeability limit for absorption. That lipid meal components might influence the extent to which lymphatic absorption of a drug would be utilized may be dictated more by lipid solubilization in the lumenal contents of the small intestine rather than permeability changes.

Clinically significant effects of high-fat meals on the absorption of poorly permeable compounds in classes III and IV have not been reported. As compared to class II compounds, this suggests that current oral formulation techniques will not improve oral bioavailability if the permeability limitation is the result of low membrane partitioning and/or diffusion through intestinal lipid membranes.

2.2.2. Carbohydrate Meals and Paracellular Absorption Pathways

Very few hydrophilic drugs or drug candidates are of small enough molecular size so that the paracellular pathway will result in class I permeability and provide for sufficient absorption for high oral bioavailability. Sodium co-transported monosaccharides have been shown to promote upper intestinal water absorption and paracellular expansion (Hu, Tse et al., 1994; Stevenson, Radulovic et al., 1997). Such nutrient effects might serve to increase the permeability of low molecular weight class III compounds. In isolated intestinal perfusions in animals, sodium co-transported monosaccharides have increased the intestinal permeability of some low-molecular weight hydrophilic drugs (Lu, Thomas et al., 1992, Lu, 1992) but significant effects on drug permeability could not be demonstrated in human intestinal perfusion studies (Fagerholm, Borgstrom et al., 1995; Lennernas, 1995). For a larger class III compound, a positive effect of oral glucose on the systemic availability of a poorly permeable octapeptide drug has been reported (Fuessl, Domin et al., 1987). Given its moderate molecular size, substantial paracellular expansion would be required to improve permeability by this pathway (Hu, Tse et al., 1994). While elevated paracellular permeability might be possible over a limited segment of upper jejunum, the restricted capacity for this effect of monosaccharides over the entire length of small intestine will limit overall enhancement of systemic availability. The glucose-enhanced availability of the oral octapeptide does not appear to be due to a nutrient effect on paracellular transport (Fricker and Drewe, 1995, Fricker, 1996). Consistent with this limitation, there is no convincing clinical evidence that carbohydrate meals will substantially and consistently improve the oral bioavailability of small class III drug candidates.

2.2.3. Protein Meals and Carrier-Mediated Absorption

Since a number of drugs have been shown in isolated animal intestinal perfusion studies to be absorbed by peptide (Friedman and Amidon, 1989) or amino acid carriers (Jezyk N, 1999), it might be expected that a protein meal would reduce their oral bioavailability by nutrient competition for carrier-mediated absorption. Surprisingly, this has not been observed in a clinical setting for either amino acid-like drugs or small peptide-like drugs. This has been the case even when dose-dependent decreases in absorption are shown to be a function of saturation of intestinal carriers (Stewart, Kugler et al., 1993). An amino acid drug example provided in the case study section of this chapter offers some explanations for these observations.

In general, nutrient effects on drug absorption pathways that have been documented in isolated experimental systems have not been shown to have significant clinical impact on the absorption of poorly permeable drugs. However, continued mechanistic studies in this regard may be warranted given the potential therapeutic value of improving the oral bioavailability of poorly permeable drug candidates through novel dosage formulation or targeted prodrug design initiated by drug discovery-development collaborations. Knowledge of the interplay of elementary nutrients with intestinal membrane lipids and proteins, including those controlling tight junction function (Berglund, Riegler et al., 2001) may someday prove useful for oral drug delivery (Ouyang, Morris-Natschke et al., 2002).

3. DRUG ADMINISTRATION FACTORS

The influence of a co-administered meal on oral drug delivery depends on drug and dosage form properties as well as the volume and content of the co-administered meal. Meal effects on drug release from modified release dosage forms have been associated with drug toxicity. Other administration factors that determine the pharmacokinetic impact of meal co-administration involve alterations in local drug concentrations influencing nonlinear pharmacokinetics. At the local concentration gradient drug delivery level, these include the saturation of an intestinal membrane carrier protein or first-pass metabolic enzyme and oral administration of a drug dose above the drug's GI solubility concentration. Local concentrations will certainly be a function of drug dose and administered fluid volume. In addition the caloric load of an administered meal and meal composition can influence GI physiology to effect variations in these local drug delivery concentrations.

3.1 Drug Dosage Form

Among the most clinically significant food effects on oral drug delivery have been in association with the administration of extended or modified release dosage formulations of narrow therapeutic window drugs (Jonkman, 1989; Abrahamsson, 1998; Schug, 2002). Since these formulations typically contain very high doses of high permeability drugs, meal component interactions with formulation components that alter the intended release rate can produce an effective under or overdose. In the extreme, meal-induced "dose dumping" as a bolus drug release process of the entire dose of a modified release formulation can result in toxicity in individual patients (Hendeles, Weinberger et al., 1985).

Less dramatic meal effects on modified release dosage forms are seen as meal-controlled delays in oral drug delivery to the systemic circulation. Since non-disintegrating particles greater than 2 mm in diameter do not empty from the stomach with the gastric liquid contents (Meyer, Dressman et al., 1985), oral drug delivery from dosage forms with these properties may be influenced by meal intake. Such dosage forms are subject to emptying with the timing of the interdigestive migrating motility complex (IMMC) under the control of the circulating gut peptide, motilin (Jadcherla and Berseth, 2002). Following administration of a meal, this complex is disrupted and gastric emptying is influenced by other gut peptides (Walsh, 1994) as regulated by caloric density and intestinal feedback control (Choe, Neudeck et al., 2001). Gastric emptying control by IMMC is not re-established until most of the meal calories have been emptied. As a result, with high caloric density input, gastric emptying of non-disintegrating dosage forms greater than 2 mm in diameter may experience substantial time lags before emptying into the intestine and absorption delays may be reflected in delayed drug plasma levels (Mojaverian, Rocci et al., 1987).

3.2 Drug Dose

Pharmacokinetic nonlinearities correspond to a decrease or increase in oral bioavailability with increasing oral dose. In the former case, either negative or positive meal effects may result (Gidal, Maly et al., 1996; Stevenson, Radulovic et al., 1997). While nutrient competition for intestinal carriers has not been observed to significantly impact clinical pharmacokinetics, decreased bioavailability with increasing dose has been reported for highly water soluble drugs which have a carrier-mediated absorption component indicative of carrier saturation at high dose (Stewart, Kugler et al., 1993). For lipophilic drugs with low aqueous solubility, increasing the oral dose may result in a decrease in oral bioavailability when drug dissolution is not complete within GI residence time at higher administered doses (Dressman, Fleisher et al., 1984). As discussed in section 2.1, a high fat meal may raise the dose range over which dose independent pharmacokinetics are operative for lipophilic drugs by increasing their concentration is solution (Perry and Noble, 1998).

In the later case, an increase in oral bioavailability with increasing dose may indicate saturation of a first-pass elimination component (Li, Stewart et al., 2000). In some cases, meals may influence drug plasma levels when saturable first-pass elimination is a pharmacokinetic characteristic (Fleisher, Li et al., 1999). Just as local drug concentrations at the intestinal mucosa dictate whether intestinal carriers mediating drug absorption are saturated or not, the magnitude of local concentrations at sites of first-pass elimination will determine whether this process is operating under first- or zero-order conditions. Local drug concentrations at sites of saturable absorption and intestinal or first-pass hepatic elimination are a function of drug delivery rates to these sites. Meals can influence these delivery rates by controlling gastric emptying rate, rate of drug exposure to intestinal export proteins and metabolic enzymes, and portal vein drug concentrations.

Drug candidate potency screening in drug discovery is often carried out in non-aqueous media. While clinical dose will likely be related to this measure of potency, aqueous and lipid solubility properties will play a role in defining oral formulation and eventual clinical dose. Drug development researchers in the early stages of drug candidate screening will have limited information on the projected oral clinical dose range of a drug candidate to guide them. However, the observation of any form of saturable phenomena in candidate screening studies may be useful in projecting the potential for a food effect. These include low aqueous solubility combined with high lipid solubility and/or the observation of concentration dependence in epithelial cell monolayer permeability, intestinal or hepatocyte microsomal metabolism or intestinal cell or hepatocyte metabolism screening studies. As more information becomes available on effective and toxic doses in animal studies, the potential for solubility limitations and local concentration-dependent nonlinearities to translate to clinical nonlinear pharmacokinetics and potential food effects may be refined.

3.3 Administered Fluid Volume and Food Effect Control Studies

In many studies of meal effects on drug absorption, a comparison of fasted versus fed conditions is not controlled for administered volume (Li, Stewart et al., 2000). This is certainly consistent with the variability in patient fluid intake with oral drug administration of prescription drugs and may therefore represent a legitimate statistical comparison. While such studies may verify whether or not there is a significant food effect on oral drug bioavailability, fasted-state pharmacokinetics may depend on administered volume for several reasons. When a drug is administered with a noncaloric aqueous liquid, the rate of human gastric emptying of liquid containing dissolved drug and small, drug particles is first-order and dependent on volume load after an initial lag period. When drug is administered with small volumes of fluid in the range of 2 fluid ounces, emptying of the gastric contents is more dependent on the inter-digestive migrating motility complex (IMMC) than is the case when larger volumes in the range of 8 fluid ounces are administered (Oberle, Chen et al., 1990). Thus, emptying of a drug contained in the gastric fluid contents will be more erratic with respect to the time of drug administration for smaller than larger co-administered fluid volumes. In addition, the first-order gastric emptying of larger volumes is more rapid than for smaller volumes for all phases of the IMMC.

Meal administration is typically high volume, but intestinal feedback control dictates that gastric emptying rate and resultant drug delivery to absorption sites in the small intestine is a function of caloric load or density (calories per volume administered). Furthermore, if fasted-state drug administration is conducted under low volume conditions compared to the typical high volumes consumed with meal administration, initial intestinal drug concentrations will be much higher in the fasted-state condition as compared to meal co-administration. In addition, certain meals may dictate significant gastric, intestinal, biliary and pancreatic secretions that can further dilute fed-state drug concentrations as compared to the fasted state. For those drugs that show nonlinear characteristics as a function of local concentrations, such differences in the volume of administration can complicate the interpretation of food-effect studies. It would be advisable to administer the same volume of noncaloric fluid in fasted state studies as the volume of meal administered in fed-state studies. While this only controls for initial conditions, since meals will influence GI fluid absorption and secretions, a more mechanistic comparison is offered when meal effect studies control for volume.

3.4 Caloric Load and GI Residence Time

Careful studies have shown that a drug's small intestinal residence time in human subjects is in the range of 200 minutes whether it is administered with or without a meal (Yu, Crison et al., 1996). However, gastric emptying rate is a function of both volume and caloric density. Caloric feedback signals from the intestine that control gastric emptying have been studied for simple carbohydrate, fat and protein meals. Triggers for these signals include sodium-monosaccharide co-transport (Hunt, 1983), peptide digestion (Jansen, Fried et al., 1994) and chylomicron formation (Glatzle, Kalogeris et al., 2002). The magnitude of the signal and the extent of gastric emptying inhibition are a function of the extent of nutrient and intestinal sensor contact down the length of intestine (Lin, Doty et al., 1989, Lin, 1990) and therefore depends on both digestion and initial caloric load. The pattern of calorie-regulated gastric emptying is different than for volume-controlled gastric emptying (Oberle, Chen et al., 1990) and has been studied in most detail for simple glucose meals (Schirra, Katschinski et al., 1996). With respect to oral drug delivery, calorie intake will result in a different volumetric input rate of gastric-liquid-containing drug into the intestine than for noncaloric liquid intake. This will, in turn, influence differences in rates of co-administered drug delivery to sites of absorption and first pass elimination in the upper intestine with nutrient versus noncaloric input. Caloric control of intestinal drug delivery rates from gastric emptying can result in less variability in oral drug pharmacokinetic profiles compared to drug administration with small volumes of noncaloric fluid. This is the case since the timing of gastric emptying with an IMMC will be highly variable with respect to the time of oral drug administration.

3.5 Meal Type

While different meal types provide a similar rate of fluid delivery form the stomach to small intestine based on caloric density (Raybould, Zittel et al., 1994), intestinal fluid volumes and resultant drug concentrations depend strongly on meal type. Simple carbohydrate meals may result in substantial water absorption in the small intestine (Lu, Thomas et al., 1992) that may, in theory, result in more concentrated drug solutions in the intestinal lumen. Protein meals promote higher intestinal fluid volumes as the result of significant pancreatic secretions (Jansen, Fried et al., 1994) which may, in theory, result in more dilute drug solutions. Even greater intestinal volumes should result from intake of high-fat meals since pancreatic and biliary secretions will be stimulated to a greater extent than with other meal types (Vidon, Pfeiffer et al., 1988). The fed-state balance between intestinal fluid secretion and intestinal water absorption is very much a function of the rate at which complex meals are converted to simple nutrients. Simple carbohydrates tend to be rapidly broken down in the upper GI tract, while protein and fat digestion are slower processes (Vidon, Pfeiffer et al., 1988). The greater extent of upper intestinal water absorption observed with simple carbohydrate meals as compared to protein meals is the result of both differences in the rate of digestion and differences in the absorption pathways of the resultant elementary nutrients. Most monosaccharides are absorbed by sodium-dependent co-transporters, which promote intestinal water absorption (Lu, Thomas et al., 1992). While a number of sodium-dependent transporters support amino acid transport, many intestinal amino acid transporters utilize sodium-independent mechanisms for mucosal absorption (Piyapolrungroj, 2001).

3.6 Meal Viscosity

While this factor is certainly related to meal type based on digestibility, the fact that meal viscosity can be studied independent of caloric input dictates consideration as an additional meal effect factor. A specific example will be provided in this regard in the case study section of this chapter. As opposed to the effect of high fluid volume intake resulting in local gastric pressure distention which speeds gastric emptying, high viscosity intake slows gastric emptying (Reppas, Meyer et al., 1991). If insufficient digestion occurs in the gastric contents to substantially reduce the solution viscosity entering the small intestine, several factors may effect drug absorption following oral administration. First, higher viscosity may increase upper intestinal residence time. In addition, based on the inverse dependence of solute diffusivity on medium viscosity, diffusion of dissolved drug from the intestinal lumen to sites of absorption at the intestinal membrane will be slowed. Finally, high viscosity can slow drug dissolution rate by decreasing solute diffusion away from the solid drug surface (Horter and Dressman, 2001).

3.7 Meal Calcium Content

There is experimental evidence that the stomach controls the rate of soluble calcium delivery to the small intestine. This element of intestinal feedback control has been verified indirectly by observations on the rate of gastric emptying of calcium chelators. The observation of feedback control appears to be an indirect effect of the capacity of calcium chelators to remove ionic calcium from the tight junctions (Hunt, 1983). In isolated intestinal tissue and cell culture, removal of calcium from the tight junctions may result in an increase in paracellular solute transport (Jezyk N, 1999). A defense mechanism to slow the delivery of calcium chelators from the stomach would thus serve a protective feedback-control function. Since a number of elementary nutrients resulting from fat digestion sequester calcium (Hunt, 1983), this may provide a parallel feedback control mechanism to that of caloric content in controlling the rate of gastric emptying. In addition, to the influence of this factor on the rate of gastric delivery to the small intestine and the availability of the paracellular pathway for absorption, calcium is known to bind a number of drugs, like tetracycline, reducing their availability for absorption in the intestine (Poiger and Schlatter, 1979).

4. BIOLOGICAL CONSIDERATIONS

The significant clinical impact of grapefruit juice on the oral bioavailability of several drugs (Ameer and Weintraub, 1997) brought meal component effects on first-pass drug elimination to the forefront of food-effect studies. This is an example of a meal component directly inhibiting the activity of first-pass elimination factors dictating an increase in oral bioavailability. Such inhibitory effects can lead to dramatic increases in oral drug delivery (Edgar, Bailey et al., 1992). Meal input can influence drug first-pass elimination elements through saturation as well as inhibition. It has been stressed that oral drug dosage form administration factors, including co-administered meals, influence drug concentration gradients that are the driving forces for drug absorption. For example, meal lipid solubilization of an orally administered drug may serve to increase class II drug concentration in the GI lumen. Oral bioavailability is further determined by intestinal and hepatic biological components with activities that may or may not be saturated as a function of local drug concentration gradients. By impacting local drug concentration gradients around first-order to zero-order transition points for saturable absorption and first-pass elimination components, meals can exert an effect on oral bioavailability independent of inhibition on first-pass elimination.

4.1 First-Pass Metabolism

Meals can affect both intestinal and hepatic first-pass metabolism. With regard to nutrient component inhibitory effects, phase I pathways have been observed to be impacted to a greater extent than phase II pathways (Chen, Mohr et al., 1996). Since grapefruit juice inhibits cytochrome P4503A4 (CYP4503A4, 3A4) which has been shown to be responsible for the intestinal metabolism of the greatest number of drugs and drug candidates, this elimination element has been the focus of drug-nutrient interaction potential. Drug candidate screening now includes human hepatocyte, microsomal or recombinant enzyme metabolism data. Since CYP4503A4 is a component of this screening, a measure of the potential for intestinal metabolism is also available. Caco-2 monolayers enhanced in 3A4 have been developed to screen drug candidate intestinal metabolism coupled to membrane transport control factors (Paine, Leung et al., 2002). Basic studies to isolate the grapefruit juice component responsible for 3A4 inhibition has generated broader investigations of elementary nutrient factors that might impact this important drug-metabolizing enzyme (Schmiedlin-Ren, Edwards et al., 1997).

Other drug oxidizing enzymes in the intestine (Lown, Bailey et al., 1997) and liver (Mohri, Uesawa et al., 2000) may be influenced by nutrient intake. In animal studies, it was reported that methionine and cysteine inhibited flavin monooxegenase (FMO)-mediated cimetidine sulfoxidation (Lu, Li et al., 1998). This interaction is less important in humans and cimetidine's safety further reduces clinical significance. In addition, the absorption of a narrow therapeutic index drug for which the FMO-mediated sulfoxide metabolite is primary, has been shown to not be influenced by meal intake (Okerholm, Chan et al., 1987). However, for a new drug entity, the screening of a battery of metabolizing enzymes and further basic investigations on elementary nutrient effects on metabolism may yet uncover meal effects on drug metabolizing enzymes other than 3A4.

4.2 Intestinal Export Permeability Limitations

Current research has implicated p-glycoprotein (p-gp) mediated drug export as a factor limiting intestinal permeability of some compounds (Kim, Fromm et al., 1998) and has led to further investigations on the effect of nutrients on this elimination pathway (Fontana, Lown et al., 1999). Inhibition of p-glycoprotein by dietary flavanoid components has been reported (Lo and Huang, 1999). Since p-gp substrates are typically hydrophobic and poorly water soluble, saturation of p-gp is difficult to achieve. However, elevated drug concentrations through meal lipid solubilization could lead to a nonlinear concentration dependence of p-gp mediated drug export (Mueller, Kovarik et al., 1994). For class II compounds that are p-gp substrates, the combined effects of increased permeability via p-gp inhibition with an increase in drug concentration through solubilization by a high fat meal might be projected to substantially increase absorptive flux. Most p-gp substrates are neutral or weak base hydrophobic compounds (Saitoh and Aungst, 1995). Some weak acid drugs are substrates for intestinal multidrug-resistance proteins (MRP) and/or multispecific organic anion transporters such as cMOAT (Guo, Marinaro et al., 2002). There may be additional intestinal membrane proteins mediating drug and/or drug metabolite export yet to be identified that could interact with the nutrient components of a meal (Piyapolrungroj, Zhou et al., 2000).

There is evidence that drug metabolites are substrates for intestinal exporters and it is proposed that intestinal metabolism and mediated mucosal export are coupled processes in intestinal drug elimination (Wacher, Salphati et al., 2001). The function of such coupling, with respect to 3A4 and p-gp, is suggested to promote efficient intestinal elimination (Hochman, Chiba et al., 2000). Since most metabolites are less hydrophobic than their parent drug, they might be weaker substrates for p-gp. Efficient intracellular metabolite production would set up a favorable metabolite-to-drug ratio minimizing potential competition for p-gp export (Li, Amidon et al., 2002). Some inhibitors of p-gp are also inhibitors of 3A4 and these include some compounds that are meal components (Eagling, Profit et al., 1999; Bhardwaj, Glaeser et al., 2002). Given the possibilities of inhibition and saturation of coupled intestinal drug elimination components, the impact of meal intake on first-pass metabolism may be mechanistically complex.

4.3 Sequential Barriers to Systemic Availability

Considering the process of drug absorption as transport across outwardly-directed radial barriers to drug permeability in series from the intestinal lumen to the systemic blood, the potential for saturation of component reaction and transport processes depends on local concentration gradients in these successive compartments. The influence of food on initial driving force for drug absorption should be greatest at the mucosal barrier of the upper small intestine. Drug and nutrient concentrations are highest in this region and nutrient solubilization interactions may occur to produce supersaturated drug solution concentrations. While the effect of a solubilizing vehicle complicates drug candidate metabolism and transport screening investigations, the capacity to study solution drug concentrations above the aqueous solubility may be valuable given the potential for lipid meal drug solubilization (Li, Amidon et al., 2002).

In addition to inhibition of first-pass elimination processes, nutrient intake can influence drug absorption by altering drug concentration gradients and the potential for saturation of these processes. At the mucosal barrier, drug concentrations determine the potential for saturating membrane proteins responsible for mediated absorption or export. This will, in turn, determine cytosolic drug concentrations presented to intercellular sites of potentially saturable metabolism. Resultant portal vein concentrations then determine the possibility of saturating sites of hepatic first-pass metabolism. While the capacity for first-pass elimination by the liver is greater than that of the intestine, intestinal drug handling dictates the rate of drug presentation to hepatic elimination following oral administration.

4.4 Region-Dependent Absorption

Many class II drugs possess sufficient lipophilicity to promote high permeability throughout the small and large intestine (Stevenson, Kim et al., 1997, Hsyu, 1994). However, for some compounds, intestinal absorption and elimination may not be homogeneous or even continuous processes throughout the entire small intestine. This is the case for some drugs that are absorbed by a carrier-mediated process (Barr, Zola et al., 1994) and is generally true for drugs of moderate lipophilicity as a function of a reduction in absorbing surface area in the lower small intestine (Li, Fleisher et al., 2001). For small hydrophilic compounds predominantly absorbed through paracellular pathways, it would be anticipated that permeability would decrease with distance down the intestine since paracellular pathways become more restricted by the tight junctions (Powell, 1981). However, this has not been confirmed with the paracellular marker compound mannitol (Krugliak, Hollander et al., 1994) and regulation of this pathway may be variable as a function of intestinal region (Kinugasa, Sakaguchi et al., 2000). What may prove to be a significant factor in regionally dependent drug absorption are differences in drug elimination as a function of intestinal region (Li, Amidon et al., 2002). Furthermore, resultant differences in the rate of absorption and elimination in different regions of the intestine can dictate variability in the rate of drug presentation to the liver.

Recent studies indicate that region-dependence in the absorption of some drugs may underlie a significant meal effect on systemic drug availability following oral administration (Pao, Zhou et al., 1998, Li, 2002). When drug absorption is better in the upper small intestine than in the mid and lower regions, meal factors that serve to reduce drug availability to the absorbing membrane may produce negative effects on systemic availability. These factors may include drug-binding interactions with meal components or any physical hindrance to drug transport provided by meal intake in the upper intestine that reduces drug availability to sites of absorption. Reduced drug absorption in the upper intestine can result in delivery of lower drug concentrations to sites of first-pass elimination. It is possible that drug administration without meals may provide intestinal concentrations sufficient to saturate first-pass metabolism, while administration with a meal results in drug concentrations below first-pass saturation levels. Based on a limited set of studies, the potential for a negative meal effect is more likely if there is region-dependent absorption.

5. PREDICTION OF DRUG-FOOD EFFECT INTERACTION

Given this information, some testing of drug candidates for region dependent absorption may be of utility in projecting the potential for a negative meal effect on systemic availability. This cannot be carried out on a high throughput basis but rather when a set of new drug candidates has been narrowed to select leads from a half dozen compounds. At this point in drug discovery/development screening, perfusion of animal intestine in situ can be carried out to determine if there is region-dependent absorption. While not a top priority in selecting a lead compound, it should be a drug development consideration since a negative meal effect cannot be easily overcome through dosage form manipulation.

The prediction of a positive food effect may be useful to formulation scientists. In addition to screening drug candidates for aqueous solubility, partition coefficient data or solubility in a lipid or oily vehicle could be obtained for a limited number of candidates. If animal studies indicate drug bioavailability increases with increasing dose, the ratio of drug oil-to-water solubility may provide some indication of the capacity of a high-fat meal to elevate oral bioavailability at a given dose. This information may allow formulators to assess the potential for a high-dose soft gel dosage form (Perry and Noble, 1998).

6. CLINICAL IMPLICATIONS

The most common meal effect observed in clinical pharmacokinetics is an increase in t_{max}, a decrease in C_{max} but no change in oral bioavailability. This is often the case since gastric emptying rate is typically decreased with meal administration resulting in a delay of gastric drug delivery to absorption sites in the small intestine. This is certainly expected for drugs with gastric emptying rate limited absorption, like acetaminophen (Nimmo, 1980), but also occurs for drugs with dissolution or permeability absorption rate limits (Welling, 1996). While not classified as a negative meal effect on oral drug delivery, there may be significant clinical consequences for patients on drug therapy where a rapid onset of action is required.

Negative meal effects (reduced oral drug bioavailability with meal administration) and positive meals effects (increased oral drug bioavailability with meal administration) are of the greatest clinical significance for drugs with a narrow therapeutic window. Previous sections offer region-dependent absorption studies and dose-dependent pharmacokinetic studies in animals as tools to project potential meal effects on oral drug delivery. Region-dependent absorption studies are difficult to carry out in humans (Buch and Barr, 1998) and are not routinely implemented in early drug development studies. However, it is quite possible that predictive region-dependent and dose-dependent studies carried out in animals may not translate to significant meal effects in clinical studies. Such projections are currently based on a very limited set of region-dependent absorption studies and differences in dose-dependent pharmacokinetics between human and animal studies are fairly common. This is not usually due to species differences in absorption. With some limited exceptions including species differences in carrier-mediated absorption (Munck and Munck, 1992) and high-capacity paracellular absorption of small hydrophilic molecules in dogs (He, Murby et al., 1998), intestinal permeability in animals is well correlated with human intestinal permeability (Amidon, Lennernas et al., 1995). In addition, some animal models have been shown to provide gastric emptying patterns similar to humans (Aoyagi, Ogata et al., 1992) permitting animal testing of oral drug formulations. Differences in GI pH between animals and humans may play a role in absorption differences for some weak acid and weak base drugs with pKa in the range of GI pH variation (Lui, Amidon et al., 1986). However, it is likely that species differences in drug metabolism and first-pass elimination underlie most observed differences in animal and human pharmacokinetics following oral drug administration. In these cases, region dependent studies in animals may not be predictive of oral drug delivery in humans resulting in a species difference with regard to a meal effect.

Regulatory recommendations for meal effect studies on oral drug delivery utilize a high-fat, high-caloric-load meal with substantial protein and carbohydrate content. This type of meal will certainly demonstrate the potential for a positive food effect. Given the spectrum of caloric content, this meal recommendation will likely pick up clinically significant negative meal effects. However, a high fat content may dictate GI physiologic response that dampens the effects of other nutrients. In these cases, a clinical study conducted in the United States utilizing a high fat meal might show a difference in food effects on oral drug delivery compared to studies in another country where typical diets are low in fat. This can be further investigated exploring additional meal types as pointed out in regulatory guidance.

7. FDA'S FOOD EFFECT GUIDANCEAND INTERPRETATIONS (FOOD EFFECT BA/BE)

A Draft FDA Guidance for Industry regarding Food-Effect Bioavailability and Fed Bioequivalence Studies was published in October 2001 (FDA, 2001). The guidance is an update of an earlier version with two major revisions. It provides for a waiver of bioequivalence (BE) studies under fed conditions when both test and reference listed drug products are rapidly dissolving, have similar dissolution profiles, and contain a drug substance that is highly soluble and highly permeable. Second, it recommends that an equivalence approach be used for fasted versus fed bioavailability (BA) and fed BE comparisons (see details below).

7.1 Food Effect Studies

7.1.1 Design

A randomized, balanced, single-dose, two-treatment (fed versus fasting for NDAs, and fed versus fed for ANDAs), two-period, two-sequence crossover design is recommended for testing whether there is a food effect. Typically the studies should be carried out in 12 subjects. The study should be tested in the highest dosage strength, unless there are some safety concerns.

7.1.2 The Meal

A high-fat and high-calorie meal is recommended for the test meal. In NDAs, other types of meals may also be chosen for exploratory or label purposes, however, one of the meals should still be a high-fat, high-calorie meal. The meal is to be consumed over 30 minutes with drug product administration immediately following the meal. A specified volume of water (8 ounces) is also to be used for the administration of the drug product. Food is then not allowed for 4 hours post dose, and water is allowed again at one hour after dosing.

7.1.3 Data Analysis

The FDA recommends that an equivalence approach for food-effect BA and fed BE studies, analyzing data using an average criterion be used. Noncompartmental pharmacokinetic exposure parameters should be obtained from the concentration time profiles for the drug product. The area under the concentration time curve (AUC0-inf, and AUC0-t) assesses total exposure to the drug product and peak (maximal) drug exposure (C_{max}) is also assessed. Other parameters to be determined are time to peak exposure (T_{max}), terminal elimination ($T_{1/2}$) and for modified release compounds lag-time (T_{lag}) should also be assessed (FDA, 2000).

Ninety percent confidence intervals should be calculated for the ratio of the population geometric means for the log-transformed C_{max} and AUC parameters between fed and fasted treatments in the case of an NDA and between the test and reference treatments for an ANDA. The parameters of the products are not considered equivalent if the 90% CI are not contained within the equivalence limits of 80%-125% (FDA, 2001). Based on the Guidance the effect of food on T_{max} for an NDA must also be assessed. For an ANDA although there is no criterion that applies to T_{max}, they should be comparable for the test and reference products.

7.2 Interpretations

The FDA Guidance for Industry regarding Food-Effect Bioavailability and Fed Bioequivalence Studies is important in that it provides recommendations for design, data analysis and product labeling and most importantly when food effect studies need to be performed. Although the Guidance recommends that a high-fat, high-calorie meal should be used for assessing if there is a food effect, it does mention that, for a new drug, other meals as well as beverages may also be assessed. Based on the examples presented below for each BCS class, assessment of other types of meals may be important to supplement the label, allowing for better dosage recommendations with regards to the type of meal ingested. These examples may provide a resource for some of the types of interactions expected. For an ANDA if there is a food effect on the reference listed product, it is also necessary to demonstrate this on the test product to determine bioequivalence. In this case if there is a specific food-drug interaction for the reference product, it may also be important to determine if the effect also occurs with the test product, although the Guidance does not specifically recommend this. A high-fat, high-caloric meal in most cases should provide the greatest effects GI physiology, so that systemic drug availability is maximally affected.

The equivalence approach using a two-way crossover study to be performed will offer an advantage to the industry as far as the ease in the study design versus the traditional three-way crossover study that was previously recommended. The parameters allow for the sponsor to make claims with regards to the effect of food on the exposure to the drug.

Regulatory guidance of food effect studies continues to be evaluated. Of interest in this regard is a recent report examining the impact of meal effects on steady state phenytoin levels when switching products (Wilder, Leper et al., 2001, Davit, 2002). While meal effect guidelines are met for both products, the importance of tighter guidelines for narrow therapeutic index drugs showing nonlinear pharmacokinetics may be appropriate in this case (FDA, 1994).

8. CASE STUDIES

Food-effect interaction case studies are provided in this section for each BCS class. These examples are provided to illustrate the principles outlined in the previous sections of this chapter on drug properties, administration conditions and biological considerations.

8.1. Case 1

The class I analgesic drug, acetaminophen, is a neutral molecule possessing high solubility and high permeability. While there are formulation differences in drug release (Ishikawa, Koizumi ET al., 2001), the drug is well absorbed throughout the small intestine (Gradate and Richter, 1994). Since this drug is fairly hydrophilic, high membrane permeability may be due in part to the drug's small molecular size resulting in significant paracellular absorption (Lu, Thomas et al., 1992). Since neither dissolution rate nor permeability limits the drug's absorption, gastric emptying controls acetaminophen's absorption rate (Hie, Hayashi et al., 1998). The drug should be most rapidly absorbed when given with a large volume of water and gastric emptying is a function of volume load. It is more erratically absorbed when taken with a small volume of water since the migrating motility complex will more strongly control the rate of gastric emptying (Oberle, Chen et al., 1990). Administration of the drug with a meal will slow the rate of gastric emptying and slow the rate of absorption of acetaminophen accordingly (Choe, Neudeck et al., 2001, Forbes, 1998). The resultant rate of drug delivery to the intestine and systemic delivery for this class I drug should be dependent on meal caloric density based on intestinal feedback control of gastric emptying. Meal viscosity, as a function of digestibility, can also impact gastric emptying and reduce acetaminophen absorption rate (Reppas, Eleftheriou et al., 1998).

Another class I drug, 5-amino salicylate (5-ASA), although zwitterionic, is predominantly negatively charged at intestinal pH. It is also of low molecular size and absorbed to a significant extent through the leaky paracellular pathway of the upper small intestine. However, lower permeability is observed in the tighter epithelia of healthy lower small intestine (Zhou, Fleisher et al., 1999). This regional absorption difference between 5-ASA and acetamionphen may be due to the fact that acetaminophen is a neutral molecule while 5-ASA is predominantly anionic at intestinal pH. The paracellular pathway is more restrictive to negatively charged small molecules as compared to neutral of positively charged small molecules (Knipp, Ho et al., 1997; Karlsson, Ungell et al., 1999). When 5-ASA is administered as a solution, it is completely and rapidly absorbed in the upper small intestine (Haagen, Nielsen and Bondesen, 1983). Although this drug possesses both high aqueous solubility and high upper intestinal permeability, a negative meal effect is observed in an unformulated administration (Yu, Elvin et al., 1990). While binding to meal components or meal viscosity could play a role since there is regional dependent absorption, gastric emptying rate and first-pass metabolism likely underlie the negative meal effect. Regardless of the mechanism of the negative meal effect, it is important to contrast 5-ASA regional-dependent absorption with acetaminophen that is well absorbed throughout human small bowel.

5-ASA is used to treat inflammatory bowel, a disease state predominantly of the lower small intestine and colon, so local targeting is a desirable drug formulation property. Since systemic delivery from oral administration is associated with toxicity, upper intestinal absorption is not desirable and controlled and delayed release oral dosage forms are utilized to minimize absorption and maximize local targeting to lower inflamed intestinal tissue. In the case of 5-ASA, meal effects on the formulation and release of the drug are more of an issue than is the gastric emptying rate. Current oral controlled release formulations have been designed to minimize significant meal effects on drug release. While critically significant food-effects due to intestinal dose dumping were observed for the narrow therapeutic index drug, theophylline in early pH-sensitive oral controlled-release preparations (Karim, 1986), this has not been observed with modified release –ASA preparations (Hardy, Harvey et al., 1993, Layer, 1995). Although a meal effect is seen with unformulated 5-ASA, likely due to its region-dependent absorption, this is not the case for the modified release dosage form.

8.2. Case 2

The class II drug phenytoin, a weak acid drug with high pK_a, does not show high first-pass metabolism or region-dependent absorption (Stevenson, Kim et al., 1997). It was recently demonstrated in a clinical study that there is no food effect on phenytoin pharmacokinetics within regulatory guidelines in subjects who have been given a 100 mg dose of Dilantin (Cook, Randinitis et al., 2001). While phenytoin possesses low aqueous solubility in the range of 100 μM at intestinal pH, complete dissolution of a 100mg dose within intestinal residence time insures complete absorption. If phenytoin required a significantly higher dose such that dissolution was not complete within intestinal residence time, incomplete absorption might be expected. Since it is a lipophilic drug, a positive effect of a high-fat meal on the absorption of much higher doses of phenytoin would also be anticipated and a greater maximum plasma concentration achieved. However, oral bioavailability is not significantly affected because the elimination phase of this long half-life drug dominates area under the plasma level versus time curve limiting the impact of the absorption phase. This is not the case in the dog, where a high capacity hydroxylation dictates a short (2-3 hour) half-life. In a canine model, a significant increase in oral bioavailability is achieved when the drug is administered with food high in fat content (Fleisher, Lippert et al., 1990). In this case, the absorption phase plays a much greater role in controlling the systemic availability. The increase in drug solubility in meal lipid and the increase in gastric emptying time allows for a greater total extent of drug dissolution in the upper intestine. While there is not significant meal impact at typical phenytoin doses in human, other drugs in this class with high doses and significant first-pass metabolism may show pronounced meal effects as detailed in case 3.

8.3. Case 3

The first HIV-protease inhibitor on the market was the class II saquinavir. The need for treatment with this drug class dictated approval in spite of a low 5% oral bioavailability. This was due to the very low intrinsic solubility and very high first-pass metabolism of this drug. While orally administered as a mesylate salt at 600-800 mg three times a day, this low pK_a weak base drug may dissolve in the acid pH of the stomach but would enter the upper small intestine at concentrations three orders of magnitude above its intrinsic solubility. Such a high level of supersaturation would promote the potential for intestinal precipitation. It was noted early on that administration with a high-fat meal increased oral bioavailability five to ten-fold (Kenyon, Brown et al., 1998) and led to the development of a lipid-melt soft gel capsule formulation that similarly increased oral bioavailability (Perry and Noble, 1998). This tremendous increase in bioavailability as a function of dosage formulation is likely due to a combination of solubilization of the drug in the intestine by lipid meal components and a saturation of some elements of first-pass elimination, particularly in the intestine. Saturation of intestinal 3A4 and p-gp should result in a faster rate of absorption and a higher rate of drug presentation in the portal vein to the liver.

The second HIV-protease inhibitor to be approved for marketing was ritonavir, which has physical-chemical properties similar to those of saquinavir. One goal of ritonavir molecular design was to reduce metabolism and maintain plasma levels. This was achieved by replacing 3A4 metabolizable moieties with thiazole (Kempf, Sham et al., 1998). The drug proved to be a mechanistic inhibitor of 3A4 (Koudriakova, Iatsimirskaia et al., 1998) which is useful in the subsequent combination HIV-protease salvage therapy employed in clinical treatment of some patients failing standard therapy (Kempf, Marsh et al., 1997). The capacity to sustain drug plasma levels permits twice a day ritonavir dosing. This drug is administered as the free base at higher single doses than for other HIV-protease inhibitors. Oral ritonavir administration provides equivalent bioavailability when administered with or without meals. In contrast to saquinavir, this may be partly due to a higher intrinsic solubility and certainly the result of limited first pass metabolism.

The third drug marketed in this class of compounds was indinavir, which showed a decrease in bioavailability when, administered with meals (Yeh, Deutsch et al., 1998). Goals in the molecular design of this drug included the addition of a weak base moiety with higher pK_a to increase its solubility. It is administered as a sulfate salt at a dosing regimen similar to that of saquinavir. As was the goal of this molecular design ploy, it is likely that indinavir achieves higher concentrations in the GI tract and a higher driving force for absorption as compared to saquinavir when administered without meals. The higher intestinal indinavir concentrations compared to saquinavir saturate elements of first-pass elimination resulting in oral indinavir bioavailability ten-fold higher than the initial saquinavir product. However, when the drug is administered with a high-caloric meal, a 60% reduction in indinavir bioavailability is observed. When the drug is administered with a light meal of low caloric density, the meal effect can be minimized (Yeh, Deutsch et al., 1998).

Possible contributions to a negative meal effect on indinavir were investigated in HIV-infected patients as a function of meal type (Carver, Fleisher et al., 1999). Indinavir plasma levels and gastric pH were simultaneously measured as a function of time after oral indinavir administration. In this clinical study, protein meals produced the greatest and most statistically consistent reduction in oral indinavir bioavailability as compared to administration with an equal volume of water. Gastric pH, as measured by radiotelemetry in these patients, showed that the protein meal caused a lengthy (4 hour) pH elevation (around pH = 6 over this time period) as compared to other meal types or drug administration with water. Only slight pH elevations of short duration were observed with the other meals since they offer little buffer capacity to gastric acid secretion. It is suggested that the protein meal will provide the greatest potential for poor dissolution and/or precipitation of indinavir in the stomach as a function of elevated pH.

However, all meal types produced a significant negative meal effect on indinavir oral bioavailability, though not to as great an extent or as consistently from patient-to-patient as the protein meal. Meal types studied in addition to the high-caloric protein meal included high-caloric carbohydrate and high caloric lipid meals as well as a noncaloric viscous meal. It is likely that high caloric density meals, as well as high viscosity meals slow gastric emptying and the rate of drug transport in the intestinal lumen to sites of first-pass elimination to an extent that they are no longer saturated.

Other contributions to the negative meal effect have been investigated in isolated animal and tissue experiments to include influences of intestinal regional differences (Li, Amidon et al., 2002). In the case of indinavir, rat intestinal perfusion studies show high permeability in the upper intestine and dramatically reduced permeability in the lower small intestine. The drug is metabolized by 3A4 in both the upper intestine and liver and the predominant intestinal metabolite is excreted into the intestinal lumen. Interestingly, no metabolism is observed in lower small intestine and metabolism is greatly reduced in the mid-jejunum as compared to the upper jejunum. Indinavir is also a substrate for intestinal p-gp and this may account for its poor permeability in the lower intestine where p-gp exports drug that is absorbed into the enterocyte back to the intestinal lumen. The fact that 3A4 metabolism dominates indinavir elimination in the upper small intestine while p-gp secretion controls its elimination in the lower small intestine permits some mechanistic studies in the rat. Reaction coupled transport in the form of cellular metabolism subsequent to cell entry increases the rate of indinavir absorption into the enterocyte by increasing the concentration gradient driving force for cellular entry. If metabolite export competes with drug export by p-gp, this could promote drug absorption across enterocytes in the upper small intestine while there would be no such competition in the lower small intestine (Li, Amidon et al., 2002).

Continued elimination as the drug moves down the intestine will depend on regional 3A4 and exporter activity as well as on changes in drug concentration down the intestinal tract. Meal effects on the rate of drug delivery to these sites of first-pass elimination might be anticipated to produce alterations in bioavailability. The potential for a positive meal effect from lipid enhanced solubility compared to negative meal effects via slowing delivery to saturable sites of first-pass elimination may also be determined by variation in these elimination factors as a function of intestinal region. Some evidence for this might be gleaned by a comparison of indinavir with nelfinavir, the fourth HIV-protease inhibitor to reach the prescription marketplace. Nelfinavir shows a positive meal effect similar to saquinavir (Li, Stewart et al., 2000). In rat intestinal perfusion of upper jejunum compared to lower ileum, nelfinavir showed no region-dependent permeability as compared to the dramatic regional permeability differences cited above for indinavir (Li, Amidon et al., 2002).

8.4. Case 4

The class III drug, disopyramide, has been on the market for a number of years to treat cardiac arrhythmia. Disopyramide is a weak base drug but in comparison to the HIV-protease inhibitors has a high pKa and is administered at a ten-fold lower dose. Given these properties, disopyramide is in predominantly ionized form over the entire range of GI pH and permeability rather than dissolution rate limits absorption. The drug's oral bioavailability is equivalent whether it is administered with or without meals. In the development of a more potent drug, bidisomide, with similar physical chemical properties to those of disopyramide, a negative meal effect on bidisomide bioavailability was noted. A negative meal effect was observed in rat, dog and human studies and an equivalent 60-70% reduction in bidisomide bioavailability was found in dogs and humans. This negative meal effect was shown to be independent of fat content in clinical studies (Pao, Zhou et al., 1998).

In mechanistic studies in a canine model, it was first observed that only semisolid dog food meals produced a reduction in bioavailability. When different meal types of equivalent calories were administered as liquids, times to reach maximum bidisomide plasma concentrations were observed but there was no reduction in oral bioavailability. This led to a hypothesis that meal viscosity might be playing a role in the negative meal effect in dogs. Given the fact that meal viscosity should be substantially reduced by the time that chyme reaches the mid-jejunum a region-dependent absorption study was carried out in dogs with upper and mid-jejunal access ports. In these studies, bidisomide plasma levels were reduced 60-70% when drug was administered into the mid-jejunal access port as compared to drug administered into the upper-jejunal access port. There was no significant difference in disopyramide plasma levels when administered into either access port (Pao, Zhou et al., 1998).

To mimic the viscosity effect, hydroxypropyl methycellulose (HPMC) meals were developed that had viscosity equivalent to homogenized semisolid dog food meals. Oral administration of bidisomide with the noncaloric HPMC meals produced reductions in bioavailability equivalent to those observed with the dog food meal as compared to drug administration with water. Oral administration of disopyramide with the HPMC meals resulted in delays in peak plasma levels with no reductions in oral bioavailability similar to observations on disopyramide administration with the dog food meals. It is suggested that bidisomide's region-dependent absorption contributes to the negative food effect in dogs. Given that neither drug bind's significantly to meal components, meal viscosity in the upper intestine could contribute to reduced upper intestinal absorption of both drugs. Disopyramide absorption is slowed but bioavailability is not reduced since it is well absorbed in the lower small intestine where the viscosity effect of a meal is no longer a factor (Pao, Zhou et al., 1998). Meal reductions in the upper intestine for bidisomide have a severe effect on oral bioavailability since it is not well absorbed in the lower small intestine. The mechanism underlying this drug difference in regional absorption was not investigated and species differences in drug metabolism limit extension of mechanistic interpretation from the canine studies.

While the experimental data supports a meal viscosity effect on bidisomide in dogs, it is difficult to make this case in humans. The clinical studies showed an equivalent meal effect to reduce bidisomide independent of fat content. These meals are remarkably different in homogenized viscosity with the light meal not much different than water. Nonetheless, the studies indicate that meal viscosity is a sufficient condition to reduce the bioavailability of a drug showing this regional-dependent absorption pattern (Pao, Zhou et al., 1998).

8.5. Case 5

LY303366 is an antifungal drug candidate of relatively large molecular size (1100 Daltons). While it is a weak acid with pKa similar to that of phenytoin, its molecular size is four times greater. Its low oral bioavailability of around 5% was initially attributed to its low aqueous solubility but formulation and soluble prodrug efforts failed to improve bioavailability in animals. In addition, a significant negative meal effect on oral bioavailability was observed. Subsequent studies in Caco-2 studies showed LY303366 possessed poor permeability making this a class IV drug. The poor permeability may be due to a combination of high molecular size and high hydrogen bonding potential. Studies in dogs with intestinal access ports showed a reduced systemic availability from lower versus upper intestinal access ports. Directionality studies across Caco-2 monolayers indicated minimal involvement of mucosal drug export so reduced absorption from lower intestinal administration is likely a result of drug exposure to reduced mucosal surface area (Li, Pao et al., 1998 (supplement)).

Mechanistic studies on the meal effect showed that both liquid fat and liquid protein meals reduced drug bioavailability compared to administration with water. However, this was not the case with simple or complex carbohydrate meals. Since both protein and lipid stimulate canine biliary excretion while carbohydrate does not, the possibility the drug binding to bile components reduced drug absorption was studied. A cholecystokinin antagonist, devazepide, which blocks canine biliary secretion, was injected prior to administering LY303366 with a fat meal. This experimental manipulation failed to reduce LY303366 bioavailability and so the mechanism of this meal effect on this class IV drug remains unresolved (Li, 2001). Reduced drug availability through an interaction with lipid-bile salt mixed micelles is not tenable since liquid protein meals, which also stimulate biliary secretion, still produced a negative meal effect. It is possible that carbohydrate stimulated water absorption promotes and higher drug concentration gradients for absorption as compared to fluid secretion and subsequent dilution of drug concentrations generated by the other two meal types.

8.6. Case 6

The amino acid-like drug, gabapentin, was originally approved for marketing as an antiepileptic drug but has since found a number of additional neurological indications. While it is a gamma amino acid drug, similar to a nutrient alpha amino acid, the drug is zwitterionic and of small molecular size. Given its structure and properties, carrier-mediated absorption through amino acid pathways as well as a parallel paracellular absorption component is expected. It has high solubility and shows sufficiently high permeability in the small intestine to be considered a Class I compound. However, its permeability in Caco-2 cell monolayers is equivalent to that of mannitol (Jezyk N, 1999) indicating a limitation for this screening tool in BCS classification. It is important to note in this discussion, that this drug is not metabolized in humans. Consistent with saturation of the carrier mediated absorption pathway, a decrease in oral bioavailability is observed with increasing oral dose. At low doses, bioavailability is typically in the range of 70% of administered dose. At higher doses often administered in clinical practice, bioavailability levels off at about 35% (Stewart, Kugler et al., 1993). This lower plateau is likely the result of paracellular absorption dominating the saturated carrier-mediated pathway at high dose.

Given this absorption pattern, a negative meal effect particularly with protein meals would be expected since nutrient amino acids would compete with the drug for the carrier pathway. Indeed, a fairly strong inhibition would be expected since isolated tissue studies show the drug to be a much weaker substrate for carrier-mediated absorption than nutrient amino acids. However, a negative meal effect is not observed even in studies where protein meals high in amino acid content are co-administered with gabapentin. In fact, these meals produce a slight but significant increase in maximum plasma concentrations of gabapentin although no significant effect on bioavailability is observed (Gidal, Maly et al., 1996). One possible explanation for this observation is that gabapentin is transported across intestinal mucosa by sodium-independent amino acid exchange transporters. Since nutrient amino acids are favorably and more rapidly absorbed by this pathway compared to the drug, trans-stimulation of gabapentin uptake by the more rapidly absorbed amino acids is possible.

A more potent gabapentin follow-up compound, isobutyl gaba, is under consideration for submission for regulatory approval. It has structure and properties similar to gabapentin, including a lack of human metabolism, and can be administered at a much lower dose than gabapentin. In isolated intestinal experiments in animal tissue, isobutyl gaba appears to be more rapidly absorbed than gabapentin (Jezyk N, 1999). Most importantly, isobutyl gaba absorption contains a significant sodium-dependent component while this is not the case for gabapentin (Piyapolrungroj, Li et al., 2001). This indicates that isobutyl gaba is also absorbed by sodium-dependent amino acid carrier pathways, which are not amino acid exchangers. Depending on drug dose, the relative contributions of sodium-dependent pathways, amino acid exchange and paracellular pathways provide a complex array for potential nutrient interactions. Even at low dose, where amino acid competition for sodium-dependent pathways should produce inhibition the other pathways may serve to reduce a potential negative meal effect.

An additional complication is the regional distribution differences in sodium-dependent versus sodium-independent amino acid absorption pathways. Rat intestinal perfusion and tissue studies indicate that isobutyl gaba is more evenly taken up by intestinal mucosa independent of intestinal region as compared to gabapentin. Since elimination half lives are similar for these two Class I drugs, different absorption patterns as a function of intestinal region may account for the fact that maximum plasma concentrations are achieved much earlier for isobutyl gaba than for gabapentin. While a delay in peak plasma levels is not observed when gabapentin is administered with meals, a food-effect to delay isobutyl gaba absorption might be expected.

9. CONCLUSIONS

Meal effects to delay peak plasma concentrations but not alter oral bioavailability are common and not generally of significant clinical consequence. However, for narrow therapeutic index drugs, reduction in drug plasma levels that remain below minimally effective levels when a drug is administered with a meal could be clinically significant even if there is no decrease in oral bioavailability. Positive meal effects are often observed when class II drugs are administered with high fat meals. These effects may go beyond enhanced absorptive flux if the meal elevates drug concentrations via lipid solubilization that are high enough to produce saturation of first-pass elimination. The potential for this to occur should be considered when a drug candidate shows a greater than proportional increase in drug plasma levels with increasing dose. Positive meal effects can lead to clinical overdose with the possibility of enhancing toxicity for marketed class II drugs with a narrow therapeutic window (Milton, Edwards et al., 1989). However, a positive meal effect on a new drug candidate can be of great utility to formulation scientists for candidates that show low oral bioavailability when administered under fasting conditions (Perry and Noble, 1998). Positive meal effects on low permeability class III and class IV drugs are rarely observed (Fuessl, Domin et al., 1987). Currently, this provides little hope that formulation can benefit the absorption of drugs in these classes. This may change with further mechanistic research of nutrient effects on solute transport through and between intestinal epithelial cells.

Of greatest concern to scientists in drug development is the potential for a negative meal effect since, currently, formulation strategies cannot overcome this problem. While appropriate labeling and health care provider supervision can serve to prevent a patient from taking the drug with a meal, patient compliance and nursing care may be compromised by this restriction. Drug precipitation in the GI tract in the presence of a meal is more likely with a high dose drug. However, any negative meal effect on a low dose drug may produce greater plasma level variability. For drug developmental scientists to project possible negative meal effects for a new drug candidate, nonlinear pharmacokinetics is a potential predictive sign.

10. OPINIONS ON FUTURE DIRECTIONS

Since absorption and first pass-metabolism may be a function of intestinal region and studies have suggested a relationship between negative meal effects and region-dependent absorption, drug development screening of intestinal region dependence may be warranted. When the field of compounds to be developed for a particular therapeutic target is narrowed to a half dozen or so based on other screening and testing, a limited study in animals to test for intestinal region dependent handling may be of value in selecting leads and backups. The extension of animal data to clinical significance is always tenuous. However, based on the minimal studies required and significance of a negative meal effect in the clinics, it would seem a worthwhile venture. The value will certainly relate to the narrowness of therapeutic range and the clinical significance of effective under or over dosage. Extension of BCS to include meal solubilization effects and first-pass elimination limitations to in vivo absorption should be of theoretical value in considering the potential of a co-administered meal to influence drug oral bioavailability.

REFERENCES

Ameer B and Weintraub RA. Drug interactions with grapefruit juice. Clin. Pharmacokinet. 1997; 33(2): 103-21.

Amidon GL, Lennernas H, Shah VP and Crison JR. A theoretical basis for a biopharmaceutic drug classification: the correlation of in vitro drug product dissolution and in vivo bioavailability. Pharm. Res. 1995; 12(3): 413-20.

Aoyagi N, Ogata H, Kaniwa N, Uchiyama M, Yasuda Y and Tanioka Y. Gastric emptying of tablets and granules in humans, dogs, pigs, and stomach-emptying-controlled rabbits. J. Pharm. Sci. 1992; 81(12): 1170-4.

Barr WH, Zola EM, Candler EL, Hwang SM, Tendolkar AV, Shamburek R, et al. Differential absorption of amoxicillin from the human small and large intestine. Clin. Pharmacol. & Ther.. 1994; 56(3): 279-85.

Berglund JJ, Riegler M, Zolotarevsky Y, Wenzl E and Turner JR. Regulation of human jejunal transmucosal resistance and MLC phosphorylation by Na(+)-glucose cotransport. Amer. J. Physiol. 2001; 281(6): G1487-1493.

Bhardwaj RK, Glaeser H, Becquemont L, Klotz U, Gupta SK and Fromm MF. Piperine, a major constituent of black pepper, inhibits human P-glycoprotein and CYP3A4. J. Pharmacol. & Exp. Ther. 2002; 302(2): 645-50.

Brasitus TA, Dudeja PK, Bolt MJ, Sitrin MD and Baum C. Dietary triacylglycerol modulates sodium-dependent D-glucose transport, fluidity and fatty acid composition of rat small intestinal brush-border membrane. Biochim. Biophys. Acta. 1989; 979(2): 177-86.

Buch A and Barr WH. Absorption of propranolol in humans following oral, jejunal, and ileal administration. Pharm. Res. 1998; 15(6): 953-7.

Carver PL, Fleisher D, Zhou SY, Kaul D, Li C and et al. Meal composition effects on the oral bioavailability of indinavir in HIV-infected patients. Pharm. Res. 1999; 16(May): 718-24.

Cass LM, Moore KH, Dallow NS, Jones AE, Sisson JR and Prince WT. The bioavailability of the novel nonnucleoside reverse transcriptase inhibitor GW420867X is unaffected by food in healthy male volunteers. J. Clin. Pharmacol. 2001; 41(5): 528-35.

Chen L, Mohr SN and Yang CS. Decrease of plasma and urinary oxidative metabolites of acetaminophen after consumption of watercress by human volunteers. Clin. Pharmacol. & Ther.. 1996; 60(6): 651-60.

Choe SY, Neudeck BL, Welage LS, Amidon GE, Barnett JL and Amidon GL. Novel method to assess gastric emptying in humans: the Pellet Gastric Emptying Test. Eur. J. Pharm. Sci. 2001; 14(4): 347-53.

Cook J, Randinitis E and Wilder BJ. Effect of food on the bioavailability of 100-mg dilantin Kapseals. Neurol.. 2001; 57(4): 698-700.

Cramer C. Emergency! Hypertensive crisis from drug-food interaction. Am. J. Nurs. 1997; 97(5): 32.

Delrat P, Paraire M and Jochemsen R. Complete bioavailability and lack of food-effect on pharmacokinetics of gliclazide 30 mg modified release in healthy volunteers. Biopharm. & Drug Dispos. 2002; 23(4): 151-7.

Dressman JB, Berardi RR, Elta GH, Gray TM, Wagner JG and et al. Absorption of flurbiprofen in the fed and fasted states. Pharm. Res. 1992; 9(7): 901-07.

Dressman JB, Fleisher D and Amidon GL. Physicochemical model for dose-dependent drug absorption. J. Pharm. Sci. 1984; 73(9): 1274-9.

Drusano GL, Yuen GJ, Morse G, Cooley TP, Seidlin M, Lambert JS, et al. Impact of bioavailability on determination of the maximal tolerated dose of 2',3'-dideoxyinosine in phase I trials. Antimicrob. Agnts & Chemother.. 1992; 36(6): 1280-3.

Eagling VA, Profit L and Back DJ. Inhibition of the CYP3A4-mediated metabolism and P-glycoprotein-mediated transport of the HIV-1 protease inhibitor saquinavir by grapefruit juice components. Br. J. Clin. Pharmacol. 1999; 48(4): 543-52.

Edgar B, Bailey D, Bergstrand R, Johnsson G and Regardh CG. Acute effects of drinking grapefruit juice on the pharmacokinetics and dynamics of felodipine--and its potential clinical relevance. Eur. J. Clin. Pharmacol. 1992; 42(3): 313-7.

Fagerholm U, Borgstrom L, Ahrenstedt O and Lennernas H. The lack of effect of induced net fluid absorption on the in vivo permeability of terbutaline in the human jejunum. J. Drug Target. 1995; 3(3): 191-200.

FDA, Interim guidance for industry: phenytoin/phenytoin sodium capsules, tablets, and suspension in vivo bioequivalence and in vitro dissolution testing. Rockville, MD, U.S. Department of Health and Human Services, Food and Drug Administration, Center for Drug Evaluation and Research. 1994.

FDA, Guidance for industry: bioavailability and bioequivalence studies for orally administered drug products: general considerations. Rockville, Md, U.S. Department of Health and Human Services, Food and Drug Administration, Center for Drug Evaluation and Research. 2000.

FDA, Guidance for industry: Food-effect bioavailability and fed bioequivalence studies: study desgin, data analysis, and labeling. Rockville, Md, U.S. Department of Health and Human Services, Food and Drug Administration, Cneter for Drug Evaluation and Research. 2001.

FDA, Guidance for industry: statistical approaches to establishing bioequivalence. Rockville, Md, U.S. Department of Health and Human Services, Food and Drug Administration, Center for Drug Evaluation and Research. 2001.

Fleisher D, Li C, Zhou Y, Pao LH and Karim A. Drug, meal and formulation interactions influencing drug absorption after oral administration. Clinical implications. Clin. Pharmacokinet. 1999; 36(3): 233-54.

Fleisher D, Lippert CL, Sheth N, Reppas C and Wlodyga J. Nutrient effects on intestinal drug absorption. J. Controlled Release. 1990; 11(1): 41-49.

Fontana RJ, Lown KS, Paine MF, Fortlage L, Santella RM, Felton JS, et al. Effects of a chargrilled meat diet on expression of CYP3A, CYP1A, and P-glycoprotein levels in healthy volunteers. Gastroent.. 1999; 117(1): 89-98.

Forbes JA, Sandberg RA and Bood-Bjorklund L. The effect of food on bromfenac, naproxen sodium, and acetaminophen in postoperative pain after orthopedic surgery. Pharmacother. 1998; 18(3): 492-503.

Fricker G and Drewe J. Enteral absorption of octreotide: modulation of intestinal permeability by distinct carbohydrates. J. Pharmacol & Exp. Ther.. 1995; 274(2): 826-32.

Friedman DI and Amidon GL. Passive and carrier-mediated intestinal absorption components of two angiotensin converting enzyme (ACE) inhibitor prodrugs in rats: enalapril and fosinopril. Pharm. Res. 1989; 6(12): 1043-7.

Fuessl HS, Domin J and Bloom SR. Oral absorption of the somatostatin analogue SMS 201-995: theoretical and practical implications. Clin. Sci. 1987; 72(2): 255-7.

Gidal BE, Maly MM, Budde J, Lensmeyer GL, Pitterle ME and Jones JC. Effect of a high-protein meal on gabapentin pharmacokinetics. Epilepsy Res. 1996; 23(1): 71-6.

Glatzle J, Kalogeris TJ, Zittel TT, Guerrini S, Tso P and Raybould HE. Chylomicron components mediate intestinal lipid-induced inhibition of gastric motor function. Amer. J. Physiol. Gastrointestinal & Liver Physiology. 2002; 282(1): G86-91.

Gramatte T and Richter K. Paracetamol absorption from different sites in the human small intestine. Br. J. Clin. Pharmacol. 1994; 37(6): 608-11.

Guo A, Marinaro W, Hu P and Sinko PJ. Delineating the contribution of secretory transporters in the efflux of etoposide using Madin-Darby canine kidney (MDCK) cells overexpressing P-glycoprotein (Pgp), multidrug resistance-associated protein (MRP1), and canalicular multispecific organic anion transporter (cMOAT). Drug Metab. & Dispos. 2002; 30(4): 457-63.

Haagen Nielsen O and Bondesen S. Kinetics of 5-aminosalicylic acid after jejunal instillation in man. Br. J. Clin. Pharmacol. 1983; 16(6): 738-40.

Hardy JG, Harvey WJ, Sparrow RA, Marshall GB, Steed KP, Macarios M, et al. Localization of drug release sites from an oral sustained-release formulation of 5-ASA (Pentasa) in the gastrointestinal tract using gamma scintigraphy. J. Clin. Pharmacol. 1993; 33(8): 712-8.

He YL, Murby S, Warhurst G, Gifford L, Walker D, Ayrton J, et al. Species differences in size discrimination in the paracellular pathway reflected by oral bioavailability of poly(ethylene glycol) and D-peptides. J. Pharm. Sci. 1998; 87(5): 626-33.

Hendeles L, Weinberger M, Milavetz G, Hill M, 3rd and Vaughan L. Food-induced "dose-dumping" from a once-a-day theophylline product as a cause of theophylline toxicity. Chest. 1985; 87(6): 758-65.

Hochman JH, Chiba M, Nishime J, Yamazaki M and Lin JH. Influence of P-glycoprotein on the transport and metabolism of indinavir in Caco-2 cells expressing cytochrome P-450 3A4. J. Pharmacol. & Exp. Ther. 2000; 292(1): 310-8.

Horter D and Dressman JB. Influence of physicochemical properties on dissolution of drugs in the gastrointestinal tract. Adv. Drug Deliv. Rev.. 2001; 46(1-3): 75-87.

Hu Z, Tse EG, Monkhouse DC, Oh CK and Fleisher D. The intestinal uptake of "enzymatically-stable" peptide drugs in rats as influenced by D-glucose in situ. Life Sci. 1994; 54(25): 1977-85.

Hunt JN. Does calcium mediate slowing of gastric emptying by fat in humans? Am. J. Physiol. 1983; 244(1): G89-94.

Ishikawa T, Koizumi N, Mukai B, Utoguchi N, Fujii M, Matsumoto M, et al. Pharmacokinetics of acetaminophen from rapidly disintegrating compressed tablet prepared using microcrystalline cellulose (PH-M-06) and spherical sugar granules. Chem. Pharm. Bul. 2001; 49(2): 230-2.

Jadcherla SR and Berseth CL. Effect of erythromycin on gastroduodenal contractile activity in developing neonates. J. Ped. Gastro. & Nutrit. 2002; 34(1): 16-22.

Jansen JB, Fried M, Hopman WP, Lamers CB and Meyer JH. Relation between gastric emptying of albumin-dextrose meals and cholecystokinin release in man. Diges. Dis. & Sci. 1994; 39(3): 571-6.

Jezyk N LC, Stewart BH, Wu X, Fleisher D. Transport of pregabalin in rat intestine and caco-2 monolayers. Pharm. Res. 1999; 16(4): 519-26.

Jonkman JH. Food interactions with sustained-release theophylline preparations. A review. Clin. Pharmacokinet. 1989; 16(3): 162-79.

Karim A. Theophylline with food: Theo-24. Am. Pharm. 1985; NS25 (3): 4-5.

Karim A. Effects of food on the bioavailability of theophylline from controlled-release products in adults. J. Allergy & Clin. Immun.. 1986; 78(4 Pt 2): 695-703.

Karim A, Burns T, Janky D and Hurwitz A. Food-induced changes in theophylline absorption from controlled-release formulations. Part II. Importance of meal composition and dosing time relative to meal intake in assessing changes in absorption. Clin. Pharmacol. & Ther.. 1985; 38(6): 642-7.

Karim A, Burns T, Wearley L, Streicher J and Palmer M. Food-induced changes in theophylline absorption from controlled-release formulations. Part I. Substantial increased and decreased absorption with Uniphyl tablets and Theo-Dur Sprinkle. Clin. Pharmacol. & Ther.. 1985; 38(1): 77-83.

Karlsson J, Ungell A, Grasjo J and Artursson P. Paracellular drug transport across intestinal epithelia: influence of charge and induced water flux. Eur. J. Pharm. Sci. 1999; 9(1): 47-56.

Kempf DJ, Marsh KC, Kumar G, Rodrigues AD, Denissen JF, McDonald E, et al. Pharmacokinetic enhancement of inhibitors of the human immunodeficiency virus protease by coadministration with ritonavir. Antimicrob. Agnts & Chemother.. 1997; 41(3): 654-60.

Kempf DJ, Sham HL, Marsh KC, Flentge CA, Betebenner D, Green BE, et al. Discovery of ritonavir, a potent inhibitor of HIV protease with high oral bioavailability and clinical efficacy. J. Med. Chem. 1998; 41(4): 602-17.

Kenyon CJ, Brown F, McClelland GR and Wilding IR. The use of pharmacoscintigraphy to elucidate food effects observed with a novel protease inhibitor (saquinavir). Pharm. Res. 1998; 15(3): 417-22.

Khoo SM, Edwards GA, Porter CJ and Charman WN. A conscious dog model for assessing the absorption, enterocyte-based metabolism, and intestinal lymphatic transport of halofantrine. J. Pharm. Sci. 1599; 90(10): 1599-607.

Kim RB, Fromm MF, Wandel C, Leake B, Wood AJ, Roden DM, et al. The drug transporter P-glycoprotein limits oral absorption and brain entry of HIV-1 protease inhibitors. J. Clin. Invest. 1998; 101(2): 289-94.

Kinugasa T, Sakaguchi T, Gu X and Reinecker HC. Claudins regulate the intestinal barrier in response to immune mediators. Gastroent. 2000; 118(6): 1001-11.

Knipp GT, Ho NF, Barsuhn CL and Borchardt RT. Paracellular diffusion in Caco-2 cell monolayers: effect of perturbation on the transport of hydrophilic compounds that vary in charge and size. J. Pharm. Sci. 1997; 86(10): 1105-10.

Koudriakova T, Iatsimirskaia E, Utkin I, Gangl E, Vouros P, Storozhuk E, et al. Metabolism of the human immunodeficiency virus protease inhibitors indinavir and ritonavir by human intestinal microsomes and expressed cytochrome P4503A4/3A5: mechanism-based inactivation of cytochrome P4503A by ritonavir. Drug Metab. & Dispos. 1998; 26(6): 552-61.

Krugliak P, Hollander D, Schlaepfer CC, Nguyen H and Ma TY. Mechanisms and sites of mannitol permeability of small and large intestine in the rat. Diges. Dis. & Sci.. 1994; 39(4): 796-801.

Lennernas H. Does fluid flow across the intestinal mucosa affect quantitative oral drug absorption? Is it time for a reevaluation? Pharm. Res. 1995; 12(11): 1573-82.

Li C, Fleisher D, Li L, Schwier JR, Stratford RE, et al. Regional-dependent intestinal absorption and meal composition effects on systemic availability of LY303366, a lipopeptide antifungal agent, in dogs. J. Pharm. Sci. 2001; 90(1): 47-57.

Li LY, Amidon GL, Kim JS, Heimbach T, Kesisoglou F, Topliss JT, et al. Intestinal metabolism promotes regional differences in apical uptake of indinavir: coupled effect of P-glycoprotein and cytochrome P450 3A on indinavir membrane permeability in rat. Journal of Pharmacology & Experimental Therapeutics. 2002; 301(2): 586-93.

Li LY, Stewart BH and Fleisher D. Oral delivery of HIV-protease inhibitors. Crit. Rev. Ther. Drug Carrier Syst. 2000; 17(2): 73-99.

Lin HC, Doty JE, Reedy TJ and Meyer JH. Inhibition of gastric emptying by glucose depends on length of intestine exposed to nutrient. Am. J. Physiol. 1989; 256(2 Pt 1): G404-11.

Lo YL and Huang JD. Comparison of effects of natural or artificial rodent diet on etoposide absorption in rats. In Vivo. 1999; 13(1): 51-5.

Lobenberg R and Amidon GL. Modern bioavailability, bioequivalence and biopharmaceutics classification system. New scientific approaches to international regulatory standards. Eur. J. Pharm. & Biopharm. 2000; 50(1): 3-12.

Lown KS, Bailey DG, Fontana RJ, Janardan SK, Adair CH, Fortlage LA, et al. Grapefruit juice increases felodipine oral availability in humans by decreasing intestinal CYP3A protein expression. J. Clin. Invest. 1997; 99(10): 2545-53.

Lu HH, Sinko PJ and Fleisher D. Fed-state effects on zidovudine absorption. AIDS. 1991; 5(7): 907-8.

Lu HH, Thomas J and Fleisher D. Influence of D-glucose-induced water absorption on rat jejunal uptake of two passively absorbed drugs. J. Pharm. Sci. 1992; 81(Jan): 21-25.

Lu HH, Thomas JD, Tukker JJ and Fleisher D. Intestinal water and solute absorption studies: comparison of in situ perfusion with chronic isolated loops in rats. Pharm. Res. 1992; 9(Jul): 894-900.

Lu X, Li C and Fleisher D. Cimetidine sulfoxidation in small intestinal microsomes. Drug Metab. & Dispos. 1998; 26(9): 940-2.

Lui CY, Amidon GL, Berardi RR, Fleisher D, Youngberg C and Dressman JB. Comparison of gastrointestinal pH in dogs and humans: implications on the use of the beagle dog as a model for oral absorption in humans J. Pharm. Sci. 1986; 75(3): 271-4.
Meyer JH, Dressman J, Fink A and Amidon G. Effect of size and density on canine gastric emptying of nondigestible solids. Gastroent. 1985; 89(4): 805-13.
Milton KA, Edwards G, Ward SA, Orme ML and Breckenridge AM. Pharmacokinetics of halofantrine in man: effects of food and dose size. Br. J. Clin. Pharmacol. 1989; 28(1): 71-7.
Mohri K, Uesawa Y and Sagawa K. Effects of long-term grapefruit juice ingestion on nifedipine pharmacokinetics: induction of rat hepatic P-450 by grapefruit juice. Drug Metab. & Dispos. 2000; 28(4): 482-6.
Mojaverian P, Rocci ML, Jr., Conner DP, Abrams WB and Vlasses PH. Effect of food on the absorption of enteric-coated aspirin: correlation with gastric residence time. Clin. Pharmacol. & Ther. 1987; 41(1): 11-7.
Mueller EA, Kovarik JM, van Bree JB, Grevel J, Lucker PW and Kutz K. Influence of a fat-rich meal on the pharmacokinetics of a new oral formulation of cyclosporine in a crossover comparison with the market formulation. Pharm. Res. 1994; 11(1): 151-5.
Munck LK and Munck BG. Variation in amino acid transport along the rabbit small intestine. Mutual jejunal carriers of leucine and lysine. Biochim. Biophys. Acta. 1992; 1116(2): 83-90.
Nimmo WS. Gastric emptying and drug absorption. Pharm. Int. 1980; 1(11): 221-23.
Oberle RL, Chen TS, Lloyd C, Barnett JL, Owyang C, Meyer J, et al. The influence of the interdigestive migrating myoelectric complex on the gastric emptying of liquids. Gastroent. 1990; 99(5): 1275-82.
Okerholm RA, Chan KY, Lang JF, Thompson GA and Ruberg SJ. Biotransformation and pharmacokinetic overview of enoximone and its sulfoxide metabolite. Am. J. Cardiol. 1987; 60(5): 21C-26C.
Oscier D, Orchard JA, Culligan D, Cunningham D, Johnson S, Parker A, et al. The bioavailability of oral fludarabine phosphate is unaffected by food. Hematol. J. 2001; 2(5): 316-21.
Ouyang H, Morris-Natschke SL, Ishaq KS, Ward P, Liu D, Leonard S, et al. Structure-activity relationship for enhancement of paracellular permeability across Caco-2 cell monolayers by 3-alkylamido-2-alkoxypropylphosphocholines. J. Med. Chem. 2002; 45(13): 2857-66.
Paine MF, Leung LY, Lim HK, Liao K, Oganesian A, Zhang MY, et al. Identification of a novel route of extraction of sirolimus in human small intestine: roles of metabolism and secretion. J. Pharmacol. & Exp. Ther. 2002; 301(1): 174-86.
Pao LH, Zhou SY, Cook C, Kararli T, Kirchhoff C, Truelove J, et al. Reduced systemic availability of an antiarrhythmic drug, bidisomide, with meal co-administration: relationship with region-dependent intestinal absorption. Pharm. Res. 1998; 15(2): 221-7.
Perry CM and Noble S. Saquinavir soft-gel capsule formulation: review of its use in patients with HIV infection. Drugs. 1998; 55(3): 461-86.
Piyapolrungroj N, Li C, Bockbrader H, Liu G and Fleisher D. Mucosal uptake of gabapentin (neurontin) vs. pregabalin in the small intestine. Pharm. Res. 1126; 18(8): 1126-30.
Piyapolrungroj N, Zhou YS, Li C, Liu G, Zimmermann E and Fleisher D. Cimetidine absorption and elimination in rat small intestine. Drug Metab. & Dispos. 2000; 28(1): 65-72.
Poiger H and Schlatter C. Interaction of cations and chelators with the intestinal absorption of tetracycline. Naunyn. Schmied. Arch. Pharmacol. 1979; 306(1): 89-92.
Porter CJ and Charman WN. Intestinal lymphatic drug transport: an update. Adv. Drug Deliv. Rev. 2001; 50(1-2): 61-80.
Powell DW. Barrier function of epithelia. Am. J. Physiol. 1981; 241(4): G275-88.
Raybould HE, Zittel TT, Holzer HH, Lloyd KC and Meyer JH. Gastroduodenal sensory mechanisms and CCK in inhibition of gastric emptying in response to a meal. Diges. Dis. & Sci. 1994; 39(12 Suppl): 41S-43S.
Reppas C, Eleftheriou G, Macheras P, Symillides M and Dressman JB. Effect of elevated viscosity in the upper gastrointestinal tract on drug absorption in dogs. Eur. J. Pharm. Sci. 1998; 6(2): 131-9.
Reppas C, Meyer JH, Sirois PJ and Dressman JB. Effect of hydroxypropylmethylcellulose on gastrointestinal transit and luminal viscosity in dogs. Gastroent. 1991; 100(5 Pt 1): 1217-23.
Rhie JK, Hayashi Y, Welage LS, Frens J, Wald RJ, Barnett JL, et al. Drug marker absorption in relation to pellet size, gastric motility and viscous meals in humans. Pharm. Res. 1998; 15(2): 233-8.
Saitoh H and Aungst BJ. Possible involvement of multiple P-glycoprotein-mediated efflux systems in the transport of verapamil and other organic cations across rat intestine. Pharm. Res. 1995; 12(9): 1304-10.
Schirra J, Katschinski M, Weidmann C, Schafer T, Wank U, Arnold R, et al. Gastric emptying and release of incretin hormones after glucose ingestion in humans. J. Clin. Invest. 1996; 97(1): 92-103.
Schmidt LaD, K. Food-drug interactions. Drugs. 2002; 62(10): 1481-502.
Schmiedlin-Ren P, Edwards DJ, Fitzsimmons ME, He K, Lown KS, Woster PM, et al. Mechanisms of enhanced oral availability of CYP3A4 substrates by grapefruit constituents. Decreased enterocyte CYP3A4

concentration and mechanism-based inactivation by furanocoumarins. Drug Metab. & Dispos. 1997; 25(11): 1228-33.

Singh BN. Effects of food on clinical pharmacokinetics. Clin. Pharmacokinet. 1999; 37(3): 213-55.

Stevenson CM, Kim J and Fleisher D. Colonic absorption of antiepileptic agents. Epilepsia. 1997; 38(1): 63-7.

Stevenson CM, Radulovic LL, Bockbrader HN and Fleisher D. Contrasting nutrient effects on the plasma levels of an amino acid-like antiepileptic agent from jejunal administration in dogs. J. Pharm. Sci. 1997; 86(8): 953-7.

Stewart BH, Kugler AR, Thompson PR and Bockbrader HN. A saturable transport mechanism in the intestinal absorption of gabapentin is the underlying cause of the lack of proportionality between increasing dose and drug levels in plasma. Pharm. Res. 1993; 10(2): 276-81.

van Hoogdalem EJ, de Boer AG and Breimer DD. Intestinal drug absorption enhancement: an overview. Pharmacol. & Ther. 1989; 44(3): 407-43.

Vidon N, Pfeiffer A, Franchisseur C, Bovet M, Rongier M and Bernier JJ. Effect of different caloric loads in human jejunum on meal-stimulated and nonstimulated biliopancreatic secretion. Am. J. Clin. Nutr. 1988; 47(3): 400-5.

Wacher VJ, Salphati L and Benet LZ. Active secretion and enterocytic drug metabolism barriers to drug absorption. Adv. Drug Deliv. Rev. 2001; 46(1-3): 89-102.

Walsh JH, Dockery GJ. Gut Peptides. 1st ed. New York: Raven Press New York; 1994.

Welling PG. Influence of food and diet on gastrointestinal drug absorption: a review. J. Pharmacokin. & Biopharm. 1977; 5(4): 291-334.

Welling PG. Effects of food on drug absorption. Ann. Rev. Nutr. 1996; 16: 383-415.

Wilder BJ, Leppik I, Hietpas TJ, Cloyd JC, Randinitis EJ and Cook J. Effect of food on absorption of Dilantin Kapseals and Mylan extended phenytoin sodium capsules. Neurol. 2001; 57(4): 582-9.

Yeh KC, Deutsch PJ, Haddix H, Hesney M, Hoagland V, Ju WD, et al. Single-dose pharmacokinetics of indinavir and the effect of food. Antimicrob. Agnts & Chemother. 1998; 42(2): 332-8.

Yu DK, Elvin AT, Morrill B, Eichmeier LS, Lanman RC, Lanman MB, et al. Effect of food coadministration on 5-aminosalicylic acid oral suspension bioavailability. Clin. Pharmacol. & Ther. 1990; 48(1): 26-33.

Yu LX, Crison JR and Amidon GL. Compartmental transit and dispersion model analysis of small intestinal transit flow in humans. Int. J. Pharm. 1996; 140(Aug 16): 111-18.

Zhou H, Khalilieh S, Lau H, Guerret M, Osborne S, Alladina L, et al. Effect of meal timing not critical for the pharmacokinetics of tegaserod (HTF 919). J. Clin. Pharmacol. 1999; 39(9): 911-9.

Zhou SY, Fleisher D, Pao LH, Li C, Winward B and Zimmermann EM. Intestinal metabolism and transport of 5-aminosalicylate. Drug Metab. & Dispos. 1999; 27(4): 479-85.

GLOBAL REGULATORY AND BIOPHARMACEUTICS STRATEGIES IN NEW DRUG DEVELOPMENT: BIOWAIVERS

Elora Gupta and Elizabeth Yamashita*

1. INTRODUCTION

To minimize cost and development time, proactive strategies and plans must be in place before embarking on pharmaceutical product process improvements, formulation changes and line extensions. With escalating drug development costs and the need for reduced development times, there has been increasing awareness within the regulatory agencies as well as the industry, of the need to develop meaningful, discriminatory *in vitro* tests that will help to assure *in vivo* bioavailability / bioequivalence (BA/BE). Assessing the risk / benefit criteria and global regulatory requirements that may include human studies for formulation or chemistry changes is a key component when determining the ultimate benefit of contemplated product changes.

The ability to request biowaivers, i.e. waiver of the requirement to conduct *in vivo* BA/BE studies, provide the opportunity to achieve the seemingly opposing regulatory goals of assuring BE while minimizing human testing. The United States Code of Federal Regulations (21CFR) provides a strong recommendation that planned BA and/or BE studies be reviewed 'to avoid the conduct of an improper study and unnecessary human research'.[1] This regulatory philosophy serves to improve the efficiency of drug development and the review process by recommending strategies for eliminating unnecessary clinical BA/BE determination. Sound company regulatory strategies utilize various Health Authority Guidances to gain global acceptance in approach and data generated for smoother review and faster approval.

The regulatory strategy for biowaivers is discussed here with a focus on immediate release (IR) oral drug products since the greatest definition and experience is available. Modified release formulations are addressed briefly, and semi solids and topical dosage forms can utilize these same strategies bearing in mind the nuances of the formulations.

* Bristol-Myers Squibb Company, Princeton, NJ, USA

2. CONCEPTS AND BACKGROUND

The concepts of BA and BE are used to better define the *in vivo* performance of a drug product. BA is defined as the rate and extent to which the active substance or therapeutic moiety is absorbed from the pharmaceutical form and becomes available at the site of action. BE is defined as the absence of a significant difference in the rate and extent to which the active ingredient or active moiety in pharmaceutical equivalents of pharmaceutical alternatives become available at the site of action when administered as the same molar dose under similar conditions in an appropriately designed study. Thus BA and BE may be viewed as product quality/performance specifications.[2]

Biopharmaceutic factors related to the oral drug absorption process can be summarized as follows:

(1) Disintegration and dissolution of the solid dosage form to release the drug substance initiates the drug absorption process. The process can be affected by formulation characteristics, pH of the gastric media, presence of physiological components such as bile, precipitation at the absorption site, degradation, and intestinal metabolism. Gastric emptying serves to present the drug in solution at the intestinal surface which is the primary site of absorption for a vast majority of drug substances.

(2) Solubilization of the drug substance is essential prior to systemic absorption and is dependent on the characteristics of the gastrointestinal fluids to which the drug substance will be exposed during transit.

(3) Permeability of the drug substance through the membranes lining the gastrointestinal tract delivers the active pharmaceutical ingredient to the systemic circulation and this process can be influenced by factors such as intestinal transit rate and variable regional permeability.

(4) For modified release formulations, which may deliver drug throughout the gastrointestinal tract, additional considerations include heterogeneity in permeability, lumenal pH, gastric contents, and gut wall metabolism.

Various regulatory Guidances that include BE considerations were available prior to 1995 providing general approaches with the possibility for waiver of *in vivo* studies without specific details of requirements. More recently, many Health Authorities have made great strides in promoting science based practice to the regulatory standards, including BA and BE (Figure 1).[2-15]

A clear evolution of testing requirements that link Chemistry, Manufacturing and Controls (CMC) and BA/BE has better defined the development aspects of new chemical entities and drug formulations from both the *in vivo* and *in vitro* testing standpoints. The continued development of additional CMC Guidances will further clarify scientific aspects relative to regulatory requirements. Thus, a proactive strategy can be applied with strategies for streamlining work while still assuring bioequivalency during API and formulation development.

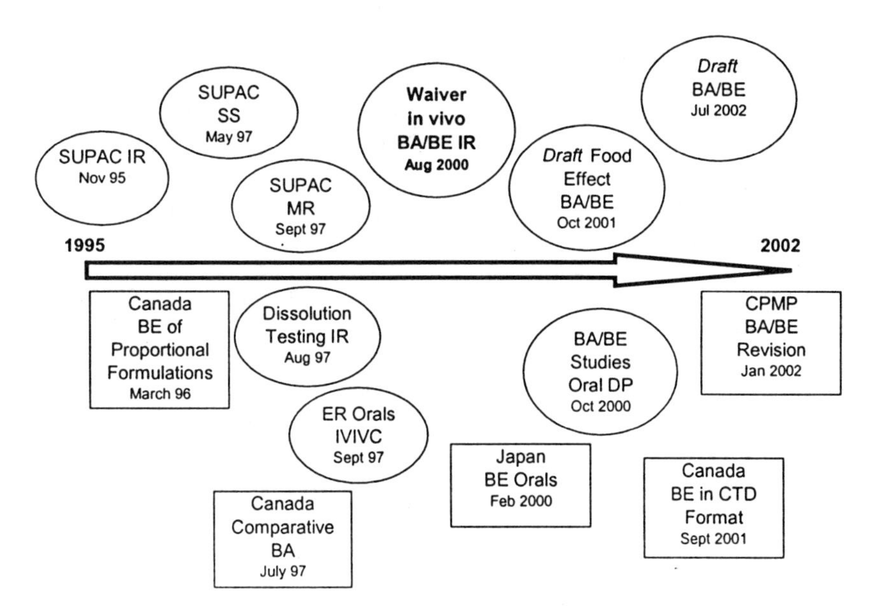

Figure 1. Evolution of regulatory guidances for biowaiver information

SUPAC Guidances (Scale-Up and Post-Approval Change) were the first step in establishing a uniform policy for CMC changes and provided greater consistency and clarity to the regulatory requirements (Figure 1). From a formulation development point-of-view, several changes in component and composition as well as process that are encountered during drug development can be qualified based on *in vitro* dissolution tests thereby reducing the requirements for BE testing. As determined by FDA's Office of Planning and Evaluation, introduction of SUPAC guidelines served to generate substantial cost savings to pharmaceutical companies by more rapid implementation of formulation changes.

Introduction of SUPAC Guidances was complemented by similar Guidances globally. For example, the European guideline on dossier requirements for Type I variations[9] describes a 'common approach to the procedures for variation to the terms of a marketing authorization' that provides assurance that qualification of these changes does not create public health concerns. According to the Australian Drug Evaluation Branch of Therapeutic Goods Administration, changes to a product could be notifiable, self-assessable, or require prior approval.[11] Similar Guidances have been issued by the Canadian and Japanese health authorities.

While SUPAC-IR represented the first application of classifying active pharmaceutical ingredients based on solubility and permeability, Amidon et al. proposed a Biopharmaceutics Classification System (BCS) based on solubility, permeability, and dissolution.[16] BCS focused on the mechanistic basis of the absorption process (relating to solubility, permeability, and dissolution) rather than empirical pharmacokinetic parameters such as Cmax and AUC for establishing BE. BCS accounted for the major determinants of drug absorption for IR drug products, namely, *in vitro* determination of solubility, permeability, and dissolution.

Since the formulation development of IR drug products is an evolutionary process, the primary application of SUPAC and BCS is the development of *in vitro* dissolution specifications that are predictive of *in vivo* performance of the drug product. The basis of BCS is that if two drug products have the same *in vivo* dissolution profile under all *in vivo* lumenal conditions, they will present the same concentration versus time profile at the intestinal surface.

The objective of BCS was to improve the efficiency of the drug development and review process by recommending a class of IR solid oral dosage forms for which BE may be assessed based solely on *in vitro* dissolution results. The FDA Guidance based on BCS was issued in August 2000 describing biowaivers for rapidly dissolving drugs that exhibit high solubility, high permeability, and wide therapeutic index.[7] The December 2000 EMEA guidance also included biowaivers based on BCS principles.[10]

Other US Guidances related to the BCS (e.g. dissolution, BA and BE studies for oral drug product[2, 4]) were also published over this time period and are in harmony with similar Guidances published by ex-US regulatory agencies.

3. STRATEGIC APPROACH

3.1 Immediate Release Solid Dosage Forms

For IR solid dosage forms, the SUPAC-IR guidance defines three levels of changes depending on the impact on formulation quality and performance (Table 1). For each level of change, SUPAC recommends CMC tests, *in vitro* dissolution tests and BE studies taking into account the solubility and permeability of the drug substance and dissolution characteristics of the drug product .[3]

Table 1. Biopharmaceutics qualification requirements for SUPAC levels for IR solid dosage forms

Formulation Changes / Other Considerations[a]	SUPAC Level	Biopharmaceutics Qualification Requirement
Components and Composition		
• Deletion of color and flavor • Excipient change with total additive effect of up to 5%	1	*In vitro* dissolution
• Change in technical grade of excipient • Excipient change with total additive effect of up to 10% • Non-NTI[c] • High permeability, high solubility drugs • Low permeability, high solubility drugs • High permeability, low solubility drugs	2	*In vitro* dissolution
• Qualitative and quantitative (>5%) excipient changes for NTI[c] and for low permeability, low solubility drugs • Excipient change with total additive effect of greater than 10% for non-NTI[c] • Addition of color	3	*In vitro* dissolution and BE study[b]
Process		
• Changes in mixing times, operating speeds within application/validation ranges	1	*In vitro* dissolution
• Changes in mixing times, operating speeds outside application/validation ranges	2	*In vitro* dissolution
• Change type of manufacturing process e.g. from wet granulation to direct compression	3	*In vitro* dissolution and BE study[b]

[a] All levels of site changes and manufacturing equipment changes can be qualified by *in vitro* dissolution
[b] BE study may be waived with an acceptable IVIVC
[c] NTI : Narrow Therapeutic Index

Level 1 changes are those that are unlikely to have any impact on formulation quality and performance. Level 2 changes are those that could have a significant impact on formulation quality and performance while Level 3 changes are those that are likely to have a significant impact on formulation quality and performance. The Guidance calls for BE evaluation for certain Level 3 changes for which biowaivers may be justified by the establishment of *in vitro* - *in vivo* correlation (IVIVC). During drug development, CMC changes that are most frequently linked to formulation development and optimization and might require a BE evaluation are component/composition and process changes. A summary of various component/composition and process changes relating to formulation development, associated SUPAC levels and strategy for qualification are summarized in the Table 1. Categorization and the biopharmaceutics qualification plan in support of the various levels of changes described by SUPAC-IR is in harmony with those described by the European Commission (as 'Type I' or 'Type II' variations) as well as by the Australian Guidelines (as 'notifiable', 'self-assessable' or 'not requiring prior notification').[9, 12]

An extension of the first biopharmaceutic qualification requirements outlined in SUPAC-IR was based on the BCS principles, which divided drugs into 4 classes on solubility and permeability and associated dissolution. High solubility can be claimed when the highest strength dissolves within 250 mL of aqueous media over a pH range of 1 - 7.5 (US) and pH 1 - 8 (EU). The underlying assumption in the solubility determination is that the highest strength corresponds to the highest clinical dose. Scientifically speaking, the highest clinical dose (if not equal to the highest strength) should be considered for solubility classification. When the extent of *in vivo* absorption is at least 90%, a drug product is considered to have high permeability. In addition, a drug product would have rapid dissolution of at least 85% dissolved within 30 minutes in 900 mL or less of simulated gastric fluid (or 0.1 N HCL), pH 4.5 buffer, and simulated intestinal fluid (or pH 6.8 buffer). When comparing test and reference products, both US and EU Guidances require a similarity factor value (F_2) of 50 or more which can be waived if greater than 85% of the drug dissolves within 15 minutes.[4,7,11] Thus, the BCS Guidance allowed for an extension of the SUPAC-IR Guidance by providing a mechanism for qualifying biowaivers for all SUPAC Level 3 changes for high permeability, high solubility drugs based on *in vitro* dissolution data.

By combining the directives provided in the BCS and SUPAC Guidances, an assessment of the risk associated with implementing a biowaiver strategy is provided below. The key component for a successful biowaiver is *in vitro* dissolution; therefore, development of discriminatory *in vitro* dissolution tests that serve as appropriate surrogates of *in vivo* dissolution, is critical for all biowaiver applications.

(A) Class 1 (High Solubility/High Permeability): The rate and extent of absorption for this class of drugs is unlikely to be dependent on drug dissolution or gastrointestinal transit time unless formulation changes affect rate and extent of absorption (e.g. using excipients that may alter gastrointestinal transit). As defined by the BCS Guidance, excipients affecting rate and extent of absorption must be excluded. Drugs in this class present little or no risk for qualification of drug product modifications using the biowaiver strategy based on the BCS guidance as long as dissolution data obtained using the three different media show similarity.

(B) Class 2 (Low Solubility/High Permeability): Dissolution rate may impact the concentration of drug at the membrane surface for this category of drugs. This category of drugs could exhibit pH sensitive dissolution and identification of *in vitro* dissolution tests that are reflective of *in vivo* dissolution is important. Drugs in this class present a moderate risk for qualification of drug product modifications using the biowaiver strategy based on the BCS guidance. An option for biowaivers for this category of drugs with modifications outside SUPAC is the identification of a discriminatory dissolution medium and method and the subsequent development of IVIVC.

(C) Class 3 (High Solubility/Low Permeability): The bioavailability of this category of compounds is primarily dependent on intestinal permeability and should be independent of dissolution unless formulation excipients affect gastrointestinal transit. This category of drugs represent low to moderate risk for qualification of drug product modifications using a biowaiver strategy.

(D) Class 4 (Low Solubility/Low Permeability): Drugs in this class present the most challenge since minor changes in the formulation may affect BA. This category of drugs represent high risk for qualification of drug product modifications using a BCS-based biowaiver strategy.

IVIVC refers to correlations between *in vitro* dissolution, assumed to be the rate limiting step, and *in vivo* input rate (derived from *in vivo* plasma concentration-time profiles) leading to the establishment of the *in vitro* dissolution rate as a measure of product quality and performance. Three primary categories of IVIVC are described in the 1997 FDA guidance[17] as follows:

Level A correlation is a point-to-point linear or non-linear relationship between *in vitro* dissolution and *in vivo* dissolution of the drug (or *in vivo* input rate). This category reflects the entire plasma concentration-time profile and is considered the most useful for regulatory purposes.

Level B correlation utilizes statistical moment analysis where the mean *in vitro* dissolution time is compared to the mean residence time or to the mean *in vivo* dissolution time. Since the entire plasma concentration-time profile is not utilized in this type of correlation, regulatory utility is limited.

Level C correlation is useful in early formulation development as this explores single point relationships between a dissolution parameter (e.g. percent dissolved in 1 h) and a pharmacokinetic parameter (e.g. Cmax or AUC). Multiple Level C correlations can be developed but have limited regulatory advantage.

Since a rate limiting dissolution step is critical for the successful development of IVIVC, Table 2 summarizes the IVIVC expectations for the four BCS classes.[20] While the FDA Guidance on IVIVC addresses modified release formulations, the basic concepts, model development, and validation techniques would be applicable to IR compounds. Overall, BCS Class 2 drugs are the optimal candidates for meaningful IVIVC development.

Table 2. IVIVC expectation based on BCS classification

BCS Class	Solubility	Permeability	IVIVC[a] Expectation
1	High	High	IVIVC may be expected if dissolution is slower than gastric emptying rate
2	Low	High	IVIVC is expected if *in vitro* dissolution rate is similar to *in vivo* dissolution rate
3	High	Low	Permeability is rate determining and limited or no IVIVC with dissolution rate
4	Low	Low	IVIVC not likely

[a] *in vitro- in vivo* correlation

Apart from the SUPAC and BCS Guidances, another option for *in vitro* BE qualification can be applied to lower strengths of drugs by establishing 'proportional similarity' between the higher and lower strengths (21CFR 320.22).[2,18] Proportional similarity between different strengths is described in the FDA BA/BE Guidance using two definitions.[2] Definition 1 is applicable when a common granulation is used in the preparation of the strengths whereby all active and inactive ingredients are in exactly the same proportion. Alternatively (definition 2), proportional similarity is achieved when that the total weight of the strengths remains nearly the same, same inactive ingredients are used and the change in any strength is obtained by altering the amount of active ingredient and one or more of the inactive ingredients. The latter definition pertains to high-potency drugs (5 mg dose). Proportional similarity justification is accepted by Canada,[13] Europe,[10] Japan,[15] and Australia[12] and therefore serves as a useful resource for lowering regulatory risk for line extensions for marketed products.

An exception to the use of any of the strategies outlined in this section is the applicability to narrow therapeutic range (NTR) or narrow therapeutic index (NTI) drugs.[7,19] These are defined as drug products 'containing certain drug substances that are subject to therapeutic drug concentration or pharmacodynamic monitoring, and/or where product labeling indicates a narrow therapeutic range (or ratio) designation'. The US guidance specifically excludes narrow therapeutic range drugs from a biowaiver based on BCS. The EU guidance has a similar stipulation for products containing approved active substances requiring that data be provided to demonstrate 'non-critical therapeutic range'.[10] Generally, *in vitro* dissolution qualification for SUPAC Level 1 changes to NTI drugs are acceptable.

Due to the co-evolution of science and regulations, the acceptance of science-based approaches enables development of biowaiver strategies for compounds that do not meet classic definitions outlined in the Guidances. The focus of Guidances has changed from being empirical to being mechanism and risk-based for regulatory decision making. Strategies can be developed to overcome specific product characteristics (e.g. NTI status) and provide continued assurance of BE by applying the scientific concepts supporting the waiving of *in vivo* studies. When unable to follow the Guidances closely, the concepts must be supported by compelling scientific data and the strategies should be presented to the appropriate Health Authority to support a biowaiver.

A decision tree for BE qualification of changes associated with formulation development is presented in Figure 2. The scheme is based on a rational combination of strategies derived from various Guidances related to BA/BE qualification. The decision tree takes into account ex-US regulatory perspectives and hence is applicable to global registration. The strategies outlined in the decision tree are applicable at all stages of drug development and marketing and are summarized below.

Phase I: Identification of BCS class is important at the early stages of drug development for prospectively evaluating qualification plans to support formulation transitions. A common strategy in the formulation development process is the initiation of early Phase I clinical studies utilizing a drug-in-bottle or drug-in-capsule presentations with the final (commercial) formulation being available prior to the proof-of-principle studies (PoP). This approach is particularly attractive for BCS Class 1 and Class 3 compounds for which dissolution is not a rate limiting step and demonstration of rapid or similar dissolution would be sufficient to link the phase I formulation to the solid dosage formulations that will be used in PoP studies. In addition, optimization and finalization of active pharmaceutical ingredient properties (e.g. polymorphism, particle size, solvates, and hydrates) at this stage would be critical to minimize the number of BE studies for poorly soluble compounds such as BCS Class 2 and 4.

Phase II/III: Changes in formulations, even after synthesis route finalization, can occur due to commercial and scale-up issues. Biowaiver approaches at this stage of drug development would be applicable to establishing links between early formulations used in phase I and formulations used in (a) pivotal clinical studies (b) long-term stability studies, and (c) launch. A combination of BCS and SUPAC Guidances could be used for qualification of subtle formulation changes as well as blinded comparators that are used in comparative clinical studies.

Phase IV: Biowaiver applications are an integral part of life cycle management options for marketed products and include qualification of new formulations, new strengths or incorporating changes to keep pace with changing standards.

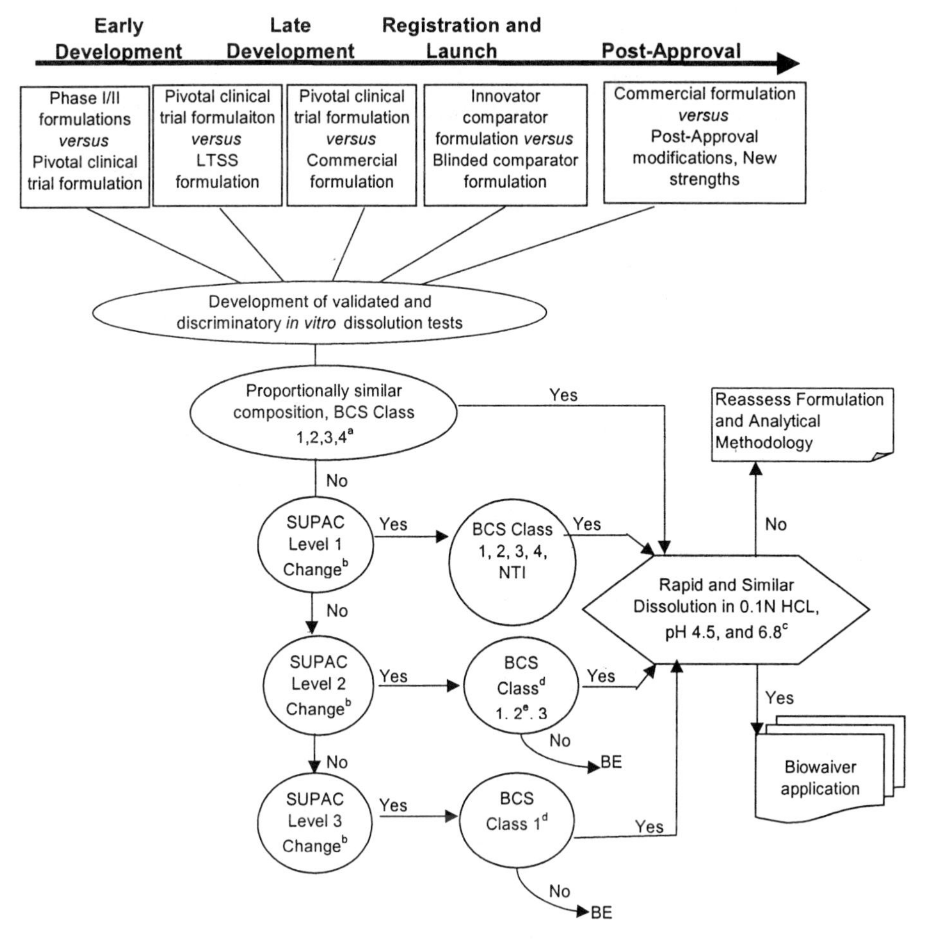

Figure 2. Decision tree for biowaiver application during new drug development and marketing

[a] This strategy is applicable to US, Canada, Europe, Japan, and Australia

[b] Justification of biowaivers based on SUPAC strategy would likely be in compliance with requirements for EU Type I variation and EMEA BA/BE Note for Guidance

[c] Dissolution similarity in only one media is sufficient to qualify a SUPAC Level 1 change

[d] BCS refers to drugs that are not in the narrow therapeutic index (NTI) category. Thorough scientific justification and consultation with regulatory agencies recommended for biowaiver justifications of NTI drugs

[e] SUPAC Level 3 changes to a BCS Class 2 drug can be qualified without a BE study if IVIVC is established

3.2 Modified Release Solid Dosage Forms

In vitro dissolution testing is also used extensively for quality control as well as biowaiver purposes for modified release dosage forms. An established basis for the utilization of *in vitro* dissolution testing as a surrogate for *in vivo* BE testing is the development of IVIVC. The value of this tool in formulation development was highlighted by the 1997 FDA Guidance where IVIVCs could be used to qualify preapproval and postapproval changes in formulation, equipment, manufacturing process or in the manufacturing site, thereby leading to reduced regulatory burden.[17] A summary of the changes that can be qualified by the use of IVIVC is provided in Table 3. It is important to note that for NTIs, use of an IVIVC regulatory strategy should be pursued only with proper consultation with the Health Authorities.

IVIVCs can serve as a basis for biowaivers for changes in the manufacturing of a drug product under the following categories:

- *Biowaivers without an IVIVC:* For SUPAC-MR Level 1 changes to lower strengths provided BE has been demonstrated for the highest strength and dose proportionality has been established.

- *Biowaivers using an IVIVC (Non-narrow therapeutic index drugs)*: For SUPAC-MR Level 3 manufacturing site and nonrelease controlling excipient changes. Also applicable to changes in highest and lower strengths if the strengths have proportional similarity, new strengths, and changes in release controlling excipients.

- *Biowaivers using an IVIVC (Narrow therapeutic index drugs)*: For SUPAC-MR Level 3 process changes and changes in release controlling excipients, changes in highest strength or lower strengths or for new strengths provided there is proportional similarity. These biowaivers require an IVIVC for which an external predictability (discussed in the following section) has been established.

- *Biowaivers when in vitro dissolution is independent of dissolution test conditions*: Biowaivers can be granted to a non-NTI or NTI drug provided IVIVC has been established with one formulation/release rate.

- *Biowaivers where IVIVC is not recommended:* For new formulations with a different release mechanism, new strengths that have not been shown to be safe and effective, formulation change that might substantially affect drug absorption.

Table 3. Biopharmaceutics qualification requirements for various SUPAC levels for MR solid dosage forms

Formulation Changes / Other Considerations	SUPAC Level	Biopharmaceutics Qualification Requirement
Components and Composition -NRC[a] Excipient		
• Deletion of color and flavor • Excipient change with total additive effect of up to 5%	1	*In vitro* dissolution
• Change in technical grade of excipient • Excipient change with total additive effect of up to 10%	2	*In vitro* dissolution
• Excipient change with total additive effect of greater than 10%	3	*In vitro* dissolution and BE study[c]
Components and Composition -RC[b] Excipient		
• Excipient change with total additive effect of up to 5%	1	*In vitro* dissolution
• Change in technical grade of excipient • Excipient change with total additive effect of up to 10%	2	Non-NTI : *In vitro* dissolution NTI : *In vitro* dissolution and BE study[c]
• Addition or deletion of RC excipients • Excipient change with total additive effect of greater than 10%	3	*In vitro* dissolution and BE study[c]
Manufacturing site		
• Changes within the same facility or change to a contiguous facility	1, 2	*In vitro* dissolution
• Change in site to a different campus	3	*In vitro* dissolution and BE study[c]
Process		
• Changes in mixing times, operating speeds within application/validation ranges	1	*In vitro* dissolution
• Changes in mixing times, operating speeds outside application/validation ranges	2	*In vitro* dissolution
• Change type of manufacturing process e.g. from wet granulation to direct compression	3	*In vitro* dissolution and BE study[c]

[a] Non-Release Controlling
[b] Release Controlling
[c] BE study may be waived with an acceptable IVIVC

4. METHODOLOGIES

The vast majority of drugs that are in development or have been approved are solid immediate release (IR) dosage forms and this section focuses on the regulatory requirements for the biopharmaceutics qualification of IR drug products.[16] The following paragraphs summarize the BCS requirements for classification (based on FDA and CPMP Guidances).

Solubility : Equilibrium solubility at 37±1°C in aqueous media with a pH range of 1-7.5 using traditional shake-flask method or acid or base titration methods can be used. A validated stability-indicating assay is required for drug quantitation. A high solubility class is attributed if the volume of an aqueous medium sufficient to dissolve the highest dose strength is ≤ 250 mL. This definition of solubility is distinct from the compendial 'high solubility' definition.

Permeability: Pharmacokinetic studies evaluating extent of absorption in humans include mass-balance studies and absolute BA studies. Intestinal permeability methods that can be used are *in vivo* intestinal perfusion studies in humans, *in vivo / in situ* intestinal perfusion studies in animals, *in vitro* permeation studies with excised human or animal intestinal tissue, and *in vitro* permeation experiments across epithelial cell monolayers (e.g. Caco-2). Suitability and validation of the permeability method have to be established based on a rank-order of test permeability values and *in vivo* extent of absorption using recommended model drugs. Following identification of a suitable method, classification of permeability can be made based on comparison with selected model drugs for high and low permeability and efflux.

Dissolution : USP apparatus I (basket) at 100 rpm or USP apparatus II (paddle) at 50 rpm using 900 mL of (1) 0.1N HCL or Simulated Gastric Fluid without enzymes (2) pH 4.5 buffer, and (3) pH 6.8 buffer or Simulated Intestinal Fluid without enzymes can be utilized. For gelatin-containing dosage forms (capsules, coated tablets) enzymes may be used with justification.

IVIVC : Since dissolution testing is a key surrogate for *in vivo* performance of a dosage form, validation of this process is critical for accurate predictions of *in vivo* performance. While validation of the analytical methods provides assurance of reliability and reproducibility of the methodology, a 'biopharmaceutical' validation of the test is the establishment of an IVIVC.

Establishment of similarity of the entire *in vitro* dissolution profile with a calculated *in vivo* dissolution profile is the basis of a Level A correlation. When absorption process is not the rate limiting step, a noncompartmental deconvolution method or compartmental methods can be used to represent *in vivo* dissolution. PK data following administration of an oral solution (or intravenous administration) and a solid oral dosage form are used for deconvolution and when performed correctly, the *in vivo* dissolution curve should asymptotically provide a point estimate of the bioavailability. Following generation of the *in vitro* dissolution profiles, the goal is then to generally establish a linear correlation between the *in vitro* and *in vivo* percent released. However, there are several instances where a nonlinear function may serve to establish a robust relationship.[21] Validation of the developed IVIVC can be achieved by convolution of the predicted *in vivo* dissolution profiles to yield predicted plasma concentration versus time profiles. Prediction error is obtained by comparison of the IVIV model predicted Cmax and AUC values to the corresponding observed values. Internal predictability is based on the initial data used for model development, while external predictability requires the use data sets not used in model development.

Numerous applications of IVIVC have been documented with modified release dosage forms and there are few examples of the application of the development of IVIVC for immediate-release solid oral dosage forms.[22-25] The United States Pharmacoepia clearly states that while Level B and Level C approaches may be possible for such dosage forms, Level A approaches are conceivable but not documented. A challenge in the development of IVIVC for these dosage forms is the different time scales for the dissolution process (~60-90 minutes) versus the *in vivo* absorption process (typically 4-6 h or more). This difference can be addressed by the use of a timescaling technique and there are reports of successful Level A IVIVC.[22] Another approach for these dosage forms is the use of convolution based IVIVC and Polli et. al have demonstrated predictable IVIVC for a BCS Class 3 drug.[26]

The following paragraphs summarize some of the recent advances in solubility, permeability, and dissolution assessments, which would be applicable in the high throughput drug discovery stage. Use of these alternate methods should be thoroughly discussed with the relevant Health Authority if the data are to be used for registrational purposes.

New technologies for solubility estimation: New computational models based on lipophilicity and non-polar surface area descriptors have been developed to predict aqueous drug solubility.[27] When combined with combinatorial chemistry and high throughput screening techniques, this method could be an important tool for identifying drug candidates. Kostewicz et al. have demonstrated the use of various biorelevant GI media to examine solubility of poorly soluble weak bases to assist in *in vivo* predictions.[28]

New technologies for permeability estimation: Computational methods of gastrointestinal simulation have been applied for the prediction of oral drug absorption and BA.[29] The ACAT model (Advanced Compartmental Absorption and Transit) has been developed to predict the rate, extent, and approximate location of drug substance liberation within the GI tract, dissolution, passive and carrier-mediated absorption, saturable metabolism, and efflux. A new rapid reduced serum culture system for Caco-2 monolayers compared to the traditional 21-day system has been developed that has been shown to effectively classify permeability according to the BCS system.[30]

New technologies for dissolution assessment: BCS Class I drugs are readily transported across the intestinal membrane. *In vitro* dissolution profiles, irrespective of the medium, demonstrate complete release is attained within a short period of time. However, the dissolution behavior of a BCS Class 2 drug is highly dependent on the dissolution medium. In addition, the presence of meals influences the rate and extent of dissolution and the small intestine may become an important site for dissolution under these circumstances. Dressman et al. have recommended suitable biorelevant media for this class of drugs for the development of IVIVC.[31] These include SGFsp plus surfactant to simulate fasted state in the stomach, Ensure or milk to simulate fed state in the stomach, FaSSIF and FeSSIF to simulate fasted and fed states in the intestine.

Dissolution of poorly soluble drugs is greatly influenced by several conditions in the gastrointestinal tract such as pH, surfactant, ionic strength, buffer capacity, and viscosity.[32] Simple additive models for the effects of pH and surfactant have been developed to predict the dissolution rate and solubility of ionizable poorly soluble compounds. The model can also serve as the basis of development of IVIVC for predicting BE of class II products. A continuous dissolution / Caco-2 system has been developed to predict dissolution-absorption relationships and can serve as a potential tool to aid formulation selection and setting dissolution specifications.[33]

Research is also being conducted by the FDA to validate *in vitro* dissolution requirement for BCS Class 3 compounds and to evaluate alternative dissolution testing apparatus.[34,35]

5. CASE STUDIES

The Bristol-Myers Squibb Experience: A common approach to the formulation development plan of an immediate release BCS Class I compound is having the initial formulation of drug-in-a-bottle evolving to a clinical capsule formulation and ultimately to the "to be marketed" tablet. Only one *in vivo* BE study may be needed linking the capsule to the tablet, even if the tablet is introduced at Phase III. This is accomplished if the formulation characteristics for the tablet are developed with the understanding of the *in vivo* impact and utilizing discriminatory *in vitro* methods. The following case studies demonstrate that biowaiver strategies have been utilized from early drug development through the marketed life of the products.

Phase I: BMS-XYZ was determined to be as permeable as metoprolol (FDA standard for highly permeable) based on *in vitro* caco-2 study results. The *in vivo* extent of absorption was > 90% thus classifying the drug substance as highly permeable. In addition, the top anticipated tablet strength was highly soluble at physiologically relevant pHs based on the BCS criteria. Overall, the BCS classification was established as 'highly soluble - highly permeable' and no BE qualification was performed for formulation transitions during development following initial pharmacokinetic assessments.

Phase II: While it may be optimal to conduct PoP in phase II with the final commercial formulation, there have been several instances where changes ranging from a SUPAC Level 1 to a Level 3 have been made between the phase II formulation and commercial formulation. Qualification of major changes made to formulations of BCS Class 1 or 3 drugs can be justified by establishing rapid and similar dissolution profiles. For BCS Class 2 drugs, where poor solubility leading to a potential for dissolution rate limited absorption exists, other strategies need to be developed.

In one instance, for a BCS Class 2 drug, phase I studies were performed using a capsule formulation, with a switch to a tablet formulation desired for commercial purposes. A pilot BE study was performed in early phase II to establish similarity in exposures between the two formulations and establish particle size limits which provided the basis for conducting the PoP studies with the tablet formulation. Subsequent minor changes made to the tablet formulation prior to launch, e.g. change in shape, color, and new strengths were qualified by establishing *in vitro* dissolution similarity in multiple media.

In another instance, an additional milling step for the tablet formulation for a BCS Class 2 compound that was used for PoP studies was not considered to be cost-effective for scale-up and launch. The solubility of the compound was pH dependent. Deconvolution of the plasma concentration versus time profiles indicated that >90% of the absorption occurred in the duodenum, limited absorption in the jejunum and negligible absorption in the other areas of the gastrointestinal tract. The strategy for biowaiver was demonstration of rapid and similar dissolution between the milled and unmilled formulations at pHs relevant to the site of absorption and in multiple media.

Phase III: Despite best efforts, scale-up issues and resultant formulation changes may sometimes arise as registration and launch approach. Biowaiver strategies have a major impact on the time to registration and launch at this stage of development. In one example, doses evaluated in clinical pharmacology studies and phase II/III trials were 2, 5, 10, 15, 20, and 30 mg, delivered using the respective tablet strengths of 1 (x 2), 5, 10, 15, 10 (x 2), and 15 (x 2) mg. The compound was intended to be marketed in tablet strengths of 2, 5, 10, 15, 20, and 30 mg. Biowaivers, based on dissolution similarity, were requested for the new lower strengths (2 and 10 mg) strengths based on both definitions provided for proportional similarity. BE evaluation was required for the new 20 and 30 mg strengths. To provide assurance of a successful BE evaluation, a time-scaled non-linear Level A IVIVC was developed which was then used for *in silico* prediction of Cmax and AUC values for the new dose strengths that provided assurance of a successful BE evaluation.

Phase IV: A classical phase IV activity is development of pediatric formulation. In this case study, a new liquid oral formulation (as powder for oral suspension) was developed for pediatric use while the initial registration was based on tablets. A biowaiver strategy was successfully applied to preclude the need for a pivotal BE study to link the commercialized tablet formulations and the powder for oral suspension. The basis of the biowaiver was (a) the active pharmaceutical ingredient was a BCS Class 1 and (b) dissolution profiles at 0.1 N HCL, pH 4.5 and 6.8 buffers of both formulations were rapid (85% within 30 minutes) and similar (F_2 values > 50). In addition, relative BA of the liquid formulation was about 95% compared to the commercial tablets thus indicating similarity in exposures following administration of the two formulations.

Marketed product: While a product can be successfully registered and launched in the US, queries on BE can surface during global registrations. In addition, changes that can potentially affect formulation quality and performance can occur over the marketed lifetime of a product. A biowaiver strategy was successfully utilized to register 75, 150, and 300 mg strengths of a drug in Australia. The pivotal clinical studies submitted for the global registration employed the 75 and 150 mg market product (tablet) which were milled and encapsulated to provide doses ranging from 75 - 300 mg. The rationale for not conducting BE studies to link the 75 and 150 mg capsules to the marketed 75 and 150 mg tablets was based on the arguments that (a) presence of the gelatin capsule encapsulating the milled market tablets was not expected to impact the formulation quality and performance as it was considered a 'self-assessable change' (based on TGA Guideline), and (b) similar dissolution profiles (F_2>50) between the 75 and 150 mg capsules and the tablets. The 300 mg commercial tablet was qualified by justifying that (a) all three strengths (75, 150, and 300 mg) of the marketed tablets were prepared from a common granulation and were thus 'proportionally similar' and (b) the pharmacokinetics was linear between the 75 and 300 mg dose range, and (c) dissolution profiles of the capsules and trade tablets were similar. Other instances where biowaivers have been successfully utilized for changes affecting formulation quality and performance of a marketed product include changes in (a) vendor for excipients that may affect disintegration and dissolution, (b) site of manufacture, (c) equipment used for processing, and (d) composition to harmonize domestic and international formulae.

6. FUTURE DIRECTIONS

Application of BCS has received strong support from the scientific community including the FDA Advisory Committee for Pharmaceutical Sciences, experts, FDA staff and the Industry. However, several concerns regarding US Guidance's conservative requirements have been raised which have hindered widespread implementation of the strategy. This has been recognized by the FDA who have begun to consider BCS revision by evaluating (a) extension of BCS based biowaivers to fed BE studies (2) solubility class boundary revision, and (c) biowaiver extensions to BCS Class 3 drugs. The rationale for the extensions is summarized in Figure 3.

The FDA draft Guidance for Industry on the study design, data analysis, and labeling associated with food-effect BA and BE studies was recently published for public comments and was discussed in a recent Advisory Committee meeting.[8] An important aspect of this Guidance is the provision of a waiver for BE studies under fed conditions for BCS Class 1 drugs. The waiver is based on the hypothesis that solubility and dissolution rates are pH and site independent and hence unlikely to be influenced by food. Therefore, the draft Guidance provides for a waiver of BE studies under fed conditions when both test and reference listed drug (RLD) products are BCS Class 1, are rapidly dissolving, and have similar dissolution profiles. The draft Guidance also proposes an equivalence limit of 80-125% for the analysis of Cmax and AUC data (90% confidence interval) in food-effect BA studies as evidence of an absence of food effects and in fed BE studies to demonstrate the BE of a test and reference product. This requirement clearly shows the emphasis on the scientific approach with clear and defined criteria.

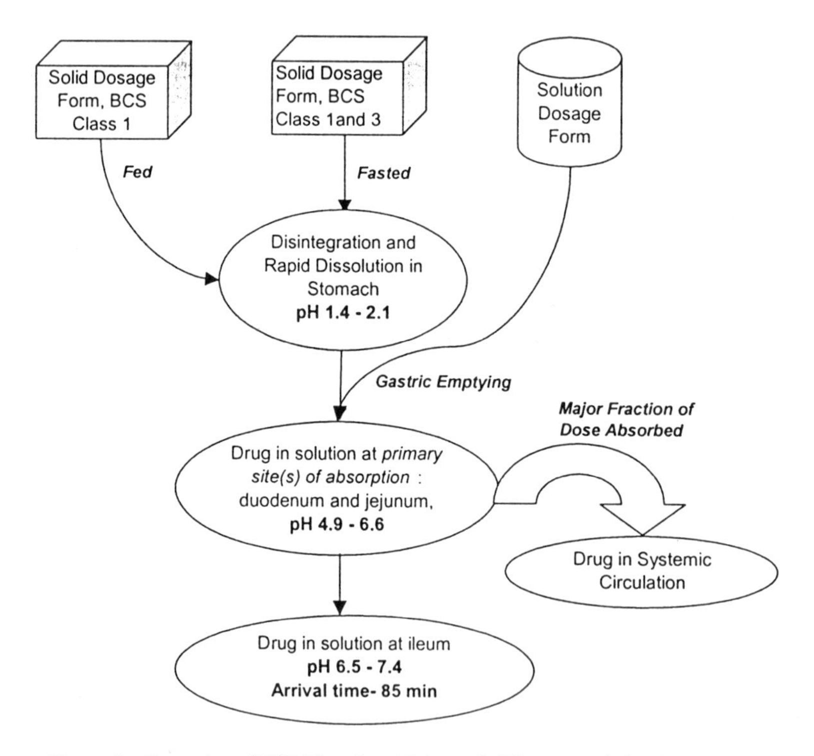

Figure 3. Absorption of BCS Class 1 and 3 drugs: Solid versus solution dosage form

In a recent FDA-authored paper on BCS extensions[36], discussions were presented to help assure that the pH range required to demonstrate solubility is over physiologically relevant pH ranges. Under fasted conditions, the pHs of the GI tract range from 1.4 to 6.6 between the stomach and the jejunum, which are the primary regions of dissolution and absorption. The pH of the ileum ranges from 6.5 to 7.4 and this distal part of the small intestine accounts for little absorption for a majority of products. In addition, it requires about 85 minutes for a drug to reach the iluem after dosage form administration during which time a major fraction of the bioavailable dose of a highly soluble drug will have been absorbed. Therefore it was felt to be reasonable to redefine the pH range for 'high solubility', from 1 - 7.5 to 1.0 - 6.8. The agency is currently gathering additional evidence to justify this revision.

Extension of biowaivers to BCS Class 3 drugs with rapid dissolution is also being considered by the FDA.[36] It is contended that for this category of drugs, the BA is determined by *in vivo* permeability and less dependent on solubility and dissolution. Hence, if the dissolution of a Class 3 compound is similar in the three recommended pH conditions (under more stringent conditions) and if excipients used in the two formulations do not affect drug absorption, it is reasonable to assume that the compound will behave as an oral solution *in vivo*. Based on the FDA's policy, BE of oral solutions is self evident and biowaivers can be granted. Proposed studies and data collection to test the hypothesis include human BE studies to compare a simple solution with solid dosage forms of at least ten model BCS Class 3 drugs, additional FDA internal studies, FDA review of other BE studies and relevant *in vitro* dissolution and cell culture studies. Potential issues identified are effect of transporters on the permeation of this class of drugs and a greater potential of excipients affecting GI motility to influence the absorption of this category of drugs.

In a recently published FDA draft Guidance on bioavailability and bioequivalnece studies, the definition of 'proportional similarity' was extended to include changes within the limits defined by SUPAC-IR Level 2.[37]

Overall, directions provided by the various Health Authority Guidances offer a broad scientific basis for biowaiver applications (Table 4) and pave the way for accelerating the drug development process without compromising quality.

Table 4. Future biowaiver opportunities

BCS Class	Solubility	Permeability	Risk of Bioinequivalence	Biowaiver
1	High	High	Minimal	√ (fasted, fed)
2	Low	High	Moderate	√ (fasted, with IVIVC)
3	High	Low	Low	√ (fasted)
4	Low	Low	High	×

7. SUMMARY

Strong scientific approaches to formulation development allow regulatory strategies to minimize unnecessary clinical testing. Scientifically founded biowaiver strategies utilize *in vitro* methodologies and serve to improve development timelines, cost, and manpower, without compromising product quality.

8. ACKNOWLEDGEMENT

We would like to convey our sincere gratitude to Robert Lipper and Ramona Lloyd for their review and helpful suggestions.

9. REFERENCES

1. 21CFR320.30. Inquiries regarding bioavailability and bioequivalence requirements and review of protocols by the Food and Drug Administration.
2. FDA Guidance for industry. Bioavailability and bioequivalence studies for orally administered drug products - General considerations. 2000.
3. FDA Guidance for industry. Immediate release solid oral dosage forms. Scale-up and postapproval changes: Chemistry, manufacturing, and controls, *in vitro* dissolution testing, and *in vivo* bioequivalence documentation. 1995
4. FDA Guidance for industry. Dissolution testing of immediate release solid oral dosage forms. 1997.
5. FDA Guidance for industry. SUPAC-MR : modified release solid oral dosage forms. Scale-up and postapproval changes : Chemistry, manufacturing, and controls, *in vitro* dissolution testing, and *in vivo* bioequivalence documentation. 1997.
6 FDA Guidance for industry. Nonsterile semisolid dosage forms. Scale-up and postapproval changes : Chemistry, manufacturing, and controls, *in vitro* dissolution testing, and *in vivo* bioequivalence documentation. 1997.
7 FDA Guidance for industry. Waiver of *in vivo* bioavailability and bioequivalence studies for immediate-release solid oral dosage forms based on a biopharmaceutics classification system. 2000.
8 FDA Draft guidance for industry. Food-effect bioavailability and fed bioequivalence studies: Study design, data analysis, and labeling. 2001
9. European Commission notice to applicants. A guideline on dossier requirements for type I variations. 1999.
10 Committee for Proprietary Medicinal Products (CPMP) Note for Guidance. The investigation of bioavailability and bioequivalence. 2001.
11. Therapeutic Goods Administration (Australia) Directive. Investigation of bioavailability and bioequivalence, 1992.
12. Australian guidelines for the registration of drugs . Changes to drug products-is notification or prior approval required ?1994
13. Therapeutic Products Directorate (Canada) Bioequivalence of proportional formulations. 1996.
14. Therapeutic Products Directorate (Canada) Preparation of drug submissions : Comparative bioavailability studies. 1997.
15. Ministry of Health, Welfare and Labor (Japan). Information on guideline for bioequivalence test on oral solid preparation for which the formulation has been changed, 2000.
16. G. L. Amidon, H. Lennernas, V.P. Shah, and J.R. Crison, A theoretical basis for a biopharmaceutic drug classification : The correlation of *in vitro* drug product dissolution and *in vivo* bioavailability, *Pharm. Res.* 12(3), 413-420 (1995).
17. FDA Guidance for Industry. Extended release oral dosage forms: Development, evaluation, and applications of *in vitro/in vivo* correlations, 1997.
18. 21CFR320.22. Criteria for waiver of evidence of *in vivo* bioavailability or bioequivalence.
19. 21CFR320.33 Criteria and evidence to assess actual or potential bioequivalence problems.
20 R. Lobenberg and G.L. Amidon. Modern bioavailability, bioequivalence and biopharmaceutics classification system. New scientific approaches to international regulatory standards. Eur. J. Pharm. Biopharm. 50(1), 5-12, 2000.
21. J. Mendell-Harary, J. Dowell, S. Bigora, D. Piscitelli, J. Butler, C. Farrell, J. Dveane, and D. Young, in : *In vitro – in vivo* correlations, edited by D. Young, J.G. Devane, and J. Butler (Plenum Press, New York and London, 1997).pp. 199-206.
22. G. Balan, P. timmins, D.S. Greene, and P.H. Marathe. In-vitro in-vivo correlation models for glibenclimide after administration of metformin/glibenclimide tablets to healthy human volunteers. J. Pharm. Pharmacol., 52 : 831-838, 2000.

23. J. Posti, K. Katila, and T. Kostianen. Dissolution rate limited bioavailability of flutamide, and *in vitro – in vivo* correlation. Eur. J. Pharmaceut. Biopharmaceut. 49: 35-39 (2000).
24. J. Radocanovic, Z. Duric, M. Jovanovic, S, Ibric, and M. Petrovic. An attempt to establish an *in vitro – in vivo* correlation : case of immediate-release tablets. Eur. J. Drug Metab. Pharmacokin. 23(1) 3-40 (1998).
25. O.A. Lake, M. Olling, and D.M. Barends. *In vitro/in vivo* correlations of dissolution data of carbamazepine immediate release tablets with pharmacokinetic data obtained in healthy volunteers. Eur. J. Pharmacokin. Biopharmaceut. 18 : 13-19 (1999).
26 J. E. Polli, in : *In vitro – in vivo* correlations, edited by D. Young, J.G. Devane, and J. Butler (Plenum Press, New York and London, 1997) pp 191-918.
27. C.A.S. Bergstrom, U. Norinder, K. Luthman, and P. Artursson. Experimental and computational screening models for prediction of aqueous drug solubility. Phar. Res. 19(2) 182-188, 2002.
28. E.S. Kostewicz, U. Braums, R. Becker, and J.B. Dressman. Forecasting the oral absorption behavior or poorly soluble weak bases using solubility and dissolution studies in biorelevant media. Pharm. Res., 19(3) 345-349 (2002).
29. B. Agoram, W.S. Woltosz, M.B. Bolger. Predicting the impact of physiological and biochemical processes on oral drug bioavailability. Adv. Drg. Del. Rev. 50 : S41-S67 (2001).
30. K.A. Lentz, J. Hayashi, L.J. Lucsiano, and J.E. Polli. Development of a more rapid, reduced serum culture system for Caco-2 monolayers and application to the biopharmaceutics classification system. Int. j. Pharmaceut. 200 ; 41-51 (2000).
31. J. B. Dressman and C. Reppas. *In vitro-in vivo* correlations for lipophilic, poorly water-soluble compounds. Eur. J. Pharm. Sci., 11(suppl.2): S73-S80 (2000).
32. J. Jinno, D-M Oh, J.R. Crison, and G.L.Amidon. Dissolution of ionizable water-insoluble drugs : The combined effect of pH and surfactant. J. Pharmaceut. Sci., 89(2) 268-274, 2000.
33. M.J. Ginski, R. Taneja, and J.E. Polli. Prediction of dissolution-absorption relationships from a continuous dissolution/Caco-2 system. AAPS Pharm.Sci. 1(3) : 1-12 (1999).
34. L.X. Yu, C.D. Ellison, D.P. Connor, L.J. Lesko, and A.S. Hussain. Influençe of drug release properties of conventional solid dosage forms on the systemic exposure of highly soluble drugs. AAPS Pharm. Sci 3(3) : 1-7 (2001).
35. L.X. Yu, J.T. Wang, and A.S. Hussain. Evaluation of USP Apparatus 3 for dissolution testing of immediate-release products. AAPS Pharm. Sci. 4(1) : 1-5 (2002).
36. L.X. Yu, G.L. Amidon, J.E. Polli, H. Zhao, M.U. Mehta, D.P. Conner, V.P. Shah, L.J. Lesko, M-L Chen, V.H.L. Lee, and A.S. Hussain. Biopharmaceutics Classification System: The scientific basis for biowaiver extensions. Pharm. Res. 19(7) : 921-925, 2002.
37 FDA Draft guidance for industry. Bioavailability and bioequivalence studies for orally administered drug products - General considerations, 2002.

SPECIAL POPULATION STUDIES IN CLINICAL DEVELOPMENT: PHARMACOKINETIC CONSIDERATIONS

John M. Kovarik*

1. APPROACHING THE INDIVIDUAL PATIENT THROUGH SPECIAL POPULATION STUDIES

Although the primary thrust of research during clinical development of a new molecular entity is to identify a dosing algorithm for the general patient population, the actual patient population in which the drug will be prescribed after marketing approval is far from homogeneous. The fields of pharmacology, pathology, and pharmacogenetics have made it increasingly clear that within the general patient population there exist subpopulations in which drug disposition as well as drug-receptor expression and activity differ from the population average. Depending on the magnitude of these differences and the width of the therapeutic window between effective and toxic blood concentrations, specific subpopulations might require a dose regimen which diverges from the population average regimen. Unmasking and exploring these special populations during clinical development is one of the focuses of clinical pharmacology.

1.1. Need for More Informative and Patient-Individualized Product Labeling

Throughout the 1990s a paradigm shift in the pharmaceutical and drug development sciences began to highlight the need for research to support individualized pharmacotherapy. This was consolidated at mid-decade in an influential editorial on "patient-oriented pharmaceutical research" which called for combining data from pharmacokinetics, pharmacodynamics, and medication compliance to provide guidelines to determine the appropriate dose for the individual patient. The author concluded that "focus on the patient as an individual is not only the duty of physicians and other health care professionals but is also becoming an increasingly important obligation of pharmaceutical scientists."[1]

In this context, the product label for a new molecular entity can serve as a barometer to assess how well drug development scientists are meeting this need. A multidisciplinary team of ten pharmaceutical scientists have assessed the quality and completeness of clinical pharmacology information in drug labeling. They reviewed 76 randomly chosen product labels for orally administered prescription drugs. A three-level scoring system (0 = absent, 1 = present, and 2 = adequate) was used to rate 31 clinical pharmacology ele-

* John M Kovarik, Clinical Pharmacology, Novartis Pharmaceuticals, Basel, Switzerland

ments in five categories. The individual categories contributed as follows to a total core information score of 50 points: mechanism of action (4 points), pharmacodynamics (12 points), metabolism (8 points), pharmacokinetics (16 points), and dose adjustment and compliance (10 points). The median percentage of core information across all categories was 31 percent. The category with the lowest median percentage of core information was metabolism with 0 percent and the highest was mechanism of action with 100 percent. The investigators concluded, "this study showed that most package inserts contain insufficient clinical pharmacology information."[2] This exercise is of particular relevance to clinical pharmacology studies in special populations because they play a central role in the dose adjustment and compliance category. The median score in this category was 40 percent, indicating again that there is a general lack of information on dose individualization in current product labeling.

1.2. The Role of Clinical Pharmacokinetics in Characterizing Special Populations

The United States Food and Drug Administration has underscored the need for special population studies within the wider framework of drug development: "...sponsors are expected to include a full range of patients in their studies, carry out appropriate analyses to evaluate potential subset differences in the patients they have studied, study possible pharmacokinetic differences in patient subsets, and carry out targeted studies to look for subset pharmacodynamic differences that are especially probable, are suggested by existing data, or that would be particularly important if present."[3]

As can be seen in the quote above, clinical pharmacokinetics, in particular, plays a pivotal role in this process often serving not only to characterize the subpopulation itself but also to establish the specific dose regimen appropriate for such patient groups. Indeed, the drug product label provides abundant testimony to the pivotal contribution of pharmacokinetics in this area of clinical development. This is outlined in Table 1 which highlights the sections and subsections in the European Summary of Product Characteristics[4] and the United States product label for prescription drugs[5] which are the bases of information for health professionals on how to use the medicinal product safely and effectively. Special populations are prominently mentioned in the "pharmacological properties" of the Summary of Product Characteristics and the "clinical pharmacology" section of the United States label. Additionally, pharmacokinetics generally underpins the dose adjustment recommendations for special populations.

With this as background, the following discussion will focus on subpopulations conventionally studied during the clinical development of a new molecular entity. These can be subsumed under two broad categories, namely, demographics and organ impairment. Conventional demographic subgroups include pediatrics, geriatrics, sex, and ethnicity. The second category includes patients with various degrees of renal or hepatic function impairment. Additional subpopulations may be identified during the development continuum depending on the drug being developed and the pathology for which it is indicated. The six subpopulations mentioned above, however, are the ones most commonly encountered in clinical development. For each subpopulation, the discussion will address pertinent regulatory guidelines, aspects of study design, methods of pharmacokinetic data analysis, derivation of the dose adjustment algorithm, and reporting of the results in the drug product label. Throughout, pertinent examples from the pharmaceutical literature

Table 1. Product Label Outlines Indicating Sections Containing Special Population Data

Summary of Product Characteristics	US Product Label
1. Name of the medicinal product	Description
2. Qualitative and quantitative composition	*Clinical pharmacology*
3. Pharmaceutical form	Mechanism of action
4. *Clinical particulars*	*Pharmacokinetics*
4.1 Therapeutic indications	*Special populations*
4.2 *Posology and method of administration*	Clinical studies
4.3 *Contraindications*	Indications and usage
4.4 *Special warning and precaution for use*	Contraindications
4.5 Interaction with other medicinal products	*Warnings*
and other forms of interaction	*Precautions*
4.6 *Pregnancy and lactation*	General
4.7 Effects on ability to drive and use machines	Information for patients
4.8 Undesirable effects	Laboratory tests
4.9 *Overdose*	Drug or drug/laboratory test interactions
5. *Pharmacological properties*	*Carcinogenesis, mutagenesis, and*
5.1 *Pharmacodynamic properties*	*impairment of fertility*
5.2 *Pharmacokinetic properties*	*Pregnancy*
5.3 Preclinical safety data	Labor and delivery
6. Pharmaceutical particulars	*Use during lactation*
7. Market authorisation holder	*Pediatric use*
8. Marketing authorisation number	*Geriatric use*
9. Date of first authorisation	Adverse reactions
10. Date of revision of the text	Drug abuse and dependence
	Overdosage
	Dosage and administration
	How supplied

Italics designate label sections which may contain data on special populations.

will be used to illustrate the application of clinical pharmacokinetics to characterizing special patient populations.

2. PEDIATRIC POPULATION

Pediatric patients have long been excluded from the benefits of drugs labeled for adult use due to exclusionary language in the product label. Accordingly, they have been termed "therapeutic or pharmaceutical orphans".[6] As such, pediatric patients are exposed to the following risks: age-specific adverse reactions which might be avoided if described properly in the product label; under or over dosing due to the lack of a pediatric dose regimen supported by pharmacokinetic and clinical trials; and poorly bioavailable oral drug formulations made extemporaneously in order to meet the administration needs of children but without proper stability and equivalence testing.[7] Through the concerted involvement of regulatory scientists, pharmaceutical industry scientists, and pediatricians, efforts are underway to "rescue the therapeutic orphan".[8, 9] The history of this effort is instructive and warrants a brief outline as it is pertinent to understanding the background to current regulatory guidances addressing pediatric studies and to appreciate the clinical need for pediatric use information in product labeling.

2.1 Overview of Efforts to Foster Pediatric Drug Development

In an attempt to encourage pediatric information in drug labeling, the United States Food and Drug Administration issued a regulation in 1979 establishing a "pediatric use" subsection in the product label and describing its contents. This regulation also required that recommendations for pediatric use be based on substantial evidence derived from adequate and well-controlled trials in the pediatric population.[10] Although some progress was made in the ensuing years, it was noted in the early 1990s that fully 70 percent of drugs used to treat pediatric patients in hospitals lacked pediatric labeling.[11]

In response, the Food and Drug Administration instituted the *Pediatric Page* in 1991 as part of the drug review process at its Center for Drug Evaluation and Research (CDER). The *Pediatric Page* documents information about the status of or need for pediatric studies at the time of drug approval. This was followed in 1994 by two additional initiatives. The first initiative was to issue a rule detailing specific requirements on the content and format of labeling and a revision of the "pediatric use" subsection.[12] This regulation states that if the course of the disease and the effect of the drug are similar in adults and pediatric patients, evidence for effectiveness can be extrapolated from adult studies to support pediatric labeling. This is usually supplemented with specific data on pharmacokinetics and safety in pediatric patients. This voluntary program was meant to ease the requirement for clinical studies in pediatrics if the above criteria could be satisfied. The second initiative was the publication of the *Pediatric Plan* designed to focus attention on pediatric patients throughout the development continuum by specifically considering the need for pediatric studies when reviewers and sponsors meet. Importantly, the plan extends into the post-marketing phase to capture the impact of adverse event reports on pediatric use of the product and any labeling changes which need to be made. A Pediatric Subcommittee, Pedicomm, was also created to oversee and track these initiatives.

In 1997, however, the Food and Drug Administration admitted that "voluntary efforts have, thus far, not substantially increased the number of products entering the market place with adequate pediatric labeling."[13] The enactment in that year of the Food and Drug Administration Modernization Act (FDAMA) and the following year of the 1998 *Pediatric Rule*[14] set the stage for the current regulatory framework for pediatric drug development.

2.2. Current Environment for Pediatric Research

2.2.1. United States Pediatric Rule and European Initiatives

The 1998 *Pediatric Rule* came into effect in April 1999 and mandates pediatric studies adequate to assess safety and effectiveness and to support pediatric use.[11] FDAMA section 505A provides the incentive by allowing a drug's original sponsor six additional months of market exclusivity for new or already-marketed drugs in exchange for conducting pediatric studies. Market exclusivity in FDAMA is not a patent extension but rather a period during which the Food and Drug Administration cannot approve an Abbreviated New Drug Application (ANDA) for the same chemical entity that relies on the safety and effectiveness data in the original sponsor's full New Drug Application (NDA). The additional six months of market protection can only be added to an already-existing period of exclusivity. Examples include: the 5-year or 3-year period of exclusivity for an

NDA or supplemental NDA, the 7-year orphan drug exclusivity period, or the existing patent term.[11] In order to qualify for exclusivity, the sponsor must conduct the pediatric study in response to a Written Request from the Food and Drug Administration. FDAMA mandated the Food and Drug Administration to report on the status of the pediatric provision by 1 January 2001. In their report, the Food and Drug Administration stated that "the pediatric exclusivity provision has done more to generate clinical studies and useful prescribing information for the pediatric population than any other regulatory or legislative process to date" and recommended that it be renewed beyond its sunset date of 1 January 2002 with modifications.[7] The resulting Best Pharmaceuticals for Children Act reauthorized the pediatric exclusivity provision and extended eligibility to October 2007.

While exclusivity and other incentives will likely change over time, the 1998 *Pediatric Rule* remains the basic regulatory anchor underlying the current requirements for sponsors to perform pediatric studies. The rule applies to a drug or biological that (1) is either used in a substantial number of pediatric patients or provides a meaningful therapeutic benefit over existing treatments and (2) whose label contains inadequate support for safe and effective use in pediatric patients for the approved indication.[14]

For marketed drugs, CDER may order the manufacturer to submit pediatric safety and effectiveness data or request for approval of a pediatric formulation. The manufacturer is given an opportunity to respond in writing and to meet with reviewers. The rule presumes that henceforth all new drugs and biologics will be studied in pediatric patients but allows sponsors to obtain a waiver of the requirement for some or all pediatric age groups under certain conditions. These include (1) the drug does not represent a meaningful therapeutic benefit for pediatric patients or will not be used in a substantial number of pediatric patients (the rule mentions 50,000 patients as substantial but does not codify it); (2) the required studies are impossible or highly impractical (for example, the population is too small or geographically dispersed); (3) the product is likely to be unsafe or ineffective in pediatric patients; or (4) reasonable efforts to develop a pediatric formulation, if one is needed, have failed.

In the European Union, complementary initiatives are forming. In addition to contributing to the International Conference on Harmonization guidance on pediatric studies in drug development (discussed below), the European Commission has issued a consultation document entitled *Better Medicines for Children*. Among the proposed regulatory actions are granting a period of marketing exclusivity as an incentive for performing pediatric studies, creation of a central database for pediatric drug information, establishing a European expert group or working party within the European Medicines Evaluation Agency with responsibility for all aspects relating to the development and follow up of pediatric medicines, and creation of a pan-European network of clinical excellence for performance of pediatric studies.[15]

2.2.2. Regulatory and Professional Guidances

A number of guidances on pediatric research are available from regulatory authorities and professional societies. An annotated listing is given below; a few entries deserve special attention. The American Academy of Pediatrics *General Considerations* contains a wealth of information on conducting studies in pediatric populations focussing primarily on issues of safety and effectiveness but touching also on pharmacokinetics. This same body's *Ethical Guidelines* are a comprehensive and seminal statement covering an essential element underlying pediatric study design and conduct. The Food and Drug

Administration draft *Pediatric Pharmacokinetic Guidance* was issued in follow-up to FDAMA and the 1998 *Pediatric Rule* and covers general aspects of data collection and analysis as they apply to product labeling. The International Conference on Harmonization *E11 Guidance* builds on the foregoing guidances and represents the most current thinking on pediatric drug development constituting a veritable development syllabus in this research field.

- *American Academy of Pediatrics. General considerations for the clinical evaluation of drugs in infants and children as excerpted by the Food and Drug Administration (1977).* This professional guidance divides the pediatric population into age categories: intrauterine, neonate, infant, child, and adolescent. In each category a useful outline of age-specific considerations for safety, efficacy, problems in drug evaluation, and ethics is given.[16]
- *American Academy of Pediatrics Committee on Drugs. Guidelines for the ethical conduct of studies to evaluate drugs in pediatric patients (1995).* This comprehensive professional guidance covers ethical aspects under the rubrics of research proposal design, the investigator, the institutional review board, drug investigations, and protection for vulnerable populations.[17]
- *Food and Drug Administration. Guidance for industry: the content and format for pediatric use supplements (1996).* This is a concise presentation of what is needed for pediatric use supplements.[18]
- *Food and Drug Administration. Regulations requiring manufacturers to assess the safety and effectiveness of new drugs and biological products in pediatric patients (1998).* This is the final *Pediatric Rule*;[14] it is also helpful to consult the proposed rule.[13] This covers required studies, waivers, pediatric use section, and legal authority,
- *Food and Drug Administration. Draft guidance for industry: general considerations for pediatric pharmacokinetic studies for drugs and biological products (1998).* This is a useful compendium discussing study design, standard versus population pharmacokinetics, biological sample collection, data analysis, labeling statements, and ethics.[19]
- *Food and Drug Administration. Qualifying for pediatric exclusivity under section 505A of the Federal Food, Drug, and Cosmetic Act (1999).* This update from the original 1998 release follows a question-answer format to guide the reader through the regulatory intricacies of pediatric exclusivity.[20]
- *International Conference on Harmonization. Guidance for industry E11: clinical investigations of medicinal products in pediatric populations (2000).* This regulatory guidance provides an outline on critical issues in pediatric development programs, pediatric drug formulations, timing of studies, age classifications, and ethics.[21]

2.2.3. Pediatric Research Facilities

Two issues that hampered pediatric drug development for many years were limited access to a sufficiently large pediatric population and difficulty identifying qualified pediatric pharmacology investigators. These valid concerns were addressed by the National Institutes of Child Health and Human Development which established a Pediatric Pharmacology Research Unit network in 1994. Today this consortium consists of 13 pediatric pharmacology units which provide a clinical research infrastructure and access to thousands of pediatric patients across the range of pediatric medical specialties. Because the units are geographically dispersed throughout the United States, the network addresses the ethical issue of distributive justice whereby benefits and risks in clinical research should be shared across the population. Members of the consortium bring specific pediatric research experience to facilitate protocol design, patient recruitment, and reporting of results. Notably, the network mission includes training of investigators dedicated to pediatric research.[22]

A number of initiatives to create networks of pediatricians have been taken up at national levels in Europe for example in France, Germany, and the United Kingdom. A multinational collaborative attempt in this context is the European Network for Drug Investigation in Children. The European Commission's *Better Medicines for Children* consultation document calls for the creation of a pan-European network which would link together existing national initiatives to foster the necessary competences at a European level, to facilitate cooperation and to avoid duplication.[15]

2.3 Influence of Maturation on Drug Disposition

Pharmacokinetic studies play a key role in pediatric clinical programs and are a central contributor to define pediatric dose adjustments specified in the product labeling. Regulatory agencies assume that a new drug will be assessed across the full range of pediatric ages. To facilitate this, guidances generally define the following age categories: *neonate* from birth to one month, *infant* from one month to two years, *child* from two to 12 years, and *adolescent* from 12 to 16 years. This categorization is a somewhat artifical construct inasmuch as biological maturation and changes in drug disposition occur over an age continuum rather than in defined stages. Indeed, both the Food and Drug Administration and the International Committee on Harmonization have more recently distanced themselves from sharply defined age categories and allow for a flexible categorization to suit the drug under development.[14, 21] Nonetheless, sponsors are expected to cover the full pediatric age range unless a specific age subgroup is unlikely to be treated with or benefit from the drug. Under these conditions, it is the burden of the sponsor to argue for any age restrictions in the pediatric development program and the subsequent product labeling.[14]

Against this background, a basic understanding of physiologic changes during maturation is essential to design a scientifically valid, safe, and ethical pharmacokinetic study in pediatric patients. An appreciation for changes in physiology which may influence drug exposure can guard against protocols that may over- or underexpose pediatric research subjects to systemic drug concentrations. The following discussion highlights general trends in drug absorption, distribution, and elimination during pediatric development. For a more comprehensive treatment, several reviews on the ontogeny of drug disposition are available.[23, 24, 25, 26]

2.3.1 Absorption

Extravascular drug administration includes topical, rectal, intramuscular, and oral routes each of which are accompanied by special age-related considerations. For instance, if properly formulated, drug absorption via the skin or rectum can be particularly efficient for neonates and infants. A thinner and more well-hydrated stratum corneum in neonates and the large surface area to volume ratio in neonates and infants compared with children and adolescents should be kept in mind for percutaneously absorbed drugs. Neonates and especially preterm infants may have erratic absorption after intramuscular administration due to reduced muscle and adipose tissue.[27] Oral administration is by far the most commonly used route of drug delivery in pediatrics. Several changes, particularly in the neonatal period and infancy, may be important. There is relative achlorhydria for the first two weeks after birth with gastric acidity reaching adult values only after two years.[26] Gastric emptying time in neonates can be irregular and prolonged reaching adult values at about six to eight months.[27] Additionally, the maturity of metabolism enzymes and transporters in enterocytes may impact on oral drug absorption and/or bioavailability.

2.3.2. Distribution

Body water distribution relative to bodyweight can influence drug distribution volumes. Between birth and puberty, total body water (expressed as a percent of bodyweight) decreases from 75 to 60 percent, extracellular fluid decreases from 44 to 20 percent, and intracellular water increases from 33 to 40 percent.[23] Total protein, serum albumin, and alpha-1 acid glycoprotein are lower at birth and reach adult concentrations by about one year of age. In addition, interfering substances such as unconjugated bilirubin in neonates can reduce drug binding sites on albumin thereby increasing the free concentration of the drug.[23]

2.3.3. Elimination

In general, it is assumed that children will form the same metabolites as adults but formation rates may be different.[19] Cytochrome P450 enzyme activity, for example, generally follows the pattern of being limited in neonates but increasing in the first year of life to levels in infants and young children which often exceed adult capacity. With the onset of puberty, oxidative drug metabolism begins to decline to adult levels. While this pattern is generally applicable, pharmacogenetic studies in children are beginning to show that there is considerable interindividual variability in these temporal patterns.[28, 29] Renal function also exhibits developmental patterns with respect to glomerular filtration and tubular secretion. A pronounced increase in glomerular filtration occurs in the first two weeks of postnatal life as adaptations in renal blood flow recruit functioning nephron mass. Proximal tubular secretion may lag behind yielding a glomerular/tubular imbalance which may persist up to six to ten months of life at which time adult competence in both glomerular and tubular function are usually reached.[23]

2.4 Case Study: Cyclosporine Absorption in Infants

Cyclosporine is an immunosuppressant indicated for prophylaxis of organ rejection after transplantation in adults and children. Given the relatively narrow window between effective and toxic blood concentrations, dosing is usually individualized based on therapeutic drug monitoring. The original oral formulation, Sandimmune, formed a crude oil-in-water emulsion in the gastrointestinal tract leading to variable absorption. In clinical practice, it was noted that especially infants and young children needed higher cyclosporine doses per bodyweight and therefore had higher apparent clearance (CL/F) compared with adolescents and adults to achieve therapeutic blood concentrations. It was assumed that pediatric patients had a higher clearance of cyclosporine due largely to higher metabolic activity.[30]

With the introduction of a microemulsion oral formulation, Neoral, cyclosporine absorption was less variable and bioavailability was higher compared with Sandimmune. A multicenter, randomized, open-label, crossover study of the two formulations in pediatric liver transplant patients revealed the relative contributions of clearance (CL) and bioavailability (F) to the age-related pattern in apparent clearance (CL/F) observed with Sandimmune.[31] In this study, 27 pediatric patients were enrolled. They were categorized as one infant (one year), 21 children (two to 11 years), and five adolescents (12 to 17 years). Patients currently receiving a maintenance cyclosporine-based immunosuppressive regimen were randomized to receive either Sandimmune or Neoral in study period 1 which lasted from day 1 to 14. In study period 2, they were converted on a mg-to-mg basis to the alternative formulation from day 15 to 28. A steady-state pharmacokinetic profile was obtained over the 12-hour morning dosing interval on days 14 and 28.

As shown in Figure 1A, the pattern of apparent clearance with age previously described for Sandimmune was confirmed in this study. By contrast, the pattern with Neoral did not exhibit a pronounced age effect. Overall, cyclosporine apparent clearance was significantly higher from Sandimmune compared with Neoral: 24.6 ± 19.6 versus 12.8 ± 4.7 mL/min/kg ($p < 0.001$). The corresponding interindividual variability from Sandimmune was nearly twice as high as for Neoral: 79.7 versus 36.7 percent.

Constructing the apparent clearance ratios of Sandimmune over Neoral (CL_S/F_S over CL_N/F_N) for each patient provided additional insight. If one assumes that the clearance of cyclosporine did not change over the month-long duration of the study ($CL_S = CL_N$) then the ratio reduces to (F_N/F_S) and can serve as an estimate for the increase in cyclosporine bioavailability upon converting from Sandimmune to Neoral. As shown in Figure 1B, 21 of the 27 patients (78 percent) had an increase in bioavailability ranging from 25 to 283 percent and the remaining five patients had relatively minor changes within ± 20 percent. The change in bioavailability with age in Figure 1B mimics the apparent clearance of cyclosporine from Sandimmune with age in Figure 1A. This indicates that especially

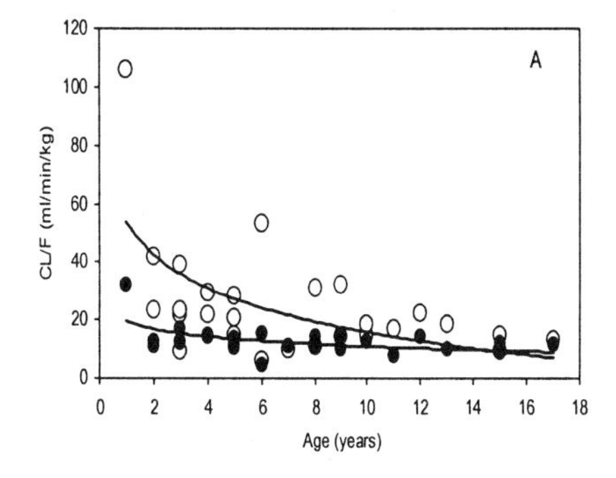

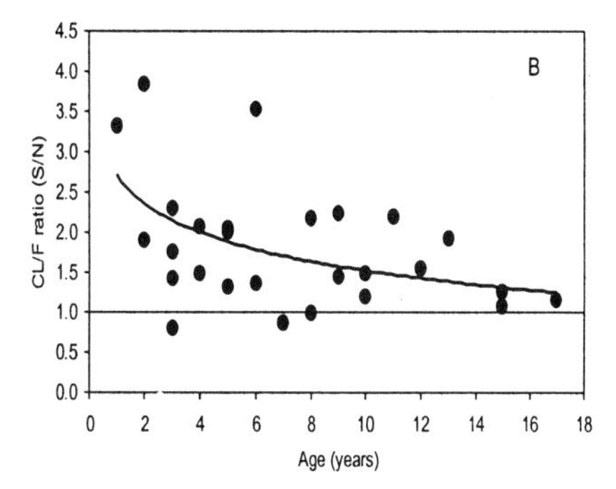

Figure 1A. A crossover comparison of cyclosporine steady-state apparent clearance (CL/F) in 27 pediatric liver transplant recipients receiving Sandimmune (*open circles*) and Neoral (*filled circles*). **B.** Cyclosporine apparent clearance ratio from Sandimmune/Neoral (S/N) versus age. The lines in both panels represent the best fit of a logarithmic function through the data to serve as a visual guide to age-related patterns. The plots were constructed by permission of Lippincott Williams and Wilkins based on data from Dunn *et al.*[31]

infants and young children have the greatest increase in bioavailability and that older children and adolescents have improvements similar to those reported in adult patients.[32] Hence, this study unmasked the predominant reason why young children undergoing liver transplantation needed higher weight-adjusted doses of Sandimmune. This was largely due to poorer bioavailability of cyclosporine from this formulation rather than a marked difference in cyclosporine clearance compared with older children and adolescents. This case study illustrates the importance of including patients over the full pediatric age range in pharmacokinetic studies and using age as a continuous variable to discover developmental changes or trends in disposition parameters.

2.5. Pediatric Programs and Ethical Considerations

2.5.1. Pediatric Drug Formulations

Before embarking on a pediatric clinical development program, a suitable pediatric drug formulation is essential. For orally administered drugs attention often focusses on liquid preparations. While these allow more accurate dose titration than tablets of fixed-dose strengths, liquids can be difficult to administered to uncooperative infants and children. Chewable tablets are an alternative which are preferred by children with chronic diseases.[33] Regardless of the oral dose form developed, stability and palatability are aspects needing attention.[34] A relative bioavailability study performed in healthy adults comparing the pediatric formulation versus the adult trial formulation is generally required before proceeding to pediatric studies.[21] For injectable drug formulations, the adult vial sizes and drug concentrations may be suboptimal for pediatric use insofar as they require substantial dilutions that may lead to dosing errors and administration of large fluid volumes in fluid-restricted neonates and infants.[33] In such cases, a pediatric vial size and concentration should be considered.[21, 27] Finally, inactive ingredients in drug formulations may be poorly tolerated by pediatric patients and should be scrutinized in this regard.[35]

2.5.2. Timing of Pediatric Development Programs

In order to avoid excessive delays between marketing approval of a drug for children relative to adults, pediatric studies should begin as soon as the product profile allows. The time to initiate pediatric clinical studies depends in general on the medicinal product category in which the drug product falls. Regulatory authorities recognize three categories.[21]

- For medicinal products to treat diseases affecting children predominantly or exclusively, testing begins in the pediatric patient population although early tolerability studies may be conducted in healthy adults if applicable.
- For medicinal products intended to treat serious or life-threatening diseases, occurring in both adults and pediatric patients, for which there are currently no or limited therapeutic options, pediatric studies are needed relatively urgently with early initiation following initial evidence of safety and effectiveness in adults. In this case pediatric study results should be part of the original marketing application database.

- For medicinal products intended to treat other diseases, pediatric studies may be delayed until sufficient safety and effectiveness data in adults are forthcoming to justify testing in pediatric patients.

Often pediatric trials can be conducted in parallel with adult trials but the submission of the adult file to regulatory authorities should not be delayed by waiting to complete the pediatric program.[14, 21]

Within a pediatric development program, pediatric studies may need to address different age groups sequentially—so-called, age de-escalation trials—with studies in neonates and infants initiated after demonstration of safety in older children and adolescents. This is particularly important for neurotropic drugs among others.[7] In addition, long-term studies may be justified to assess effects on behavioral, cognitive, sexual, and other developmental processes.[36]

2.5.3. Pharmacokinetic Study Designs and Sample Collection

The full pediatric program may consist of several studies including clinical studies to generate safety data and pharmacokinetic studies to determine dosing strategies. At least one pharmacokinetic study is likely to be included in all pediatric programs. This may be a dedicated study in which pharmacokinetics is the main objective or a clinical trial assessing safety and effectiveness in which pharmacokinetics is a secondary objective. In either case, the goals of pharmacokinetic studies are to evaluate the influence of maturation on drug disposition and to design an appropriate pediatric dose regimen to obtain similar systemic exposure to the drug as is safe and effective in adults. Accordingly, pediatric safety and effectiveness are also key components in the latter objective. There are two basic approaches to gathering pharmacokinetic data: standard and population approaches.[14, 21]

The standard approach involves studying a relatively limited number of pediatric subjects who receive either a single dose or multiple doses of drug with frequent blood sampling. Additional biological matrices sampled may include urine or saliva, depending on the drug. Single-dose studies may be appropriate for drugs that obey linear pharmacokinetics in adults. For drugs with time-dependent or nonlinear pharmacokinetics, multiple-dose studies are essential for a proper interpretation of the data. Timing of blood samples should be based on what is known from adult pharmacokinetic studies. In order to reduce the number of samples, optimal sampling theory may be useful to identify those sampling time points which are information-rich.[37] The number of patients to study depends on the breadth of the age range to be covered and the interindividual variability known from adults. Prompt assay of biological samples, preferably during the course of the trial, will alert investigators if sample timing or the number of patients enrolled into the study need to be amended. If a single-dose study is undertaken, a subsequent assessment of steady-state blood levels in pediatric patients on chronic treatment should be considered to corroborated that the desired exposure has been obtained based on single-dose pharmacokinetic extrapolations.[21] This could be accomplished by obtaining predose blood levels in a limited number of patients.

The population approach consists of obtaining two to four blood samples per patient usually in patients receiving the drug for therapeutic purposes. Sample timing can be random or the protocol can specify restricted postdose time windows within which a random sample is obtained. The latter can be guided by optimal sampling theory or based intuitively on what is known from adult pharmacokinetics. In either case, the time of dose administration and blood sampling should be accurately recorded. The population approach is usually applied in a larger cohort of patients than the standard approach. If samples are obtained after more than one dosing occasion, intraindividual interoccasion variability can be estimated.[19, 38]

Given the emotional trauma blood drawing may cause for infants and children, every effort should be made to minimize the occasions for venous access. Sparse and optimal sampling strategies are warranted as are the use of micro-volume drug assays and an indwelling venous cannula if these approaches do not compromise the scientific integrity and rigor of the data collected.[37, 39] The volume of blood taken should also be minimized; this is especially important in neonates whose total circulatory blood volume averages about 40 mL. The total amount of blood withdrawn in a 24-hour period in pediatric patients is generally recommended not to exceed ten percent of circulating blood volume.[39] The use of less invasive sample collection of urine or saliva may be appropriate if the correlation with blood concentrations has been established. Accurate, timed urine collections can be very difficult in infants and young children but provide important information for renally cleared drugs.[39]

2.5.4. Ethical Considerations in Pediatric Research

The ethical guidelines published by the American Academy of Pediatrics Committee on Drugs states that "there is a moral imperative to formally study drugs in children so that they can enjoy equal access to existing as well as new therapeutic agents." [17] Naturally, accompanying the benefits come various risks which must be minimized given the special vulnerability of the pediatric population. Minimal risk is defined as research activities with a level of risk similar to that encountered in the child's usual daily activity. Minimal risk would include procedures such as physical examinations, venipuncture, or urine sample collection. More than minimal risk may be justified if it is part of the treatment for the child's disease in the context of therapeutic research studies which offer the possibility of direct benefit to the individual research subject.[17, 39]

Investigational review boards which review pediatric study protocols must include or seek advice from health care professionals aware of the needs of pediatric research subjects. The adult legally acting on behalf of the child should be asked first for permission to approach the child. The study should be explained in terms suitable to the child's understanding and informed *assent* should be solicited from children with an intellectual age of seven years or older. Formal written *consent* should be solicited from the parent or legal guardian and children aged 13 and older should participate in the consent decision. Investigators must be respectful of the withdrawal of assent on the part of the research subject or of consent on the part of parents or guardian at any time during the course of the study.[17, 39] The principle of distributive justice requires that subjects enrolled in clinical investigations should represent a cross-section of society unless the disease for which the drug is being studied is limited to a particular subgroup.

2.6. Case Study: Basiliximab in Pediatric Kidney Transplantation

Basiliximab is a chimeric human/mouse monoclonal antibody directed to the interleukin-2 receptor (CD25) on activated T-lymphocytes which play a key role in rejection episodes after organ transplantation. It is indicated for the prophylaxis of acute organ rejection in patients receiving renal transplantation. Adult pharmacokinetic studies indicated that the elimination half-life is 7.4 ± 3.0 days with a total body clearance of 37 ± 15 mL/h. The adult dose is 20 mg administered by bolus intravenous injection or 20-minute infusion pretransplant and on day 4 posttransplant to provide 30 to 45 days of CD25 receptor saturation.[40]

A pediatric safety, pharmacokinetic, and pharmacodynamic trial employing an adaptive study design was initiated while adult phase 3 kidney transplant trials were in progress.[41, 42] The pediatric study planned to enrolled 40 newly transplanted patients over the age range one to 16 years. The relatively long elimination half-life and two-dose regimen of this drug in adults were conducive to the standard approach for pharmacokinetic data collection and analysis in the pediatric population. Sixteen 1-mL blood samples were collected from each patient over a three-month period posttransplant. Blood samples were drawn daily in the first week while the patient was hospitalized and had a intravenous catheter in place and then weekly to biweekly at clinic visits up to month three posttransplant at the same time that blood was scheduled to be taken for clinical laboratory tests. A validated enzyme-linked immunosorbent assay method was used to analyze basiliximab in serum. Individual pharmacokinetic parameters were derived by noncompartmental methods as well as nonlinear least squares fitting of a two-compartment pharmacokinetic model to each patient's concentration-time data. In a subset of patients, flow cytometric analysis was performed on blood samples to determine if the CD25 receptor on T lymphocytes was saturated by basiliximab as a biomarker of immunosuppressive activity.

The study was divided into two parts. In part 1, patients received the standard adult dose scaled to body surface area, namely 1.2 mg/m^2 (maximum 20 mg) on days 0 and 4. A protocol-specified interim pharmacokinetic analysis was performed to assess the systemic exposure in children relative to adults and to adjust the pediatric dosing algorithm if warranted. The interim evaluation in the first 12 patients revealed that clearance in infants and children 1 to 11 years of age (17 ± 6 mL/h, n = 8) was reduced by about half relative to adults but appeared age-independent over this range. Clearance in adolescents 12 to 16 years of age (45 ± 25 mL/h, n = 4) began to approach and attain adult values. A similar relationship was observed for weight versus clearance with an inflection point of about 40 kg. Flow cytometry in a 2-year-old and 11-year-old indicated that basiliximab

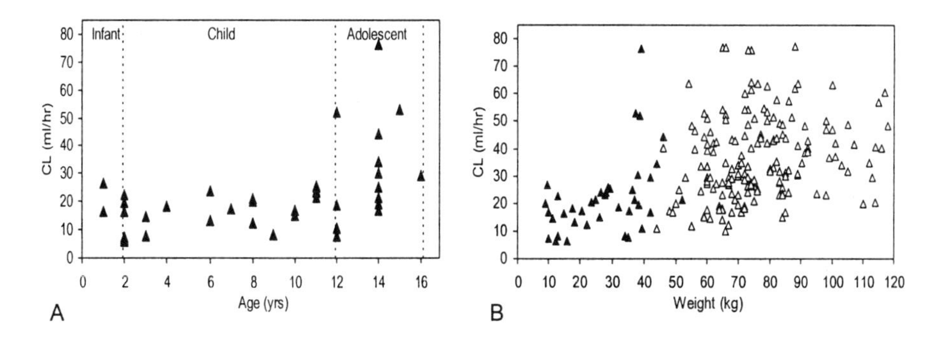

Figure 2A. Age-clearance (CL) relationship for basiliximab in pediatric kidney transplant patients. **B.** Weight-clearance relationship of basiliximab in pediatric (*filled symbols*) and adult (*open symbols*) kidney transplant patients. Reproduced by permission of Lippincott Williams and Wilkins from Kovarik et al.[42]

concentrations in excess of 0.2 μg/mL were sufficient to saturate CD25 and the corresponding duration of CD25 saturation was 26 and 46 days in these two children. Both the threshold receptor-saturating concentration and the durations of saturation were similar to adult values. Based on these interim data, the dosing algorithm was simplified in study part 2 whereby infants and children received two 10-mg doses and adolescents received two 20-mg doses using a weight cut-off of 40 kg for transferring between the two regimens. It was the intention in study part 2 to apply this regimen prospectively and to reexamine this weight cutoff from a fuller database.

The completed study in a total of 41 patients demonstrated that basiliximab was well tolerated and safe in pediatric de novo kidney allograft recipients.[41] As illustrated in Figures 2A and 2B, the full study confirmed the overall relationships between age and weight versus clearance observed in study part 1, and suggested 35 kg as an optimal weight cutoff for choosing the 10-mg versus 20-mg regimen in the pediatric population.[42] The results of this study were submitted to the Food and Drug Administration and the European Medicines Evaluation Agency for pediatric labeling. Marketing approval for use of basiliximab in pediatric kidney transplantation was granted in the United States and in the European Union. The pharmacokinetic data from this study appear in the "clinical pharmacology" section, the safety and clinical data are described in the "clinical studies" section, and the pediatric dose regimen identified in study part 2 appears in the "dose administration" section of the United States label[43] and corresponding sections of the European Summary of Product Characteristics.

2.7. Data Analysis and Product Labeling

2.7.1. Pharmacokinetic Data Analysis

In general, pharmacokinetic data are collected in pediatric patients to determine if biological maturation influences drug disposition and to use this information to derive a dosing algorithm which achieves similar exposure to that in adults. The standard approach should generate concentration-time data sufficient to derive individual pharmacokinetic parameters by noncompartmental or compartmental analysis including Cmax, AUC, half-life, and clearance for each patient. The population approach is generally supported by an underlying pharmacokinetic structural model such as a compartmental disposition model which is used to generate population typical values and their variances for pharmacokinetic parameters of interest. Subsequent Bayesian methods may be used to generate individual patient parameter estimates.[19, 44]

Both standard and population approaches are conducive to exploring the influence of covariates on pharmacokinetic parameters. The draft *Pediatric Pharmacokinetic Guidance* lists the following covariates which should be recorded and explored: height, weight, body surface area, gestational age and birth weight for neonates, relevant laboratory tests that reflect the function of organs responsible for drug elimination, and concomitant drug therapy. Scatterplots of covariates versus a relevant pharmacokinetic parameter such as clearance can be used to identify "developmental breakpoints" of age or weight at which pharmacokinetics change or to establish linear or nonlinear correlations.[14, 21] These relationships can be decisive in supporting a pediatric dosing regimen along with safety information. If sufficient numbers of patients are treated with a given comedication, the population approach may be used to explore for the influence of selected concomitant medications on the pharmacokinetics of the investigational drug.

2.7.2. Labeling Statements

In the United States product label, pediatric pharmacokinetic parameters should be listed in the "clinical pharmacology" section clearly indicating to which age ranges they pertain and the number of patients from which they were derived. If applicable, differences from adult pharmacokinetics should be stated. Safety and drug activity information in pediatrics should be described in the "precautions—pediatric use" section and in the "clinical studies" section, as appropriate. Pediatric dose regimens are given in the "dosage and administration" section. This should specify the age range to which the regimen applies and the dose should be expressed on a mg/kg or mg/m^2 basis, if possible, being mindful that accurately determining the body surface area of a neonate or infant can be difficult.[21] If there are special administration conditions for pediatric patients, these should be stated here. Pediatric formulations may be mentioned in the "preparations available" section. In the European Summary of Product Characteristics, the parallel sections to those mentioned above are "pharmacological properties", "special warnings and special precautions for use", "posology and method of administration", and "pharmaceutical form".

2.8. Case Study: Pediatric Labeling of Midazolam

Midazolam is one of the most commonly used benzodiazepines in pediatrics. Its rapid onset of activity and relatively short half-life are advantageous for short-term sedation in children undergoing diagnostic or therapeutic procedures. Until recently, it was available only as an injectable solution for intravenous and intramuscular administration. Oral syrup formulations were prepared extemporaneously by mixing the parenteral formulation with sweetener or juice.[45] As part of its pediatric program for this drug, the sponsor produced a new oral formulation, conducted pharmacodynamic and pharmacokinetic studies which yielded pediatric dosing information, and identified a subpopulation of pediatric patients at higher risk for adverse events.[7]

Pediatric patients who were scheduled to receive sedative premedication before minor in-hospital or day-stay procedures were enrolled in a multicenter, randomized, prospective study.[46] Seven of the eight participating centers were members of the Pediatric Pharmacology Research Unit network. The patients were stratified into three age groups: 6 months to < 2 years; 2 to < 12 years; and 12 to < 16 years Midazolam was administered intravenously as a 0.15 mg/kg bolus injection over two to three minutes or orally at a dose of 0.25, 0.5, or 1 mg/kg (maximum 40 mg) cherry-flavored syrup containing the equivalent of 2 mg/mL midazolam. The study randomized patients to one of three dosing arms for pharmacokinetic evaluations. In arm 1, midazolam was administered at one of the three oral dose levels under blinded conditions. In arm 2, a single intravenous and a single oral dose were administered in a open-label, randomized, crossover design with doses separated by at least 24 hours. In arm 3, a single intravenous dose was administered. Twelve 2.5-mL blood samples were obtained after each administration over a total period of ten hours. In patients weighing less than 12 kg, the sampling strategy was modified to limit total blood removal to less than three percent of the patient's estimated blood volume (75 mL blood per kg). Midazolam and alpha-hydroxymidazolam were quantified by a validated chromatographic method.

Of the 111 patients enrolled, 87 were evaluable for pharmacokinetics. Both midazolam C_{max} and AUC were dose proportional over the oral dose range applied. There was a trend of increasing half-life and a corresponding decrease in total body clearance with rising age. The metabolite-to-parent AUC-ratios suggested that patients 2 to < 12 years old have a slightly higher metabolism capacity or slower clearance of the metabolite than adolescents or adults.

The absolute bioavailability of midazolam from the syrup could only be evaluated in six patients using standard noncompartmental analysis and averaged 36 percent. To supplement this estimate, a population two-stage approach was applied to data from all three study arms along with published data. In stage 1, individual estimates of CL and CL/F after intravenous and oral administration, respectively, were generated by noncompartmental methods. In stage 2, these intravenous and oral clearance estimates were simultaneously modeled using nonlinear mixed-effects analysis to derive population estimates of midazolam clearance and bioavailability across age groups. As applied to a total of 47 patients, the population estimate of bioavailability was 35 percent. This analysis also revealed that midazolam CL/F changed with weight or age due to changes in clearance rather than bioavailability.

The effectiveness of midazolam syrup as a premedicant to sedate and calm pediatric patients prior to induction of general anesthesia was compared among three dose levels (0.25, 0.5, and 1.0 mg/kg up to a maximum dose of 20 mg) in a randomized, double-blind, parallel-group study.[47, 48] More than 90 percent of treated patients achieved satisfactory sedation and anxiolysis at at least one timepoint within 30 minutes posttreatment. A significantly higher proportion of patients in the 1.0 mg/kg dose group exhibited satisfactory sedation and anxiolysis as compared to the 0.25 mg/kg group.[48] A study in 22 children undergoing elective repair of congenital cardiac defects suggested that hypercarbia or hypoxia following premedication with oral midazolam might pose a risk to children with congenital heart disease and pulmonary hypertension, although there are no known reports of pulmonary hypertensive crises that had been triggered by premedication.[48]

On the basis of these data, midazolam syrup was approved for marketing in pediatric patients as a single dose for preprocedural sedation and anxiolysis. The recommended dose for pediatric patients is a single dose of 0.25 to 0.5 mg/kg, depending on the status of the patient and desired effect, up to a maximum dose of 20 mg. Younger (6 months to < 6 years of age) and less cooperative patients may require a higher than usual dose up to 1.0 mg/kg. For all pediatric patients, a dose of 0.25 mg/kg should be considered for patients with cardiac or respiratory compromise, other higher-risk surgical patients, and patients who have received concomitant narcotics or other central nervous system depressants. It is supplied as a cherry-flavored syrup containing the equivalent of 2 mg/mL of midazolam. This information is described in the "dosage and administration" and "how supplied" sections of the product label.[48]

3. GERIATRIC POPULATION

Elderly or geriatric patients are generally defined as those 65 years of age or older. The frail elderly are defined as those persons over 65 years of age who depend on others for activities of daily living.[49] Not only is the proportion of the world's population which is elderly increasing, but they are also becoming more vocal that their medical needs be better recognized in drug development and product labeling. The Alliance for Aging Research has proposed that incentives to the pharmaceutical industry similar to those operative in pediatrics be extended to research in geriatrics.[50]

3.1. Need for Pharmacokinetic Research in Geriatrics

The elderly may be considered a vulnerable population insofar as decline in organ function, changes in receptor activity, and prevalence of polypharmacy increase the chances that they may suffer from adverse events secondary to drug overexposure and to unrecognized drug-drug interactions. Characterizing age-related changes in pharmacokinetics is the key to uncovering potential problems and to designing safe and effective drug regimens in this population.[51]

Age-related changes in pharmacokinetics have been extensively reviewed.[52, 53] General pharmacokinetic trends in the elderly may be highlighted as follows. Decreased salivation and esophageal motility disorders are common in the elderly which can make administration of solid oral dose formulations difficult. Although several changes in gastrointestinal function have been documented in the elderly, there is scant evidence that these changes cause clinically significant impairment of drug absorption. Body fat increases and total body water decreases with advancing age and these changes have had documented effects on the distribution volume for some drugs. In general, cytochrome P450 enzyme activity is similar in the elderly compared with nonelderly adults; hence, the observation of decreased clearance in the elderly of some drugs undergoing extensive biotransformation is probably attributable to age-related reductions in liver size and blood flow. A measurable and predictable decline in renal function with increasing age can have a clinically significant effect on drugs which are renally excreted.[49, 52, 53] Alone or in aggregate, these trends can alter the disposition of drugs in the elderly.

In addition to potential changes in drug disposition, pharmacokinetic parameters appear to have higher interindividual variability in the elderly and the elderly appear more prone to drug-related adverse events. Underlying mechanisms may include wide heterogeneity in the rate of decline in hepatic and renal function with ageing, increased sensitivity of drug receptors, and occult drug interactions due to multiple comedications. There is an attempt to combine these multiple factors under a unifying concept of "frailty".[49] Whether this emerging concept will have scientific merit remains to be seen.

3.2. Case Study: Geriatric Product Labeling for Benzodiazepines

The elderly are almost twice as likely to be prescribed benzodiazepine hypnosedatives as the nonelderly. The incidence of use is even higher in institutionalized and frail elderly.[52] The elderly appear to be generally more sensitive to the effects of benzodiazepines based on pharmacokinetic and pharmacodynamic data.[52, 54] Epidemiological studies suggest an association between benzodiazepine use and cognitive impairment, daytime sedation, falls, and hip fractures.[52] Accordingly, it is generally recommended that benzodiazepines be dosed cautiously in the elderly initiating therapy at low doses and titrating to effect with small dose increments. Likewise, discontinuation of benzodiazepines should be carried out with small dose decreases over a suitable time period relative to the half-life of the drug in the elderly.[54]

The author consulted the 2001 edition of the *Physician's Desk Reference* to determine whether a geriatric mention was included in benzodiazepine product labeling. Chosen were the 13 benzodiazepines most widely prescribed in elderly outpatients in an oral dose formulation (tablet, capsule, or solution) as tabulated by Maletta *et al.*[54] Three sections in the product label were assessed for geriatric-specific data: "clinical pharmacology", "precautions" or "warnings", and "dosage and administration".

Nine of the thirteen drugs had full labeling represented. The median time since last label update was 3.0 years with a range of 0.9 to 6.0 years. The "clinical pharmacology" section either cited geriatric-specific pharmacokinetic parameters or stated that there was no difference in pharmacokinetics with age in only two labels (22.2 percent). All nine labels included a geriatric mention under either "precautions" or "warnings" generally to the effect that "the elderly may be more sensitive to the effects of benzodiazepines". This conforms to the recommendation of the Food and Drug Administration.[5] Likewise, all nine labels included a geriatric-specific dose recommendation in the "dosage and administration" section, generally to the effect that "in elderly and debilitated patients it is recommended that dosage be limited to the smallest effective amount to preclude the development of ataxia, oversedation, or confusion and that the dose be increased gradually as needed and tolerated."

Hence, this informal survey of geriatric labeling for benzodiazepines indicates that while geriatric-specific pharmacokinetic data are seldom cited, the implications of the greater sensitivity to benzodiazepines in the elderly for dosing are captured in current product labeling.

3.3. Geriatric Drug Development Programs

The International Conference on Harmonization issued *Industry Guidance E7* on geriatric studies in 1993.[55] The frequent occurrence of underlying diseases, concomitant drug therapy, and the consequent risk of drug interactions in this population are cited as justification for special consideration of this age group during drug development. The *Geriatric Guidance* is directed mainly toward new molecular entities likely to have significant use in the elderly. In addition, it is directed toward new formulations or combinations of existing products for which conditions common in the elderly may affect the pharmacokinetics or response to the drug if not already dealt with in current labeling. While the common definition for elderly as persons aged 65 and over is accepted, it is recommended that efforts be made to enroll patients 75 and older in drug development studies to the extent possible. Geriatric patients should be included in phase 3 trials in meaningful numbers with a minimum of 100 patients considered necessary to detect clinically important differences in this population. For drugs intended to treat diseases

Table 2. Geriatric Components in Drug Development Programs

1	*Basic pharmacokinetics in the elderly* Formal pharmacokinetic study or population screen
2	*Drug disposition under conditions of organ impairment* Conventional renal and hepatic impairment studies either in elderly or nonelderly adults
3	*Exposure-response relationships in the elderly* Not routinely performed unless the drug is to be used exclusively to treat the elderly or there is an indication that the relationship may be different in elderly versus nonelderly
4	*Drug-drug interaction studies* If the conventional drug interaction program does not include drugs likely to be coadministered in the elderly, specific studies may be added based on a mechanistic rationale. These studies may be performed either in elderly or nonelderly subjects.

characteristically associated with ageing, geriatric patients are expected to constitute the major portion of the clinical database.

The cardinal importance of characterizing drug disposition in the elderly is recognized inasmuch as fully two-thirds of the *Geriatric Guidance* is devoted to this topic. The *Geriatric Guidance* specifically states: "Most of the recognized important differences between younger and older patients have been pharmacokinetic differences, often related to impairment of excretory (renal and hepatic) function or to drug-drug interactions." Table 2 lists the four components which should be addressed in a comprehensive geriatric pharmacokinetic program.

3.3.1. Basic Pharmacokinetics in the Elderly

This component is addressed either in a formal pharmacokinetic study or a pharmacokinetic screen. The former is a dedicated study either in healthy elderly subjects or in elderly patients with the disease to be treated by the drug. An initial or pilot study may be conducted under single-dose or steady-state conditions in a sufficient number of elderly persons to allow detection of a sizable difference in drug exposure between elderly and nonelderly adults. If a clinically relevant age-related difference is found, the pilot study may need to be confirmed in a properly powered comparative study in nonelderly and geriatric subjects at steady state. The pharmacokinetic screen involves collecting one or two blood concentrations either predose (trough) or at other specified times during a steady-state dose interval from sufficient numbers of geriatric patients to detect age-related differences in drug exposure compared with younger adults. To properly interpret the pharmacokinetic data, demographic and disease characteristics should also be recorded such as sex, weight, ethnicity, measures of renal and hepatic function, and concomitant illnesses. Regardless of the approach taken, it is prudent—but not mandated—to include nonelderly subjects in the study to facilitate the comparison between age groups and to avoid the potential difficulties of a cross-study evaluation.

3.3.2. Drug Disposition under Conditions of Organ Impairment

The elderly are more likely to suffer from compromised renal and hepatic function than the nonelderly. Because these two organs are the primary routes for xenobiotic elimination, age-related organ impairment may lead to altered drug exposure in this population. It is therefore important to quantify the influence of organ impairment on the pharmacokinetics of a drug to alert the clinician when caution should be exercised in the use of the drug and to derive guidelines if dose adjustment is warranted. Conventional hepatic and renal impairment studies performed either in nonelderly or elderly subjects can satisfy this requirement. These studies are described more fully below in sections 6 and 7.

3.3.3. Exposure-Response Relationships in the Elderly

If the investigational drug is intended to treat a disease which primarily or exclusively occurs in the elderly, then studies to characterize exposure-response relationships in patients during treatment will likely include geriatric patients. Outside of this drug category, a study of exposure-response relationships in the elderly is not routinely performed as there are few examples of clinically significant age-related pharmacodynamic differences. Nonetheless, such a study should be considered for drugs with central nervous system effects to which the elderly may be more sensitive and for drugs with age-associated differences in effectiveness or adverse reactions noted during development which are not explained by pharmacokinetics alone.

3.3.4. Drug Interactions with Coadministered Drugs Likely to be Used in the Elderly

This component will generally be covered in the overall clinical pharmacology development program but there may be specific drugs commonly taken by the elderly not included in drug-drug interaction studies to support a therapeutic indication directed toward the general adult patient population. If additional drug interaction studies of this type are needed they could be addressed with either a formal pharmacokinetic study in nonelderly or elderly subjects or a pharmacokinetic screen in phase 3. The latter approach has some restrictions, however. Firstly, a sufficient number of patients must have received the concomitant drug in order to detect a relevant difference in exposure if one exists. Secondly, full dosing histories of concomitant drugs need to be captured in the case report forms for proper documentation of coadministration and interpretation of the data. Thirdly, lack of an interaction cannot generally be claimed in product labeling by this approach, only that an interaction was not detected.[56] Finally, this approach is usually applied to evaluate the influence of comedications on the investigational drug. The study conditions and blood sampling scheme may not be optimal to retrospectively assess the interaction in the reverse direction.

The *Geriatric Guidance* acknowledges that the drug interaction study program must be designed on a case-by-case basis but generally recommends the following drugs be included in the drug interaction study package: (1) digoxin and oral anticoagulants because of the frequency of use in the elderly and their narrow therapeutic ranges; (2) a prototype hepatic-enzyme inducer and inhibitor for investigational drugs undergoing extensive metabolism; and (3) a prototype cytochrome P450 inhibitor if the investigational drug is metabolized via known isozymes in this family.[55]

3.4. Case Study: Rivastigmine Drug Development Program

Rivastigmine is a reversible cholinesterase inhibitor indicated for the treatment of mild to moderate dementia of the Alzheimer's type. It is available as capsules for twice daily oral administration.[57] As a drug used in a disease which primarily occurs in the elderly, its development program illustrates several of the components outlined in the *Geriatric Guidance* for studies in the elderly.

The basic pharmacokinetic properties of rivastigmine such as absolute bioavailability, food effect, radiolabeled study, and penetration into the cerebrospinal fluid were conducted in nonelderly subjects. These data appear in the "clinical pharmacology—pharmacokinetics" section of the product label. A dedicated parallel-group single-dose study compared the pharmacokinetics of rivastigmine in 24 elderly versus 24 nonelderly subjects using a standard design with intensive blood sampling and demonstrated a minor 30 percent lower oral clearance in the elderly. This is described in the "clinical pharmacology—special populations" section. A population pharmacokinetic evaluation in 625 patients enrolled in phase 3 trials provided further pharmacokinetic data in elderly but are not specifically mentioned in this regard in the "clinical pharmacology" section of the product label.

Dedicated studies were performed in subjects with hepatic and renal impairment to assess whether dose adjustments are needed under these conditions. Rivastigmine is primarily metabolized through hydrolysis by esterases with minimal metabolism via the major cytochrome P450 isoenzymes. Single-dose and multiple-dose studies demonstrated 60 to 65 percent lower oral clearance in hepatic impaired subjects versus healthy subjects. The dosing recommendation in the "clinical pharmacology—special populations" section states "Dosage adjustment is not necessary in hepatically impaired patients as the dose of drug is individually titrated to tolerability." The major elimination pathway of rivastigmine is via the kidneys. A single-dose study demonstrated 64 percent lower oral clearance in moderate renal impairment and 43 percent higher oral clearance in severe renal impairment versus healthy subjects. The dosing recommendation in the "clinical pharmacology—special populations" section states "For unexplained reasons, the severely impaired renal patients had a higher clearance of rivastigmine than moderately impaired patients. However, dosage adjustment may not be necessary in renally impaired patients as the dose of the drug is individually titrated to tolerability". All three studies were performed in nonelderly subjects.

Exposure-response relationships mentioned in the "clinical studies" section of the label consist of dose-response observations from the two randomized, double-blind, placebo-controlled clinical trials which demonstrated the effectiveness of rivastigmine in the treatment of patients with Alzheimer's Disease. Dose levels were placebo, 1 to 4 mg/day, and 6 to 12 mg/day given in divided doses twice daily in a forced titration followed by a maintenance phase. The response was the cognitive subscale of the Alzheimer's Disease Assessment Scale which examines selected aspects of cognitive performance including elements of memory, orientation, attention, reasoning, language, and praxis. In one trial, both treatments were statistically significantly superior to placebo and the higher dose was significantly superior to the lower dose. In the other trial, the upper dose was statistically significantly superior to placebo, as well as to the lower dose. The mean age of patients participating in these trials was 73 years with a range of 41 to 95 years. A subanalysis of clinical outcome did not show any age-effect. Because the elderly were well represented in these trials and there was no age effect, a dedicated exposure-response study in elderly was not necessary.

Drugs that induce or inhibit cytochrome P450 metabolism are not expected to alter the metabolism of rivastigmine. This was supported by single-dose pharmacokinetic studies in nonelderly subjects with digoxin, warfarin, diazepam, and fluoxetine as stated in the "clinical-pharmacology—drug interactions" section of the label. The population pharmacokinetic analysis in 625 patients including elderly patients showed that the pharmacokinetics of rivastigmine were not influenced by commonly prescribed medications such as antacids (n = 77), antihypertensives (n = 72), beta-blockers (n = 42), calcium channel blockers (n = 75), antidiabetics (n = 21), nonsteroidal antiinflammatory drugs (n = 79), estrogens (n = 70), salicylate analgesics (n = 177), antianginals (n = 35), and antihistamines (n = 15).

Hence, all four components mentioned in the *Geriatric Guidance* were included in the rivastigmine drug development program. Because there were only minor differences between elderly and nonelderly subjects, the basic pharmacokinetics, disposition in organ impairment, and single-dose drug interactions could be assessed in studies in nonelderly subjects. Inasmuch as the elderly constituted a considerable part of the clinical trial population, they were well represented in the clinically-derived dose-response relationships and in the associated screen for drug interactions using the population pharmacokinetic approach.

3.5. Geriatric Use Labeling

The Food and Drug Administration has issued a guidance for industry on the content and format for geriatric labeling[58] which outlines how to submit and include in the product label information on the use of a drug in the elderly. Data which can be submitted for geriatric labeling include effectiveness, pharmacokinetics, exposure-response relationships, and safety based on dedicated studies, database analyses, or literature searches. The geriatric labeling regulation[5] and the *Geriatric Labeling Guidance*[58] outlines various options for statements in the "precautions—geriatric use" subsection and other sections of the label depending on the type of information available and its interpretation. The five information categories are outlined below; of particular note for special population studies are options 3 and 5.

1. *Specific geriatric indication.* If an adequate and well-controlled study in the elderly supports a specific geriatric use, this is stated in the "indications and usage" and "dosage and administration" sections. Any limitations or precautions with respect to the geriatric indication are cited in the "precautions—geriatric use" subsection.
2. *Geriatric use for the general indication.* A safety/effectiveness statement on geriatric use for the same indication as for the general population is included in the "precautions—geriatric use" subsection. Three wording options are suggested in the *Geriatric Labeling Guidance* based on whether sufficient numbers of elderly have been studied—more than 100 patients is generally regarded as sufficient—and whether any difference in response to the drug was noted.
3. *Clinical pharmacology information.* If a dedicated pharmacokinetic or exposure-response study has been performed in the elderly this is briefly mentioned in the "precautions—geriatric use" subsection and details given in the "clinical pharmacology—special populations" subsection or the "precautions—drug interactions" subsection depending on the nature of the study. If the drug is substantially excreted by the kidneys, a statement must be made in the "precautions—geriatric use" subsection that care should be taken in dose selection and monitoring of renal function may be useful during use of the drug in the elderly.
4. *Specific hazards in geriatric use.* These are described as appropriate in "contra-indications", "warnings" or "precautions" sections.
5. *Enhancing safe geriatric use.* Statements reflecting good clinical practice or past experience in a particular situation which may promote safe use in the elderly can be made in the "precautions—geriatric use" subsection. An example is initiating sedative drug therapy at the lower end of the therapeutic dose range with slow titration to response in the elderly because they may be more sensitive to these agents.

If none of the above options is relevant to the drug, the sponsor must provide reasons for omission of the statements, for example, if the drug is unlikely to be used in the elderly.

4. SEX-RELATED DIFFERENCES, PREGNANCY, AND LACTATION

4.1. Addressing the Needs of Women During Drug Development

The pharmaceutical literature of the last few years is replete with studies assessing whether the pharmacokinetics of a drug differ between the sexes either in a post hoc exploratory analysis or in a prospectively designed clinical pharmacology study. This was not always so. Just 25 years ago, women of childbearing potential were categorically excluded from early drug development studies out of concern for the possible effects of an investigational drug on the fetus should the research subject become pregnant.[59] By the late 1980s the exclusion of women from early development trials was interpreted not only as rigid, overprotective, and paternalistic but also as a hindrance to providing women with the full benefits of new medicines. Various interest groups stressed that women themselves should decide if they wish to participate in a clinical trial after being fully informed about the risks and benefits of the study.

A survey by the Food and Drug Administration of the relative numbers of men and women in the databases of approved drugs in 1988 revealed that women were generally adequately represented in the later-phase development trials with a few exceptions. Nonetheless, in about half the cases, the sponsor had not analyzed the databases to assess potential differences in drug response between men and women.[3] Contemporary with this survey was the issuance by the Food and Drug Administration of a *Guideline for the format and content of the clinical and statistics sections of new drug applications* which called for a "by-gender analysis" of differences in response to the drug between men and women.[60] The lack of women in early phase development trials was specifically addressed in the *Guideline for the study and evaluation of gender differences in the clinical evaluation of drugs* issued by the Food and Drug Administration in 1993.[3] This guideline delineates the Food and Drug Administration's expectations regarding inclusion of men and women in drug development, analyses of clinical data by sex, assessment of potential pharmacokinetic differences between the sexes, and the conduct of specific additional studies in women if indicated.

Before addressing each of these issues in more detail, a note on terminology is warranted. The pharmaceutical literature, regulatory guidances, and product labeling use the terms *gender* and/or *sex* when referring to differences in drug disposition or response between men and women. These terms are not interchangable. *Sex* refers to biological differences between men and women, whereas, *gender* refers to differences attributable to culture or environment.[61] By way of example, when hormones or hormonal patterns over time influence drug disposition, they may lead to *sex*-related pharmacokinetic differences between men and women. On the other hand, when men and women in specific cultures consume different foods or amounts of alcohol or nicotine, drug disposition may be differentially affected. Such pharmacokinetic differences are *gender*-related (although, in this example, they would more appropriately be described as differences due to food or drug interactions which cosegregate with gender). Hence, *sex* rather than *gender* is generally—but not exclusively—the correct term to describe differences in drug disposition between men and women.[61]

4.2. Participation of Women in Drug Development Studies

In the *Gender Guideline*, the Food and Drug Administration jettisoned the restriction on the participation of women of childbearing potential in early clinical trials including clinical pharmacology studies and early therapeutic studies.[3] It further encouraged the enrollment of men and women in later phase 2 and phase 3 trials in numbers adequate to detect clinically-relevant differences in their response to the drug. If pharmacokinetic data are also collected, this is an added source of information to confirm or exclude sex-related differences. However, the *Gender Guideline* points out that data in later drug development phases is often too late to be used to guide pivotal study design and dose selection. Therefore, the inclusion of women in earlier phase trials is encouraged. If a disease under study is serious and affects women, "a case could be made for requiring that women participate in clinical studies at an early stage".[3]

It is expected that reproductive toxicology studies will usually be completed by the end of phase 2 before large-scale exposure of women of childbearing potential to the drug.[3] Proper precautions to guard against inadvertent exposure of fetuses to potentially toxic drugs should be taken in all development phases. These include providing women volunteers with full information on potential risks, use of routine pregnancy testing (beta-human chorionic gonadotropin), timing of short-term studies to coincide with or immediately follow menstruation, and abstinence or use of reliable contraception for the duration of drug exposure.[3, 62] For drugs with a toxic potential for the fetus, inclusion of women who are surgically sterilized or postmenopausal should be considered until the risk is better defined.[62]

The *Gender Guideline* acknowledges that possible hormonally-mediated differences in metabolism or shifts in fluid balance accompanying the menstrual cycle could influence the disposition or pharmacokinetic variability of some drugs. This has given rise to concern from sponsors about including women in bioequivalence trials the outcome of which may be particularly sensitive to pharmacokinetic variability. The *Gender Guideline* counters this concern by clearly stating, "there is no regulatory or scientific rational for the routine exclusion of women from bioequivalence trials".[3] Recommended measures to counteract such influences are to administer the drug at the same phase of the menstrual cycle in a crossover study or to increase the number of subjects in the study.[63]

4.3. Case Study: Women in Bioequivalence Studies

The *Gender Guideline* on sex-related analyses in drug applications acknowledged a general hesitance to enroll women in bioequivalence trials. Concerns included practicality, liability for drug effects on the fetus should a subject become pregnant, and increased pharmacokinetic variability due to hormonal fluctuation and concomitant use of oral contraceptives. In an effort to resolve the pharmacokinetic concerns, the Food and Drug Administration undertook two initiatives: (1) to review the medical literature for the effect of the menstrual cycle on intrasubject pharmacokinetic variability and (2) to perform a retrospective by-sex analysis of bioequivalence trials that studied both men and women.

Their comprehensive literature search from 1977 to 1994 yielded 20 studies that assessed the variance of pharmacokinetic parameters at different phases of the menstrual cycle on 19 drugs. Fifteen of the 20 studies (75 percent) either did not show any significant ($p > 0.1$) pharmacokinetic differences during the cycle or exhibited differences of a magnitude unlikely to substantially influence bioavailability or bioequivalence. Most of these drugs were biotransformed by conjugation, hydroxylation, and/or oxidation. Four studies (25 percent) detected pharmacokinetic changes during the menstrual cycle which might influence bioequivalence study outcomes. The drugs involved were theophylline, caffeine, xylose, and methaqualone. Fluctuations in exposure for the first three were within the normal ranges seen in bioequivalence trials. Only methaqualone provided well documented evidence for an influence from the menstrual cycle. The investigators concluded that "Overall, the literature review revealed limited evidence to suggest a significant or important impact of the menstrual cycle on the pharmacokinetics of drugs....Although specific data are lacking, participation of women in bioequivalence trials does not appear likely to compromise the results of a bioequivalence study." [63]

The second initiative was an exploratory analysis of 26 bioequivalence studies involving men and women that were submitted to the Food and Drug Administration. Three areas were addressed: intrasubject variability, sex-by-formulation interactions, and pharmacokinetic differences between the sexes. There were no statistically significant differences in intrasubject variability in 87 percent of the Cmax and AUC datasets. Significant by-sex differences were detected in six cases each for Cmax and AUC. For either Cmax or AUC, men exhibited higher variability in only one of these six instances. A signal for a sex-by-formulation interaction was defined as a 20 percentage-point difference in the test/reference mean parameter ratio between sexes. Such an interaction was infrequent inasmuch as only 13 percent of Cmax or AUC datasets met this criteria. A 20 percent or greater difference in pharmacokinetic parameters from the reference product between men and women was defined as a sex-related difference. This was observed in 39 percent of the datasets; in 29 percent of the datasets this difference was statistically significant. After correction for bodyweight, this percentage dropped to 15 percent. The investigators concluded: "Although exploratory, the results of this study support recommendations of the 1993 Food and Drug Administration gender guideline that women not be excluded from bioequivalence trials." [64]

4.4. Sex-related Pharmacokinetic Differences

The *Gender Guideline* emphasizes that for practical and theoretical considerations, the evaluation of possible sex-related differences in drug response should initially focus on pharmacokinetics. This can be accomplished either in a dedicated study or via a pharmacokinetic screen in the context of phase 2 and 3 trials. A screen involves obtaining a small number of blood samples in the treatment population to detect differences in drug exposure in subpopulations based, in this case, on sex. When such blood level data are available in addition to the by-sex analysis of clinical safety and effectiveness data, the combination can be a powerful screening tool. Should differences between men and women be found, dedicated studies to more carefully explore the magnitude of these differences can be conducted and the impact on the dose regimen for men and women quantified. It is essential when analyzing by-sex pharmacokinetic data, that the pharmacokinetic parameters be normalized to bodyweight or body surface area to better distinguish between true sex-related differences and differences due merely to body size.

Sex-related differences in pharmacokinetics have been extensively reviewed.[65, 66, 67] The following generalizations can be made. Women tend to have smaller body size, and a proportionally higher body fat content and lower body water volume, compared with men.[68] Protein binding exhibits minor sex differences. Data on cytochrome P450 activity suggests that CYP3A activity in women is about 1.4 times that in men, whereas,

CYP2C19 activity may be higher in men than women. Currently, no definite conclusions about inter-sex differences in CYP2D6 or CYP1A2 can be made.[65] Glomerular filtration rate is, on average, higher in men than in women, but this is due to differences in bodyweight rather than a true sex difference. Overall, these pharmacokinetic differences are generally of a small magnitude relative to other sources of total interindividual variability in exposure. Consequently, sex-related pharmacokinetic differences seldom translate into different dose regimens in men and women.[65, 67]

On this basis, screening for differences in exposure between men and women in an early therapeutic trial or clinical pharmacology studies is generally sufficient to assess whether a sex-based difference in exposure of a magnitude to be of clinical importance is present or not. Pooling clinical pharmacokinetic studies from throughout the development program and exploring for a sex difference in exposure with population pharmacokinetic methods can provide a firmer basis from which to make conclusions. The presence or lack of an influence of sex on drug exposure should be complemented with a by-sex analysis of phase 2 and 3 data with respect to drug response. Pooling of databases from several development studies and applying meta-analysis methods can be helpful in this regard.

4.5. Case Study: Fluvoxamine in Male and Female Pediatric Patients

Fluvoxamine maleate is a selective serotonin reuptake inhibitor indicated for the treatment of obsessions and compulsions in patients with Obsessive Compulsive Disorder. Fluvoxamine exhibits nonlinear pharmacokinetics. Although a sex-related difference in the fluvoxamine dose-concentration relationship has been reported in adults,[69] exploratory analyses for effects on outcomes in adult clinical trials did not suggest any differential responsiveness on the basis of sex.[70] The fluvoxamine product label does not distinguish adult men from women in the "dosage administration" section.

The multiple-dose pharmacokinetics of fluvoxamine were characterized in children, ages six to 11, and adolescents, ages 12 to 17. As shown in Table 3, weight-normalized AUC and Cmax in children were 1.5- to 2.7-fold higher than in adolescents; whereas, exposure in adolescents was similar to that reported in adults. When examined by sex, female children showed significantly higher AUC and Cmax compared with male children. No sex-related differences in pharmacokinetics were observed among adolescents.[70]

The effectiveness of fluvoxamine for the treatment of moderate-to-severe Obsessive Compulsive Disorder was demonstrated in a 10-week multicenter, parallel group study in a pediatric outpatient population consisting of children and adolescents eight to 17 years old. Patients were titrated to a total daily dose of approximately 100 mg/day over the first

Table 3. Fluvoxamine pharmacokinetics in pediatric patients.

Parameter	Male children (n = 7)	Female children (n = 3)	Adolescents (n = 17)
Cmin (ng/mL/kg)	6.6 ± 6.1	21.2 ± 17.6	2.9 ± 2.0
Cmax (ng/mL/kg)	9.1 ± 7.6	28.1 ± 21.1	4.2 ± 2.6
AUC (ng·h/mL/kg)	96 ± 84	294 ± 233	44 ± 28

Patients received fluvoxamine 100 mg twice daily orally.
Data are mean ± sd. Cmin, minimum concentration; Cmax, maximum concentration; AUC, area under the concentration-time curve.
Data are from the Luvox (fluvoxamine) product label.[70]

two weeks of the trial, following which the dose was adjusted within a range of 25 to 100 mg twice daily based on response and tolerability. The mean baseline rating on the Children's Yale-Brown Obsessive Compulsive Scale total score was 24. Patients receiving fluvoxamine experienced mean reductions of approximately six units compared with three units for placebo patients. Post hoc exploratory analyses for sex-related effects on outcomes did not suggest any differential responsiveness. Further exploratory analyses however revealed a prominent treatment effect in children and essentially no effect in adolescents. The higher drug exposure in children compared with adolescents was suggestive that decreased exposure in adolescents may have been a factor.[70, 71]

The "dosage and administration—pediatrics" subsection of the product label reflects the differential data on both pharmacokinetics and outcomes in this population: "Physicians should consider age and gender differences when dosing pediatric patients....Therapeutic effect in female children may be achieved with lower doses. Dose adjustment in adolescents...may be indicated to achieve therapeutic benefit." [70]

4.6. Influence of Endogenous Gonadal Steroids on Pharmacokinetics

Temporal patterns of the sex hormones testosterone, estrogen, and progesterone over the life cycle may differentially influence drug disposition in men and women. In men testosterone levels gradually decline with age such that a pronounced effect on drug pharmacokinetics or variability is generally not expected. In women, by contrast, estrogen and progesterone temporal patterns define various stages in the female life cycle which need to be considered in fully addressing optimal drug use in women. Specific issues in this context are the influence of the menstrual cycle and menopause on drug disposition.

Hormone surges over the course of the menstrual cycle may theoretically influence the disposition of drugs. Briefly, estrogen and progesterone levels are at a minimum at menses in the early follicular stage and are elevated prior to ovulation in the late follicular stage. In the mid-luteal phase, estrogen concentrations rise again while progesterone concentrations plateau. Monthly changes in fluid volume could affect drug distribution and clearance and estrogen levels could effect the activity of cytochrome P450 enzymes. A number of studies have addressed the influence of the menstrual cycle on a handful of drugs, but in general, influences of a magnitude justifying a dose alteration have not been found.[65, 72] The fluctuation in pharmacokinetic parameters which may occur over the course of the menstrual cycle are generally narrower than interindividual variability in the population as a whole.

When assessing the impact of the menstrual cycle on the pharmacokinetics of a drug, both the follicular and luteal phases need to be investigated, ovulation should be documented by plasma or urine luteinizing hormone concentrations, and the study needs to be properly powered. If possible, assessing more than one menstrual cycle can provide information on the within-subject consistency of results.[72]

Drug disposition after menopause is a generally neglected area of research although a few examples of such studies have been reported.[73] Speculation on the mechanism of postmenopausal changes in pharmacokinetics centers on the role of estrogen and progesterone levels and on decreased enzyme activity after menopause.[73]

4.7. Drug Interactions with Exogenous Gonadal Steroids

Commonly used exogenous gonadal steroids include oral contraceptives in women of childbearing potential, estrogen replacement therapy in postmenopausal women, androgens for women with endometriosis, and estrogen therapy for men with prostate cancer. The need for drug interaction studies with exogenous steroids should be based on knowledge of the pharmacokinetics of both the exogenous steroid and the drug under development coupled with a mechanistic understanding of their actions. Preclinical and in vitro models that assess the effect of hormones on drug metabolism can help to guide the design of drug interactions studies.

By far the most widely-performed interaction study in this regard is with oral contraceptives. The most commonly prescribed oral contraceptive agents are combination products containing estrogen, such as ethinyl estradiol, and a progestin. Briefly, the metabolic and pharmacokinetic characteristics of these steroids are as follows. Ethinyl estradiol is metabolized primarily via CYP3A, glucuronosyltransferase, and sulfotransferase. Progestin can inhibit the activity of CYP1A2, xanthine oxidase, and N-acetyltransferase.[74, 75, 76] Both ethinyl estradiol and progestins containing the 17α-acetylenic moiety can undergo oxidation resulting in mechanism-based inhibition of CYP3A. Finally, these hormones undergo extensive enterohepatic circulation in which intestinal bacteria play a central role in hydrolyzing the conjugated hormones and releasing them for reabsorption. Depending on the contribution of these various pathways to the disposition of the sponsor's drug, a drug interaction study may need to be performed. These studies generally use a randomized crossover design coadministering the sponsor's drug and placebo with either single-dose or multiple-dose oral contraceptives. In addition to measuring the estrogen and progestin components of the oral contraceptive, some studies also assess sex hormone binding globulin and albumin. By way of reference, increases or decreases in ethinyl estradiol AUC by 20 to 30 percent appear to warrant mention in the product label.[77] Results of the study can be described in the "precautions—drug interactions" section of the product label.

4.8. Pregnancy and Lactation

4.8.1. Need for Better Drug Labeling in Pregnancy

Most prescription drug labels do not contain clinical or pharmacokinetic data in pregnant women. Yet drug exposure in women during pregnancy is surprisingly common. A study conducted by the Food and Drug Administration on drug use in pregnancy revealed that the average expectant mother under 35 years of age took three prescription drugs during the course of her pregnancy and those over 35 years took five prescription drugs.[78] The Food and Drug Administration has undertaken a number of initiatives to improve product labeling for pregnant women and to expand recognition of them as a special population in the context of drug development. In 1996 a Pregnancy Labeling Task Force was established and charged with ensuring that pregnancy labeling of medical products is based on sound scientific information and provides better guidance for healthcare professionals in making therapeutic decisions. In 1999 a draft guidance for industry was produced on the topic of establishing a pregnancy registry to systematically collect and periodically report pregnancy outcomes from exposure to medical products.[79] Registry data may be included in the product label to guide clinicians and patients on assessing maternal and fetal risks associated with exposure to the drug. In 2000 the Food and Drug Administration and the National Institutes of Health cosponsored a conference on clinical pharmacology during pregnancy addressing ethical, legal, and clinical pharmacology issues.[78]

Currently the United States drug label includes a subsection describing a drug's ability to cause birth defects and other effects on reproduction and pregnancy. Products must also be classified under a letter system (A, B, C, D, X) to assess risk based on human and animal data. If adequate, well-controlled studies in pregnant women have not shown an increased risk of fetal abnormalities, the drug would be classified as "A". Risk-to-benefit ratios increase with progression through the letters. Fully two-thirds of prescription drugs are classified as "C" because either no animal or human studies have been performed or animal studies have shown an adverse effect. This system has been criticized as overly simplistic and minimally informative to clinicians and patients. The Pregnancy Labeling Task Force is considering a three-component pregnancy label containing "clinical considerations", "risk assessment", and "description of data" subsections. These would provide practical and data-driven information in narrative format needed by the clinician for risk assessment and therapeutic decision making.[78]

The "pregnancy and lactation" section of the European Summary of Product Characteristics consists of three subsections. Information for the "pregnancy" subsection may include facts on human experience and conclusions from preclinical toxicology studies relevant to risk assessments; recommendations on use at different times during pregnancy; and management of inadvertent exposure. Suggested wording is given in the associated guideline.[4] In the "women of child-bearing potential" subsection recommendations on use may be described if appropriate. In the "lactation" subsection data on transfer of active drug and/or metabolites into breast milk and recommendations whether to stop or continue breastfeeding should be given.[4]

4.8.2. Pharmacokinetic Studies in Pregnancy

There are dramatic physiological changes during pregnancy such as plasma volume expansion, changes in regional blood flow, decreases in albumin concentrations, and increases in glomerular filtration which could affect drug disposition.[75, 80] In addition the fetus and the placenta may also be involved in drug disposition and their influence can change over the course of pregnancy. Safety and clinical pharmacology studies in pregnancy should be considered for (1) drugs used to treat medical conditions that are caused or exacerbated by pregnancy such as asthma, diabetes, and hypertension; (2) drugs used to treat conditions with a high incidence of morbidity or mortality for which treatment of the mother cannot be discontinued during pregnancy such as anti-infectives, antiepileptics, and antidepressants; and (3) drugs with a high-use pattern in women of childbearing age.[79]

Rational drug dosing in pregnancy can be supported by dedicated clinical pharmacology studies or pharmacokinetic sampling in the context of a safety study in pregnant women receiving the drug for therapeutic purposes. Generally such studies would not be performed until effectiveness and safety of the drug have been demonstrated in the general patient population and use in pregnant women is likely. Especially informative are serial pharmacokinetic studies using each subject as her own control by collecting data longitudinally at appropriate stages during pregnancy and postpartum. An alternative study design uses a parallel-group of weight-matched nonpregnant women undergoing treatment for the same medical condition. The goal of such pharmacokinetic assessments is to design a dosing regimen that yields average drug exposure in pregnant women comparable to that which is safe and effective in the general population.

Pharmacokinetic data in pregnancy can be described in the "clinical pharmacology" and "use in pregnancy" sections of the label and dosing recommendations in pregnant women in the "dosage and administration" section.

4.8.3. Drugs in Lactation

Drugs enter breast milk predominantly via passive diffusion and, generally to a lesser extent, via reverse pinocytosis or apocrine secretion. Factors influencing transfer include protein binding, lipid solubility, pKa, and molecular weight.[80] Quantification is usually expressed as the milk to maternal plasma ratio. These generally range from 0.5 to 1.0. The dose of drug actually received by the neonate can be estimated from the volume of milk consumed, the concurrent maternal plasma concentration, and the milk-to-plasma ratio. For general calculations, the volume of milk consumed averages 150 mL/kg/day and the maternal plasma concentration is the average concentration over the dosing interval, Cavg. Using Cmax for the maternal plasma concentration provides a worst case assessment.[81] For most drugs, less than one percent of the maternal dose is available for absorption by the neonate. Clinically, the dose received by the neonate is minimized by planning nursing periods to coincide with the time of low maternal plasma drug concentrations such as just before the maternal dose.[80] A reference listing of drugs and other chemicals transferred into human milk is provided by the American Academy of Pediatrics Committee on Drugs.[82]

Drug excretion into breast milk may be quantified in a clinical pharmacology or clinical study. The milk-to-plasma concentration ratio can be derived from simultaneous milk and plasma samples from a single point in time or from the ratio of the areas-under-the-curve based on multiple paired milk and plasma samples generally after a single dose of drug.[83] For some drugs there is a discrepancy in the single verus multiple dose milk-to-plasma concentration ratios. This tends to be the case for drugs with milk-to-plasma ratios greater than unity and steady-state studies may be justified under these circumstances.[81]

Pharmacokinetic data on passage of a drug into breast milk and the milk-to-plasma ratio can be cited in the "use in pregnancy" section of the United States product label. The current format and content of the label have been criticized as relatively uninformative for clinicians and patients.[84] The Food and Drug Administration is considering revising this label section to conform with the proposed narrative model for the pregnancy section.[78] The "pregnancy and lactation" section of the European Summary of Product Characteristics can contain information on the excretion of drugs into milk and a recommendation as to whether the patient should stop or continue breastfeeding.[4]

4.9. Case Study: Disposition of Cefprozil in Lactating Women

Cefprozil is a semisynthetic broad-spectrum cephalosporin antibiotic indicated for the treatment of mild to moderate infections caused by susceptible strains of microorganisms. The transfer of cefprozil into breast milk was characterized in nine healthy lactating women who had decided to discontinue breast feeding.[85] Each subject received a single 1 gram oral dose of cefprozil consisting of cis- and trans-isomers in an approximately

90:10 ratio. Fifteen serial blood samples were collected over a 12-hour interval postdose, six fractional urine collections were made over a 24-hour period postdose, and breast milk was collected at seven timepoints up to 24-hour postdose. Samples were analyzed for the concentrations of both isomers by a specific high-pressure liquid chromatography assay.

The serum Cmax of the cis-isomer was 14.8 ± 3.2 μg/mL occurring at 2.0 hours postdose. The corresponding AUC was 54.8 ± 10.9 μg·h/mL and the elimination half-life was 1.7 ± 0.4 hours. Renal clearance of the cis-isomer was 164 ± 33 mL/min and urinary excretion was 60 ± 7 percent. These parameters were comparable to those of young non-pregnant women characterized in independent studies. The mean concentrations in milk of the cis-isomer over a 24-h period ranged from 0.25 to 3.36 μg/mL, with the maximum concentration in milk appearing at 6 hours after dosing. Because of this offset in time to reach the peak in plasma (2 hours) versus milk (6 hours), plasma concentrations were greater than milk concentrations up to 6 hours postdose yielding milk-to-plasma ratios of 0.05 ± 0.02 to 0.31 ± 0.13 and milk concentrations were greater than plasma concentrations from 6 hours and thereafter yielding milk-to-plasma ratios of 1.26 ± 0.65 to 5.67 ± 1.86. The overall milk-to-plasma ratio based on AUCs was 0.6 ± 0.2. The concentrations of the trans-isomer in plasma and in breast milk were about a tenth of those for the cis-isomer. Less than 0.3 percent of the dose was excreted in breast milk for both isomers. The investigators concluded, "[e]ven if one assumes that the concentration of cefprozil in milk remains constant at 3.36 μg/mL (the highest concentration of cefprozil observed in breast milk), an infant ingesting an average of 800 mL of milk per day will be exposed to a maximum amount of about 3 mg of cefprozil per day."[85]

The results of this study are presented in the product label in the "precautions—nursing mothers" subsection: "Small amounts of cefprozil (< 0.3% of dose) have been detected in human milk following administration of a single 1 gram dose to lactating women. The average levels over 24 hours ranged from 0.25 to 3.3 μg/mL. Caution should be exercised when Cefzil is administered to a nursing woman, since the effect of cefprozil on nursing infants is unknown." [86]

5. ETHNICITY

A corollary to the increasing focus on global drug development is the likelihood that regulatory dossiers will be submitted in many regions of the world for marketing approval. Consequently, efficiency demands that registration dossiers fulfill as broadly as possible the filing requirements of regulatory agencies serving the health needs of various ethnic regions and populations. By characterizing the ethnic sensitivity of a drug early in development, the sponsor can better determine the need for bridging studies, reduce duplication of clinical trials, and—most importantly—expedite the availability of innovative medicinal products to patients worldwide. To facilitate global development of medicines, the Intentional Conference on Harmonization has issued Industry Guidance E5 entitled *Ethnic factors in the acceptability of foreign clinical data.*[87] The *Ethnicity Guidance* addresses regulatory strategies to facilitate acceptance of foreign data in a new region, characterization of a new molecular entity's ethnic sensitivity, and use of bridging studies to allow extrapolation of foreign clinical data to a new region.

5.1. Multiregional Regulatory Submissions

The *Ethnicity Guidance* acknowledges that, from the viewpoint of any regional regulatory agency, a global registration dossier will often contain clinical data generated in a foreign region. Before the foreign data can be extrapolated to the new region, the complete clinical data package needs to be assessed by the regional regulatory authority for compliance with local regulatory standards along with adequate information in the following three areas:

- Characterization of the pharmacokinetics, exposure-response relationships, safety, and effectiveness in the population of the foreign region
- Support for a claim of safety and effectiveness based on adequate and well-controlled clinical trials conducted according to standards acceptable in the new region
- Evaluation of the pharmacokinetics and, where possible, exposure-response relationships in a population relevant to the new region

The *Ethnicity Guidance* outlines a decision tree for assessing the acceptability of the complete clinical data package in the new region. If there are shortcomings which do not allow the extrapolation of the foreign-generated data to the new region, a bridging study or studies may need to be conducted to complete the regulatory dossier for the new region. Dedicated clinical pharmacology studies or auxiliary pharmacokinetic data can play a decisive role in at least two elements in this process: characterizing the ethnic sensitivity of a new molecular entity and contributing to the bridging strategy.

5.2. Characterizing the Sensitivity of a Medicine to Ethnic Factors

5.2.1. Defining Ethnic Factors

This special population category in clinical pharmacology has historically been dominated by the concept of *race* with prospective studies or retrospective data explorations seeking to compare drug disposition between groups of subjects classified as white, black, Asian, Hispanic, and so forth. In this paradigm, *race* meant a group of subjects sharing to a high degree phenotypical features which are passed on between generations.[88, 89] Biological and anthropological research, however, has concluded that this categorization is no longer tenable "for, although it must be true that phenotype variation has a genetic basis, the point is that there is no consistent categorization across characteristics." [89] Furthermore, it has been shown that more human genetic variation exists within a racial group than between racial groups.[89, 90] Such a concept may also be misused to support stereotyping and discriminatory practices in disease diagnosis, treatment, and access to medical therapies.[88, 89] Concisely put, *race* is an inadequate marker for genetic-related variation in drug disposition and should be abandoned in clinical research.[89, 91]

The International Conference on Harmonization has taken a fresh look at this special population category as embodied in their *Ethnicity Guidance*. They have created a broader construct under the rubric of *ethnicity* and have charged it with a practical mission—namely, to assist in the transfer of clinical research data among different regions of the world thereby facilitating global registration of innovative medicines.[87]

While the *Ethnicity Guidance* acknowledges the conventional concept of race as a biologically-related population, it is only one element among a host of ethnic factors

Table 4. Ethnic Factors Influencing Acceptability of Foreign Clinical Data

Intrinsic	Genetic	sex, polymorphism of drug metabolism, genetic disease, genetically-based responsiveness to a drug, membership in a biologically-related subpopulation ("race")
	Physiological	age, body size and composition, smoking, alcohol consumption
	Pathological	receptor sensitivity, liver function, kidney function, cardiovascular function, systemic disease, stress
Extrinsic	Environmental	climate, sunlight, pollution
	Cultural	socioeconomic status, educational status, language, diet, smoking, alcohol consumption
	Medical practice	disease definition, diagnosis, therapeutic approach, drug regimen compliance
	Research practice	regulatory standards, good clinical practices (GCP), trial methodology, trial endpoints

Compiled from references 87, 88, 92.

which may influence the acceptability of foreign clinical data in a new region. As outlined in Table 4, ethnic factors can be divided between those that are more biological or intrinsic—including genetic, physiological, and pathological factors—and those that are more sociocultural or extrinsic—including environmental, cultural, medical practice, and research practice factors. Below are selected examples which underscore the breadth and complexity of the ethnicity construct and illustrate some of the less conventional components which might escape consideration in traditional or single-region drug development programs.

One of the best researched ethnic factors is genetic polymorphism of drug metabolism. Among the cytochrome P450 family of oxidative metabolic enzymes, for example, genetic variations of clinical significance have been found in CYP2C9, CYP2C19, and CYP2D6.[93, 94] In each subfamily at least one variant has been identified which is associated with decreased metabolism of a clinically-used substrate drug in heterozygotes and/or homozygotes. Furthermore, the variants have demonstrated differential distribution when members of various ethnic groups are genotyped. However, it is not always the case that an ethnic group in which the variant frequency is higher needs a different average dose of the substrate under clinical conditions. CYP2C9 can serve as an illustration. At least two variants occur for CYP2C9 (*2 and *3). Individuals homozygous for CYP2C9*3 are poor metabolizers of *S*-warfarin, phenytoin, tolbutamide, glipizide, and losartan.[94] These variants exhibit some association with ethnicity such that white subjects have a significantly higher frequency (~10 percent) compared with Asians (~1 to 2 percent) or blacks (~1 percent). However, counter to what would be expected, whites need higher warfarin doses than Asians to achieve comparable anticoagulation, they do not appear to have a higher incidence of warfarin-induced bleeding complications, nor a higher incidence of hypoglycemia associated with tolbutamide or glipizide.[94] On the other hand, genetic polymorphism of CYP2C19 appears to exhibit the expected ethnic pattern with regard to clinical response to omeprazole as detailed below in the case study in section 5.3. These examples suggest that membership in a specific ethnic subpopulation is a poor marker for dose regimen adjustment but can serve as a general signal for potential ethnic sensitivity for a drug in development.

Environmental factors may have implications for registration of topical drug products in different regions of the world. A dermatology product for topical application with a potential for photosensitivity, for example, may undergo pivotal trials in countries in the northern latitudes with a low incidence of adverse events in patients. Submission of the dossier to a regulatory agency in an equatorial country with more intense sunlight and longer daylight periods throughout the year might require an additional study to characterize safety under these new environmental condition not adequately reflected in the foreign-conducted clinical trials.

Interregional differences in medical practices could give rise to the likelihood that a new molecular entity would be used in addition to an existing therapy in the new region which was not coadministered in the foreign-conducted clinical trials. The regulatory agency in the new region might request a drug-drug interaction study to address the appropriateness of using the new drug as an add-on therapy with an existing treatment. The results of such a study might have implications for the regional product label or the ability to extrapolate clinical data to the new region.

5.2.2. Drug Characteristics Predisposing to Ethnic Sensitivity

The *Ethnicity Guidance* acknowledges that "[t]he impact of ethnic factors upon a medicine's effect will vary depending upon the drug's pharmacological class and indication and the age and gender of the patient. No one property of the medicine is predictive of the compound's relative sensitivity to ethnic factors." [87] In assessing the ethnic sensitivity of a new molecular entity, a basic knowledge of its pharmacokinetic properties and the relationships between exposure versus effectiveness and safety are essential. The *Ethnicity Guidance* provides an extensive list of drug characteristics which may signal a greater likelihood that a new molecular entity is sensitive to ethnic factors. These can be grouped as follows:

- *Bioavailability:* low bioavailability (more susceptible to dietary influences); high intersubject variation in bioavailability
- *Metabolism:* highly metabolized (especially through a single pathway with greater potential for drug interactions); metabolism by enzymes exhibiting genetic polymorphism; administration as a prodrug (potentially requiring enzymatic conversion via an ethnically sensitive system)
- *Exposure-response relationships:* nonlinear pharmacokinetic disposition; steep efficacy and safety pharmacodynamic curves in the therapeutic dose range; narrow therapeutic range
- *Use patterns:* high likelihood of use with other comedications; high likelihood for inappropriate use or abuse

Particular pharmacokinetic properties which predispose to ethnic sensitivity have been reviewed[94, 95] and underscore several of the aspects mentioned in the *Ethnicity Guidance*. Briefly, drugs which are absorbed by active processes appear more likely to show ethnic differences than drugs absorbed passively, however, examples are yet too sparse to make categorical conclusions.[95] Ethnic differences in protein binding appear unlikely for drugs that bind exclusively to albumin. By contrast, several studies have demonstrated that plasma concentrations of alpha-1 acid glycoprotein are higher in whites leading to lower plasma free fractions of drugs bound to this protein compared with Asians and blacks.[95]

By far the most important route of drug biotransformation is via the cytochrome P450 enzyme family. It has been estimated that about 40 percent of CYP-mediated drug metabolism is catalyzed by polymorphic enzymes, particularly via CYP2C and CYP2D6. Some general trends in the ethnic distribution of polymorphic variants associated with decreased metabolism of CYP substrates are emerging from pharmacogenetic research. The frequency of CYP2C9*2 and *3 appear higher in whites (~10 percent) compared with Asians and blacks (~0-2 percent). CYP2C19*2 is twice as frequent in Chinese (~30 percent) as in blacks (~17 percent) or whites (~15 percent); likewise, CYP2C19*3 is found in 5 percent of Chinese versus less than 1 percent in blacks or whites. Several variant alleles of CYP2D6 are known and produce a poor metabolizer phenotype. This phenotype appears to be rare in Asians (~1 percent), higher in whites (~5 to 10 percent) and variable in African populations (~0 to 19 percent).[94] Importantly, however, these ethnic distributions in metabolism do not always lead to predictable ethnic patterns in drug pharmacokinetics.[95] Finally, ethnic differences are more likely for drugs excreted via renal tubular secretion as this is an active process than for drugs excreted by filtration which is a passive process.[95]

5.2.3. Clinical Pharmacokinetic Considerations

The *Ethnicity Guidance* stresses the importance of clinical pharmacology studies in determining the acceptability of foreign clinical data in a new region: "Evaluation of the pharmacokinetics and pharmacodynamics, and their comparability, in the three major racial groups most relevant to the ICH regions (Asian, Black, and Caucasian) is critical to the registration of medicines in the ICH regions."[87] While pharmacokinetic studies tend to identify those ethnic factors which are intrinsic in nature, pharmacodynamics and clinical response are often influenced by both intrinsic and extrinsic factors. In this context, comparative pharmacokinetic studies serve two purposes. Firstly, they constitute an initial signal whether the new molecular entity is likely to be ethnically sensitive. Secondly, they can contribute to an informed decision of what kinds of pharmacodynamic or clinical bridging studies are needed in the new region. A new molecular entity which is shown to be relatively insensitive to ethnic factors would make it easier to extrapolate clinical data interregionally. Conversely, an ethnically sensitive drug may need a bridging data package to make the complete clinical data package acceptable for review in a new region.[87]

5.3. Case Study: Ethnic Sensitivity of Omeprazole in Asians

Omeprazole is a benzimidazole proton pump blocker that inhibits gastric acid secretion by parietal gastric mucosa cells. It is indicated for the treatment of duodenal ulcers, peptic ulcers, reflux esophagitis, and other hyperacidic conditions. Omeprazole is metabolized in the liver primarily to hydroxyomeprazole and omeprazole sulfone. The hydroxylation of omeprazole is mediated primarily by CYP2C19 and omeprazole clearance cosegregates with the polymorphic mephenytoin hydroxylase system (CYP2C19). Omeprazole AUCs were significantly higher in white and Chinese poor metabolizers of mephenytoin compared with white and Chinese extensive metabolizers, respectively. Furthermore, Chinese extensive metabolizers had 2.5-fold higher omeprazole AUCs than white extensive metabolizers.[96]

Given the generally higher frequency of the poor metabolizer phenotype in Asian compared with white populations, investigators have characterized the pharmacokinetics of omeprazole and its effect on gastric acid secretion after multiple doses of 40 mg/day in seven Chinese and eight white extensive metabolizers.[97] Mephenytoin hydroxylation capacity was evaluated by calculating the urinary enantiomeric ratio between *S*- and *R*-mephenytoin determined before the study. Plasma omeprazole and gastrin concentrations were measured for 12 and 24 hours, respectively, after the omeprazole dose on day 8 of the study. The oral clearance of omeprazole was significantly higher in white versus Chinese subjects: 319 ± 60 versus 183 ± 35 mL/min ($p < 0.05$). The 24-hour plasma gastrin AUC, used as a marker for gastric acid inhibition, was significantly lower in whites versus Chinese: 747 ± 99 versus 1414 ± 228 pmol·h/L ($p < 0.004$). Both the *S/R* mephenytoin ratio and omeprazole AUC correlated with the extent of omeprazole-induced hypergastrinemia. The investigators concluded that "[t]he need for dosage adjustment in patients with diminished CYP2C19 activity and in patients with Chinese ancestry deserves further evaluation."[97] Other researchers have subsequently shown that clinical outcomes in the treatment of *Helicobacter pylori* infection and peptic ulcer disease in patients who receive omeprazole and amoxicillin have been associated with metabolic genotypes.[98, 99]

While it is not known which specific studies were submitted by the sponsor to support product labeling statements, the above research is reflected in the omeprazole label. The "clinical pharmacology" section states "[i]n pharmacokinetic studies of single 20 mg omeprazole doses, an increase in AUC of approximately four-fold was noted in Asian subjects compared to Caucasians. Dose adjustment, particularly where maintenance of healing of erosive esophagitis is indicated, for the hepatically impaired and Asian subjects should be considered."[100]

5.4. Strategies to Bridge Foreign Data to New Regions

The decision whether a complete clinical data package is acceptable for review in a new region is made on the basis of a "bridging data package". This consists of (1) selected information from the complete clinical data package that is relevant to the new region including pharmacokinetic data and, if available, exposure-response data and (2) bridging studies, if needed, to extrapolate efficacy or safety data to the new region. The first element in the bridging data package can be based on data generated either in the new region or in a foreign region in a population of relevance to the new region. The second element, or bridging study, is a trial specifically conducted in the new region. The *Ethnicity Guidance* outlines conditions in which no bridging study may be required and in which efficacy or safety bridging studies may be needed. In all three cases, clinical pharmacology and/or pharmacokinetics may contribute to the study objections.

No bridging study may be necessary (1) if the medicine is ethnically insensitive and extrinsic factors are similar between the two regions or (2) if the medicine is ethnically sensitive but the two regions are ethnically similar and there is clinical experience with similar compounds in the new region to show that the class behaves comparably in patients in the two regions.

An efficacy bridging study may be needed if the following conditions hold: the medicine is ethnically sensitive, the regions are ethnically dissimilar, extrinsic factors are comparable, and the drug class is familiar in the new region. The bridging study could be a "controlled pharmacodynamic study in the new region, using a pharmacologic endpoint that is thought to reflect relevant drug activity...[to] provide assurance that the efficacy, safety, dose and dose regimen data developed in the first region are applicable in the new region."[87] In the absence of these conditions, a controlled clinical trial would be needed to replicate the foreign clinical trial. In either case, pharmacokinetic measurements could enhance the interpretation of the study outcome.

A safety bridging study may be needed if there is a safety concern in the new region. In the case where an efficacy bridging study is already being performed, the study may also be powered to assess the rates of common adverse events. Otherwise, a separate safety bridging study may be called for.

The *Ethnicity Guidance* outlines the main interpretations that may result from a bridging data package.

- If the bridging study demonstrates that all three elements—dose response, efficacy, and safety—are similar in the new region, then extrapolation of the complete clinical data package to the new region is supported.
- If the bridging study indicates that a different dose is needed in the new region to yield similar efficacy and safety, then extrapolation of the foreign data will often be possible with appropriate dose adjustment.
- If the bridging study is too small to verify safety data or if the bridging study does not verify similar efficacy and safety in the new region, then additional safety or clinical trials will be necessary.

5.5. Case Study: Interethnic Study of Nitrazepam in Europeans and Japanese

In two double-blind, randomized, placebo-controlled, crossover studies the pharmacokinetics and pharmacodynamics of the benzodiazepine nitrazepam were assessed separately in eight Japanese and eight white European healthy subjects matched between studies for sex, age (± 6 years), weight (± 6.5 percent), and height (± 10 percent).[101] The combined data were not a formal bridging study for registration purposes but rather designed "to test the feasibility and informativeness of interregional, interethnic pharmacokinetic and pharmacodynamic studies."[101] Subjects received a single 5 mg dose of nitrazepam or placebo in a randomized sequence. In addition to nitrazepam pharmacokinetic measurements, pharmacodynamic data included saccadic eye movements, smooth pursuit eye movements, and visual analog scales of sedation and unsteadiness.

Nitrazepam AUC and Cmax were nonsignificantly higher in the Japanese subjects by 26.3 and 15.9 percent, respectively. Placebo-corrected saccadic eye movement peak velocity was comparable in the two groups. Nitrazepam did not affect smooth eye movements in either group. Nitrazepam caused clear sedative effects in whites based on the visual analog scale but nonsignificant, and often reverse, effects in Japanese.

Among their observations, the investigators noted that intersubject variability in exposure-response relationships was much larger than interethnic variation and that validating objective measurements such as saccadic and smooth pursuit eye movements was easier than subjective measurements such as the visual analog scales. Indeed, the different results in Japanese and Europeans from the sedation scale were likely due to different interpretation of the scales by the two groups. The investigators concluded that "[t]he results of this study show that interethnic comparative pharmacokinetic and pharmacodynamic investigations with rigorous similarity in design, subjects, and methodology are feasible across different regions. Such studies can make a useful addition to in vitro assessment of ethnically sensitive pharmacokinetic differences, not only for global drug development but also in multiethnic societies."[101]

6. RENAL IMPAIRMENT

For many drugs, renal excretion is a significant route of elimination for the parent compound and/or metabolites. Consequently, impairment of renal function can affect the disposition of a drug to an extent that warrants a dose adjustment for patients with decreased renal function or with end-stage renal disease on dialysis. It is the object of renal impairment studies to quantify this effect and translate this, when justified, into dosing guidelines in the product label. Intuitively, impaired kidney function would be expected to lead to decreased urinary excretion and/or renal metabolism of a drug that depends on this route for elimination. However, numerous studies have identified additional factors in renal impairment which can influence pharmacokinetics. These include changes in absorption, drug distribution, plasma protein binding, and hepatic metabolism. It is important to be mindful of both the parent drug and active or toxic metabolites when considering the impact of renal impairment on drug disposition. Drug disposition in renal impairment has been the subject of a comprehensive review[102] which serves to underscore the need for detailed pharmacokinetic studies in this special population during drug development. In order to provide guidance on this topic, the Food and Drug Administration conducted a survey on how these studies were performed in the past[103] and used this survey, in part, to develop a guidance for industry.[104]

The information sources for the survey were renal impairment studies submitted to the agency as part of a new drug application or a supplemental new drug application over a one-year period from 1996 to 1997 and data contained in the sponsor's product label. Data elements collected for the survey included the percent of unchanged drug excreted in the urine in subjects with normal renal function, drug dose applied in the study, study design, method for estimating renal function, renal function stratification, number of subjects per stratification group, and data analysis methods. There were a total of 71 drug reviews over this period of which 40 (56 percent) contained a renal impairment study.[103] The impact on product labeling was garnered from the "clinical pharmacology", "precautions", "contraindications", and "dosage and administration" sections.[103]

The subsequent *Renal Guidance* for industry was released in 1998 and covers the following topics: deciding whether a renal impairment study is necessary, designing the study, analyzing the data, and deriving dosing recommendations and labeling statements.[104] Pertinent elements of renal impairment studies are discussed below using the findings of the survey and the *Renal Guidance* as a leitmotif.

6.1. Determining the Need for a Renal Impairment Study

The Food and Drug Administration survey revealed that the decision to perform a renal impairment study did not appear to depend consistently on any particular pharmacokinetic property of the compound including fraction of drug excreted unchanged in the urine, percent protein binding, or extent of metabolism.[103] In an attempt to standardize and direct decision-making in this regard, the *Renal Guidance* describes three levels of advice—recommendation, consideration, and desirability— for deciding whether to perform a study in patients with impaired renal function.

A renal impairment study is *recommended* if a drug is likely to be used in such patients and (1) renal impairment is anticipated to significantly alter the pharmacokinetics of the parent or its active/toxic metabolites, (2) a dose adjustment is likely to be necessary for safe and effective use, (3) the parent drug or active metabolite has a narrow therapeutic index, or (4) the drug or its active metabolite is primarily excreted and/or metabolized via renal mechanisms.

A renal impairment study should be *considered* under two further conditions: (1) If the parent drug or an active metabolite has high hepatic clearance relative to hepatic blood flow and significant protein binding, renal impairment could cause an increase in unbound concentrations after parenteral administration due to decreased protein binding coupled with minimal change in total clearance. (2) Even if renal impairment is not expected to influence the drug's pharmacokinetics, the impact of dialysis may warrant study.

Finally, the *Renal Guidance* recognizes that controversy exists regarding the impact of severe renal impairment on hepatic metabolism. Hence, even if the drug is primarily metabolized, a renal impairment study may be *desirable* unless the therapeutic index of the drug is wide.

6.2. Measuring and Stratifying Renal Function

The survey indicated that nearly all studies (39 of 40) used measured creatinine clearance based on a 24-hour urine collection to quantify renal function. Only one study used an isotope method.[103] The *Renal Guidance* acknowledges the primacy of creatinine clearance to assess renal function adding that because of its practicality and universality, it is also a good index upon which to base dose adjustment recommendations. Hence, it should be used in all renal impairment studies.[104] Two options exist for deriving creatinine clearance, namely, measurement and estimation. The standard method for measurement involves collecting total urine output for 24 hours for urine creatinine measurement and obtaining a serum creatinine concentration at the midpoint of the collection interval. Creatinine clearance (CLcr) is then calculated as:

$$CLcr = (Ucr \times Vur) / (Scr \times t)$$

where Ucr and Scr are the urine and serum concentrations of creatinine (mg/dL), Vur is the urine volume (mL), and t is the collection time interval (min). This yields creatinine clearance in mL/min which is commonly standardized to body surface area as mL/min/1.73 m^2. The alternative is to estimate creatinine clearance based on a determination of serum creatinine and anthropometric data for the subject which is the preferred method as this is what is commonly applied clinically.[103] The *Renal Guidance* mentions the equation of Cockroft and Gault for adults and adolescents which uses age, bodyweight, and sex as inputs.[105] Formulas for infants less than one year[106] and for children from one to 12 years[107] are cited in the *Renal Guidance* both of which use length or height as input. These and alternative formulas for estimating creatinine clearance in subjects with stable as well as unstable renal function have been assembled.[102]

Finally, the *Renal Guidance* acknowledges that other methods to assess renal function exist which may yield additional insight into the renal handling of a drug such as glomerular filtration, tubular secretion and reabsorption, and renal blood flow. Such methods are encouraged as a potentially useful addition but should not be used to the exclusion of creatinine clearance.[104]

The survey revealed that about a quarter of the studies grouped the subjects into two renal impairment categories (normal and severe), a quarter into three categories (normal, moderate, severe), a quarter into four categories (normal, mild, moderate, severe), and a quarter into other groupings.[103] The *Renal Guidance* recommends as a comprehensive stratification, a five-level grouping based on creatinine clearance with the following boundaries:

1. normal renal function, CLcr > 80 mL/min
2. mild renal impairment, CLcr 50 to 80 mL/min
3. moderate renal impairment, CLcr 30 to 50 mL/min
4. severe renal impairment, CLcr < 30 mL/min
5. end-stage renal disease, requiring dialysis

6.3. Study Designs and Data Collection

The survey indicated that 72.5 percent of renal impairment studies were single-dose, 10 percent were multiple-dose, 12.5 percent were combined single- and multiple-dose, and 5 percent used population pharmacokinetics. The choice of study design was independent of whether the drug exhibited linear or nonlinear disposition characteristics. Three-quarters of the studies used doses in the therapeutic range and generally included six to eight subjects per renal function group with a range from three to 18 subjects per group.[103] The *Renal Guidance* acknowledges this diversity of approaches by proposing several options to characterize the renal impairment subpopulation.

6.3.1. Comprehensive Study Design

The *Renal Guidance* describes a "full" study design that could be applied to drugs whose pharmacokinetics are likely to change in renal impairment in order to comprehensively characterize the impact of renal function on drug disposition. It is meant to be applied to drugs whose exposure-response relationships are known not to be influenced by renal impairment or whose therapeutic index is wide enough to preclude safety concerns for the participants. It enrolls subjects in all five renal function categories defined above. The *Renal Guidance* stresses that renal impaired subjects should be compared with patients whose renal function is typical of the usual patient population as a control group. Dose adjustment recommendations should then use this control group as the comparator. A group of subjects with renal function greater than the control group—for example, healthy young volunteers—can also be included to expand the ability to detect the impact of renal function on drug pharmacokinetics.

Approximately equal numbers of subjects should be recruited into the renal function groups. Although an exact number of subjects is not specified, the *Renal Guidance* recommends that enough subjects be enrolled to allow detection of a pharmacokinetic difference large enough to justify a dose adjustment if one exists. Furthermore, the groups should be comparable with respect to age, sex, weight, and other characteristics potentially influencing the drug's disposition.

It is important to consider both the parent drug and therapeutic or toxic metabolites in selecting single-dose or multiple-dose application for a renal impairment study. If all analytes of interest have linear, time-independent pharmacokinetics, then a single-dose study can be justified and results extrapolated to steady-state. If these conditions do not hold, a steady-state study is generally warranted. When possible, the dose selected should be in the projected therapeutic range keeping in mind that in multiple-dose studies, a lower dose or less frequent administration may be necessary to prevent drug or metabolite accumulation with attendant safety concerns.

6.3.2. Alternative Study Designs

The survey indicated that less than half of studies used a full design with four or five renal function groups.[103] In acknowledgment, the *Renal Guidance* allows for two alternatives: a staged approach and a population pharmacokinetic study. A staged approach may be applicable for drugs unlikely to be influenced by renal impairment. In stage 1, the extremes of renal function are studied by enrolling subjects with normal (group 1) and severely impaired (group 4) renal function. If results confirm that drug disposition does not differ in the two groups to an extent warranting a dose adjustment, no further studies are necessary. If this conclusion is not supported, stage 2 assesses subjects with mild and moderate renal impairment (groups 2 and 3). For purposes of analysis, data from both stages can be combined.

Alternatively, population pharmacokinetic data from phase 2 or 3 trials can be evaluated with nonlinear mixed effects modeling to characterize the influence of renal function on drug disposition. Generally, calculated creatinine clearance is the covariate and drug clearance is the exposure parameter for such an evaluation. This approach is only justifiable if the design and data collection retain the key features of the conventional renal impairment study design, namely, sufficient patients representing an appropriate range of renal function and measurement of protein binding and metabolites in addition to the parent drug, if needed. A potential limitation is that patients with severe renal impairment may be excluded from phase 2 and 3 trials and therefore not represented in a population pharmacokinetic database. If the drug is likely to be used in such patients, a separate dedicated study in this group may be conducted in parallel with the clinical trials and data from both sources combined for a broader assessment.[44, 104]

6.3.3. Sample Collection and Bioanalysis

While dose adjustments are usually made on the basis of drug exposure in blood, collecting urine samples for drug bioanalysis provides additional mechanistic information in a renal impairment study. In single-dose studies the duration of blood sampling may need to be extended to accommodate prolonged elimination with progressive renal impairment relative to the usual sampling duration in healthy volunteer studies. Multiple-dose studies may also need to be extended to allow for a slower rise to steady state due to reduced clearance or for full accumulation of metabolites in renal impairment. In these situations, a loading dose may be considered if clinical safety data support such an approach. For accurate urine collection, the subjects should void their bladder completely just before dose administration. To make complete urine collection over the full period of pharmacokinetic sampling more practical, urine may be collected in serial intervals of 2 to 8 hours duration. Fluid intake should be standardized to promote proper hydration and thereby facilitate urine production over the full collection period. Intermediate voidings during a collection interval may need to be refrigerated depending on drug or metabolite stability. The subject should void completely at the end of each collection interval, the full urine volume measured, and a portion frozen for analysis. Renal clearance is then calculated from the drug concentration in urine (Cur), the urine volume (Vur), and the systemic drug exposure over the urine collection interval:

$$CLr = (Cur \times Vur) / AUC = Ae / AUC$$

The numerator in this equation is the amount of drug eliminated in urine (Ae).

Measurement of total drug and metabolite concentrations should be adequate for drugs less than 80 percent bound to plasma proteins. For more highly protein-bound drugs, renal impairment may lead to alteration in binding due to hypoalbuminemia, accumulation of competitive endogenous substances or drug metabolites for binding sites, or conformational changes in binding sites in uremia.[102] The *Renal Guidance* recommends that unbound drug concentrations should be measured in each plasma sample. If the binding of drug and metabolites is concentration independent and time-invariant, the fraction of unbound drug may be measured in a single plasma sample per subject and the total concentration in all other samples for that subject multiplied by this factor to estimate the unbound analyte concentration.

6.3.4. Influence of Dialysis on Pharmacokinetics

Pharmacokinetic studies to assess the dialyzability of drugs have two potential applications: they can be used to quantify a dose adjustment or the need for a dose supplement after dialysis in patients with end-stage renal disease and to determine the utility of dialysis to manage a drug overdose. Dialysis studies should be performed if the drug is likely to be used in patients with end-stage renal disease and the drug has pharmaceutical characteristics which favor removal by dialysis. These include molecular weight less than 500, high water solubility, low plasma protein binding, small unbound distribution volume, and large unbound renal clearance.[102] Conversely, it may be justified to omit a dialysis study from the development program for drugs with a large unbound volume of distribution or a large unbound nonrenal clearance.

The most commonly used dialysis technique in patients with end-stage renal disease is intermittent hemodialysis. Hemodialysis removes substances from the blood by adjusting blood and dialysate flow rates to maintain a transmembrane diffusion gradient. The *Renal Guidance* recommends that the pharmacokinetics of the drug be characterized when the subject is both on and off dialysis. The off-dialysis study is performed after a single-dose or over the multiple-dose interval between dialysis runs similar to the pharmacokinetic sampling performed in nondialyzed subjects with renal impairment. The clearance of drug by dialysis (CLhd) is derived from paired drug concentrations in arterial blood entering the dialyzer (Ca) and venous blood leaving the dialyzer (Cv) and the rate of blood flow through the dialyzer (Q):

$$CLhd = [(Ca - Cv) / Ca] \times Q = ER / Q$$

The quotient in brackets in the above equation is also known as the extraction ratio across the dialyzer (ER).[108] To determine the change in systemic drug or metabolite exposure due to dialysis, a minimal approach is to obtain a venous blood sample before and after dialysis. Some investigators recommend drawing the postdialysis blood sample approximately 2 hours after the end of the dialysis run to allow any postdialysis rebound or redistribution in blood concentrations to occur.[102] This minimal approach can be supplemented with additional timed blood samples during the dialysis run to more fully characterize the time course of drug removal. The total amount of drug removed during a dialysis run can be calculated by collecting all dialysis fluid and measuring the drug or metabolite concentration. The product of this concentration and the total dialysate volume yields the amount of analyte removed. Dialysis studies should report the type of dialyzer used, the surface area of the dialysis membrane, and the blood and dialyzer flow rates.[108]

The *Renal Guidance* recommends that a pharmacokinetic study in patients treated with chronic ambulatory peritoneal dialysis be conducted if the drug is likely to be used in this patient population and likely to be cleared by this renal replacement therapy. Peritoneal dialysis removes substances from the blood by diffusion and osmosis across the peritoneal membrane into the dialysis fluid in the peritoneal cavity. To quantify the clearance of drug from the body by peritoneal dialysis, peritoneal dialysis effluent is collected after a timed dwell. Venous blood samples are also collected over this time interval to determine the systemic AUC. Drug clearance via peritoneal dialysis (CLpd) is then derived from the drug concentration in the effluent (Cpd) and its volume (Vpd) and the systemic AUC:

$$CLpd = (Cpd \times Vpd) / AUC.$$

Peritoneal drug clearance cannot exceed the dialysate outflow rate.[109] The numerator in this equation yields the amount of drug removed over the dialysis dwell time. This can be used as a basis to recommend supplemental doses if warranted. These and other aspects of drug therapy in peritoneal dialysis have been reviewed.[109]

6.4. Data Analysis

The intent of data analysis is to determine if renal impairment affects drug disposition to an extent requiring a dose adjustment. Standard pharmacokinetic parameters of the parent drug and active metabolites in plasma and urine should be determined by noncompartmental and/or compartmental approaches. Parameters for parent drug should include Cmax, tmax, AUC, CL/F, CLr, Vd/F, and half-life and those for metabolites should include Cmax, tmax, AUC, and CLr. If protein binding of parent drug and metabolites is determined, parameters should be expressed in terms of unbound concentrations.

A mathematical relationship should be sought between pertinent pharmacokinetic parameters and clinically practical measures of renal function to yield a model which successfully predicts drug exposure based on renal function. The most informative parameters to explore are CL/F (alternatively, dose-normalized AUC) and dose-normalized Cmax as measures of exposure on the one hand and creatinine clearance as a measure of renal function on the other. Both the *Renal Guidance* and a perusal of the clinical pharmacology literature indicate that linear regression between drug clearance and creatinine clearance are most commonly used in this context. This basic model should be explored in all studies although alternative mathematical models may be justified if they adequately describe the data and are mechanistically reasonable. The simplest relationship between exposure and renal function is generally preferred keeping in mind the goal of producing a model that can be easily applied in the clinic to derive dose adjustments for patients.

6.5. Developing Dose Recommendations and Product Labeling Statements

6.5.1. Deciding Whether a Dose Adjustment is Needed

The survey showed that of the 40 drug applications which included a renal impairment study, 10 studies did not derive any dose adjustment recommendation even though four of these studies found an influence of renal impairment on the drug. The remaining 30 studies either made a dose adjustment recommendation (12 studies) or stated that an adjustment was not needed because of a wide therapeutic index or a general recommendation to titrate to clinical response (18 studies). Twenty-nine of these 30 drugs could be categorized by the fraction of drug excreted unchanged in urine into the following groups: < 10% (n = 21), 10-50% (n = 6), and > 50% (n = 2). There was a positive association between the need for a dose adjustment and increasing fe of 19, 83, and 100 percent of drugs in the respective goups.[103] Hence, the fraction of drug excreted via renal mechanisms had some predictive capacity of the need for dose adjustment.

The overall goal of a renal impairment study is to derive a dose algorithm for renally impaired patients which yields systemic exposure to unbound drug similar to that in patients with renal function typical for the treatment population. The *Renal Guidance* encourages a categorical approach to dose adjustment from both a statistical and product labeling perspective. The ratios and 90 percent confidence intervals should be constructed for the unbound Cmax and AUC in patients with severe renal impairment to those in the control group (group 4 / group 1). The sponsor can either prespecify no-effect boundaries before performing the study or use 70 to 143 percent for Cmax and 80 to 125 percent for AUC without further justification. If the 90 percent confidence interval is within the respective boundary, the sponsor could claim no dose adjustment is needed for patients with renal impairment.

If an effect is observed when comparing the extreme renal function groups (1 versus 4), then the parameter ratios for each renal function group to the control group should be constructed. Parameters of interest may include Cmax, AUC, CL/F, and half-life. The decision whether a dose adjustment is warranted in each renal impairment group is based on the magnitude of the parameter ratios and the therapeutic index of the drug. The general dose adjustment approaches are (1) maintaining the usual dose but lengthening the dose interval, (2) reducing the dose but maintaining the usual dose interval, or (3) a combination of modifying both the dose and interval. Factors to consider when choosing the approach are the class of drug, the amplitude of the peak-trough fluctuation relative to the therapeutic index, the magnitude of the dose with respect to the dose strengths to be marketed, and the practicality of the calculated dosing interval.[102, 110] Pharmacokinetic simulations can be especially helpful in visualizing the impact of various dose and interval changes on the concentration-time profile at steady state. A helpful relationship for calculating the average change in the rate of drug administration in a renal impairment group relative to the control group is:

$$(D/\tau)_{impaired} = [CLu_{impaired} / CLu_{control}] \times (D/\tau)_{control}$$

where the quotient dose/dose interval (D/τ) is the dosing rate and CLu is the unbound drug clearance.[111] A number of assumptions underlie the use of this equation including unchanged bioavailability and metabolism of the drug in renal impairment, linear pharmacokinetics, and the absence of therapeutic or toxic metabolites.[110]

The need for a supplemental dose after a dialysis procedure, depends on the change in unbound drug concentration during dialysis relative to the therapeutic range or on the fraction of the dose removed during dialysis. One general recommendation is that dose adjustment is needed only if hemodialysis clearance increases total clearance by 30 percent or more.[102, 112]

6.5.2. Product Labeling Statements

The survey did not indicate any consistent method for presenting the results of renal impairment studies in product labeling.[103] The *Renal Guidance* seeks to redress this with proposed wording for several sections of the product label.

- *Clinical pharmacology—pharmacokinetics.* This subsection can contain information on the percentage of drug that is eliminated by renal excretion unchanged or as metabolites, the disposition of metabolites in patients with renal impairment, the effects of renal impairment on protein binding of parent drug and metabolites, and the effects of changes in urine pH or other special situations of clinical relevance.
- *Clinical pharmacology—special populations.* This subsection should briefly summarize the impact of renal impairment on the drug's pharmacokinetics stating either that no influence occurs or describing the results of a renal impairment study performed in accordance with the *Renal Guidance*. In the latter case, the relationship between renal function and pharmacokinetic parameters can be given as an equation (for example, linear regression) or by describing the average percent change in pertinent pharmacokinetic parameters in each renal function group relative to the control group. Removal of drug by dialysis can also be mentioned here.
- *Precautions, warnings.* These sections should describe clinically important changes in drug pharmacokinetics to which prescribers should be alert.
- *Dose and administration.* This section should clearly state that either that no dosing adjustment is needed in renal impairment or that the dose needs to be individualized. In the latter case the relationship between drug clearance and endogenous clearance should be stated, equations given for estimating patient's creatinine clearance based on creatinine and anthropometric data,[105-107] and a tabular presentation of dose (mg) and frequency (every __ hours) in each creatinine clearance range to which this applies. A supplemental dose after dialysis can be included in the table if applicable for end-stage renal disease patients.
- *Overdose.* Mention can be made as to the value of hemodialysis to remove drug in cases of overdosing.

6.6. Case Study: Ciprofloxacin Dosing in Renal Impairment

Ciprofloxacin is a fluoroquinolone antimicrobial agent indicated for the treatment of infections caused by susceptible strains of gram-negative and gram-positive bacteria. Approximately 40 to 50 percent of an oral dose is excreted unchanged in the urine. Several renal impairment studies have been published using oral or intravenous ciprofloxacin administration.[113, 114, 115, 116] All studies observed a reduction in renal drug clearance with decreasing creatinine clearance. Some also reported a similar pattern for total drug clear-

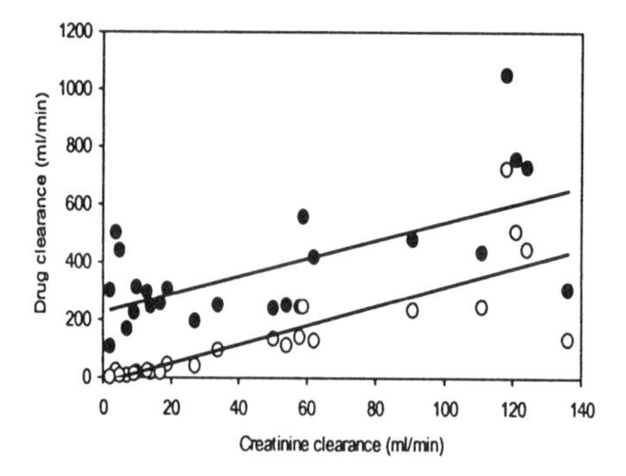

Figure 3. Linear regression of ciprofloxacin total body clearance versus creatinine clearance (*filled circles*) and ciprofloxacin renal clearance versus creatinine clearance (*open circles*) in 24 subjects with various degrees of renal impairment. Plot constructed by permission of the British Society of Antimicrobial Chemotherapy using data from Webb *et al.*[114]

ance. Figure 3 was constructed from the individual pharmacokinetic data reported in a study in which 18 subjects with various degrees of renal impairment received 100 mg ciprofloxacin intravenously.[114] Plasma samples were collected to 48 hours post-infusion with an extension to 72 hours for subjects with creatinine clearances less than 30 mL/min. Fractional urine collections were made over the same intervals. Ciprofloxacin was quantified in plasma and blood by a validated high performance liquid chromatography method. Both renal and total drug clearance were positively correlated with creatinine clearance yielding significant and parallel regressions as illustrated in Figure 3. The impact of hemodialysis on ciprofloxacin disposition has also been reported.[113, 115, 116] In one representative study the dialysis extraction ratio was 23 percent and dialysis clearance was 40 mL/min. Dialysis clearance accounted for 13 percent of total body clearance.[115]

Although the specific studies and data used by the sponsor to support the product labeling of ciprofloxacin in renal impairment are not known, the general findings reported in published studies are reflected in the ciprofloxacin label. In the "dosage and administration" section specific recommendations on dosing in renal impairment are given in text and tabular formats. "Ciprofloxacin is eliminated primarily by renal excretion; however, the drug is also metabolized and partially cleared through the biliary system of the liver and through the intestine. These alternate pathways of drug elimination appear to compensate for the reduced renal excretion in patients with renal impairment. Nonetheless, some modification of dosage is recommended, particularly for patients with severe renal dysfunction." Guidelines are then given as shown in Table 5. The product label also mentions the use of serum drug level monitoring especially in patients with severe infections and severe renal impairment. The "overdose" section of the label, states "Only a small amount of ciprofloxacin (<10%) is removed from the body after hemodialysis or peritoneal dialysis."[117]

Table 5. Recommended ciprofloxacin doses in renal impairment

Creatinine clearance	Dose
> 50 mL/min	Usual dosage
30 - 50 mL/min	250 - 500 mg q 12 h
5 - 29 mL/min	250 - 500 mg q 18 h
Patients on hemodialysis or peritoneal dialysis	250 - 500 mg q 24 h (after dialysis)

Usual dosage ranges from 250-750 mg q 12 h depending on seriousness and site of infection.
Reproduced by permission of Bayer Corporation from the Cipro (ciprofloxacin) product label.[117]

7. HEPATIC IMPAIRMENT

In contrast to renal impairment, hepatic impairment appears to be a special population seeking a drug development paradigm. Whereas the degree of renal impairment can be relatively easily quantified by creatinine clearance in the clinical setting, there is no widely accepted method to apply in the clinic to quantify the degree of hepatic impairment. In addition, changes in renal function have a generally consistent and predictable influence on drug elimination; whereas, hepatic impairment is an umbrella term covering a broad variety of pathologies having widely different impacts on drug behavior. Consequently, unlike renal impairment studies which use creatinine clearance as a relatively good predictor for individual dose adjustment, hepatic impairment studies must have a more modest and restricted aim. Their goal is to recommend an *initial* dose adjustment recognizing that individual dose titration based on clinical assessments and possibly therapeutic drug monitoring will frequently be needed in this special population. The Food and Drug Administration has released a draft guidance on pharmacokinetic studies in patients with hepatic impairment which attempts to apply the renal impairment paradigm to hepatic impairment studies. The guidance proposes the use of a clinical scoring system for staging the degree of hepatic impairment and classifying patients into various categories with initial dose adjustment recommendations in product labeling.[118]

7.1. Pharmacokinetics in Hepatic Impairment

The hepatic clearance of a drug (CLh) can be quantified based on blood flow through the liver (Q) and drug concentrations in arterial (Ca) and venous (Cv) blood.

$$CLh = Q \times [(Ca - Cv) / Ca] = Q \times E$$

The difference of drug concentrations in the inflowing and outflowing blood is the extraction ratio across the liver (E). Drugs with a high extraction ratio (greater than 0.7), are said to be *flow-limited* because their clearance is largely determined by hepatic blood flow. Drugs with low extraction ratios (less than 0.3) are termed *capacity-limited* because their clearance is largely determined by the rate at which the liver metabolizes them. Protein binding can also influence hepatic clearance. If only unbound drug is available for entry into the hepatocyte, the drug is classified as *restrictive* or *binding-sensitive*; whereas, if both bound and unbound drug are extracted into the hepatocyte, the drug is termed *nonrestrictive* or *binding-insensitive*.[119]

These pharmacokinetic concepts are important for understanding how liver impairment can influence drug clearance. The are three pathological changes which may accompany liver dysfunction and have a significant impact on drug clearance. Firstly, liver blood flow may change due to changes in intrahepatic vascular resistance and/or shunting of blood from the portal to the systemic circulation bypassing the liver. These alterations could reduce the clearance of flow-limited drugs or increase the bioavailability of orally administered drugs ordinarily exposed to first-pass elimination. Secondly, metabolic function may be reduced either due to a decreased number of viable hepatocytes or to decreased enzyme activity in the remaining cells. In theory this could decrease the clearance of capacity-limited drugs. Finally, drug protein binding may be decreased due to a reduction in the concentration of plasma proteins or changes in binding sites.[119, 120, 121] It has been argued that only rarely would protein binding changes warrant dose adjustment except in cases of a parenteral drug with a high extraction ratio and narrow therapeutic index or an oral drug with a narrow therapeutic index that has a very rapid exposure-response equilibration time.[122]

Given this complex interplay of pathologic and pharmacokinetic factors, predicting the impact of liver impairment on systemic drug exposure is extremely difficult. A few broad generalizations have been offered based on clinical observations and dedicated studies. The presence of cirrhosis appears to be the principal determinant associated with clinically significant impairment of drug metabolism. The available evidence from a variety of hepatic diseases suggests that, unless cirrhosis is present, there is usually not a clinically relevant reduction in drug elimination.[119, 121, 123] While conjugative metabolism appears to be relatively spared even in severe chronic liver disease, oxidative metabolism appears to be differentially affected by liver impairment.[124] This is illustrated in a study of cytochrome P450 levels, catalytic activity, and corresponding messenger RNA from livers removed at transplantation from 50 patients with end-stage cirrhosis.[125] The severity of liver disease was classified according to Child-Pugh[126] as mild (class A) for four livers, moderate (class B) for 17 livers, and severe (class C) for 29 livers. CYP1A expression was substantially reduced, CYP3A and CYP2E1 were differentially reduced depending whether cholestasis was present, and CYP2C was minimally changed.[125] In addition to the direct effects of liver disease on hepatic function, a further complicating factor is that the renal elimination of drugs can also be affected even when the liver is not primarily responsible for the drug's elimination.[123] These manifold considerations underscore the need to perform dedicated hepatic impairment studies for drugs likely to be used in patients with decreased hepatic function.

7.2. Determining the Need for a Hepatic Impairment Study

The foregoing pharmacokinetic and clinical considerations and the draft *Hepatic Guidance* suggest three general drug categories for which a hepatic impairment study should be performed: (1) drugs for which hepatic metabolism or biliary excretion accounts for a substantial portion of parent drug or active metabolite elimination; (2) drugs with narrow therapeutic windows even if hepatic clearance is minor; and (3) drugs for which hepatic elimination could become important in the event of renal impairment. Conversely, the draft *Hepatic Guidance* considers the need for a hepatic impairment study less important if the drug is entirely eliminated by the kidney, hepatic metabolism is minor and the therapeutic window is wide, or the drug is meant for single-dose administration.[118]

7.3. Measuring and Stratifying Hepatic Function

Currently there is no universally accepted and clinically applicable parameter or composite scoring system to quantify the extent of hepatic impairment. Attempts have

Table 6. Child-Pugh Classification of Hepatic Impairment

Parameter	1 point	2 points	3 points
Serum bilirubin (mg/dL)	< 2	2 to 3	> 3
Serum albumin (g/dL)	> 3.5	2.8 to 3.5	< 2.8
Prothrombin time (sec prolonged)	< 4	4 to 6	> 6
Encephalopathy grade[a]	none	1 or 2	3 or 4
Ascites	absent	slight	moderate

[a] Grade 0: normal consciousness, personality, neurological examination, electroencephalogram
Grade 1: restless, sleep disturbed, irritable/agitated, tremor, impaired handwriting, 5 cps waves
Grade 2: lethargic, time-disoriented, inappropriate, asterixis, ataxia, slow triphasic waves
Grade 3: somnolent, stuporous, place-disoriented, hyperactive reflexes, rigidity, slower waves
Grade 4: unrousable coma, no personality/behavior, decerebrate, slow 2-3 cps delta activity
Table modified by permission of Blackwell Science Limited from Pugh *et al.*[126]

been made to use biochemical parameters, clinical signs, or elimination of marker drugs.[119, 127] Confounding these attempts is the fact that hepatic impairment includes a complex and highly variable set of pathological changes whereas any such classifying system needs to be simple enough for routine clinical application in order to be useful as a dose adjustment guide. The Food and Drug Administration surveyed 57 hepatic impairment studies submitted between 1995 and 1998 and found that nearly all (55 studies) employed the Child-Pugh classification[126] to assess hepatic impairment.[118] In an attempt to standardize the conduct of hepatic impairment studies, the draft *Hepatic Guidance* recommends the use of this scale.[118]

A classification originally proposed nearly three decades ago by Child was subsequently modified by Pugh to produce a quantitative system for surgical evaluation of alcoholic cirrhotics.[126] It is a semiquantitative system based on three biochemical markers and two clinical stigmata of hepatic dysfunction: bilirubin, albumin, prothrombin time, encephalopathy, and ascites. Points are scored for each of the five items and summed to yield the Child-Pugh score as outlined in Table 6. A score of 5 to 6 is *mild*, 7 to 9 *moderate*, and 10 to 15 *severe* hepatic impairment. Although this classification depends in part on subjective clinical observations, it correlates well with more invasive approaches to assess liver blood flow (indocyanine green) and general metabolic capacity (antipyrine) and provides a useful staging of liver disease severity in the clinical setting.[128]

The draft *Hepatic Guidance* also mentions other scoring systems to classify the severity of liver impairment. While it acknowledges that alternative approaches may be valuable in certain drug development programs, it urges that a Child-Pugh score be derived and reported for each subject in a hepatic impairment study.[118] The use of marker substances such as indocyanine green to estimate liver blood flow or other probe drugs may provide additional insight into mechanisms of drug handling in hepatic impairment and can be included in the study assessments. If the drug is biotransformed via pathways known to exhibit genetic polymorphism, such as CYP2D6, CYP2C9, or CYP2C19, genotyping the subjects should be considered for a more informative data analysis.

7.4. Study Designs and Data Collection

The Food and Drug Administration survey of hepatic impairment studies revealed that 19 of the 57 studies contained oral drug clearance data in subjects with normal hepatic function and one or more of the Child-Pugh classification groups. Sixteen of these studies demonstrated reduced metabolism in subjects with Child-Pugh moderate hepatic impairment.[118] This survey and the renal impairment study paradigm are then reflected in the proposed hepatic study designs. These include full, reduced, and population approaches.[118]

7.4.1. Full Study Design

This study design attempts to characterize the drug's pharmacokinetics across the full range of hepatic impairment by studying all three Child-Pugh classes—mild, moderate, severe—and a control group. The control group enrolled in the study should have hepatic function representative of that in the population for which the drug is intended, not necessarily subjects with normal hepatic function. To the extent possible, the control subjects should be similar in common demographic characteristics such as age, weight, and sex distribution, as the hepatic impaired subjects. Sufficient numbers of subjects should be enrolled to yield evaluable data in six subjects in each study group.

Drug application could be single dose if both the parent drug and therapeutic/toxic metabolites obey time-independent linear pharmacokinetics for which single-dose information is predictive of steady-state exposure. A multiple-dose study is warranted if either the parent or metabolites have nonlinear or time-dependent disposition. Generally, the dose studied should be the therapeutic dose envisioned for product labeling, however, if there is concern about higher drug exposure in hepatic impaired subjects, a reduced dose may be chosen.[118]

7.4.2. Alternative Study Designs

The survey finding that 16 of 19 studies with oral clearance data (84 percent) showed reduced metabolism in subjects with moderate Child-Pugh hepatic impairment led, in part, to the proposal of a reduced study design in the draft *Hepatic Guidance*. This approach requires sufficient subjects to be enrolled to yield evaluable data in at least eight subject with moderate hepatic impairment (Child-Pugh score 7 to 9) and eight control subjects. The two groups should be similar demographically and in other extrinsic characteristics which might influence drug disposition such as diet, alcohol consumption, and smoking, if warranted.[118] Enrollment of control and hepatic impaired subjects as demographically-matched pairs is a particularly effective way to conduct a rigorously controlled assessment. It is noteworthy that choosing a reduced study design has restrictive implications on product labeling as discussed below in section 7.7.2.

A third approach is to assess the impact of hepatic impairment on drug pharmacokinetics using a population screen of data collected prospectively in phase 2 or 3 clinical trials. The draft *Hepatic Guidance* urges that this be prospectively planned so that the appropriate information can be gathered especially with reference to the five biochemical and clinical components of the Child-Pugh scoring system. It is also important to demonstrate that a sufficient number of hepatic impaired patients were available in the screen to adequately detect a difference in drug exposure of clinical significance.[44, 118]

7.4.3. Sample Collection and Bioanalysis

In single-dose studies, blood sampling may need to be extended in hepatic impaired subjects in order to appropriately characterize the terminal portion of the concentration-time profile if a prolongation in half-life accompanies hepatic dysfunction. For multiple-dose studies, the draft *Hepatic Guidance* recommends a concentration profile both after the first dose and at steady state. It furthermore recommends that both total and unbound drug and metabolite concentrations be determined for drugs with a high extraction ratio (greater than 0.7) and high protein binding (fraction bound greater than 80 percent). Protein binding should be determined at least from blood samples corresponding to maximum and trough concentrations.[118]

7.5. Data Analysis

Noncompartmental or compartmental analysis should be applied to the concentration-time data to derive Cmax, AUC, CL/F, Vd/F, and half-life of the parent drug. If urine drug concentrations are determined, total clearance can be apportioned into renal and nonrenal components (CLr and CLnr) providing additional insight into elimination mechanisms as both routes can be affected by hepatic impairment.[120] Therapeutic or toxic metabolite Cmax and AUC are also informative parameters. Where warranted, parameters from unbound concentration data should be derived, for example, CLu/F or AUCu.

An appeal has been made to drug development and regulatory scientists to more fully explore pharmacokinetic and clinical data generated in hepatic impairment studies.[129] The authors urged analysts to explore relationships not only between drug exposure and the Child-Pugh class but also between drug exposure and individual biochemical markers for hepatic impairment. They noted that, in general, aspartate aminotransferase and alanine aminotransferase serum levels poorly correlated with drug exposure but that biochemical components of the Child-Pugh classification—serum bilirubin, albumin, prothrombin time—were potentially more informative markers of drug exposure in hepatic impaired subjects.[129]

The draft *Hepatic Guidance* also recommends linear and nonlinear regression methods to seek for associations between exposure and indices of hepatic function. The former may include AUC, CL/F, and Vd/F and the latter may be categorical such as the Child-Pugh score or continuous such as biochemical markers or liver blood flow. Model parameter estimates and their precision should be reported as well as prediction error estimates to demonstrate the appropriateness of the chosen model. Exposure parameters should also be compared between Child-Pugh groups and the control group by calculating parameter ratios and confidence intervals.[118] In this manner, the average magnitude of change in exposure can be quantified in each Child-Pugh class which can be helpful in developing group-specific dose recommendations.

7.6. Case Study: Reduced Study Design for Everolimus in Hepatic Impairment

Everolimus is an investigational macrolide immunosuppressant for prophylaxis of allograft rejection after organ transplantation. Everolimus is extensively metabolized by CYP3A with elimination of metabolites in bile. Consequently, hepatic impairment would be expected to influence everolimus disposition. Use of everolimus in kidney transplant patients is occasionally associated with a decrease in platelet count which generally returns to normal without the need for dose reduction.[130] Given the above information, performing a comprehensive hepatic impairment study across the full range of Child-Pugh impairment classes could raise concern for subjects with severe hepatic impairment who may be more sensitive to the thrombocytopenic effects of this class of drugs. Accordingly

Table 7. Demographic, laboratory, and pharmacokinetic parameters in healthy and hepatic impaired subjects receiving everolimus.

Parameter	Healthy subjects	Hepatic impaired
Demographics:		
Age (years)	50.3 ± 5.7	50.3 ± 6.0
Weight (kg)	83.9 ± 9.7	86.4 ± 11.7
BSA (m^2)	2.04 ± 0.16	2.06 ± 0.19
Laboratory parameters:		
Bilirubin (mg/dL)	0.6 ± 0.2	2.4 ± 1.7
Albumin (g/dL)	4.3 ± 0.1	3.9 ± 0.5
Prothrombin time (s)	10.6 ± 0.5	14.1 ± 1.8
Pharmacokinetics:		
tmax (h)	0.8 ± 0.5	0.7 ± 0.3
Cmax (ng/mL)	15.4 ± 8.6	11.7 ± 4.3
AUC (ng·h/mL)	114 ± 45	245 ± 91
CL/F (L/h)	19.4 ± 5.8	9.1 ± 3.1
t1/2 (h)	43 ± 18	79 ± 42
Protein binding (%)	73.5 ± 2.4	73.8 ± 3.6

BSA, body surface area; tmax, time to reach Cmax; Cmax, maximum blood concentration; AUC, area under the concentration-time curve; CL/F, apparent clearance; t1/2, half-life.
Data reproduced by permission of Elsevier Science Limited from Kovarik *et al.*[131]

a reduced study was performed[131] which illustrates several aspects of study conduct and analysis pertinent to hepatic impairment studies in general.

In an open-label, single-dose, case-controlled study, eight subjects with liver cirrhosis classed as moderate hepatic impairment (Child-Pugh score 7 to 9) and eight healthy control subjects with normal hepatic function were enrolled. Each hepatic impaired subject was matched to a healthy control subject by sex, age (within five years), bodyweight (within ten percent), and height (within five cm). Each subject received a single oral 2-mg dose of everolimus. The period of blood sampling usually adequate for healthy subject pharmacokinetic studies at this dose level was extended to better characterize a prolonged elimination half-life should this occur in the hepatic impaired subjects. Although everolimus is not highly protein bound, blood samples were obtained to quantify protein binding. Pharmacokinetic parameters were derived by standard noncompartmental methods. Cmax, AUC, CL/F, and half-life were log transformed and compared in an analysis of variance based on the model: response = pair + condition + residual, where *pair* referred to the matched subject pairs and *condition* referred to the presence or absence of hepatic impairment. Scatterplots were constructed and evaluated by linear regression for relationships between everolimus AUC and each biochemical component of the Child-Pugh score—bilirubin, albumin, and prothrombin time.

Table 7 compares demographic, clinical, and pharmacokinetic data between the two study groups. Figure 4 shows the mean concentration-time profiles and the scatterplots of everolimus AUC versus various biochemistry parameters. The apparent clearance of everolimus was significantly reduced by 53% in subjects with moderate hepatic impairment compared with healthy subjects. This was manifested by a 115% higher AUC and 84% prolonged half-life in hepatic impaired subjects. The rate of absorption of everolimus was not altered by hepatic impairment based on Cmax and tmax. Likewise, protein binding was similar in the two groups. A significant positive correlation of everolimus

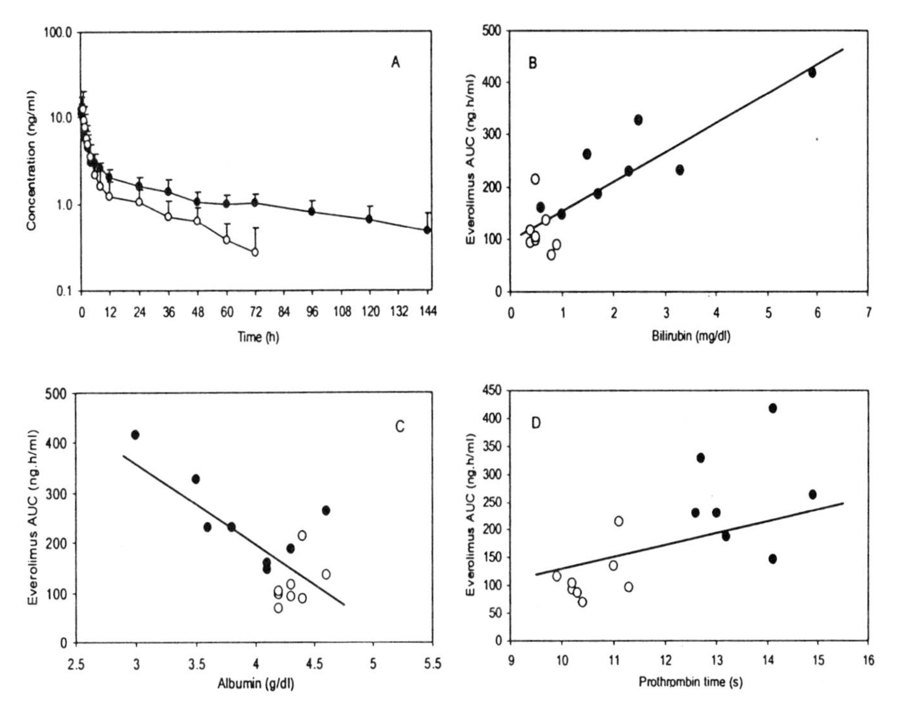

Figure 4. Mean concentration-time profiles of everolimus *(A)* and linear regression of everolimus AUC versus bilirubin *(B)*, albumin *(C)*, and prothrombin time *(D)* in healthy subjects (*open circles*) and subjects with moderate hepatic impairment (*filled circles*). Reproduced by permission of Elsevier Science Limited from Kovarik *et al.*[131]

AUC with bilirubin level (r = 0.86) and a significant negative correlation with albumin concentration (r = 0.72) was observed. There was a borderline significant positive correlation with prothrombin time (r = 0.49). The investigators pointed out that a bilirubin level greater than 1.5 mg/dL could serve as a general signal for patients who may need an everolimus dose reduction, however, data in a broader population would be needed to confirm this initial observation.

Based on this study and in accordance with the draft *Hepatic Guidance*, the investigators recommended that "the dose of everolimus should initially be reduced by half in patients with mild and moderate hepatic impairment based on the Child-Pugh classification. Therapeutic monitoring would be a helpful adjunct to subsequently titrate everolimus exposure in this subpopulation". They furthermore cautioned that everolimus has not been assessed in patients with severe hepatic impairment.[131]

7.7. Developing Dose Recommendations and Product Labeling Statements

7.7.1. Deciding Whether a Dose Adjustment is Needed

Recognizing the wide variety of pathologies assembled under the term *hepatic impairment* and the variable influence hepatic impairment has on drug disposition from patient to patient, the goal of a hepatic impairment study is somewhat restricted. As stated in the draft *Hepatic Guidance*, hepatic impairment studies should be designed to "provide information that can help guide initial dosing in patients. This information should be used with the understanding that careful observation and dose titration could be necessary to achieve the optimal dose in any given patient." [118]

The draft *Hepatic Guidance* proposes using the pharmacokinetic parameter ratios and confidence intervals for each Child-Pugh impairment class that was studied (mild, moderate, severe) compared with the control group. If the sponsor wishes to claim lack of an influence of hepatic impairment from a given Child-Pugh class, either a no-effect boundary for the 90 percent confidence interval should be prespecified in the study protocol or 70 to 143 percent for Cmax and 80 to 125 percent for AUC could be used without further justification. However, given the small group sample sizes relative to the wide confidence intervals expected in such studies, lack of an influence may be difficult to demonstrate statistically.[118]

On the other hand, if the parameter ratio is two or greater, necessary dose adjustments for a given Child-Pugh group should be made in the product label. The draft *Hepatic Guidance* foresees that a significant reduction in clearance in moderate hepatic impairment will generally lead to a similar dose reduction recommendation for mild impairment. Depending on the therapeutic index of the drug, either caution or a contraindication of use in severe impairment would be made. This is the implied regulatory approach in the case of a reduced study design in which only moderate impairment is studied. In the case of a full study design, this approach would be generally applied unless specific data generated in mild and severe impairment was strong enough to support other dose recommendations. As in renal impairment, so too in hepatic impairment, either the dose may be reduced while maintaining the usual dose interval, or the interval may be lengthened with a usual dose, or a combination of a dose and interval change may be made. In any case, the general aim is to design a dose regimen to yield similar exposure in hepatic impaired patients relative to the safe and effective exposure in the general patient population.[121]

7.7.2. Product Labeling Statements

The draft *Hepatic Guidance* attempts to standardize hepatic impairment labeling with proposed wording for several sections of the product label.

- *Clinical pharmacology—pharmacokinetics.* This subsection can contain information on the percent of drug eliminated by the liver, the mechanism of hepatic elimination with respect to the enzyme and conjugation pathways involved or the role of biliary excretion, and the effects of hepatic impairment on protein binding.
- *Clinical pharmacology—special populations.* This subsection should briefly summarize the impact of hepatic impairment on the drug's pharmacokinetics by describing the results of the hepatic impairment study and the specific Child-

Pugh groups assessed. The average percent change in pertinent pharmacokinetic parameters in each hepatic impairment group relative to the control group should be provided. If a clinically useful relationship between a marker of hepatic function and drug exposure was characterized, it could be described here. The draft *Hepatic Guidance* makes the following wording recommendations based on the outcome of the hepatic impairment study:

(1) If no influence of hepatic impairment was found: "No dosing adjustment is required with mild and moderate hepatic impairment."

(2) If an influence was found: "The dosage should be reduced in patients with mild and moderate hepatic impairment receiving (drug name). (Drug name) should be contraindicated or used with great caution in severe hepatic impairment."

(3) If no hepatic impairment study was performed, this must be stated and a recommendation based on the percent of drug eliminated by metabolism is suggested.

(a) <10% hepatic elimination: "hepatic impairment is not expected to have a significant effect on blood levels".

(b) <20% hepatic elimination: "hepatic impairment is not expected to have significant safety effects on blood levels" for wide therapeutic index drugs or "hepatic impairment could lead to an increased rate of adverse effects" for narrow therapeutic index drugs.

(c) >20% hepatic elimination: "hepatic impairment would be expected to have significant pharmacokinetic effects". For wide therapeutic index drugs "patients with impaired liver function would require reduced initial and maintenance doses" or for narrow therapeutic index drugs "should be contraindicated or used with great caution in this patient population".

- *Precautions, warnings,* and *contraindications.* If there are clinically relevant changes in drug pharmacokinetics or activity in hepatic impairment, this should be stated in the "precautions" section. If the drug additionally has a narrow therapeutic index, a hepatic impairment statement may also be included in the "warnings" or "contraindications" sections.
- *Dose and administration.* This section should clearly state whether a dose adjustment is needed in hepatic impairment, and if so, the dose recommendation should be specified.

8. PROSPECTUS

The foregoing has demonstrated that the rationale for exploring the six conventional special populations during drug development has a firm basis in decades of pharmacokinetic research and clinical experience. The many regulatory guidances for industry in these research areas testify to the importance of these patient subgroups. Furthermore, there are myriad examples of how special population studies in these six categories have improved the use of drugs by providing clinicians with guidelines how to tailor exposure in patients who are members of these clinical subgroups and thereby to improve the individual benefit/risk balance. Using the impact which these studies have on the product label as a success criteria, however, reveals that there is an uneven impact range across the six special populations. At one end of the spectrum is renal impairment for which clinically-practical measurement methods exist and predictable effects on drug exposure yield relatively precise dose adjustment recommendations. At the other end of the spectrum are subgroups such as geriatrics or hepatic impairment which exist more as umbrella terms capturing heterogenous groups of individuals and supporting only general trends or broad dosing considerations for product labeling. The lack of scientific rigor in deriving dose regimens for such subpopulations likely occurs because the group label is a poor marker for the underlying reason why the pharmacokinetics of a drug differ between some members in the subgroup and the general population. Two developments in clinical pharmacology may help to redress these shortcomings in the future. One, in its adolescence, is population pharmacokinetics and the other, in its infancy, is pharmacogenomics.

Population pharmacokinetics is a recognized tool in drug development[44] which is beginning to demonstrate its power to unmask pharmacokinetic subpopulations in clinical trials. Results from this approach are more frequently gaining a place in product labeling. Population pharmacokinetic modeling can incorporate both categorical and continuous variables. However, during the modeling procedure, each covariate must compete in a statistically rigorous process against other covariates to survive in the final model. This will tend to eliminate diffuse group markers—which are often categorical variables—in favor of more precise, science-driven covariates—which are more often numerical, continuous variables. As a result, population pharmacokinetics has the potential to discard uninformative group labels in favor of more objectively and precisely defined subpopulations truly needing a dose adjustment. In order for this analysis method to reach its full potential, critical study design factors that influence the power of a population pharmacokinetic study to identify subpopulations need to be better defined ánd implemented in the trial protocol. Research in this area is underway.[132]

Pharmacogenomics is the science of using genetic information from an individual or population to explain differences in drug disposition, to identify responders to a drug, or to predict efficacy or toxicity from a drug. In a regulatory position paper on pharmacogenomic-guided drug development, the authors cautioned that several scientific and ethical implications of this science have yet to be worked out and that the use of such data to support labeling claims needs to be addressed.[133] Nonetheless, they hold out the hope that pharmacogenomics may approach the yet-elusive goal of true individualization of therapy by providing more precise diagnoses based on underlying genotype or gene expression and by identifying variability in drug exposure which impacts on efficacy and toxicity. The authors foresee that "[t]he ability to predict and account for such differences could markedly improve the therapeutic index of many drug interventions." [133]

If the potential of these and other trends in clinical pharmacology are realized, they would not mean the demise of special population studies nor the removal of such information from its prominent place in product labeling. Rather they would help to refine and focus this area of drug development to ultimately better serve the individual patient.

9. REFERENCES

1. Levy G. Patient-oriented pharmaceutical research: focus on the individual. Pharm Res. 1995;12:943-944.
2. Spyker DA, Harvey ED, Harvey BE, Harvey AM, Rumack BH, Peck CC, et al. Assessment and reporting of clinical pharmacology information in drug labeling. Clin Pharmacol Ther. 2000;67:196-200.
3. Food and Drug Administration. Guideline for the study and evaluation of gender differences in the clinical evaluation of drugs. Fed Register. 1993;58:39406-39416.

4. European Commission, Enterprise Directorate-General. Notice to applicants: a guideline on summary of product characteristics. 1999.
5. General requirements on content and format of labeling for human prescription drugs (revised 1 April 2001). US Code Fed Regul. 21:201.56.
6. Shirkey H. Editorial comment: therapeutic orphans. J Pediatrics. 1968;72:119-120.
7. Food and Drug Administration. Pediatric exclusivity provision: status report to Congress. 2001. http://www.fda.gov/cder/pediatric/reportcong01.pdf
8. Kearns GL. Introduction: drug development for infants and children: rescuing the therapeutic orphan. Drug Inform J. 1996;30:1121-1123.
9. Wilson JT. An update on the therapeutic orphan. Pediatrics. 1999;104:585-590.
10. Roberts R, Maldonado S. FDA center for drug evaluation and research (CDER) pediatric plan and new regulations. Drug Inform J. 1996;30:1125-1127.
11. Milne CP. Pediatric research: coming of age in the new millennium. Am J Ther. 1999;6:263-282.
12. Food and Drug Administration. Specific requirements on content and format of labeling for human prescription drugs; revision of "pediatric use" subsection in the labeling: final rule. Fed Register. 1994;59:64240-64250.
13. Food and Drug Administration. Pediatric patients: regulations requiring manufacturers to assess the safety and effectiveness of new drugs and biological products: proposed rule. Fed Register. 1997;62:43899-43916.
14. Food and Drug Administration. Regulations requiring manufacturers to assess the safety and effectiveness of new drugs and biological products in pediatric patients: final rule. Fed Register. 1998;63:66632-66671.
15. European Commission, Enterprise Directorate-General. Better medicines for children: proposed regulatory actions on paediatric medicinal products, consultation document. 2002.
16. Food and Drug Administration. General considerations for the clinical evaluation of drugs in infants and children. 1977.
17. American Academy of Pediatrics Committee on Drugs. Guidelines for the ethical conduct of studies to evaluate drugs in pediatric patients. Pediatrics. 1995;95(2):286-294.
18. Food and Drug Administration. Guidance for industry: the content and format for pediatric use supplements. 1996.
19. Food and Drug Administration. Draft guidance for industry: general considerations for pediatric pharmacokinetic studies for drugs and biological products. 1998.
20. Food and Drug Administration. Qualifying for pediatric exclusivity under section 505A of the Federal Food, Drug, and Cosmetic Act. 1999.
21. International Conference on Harmonization. Guidance for industry E11: clinical investigations of medicinal products in pediatric populations. 2000.
22. Cohen SN. Pediatric pharmacology research unit (PPRU) network and its role in meeting pediatric labeling needs. Pediatrics. 1999;104:644-645.
23. Kearns GL, Reed MD. Clinical pharmacokinetics in infants and children. Clin Pharmacokinet. 1989;17(Suppl 1):29-67.
24. Kearns GL. Impact of developmental pharmacology on pediatric study design: overcoming the challenges. J Allergy Clin Immunol. 2000;106:S128-S138.
25. Reed MD. Ontogeny of drug disposition: focus on drug absorption, distribution, and excretion. Drug Inform J. 1996;30:1129-1134.
26. Morselli PL. Clinical pharmacology of the perinatal period and early infancy. Clin Pharmacokinet. 1989;17(Suppl 1):13-28.
27. Zenk KE. Challenges in providing pharmaceutical care to pediatric patients. Am J Hosp Pharm. 1994;51:688-694.
28. Leeder JS, Kearns GL. Pharmacogenetics in pediatrics: implications for practice. Ped Clin North Am. 1997;44:55-77.
29. Rane A. Phenotyping of drug metabolism in infants and children: potentials and problems. Pediatrics. 1999;104:640-643.
30. Ptachcinski RJ, Burckart GJ, Rosenthal JT. Cyclosporine pharmacokinetics in children following cadaveric renal transplantation. Transplant Proc. 1986;18:766.
31. Dunn S, Cooney G, Sommeraurer J, Lindsay C, McDiarmid S, Wong RL, et al. Pharmacokinetics of an oral solution of the microemulsion formulation of cyclosporine in maintenance pediatric liver transplant recipients. Transplantation. 1997;63:1762-1767.
32. Kovarik JM, Mueller EA, Niese D. Clinical development of a cyclosporine microemulsion in transplantation. Ther Drug Monit. 1996;18:429-434.

33. McRorie T. Quality drug therapy in children: formulations and delivery. Drug Inform J. 1996;30:1173-1177.
34. Nahata MC. Lack of pediatric drug formulations. Pediatrics. 1999;104:607-609.
35. American Academy of Pediatrics Committee on Drugs. "Inactive" ingredients in pharmaceutical products: update. Pediatrics. 1997;99:268-278.
36. Autret E. European regulatory authorities and pediatric labeling. Pediatrics. 1999;104:614-618.
37. Reed MD. Optimal sampling theory: an overview of its application to pharmacokinetic studies in infants and children. Pediatrics. 1999;104:627-632.
38. Collart L, Blaschke TF, Boucher F, Prober CG. Potential of population pharmacokinetics to reduce the frequency of blood sampling required for estimating kinetic parameters in neonates. Devel Pharmacol Ther. 1992;18:71-80.
39. Kauffman RE, Kearns GL. Pharmacokinetic studies in paediatric patients: clinical and ethical considerations. Clin Pharmacokinet. 1992;23:10-29.
40. Kovarik JM, Kahan BD, Rajagopalan PR, Bennett W, Mulloy LL, Gerbeau C, et al. Population pharmacokinetics and exposure-response relationships for basiliximab in kidney transplantation. Transplantation. 1999;68:1288-1294.
41. Offner G, Broyer M, Niaudet P, Loirat C, Mentser M, Lemire J, et al. A multicenter, open-label, pharmacokinetic/pharmacodynamic safety and tolerability study of basiliximab (Simulect) in pediatric de novo renal transplant recipients. Transplantation. (in press).
42. Kovarik JM, Offner G, Broyer M, Niaudet P, Loirat C, Mentser M, et al. A rational dosing algorithm for basiliximab (Simulect) in pediatric renal transplantation based on pharmacokinetic-dynamic evaluations. Transplantation. (in press).
43. Simulect (basiliximab). In: Physicians' desk reference. 55th ed. Montvale, NJ: Medical Economics Company; 2001. p. 2218-2220.
44. Food and Drug Administration. Guidance for industry: population pharmacokinetics. 1999.
45. Parnis SJ, Foate JA, van der Walt JH, Short T, Crowe CE. Oral midazolam is an effective premedication for children having day-stay anaesthesia. Anaesth Intensive Care. 1992;20:9-14.
46. Reed MJ, Rodarte A, Blumer JL, Khoo KC, Akbari B, Pou S, et al. Single-dose pharmacokinetics of midazolam and its primary metabolite in pediatric patients after oral and intravenous administration. J Clin Pharmacol. 2001;41:1359-1369.
47. Marshall J, Rodarte A, Blumer JL, Khoo KC, Akbari B, Kearns GL and the Pediatric Pharmacology Research Unit Network. Pediatric pharmacodynamics of midazolam. J Clin Pharmacol. 2000;40:578-589.
48. Versed (midazolam hydrochloride). In: Physicians' desk reference. 55th ed. Montvale, NJ: Medical Economics Company; 2001. p. 2800-2804.
49. Kinirons MT, Crome P. Clinical pharmacokinetic considerations in the elderly: an update. Clin Pharmacokinet. 1997;33:302-312.
50. Anonymous. Geriatric study incentives modeled on pediatric patent extensions suggested. The Pink Sheet. 1998;60:11.
51. Crome P, Flanagan RJ. Pharmacokinetic studies in elderly people: are they necessary? Clin Pharmacokinet. 1994;26:243-247.
52. Ozdemir V, Fourie J, Busto U, Naranjo CA. Pharmacokinetic changes in the elderly: do they contribute to drug abuse and dependence? Clin Pharmacokinet. 1996;31:372-385.
53. Tregaskis BF, Stevenson IH. Pharmacokinetics in old age. British Med Bull. 1990;46:9-21.
54. Maletta G, Mattox KA, Dysken M. Guidelines for prescribing psychoactive drugs in the elderly: part 1. Geriatrics. 1991;46:40-47.
55. International Conference on Harmonization. Note for guidance on studies in support of special population: geriatrics. 1994.
56. Food and Drug Administration. Guidance for industry: in vivo drug metabolism/drug interaction studies—study design, data analysis, and recommendations for dosing and labeling. 1999.
57. Exelon (rivastigmine tartrate). In: Physicians' desk reference. 55th ed. Montvale, NJ: Medical Economics Company; 2001. p. 2171-2174.
58. Food and Drug Administration. Guidance for industry: content and format for geriatric labeling. 2001.
59. Food and Drug Administration. General considerations for the clinical evaluation of drugs. Publication no. HEW (FDA) 77-3040. Washington, DC: Government Printing Office; 1977.
60. Food and Drug Administration. Guideline for the format and content of the clinical and statistics sections of new drug applications. 1988.
61. Kim JS, Nafziger AN. Is it sex or is it gender? Clin Pharmacol Ther. 2000;68:1-3.
62. Bush JK. Industry perspective on the inclusion of women in clinical trials. Acad Med. 1994;69:708-715.

63. Chen ML, Williams RL. Women in bioavailability/bioequivalence trials: a regulatory perspective. Drug Inform J. 1995;29:813-820.
64. Chen ML, Lee SC, Ng MJ, Schuirmann DJ, Lesko LJ, Williams RL. Pharmacokinetic analysis of bioequivalence trials: implications for sex-related issues in clinical pharmacology and biopharmaceutics. Clin Pharmacol Ther. 2000;68:510-521.
65. Harris RZ, Benet LZ, Schwartz JB. Gender effects in pharmacokinetics and pharmacodynamics. Clin Pharmacokinet. 1995;50:222-239.
66. Beierle I, Meibohm B, Derendorf H. Gender differences in pharmacokinetics and pharmacodynamics. Int J Clin Pharmacol Ther. 1999;37:529-547.
67. Tanaka E. Gender-related differences in pharmacokinetics and their clinical significance. J Clin Pharmacol Ther. 1999;24:339-346.
68. Merkatz RB, Temple R, Sobel S, Feiden K, Kessler DA. Women in clinical trials of new drugs: a change in Food and Drug Administration policy. N Engl J Med. 1993;329:292-296.
69. Hartter S, Wetzel H, Hammes E, Torkzadeh M, Hiemke C. Nonlinear pharmacokinetics of fluvoxamine and gender differences. Ther Drug Monit. 1998;20:446-449.
70. Luvox (fluvoxamine maleate). In: Physicians' desk reference. 55th ed. Montvale, NJ: Medical Economics Company; 2001. p. 3153-3156.
71. Sekar V, Gobburu J, Chang J, ZumBrunnen T. Population pharmacokinetic modeling to support pediatric use of fluvoxamine. Abstract. J Clin Pharmacol. 2001;41:1021.
72. Kashuba ADM, Nafziger AN. Physiological changes during the menstrual cycle and their effects on the pharmacokinetics and pharmacodynamics of drugs. Clin Pharmacokinet. 1998;34:203-218.
73. Gustavson LE, Benet LZ. Menopause: pharmacodynamics and pharmacokinetics. Exp Gerontol. 1994;29:437-444.
74. Belle DJ, Callaghan JT, Gorski JC, Maya JF, Mousa O, Wrighton SA, et al. Effects of an oral contraceptive containing ethinyloestradiol and norgestrel on CYP3A activity. Br J Clin Pharmacol. 2002;53:67-74.
75. Xie CX, Piecoro LT, Wermeling DP. Gender-related consideration in clinical pharmacology and drug therapeutics. Crit Care Nurs Clinics of North Am. 1997;9:459-468.
76. Back DJ, Orme ML. Pharmacokinetic drug interactions with oral contraceptives. Clin Pharmacokinet. 1990;18:472-484.
77. Food and Drug Administration. Draft guidance for industry: combined oral contraceptives—labeling for healthcare providers and patients. 2000.
78. Food and Drug Administration and the National Institutes of Health Conference. Clinical pharmacology during pregnancy: addressing clinical needs through science. 2000. Conference transcript at http://www.fda.gov/cder/present/clinpharm2000/1204preg.txt.
79. Food and Drug Administration. Draft guidance for industry: establishing pregnancy registries. 1999.
80. Murray L, Seger D. Drug therapy during pregnancy and lactation. Emerg Clin North Am. 1994;12:129-149.
81. Wilson JT, Brown RD, Hinson JL, Dailey JW. Pharmacokinetic pitfalls in the estimation of the breast milk/plasma ratio from drugs. Ann Rev Pharmacol Toxicol. 1985;25:667-689.
82. American Academy of Pediatrics Committee on Drugs. Transfer of drugs and other chemicals into human milk. Pediatrics. 1994;93:137-150.
83. Breitzka RL, Sandritter TL, Hatzopoulos FK. Principles of drug transfer into breast milk and drug disposition in the nursing infant. J Human Lact. 1997;13:155-158.
84. Scialli AR. Drugs and lactation: another failure of product labeling. Reproductive Toxicol. 1996;10:91-92.
85. Shyu WC, Shah VR, Campbell DA, Venitz J, Jaganathan V, Pittman KA, et al. Excretion of cefprozil into human breast milk. Antimicrob Agents Chemother. 1992;36:938-941.
86. Cefzil (cefprozil). In: Physicians' desk reference. 55th ed. Montvale, NJ: Medical Economics Company; 2001. p. 998-1001.
87. International Committee on Harmonization. Guidance for industry E5: ethnic factors in the acceptability of foreign clinical data. 1998.
88. Foster MW, Sharp RR, Mulvihill JJ. Pharmacogenetics, race, and ethnicity: social identities and individualized medical care. Ther Drug Monitor. 2001;23:232-238.
89. Sheldon TA, Parker H. Race and ethnicity in health research. J Public Health Med. 1992;14:104-110.
90. Anand SS. Using ethnicity as a classification variable in health research: perpetuating the myth of biological determinism, serving socio-political agendas, or making valuable contributions to medical sciences? Ethnic Health. 1999;4:241-244.
91. Freeman HP. The meaning of race in science—considerations for cancer research. Cancer. 1998;82:219-225.
92. Kitler ME. Clinical trials and transethnic pharmacology. Drug Safety. 1994;11:378-391.

93. Wood AJJ. Ethnic differences in drug disposition. Ther Drug Monit. 1998;20:525-526.
94. Xie HG, Kim RB, Wood AJJ, Stein CM. Molecular basis of ethnic differences in drug disposition and response. Ann Rev Pharmacol Toxicol. 2001;41:815-850.
95. Johnston JA. Predictability of the effects of race or ethnicity on pharmacokinetics of drugs. Int J Clin Pharmacol Ther. 2000;38:53-60.
96. Andersson T, Regårdh CG, Lou YC, Zhang Y, Dahl ML, Bertilsson L. Polymorphic hydroxylation of S-mephenytoin and omeprazole metabolism in Caucasian and Chinese subjects. Pharmacogenetics. 1992;2:25-31.
97. Caraco Y, Lagerstrom PO, Wood AJJ. Ethnic and genetic determinants of omeprazole disposition and effect. Clin Pharmacol Ther. 1996;60:157-167.
98. Furuta T, Ohashi K, Kamata T, Takashima M, Kosuge K, Kawasaki T, et al. Effect of genetic differences in omeprazole metabolism on cure rates for *Helicobacter pylori* infection and peptic ulcer. Ann Intern Med. 1998;129:1027-1030.
99. Goldstein JA. Clinical relevance of genetic polymorphisms in the human CYP2C subfamily. Br J Clin Pharmacol. 2001;52:349-355.
100. Prilosec (omeprazole). In: Physicians' desk reference. 55th ed. Montvale, NJ: Medical Economics Company; 2001. p. 587-591.
101. Van Geerven JMA, Uchida E, Uchida N, Pieters MSM, Meinders AJ, Schoemaker RC, et al. Pharmacodynamics and pharmacokinetics of a single oral dose of nitrazepam in healthy volunteers: an interethnic comparative study between Japanese and European volunteers. J Clin Pharmacol. 1998;38:1129-1136.
102. Lam YWF, Banerji S, Hatfield C, Talbert RL. Principles of drug administration in renal insufficiency. Clin Pharmacokinet. 1997;32:30-57.
103. Ibrahim S, Honig P, Huang SM, Gillespie W, Lesko LJ, Williams RL. Clinical pharmacology studies in patients with renal impairment: past experience and regulatory perspectives. J Clin Pharmacol. 2000;40:31-38.
104. Food and Drug Administration. Guidance for industry: pharmacokinetics in patients with impaired renal function—study design, data analysis, and impact on dosing and labeling. 1998.
105. Cockcroft DW, Gault MH. Prediction of creatinine clearance from serum creatinine. Nephron. 1976;16:31-41.
106. Schwartz GJ. A simple estimate of glomerular filtration rate in full-term infants during the first year of life. J Pediatr. 1984;104:849-854.
107. Schwartz GJ. A simple estimate of glomerular filtration rate in children derived from body length and plasma creatinine. Pediatrics. 1978;35:53-62.
108. Lee CS, Marbury TC, Benet LZ. Clearance calculations in hemodialysis: application to blood, plasma, and dialysate measurements for ethambutol. J Pharmacokinet Biopharm. 1980;8:69-81.
109. Keller E, Reetze P, Schollmeyer P. Drug therapy in patients undergoing continuous ambulatory peritoneal dialysis: clinical pharmacokinetic considerations. Clin Pharmacokinet. 1990;18:104-117.
110. Gambertoglio JG. Drug use in renal disease. In: Knoben JE, Anderson PO, eds. Handbook of clinical drug data. Hamilton, IL: Drug Intelligence Publications. 1993. p. 161-170.
111. Rowland M, Tozer TN. Clinical pharmacokinetics: concepts and applications. 2nd ed. Philadelphia: Lee and Febiger; 1989. p. 238-254.
112. Lee CC, Marbury TC. Drug therapy in patients undergoing hemodialysis: clinical pharmacokinetic considerations. Clin Pharmacokinet. 1984;9:42-66.
113. Boelaert J, Valcke Y, Schurgers M, Daneels R, Rosseneu M, Rosseel MT, Bogeart MG. Pharmacokinetics of ciprofloxacin in patients with impaired renal function. J Antimicrob Chemother. 1985;16:87-93.
114. Webb DB, Roberts DE, Williams JD, Asscher AW. Pharmacokinetics of ciprofloxacin in healthy volunteers and patients with impaired kidney function. J Antimicrob Chemother. 1986;18(Suppl D):83-87.
115. Singlas E, Taburet AM, Landru I, Albin H, Ryckelinck JP. Pharmacokinetics of ciprofloxacin tablets in renal failure: influence of haemodialysis. Eur J Clin Pharmacol. 1987;31:589-593.
116. Bergan T, Thorsteinsson SB, Rohwedder R, Scholl H. Elimination of ciprofloxacin and three major metabolites and consequences of reduced renal function. Chemother. 1989;35:393-405.
117. Cipro (ciprofloxacin hydrochloride). In: Physicians' desk reference. 55th ed. Montvale, NJ: Medical Economics Company; 2001. p. 847-852.
118. Food and Drug Administration. Draft guidance for industry: pharmacokinetics in patients with impaired hepatic function—study design, data analysis, and impact on dosing and labeling. 1999.
119. Howden CW, Birnie GG, Brodie MJ. Drug metabolism in liver disease. Pharmacol Ther. 1989;40:439-474.
120. McLean AJ, Morgan DJ. Clinical pharmacokinetics in patients with liver disease. Clin Pharmacokinet. 1991;21:42-69.

121. Rodighiero V. Effects of liver disease on pharmacokinetics: an update. Clin Pharmacokinet. 1999;37:399-431.
122. Benet LZ, Hoener B. Changes in plasma protein binding have little clinical relevance. Clin Pharmacol Ther. 2002;71:115-121.
123. Morgan DJ, McLean AJ. Clinical pharmacokinetic and pharmacodynamic considerations in patients with liver disease: an update. Clin Pharmacokinet. 1995;29:370-391.
124. Westphal JF, Brogard JM. Drug administration in chronic liver disease. Drug Safety. 1997;17:47-73.
125. George J, Murray M, Byth K, Farrell GC. Differential alterations of cytochrome P450 proteins in livers from patients with severe chronic liver disease. Hepatology. 1995;21:120-128.
126. Pugh RNH, Murrary-Lyon IM, Dawson JL, Pietroni MC, Williams R. Transection of the oesophagus for bleeding oesophageal varices. Brit J Surg. 1973;60:646-649.
127. Brockmoller J, Roots I. Assessment of liver metabolic function: clinical implications. Clin Pharmacokinet. 1994;27:216-248.
128. Figg WD, Dukes GE, Lesesne HR, Carson SW, Songer SS, Pritchard F, et al. Comparison of quantitative methods to assess hepatic function: Pugh's classification, indocyanine green, antipyrine, and dextromethorphan. Pharmacother. 1995;15:693-700.
129. Bergquist C, Lindegard J, Salmonson T. Dosing recommendations in liver disease (Letter to the editor). Clin Pharmacol Ther. 1999;66:201-204.
130. Kovarik JM, Kaplan B, Tedesco Silva H, Kahan BD, Dantal J, Vitko S, et al. Exposure-response relationships for everolimus in de novo kidney transplantation: defining a therapeutic range. Transplantation. 2002;73:920-925.
131. Kovarik JM, Sabia HD, Figueiredo J, Zimmermann H, Reynolds C, Dilzer S, et al. Influence of hepatic impairment on everolimus pharmacokinetics: implications for dose adjustment. Clin Pharmacol Ther. 2001;70:425-430.
132. Lee PID. Design and power of a population pharmacokinetic study. Pharm Res. 2001;18:75-82.
133. Lesko LJ, Woodcock J. Pharmacogenomic-guided drug development: regulatory perspective. Pharmacogenomics J. 2002;2:20-24.

DRUG-DRUG INTERACTIONS

Shiew-Mei Huang*

1. INTRODUCTION

FDA evidence standards for demonstrating the effectiveness and safety of a new molecular entity (NME) are well evolved and defined by current laws and regulations. However, because only a limited number of patients can be studied during the drug development process, unpredictable adverse events can occur following approval. While serious post-marketing adverse events are rare, their occurrence can lead to market withdrawal of a drug. Five of 12 drugs that have been withdrawn from the U.S. market from 1997 to 2002 were prone to metabolic drug-drug interactions[1]. Three of these drugs, terfenadine, astemizole, and cisapride are cytochrome P450 (CYP) isozyme 3A4 (CYP3A4) substrates and can induce alterations in cardiac repolarization at high exposure levels in sensitive individuals. Inhibition of the intestinal and hepatic CYP3A4 by concomitant drug therapy can lead to unanticipated increases in drug exposure and to clinically significant QT prolongation or, in some cases, serious arrhythmias. Mibefradil, also a CYP3A4 substrate, is a potent inhibitor of CYP3A4 and P-gp transporter. Concomitant administration of mibefradil with other CYP3A4 and P-gp substrates can lead to increased adverse events of these other drugs by significantly elevating their systemic exposure. The fifth drug, cerivastatin, is a substrate of CYP3A, CYP2C8, and glucuronosyl transferases[2]. Inhibition of these enzymes may have contributed to the interactions of other co-administered drugs with cerivastatin, which may result in muscle toxicity, or in some cases, rhabdomyolysis.

How to determine if an NME being reviewed by the Agency is a potent inhibitor (like mibefradil) or a substrate (like terfenadine, astemizole, cisapride or cerivastatin) of metabolizing enzymes and/or transporters, is an important task faced by the Agency day-to-day. In this chapter, methods to evaluate an NME's interaction potential and the impact of the evaluation results on regulatory decision-making and labeling languages will be discussed.

---

* Shiew-Mei Huang, Office of Clinical Pharmacology and Biopharmaceutics, HFD-850, Center for Drug Evaluation and Research, Food and Drug Administration, Parklawn 6A/19, 5600 Fishers Lane, Rockville, MD, USA

2. AN INTEGRATED APPROACH

Drug interaction potential needs to be assessed early in drug development and thoroughly evaluated throughout the development process employing an integrated approach. Figure 1 depicts one algorithm how this investigation can take place. As CYP enzymes have been involved in many clinically important drug interactions, *in vitro* evaluation of the drug's metabolic clearance via, and modulating effects of, key cytochrome P450 enzymes (e.g., CYP1A2, CYP2D6, CYP2C9, CYP2C19, CYP3A) appears to be a critical first step in this evaluation[3-9]. Depending on the outcome of this *in vitro* evaluation using human tissues or expressed human enzymes, subsequent *in vivo* evaluation in human subjects may be designed to address specific questions. In addition to CYP enzymes, various other metabolic enzymes (e.g., glucuronosyl transferases) and transporters (e.g., P-glycoprotein (P-gp) and organic anion transporting polypeptides (OATP)) also play an important role in the absorption, distribution and clearance of drugs. The use of other complementary approaches (e.g., population pharmacokinetic assessments) can provide additional opportunities to uncover unexpected pharmacokinetic drug interactions and, with proper design, pharmacodynamic interactions.

In addition to literature articles, the following documents[9-11] are available to provide guidance to industry and Agency reviewers regarding the use of various methodologies to address drug-drug interaction issues.

- Guidance for industry: drug metabolism/drug interactions in the drug development process: studies *in vitro*.
- Guidance for industry: *in vivo* metabolism/drug interactions: study design, data analysis and recommendation for dosing and labeling.
- Guidance for industry: population pharmacokinetics

3. *IN VITRO* TECHNOLOGIES

3.1. CYP Inhibition

An NME's potential to inhibit CYP enzymes can be assessed using various *in vitro* systems, such as human liver microsomes, expressed enzymes, hepatocytes, and liver slices. Considering the known genetic polymorphism of many CYP enzymes and the well-recognized large inter-subject variability in drug metabolism, liver tissues derived from more than one individual should be used in *in vitro* evaluation. A specific pool with known enzyme activities can be used for this type of study.

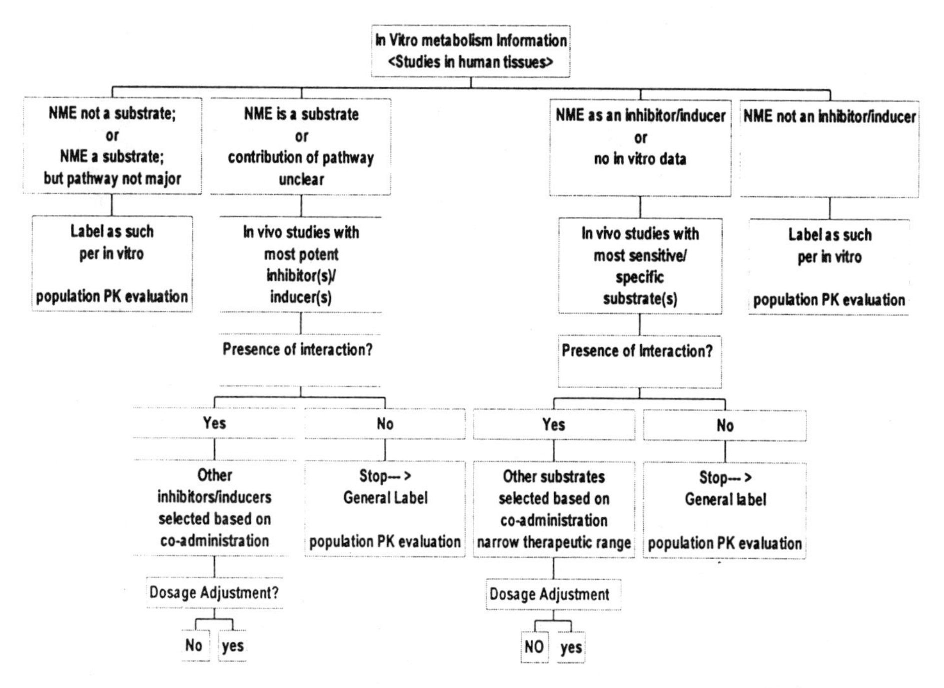

Figure 1. An algorithm for evaluating drug-drug interactions[12,13]

An NME's ability to inhibit CYP enzymes *in vitro* can be evaluated with the following considerations.

- Specific probe substrates are used (Table 1 lists preferred *in vitro* probe substrates)
- To determine the Ki values, at least 5-6 probe substrate concentrations and 2-3 NME concentrations should be used in the assays
- Probe substrate concentrations should cover at least 10-20 fold range with the number of concentrations evenly distributed below and above the Km value
- Protein concentrations and incubation times are established to be linear with metabolite formation
- Avoid using organic solvents or solvent concentrations be kept low to avoid solvent inhibition or activation effects[15]

Table 1. Recommended *in vitro* probe substrates and inhibitors for CYPs[14]

CYP	Substrates		Inhibitors	
	'Preferred'	'Acceptable'	'Preferred'	'Acceptable'
1A2	Ethoxyresorufin Phenacetin	Caffeine (low turnover) Theophylline (low turnover) Acetanilide (mostly applied in hepatocytes) Methoxyresorufin	Furafylline	α-naphthoflavone (but can also activate and inhibit CYP3A4)
2A6	Coumarin			Coumarin (but high turnover)
2B6	S-Mephenytoin (N-desmethyl metabolite)	Bupropion (availability of metabolite standards?)		Sertraline (but also inhibits CYP2D6)
2C8	Paclitaxel (availability of standards?)		('glitazones' – availability of standards?)	
2C9	S-Warfarin Diclofenac	Tolbutamide (low turnover)	Sulphaphenazole	
2C19	S-Mephenytoin (4-hydroxy metabolite) Omeprazole			Ticlopidine (but also inhibits CYP2D6) Nootkatone (but also inhibits CYP2A6)
2D6	Bufuralol Dextromethorphan	Metoprolol Debrisoquine Codeine (all with no problems, but less commonly used)	Quinidine	
2E1	Chlorzoxazone	4-nitrophenol Lauric acid		4-methyl pyrazole
3A4	Midazolam Testosterone (strongly recommended to use at least two structurally unrelated substrates)	Nifedipine Felodipine Cyclosporin Terfenadine Erythromycin Simvastatin	Ketoconazole (but recent evidence indicates that it is also a potent inhibitor of CYP2C8) Troleandomycin	Cyclosporin

- Metabolite formation rate should be measured; preferably no more than 10% substrate or inhibitor depletion should occur
- 2 or more substrates may be needed when evaluating inhibitors of CYP3A using *in vitro* methods[14-16], as substrate-dependence in inhibitor potency has been reported earlier[16,17].
- If major metabolites are found *in vitro* or *in vivo*; their possible inhibition effects should be taken into consideration
- Time-dependent inhibition is evaluated as part of the protocol to determine if the NME is a mechanism-based inhibitor. NME, at various concentrations (covering a 10- 20 fold range), is pre-incubated with human liver microsomes with and without NADPH for various lengths of time to allow the generation of reactive metabolites that inhibit cytochrome P450 activity irreversibly or quasi-irreversibly and to determine the Ki and Kinact values[18,19].

When the *in vitro* data indicate that the NME is an inhibitor, one will then need to design further *in vivo* studies to evaluate the extent of interaction, as the quantitative projection of *in vivo* interactions from *in vitro* data is limited. When the *in vitro* data indicate that the NME is not an inhibitor, no further *in vivo* investigation is needed and the information can be used for the labeling (Figure 1). The following criteria have often been used for the determination of inhibition potential[14,66].

Impact of $[I]/K_i$ ratio on the risk of great change in AUC of substrate ($[S] < K_m$) in presence and absence of a competitive inhibitor	
I/Ki 0.01 to 0.1	low risk
I/Ki 0.1 to 1	medium risk
I/Ki >1	high risk

Where "Ki" is the inhibitory constant and "I" may be estimated from the steady state total (free and protein-bound) maximum plasma concentrations at the highest proposed clinical dose.

When the I/Ki ratio fell between the gray area, one practice is to rank order the I/Ki values for various CYP enzymes and conduct an *in vivo* study in humans with probes of the CYP enzyme with the highest I/Ki (or lowest Ki) value. A negative result from this *in vivo* study would obviate the need to further evaluate other CYP enzymes that the NME inhibits with less potency.

3.2. CYP Induction

An NME's potential to induce CYP enzymes can be assessed using various *in vitro* systems, depending on its stage of development and the purpose of the study[20-23]. An NME's ability to induce CYP enzymes *in vitro* can be evaluated with the following considerations.

- Freshly prepared human hepatocyte system is used
- The hepatocyte system is incubated with the NME for 2-5 days
- The NME concentrations should be relevant to its therapeutic range or, if unknown, covering two to three orders of magnitude
- At least 3 individual donor livers are used for the study
- A positive control is provided using known inducers (e.g., rifampin for CYP3A, CYP2C9, CYP2C19 and CYP2B6; omeprazole for CYP1A2)
- Enzyme activity assays using probe substrates (see Table 1 for examples) are preferred; other measures such as protein or mRNA levels can be used to provide for additional mechanistic evaluation.
- Other conditions (protein concentration, incubation time, solvent concentration, substrate depletion, etc) for the activity assay should follow those for inhibition studies described earlier
- Other methods such as reporter systems and binding assays have been used in the early screening for NMEs' induction potential[20,21,24].

When the *in vitro* data is positive (e.g., 2-fold increase over the baseline control), proper clinical trials can be designed to determine if the induction is likely to occur at clinical doses and if the extent of induction may result in significant drug-drug interactions.

3.3. CYP Reaction Phenotyping

Cytochrome P450 reaction phenotyping is generally carried out with human liver microsomes and recombinant cytochrome P450s using a combination of several basic approaches[18]. An NME's metabolic pathways *in vitro* can be evaluated with the following considerations.

- Initial reaction rates are measured in the absence and the presence of chemical inhibitors or antibodies, or with a panel of human liver microsomes for correlation analysis with various cytochrome P450 probe substrates
- metabolic formation rates are linear with respect to enzyme concentration and incubation time
- Specific probe substrates are used when a correlation analysis is used (Table 1 lists preferred *in vitro* probes substrates); the incubation usually includes a panel of more than 10 human liver microsomal preparations
- Highly selective chemical inhibitors are generally available. Table 1 lists preferred chemical inhibitors for individual cytochrome P450 enzymes
- If there is an indication for the involvement of more than one cytochrome P450 in the metabolism of the NME, several drug concentrations spanning approximately two orders of magnitude should be used for inhibition studies

When the *in vitro* data indicate that the NME is a substrate of CYP enzymes, additional *in vivo* data are reviewed to determine the importance of the CYP pathways. If the contribution of a particular CYP pathway to the overall clearance process is important, additional clinical studies can be designed (see section 4).

3.4. Other Metabolic Enzymes

Analytical tools are generally not available for enzymes such as flavin-containing monooxygenases, monoamine oxidases, epoxide hydrolases, glucuronosyl transferases (UGT), sulfotransferases (SULT), methyltransferases, acetyltransferases and glutathione-S- transferases. Recently published review articles discussed UGT reaction phenotyping and its role in drug-drug interactions[25,26]. Similar to CYP enzymes, multiple forms of UGT exist in human liver and extrahepatic tissues such as kidneys, intestine, etc and interact with substrates with diverse molecular structures. However, unlike CYP enzymes, specific probe substrates and inhibitors for various UGT isoforms are generally not available. Therefore, *in vitro* studies of UGT-related metabolism and interactions are often conducted with multiple probe substrates and inhibitors or with recombinant UGT enzymes. A list of substrates, inhibitors, and inducers is available[25]. Ethinylestradiol, an important component in various oral contraceptive products is a substrate of UGT1A1 (as well as SULT1E1) and appears to be an important drug to evaluate for interactions involving UGT *in vitro* and *in vivo*.

3.5. Transporters

Various *in vitro* models have been used to screen for of P-gp substrates or inhibitors[27,28]. Whether an NME is a P-gp substrate or inhibitor may be evaluated *in vitro* with the following considerations.

- Caco-2 cells, cDNA transfected Madine-Darby canine kidney cells (MDR1-MDCK) and LLC-PK1 pig kidney cells and derivative cells containing MDR1 (L-MDR1), may be used.
- A positive control (e.g., digoxin) is included when studying an NME as a substrate
- Digoxin and vinblastine are often used as probe substrates when studying an NME as an inhibitor
- The experiments are usually carried out under linear condition and the substrate concentrations at or below their Km values.
- The bi-directional efflux assay is the method of choice for evaluating NME's or probe substrate's transport
- Both ATPase and calcein-AM assays have been used in high throughput assays

Although many P-gp inhibitors are also CYP3A4 inhibitors, various degrees of selectivity exist[29-31]. The comparison of IC50 or Ki values for inhibition of P-gp and CYP3A may suggest their relative contribution to a systemic drug interaction. Wandel et al[30] estimate the ratios of IC50 between CYP3A and P-gp to vary widely, from 1.1 to 125, for 14 CYP3A inhibitors. Yasuda's recent work[32] suggest that although some systemic drug interactions involve both CYP3A4 and P-gp locus (e.g., ketoconazole, erythromycin), other interactions (e.g., fluconazole and troendoandromycine) may predominantly involve CYP3A4. Table 2 shows a list of substrates, inhibitors and inducers of P-gp[33]. Similar to *in vitro* CYP3A inhibition, there appears to be substrate-dependence in P-gp interactions *in vitro*, more than two P-gp substrates may be needed when screening for drug interactions[32].

Table 2. Substrates, Inhibitors, and Inducers of P-glycoprotein[33]

Substrate	Inhibitor	Inducer
Aldosterone	Atorvastatin	Amprenavir
Amprenavir	Bromocriptine	Clotrimazole
Bilirubin	Carvedilol	Dexamethasone
Cimetidine	Cyclosporine	Indinavir
Colchicine	Erythromycin	Morphine
Cortisol	GF120918	Nelfinavir
CPT-11 (Irinotecan)	Itraconazole	Phenothiazine
Cyclosporine	Ketoconazole	Retinoic acid
Dexamethasone	LY335979	Rifampin
Digoxin	Meperidine	Ritonavir
Diltiazem	Methadone	Saquinavir
Domeperidone	Nelfinavir	St John's Wort
Doxorubicin	Pentazocine	
Erythromycin	Progesterone	
Estradiol-17B-D-glucuronide	Quinidine	
Etoposide	Ritonavir	
Fexofenadine	Saquinavir	
Indinavir	Tamoxifen	
Itraconazole	Valspodar (PSC-833)	
Ivermectin	Verapamil	
Loperamide		
Methylprednisolone		
Morphine		
Nelfinavir		
Paclitaxel		
Quinidine		
Ranitidine		
Rhodamine		
Saquinavir		
Sparfloxacin		
Terfenadine		
Tetracycline		
Vecuronium		
Verapamil		
Vinblastine		

3.6. GLP vs. Non-GLP Studies

There have been discussions on whether *in vitro* metabolism ·studies should be conducted under GLP conditions[14]. Currently FDA has no GLP requirements for *in vitro* metabolism or drug interaction studies. Proper standard operating procedures, data tracking process and analytical method validation should be established. The validity of the *in vitro* data (especially for labeling purposes) will be reviewed based on scientific principles and study conditions as discussed previously in this section.

4. *IN VIVO* APPROACHES

In vivo studies may be designed to address specific questions. For example, when *in vitro* metabolism information indicates that CYP1A2, CYP2D6, CYP2C9, CYP2C19, or CYP3A may be responsible for the metabolism of an NME and *in vivo* clearance data indicate that the metabolic pathway is a major clearance pathway (or contribution of the metabolic pathway unclear), an *in vivo* study using a selective inhibitor or inducer to evaluate the effect of alteration in the specific enzyme activity on the clearance of the NME is indicated (Figure 1). Initially, the most potent inhibitors or inducers for respective CYP enzymes (e.g., ketoconazole for CYP3A inhibition and rifampin for CYP3A induction) may be used to maximize the opportunity to see an interaction. Similarly, if an NME is an inhibitor or an inducer, based on *in vitro* evaluations, a study using the most sensitive substrates (such as midazolam for CYP3A, warfarin for CYP2C9, desipramine for CYP2D6, theophylline for CYP1A2, and omeprazole for CYP2C19; or other probe substrates as listed in Table 3) to evaluate the effect *in vivo* is suggested.

Table 3. Recommended *in vivo* probe substrates for CYPs[14]

CYP	Probe substrate	Comments
1A2	Caffeine 2B6	Alternative: Theophylline (clinical relevance, but concern about selectivity?)
2B6	Bupropion	More validation required
2C8	Unclear	Paclitaxel cannot be given to healthy subjects
2C9	Tolbutamide	Alternatives: Flurbiprofen, diclofenac, phenytoin, warfarin (all clinically relevant; safety issue with warfarin?)
2C19	Mephenytoin	Availability?
	Omeprazole	Potential contamination from 3A4 pathway?
2D6	Debrisoquine	Availability? Alternatives: Dextromethorphan (urine pH-dependent renal excretion; potential contamination from downstream 3A4 pathway?); metoprolol (urine pH-dependent renal excretion); desipramine (clinically relevant)
2E1	Chlorzoxazone	
3A4	Midazolam (oral)	Not selective for 3A4 versus 3A5
	Midazolam (oral and intravenous)	Separates liver versus gut contributions; need for stable-isotope labeling for concurrent oral and intravenous administration; staggered oral and intravenous dosing may avoid use of labeled drug?
	Midazolam (oral) + erythromycin (iv)	Liver versus gut; erythromycin marks 3A4 preferentially to 3A5, but precise mechanistic interpretation is confounded by P-glycoprotein transport, and use of radioactive compound (breath test) may be an issue in some countries
	Simva-/atorva-statin	Availability of metabolite standards

If no interactions are detected under these "stressed" systems, then no further *in vivo* evaluations on metabolic interactions are needed. However, if there is an interaction, subsequent evaluations using other substrates or inhibitors/inducers may be desirable. Results from these *in vivo* investigations and previous *in vitro* evaluation form the primary basis of labeling language (see section 7). For example, rifampin has been shown to decrease amprenavir AUC, Cmax, and Cmin values by 82, 70, and 92%, respectively. A separate study with rifabutin, a less potent inducer, indicates acceptable changes in the exposure of amprenavir[34]. The labeling indicates that rifampin not be used with amprenavir and that rifabutin be given at lower doses (as rifabutin levels were increased by co-administration of amprenavir).

Population pharmacokinetics analyses are useful to further support the absence of metabolic interactions when indicated by *in vitro* studies. It is, however, more difficult to use population analyses to prove the absence of an interaction when strongly indicated by *in vitro* evaluations. Carefully designed population studies (where there is documented information on the dose, dosing time and other relevant information of the interacting drugs) can be helpful in defining the clinical significance of known or newly identified interactions. When skillfully applied, population studies may be able to detect unsuspected drug-drug interactions[10,11].

4.1. Study Design

Various study designs can be used. Depending on the following factors, single or multiple doses of the substrate (S) and the interacting drug (I) may be studied.

- The pharmacokinetics characteristics of S and I; if the drug does not accumulate or change over time (such as autoinduction), then single dose may be sufficient to study.
- Whether the metabolites of I are also inhibitors; it may be more appropriate to study multiple doses of I if its metabolites are also inhibitors and accumulate over time, as multiple doses may result in higher degree of inhibition
- Whether pharmacodynamic measurements will be made; if so, multiple doses may enhance the probability to see changes
- If induction is being studied, the inducer will need to be given in multiple doses

The highest dose and the shortest dosing interval should be used. For example, if a drug is given q.i.d. for one indication and b.i.d. for another indication, q.i.d. should be the dosing regimen when a multiple-dose study is indicated. Lower doses of the substrate may be used when there are safety concerns of increased exposure due to drug interactions.

Crossover study designs are generally preferred as each subject provides his/her own control. Both randominzed cross-over and one-way fixed sequence crossover designs have been used by different investigators. When one of the drugs being evaluated has a long half-life, it may be more feasible to use a one-way fixed sequence or a parallel design.

Measurements of substrate plasma concentrations in the presence or absence of the interacting drugs provide the basis of quantitative evaluation of the extent of interactions. It is, at times, critical to measure the plasma concentrations of inhibitors or inducers. This can be helpful in uncovering bi-directional interactions. It may also uncover other intrinsic (genetics, age, race, gender, etc) or extrinsic factors (concomitant antacids, etc) that may have affected the exposure levels of the inducers or inhibitors resulting in varied extent of interactions.

In many cases, the interaction studies are carried out in healthy volunteers. This is acceptable when pharmacokinetic endpoints are the focus of the study. However, when pharmacodynamic parameters that are unique to patients are measured or when there are safety concerns, patients may be more appropriate to study. When studying interactions involving enzymes with polymorphic distribution (such as CYP2D6, CYP2C9, CYP2C19), genotype information is important to properly interpret the outcome. Recent publications show differential extent of interactions between extensive and poor metabolizers of CYP2D6 substrates, such as metoprolol[35]. The study population should usually include both genders unless the intended use of the drug being studied is gender-specific. Recent publications and NDA (new Drug Applications) reviews[1, 36] suggest that female and male subjects may show differences in the extent of inhibition or induction of CYP3A substrates.

4.2. Data Analysis and Sample Size Consideration

To determine if there is a pharmacokinetic (PK) interaction, 90% confidence interval (90%CI) of the geometric mean ratio of PK parameters (e.g., AUC, Cmax) of the substrate in the presence and absence of the interacting drug is estimated. If the 90%CI is within 80 and 125%, no interactions are indicated. If, however, an exposure-response relationship has been established for the substrate drug with either a safety or efficacy measure, a «no effect boundary» other than the 80-125% range, where changes in the ratio do not translate to clinically relevant effects, can be defined[10]. A draft guidance entitled « Exposure-Response Relationships: Study Design, Data Analysis, and Regulatory Applications » discusses how the exposure-response relationship may be used in determining if a change in an exposure parameter due to intrinsic (age, gender, race, genetics, hepatic and renal impairment, etc.) or extrinsic factors (drug interaction, food, etc.) is of clinical significance[37]. The guidance also discusses study design and data analysis considerations[37].

With this type of data analysis, if there are strong suggestions from *in vitro* that an interaction is likely to occur, a small number of subjects may be sufficient to demonstrate an *in vivo* difference and to provide an estimate of the extent of interaction. If the interaction is not likely based on *in vitro* data, it may take a larger number of subjects to show no interactions, depending on the size of the intra- or inter-subject variation of the parameters being evaluated.

4.3. Classification of Inhibitors

A system for classifying CYP enzyme inhibitors has been suggested as a way for labeling[38-40]. For example, the following have been suggested for classifying CYP3A inhibitors.

Classification of CYP3A inhibitors based on the fold-change in AUC of oral midazolam in the presence and absence of the inhibitor		
Potent	Moderate	Mild
AUC ratio ≥5	AUC ratio ≥2 and <5	AUC ratio <2
Ketoconazole Itraconazole....	Verapamil Diltiazem....	Ranitidine Azithromycin....

Some limitations of this system may include the following.

- The data may not be extrapolated to other substrates. Midazolam is a substrate for both CYP3A4 and CYP3A5; many of the inhibitors may have differential activities toward these enzymes; depending on the relative contribution of either enzyme in the metabolism of various substrates and relative inhibitory potency of the inhibitor on these enzymes; the classification may vary
- Concerns that the effect of « mild » or « moderate » inhibitors may be understated. For example, for a patient on a CYP3A substrate drug who is also taking multiple other drugs or food/juice products and dietary supplements that may be classified as mild or moderate inhibitors individually but may collectively exert potent inhibition effect, the clinical outcome may manifest as if he/she was given a potent inhibitor
- Many CYP3A inhibitors affect more than one enzyme and/or transporter system; their effects on drugs that are substrates of CYP3A only and on drugs that are substrates of CYP3A and other enzymes/transporters may differ
- Some inhibitors also act as inducers; when a CYP3A substrate is also a substrate of that induced enzyme, net results may not be determined by the potency of CYP3A inhibition
- Inhibitors may be classified differently depending on the available human data that may be generated using various study designs or under varied study conditions. For example, ketoconazole increased the AUC of co-administered oral midazolam to various extent, ranging from 5- to 16- fold (Figure 2a), depending on the study conditions that varied in ketoconazole dose, dosing regimen, dosing length, and in midazolam dose, dosing time relative to ketoconazole, etc. Ketoconazole increased the AUC of various CYP3A substrates to various degrees, ranging from 1.3- to 16- fold (Figure 2b), depending on the study conditions and the importance of CYP3A in the substrates' clearance pathway.

However, the classification system may provide a useful guide in prioritizing studies to evaluate *in vivo*. For example, the list is helpful in evaluating the degree of increased exposure of a CYP3A substrate with various inhibitors. Initially, an inhibitor from the potent list can be studied. If the results are negative, no studies with the drugs in the « moderae » or « mild » list will be needed. However, if the results with the potent inhibitor is positive and the interaction is of such an extent that contraindication may be suggested to avoid a serious adverse event, further studies with inhibitors in the other two classes may be warranted to determine if drugs in the other two classes can be co-administered (with or without dosage adjustments) and to determine proper labeling language for co-administion with these other inhibitors.

(A)

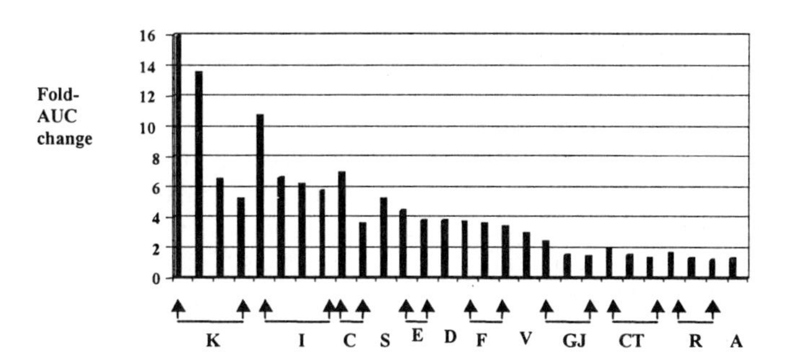

(B)

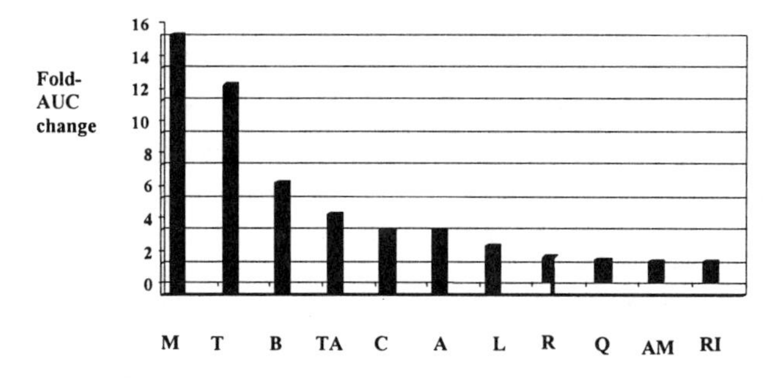

Figure 2. (A) Fold-changes of oral midazolam AUC in the presence of various CYP3A inhibitors under varied study conditions for either drug. Data obtained from PubMed via University of Washington Metabolism and Transport Drug Interaction Database [http://depts.washington.edu/didbase/] – searched until January 2000. (K, ketoconazole; I, itraconazole; C, clarithromycin; S, saquinavir; E, erythromycin; D, diltiazem; F, fluconazole; V, verapamil; GJ, grapefruit juice; CT, cimetidine; R, ranitidine; A, azithromycin). (B) Fold-changes in AUC of various CYP3A substrates in the presence of ketoconazole. (M, midazolam; T, triazolam; B, budesonide; TA, tacrolimus; C, cyclosporin; A, alprazolam; L, loratadine; R, reboxetine; Q, quinine; AM, amprenavir; RI, ritonavir)

4.4. Cocktail Approaches

Investigators have used various cocktail approaches to evaluate the effect of drugs or herbal products on several CYP enzymes simultaneously[41-44]. For example, Wang et al[41] and Gorski et al[42] used caffeine, tolbutamide, dextromethorphan, and midazolam (both IV and PO) to evaluate the changes of CYP1A2, CYP2C9, CYP2D6, and CYP3A (both hepatic and intestinal) enzyme activities, respectively, after acute or chronic dosing of St. John's Wort or Echinacea. The statistical approaches and sample size requirements are similar to those described earlier (section 4.2).

4.5. P-gp and Other Transporters

Recent publications[49,51] or NDA submissions have included evaluation of P-gp based interactions. For example, when *in vitro* data suggest that an NME is either a substrate or inhibitor of P-gp; *in vivo* studies are often carried out using digoxin or fexofenadine as probe drugs. Recent findings suggest that other transporters may also be involved in the transport of these drugs. Work from Cvetkovic et al suggests that members of OATP (organic anion transporting polypeptides), in addition to P-gp, appear to be involved in fexofenadine's disposition.[57]

5. CLINICAL CASES

Serious drug-drug interactions have contributed to recent withdrawal of drugs from the US market[1,45]. When FDA's adverse reaction reports in the Adverse Event Reporting System were searched for interactions between mibefradil and others, mibefradil was found to be a possible suspect drug in various case reports as interacting with cyclosporine, statins (lovastatin, simvastatin, cerivastatin and atorvastatin) , and cisapride, in cases when renal failure, rhabdomyolysis, and torsads de pointes, respectively, were reported[1,46]. Similarly, cases of torsades de pointes have been reported when terfenadine was given with ketoconazole or erythromycin[1].

A search of the MedWatch reports up to July 2002 in the Adverse Events Report Databases in both Center for Drug Evaluation and Research and Center for Food Safety and Applied Nutrition on St John's Wort resulted in 498 reports. Of these, 42 implicated St. John's Wort's possible role in the varied clinical responses of cyclosporine (rejection, lowered plasma levels), oral contraceptives (breakthrough bleedings, pregnancy), selective serotonin reuptake inhibitors (serotonin syndrome, hypertension), sildenafil (loss of efficacy), etc[47,48]. Recent studies by various investigators suggest that St. John's Wort induces CYP3A and P-gp transporter[41,49-51].

Another recent survey of MedWatch reports for grapefruit juice interactions yielded 167 reports[52]. Of these, 39 appear to implicate grapefruit juice as an interacting component. Grapefruit juice appears to have contributed to the exaggerated pharmacologic effects of various drugs, such as calcium channel blockers (hypotension, dizziness, syncope, etc), statins (myalgia, rhabdomyolysis) and others. Literature data suggest that grapefruit juice inhibits intestinal CYP3A and other enzymes, in addition to

P-gp transporter[53-54]. Grapefruit juice and other citrus juices may be affecting additional transporters, such as OATP[58].

Many CYP (CYP2D6, CYP2C9, CYP2C19, CYP3A5, etc), non-CYP enzymes (UGT, NAT, etc), and tansporters such as P-gp and OATP also exhibit genetic polymorphism. The genetic variation may have contributed to large inter-individual variations in pharmacokinetics[59] and clinical outcomes for efficacy (e.g., lansoprazole cure rate in GERD and CYP2C19 polymorphism)[60] and safety (e.g., irinotecan induced gastrointestinal and bone marrow toxicity and UGT1A1 polymorphism)[61] for substrates of these enzymes or transporters. As indicated earlier, genetic variation can contribute to differential extent of interactions among subjects with different genetic makeups[35].

Drug interactions have often been linked to adverse events or loss of efficacy as indicated in various clinical cases. However, drug interactions have also been explored and used to increase exposoure and to affect clinical outcomes. For example, ketoconazole and other CYP inhibitors have been used for « dose sparing» (e.g., diltiazem and cyclosporine)[55] or to increase the bioavailability of a poorly available drug that may not be a viable drug product if administered alone. An example of the latter is the use of ritonavir in the recently approved product for AIDS therapy, KALETRA, which contains both ritonavir and lopinavir[56].

6. REGULATORY CONSIDERATIONS

Information on a drug's metabolism/disposition pathways and its interaction potential is critical in the decision-making in drug discovery and development. A drug that is subject to drug interaction as a strong inducer or as a substrate that is primarily metabolized by a single enzyme (e.g., CYP2D6, CYP3A) may be screened out early, if there are other similar leads with a more diverse metabolic profile (i.e., metabolized by multiple metabolic and other elimination pathways). Similarly, drug metabolism and interaction information is critical in regulatory decision-making on approval and on labeling languages. Figure 2 shows that drug interaction is one of the critical questions to ask during regulatory review of IND/NDA data and/or post-marketing case reports[45].

1. <u>Is drug interaction an issue?</u> Drug interaction is one of the key clinical pharmacology questions to ask during the FDA review[62]. This question can be addressed with the methods discussed in earlier sections of this chapter.
   - *In vitro* metabolism and interaction evaluation
   - *In vivo* interaction evaluation using a tiered approach by studying specific inhibitors/inducers with various potencies or substrates with varied degrees of sensitivity
   - Complementary population pharmacokinetic and pharmacodynamic evaluation
   - Post-marketing monitoring of adverse event reports and other literature information

   If drug interaction is an issue, the second question is posed.

2. Does the benefit outweigh the risk? Whether the treatment is for unmet medical needs, for life-threatening diseases, or where alternative therapies are not available are important considerations for risk-benefit assessment. If there are other therapies available for a non-life threatening diseases or symptoms and the potential of drug interactions is high (i.e., the drug is a strong inhibitor, a strong inducer, or its clearance is subject to a single enzyme metabolism) and the consequences of drug interactions are severe, the drug may not be approved or it may be removed from the market. If, however, it is deemed that the benefit of the drug outweigh the risk, then we need to manage the risk due to drug interactions.
3. Can we manage through labeling? From past experiences, we learned that labeling may not be effective in preventing practitioners to prescribe, dispense or patients to use drugs that are contraindicated[63]. Various measures to improve the communication of risk of drug interactions continue to be developed and implemented.
   - New labeling initiative[64], when finalized, would require the drug labeling to have a highlighted section on drug interactions
   - Medication guides, implemented for various drug products, such as RU486, lotronex, can provide additional information on risk management to patients
   - Measures to educate the health care providers and consumers are ongoing and need to continue to evolve
   - Other systems such as the classification of inhibitors discussed earlier may be considered

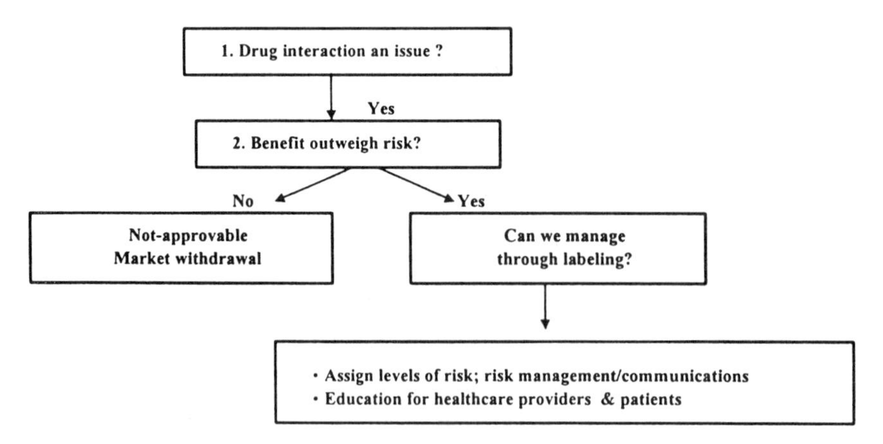

Figure 3. Questions to ask when reviewing NDA/post-marketing data[45]

7. LABELING

In a proposed revision of physician labeling format and content, significant (or evidence of no) drug-drug interactions would appear in the Highlights section, in addition to having this information in the main body of the labeling[64]. *In vitro* and *in vivo* information on the metabolic pathways and metabolites, including contribution of specific enzymes, and known or expected effects of inducers or inhibitors of the pathway, is described in the CLINICAL PHARMACOLOGY section of the labeling. Any information on pathways or interactions that have been ruled out by *in vitro* data is also included in this section. Important clinical consequences of this information would be placed in DRUG INTERACTIONS, WARNINGS, PRECAUTIONS, BOXED WARNINGS, CONTRAINDICATIONS and DOSAGE AND ADMINISTRATION sections of the main labeling, as appropriate. Examples of appropriate labeling language are provided in italic below.

[Case 1] *In vitro* interaction has been studied for (the new drug) and showed interaction; no *in vivo* studies have been conducted to confirm or refute the *in vitro* finding.

In vitro drug metabolism studies reveal that the metabolism of <u>(the new drug)</u> is by CYP3A4 and can be inhibited by the CYP3A4 inhibitor ketoconazole. No clinical studies have been performed to evaluate this finding. Based on the in vitro findings, it is likely that ketoconazole, itraconazole, and other CYP3A4 inhibitors will lead to substantial increase in blood concentrations of <u>(the new drug)</u>.

[Case 2] *In vivo* interaction has been studied for (the new drug) and showed interaction

> *The effect of <u>(the new drug)</u> on the pharmacokinetics of (the prototypical CYP3A substrate) has been studied in ___ patients/healthy subjects. The Cmax, AUC, half-life, and clearances of (the prototypical CYP3A substrate) were increased/ decreased by _% (90% confidence interval : ___ to ___ %) in the presence of (the new drug). This indicates that (the new drug) can inhibit/induce the metabolism of drugs metabolized by CYP3A and can increase/decrease blood concentrations of such drugs.*

This above information is important if the clinical consequences of such an interaction are significant and/or the likelihood of coadministration is high. As appropriate, this information and the extrapolated conclusions (to drugs metabolized by the same class of enzymes) should be represented in the CONTRAINDICATIONS, WARNINGS, PRECAUTIONS, and DOSAGE AND ADMINISTRATION sections. Tables 4 and 5 list examples of labeling information in the Physicians Desk Reference based on *in vitro* and *in vivo* evaluation, respectively. Table 6 gaives an exmaple of labeling based on population analyses.

Table 4. Labeling examples of *in vitro* metabolism and drug-drug interaction information

Drug Name (Brand Name)	Labeling section	Labeling Statement
Valdecoxib (BEXTRA)- April 2002 labeling	CLINICAL PHARMACOLOGY	Metabolism In humans, valdecoxib undergoes extensive hepatic metabolism involving both P450 isoenzymes (3A4 and 2C9) and non-P450 dependent pathways (i.e., glucuronidation). Concomitant administration of BEXTRA with known CYP 3A4 and 2C9 inhibitors (e.g., fluconazole and ketoconazole) can result in increased plasma exposure of valdecoxib.
	DRUG INTERACTIONS	*In vitro* studies indicate that valdecoxib is a moderate inhibitor of CYP 2C19 (IC50 = 6 mcg/mL), and a weak inhibitor of both 3A4 (IC50 = 44 mcg/mL) and 2C9 (IC50 = 13 mcg/mL). In view of the limitations of *in vitro* studies and the high valdecoxib IC50 values, the potential for such metabolic inhibitory effects *in vivo* at therapeutic doses of valdecoxib is low.
Pioglitazone (ACTOS)- July 2002 labeling	CLINICAL PHARMACOLOGY	Metabolism The major cytochrome P450 isoforms involved in the hepatic metabolism of pioglitazone are CYP2C8 and CYP3A4 with contributions from a variety of other isoforms including the mainly extrahepatic CYP1A1. Ketoconazole inhibited up to 85% of hepatic pioglitazone metabolism *in vitro* at a concentration equal molar to pioglitazone. Pioglitazone did not inhibit P450 activity when incubated with human P450 liver microsomes. *In vivo* human studies have not been performed to investigate any induction of CYP3A4 by pioglitazone.
	DRUG INTERACTIONS	The cytochrome P450 isoform CYP3A4 is partially responsible for the metabolism of pioglitazone. Specific formal pharmacokinetic interaction studies have not been conducted with ACTOS and other drugs metabolized by this enzyme such as: erythromycin, astemizole, calcium channel blockers, cisapride, corticosteroids, cyclosporine, HMG-CoA reductase inhibitors, tacrolimus, triazolam, and trimetrexate, as well as inhibitory drugs such as ketoconazole and itraconazole. *In vitro*, ketoconazole appears to significantly inhibit the metabolism of pioglitazone. Pending the availability of additional data, patients receiving ketoconazole concomitantly with ACTOS should be evaluated more frequently with respect to glycemic control.

Table 5. Labeling examples of *in vivo* metabolism and drug-drug interaction information

Drug Name (Brand Name)	Labeling section	Labeling Statement
Nefazodone (SERZONE®)-February 2002 labeling	WARNING	*Triazolam* When a single oral 0.25-mg dose of triazolam was coadministered with nefazodone (200 mg BID) at steady state, triazolam half-life and AUC increased 4-fold and peak concentrations increased 1.7-fold. Nefazodone plasma concentrations were unaffected by triazolam.
	CONTRA-INDICATIONS	**Coadministration of terfenadine, astemizole, cisapride, pimozide, or carbamazepine with SERZONE (nefazodone hydrochloride) is contraindicated (see WARNINGS and PRECAUTIONS).** SERZONE tablets are contraindicated in patients who were withdrawn from SERZONE because of evidence of liver injury (see **BOXED WARNING**). SERZONE tablets are also contraindicated in patients who have demonstrated hypersensitivity to nefazodone hydrochloride, its inactive ingredients, or other phenylpiperazine antidepressants. The coadministration of triazolam and nefazodone causes a significant increase in the plasma level of triazolam (see **WARNINGS** and **PRECAUTIONS**), and a 75% reduction in the initial triazolam dosage is recommended if the two drugs are to be given together. Because not all commercially available dosage forms of triazolam permit a sufficient dosage reduction, the coadministration of triazolam and SERZONE should be avoided for most patients, including the elderly
Sildenafil (VIAGRA)-August 2002 labeling	CLINICAL PHARMACOLOGY	It is eliminated predominantly by hepatic metabolism (mainly cytochrome P450 3A4) and is converted to an active metabolite with properties similar to the parent, sildenafil. The concomitant use of potent cytochrome P450 3A4 inhibitors (e.g., erythromycin, ketoconazole, itraconazole) as well as the nonspecific CYP inhibitor, cimetidine, is associated with increased plasma levels of sildenafil (see **DOSAGE and ADMINISTRATION**).
	WARNINGS	The concomitant administration of the protease inhibitor ritonavir substantially increases serum concentrations of sildenafil (**11-fold increase in AUC**). If VIAGRA is prescribed to patients taking ritonavir, caution should be used. Data from subjects exposed to high systemic levels of sildenafil are limited. Visual disturbances occurred more commonly at higher levels of sildenafil exposure. Decreased blood pressure, syncope, and prolonged erection were reported in some healthy volunteers exposed to high doses of sildenafil (200-800 mg). To decrease the chance of adverse events in patients taking ritonavir, a decrease in sildenafil dosage is recommended (see **Drug Interactions**, **ADVERSE REACTIONS**, and **DOSAGE and ADMINISTRATION**).

Table 6. Labeling examples of drug-drug interaction information based on population pharmacokinetic approach

Drug Name (Brand Name)	Labeling section	Labeling Statement
Dofetilide (TIKOSYN)- December 1999 labeling	DRUG INTERACTIONS	Population pharmacokinetic analyses were conducted on plasma concentration data from 1445 patients in clinical trials to examine the effects of concomitant medications on clearance or volume of distribution of dofetilide. Concomitant medications were grouped as ACE inhibitors, oral anticoagulants, calcium channel blockers, beta blockers, cardiac glycosides, inducers of CYP3A4, substrates and inhibitors of CYP3A4, substrates and inhibitors of P-glycoprotein, nitrates, sulphonylureas, loop diuretics, potassium sparing diuretics, thiazide diuretics, substrates and inhibitors of tubular organic cation transport, and QTc-prolonging drugs. Differences in clearance between patients on these medications (at any occasion in the study) and those off medications varied between -16% and +3%. The mean clearances of dofetilide were 16% and 15% lower in patients on thiazide diuretics and inhibitors of tubular organic cation transport, respectively.

When a drug's disposition pathway is defined, labeling with St John's Wort products and grapefruit juice is possible without conducting actual *in vivo* studies. As the components in St. John's Wort and grapefruit juice products that are responsible for their interactions are still being defined[53, 65], and the contents of these components in various products remain uncertain, it becomes difficult to interpret negative *in vivo* studies when *in vitro* mechanistic studies and clinical cases strongly suggest an interaction. For example, St. John's Wort products are standardized by their hypericin content when the CYP3A induction activity may be attributed to other components including hyperforin[65]. Although the plasma levels of hypericin and hyperforin have been reported in recent studies[49, 51], the systemic availability of these components and factors affecting their systemic availability have yet to be elucidated.

Labeling decisions for interactions with St. John's Wort or grapefruit juice (see (examples in Tables 7 and 8, respectively) can be based on the metabolic and disposition characteristics of the drugs being labeled. If a drug is a CYP3A substrate and/or P-gp substrate and interaction of these pathways may significantly change their pharmacokineitcs (mostly lower systemic concentrations) or pharmacodynamics and affect the efficacy measures, St. John's Wort is added to the labeling. In addition to being a CYP3A or P-gp substrate, if the drug also has low bioavailability, grapefruit juice may be added to the labeling.

Table 7. Labeling examples of St John's Wort-drug interaction information

Drug Name (Brand Name)	Labeling section	Labeling Statement
Mifepristone (MIFEPREX™)	DRUG INTERACTIONS	rifampin, dexamethasone, St. John's Wort, and certain anticonvulsants (phenytoin, phenobarbital, carbamazepine) may induce mifepristone metabolism (lowering serum levels of mifepristone).
Ritonavir & Lopinavir (KALETRA)- January 2002 labeling	CLINICAL PHARMACOLOGY	Concomitant use of KALETRA and St. John's wort (hypericum perforatum), or products containing St. John's wort, is not recommended.
Imatinib (GLEEVEC)- January 2002 labeling	DRUG INTERACTIONS	Drugs that may decrease imatinib plasma concentrations.... Co--medications that induce CYP3A4 (e.g., dexa-methasone, phenytoin, carbamazepine, rifampicin, phenobarbital or St. John's Wort) may reduce exposure to Gleevec. No specific studies have been performed and caution is recommended.
Cyclosporine (NEORAL)- January 2001 labeling	DRUG INTERACTIONS	There have been reports of a serious drug interaction between cyclosporine and the herbal dietary supplement, St. John's Wort. This interaction has been reported to produce marked reduction in the blood concentrations of cyclosporine, resulting in subtherapeutic levels, rejection of transplanted organs, and graft loss.

Table 8. Labeling examples of grapefruit juice-drug interaction information

Drug Name (Brand Name)	Labeling section	Labeling Statement
Cyclosporine (NEORAL)- January 2001 labeling	DOSAGE AND ADMINSTRATION	Grapefruit and grapefruit juice affect metabolism, increasing blood concentration of cyclosporine, thus should be avoided
Simvastatin (ZOCOR)- May 2002 labeling	CLINICAL PHARMACOLOGY	To avoid possible serious side effects, avoid drinking large quantities of grapefruit juice (more than on quart daily) while on ZOCOR (see WARNING, Muscle)

When the drug interaction information becomes too complex to describe sufficiently in paragraphs, tables may be helpful. Tables 9 and 10 illustrate how results of interactions with KALETRA are listed in the « CLINICAL PHARMACOLOGY » section and the recommendations in the « CONTRAINDICATION » section in a table format.

Table 9. Drug interactions: pharmacokinetic parameters for lopinavir in the presence of the co-administered drug

Co-administered Drug	Dose of Co-administered Drug (mg)	Dose of KALETRA (mg)	n	Ratio (with/without co-administered drug) of Lopinavir Pharmacokinetic Parameters (90% CI); No Effect = 1.00		
				C_{max}	AUC	C_{min}
Amprenavir [1]	450 BID, 5 d	400/100 BID,	12	0.89	0.85	0.81
	750 BID, 5 d	22 d	10	(0.83, 0.95)	(0.81, 0.90)	(0.74, 0.89)
Atorvastatin	20 QD, 4 d	400/100 BID,	12	0.90	0.90	0.92
		14 d		(0.78, 1.06)	(0.79, 1.02)	(0.78, 1.10)
Efavirenz [2]	600 QHS, 9 d	400/100 BID,	11, 7 *	0.97	0.81	0.61
		9 d		(0.78, 1.22)	(0.64, 1.03)	(0.38, 0.97)
Ketoconazole	200 single dose	400/100 BID,	12	0.89	0.87	0.75
		16 d		(0.80, 0.99)	(0.75, 1.00)	(0.55, 1.00)
Nevirapine	200 QD, 14 days;	400/100 BID,	5, 9 *	0.95	0.99	1.02
	BID, 6 days	20 d		(0.73, 1.25)	(0.74, 1.32)	(0.68, 1.53)
	7 mg/kg or 4 mg/kg	300/75 mg/m [2]	12, 15 *	0.86	0.78	0.45
	QD, 2 wk; BID 1 wk [3]	BID, 3 wk		(0.64, 1.16)	(0.56, 1.09)	(0.25, 0.81)
Pravastatin	20 QD, 4 d	400/100 BID,	12	0.98	0.95	0.88
		14 d		(0.89, 1.08)	(0.85, 1.05)	(0.77, 1.02)
Rifabutin	150 QD, 10 d	400/100 BID,	14	1.08	1.17	1.20
		20 d		(0.97, 1.19)	(1.04, 1.31)	(0.96, 1.65)
Rifampin	600 QD, 10 d	400/100 BID,	22	0.45	0.25	0.01
		20 d		(0.40, 0.51)	(0.21, 0.29)	(0.01, 0.02)
Ritonavir [4]	100 BID, 3-4 wk	400/100 BID,	8, 21 *	1.28	1.46	2.16
		3-4 wk		(0.94, 1.76)	(1.04, 2.06)	(1.29, 3.62)

All interaction studies conducted in healthy, HIV-negative subjects unless otherwise indicated.
[1] Composite effect of amprenavir 450 and 750 mg Q12h regimens on lopinavir pharmacokinetics.
[2] The pharmacokinetics of ritonavir are unaffected by concurrent efavirenz.
[3] Study conducted in HIV-positive pediatric subjects ranging in age from 6 months to 12 years.
[4] Study conducted in HIV-positive adult subjects.
* Parallel group design; n for KALETRA + co-administered drug, n for KALETRA alone.

Table 10. Drugs that are contraindicated with KALETRA

Drug Class	Drugs Within Class That Are Contraindicated With KALETRA
Antiarrhythmics	Flecainide, Propafenone
Antihistamines	Astemizole, Terfenadine
Ergot Derivatives	Dihydroergotamine, Ergonovine, Ergotamine, Methylergonovine
GI motility agent	Cisapride
Neuroleptic	Pimozide
Sedative/hypnotics	Midazolam, Triazolam

When there are serious consequences of the interactions, direct labeling of the product has also been attempted. For example, as described in the product labeling of KALETRA, the following statement is included on the product's bottle label : **ALERT: Find out about medicines that should NOT be taken with KALETRA.**

8. SUMMARY

Intrinsic factors (e.g., age, gender, race, genetics, hepatic, renal and other diseases) and extrinsic factors (e.g., concomitantly administered drugs, biologics, food and juice products, dietary supplements) contribute to the intra- and inter-subject variations in drug responses. Of these factors, drug interactions can cause abrupt changes in pharmacokinetics and pharmacodynamics of the co-administered drugs and exert unwanted adverse events or loss of efficacy. When a drug's clearance pathway and its enzyme- or transporter- modulating effects are well understood, potential drug-drug, drug-food, drug-juice or drug-dietary supplement interactions may be projected and managed. The following steps are critical to assess and manage risks due to drug interactions.

- Drug's metabolism and transport mechanisms well studied during drug development
- Clinical significance of drug interactions be interpreted based on well defined exposure-response data and analyses
- Drug interaction information well placed in the labeling; additional warning placed physically on the product label, when necessary
- Risk management tools (e.g., medication guides, continued post-marketing surveillance, etc) are in place and additional effective communication measures continue to be developed
- Useful drug information readily accessible to health care providers and patients

With improved understanding of the molecular bases of drug interactions and of the interplay of various intrinsic factors (including genetics) affecting drug interactions, the risks associated with drug interactions can be managed to minimize untoward effects. When used in a controlled environment, drug interactions can also be utilized to improve therapy.

9. REFERENCES

1. S.-M. Huang, M. Miller, T. Toigo, M.C. Chen, C. Sahajwalla, L.J. Lesko, R. Temple, in: Section 11, Drug Metabolism/Clinical Pharmacology (section editor: Schwartz, J), in "Principles of Gender-Specific Medicine", Ed., Legato M, Academic Press, in press.
2. T. Prueksaritanont, J.J. Zhao, B. Ma, B.A. Roadcap, C. Tang, Y. Qiu, L. Liu, J.H. Lin, P.G. Pearson, T.A. Baillie. *J Pharmacol Exp Ther*. **301**,1042 (2002)
3. J.H. Lin, and A.D. Rodrigues, in "pharmacokinetic optimization in drug research: biological, physicochemical and computational strategies," edited by B. Testa, H. Vander Waterbeemed, G. Folkes and R. Guy, Wiley-Verlag, pp. 217-243 (2001).
4. A.D. Rodrigues and J.H. Lin, *Current opinion in chemical Biology* **5**, 396 (2001).
5. A.Y.H. Lu, and S.-M. Huang, in: New Drug Development : Clinical Pharmacology and Biopharmaceutics, Ed., Sahajwalla, Marcell Dekker, in press.
6. B. Davit, K. Reynolds, R. Yuan, F. Ajayi, D. Conner, E. Fadiran, B.Gillespie, C. Sahajwalla, S.-M. Huang, and L.J. Lesko, *J. clin. Pharmacol*. **39**,899 (1999).
7. K. Ito, Y. Iwatsudo, S. Kamamitsu, K. Ueda, H. Suzuki, and, Y. Sugiyama, *Pharmacol*.Rev. **50**, 387 (1998).
8. L.L. Von Moltke, D.J. Greenblatt, J. Schmider, C.E. Wrigh, J.S. Harmatz, and R.I. Shader, *biochem. pharmacol*. **55**,113 (1998).
9. Guidance for Industry: Drug Metabolism/Drug Interactions in the Drug Development Process: Studies *in vitro*. Internet: http://www.fda.gov/cder, April, 1997
10. Guidance for industry: *in vivo* metabolism/drug interactions: study design, data analysis and recommendation for dosing and labeling, Internet: http://www.fda.gov/cder, December, 1999.
11. Guidance for industry: population pharmacokinetics. Internet: http://www.fda.gov/cder, February, 1999.
12. S.-M. Huang, P. Honig, L.J. Lesko, R. Temple and R.L.Williams, in "Drug-Drug Interactions", edited by A.D. Rodrigues, Marcel Dekker, N.Y., pp. 605-632, 2001
13. S.-M. Huang, L.J. Lesko and R.L. Williams, *J Clin Pharmacol*. **39**, 1006, (1999).
14. T. Tucker, J.B. Houston, S.-M. Huang, *Clin Pharmacol. Ther*. **70**, 103, *Br J Clin Pharmacol*. **52**, 107, *Eur J Pharm Sci*. **13**, 417 and *Pharm Res*. **18**, 1071 (2001).
15. R. Yuan, R, S. Madani, S. Wei, K. Reynolds, S.-M. Huang, *Drug Metab Disp*, **30**, 1311 (2002)
16. K.E. Kentworthy, J.C. Bloomer, S.E.Clarke, J.B. Houston, *Br J clin Pharmacol* **48**: 716 (1999).
17. R.W. Wang, and A.Y.H. Lu, *Drug Metab. Dispos*. **25**: 762 (1997).
18. A. Madam, E. Usuki, E., L.A. Burton, B.W. Ogilive, and A. Parkinson, In "Drug-Drug Interaction". Edited by A.D. Rodrigues, Marcel Dekker, N.Y. pp. 217-294, 2001.
19. D.R. Jones, and S.D. Hall, In " Drug-Drug Interactions". Edited by A.D. Rodrigues, Marcel Dekker, N.Y. pp. 387-413, 2001.
20. J.M. Silva, and D.A. Nicoll-Griffith, In "Drug-Drug Interactions", edited by A.D. Rodrigues, Marcel Dekker, N.Y. pp. 189-216, 2001.
21. E. Lecluyse, *Eur J Pharm Sci* **4**:343 (2001).
22. A.P. Li., M.K. Reith, A. Rasmussen, J.C. Gorski, S.D. Hall, L. Xu, D.L. Kaminski, and K.L. Cheng, *Chem Biol*. **107**,17 (1997).
23. A.P. Li, P.D. Gerycki, J.G. Hengstler, G.L. Kedderis, H.G. Keebe, R. Rahman, G. de Sousas, J.M. Silva, and P. Skett, *Chem Biol Interact*. **121**, 117 (1999)
24. B. Goodwing, M.R. Redinbo, S.A. and S.A. Kliewer, *Annu. Rev. pharmacol toxicol*. **42**, 1 (2002).
25. R.P. Remmel, in "Drug-Drug Interactions", edited by A.D. Rodrigues, Marcel Dekker, N.Y. pp. 89-114, (2001).
26 M.D.Green, and T.R. Tephly, *Drug Metab Dispos* **26**:860 (1998).
27. J.W. Polli, S.A. Wring, J.E. Humpreys, L. Huang, J.B. Morgan, L.O. Webster, and C.S. Serabjit-Singh, *J. Pharmacol. Exp. Ther*. **299**, 620 (2001).
28. J.H. Hochman, M. Yamazaki, T. Ohe, and J.H. Lin, *Current Drug Metabolism*, **3**, 257 (2002).
29. R.B. Kim, C. Wandel, B. Leake, M. Cvetkovic, M.F. Fromm, P.J. Dempsey, M.M. Roden, F. Belas, A.K. Chaudhary, D.M. Roden, A.J. Wood, nad G.R. Wilkinson, *Pharm Res* **16**, 3944 (1999).
30. C. Wandel, R.B. Kim, S. Kajiji, F.P. Guengerich, G.R. Wilkinson, A.J.J. Wood, *Cancer Res*. **59**, 3944 (1999)
31. C.L. Cummins, W. Jacobsen, L.Z. Benet, *J Pharmacol Exp Ther* **300**, 1036 (2002).
32. K. Yasuda, L.B. Lan, D. Sanglard, K. Furuya, J.D.Schuetz, E.G. Schuetz, *J Pharmacol Exp Ther* **303**, 323 (2002)
33. R.B. Kim RB, *Drug Metab. Rev*. **34**,47 (2002)
34. Pysician's Desk Reference for AGENERASE® (amprenavir), February 2002 labeling

35. B.A. Hamelin, A. Bouayad, J. Methot, J. Jobin, P. Desgagnes, P. Poirier, J. Allaire, J. Dumesnil, J. Turgeon, *Clin Pharmacol Ther* **67**,466 (2000)
36. J.C. Gorski, D.R. Jones, B.D. Haehner-Daniels, M.A. Hamman, E.M. O'Mara Jr, and S.D. Hall, *Clin Pharmacol Ther* **64**,133 (1998)
37. Guidance for Industry: Exposure-Response Relationships: Study Design, Data Analysis, and Regulatory Applications Internet: http://www.fda.gov/cder, Draft published in April, 2002
38. T.A. Baillie, presentation at the AAPS/ACCP/ASCPT/EUFEPS/FDA co-sponsored workshop on drug-drug interactions, Crystal City, Virginia, December, 1999
39. D. Rodrigues, presentation at the annual ASCPT meeting, Atlanta, Georgia, March 24, 2002,
40. T.D. Bjornsson, J.T. Callaghan, H.J. Einolf, V. Fischer, L. Gan, S. Grimm, J. Kao, P. King, G. Miwa, G. Kumar, J. McLeod, L. Ni, S. Obach, S. Roberts, A. Roe, A. Shah, F. Snikeris, J. Sullivan, D. Tweedie, J.M. Vega, J. Walsh, S.A. Wrighton, *Drug Metab Disp* (in press)
41. Z. Wang, J.C. Gorski, M.A. Hamman, S.-M. Huang, L.J. Lesko, S.D. Hall. *Clin Pharmacol Ther* **70**,317 (2001)
42. J.C. Gorski, S.-M. Huang, N.A. Zaheer, M. Desai, A. Pinto, M. Miller, S.D. Hall, presentation at the Annual Meeting of American Society of Clinical Pharmacology and Therapeutics. April 2-5, 2003, Washington, D.C., abstract in Clin Pharmacol Ther
43. D.S. Streetman, J.F. Bleakley, J.S. Kim, A.N. Nafziger, J.S. Leeder, A. Gaedigk, R. Gotschall, G.L. Kearns, J.S. Bertino Jr., *Clin Pharmacol Ther* **68**,375 (2000)
44. R.A. Branch, A. Adedoyin, R.F. Frye, J.W. Wilson, M. Romkes, *Clin Pharmacol Ther* **68**,401 (2000)
45. S.-M. Huang, B. Booth, E. Fadiran, R.S. Uppoor, S. Doddapaneni, M.C. Chen, F. Ajayi, T. Martin, and L.J. Lesko, presentation at the 101 Annual Meeting of American Society of Clinical Pharmacology and Therapeutics. March 15-17, 2000, Beverly Hills, CA, abstract in Clin Pharmacol Ther
46. D.K. Wysowski, A. Corken, H. Gallo-Torres, L. Talarico, E.M. Rodriguez, *Am J Gastroenterol* **96**,1698 (2001)
47. M.C. Chen, S-M. Huang, R. Mozersky, J. Beitz and P. Honig, presented at the AAPS meeting, Denver, Colorado, October 2001
48. S-M. Huang, presented at the ASCPT workshop on herb-drug interactions, Bethesda, Maryland, July 23, 2002
49. Z. Wang, M.A. Hamman, S.M. Huang, L.J. Lesko, S.D. Hall, *Clin Pharmacol Ther* **71**,414 (2002)
50. S.C. Piscitelli, A.H. Burstein, D. Chaitt, R.M. Alfaro, J. Falloon, *Lancet* **355**,547 (2000)
51. A. Johne, J. Brockmoller, S. Bauer, A. Maurer, M. Langheinrich, I. Roots, *Clin Pharmacol Ther* **66**, 338 (1999)
52. T. Piazza, S.-M. Huang, and P. Hepp, presented at the ACCP annual meeting, San Francisco, Calofornia, September 2002
53. S. Malhotra, D.G. Bailey, M.F. Paine, P.B. Watkins, *Clin Pharmacol Ther* **69**,14 (2001)
54. K. He, K.R. Iyer, R.N. Hayes, M.W. Sinz, T.F. Woolf, P.F. Hollenberg, *Chem Res Toxicol* **11**,252 (1998)
55. S.P. McDonald and G.R. Russ, *Kidney Int* **61**,2259 (2002)
56. Pysician's Desk Reference for KALETRA ® (ritonavir and lopinavir), January 2002 labeling,
57. Cvetkovic M, Leake B, Fromm MF, Wilkinson GR, Kim RB. *Drug Metab Dispos.* **27**, 866 (1999)
58. G.K. Dresser, D.G. Bailey, B.F. Leake, U.I. Schwarz, P.A. Dawson, D.J. Freeman, and R.B. Kim. *Clin Pharmacol Ther* **71**,11 (2002)
59. Y. Kurata, I. Ieiri, M. Kimura, T. Morita, S. Irie, A. Urae, S. Ohdo, H. Ohtani, Y. Sawada, S. Higuchi, K. Otsubo. *Clin Pharmacol Ther* **72**, 209 (2002)
60. T. Furuta, N. Shirai, F. Watanabe, S. Honda, K. Takeuchi, T. Iida, Y. Sato, M. Kajimura, H. Futami, S. Takayanagi, M. Yamada, K. Ohashi, T. Ishizaki, H. Hanai. *Clin Pharmacol Ther* **72**, 453 (2002)
61. L. Iyer, S. Das, L. Janisch, M. Wen, J. Ramirez, T. Karrison, G.F. Fleming, E.E. Vokes, R.L. Schilsky, M.J. Ratain, *Pharmacogenomics J* **2**, 43 (2002)
62. L.J. Lesko and R.L. Williams, *Appl Clin Trials* **8**,56 (1999)
63. W. Smalley, D. Shatin, D.K.Wysowski, J. Gurwitz, S.E. Andrade, M. Goodman, K.A. Chan, R. Platt, S.D. Schech, W.A. Ray. *JAMA* **284**, 3036 (2000)
64. FR notice (2000): Labeling guideline (Federal Register 65: 247; 81082-81131; December 22, 2000
65. L.B. Moore, B. Goodwin, S.A. Jones, G.B. Wisely, C.J. Serabjit-Singh, T.M. Willson, J.L. Collins, S.A. Kliewer. *Proc Natl Acad Sci* **97**, 7500 (2000)
66. S.A. Wrighton, E.G. Schuetz, K.E. Thummel, D.D. Shen, K.R. Korzekwa, P.B. Watkins, *Drug Metab Rev* **32**, 339 (2000)

PHARMACOKINETIC/PHARMACODYNAMIC MODELING IN DRUG DEVELOPMENT

Shashank Rohatagi*, Nancy E. Martin and Jeffrey S. Barrett

1. INTRODUCTION

Pharmacokinetic (PK) and pharmacodynamic (PD) modeling is employed to establish correlation of the concentration-time relationship (PK) with effect-concentration relationship (PD) in order to provide a better understanding of the time course of an effect (PK/PD) after administration of drug[1, 2, 3, 4].

Pharmacokinetics is the study of what the body does to the drug i.e. the absorption, distribution, metabolism and excretion of the drug. Pharmacokinetic modeling characterizes the blood or plasma concentration-time profiles following administration of the drug via various routes. Pharmacokinetic characterization *via* parameterization (i.e. volume of distribution, clearance and half-life) can be used to determine desired blood or plasma concentrations, optimize doses or select the route of administration[1, 2, 3, 4]. These parameters allow simulation of drug-concentrations in different situations, identification of important metabolites and characterization of the pharmacokinetics in various disease states.

Pharmacodynamics (PD), in general terms, seeks to define what the drug does to the body (i.e. the effects). Pharmacodynamic modeling attempts to characterize measured physiological parameters before and after drug administration with the effect defined as the change in a physiological parameter relative to its pre-dose or baseline value. Baseline is defined as the physiological parameter without drug dosing and may be complicated in certain situations due to diurnal variations. Efficacy is the expected sum of all beneficial effects following treatment. While this characterization implies a human / patient stage of development, many of the assumptions under which PK/PD studies are designed are built from *in vitro* data, animal pharmacology studies, *in vitro* and *in vivo* PK assessment in animals and healthy volunteers and ultimately challenged in patients. There is an inherent evolutionary nature to the process by which PK/PD correlations are

* Shashank Rohatagi, Nancy E. Martin, and Jeffrey S. Barrett, Aventis Pharmaceuticals, Inc., Drug Metabolism and Pharmacokinetics, 1041 Route 202-206, Bridgewater, NJ, USA.

Applications of Pharmacokinetic Principles in Drug Development
Edited by Krishna, Kluwer Academic/Plenum Publishers, 2004

developed. PK/PD modeling is a tool by which this knowledge is quantified and ultimately bridged.

Pharmacokinetic and pharmacodynamic modeling (PK/PD) currently receives a great deal of attention in the regulatory agencies, industry and the academia since it aims to link the dose to the causative effect taking advantage of traditional pharmacokinetic and pharmacodynamic strategies (Table 1). And yet, changes in regulations and extra attention given by industry and academia have not guaranteed that the quality of the data and ultimate prescription guidance is aligned with the financial investment. The 2002 edition of Physician's Desk Reference, reports for several drugs that "the precise mechanisms of action are unknown[5]." This statement may be true despite substantial efforts to propose and challenge various mechanisms of action. It is likely however, that it also represents the failure of medical and pharmaceutical science to optimally study and evaluate new chemical entities. There are variety of reasons for such failures, but a popular belief is that disparate efforts to characterize drug actions and monitor and characterize drug exposure are not joined on the appropriate occasions[6]. Simply put, pharmacokinetics and pharmacodynamics (PK/PD) are not often embraced in an organized, evolving effort. The end result is poorly defined safety-efficacy windows, little or no support for dose selection and occasionally, no understanding of drug mechanism.

Table 1. Milestones in PK/PD research: a partial list

Year	Milestone	Reference
1937	Teorell lays foundation of modern pharmacokinetics.	7, 8
1938	Adoption of US Food, Drug and Cosmetic Act.	
1950	First PK/PD paper published by De Jongh and Wijnans where using one compartment body model linked to log-linear/linear effect model predicted the dose sparing effect of divided doses.	9
1960	Garrett applied analog computing to pharmacokinetics.	10
1966	Levy describes the kinetics of pharmacological effects.	11
1968	Wagner uses the E-max model to evaluate different dosage regimen with respect to pharmacological outcome.	12
1968	Sheiner describes an indirect-effect model to account for hysteresis behavior of warfarin.	13
1969	Metzler introduced NONLIN, the first digital pharmacokinetic program.	
1975	Gibaldi and Perrier publish the first edition of Pharmacokinetics.	14
1979	Beal and Sheiner form the NONMEM Project Group.	
1981	Holford and Sheiner describes the use of effect compartment model to collapse the hysteresis loop.	15
1986	Boudinot and Jusko linked mechanistic/biochemical approaches to PK/PD	16
1987	Temple from US FDA supported the idea of using population PK/PD in special population thus encouraging industrial scientist and academics to include PK/PD during drug development.	
1992	Peck introduced the concept of concentration-controlled clinical trials and importance of PK/PD in drug development.	17
1992	Sheiner and Ludden introduced population PK/PD.	18
1994	Derendorf and Gabrielsson publish books on PK/PD correlations.	3, 19
	Derendorf and Hochhaus introduced dose optimization using PK/PD in therapeutic drug monitoring.	
1996	Bruno used population PK/PD rigorously for the anticancer drug, Taxotere, allowing the approval of the drug on the basis of this analysis.	20
1999	US FDA issues a guidance on population pharmacokinetics.	21
2002	US FDA issues the draft guidance on exposure-effect relationship.	22

This chapter defines the terminology around and stages of PK/PD modeling with associated case studies. Concepts for a variety of PK/PD models are described, several examples of many are presented, along with general recommendations for PK/PD analyses. Examples pertaining to specific therapeutic areas specific examples are also presented along with the current associated draft guidance provided by the US Food and Drug Administration (FDA). Finally, a proposal for PK/PD workflow is presented along with guidance on the relevant assumptions and assumption testing.

2. PHARMACOKINETIC MODELS

Pharmacokinetics of a drug can be described by using both compartmental and noncompartmental analysis[3, 4, 23]. Compartmental methods divide the body into well-stirred hypothetical areas (compartments) for a drug and its metabolite and describe the disposition of the drug using the appropriate mathematical equations. Noncompartmental analysis uses the calculation of areas under the concentration-time curve (AUC), concentration·time-time curve (AUMC), terminal slopes of the concentration-time curve (k_e) to determine the pharmacokinetic parameters[24].

Specifically, non-compartmental analysis typically proceeds *via* the partitioning of concentration-time profiles into regions within which certain physiologic processes may be estimated as described previously. Assuming an extravascular administration, concentrations up to the maximum observed concentration (C_{max}) are used to estimate absorption parameters (e.g. C_{max}, time of maximum concentration, T_{max}, mean absorption time, MAT, etc.), post-T_{max} concentrations contribute to parameters that define drug distribution (e.g. mean residence time, MRT, volume of distribution, V_d) and concentrations in the terminal phase are used to estimate drug elimination and ultimately provide information about drug half-life. The full-profile is interpolated with the areas-under-the-curve (AUC) to the last time point and extrapolated to infinity used as measures of drug exposure. Other physiologically-based parameters such as drug clearance (CL) are derived from essentially profile-based parameters. These parameters such as clearance are compared to liver blood flow (90L/h) to assess whether drug has a high or low clearance. Similarly, V_d can be compared to plasma volume and total body volume to assess extent of volume of distribution[24]. The effect of extrinsic and intrinsic factors such as food, gender, age, disease state and concomitant therapy may be tested on these non-compartment parameters to provide dosing guidance. Other parameters such as urinary and biliary clearances can also be calculated.

During noncompartmental analysis, the accuracy of these individual parameter estimates is dependent on the study conduct (accurate dosing and collection time information) and appropriate data density (concentration-time data within a curve region where certain parameters are derived). It is often observed that parameters are not estimable when adequate data density is not achieved as would be the case when only two observations are available in the terminal phase (precludes estimation of half-life, AUC, CL or V_d) or the first measured concentration is the maximum observation (precludes estimation of C_{max} or T_{max} accurately). Hence, missing values are generated in the study data set despite having administered drug and measured drug concentrations. Another limitation of the non-compartmental approach is the fact that the analysis of metabolite data proceeds in a similar manner completely independent of the parent compound data. Hence, the within-subject information is ignored and parent exposure as an input to metabolite concentrations is not considered.

Non-compartmental or equivalent empirical approaches also are implicitly two-stage in nature, relying on descriptive statistics based primarily on normal distribution assumptions in order to summarize data within and across groups. While the normality assumptions are fairly robust especially within the relatively homogenous distributions of a phase I population, the accuracy of such normal-theory-based estimates can be in jeopardy when sample sizes are small[3, 25].

A compartmental model consists of inter-connected compartments that are homogenous with respect to the analyte of interest[3, 4, 23]. The transfer rate between these compartments can be expressed by intercompartmental clearances (CL) or rate constants (k_i) which are usually linear but can also be non linear. The central compartment is traditionally where the samples are collected and the matrix is usually the plasma or blood. The peripheral compartments refer to the deeper compartments. There may be inflows to a compartment (input or dose) and/or outflows from a compartment (output, excretion, elimination), but mass-balance across all compartments is maintained. Most of the pharmacokinetic models are mammillary system models where the central compartment is attached to all other compartments and exchange occurs only through the central compartment, not between the peripheral compartments. The elimination is mainly through the central compartment described by elimination rate constant (k_e). For intravenous dosing, the dose administration is directly in the blood or central compartment, while for oral dosing, the dose administration is in the gut compartment and an absorption constant (k_a) describes the absorption of the drug in the blood stream. Depending on the distribution properties of the drug, the drug can be distributed in to one, two or more peripheral or tissue compartments leading to two, three or more compartment body models. For example, the concentrations (C) of drug following one-compartment body model with oral absorption can be described by $C=A*exp(-k_a*t)-exp(-k_e*t))$, where A is the hybrid constant, k_a is the absorption constant, k_e is the elimination constant and t is time after dose administration.

Compartmental models provide a system to describe the underlying mechanism of drug disposition with mathematical equations and hence compartmental analysis is essential for quantitive predictions and PK/PD modeling. Noncompartmental analysis is of limited use in PK/PD analysis. Detailed understanding regarding mathematical compartmental modeling can be found in several text-books[3, 4, 23].

3. PHARMACODYNAMIC AND PHARMACOKINETIC/ PHARMACODYNAMIC MODELS

While PK and PD can be quantified independently *via* non-compartmental or equivalent empirical approaches, the ability to perform PK/PD analysis relies heavily on modeling. Noncompartmental pharmacodynamic parameters can be calculated for the effect-time curves to generate AUC_E (area under the effect-time curve), T_{maxE} (Time of maximal effect) and C_{maxE} (Maximal effect). Also, duration of effect can be calculated using the AUC_E or the concentration above the *in vitro* EC_{50} (concentration that gives 50% of the maximal effect) or the MIC (minimum inhibitory concentrations such as for an antibiotic)[19].

A PK/PD modeling approach implies that functional expressions can be defined to describe the concentration-response to a given dose of a drug. While still dependent on adequate data density, parameter estimates are more easily attained with a greater appreciation for the error (uncertainty) in their prediction. A modeling approach has the advantage of being able to scale the concentration response to dose based on linear or nonlinear assumptions and pulling the data across dose (within subject) in order to estimate such dose scaling relationships. Both parametric and non-parametric approaches are possible; the chapter focuses on parametric techniques and applications herein.

The definition of structural model components requires that the underlying PK properties including influences of drug input, dosing duration, drug interactions along with concomitant medications, chronopharmacokinetics and possible population differences (pharmacogenomics or differences in special populations) be specified. Simple structural models can easily be extended to accommodate more complex pharmacokinetic behavior.

Likewise, PD parameterization (effect-time relationship) should be defined for all relevant response variables with appreciation for input, lifestyle (such as alcohol-consumption, cigarette smoking etc) and special population differences. Given the sequential nature of drug development, many of these potential effects are assumed based on prior (pre-clinical) experience or from historical evidence with related agents. As drug candidates proceed to later stages of drug development, these assumptions should be challenged via appropriately designed, prospective trials. As needed, models are constructed and/or modified to reflect fundamental PK and PD behavior with the critical dependencies incorporated as model parameters.

PK/PD parameterization (concentration-effect) reflects a separate modeling effort, which, while dependent on PK and PD data, proceeds in an independent manner. The environment is inherently multivariate as multiple drug moieties (i.e., parent compound and metabolites) may be responsible for multiple drug actions. There are a variety of ways in which this apparently complex situation can be solved. Some of these approaches are demonstrated in subsequent sections of this chapter. In general, PK/PD modeling is based on the following relationships.

Pharmacokinetics	→	Concentration=f_n(Dose, Time)
Pharmacodynamics	→	Activity or Effect = f_n(Concentration)
Pharmacokinetic/Pharmacodynamic Modeling	→	Activity or Effect = f_n(Dose, Time)

A common problem in choosing a parameter is whether it quantifies the drug effect. Hence, it is important to differentiate between certain terms such as exposure, surrogate endpoints and clinical outcomes. Biomarkers and surrogate endpoints are critical to the future of efficient drug development. Poor correlation between a biomarker and its clinical endpoint may result from poor measurement, selection of an inappropriate biomarker, or use of an inappropriate clinical endpoint. Successful pharmacokinetic/pharmacodynamic (PK/PD) modeling depends on the biomarkers or surrogate endpoints used for the modeling[26]. Hence, the understanding of correlation between biomarkers, surrogate markers, mechanistic correlation with clinical outcome will improve the predictive power of PK/PD modeling. These terms are defined below:

- **Exposure.** *Exposure* refers to dose (drug input to the body) and various measures of acute or integrated drug concentrations in plasma and other biological fluid such as maximum concentration (C_{max}), minimum concentration (C_{min}), steady-state concentration (C_{ss}) or area under the plasma concentration – time curve (AUC)[22].

- **Response.** *Response* refers to a direct measure of the pharmacologic effect of the drug. Response includes a broad range of endpoints, including biomarkers and surrogate endpoints[22].
- **Biomarker.** *Biomarker* is a characteristic that is measured and evaluated as an indicator of normal biologic process, pathophysiological processes, or pharmacologic processes to a therapeutic intervention[26]. Biomarkers can also be defined as quantifiable physiological or biochemical-marker which is sensitive to intervention (drug treatment)[22, 26, 27]. A biomarker may or may not be relevant for monitoring clinical outcome but may be useful in early drug development for dose selection / discrimination. This situation is exemplified with RGD 891, where the inhibition of platelet aggregation by this GPIIb/IIIa receptor antagonist was used as a biomarker for selection of doses in animals and humans[28]. In certain cases, a drug can have an effect on the disease state but it is only partially mediated through the biomarker similar to the relationship between cholesterol lowering effect of statins and mortality[26]. In addition, for some biomarkers there may be a negative-link, or a false positive-link that may not allow the marker to be used as a surrogate endpoint[26, 29]. For example, VPC displays a false positive link as VPC suppression decreases sudden death in CAST trial but Flecanide suppresses VPC, however, Flecanide shows increased sudden death over placebo[26, 29]. Interferon-gamma displays a negative link as it appears to stimulate production of superoxide and bacterial killing[26, 30]. However, in patients with chronic granulomatous disease, interferon increases bacterial killing, but interferon administration in these patients did not effect superoxide production.
- **Surrogate Marker (SM).** S*urrogate marker* is a biomarker that substitutes for a clinical endpoint [22, 26]. Biomarkers need to be validated and converted into surrogate markers i.e. if a biomarker has been shown to reflect clinical outcome, it can be called surrogate marker. Examples include HIV viral load in HIV+ patients, blood glucose in diabetic patients, effects on blood pressure and lipids in cardiovascular disease and FEV_1 in asthma patients[1, 2, 26, 27].
- **Clinical Endpoint.** *Clinical endpoint* is a characteristic or variable that measures how a patient feels, functions or survives. However, the assessment is often difficult to perform[1, 2, 26, 27].

Various pharmacodynamic models have been proposed to characterize a wide range of pharmacologic effects. Representative examples are described in Table 2. The general premise for any of these models is to capture the functional manner in which effect varies with measured concentration in some biological matrix. An inherent assumption in this regard is that the concentration referenced is that at the site of action. In many cases, there is the assumption of equilibrium between site of action and systemic concentration. This is not mandatory as modeling strategies have evolved in order to accommodate situations in which systemic concentration cannot be used directly in modeling expressions (e.g. link model, indirect response etc).

Table 2. Pharmacokinetic/pharmacodynamic (PK/PD) models

PK/PD-Models	Salient Features	Examples
Fixed Effect Model	• Quantal Effect Model[3, 4, 19]. • Threshold model. • E=Fixed, $C \geq C_{threshold}$.	• Hydromorphone concentrations> 4 ng/mL provided good pain control in patients with bone or soft tissue injury but not effective in neuropathic pain[31]. • Ototoxicity with gentamycin at concentrations greater than 4 mcg/mL[32].
Linear Model	$Effect = S \bullet Ce + E_0$ • Assumes baseline effect is constant[3, 4, 19]. • Assumes an effect will continuously increase as concentration increases giving erroneous results at low and high concentrations. 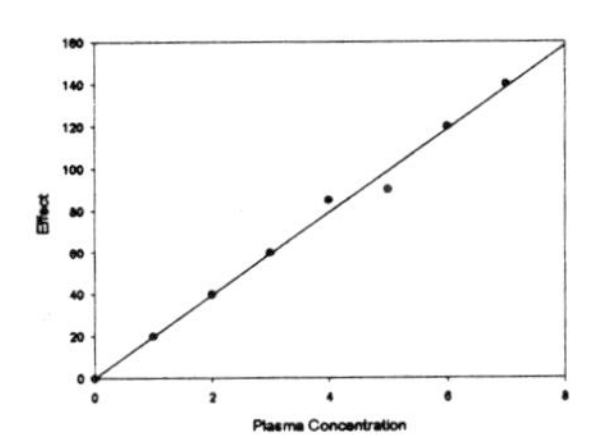	• Relationship between pilocarpine plasma level and saliva flow[33]. 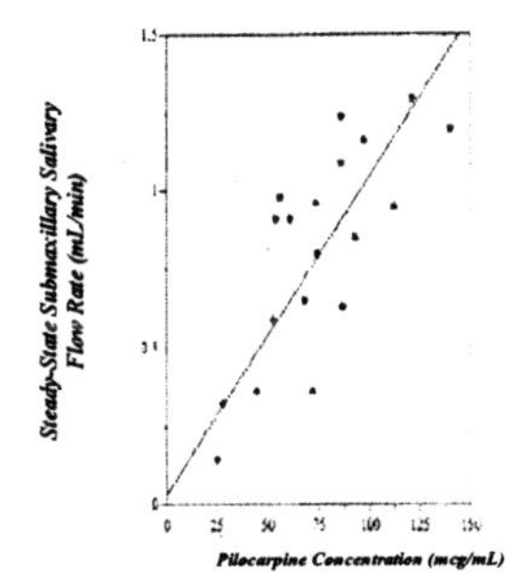
Log-Linear Model	$Effect = S \bullet \log Ce + E_0$ • E_0 has no physiological meaning[3, 4, 19]. • Valid only over 20-80% of the maximal effect.	• Relationship between % reduction in exercise-induced tachycardia and propranolol concentration in plasma[34].

PK/PD-Models	Salient Features 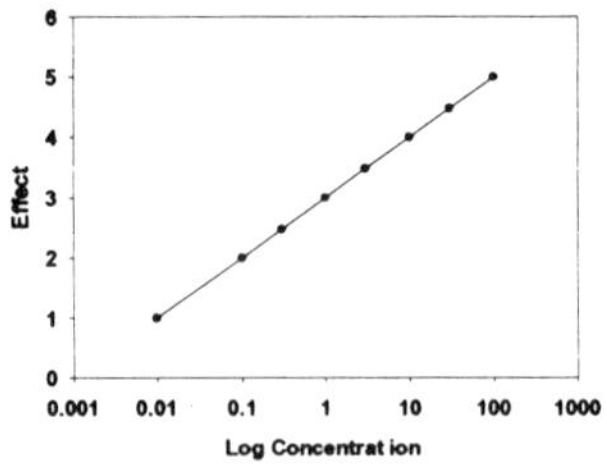	Examples 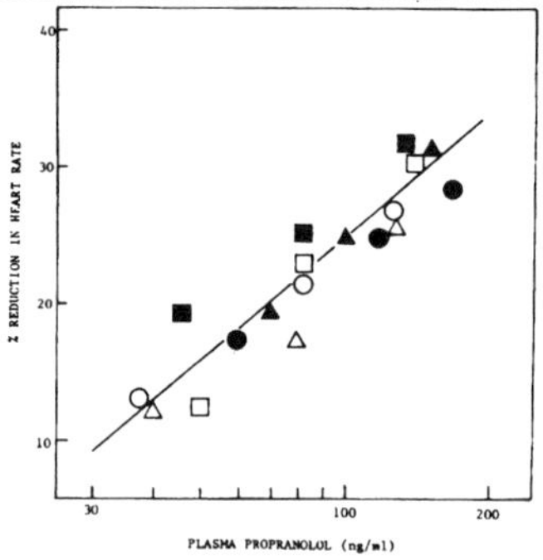
E_{max}-Model	$$Effect = \frac{E_{max} \bullet C_e}{EC_{50} + C_e} + E_0$$ • Valid over the entire range of effect and can describe maximal and baseline effects[3, 4, 19]. 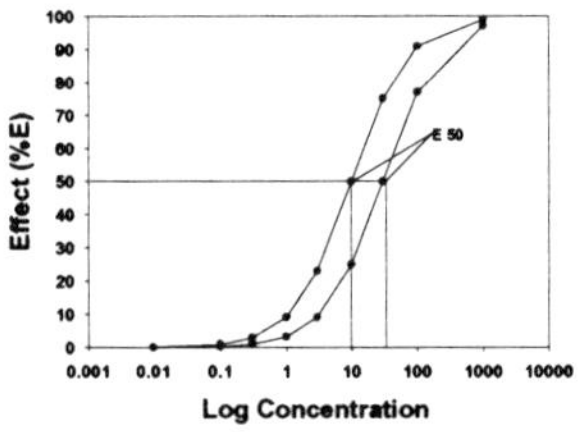	• Relationship between propanolol and heat rate[35]. 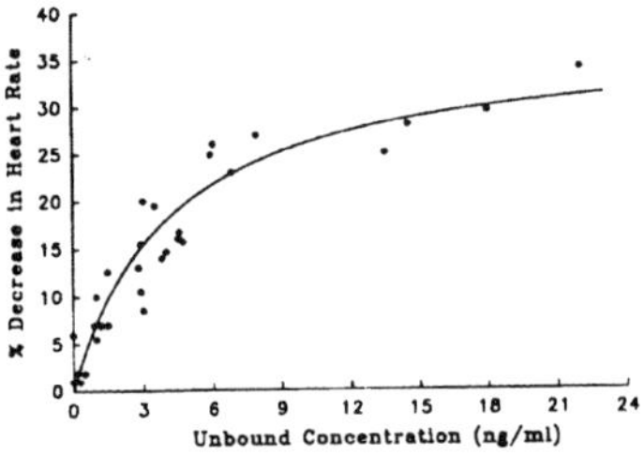
Sigmoid E_{max}-Model	$$Effect = \frac{E_{max} \bullet C_e^n}{EC_{50}^n + C_e^n} + E_0$$ • Valid over the entire range of effect and can describe maximal and baseline effects. • n describes curve shapes[3, 4, 19]. 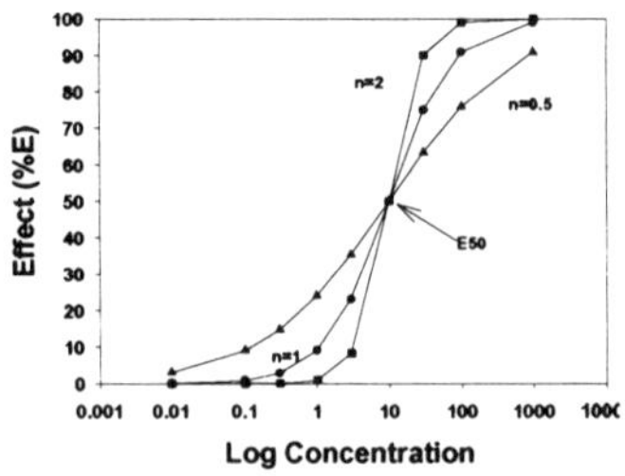	• Relationship between phenobarbital concentration and the anticonvulsant effect as defined by change in pentylenetetrazole (△PTZ)[36]. 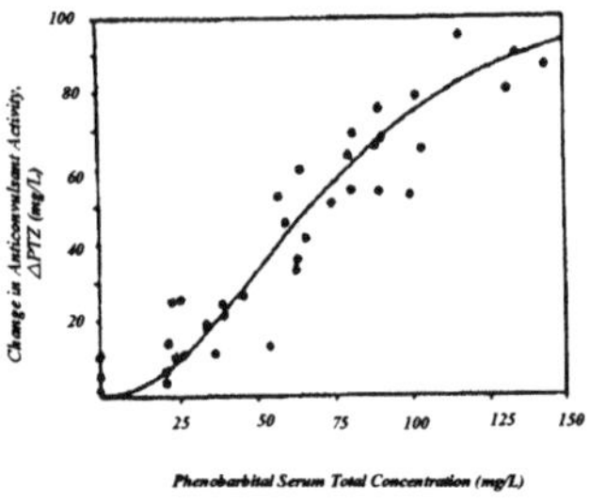

PK/PD-Models	Salient Features	Examples
Direct Link	• Effect is directly correlated to plasma concentrations[3, 4, 19].	• Effect of oral nisoldipine on diastolic blood pressure[37].
	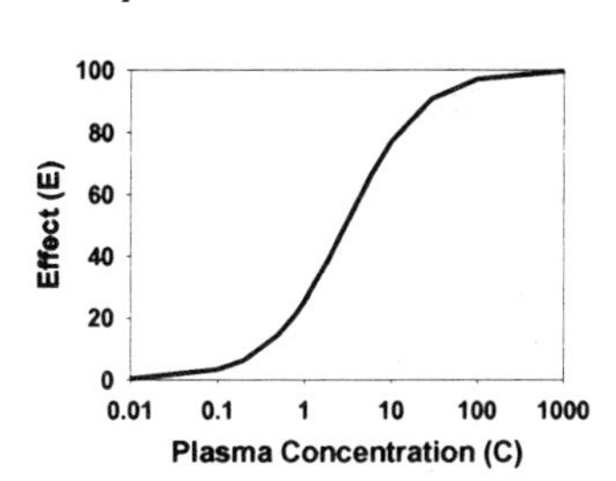	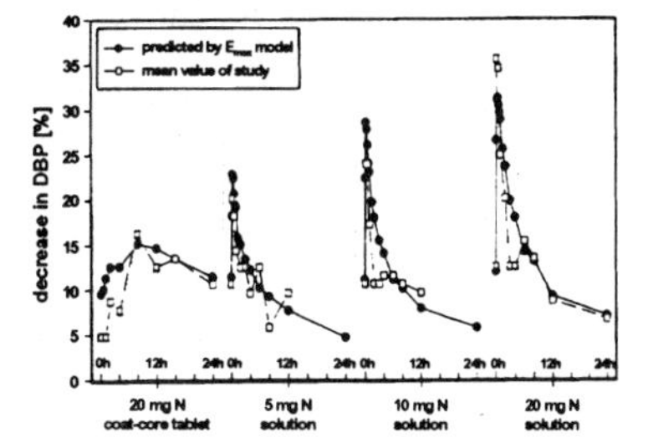
Indirect Link	$$Effect = \frac{E\max \bullet C_e^n}{EC_{50}^n + C_e^n} + E_0$$ $$C_e = \frac{D{\cdot}Ka{\cdot}Keo}{Vd}\left[\frac{e^{-KT}}{(Ka-K)(Keo-K)} + \frac{e^{-KaT}}{(K-Ka)(Keo-Ka)} + \frac{e^{-KeoT}}{(K-Keo)(Ka-Keo)}\right]$$	Input Dose k_a Central Compartment 1 Hypothetical Effect Compartment k_{eo} Effect E_{max} Model k_e Pharmacokinetics Pharmacodynamics
	• Time courses of plasma concentration and effect are dissociated; temporal delay caused by distribution process whereas the effect site concentration is directly related to the o observed effect [3, 4, 19].	• Analgesic effect of 400 mg oral ibuprofen quantified by subjective pain intensity rating [38, 39].
	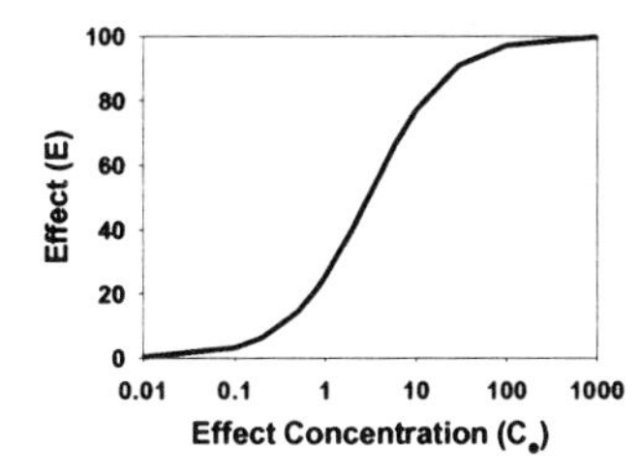	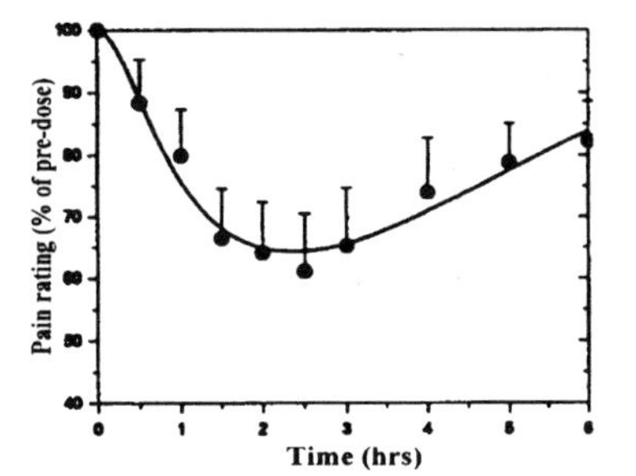
Direct Response	• Response can be directly linked or indirectly linked; direct correlation between effect site concentration and effect[1, 2, 26, 27].	

PK/PD-Models	Salient Features	Examples
Indirect Response	• Inhibition or induction of a physiological process to characterize the PD response[1, 2, 26, 27]. • Temporal dissociation between the concentration time course and the effect-time course (hysteresis) due to synthesis of a protein, reduction in a synthesis rate (reduction in hormonal levels). 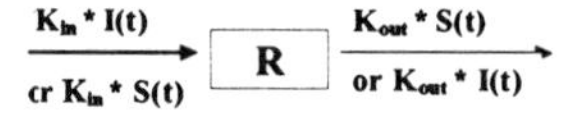 $K_{in} * I(t)$ or $K_{in} * S(t)$ → R → $K_{out} * S(t)$ or $K_{out} * I(t)$	
Indirect Response: Stimulation of input	$\frac{dR}{dt} = k_{in} * \left(1 + \frac{E\max * C}{EC_{50} + C}\right) - k_{out} * R$ 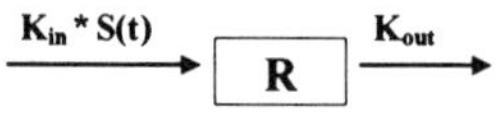 $K_{in} * S(t)$ → R → K_{out}	• Granulocyte induction after administration of methylprednisolone[40]. 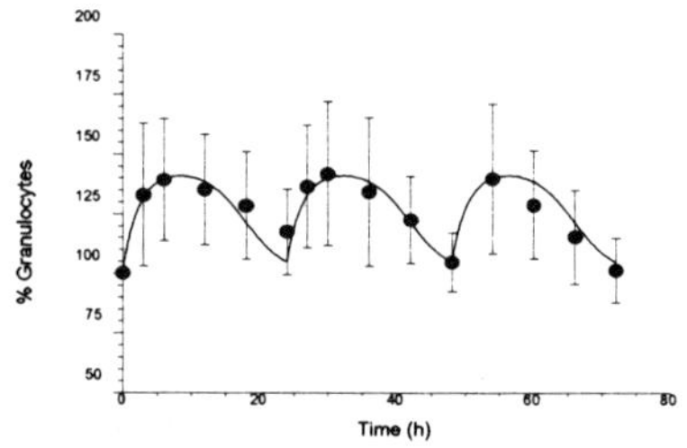
Indirect Response: Stimulation of output	$\frac{dR}{dt} = k_{in} - k_{out} * \left(1 + \frac{E\max * C}{EC_{50} + C}\right) * R$ 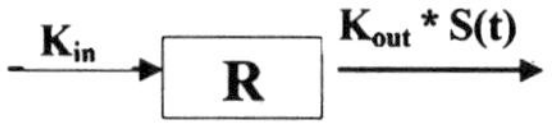 K_{in} → R → $K_{out} * S(t)$	• Lymphocyte inhibition after administration of methylprednisolone[40]. 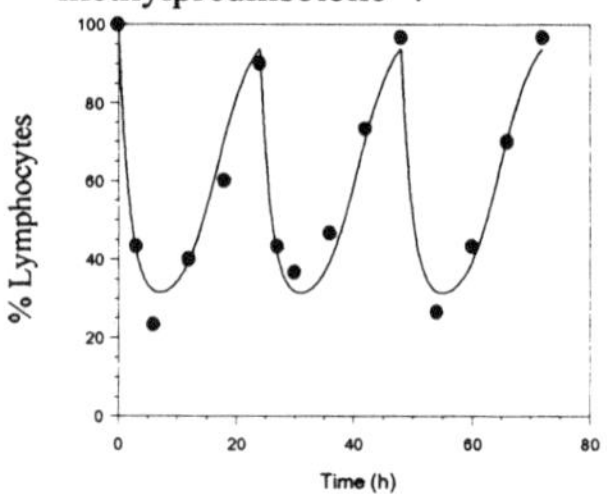

PK/PD-Models	Salient Features	Examples
Indirect Response: Inhibition of output	$\frac{dR}{dt} = k_{in} - k_{out} * \left(1 - \frac{C}{C + EC_{50}}\right) * R$ 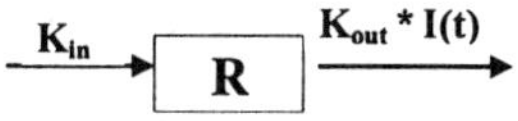 K_{in} → R → $K_{out} * I(t)$	• Granulocyte induction after administration of methylprednisolone[40]. 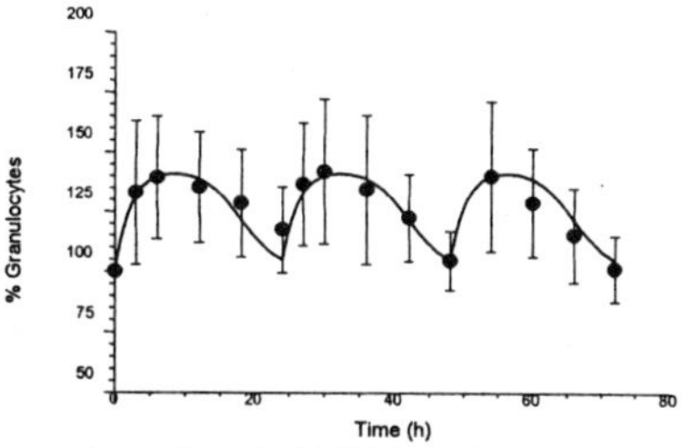
Indirect Response: Inhibition of input	$\frac{dR}{dt} = k_{in} * \left(1 - \frac{C}{C + EC_{50}}\right) - k_{out} * R$  $K_{in} * I(t)$ → R → K_{out}	• Lymphocyte inhibition after administration of methylprednisolone[40]. 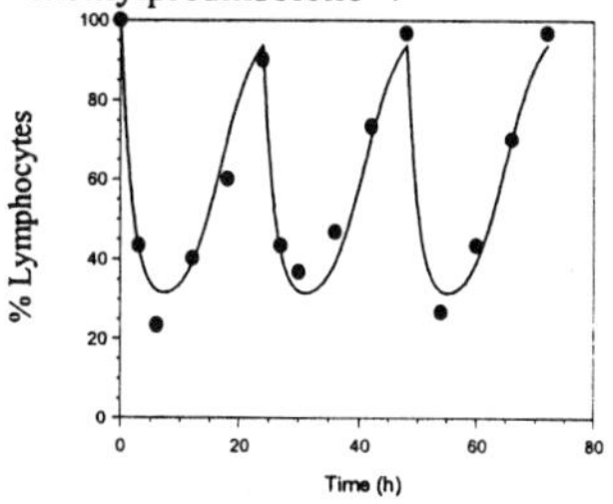
Soft Link	• PK and PD data used to determine the link[2]. 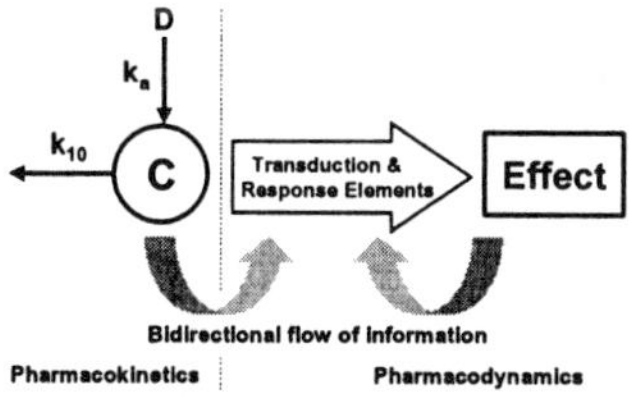	
Hard Link	• Pharmacokinetic data is combined with additional information from in vitro experiments such as receptor binding affinities to predict PD[2]. 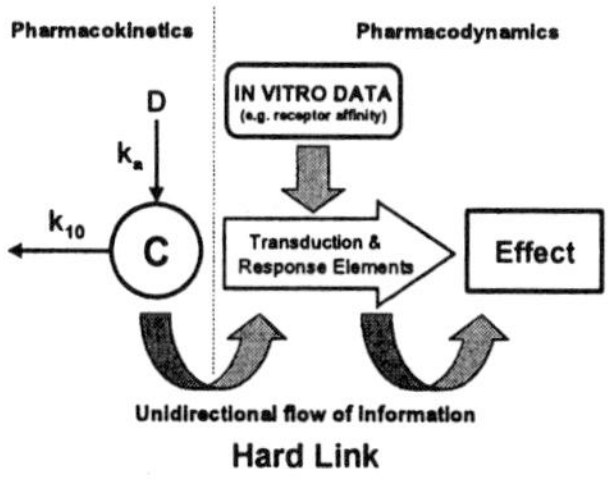	• *In vitro* measurements in PK/PD models to predict effect of tazobactam on antimicrobial activity of piperacillin[2, 41].

PK/PD-Models	Salient Features	Examples
Time Variant	• Tolerance or sensitization.	
Tolerance Models	• Functional vs. metabolic tolerance. • Reduction in effect intensity at concentrations that earlier produced a greater effect, or a decrease in drug effect over time despite constant effect site concentrations[4, 19]. • Diminishing response with re-challenging stimulus may be caused by a decrease in receptor number or receptor affinity at the effect site. • Decrease in pharmacological response after exposure to drug leading to clockwise hystersis or proteresis. Could be due to: • Down-regulation of receptors (replace E_{max} with E_{max}*exp(-k*t) in the E_{max} model where k is the constant defining tolerance and t is the time of exposure to drug. • Decrease in affinity to receptors (replace EC_{50} with E_{50}*exp(k*t) in the E_{max} model where k is the constant defining tolerance and t is the time of exposure to drug. • Decrease in receptor-generated response[4, 19].	• Hypokalemia resulting from the β2-agonist terbutaline[42]. • Clockwise hysteresis loop in a plot of effect vs. concentration. 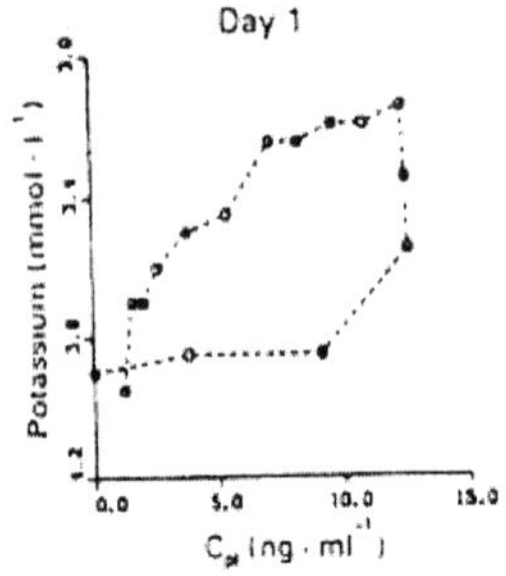
Sensitization Models	• Increase in pharmacological response[4, 19]. • Upregulation of receptors (replace E_{max} with E_{max}*exp(k*t) in the E_{max} model where k is the constant defining tolerance and t is the time of exposure to drug.	• Upregulation of beta-adrenoreceptors during chronic administration of beta-blockers[43].
Time Invariant	• Most of the PK/PD models are Time Invariant[19].	

<table>
<tr><th>PK/PD-Models</th><th>Salient Features</th><th>Examples</th></tr>
<tr><td>Competitive Agonists</td><td>

$$E = \frac{\sum_{i=1}^{n} \frac{E_{max,i} \cdot C_i}{EC_{50,i}}}{1 + \sum_{i=1}^{n} \frac{C_i}{EC_{50,i}}}$$

• Do not alter E_{max} or EC_{50} but change the location of the inflection point of the curve[3].</td><td>

• PK/PD relationship describing the effect of RGD and its competitive active metabolite on the inhibition of platelet aggregation induced by collagen , TRAP or ADP[28].

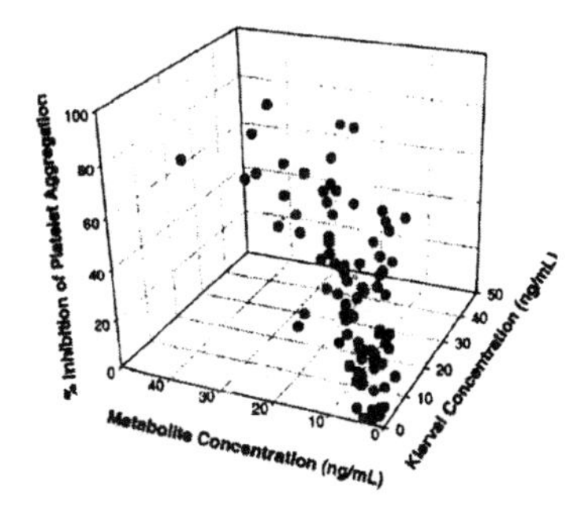

</td></tr>
<tr><td>Noncompetitive Agonists</td><td>

$$E = \sum_{i=1}^{n} \frac{E_{max,i} \cdot C_i}{EC_{50,i} + C_i}$$

• Shifts the maximal effect [3, 4].

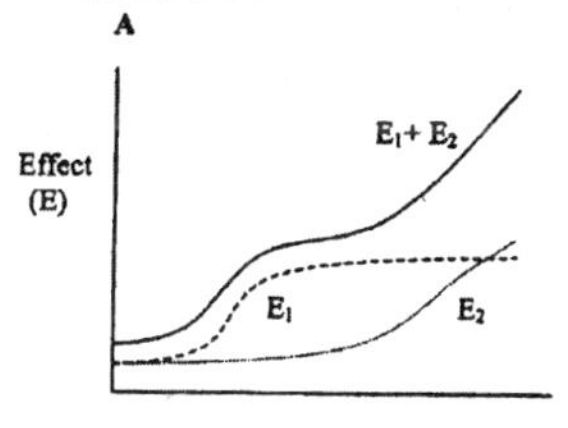

</td><td></td></tr>
<tr><td>Logistic Regression</td><td>

• Relationship between degree or probability of response and concentrations[3, 4].

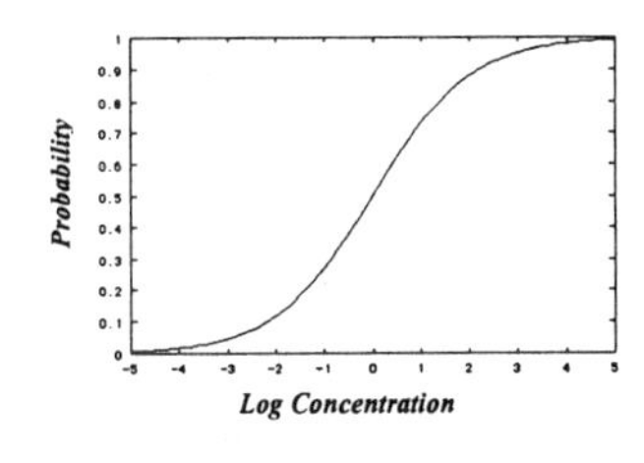

</td><td>

• Probability of patient being classified as major-, moderate- or non-responder in treatment of panic disorder by alprazolam [44].

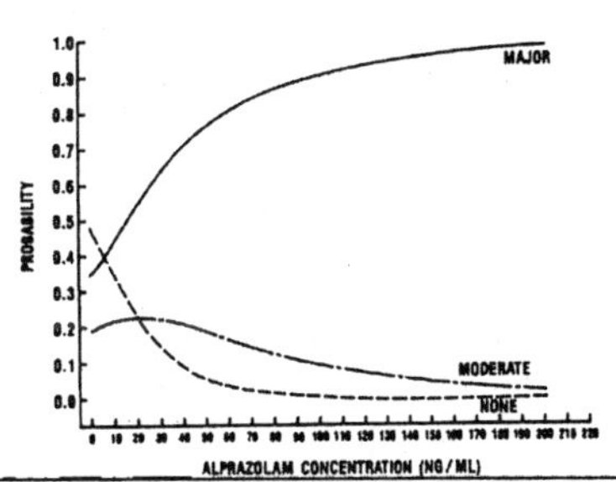

</td></tr>
</table>

S = Slope, C_e = Concentration at the Effect Site, E_o = Baseline Effect E_{max} = Maximum Effect, EC_{50} = Concentration that Produces 50% of the Effect, n=Sigmoidity Factor that Determines the Shape of the Curve or shape factor, D = Dose, K_a = Absorption Rate Constant, K = Elimination Rate Constant from the Central Compartment, K_{eo} = Elimination Rate from the Effect Compartment, T = Time, k_{in} = Input Constant , k_{out} = Output Constant, R= Response in Relation to Time, 1 = $Drug_1$ and 2 =$Drug_2$, I(t) = Inhibition Function, S(t) = Stimulation Function.

The simplistic model that can correlate the effect to concentration is the fixed effect model followed by the linear or log-linear model as described in Table 1. However, most drug targets are either functional receptors or enzymes that can be described by E_{max}-or the sigmoid E_{max}-model and hence these models are the most frequently used. This model is discussed in more detail in this chapter. The model can be described by the expression below:

$$E = \frac{E_{max} \cdot C^n}{E_{50} + C^n} + E_0 \qquad (1)$$

E is intensity of the effect, C is concentration, E_{50} (or EC_{50}) is concentration at which 50% of the effect is observed, E_{max} is maximum effect, E_0 is baseline effect and n is shape factor. The model is consistent with an effect caused by an interaction of a large and a small molecule, e.g. drug-receptor interaction. It is linear from 20%-80% of maximal effect range where it is identical to the log-linear model[45]. If the value of the shape factor (n) is 1 then it implies that one drug molecule interacts with one site on the receptor (which is the most probable case). Non-integer values actually are used as correction factors to give a better curve fit. E_0 is the baseline value, which is usually a constant but it may be a complex function as in the case of cortisol. Corticosteroid effects are mainly receptor mediated and can be expressed by using the E_{max} model. Using lag-time or introducing a hypothetical effect compartment can solve the discrepancy between maximal effect and maximum time shown by a hysteresis loop. Measurement of the drug at the biophase (hypothetical effect compartment) or free plasma drug concentrations have been used in these models.

3.1. Derivation of the E_{max}-model

As discussed earlier, drug targets are often either functional receptors or enzymes. These interactions can be described by interaction between small molecules such as drugs (D_g) in most cases and large molecules such as receptors or enzymes (R_c). The equilibrium between the drug (D_g), receptor (R_c) and the drug-receptor complex can be described the equation below[19, 46]:

$$D_g + R_c \Leftrightarrow DR \qquad (2)$$

If K_d is the equilibrium constant between the drug, receptor and the drug-receptor complex, then the value of K_d may be expressed mathematically by the following equation:

$$K_d = \frac{\lfloor D_g \rfloor \bullet [R_c]}{[DR]} \qquad (3)$$

Free receptors (R_c) are equal to the difference between the total number of receptors (R_{tot}) and the drug-receptor complex. Substituting this value, the following expression is achieved:

$$K_d = \frac{\lfloor D_g \rfloor \bullet ([R]_{tot} - [DR])}{[DR]} \quad (4)$$

Dividing both sides by the drug concentrations, following expression is achieved:

$$\frac{K_d}{[D_g]} = \frac{[R]_{tot}}{[DR]} - 1 \quad (5)$$

Adding 1 on both sides and taking the reciprocal the following equation is achieved:

$$\frac{[DR]}{[R]_{tot}} = \frac{\lfloor D_g \rfloor}{[D_g] + K_d} \quad (6)$$

Multiplying both sides by R_{tot}, the following expression is achieved:

$$[DR] = \frac{\lfloor D_g \rfloor \bullet [R]_{tot}}{[D_g] + K_g} \quad (7)$$

As DR is correlated to the effect as it reflect the drug binding to the receptor. D_g reflects the concentration and R_{tot} reflects the maximal total effect or E_{max} and K_d reflect the concentration that can produce 50% of the drug effect or EC_{50}. The equation can be re-written as:

$$E = \frac{C \bullet E_{max}}{C + EC_{50}} \quad (8)$$

3.2. Explanation of E_{max}-Model

The E_{max} model can be divided into 3 distinct regions, E < 20% of E_{max}, 20% ≤ E ≤ 80% of E_{max} and E > 80% of E_{max} (Figure 1)[19]. At very low doses and at high doses, small changes in concentration (Region 1 and Region 3) elicit no discernable change in the effect. Only in the steep part of the curve (Region 2), effect changes with respect to concentrations, but the effect has no constant half-life and hence the concept of constant biological half-life is inappropriate[19].

Region 1: E < 20% of E_{max}. In this region, concentrations are much lower than EC_{50} and the effect is nearly directly proportional to the concentration, i.e. follows an exponential decline. This is described in the equation below.

$$E = \frac{C \bullet E_{max}}{C + EC_{50}} \approx \frac{E_{max}}{EC_{50}} \bullet C \quad (9)$$

Region 2: 20% ≤ E ≤ 80% of E_{max}. In this region, concentrations are in a similar range than EC_{50} and the effect intensity declines approximately linearly with time. Linear decline in effect in contrast to the exponential decay in concentration, i.e. drug effect disappears with a zero-order rate. The slope of the linear phase is described below.

$$m = Slope = -\frac{n \bullet E_{max} \bullet K}{4} \quad (10)$$

Where n is the Sigmoidity Factor that determines the shape of the curve. Hence, the effect decreases by a constant 17% of E_{max} during one pharmacokinetic half-life. Duration of the linear phase in a sigmoid E_{max}-model is described below.

$$0.8 \bullet E_{max} = Intercept - m \bullet t_{80} \tag{11}$$

$$0.2 \bullet E_{max} = Intercept - m \bullet t_{20} \tag{12}$$

Subtracting the two equations the following result is achieved.

$$t_{20} - t_{80} = \frac{0.6 \bullet E_{max}}{m} \tag{13}$$

Substituting the value of m, the following relationship is achieved:

$$t_{20} - t_{80} = \frac{0.6 \bullet 4}{n \bullet k} \approx 3.5 \bullet t_{1/2} \tag{14}$$

Hence, it takes 3.5 pharmacokinetic half-lives for the effect to drop from 80% to 20% of E_{max} (for n=1, i.e. a simple E_{max}-model). Thus, the effect has no constant half-life and the concept of biological half-life is frequently used inappropriately.

Region 3: E > 80% of E_{max}. In this phase, the concentration always remains much higher than EC_{50}. Thus, response remains almost maximal despite a dramatic fall in the concentration. There is a shallow relationship between concentration and effect. Only a 9% increase in effect (i.e. 90 to 99%) requires a 90-fold increase in drug concentration. Drug concentration appears to have little influence on drug effect as illustrated in the equation below:

$$E = \frac{C \bullet E_{max}}{C + EC_{50}} \approx E_{max} \tag{15}$$

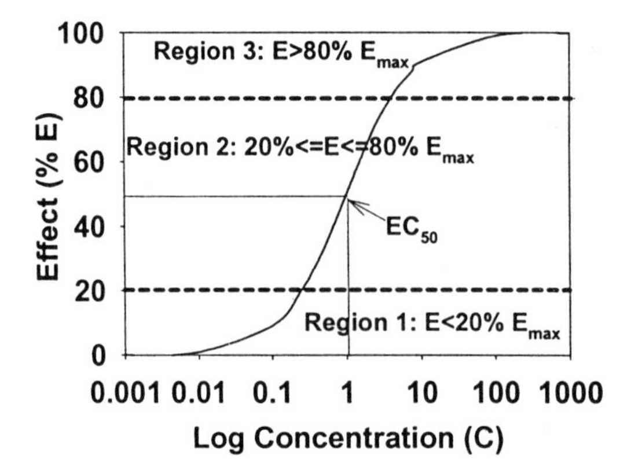

Figure 1. Behavior of the E_{max}-Model.

3.3. Estimating Doses Mathematically Using PK/PD Models

Total drug concentrations can be used to derive PK/PD relationships, but free drug concentrations (not bound to proteins) are more appropriate, as only the free drug is typically available to the receptors at the site of action. Certain parameters like AUC_E (area under the effect time curve), D_{50}, and DR_{50} can be used to characterize the efficacy of the drug. D_{50} is the dose necessary to produce and maintain 50% of the maximum effect while DR_{50} is the dosing rate necessary to produce and maintain 50% of the maximum effect. They can be expressed mathematically by the following equations:

$$D_{50} = \frac{E_{50} \cdot V_d}{1 - f_b} \qquad (16)$$

$$DR_{50} = \frac{E_{50} \cdot CL}{1 - f_b} \qquad (17)$$

It is evident from the above equations that $DR_{50}=k_e \cdot D_{50}$ where k_e is the elimination constant of the drug as $CL=k_e \cdot V_d$. E_{50} is concentration at 50% of maximal effect, V_d is volume of distribution, $(1-f_b)$ is free fraction of drug and CL is clearance. The PD parameters generally chosen for PK/PD modeling are usually easy to measure. However, it is necessary to correlate the parameters to the therapeutic outcome. Some effort has been made in this area. Parameters such as DR_{50} and D_{50}, which include pharmacokinetic (CL, V_d), pharmacodynamic (E_{50}) and protein binding factors (f_b) are a good estimate of clinical potency as illustrated in Table 2[47].

Table 3. Calculated values of DR_{50} for representative corticosteroids (adapted from 47)

Corticosteroid	E_{50} (ng/mL)	CL (L/h)	$1-f_b$	DR_{50} (mg/day)	Relative Potency (HC=1)	Clinical Potency (HC =1)
Betamethasone	0.70	9.00	0.36	0.40	24.3	25.0
Dexamethasone	0.40	17.0	0.32	0.50	20.0	25.0
Triamcinolone Acetonide	0.20	21.0	0.20	1.70	6.00	6.00
Methylprednisolone	1.00	10.0	0.23	2.20	4.60	5.00
Prednisolone	2.60	10.0	0.25	2.50	4.10	4.00
Fluocortolone	0.60	32.0	0.13	3.50	2.90	5.00
Cortisol	4.70	18.0	0.20	10.2	1.00	1.00

4. FDA EXPOSURE EFFECT GUIDANCE-SALIENT FEATURES

FDA has been concerned with the number of applications for which a satisfactory dose rationale can be defined and hence a guidance was introduced in 2002 pertaining to concentration-effect relationships[22]. This Food and Drug Administration's (FDA) draft guidance on concentration-effect relationships describes the following steps in detail:

- The uses of exposure-response studies in regulatory decision-making,
- The important considerations in exposure-response study designs to ensure valid information,
- The strategy for prospective planning and data analyses in the exposure-response modeling
- The integration of assessment of exposure-response relationships into all phases of drug development, and
- The format and content for reports of exposure-response studies.

The FDA guidance defines PK/PD information as data that corroborates the safety and effectiveness of drugs. A drug can be considered to have an appropriate benefit-risk ratio when its relationship to concentration is determined. Except in the cases of very toxic or very safe drugs, it is important to develop PK/PD relationships for favorable and unfavorable effects as well as information assessing whether, exposure should be adjusted for various subsets of the population. Historically, there has been success in establishing the relationship of dose to blood levels in various populations, thus providing a basis for adjustment of dose for pharmacokinetic differences among demographic subgroups or subgroups with impaired elimination hepatic or renal disease), assuming systemic concentration-response relationships are unaltered. Far less attention has been paid to establishing the relationship between blood levels and pharmacodynamic (PD) responses and possible differences among population subsets in these PK/PD relationships.

The guidance provides a description of basic terminology, defines the process of PK/PD modeling, selection of models and validation of models. It also describes the expectations, format and data to be included in a PK/PD analysis report. The guidance expects the following PK/PD information during the drug development process as described in Table 4. Further details can be found in the FDA Guidance[22].

Table 4. PK/PD requirements at various stages of drug development as defined in FDA concentration-effect draft guidance[22].

Drug Development Stage	PK/PD Requirements
Drug Discovery and Development Processes	• In Phase I and Phase II studies, along with safety and pharmacokinetics, relationship of exposure to response (biomarkers, surrogate endpoints, or short- term clinical effects) should be explored. • PK/PD relationship in animals should be linked to humans. • Provide that the proof that hypothesized mechanism is affected by the drug and leads to desired short-term clinical outcome (*proof of concept*). • Provide guidance for designing initial clinical endpoint trials. • Select an appropriate dose range, doing interval and dosage form that provides appropriate magnitude and time course of an effect, dosing interval, and monitoring procedures. • Exposure- response and PK data can also define the changes in dose and dosing regimens that account for intrinsic and extrinsic patient factors.
Support Determination of Safety and Efficacy	• PK/PD studies facilitate the designing of well-controlled studies that will establish the effectiveness of a drug. • Depending on study design and endpoints well characterized PK/PD relationship can provide the following advantages: ▪ Represent a well-controlled clinical study, in some cases a particularly persuasive one, contributing to substantial evidence of effectiveness (where clinical endpoints or accepted surrogates are studied). ▪ Add supporting value to efficacy of the drug where mechanism of action is well understood (e.g., when an effect on a reasonably well-established biomarker/surrogate is used as an endpoint). ▪ Support or even provide, primary evidence for approval of different doses, dosing regimens, or dosage forms, or use of a drug in different populations, when effectiveness is already well-established in other settings and the study demonstrates a PK-PD relationship that is similar to, or different in an interpretable way from the established setting.

5. PK/PD MODELING PROCESS

The process of PK/PD modeling contains several iterative steps (Figure 2)[3]. The first step is to state the problem and based on assumptions and prior knowledge, propose a tentative model. Based on this model, an experiment with PK/PD observations should be designed and performed. The data should be explored, plotted and fit to the proposed models. The output should be analyzed and optimized leading to proposal of a new model and the cycle can be repeated, if needed, or until an optimal result is achieved. These steps are described in detail below.

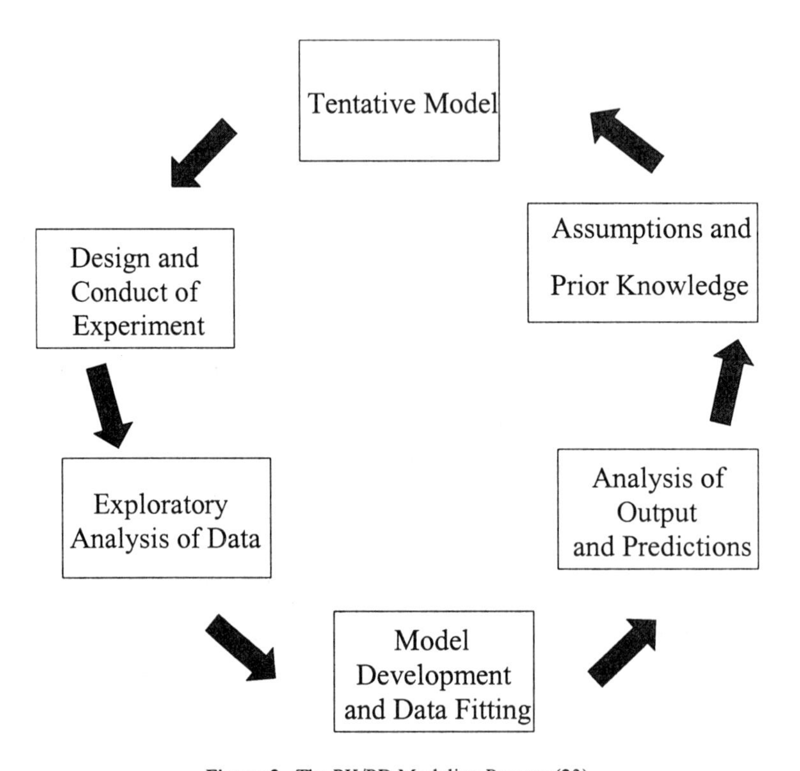

Figure 2. The PK/PD Modeling Process (23).

1. Statement of the Problem. The objectives of the modeling, the study design, and the available PK and PD data should be clearly identified. For example the statement of a problem could read as follows:

"The objective of the present analysis was to correlate the concentrations of a new corticosteroid to its potential effects on endogenous cortisol, which is considered a biomarker for safety. The concentration-time data for the drug and cortisol will be pooled from all Phase I, Phase II and Phase III trials to evaluate this concentration-effect relationship".

2. Statement of Assumptions. The assumptions of the model should be clearly articulated. The assumptions can be related to dose-response, PK, PD, and/or one of the following:

- The mechanism of the drug actions for efficacy and adverse effects
- Immediate or cumulative clinical effects
- Development of tolerance or absence of tolerance
- Drug-induced inhibition or induction of PK processes
- Disease state progression
- Circadian variations in basal conditions
- Influential covariates
- Absence or presence of an *effect compartment*
- Presence or absence of active metabolites and their contribution to clinical effects
- The PK model of absorption and disposition and the parameters to be estimated
- The PD model of effect and the parameters to be estimated
- Distribution of PK and PD measures and parameters
- Distributions of intra- and inter-individual variability in parameters
- Inclusion and/or exclusion of specific patient data

3. Plot and Explore Data. The most important step is to plot the data. This may include, concentration-time curve, response-time curve, response-concentration curve, linear and log-transformed curve, individual and pooled curve. This first step assists the analyst with the assumption regarding what models to consider. The composite plots also help in identification of outliers. The plotting of output data with covariates also assists in the identification of important covariates, as well. Scatter plots, residual plots, correlation plots and distributional plots help to identify goodness of fit, examination of residual error model, evaluation of structural model and evaluation of functional form of covariates, respectively.

4. Questions to Ask. During the modeling process several questions should be asked such as, is the behavior linear, nonlinear or mixed? Does the plasma concentration curve decline in a mono- or multi-exponential manner? Is the data highly variable? Are there single or multiple peaks or other forms of unconventional behavior in the extravascular data? After dose-normalizing the data, do the curves superimpose? Is there a baseline concentration value? If so, is it constant or circadian or is it feedback regulated? There are more questions that can be asked in order to understand the data.

5. Pooling the Data. Ideally, a full structural model may be developed if there is a data rich situation. However, when data is sparse and/or highly variable the analyst can conduct naïve average data analysis (NAD, i.e. concentrations at different time points are averaged and mean data is fitted to the model) or create naïve pooled data sets (NPD, i.e. all observations are treated as if they are collected from one subject and this data is fit to the model). This may be good for a general trend but there is a major risk of masking individual behavior.

6. Data Transformation. In certain situations, there may be a need for data transformation i.e. convert curvilinear relationships to straight lines and make use of linear regression, or to fit percentage change from baseline versus actual data. However, this may be associated with risk of distorting the error distribution. For example, in order to evaluate the effect of a drug on the heart rate, the modeler may decide that it is more meaningful to correlate the concentrations to the percentage change in heart-rate from

baseline. Alternatively, it may be decided to mathematically characterize the baseline, and then model the effect of the drug on heart-rate without any data transformation.

7. Parameter Estimability. Estimability is a statistical property that requires a statement concerning the degree of uncertainty that the modeler is willing to accept and is dependent on the variation in the data, the study design, the model and the algorithm. While it is always faster to work with integrated equations, differential equations may also be used.

8. Initial Estimates. Initial estimates are very important when the data have a lot of scatter or range of independent variable collected such as concentrations may not cover the entire range of potential dependent variables such as effects. The initial estimates are also important when the data do not really fit a model and there are multiple minima or local minima in the sum of squares surface. To get the initial estimates the analyst may use graphic methods and linear regression or use data from non-compartmental analysis, refer to the literature value of relevant compounds, make an educated guess, specify reasonable lower and upper boundaries and try the physiological range.

9. Parameter Estimation. Various methods can be used for parameter estimation including linear and non-linear methods. The non-linear models include grid search, simplex procedure (Nelder-Mead), Gauss-Newton method (with various modifications), and the Marquardt procedure. For population PK/PD analysis, modeler may use maximum likelihood functions with first or second order expansions about the conditional estimates (NONMEM), generalized least square procedure with Taylor series expansion of condition estimates. Other procedures include estimated generalized least squares, iteratively reweighted generalized least squares, pseudo-maximum likelihood and pseudo-restricted maximum likelihood (MIXLIN, SAS).

10. A Check for Goodness of Fit. For goodness of fit, the following question should be answered. Does the model have biological relevance? Does the fitted curve mimic trends in the data? Are the parameters estimated with adequate precision? Do the residuals show a lack of systematic deviation? Do the residual plots display a random scatter?

6. SELECTION OF THE PK/PD MODEL

Selecting a PK/PD model is not an easy process. Problem may arise due the following reasons:

- "Model building is as much an art as it is a science. It involves intuition, imagination and skill[48]."
- "Models are, for the most part, caricatures of reality, but if they are good, then like good caricatures, they portray, though perhaps in distorted manner, some of the features of the real world" i.e. all models are wrong but some are useful[3, 49].

The answer to the question of what constitutes an appropriate model is complex. The model selected should be based on the assumptions made and the intended use of the model in decision-making. If the assumptions do not lead to a mechanistic model, an empirical model can be selected. In this case, the validation of the model predictability becomes especially important. The available data can also govern the types of models that can be used. The model selection process can be a series of trial and error steps. However, some of the features to consider in selection of the model should include the need of estimating the particular parameters such as E_{max}, E_{50}, E_0 and the ability of the sufficient data to provide those estimates. Different model structures or newly added or dropped components to an existing model can be assessed by visual inspection and tested using one of several objective criteria. New assumptions can be added when emerging data indicates that this is appropriate[4, 22]. The final selection of the model should be the simplest possible (parsimonious), have reasonable goodness of fit, and provide a level of predictability appropriate for its use in decision-making. Some of the goodness of fit statistics could include smaller sum of squares of residuals, small asymptotic standard deviation, smaller 95% confidence intervals, smaller Akaike's Information criteria (AIC), smaller Schwartz criteria, larger model selection criteria. Selection of models should always include plotting the data, i.e. plotting concentration (mean or composite) concentration versus time or effect versus time.

6.1 Clinical PK/PD Case Studies

Warfarin: Warfarin sodium is administered as a R,S racemate, belonging to family of coumarins used in the treatment of thrombotic disorders such as deep vein thrombosis, pulmonary embolism and myocardial infarction[19, 50]. Warfarin is completely absorbed, reaching a maximum plasma concentration between 2 and 6 hours. It distributes into a small volume of distribution (10 L/70kg) and is eliminated by hepatic metabolism with a very low clearance (0.2 L/h/70kg). The elimination half-life is about 35 hours. Warfarin inhibits Vitamin K epoxide reductase leading to a decrease of Vitamin K_1 thus inhibiting clotting factors such as Factor II, VII, IX and X. The pharmacodynamic endpoint for warfarin is prolongation of prothrombin time (PT) and the international normalized ration (INR)[19, 50]. It is difficult to correlate the warfarin dose to the PT because the effect is not immediate to dose administration. The reason is that warfarin inhibits the synthesis of Vitamin K-dependent clotting factors, but not on their degradation. There seems to be delay in the response that reflects the rate of metabolism of Factor II, VII, IX and X. Prothrombin complex synthesis is inhibited 50% at a warfarin concentration of about 1.5 mg/L[50]. Warfarin concentrations associated with therapeutic anticoagulation are of similar magnitude. There is a hyperbolic relationship between the activity of prothrombin complex and the prothrombin time. Four distinct pharmacodynamic models have been proposed: linear, log-linear, power and E_{max} for predicting the effect of warfarin and comparing them to empirical dose[50]. $F(W_c)$ is assumed to be a function of the warfarin concentration (W_c) that predicts the synthesis rate of clotting factors as a percentage of baseline value.

Linear model:
$$f_n(W_c) = 100 - S \bullet W_c \quad (18)$$

Power function:
$$f_n(W_c) = 100 - \left(1 - \frac{1}{S \bullet W_c^n}\right) \quad (19)$$

Log-Linear Function:
$$f_n(W_c) = 100 - S \bullet \left(\log(W_c) - \log(C_{\min})\right) \quad (20)$$

E_{max}:
$$f_n(W_c) = 100 - \left(1 - \frac{1}{IC_{50} \bullet W_c^n}\right) \quad (21)$$

$$E_{max}: \; f_n(W_c) = 100 - \left(1 - \left(1 + \frac{W_r}{IC_{50r}}\right)^n + \left(1 + \frac{W_s}{IC_{50s}}\right)^n\right) \qquad (22)$$

Where S is the slope, C_{min} is the intercept, IC_{50} is the concentration that gives 50% of the effect, and subscripts r and s represent the r and s enantiomers. Although the E_{max} model was superior on a theoretical basis, it was similar to other models based on performance.

Methlprednisolone. Data from a clinical study was used to simulate the concentrations of 20 mg of methylprednisolone in humans[51]. The study was performed in a randomized, cross-over design in eight healthy subjects (5 male, 3 female). The average age was 29 years, the average weight 71 kg. The purpose of the study was to evaluate the pharmacokinetics and pharmacodynamics of three corticosteroids, deflazacort after oral administration (30 mg) and compare the results with those after oral administration of 20 mg of methylprednisolone and 25 mg of prednisolone. Blood samples were obtained prior to drug administration and after 15, 30, 45, 60, 90, 120 and 150 minutes as well as 3, 4, 5, 6, 8, 10, 12, 14, 16, 18, 20 and 24 hours after drug administration. The samples were centrifuged; the harvested plasma was frozen immediately and stored at -20°C until analyzed. For the pharmacodynamic evaluation, differential white blood cell counts were performed before drug administration and 2, 4, 6, 8, 10, 12, 14, 16, 20 and 24 hours post dose. The free methylprednisolone concentrations (C_m) were simulated using a one-compartment body model equation with oral absorption (C_m=A*exp(-k_a*t)-exp(-k_e*t)). The estimates for A, k_a and k_e were 201.5 ng/mL, 0.55 h^{-1} and 0.366 h^{-1} for unbound methylprednisolone, where A is the hybrid constant, k_a is the absorption constant, k_e is the elimination constant and t is time after dose administration. The pharmacodynamics was also simulated and fitted to the various pharmacodynamic models.

The PK/PD data was fitted in a sequential manner and the concentrations were correlated to the effect using a linear relationship, followed by E_{max} and sigmoid E-$_{max}$ model. The table below shows that goodness of fit statistics were not appropriate.

The concentration versus effect relationship shows a clockwise hystersis loop and hence an indirect E model was explored. The goodness of fit statistics improved but the visual plots still indicated an improvement of the model was desirable. Next, the indirect response model with stimulation of output and inhibition of input were explored and the latter appeared to be the most appropriate model.

Table 5. Results of sequential fitting of PK/PD data for methylprednisolone.

Model	Parameters, r^2, Goodness of Fit	Pictorial Representation	Comments
Linear Model $Effect = S \bullet Ce + E_0$	• Effect=-1.1*C_e+82.1 • r^2=0.21		• Not a good fit. • Model misspecificat ion.

Continued on the next page

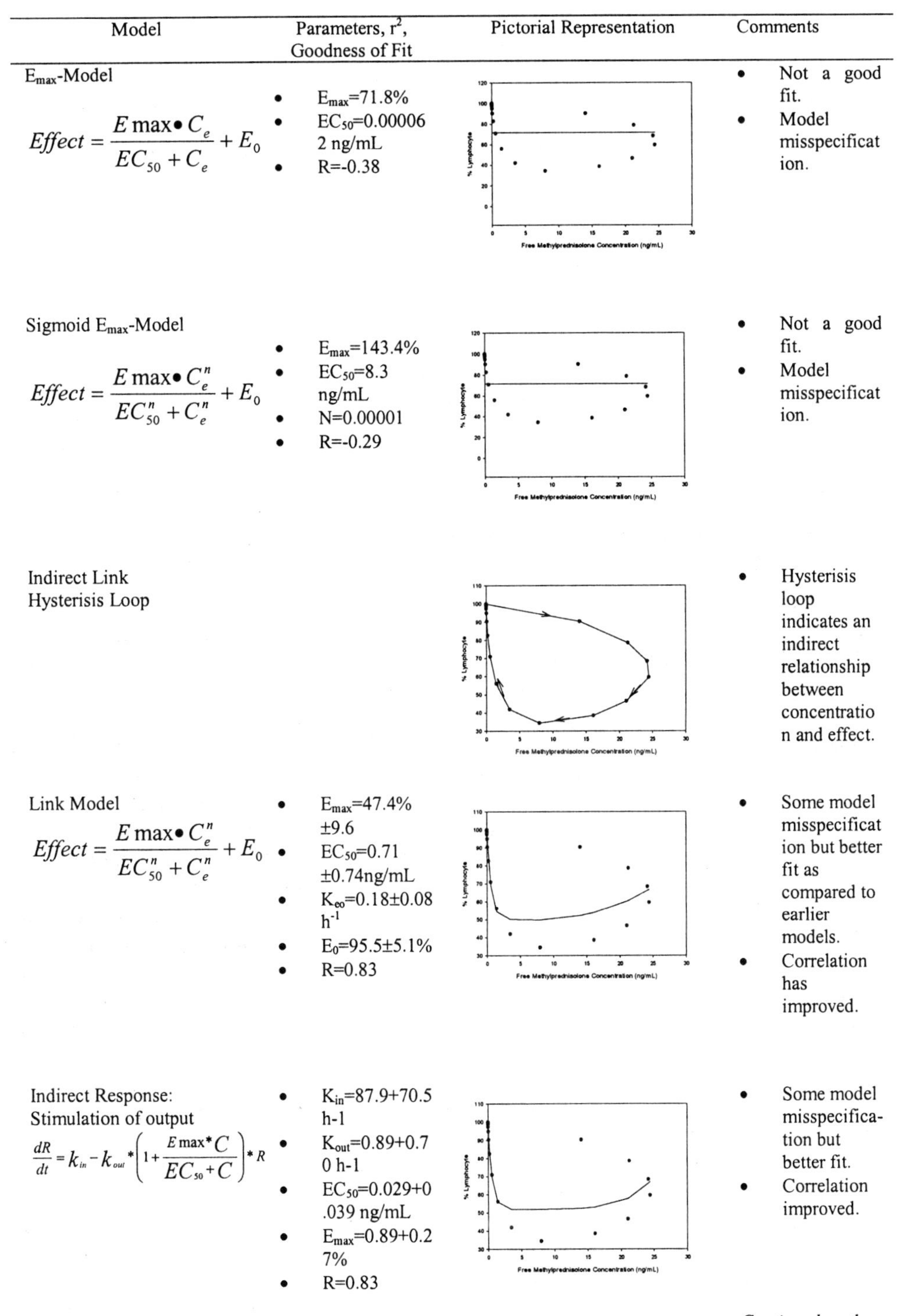

Model	Parameters, r^2, Goodness of Fit	Pictorial Representation	Comments
E_{max}-Model $Effect = \frac{E\max \bullet C_e}{EC_{50} + C_e} + E_0$	• E_{max}=71.8% • EC_{50}=0.00006 2 ng/mL • R=-0.38		• Not a good fit. • Model misspecificat ion.
Sigmoid E_{max}-Model $Effect = \frac{E\max \bullet C_e^n}{EC_{50}^n + C_e^n} + E_0$	• E_{max}=143.4% • EC_{50}=8.3 ng/mL • N=0.00001 • R=-0.29		• Not a good fit. • Model misspecificat ion.
Indirect Link Hysterisis Loop			• Hysterisis loop indicates an indirect relationship between concentratio n and effect.
Link Model $Effect = \frac{E\max \bullet C_e^n}{EC_{50}^n + C_e^n} + E_0$	• E_{max}=47.4% ±9.6 • EC_{50}=0.71 ±0.74ng/mL • K_{eo}=0.18±0.08 h^{-1} • E_0=95.5±5.1% • R=0.83		• Some model misspecificat ion but better fit as compared to earlier models. • Correlation has improved.
Indirect Response: Stimulation of output $\frac{dR}{dt} = k_{in} - k_{out} * \left(1 + \frac{E\max * C}{EC_{50} + C}\right) * R$	• K_{in}=87.9+70.5 h-1 • K_{out}=0.89+0.7 0 h-1 • EC_{50}=0.029+0 .039 ng/mL • E_{max}=0.89+0.2 7% • R=0.83		• Some model misspecifica- tion but better fit. • Correlation improved.

Continued on the next page

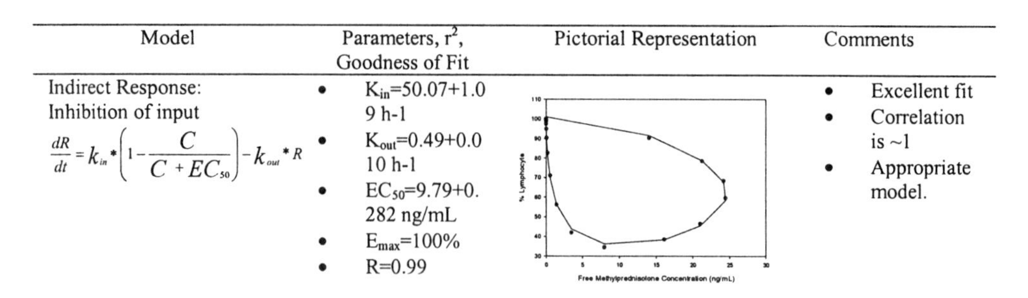

Model	Parameters, r^2, Goodness of Fit	Pictorial Representation	Comments
Indirect Response: Inhibition of input $\frac{dR}{dt} = k_{in} * \left(1 - \frac{C}{C + EC_{50}}\right) - k_{out} * R$	• K_{in}=50.07+1.09 h-1 • K_{out}=0.49+0.010 h-1 • EC_{50}=9.79+0.282 ng/mL • E_{max}=100% • R=0.99		• Excellent fit • Correlation is ~1 • Appropriate model.

S = Slope, C_e = Concentration at the Effect Site , E_o = Baseline Effect E_{max} = Maximum Effect, EC_{50} = Concentration that Produces 50% of the Effect, n = Sigmoidity Factor that Determines the Shape of the Curve or shape factor, D = Dose, K_a = Absorption Rate Constant, K = Elimination Rate Constant from the Central Compartment, K_{eo} = Elimination Rate from the Effect Compartment, T = Time, k_{in} = Input Constant, k_{out} = Output Constant, R= Response in Relation to Time, 1 = $Drug_1$ and 2 = Drug $_2$, I(t) = Inhibition Function, S(t) = Stimulation Function.

7. VALIDATION OF THE PK/PD MODEL

With regards to the validation of a PK/PD model, the FDA guidance provides a comprehensive approach[22]. The guidance states "the issue of model validation is not totally resolved. Models are never completely valid, but the application of a model to a specific purpose can be checked or evaluated. The main role of models is not so much to explain and predict -- though ultimately these are the main functions of science -- as to polarize thinking and to pose sharp questions. The answer to the question of what constitutes an appropriate model is complex[22]".

It is important to note that predictive power of the final models derived from the study results is a function of study design factors, number of subjects and data sampling plan. Hence, the studies should be designed robustly to allow proper PK/PD modeling of data and to obtain a predictive model with accurate and precise model parameter estimations that are insensitive to model assumptions. In order to facilitate adequate PK/PD data acquisition from a study, models can be identified based on prior knowledge of the drug or the class of drugs and various assumptions. The predictive power can be estimated through simulation of study outcomes, using the estimates and distributions of pharmacokinetic, pharmacodynamic, and study design parameters. During the analysis stage of a study, models can be validated based on internal and/or external data and also estimate the predictive power of the model. The common method for estimating predictability is to split the data set into two parts. One set of the data should be used to build the model and the second set of data should test the predictability of the resulting model. The predictability is especially important when the model is intended to provide supportive evidence for primary efficacy studies, address safety issues, support new doses and dosing regimens in new target populations or sub-populations defined by intrinsic and extrinsic factors or when there is a change in dosage form and/or route of administration.

While predictive performance is perhaps best estimated via the projection of modeling results from one dataset onto another via data splitting techniques or an external validation dataset (i.e., separate study), resampling techniques are valuable when such data is not available[52, 53]. As mentioned above, these techniques, coupled with simulations to test design constraints, parameter identifiability and sensitivity and dose regimen manipulations are extremely important in defining the space across which the proposed model may be further explored. The combination of resampling, simulation and prediction onto new datasets help describe space within which the model can be applied and ultimately define the validity of a model.

8. PK/PD DESIGN FEATURES

Despite the evolution of sophisticated analysis techniques and models, the design of trials hoping to develop PK/PD correlations has not always been optimal. Much of this can be attributed to the lack of *a priori* defined objectives defining the PK/PD analysis. Hence, it has often been lumped into "exploratory data analyses" which carry less lofty importance with respect to study outcomes. In addition, the collection of measured responses is not always in alignment with the relevant exposure metric or often the desired time course of pharmacologic/pharmacodynamic effect study. Hence, at the protocol creation phase, there are four general principles, which should guide an effective PK/PD study design.

1. Define the PK/PD assumptions as described previously. These should be incorporated into the background section of any protocol or incorporated into the Investigator's Brochure in a section describing the drug actions.
2. Define study hypotheses and objectives pertaining to the PK/PD approach. These also should be stated in the protocol along with the appropriate statistical analysis plan.
3. Define an event schedule that captures the most meaningful data (i.e., consistent with the time course of exposure and effect response stated in your assumptions and hypotheses). Do not obligate an analysis plan when appropriate samples cannot be collected. Study conduct difficulties (clinical support during off hours, sampling time or frequency, other logistical issues) are not an excuse to construct or conduct a poor study that will not fulfill its objectives.
4. Simulate expected results based on assumptions. If possible, these simulations should also be included in the protocol, especially if the response measure has been utilized for other agents where prior knowledge can be incorporated into the simulation. Use a stochastic approach where possible (make assumptions and challenge them).

9. DIVERSITY IN PK/PD ACROSS THERAPEUTIC AREAS / DRUG CLASS

The role of PK/PD modeling is important and this tool has been applied to a variety of therapeutic areas to either correlate plasma concentrations to effects, side-effects, biomarkers or surrogate end-points. In the case of corticosteroids, indirect response models have been used to predict the systemic side effects of these drugs such as lymphocyte suppression and granulocyte induction[40, 51]. Cortisol suppression after administration of exogenous corticosteroids has also been modeled after mathematically modeling the complex circadian rhythm of cortisol release using cosine function or linear release-rate model and using the indirect response model to express cortisol suppression[40, 54]. Attempts have also been made to correlate the systemic EC_{50} for these effects with relative glucocorticoid receptor affinity[40]. Similarly, Hochhaus et al have correlated the effect of beta-agonists on heart rate (side-effect) and improvement in forced expiratory volume in 1 second (FEV_1, desired effect)[19]. For antibiotics, Dalla Costa et al used hard-linked PK/PD models to predict the effect of tazobactam on antimicrobial actvity of piperacillin[41]. For analgesics and anesthetics, sigmoid E_{max}-models have been used to predict the respective effects using EEG outputs[19]. In the field of cardiovascular disease, area under the effect time curve (AUE) for inhibition of exercise-induced tachycardia after administration of RR-labetalol has been modeled using the E_{max}-model[55]. A modified E_{max}-model has been used to predict the inhibition of platelet aggregation by GPIIb/IIIa antagonists[56]. Although it is almost impossible to summarize all the examples in the literature, a more exhaustive list of attempts of PK/PD modeling in various therapeutic areas is summarized in Table A.1.

10. ROLE OF PK/PD BY PHASE AND ADVANTAGES TO DRUG DEVELOPMENT

The cost and time of developing new drugs is increasing with the latest estimates claiming that it takes 10-12 years and $600 million to develop a drug[57, 58]. These figures may be explained in part, by the highly regulated nature of the pharmaceutical industry that requires additional and well-designed safety and efficacy data. This has lead to increased number of studies, repetition of studies and greater number of patients tested prior to final marketing. An important benefit to the patients who will ultimately receive such new medicines is the reduction in consumer risk afforded them with the increased attention to safety and efficacy. However, there is a real need to transform and accelerate the drug development process within the pharmaceutical industry. Also, there is a huge failure rate with approximately 50,000 compounds screened in one day but resulting in only 1 drug per year in an average big pharmaceutical company. Furthermore, with the explosion of genomics, proteomics, high-throughput screening and combinatorial chemistry, there are even more interesting targets and there is a need for efficiently selecting these targets.

Traditionally the pharmaceutical industry performs sequential testing of drug candidates by screening and selecting the best drug candidates at the preclinical, Phase I /Phase IIa (proof of concept stage) and Phase IIb /III (confirmation stage). The objective is to minimize the failures in the expensive Phase IIb/III part of drug development. Pharmacokinetic (PK) and efficacy data obtained during the development of most drugs can be used together with pharmacodynamic (PD) data to provide insight into the critical factors influencing the disposition and effects of the drug. In an attempt to expedite and scientifically support drug development process, PK/PD concepts can be applied at every development stage including preclinical screening, preclinical development, clinical development through post marketing, and line extension strategies (See Figure 3). At the preclinical stage, in vivo potency, intrinsic activity, identification of biological or surrogate markers, identification of dose and dosage form can be optimized using PK/PD studies. At clinical stage, dose-concentration-effect and side effect relationships can be explored and or established, along with evaluation of sources of PK/PD variability, i.e. effect of food, age, gender, drug-drug interaction and tolerance development. Predictive PK/PD can also provide simulations of drug responses and thus necessitate fewer and more-focused studies. The increased application of PK/PD concepts in all stages of drug development should improve quality of information, create efficient decision-making, and cost effectiveness of current development programs. As stated in an section earlier, based on the new draft guidance by the FDA, there is an expectation for PK/PD modeling in various stages of drug development as summarized in Table 4. Table 6 describes the utility of PK/PD in various parts of drug development.

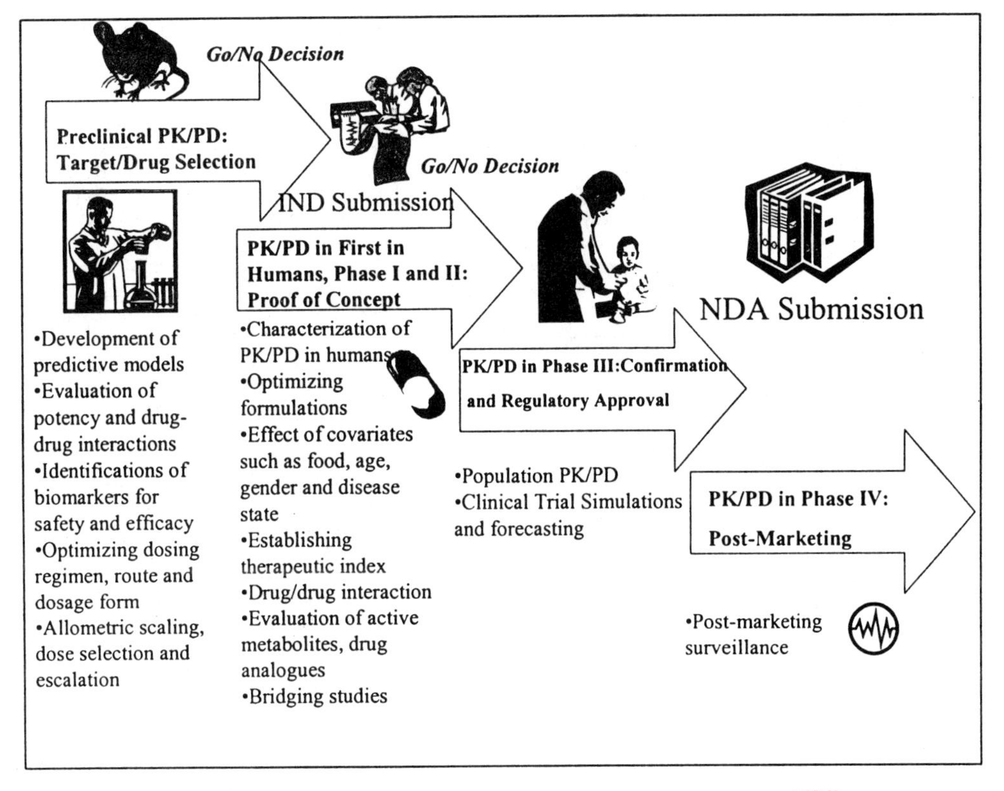

Figure 3. Scheme for incorporation of PK/PD studies in drug development[1, 27, 59].

Table 6. PK/PD studies performed during the course of drug development

Study	Objective	Example
	PK/PD in Preclinical Development (Target/Drug Selection)	
• Identification of potential surrogates and animal models for efficacy and toxicity	• Identification of relevant endpoints and thus biomarkers or receptor occupancy to be tested early in development for potential use and surrogates or endpoints.	• For antisicling agent Tucaresol, proportion of hemoglobin modified to high affinity for was used as PD marker[60]. • Use of inhibition of *ex-vivo* platelet aggregation as surrogate for PD activity for GPIIb/IIIa antagonists[61]. • Identification of guinea pigs as appropriate *in vivo* species to investigate PK/PD of GPIIb/IIIa antagonists[62].
• Development of mechanism based models for efficacy and toxicity	• Appreciate underlying principles. • Accurate characterization of dose-concentration-effect relationship. • Gives measure of potency and intrinsic activity based on concentration rather than dose. • Allows modulation of physiological, pathophysiological parameters. • Explore tolerance and/or sensitization.	• Physiological correlation identifying major determinants of effects of humanized anti-factor IX monoclonal antibody (SB249417) in monkeys, thus providing guidance for new drug in development for anti-IX activity in humans[63]. • Physiologically based immunodynamic model for monoclonal antibody (mAB5c8) in monkeys to provide guidance for development of drugs in humans[64]. • Mechanism-based model for agonism for evaluating adenosine A1 receptor-mediated effect of N^6-cyclopentyl-adenosine analogs on heart rate in rats[65].
• Evaluation and prediction of *in vivo* potency and intrinsic activity	• Unbound effective concentrations is usually similar in animals and humans, although inter-individual variability is large in PD.	• Similar EC_{50} for opioids in rats and humans. A PK/PD model for EEG effect of active metabolite of remifentanil in rats displayed lack of PD interaction with parent compound and data could be extended to humans[66]. • EC_{50} of corticosteroids based on cortisol suppression, granulocyte induction and lymphocyte suppression can be correlated to relative in vitro glucocorticoid receptor affinity[40].
• Evaluation of Drug Interaction	• An established PK/PD relationship can help in evaluation of drug-drug, drug metabolite or drug protein binding interactions.	• Prediction of EEG effect of multiple combination of competitive antagonists benzodiazepine and flumazenil in rats[67]. • Using *in vitro* measurements in PK/PD models to predict effect of tazobactam on antimicrobial actvity of piperacillin[41].
• Dosage form and dosing regimen optimization	• Previously developed PK/PD relationships may help in optimizing dosage forms and drug delivery systems.	• Oral selegiline gave low and variable concentrations but higher concentrations were needed to obtain anti-parkinson effect, leading to development of transdermal selegiline patch that by-passed first-pass metabolism to give high concentrations of selegiline[68]. • Using PK/PD model to design dosage form that can overcome hemodynamic tolerance following continuous nitroglycerin administration[69].

Continued on the next page

Study	Objective	Example
• Supporting decision making process	• Provide real-time implementation of PK/PD modeling in preclinical drug development to provide sufficient information for go/no go decision thus optimizing resource allocation.	
• PK/PD at transition from preclinical to clinical studies	• Selection of doses and dose escalation for First in Human studies. • Effective concentrations may be similar across species. • Pharmacokinetic parameters vary across species and can be predicted using physiological based PK models, allometric scaling and with adjustment such as the Campbell method.	• For a GPIIb/IIIa antagonist, the target dose was predicted as a product of clearance, in vitro EC_{50} and dosing interval. Clearance was scaled using allometric scaling[28]. • Allometric scaling and PK/PD modeling was used to select doses for Phase I studies with recombinant P-selecting antagonist (rPSGL-Ig)[70].
	PK/PD During Phase I, Phase II (Proof of Concept)	
• Evaluation of dosage form and routes of administrat-ion	• Assessment of impact on route of administration and release rates of drug on effect *in vivo* in humans.	• PK/PD analysis to compare the suppressive effect of cetrorelix (luteinizing hormone-releasing antagonist) on testosterone and luteinizing hormone after iv and s. administration [71]. • Interluking-10 had better delivery after subcutaneous administration as compared to intravenous administration [72].
• Food effect	• Food may increase decrease or keep the effect the same for any drug. • Food may physically or chemically interact with a dosage form and may change the solubility of a drug substance or drug product. • Food may delay gastric empty and change gastrointestinal (GI) pH • Food may increase bile flow and splanchnic blood flow. • Change GI metabolism of a drug. • Have physiological changes in the GI.	• PK/PD model evaluated that food decreased blood pressure for controlled release formulation of nisoldipine[73]. • Food decreased the C_{max} of phosphodiestrase IV inhibitor, RP73401 by ~51%[74].
• Gender effect	• Gender effects may be true. • Some gender effect may be attributed to differences in weight • Gender effect may be seen in drug with inhaled administration also where there are differences in inspiration rate and lung capacity.	• Females and asthmatic subjects had lower cyclosporine concentrations after inhalation probably due to lower inspiratory flow rates[75]. • No effect of gender on the PK of phosphodiestrase IV inhibitor, RP73401[74]. • PK, not PD of pyridostigmine had gender effect on acetylcholinestrase activity[76]. • Women have higher baseline QTc than males[77].

Continued on the next page

Study	Objective	Example
• Special population	• PK/PD mode to identify pharmacological differences and their underlying cause in sub-populations	• Use PK/PD model to determine the required acetaminophen concentration for analgesic effect in children after tonsillectomy[78].
• Therapeutic index	• Simultaneous characterization of dose-concentration-effect relationship for therapeutic as well as adverse effect by using PK/PD model to allow adequate dose selection.	• Therapeutic index of inhaled corticosteroids can be determined by identifying the dose that gives adequate increase in asthma control as determined by increase in FEV_1 with minimal adverse effect as determined by cortisol suppression[40, 79, 80].
• Characterization of active metabolites	• PK/PD modeling may quantify the *in vivo* potency of metabolites and their contribution to overall effect. • Suggest presence of previously unknown metabolites	• Klerval, a GPIIb/IIIa antagonist, and its active metabolite were modeled using a competitive E_{max} model to describe the inhibition of platelet aggregation[56]. • PK/PD modeling described that direct sinus node inhibitor ivabradine and its N-dealkylated metabolite contribute to the bradycardiac effect of the drug[81].
• Drug-drug interaction	• Using PK/PD models to evaluate potential drug-drug interactions and the PK and PD parameters affected.	• No PK interaction between formoterol and theophylline, but noncompetitive PD interaction for eosinopenic and hypokalemic effects[82]. • No PK interaction between paclitaxel and carboplatin but paclitaxel has platelet sparing effect on thrombocytopenia of carboplatin[83].
• Drug-disease interaction	• PK/PD techniques can elucidate the changes in various parameters caused by disease states	• Females and asthmatic subjects had lower cyclosporin concentrations after inhalation probably due to lower inspiratory flow rates[75]. • Renal impairment decreased clearance of GPIIb/IIIa inhibitor, lamifiban but patients were sensitized to its anti-platelet effect[84].
• Tolerance or sensitization	• PK/PD model can characterize the time course and extent of tolerance or sensitization	• Tolerance development to acid inhibitory effect of ranitidine increase EC_{50} by 100% within 6-10 h after prolonged intravenous therapy[85].
• Discrimination of lead candidate	• Use established PK/PD models for a class of compounds with same mechanism of action to define the therapeutic index of a new compound based on PK and in vitro PD properties.	• Using integrated PK/PD models to the systemic PD effects of new corticosteroids using PK and in vitro receptor binding data[47]. • Opioid trefentanil had faster recovery from opioid effects as compared to fentanyl and alfentanil using EEG changes as PD markers[86].
• Bridging studies	• During global drug development bridging PK/PD studies may be performed to allow switch from formulation A to formulation B, adult population to pediatric population, Japanese to caucasian population and one route of elimination to another.	• Using PK/PD model it was established that same dosing principle should be used for treatment of type-2 diabetes in Chinese and Caucasian populations with intravenous glibenclamide[87].
• Bioequivalence of drugs	• Bioequivalence is defined as the absence of a significant difference in the rate and extent to which the	

Continued on the next page

Study	Objective	Example
	active ingredient or active moiety in pharmaceutical equivalents or pharmaceutical alternatives becomes available at the site of drug action when administered at the same molar dose under similar conditions in an appropriately designed study. • After formulation is changed then bioequivalence may be established using pharmacokinetic data with AUC and C_{max} as major endpoints. • Bioequivalence may be established using in vitro metrics, pharmacokinetics, clinical/therapeutic endpoints or pharmacodynamic markers.	• Pharmacodynamic studies are not recommended for orally administered drug products when the drug is absorbed into the systemic circulation and a pharmacokinetic approach can be used to assess systemic exposure and establish BE[22, 88, 89, 90, 91]. • FDA has special bioequivalence guidance for orally administered drugs such as clozapine tablets, potassium chloride slow release tablets and capsule and phenytoin tablets[92, 93, 94].
	PK/PD in Phase III (Confirmation and Regulatory Approval)	
• Population PK/PD analysis	• Applied to clinical and preclinical data but mainly to Phase IIb, Phase III data where sparse PK/PD sampling is available in a large, diverse population with efficacy and safety endpoints.	• PK/PD model was used to evaluate the population PK parameters of inhaled corticosteroid ciclesonide and its lack of cortisol suppression[95]. • Population PK of GPIIb/IIIa antagonist was evaluated in several patient populations and the effect on inhibition of platelet aggregation was also evaluated [28, 96].
• Clinical trial simulation	• PK/PD model along with Monte Carlo simulations used as predictive tools to predict outcomes of future clinical trials.	• PK/PD based simulation were used to design a trial for oral anticancer drug and appropriate does were selected[97]. • PK/PD model including cytokinetic processes and disease progression that described development of drug resistance towards reverse transcriptase inhibitors and aspartyl protease inhibitors during therapy of HIV inspection was used to optimize the ways administering this drug[98].
• Preparation and review of regulatory documents	• PK/PD integrates information from preclinical, Clinical Phase I-III along with sub-population and other demographic information to perform simulations for various scenarios and helps reviewer to gain a better understanding of dose selection and characteristics of the drug.	
	PK/PD in Post Marketing	
• Line extensions and new indications	• If underlying PK/PD model is known it is easier to predict the effect after formulation change or to select doses for a new indication.	

Continued on the next page

Study	Objective	Example
		• Oral selegiline used as an adjunct with levodopa for treatment of Parkinson's disease gave low and variable concentrations but higher concentrations were needed for treatment as mono-therapy for Parkinson's Disease and Alzhiemer's Disease, leading to development of transdermal selegiline patch that by-passed first-pass metabolism to give high concentrations of selegiline [68].
• Post-marketing surveillance	• Integration of PK/PD modeling with post marketing surveillance should be encouraged.	• Population PK parameters for tianeptine obtained in developmental studies did not accurately predict drug concentrations obtained in-patients receiving drug in post marketing[99].

11. CHALLENGES IN PK/PD MODELING

There are several challenges with conduct and implementation of PK/PD. The first problem is that plasma concentrations may not reflect the concentrations at the effect site. This leads to disconnect between plasma concentrations and the effect. Usually these types of effects leads to a hysteresis or a proteresis loops. These can be resolved by using indirect link models.

Another major problem is the validation of pharmacodynamic markers[26]. Even though great care is taken in validating the bioanalytical assays for the drug or metabolite concentrations, very little effort is often made in validating pharmacodynamic markers. Complicating matters is the variety of pharmacodynamic markers ranging from endogenous substances, laboratory evaluations to adverse events of efficacy end-points. Many of these endpoints may not be reproducible due to either the inter- and intra-subject variability or due to the variability in the measurements by various techniques and even lack of standardization.

In order to adequately describe the concentration-effect relationships, complicated PK/PD models may have to be developed that require the understanding of multiple receptors, concentration-dependent effects, presence of active metabolites, partial agonists, agonist/antagonist relationships, presence of endogenous agonists or antagonists, baseline effects and baseline characterization (e.g. circadian rhythm of cortisol)[54]. There is always a problem of development of tolerance that may be related to down-regulation of receptors or change in the sensitivity of the receptors. This may lead to a proteresis loop. This kind of tolerance can be observed with morphine[4]. On the other side there may be a problem of sensitization or up-regulation of receptors. Feedback mechanism such as those seen with cortisol and ACTH may require more sophisticated modeling techniques[54]. Other factors such as genetic and environmental differences in response may increase pharmacodynamic variability. Even the status or disease state may modify pharmacodynamic outcome. In many disease states such as drug controlling blood pressure as in case of ACE inhibitors, there the pharmacodynamic effect can only be observed in disease state and not in healthy volunteers[4, 19]. In cases where there are difficulties in distinguishing the variability associated with the pharmacokinetics or pharmacodynamics, population approaches to dissect these sources of variability may be used.

The most challenging aspect of PK/PD modeling is the correlation of the PK/PD model with safety or efficacy outcomes. This involves a detailed knowledge of pathophysiology of disease and modeling of disease progression. Most of the advances in disease stage progression are in form of epidemiological studies but they have not been correlated to specific modulations in any of the underlying factors. Only in the case of cardioavasuclar disease have the mortality outcomes been correlated to decrease in cholesterol levels or decrease in systolic or diastolic blood pressure. In the case of HIV anti-viral agents, there is a correlation between viral load and efficacy. Jackson described a PK/PD model including cytokinetic processes and disease progression that characterized development of drug resistance towards reverse transcriptase inhibitors and aspartyl protease inhibitors during therapy of HIV inspection that was used to optimize the ways administering this agent[98]. However, the modeling of disease progression is still at the stage of infancy and not having adequate disease models in animals also adds to the challenge. This has lead to the unfortunate outcome of many agents that have failed in humans that worked very well in animal models.

12. SUMMARY AND DISCUSSION

With the advent of sophisticated software, training of scientists, evolution of population PK/PD and simulation, it is evident that the field of PK/PD has come of age. However, it still needs to evolve and grow into a major field of study. Integrating knowledge generated from improvements in pharmacogenomics, pharmacogenetics, disease modeling, clinical trial designs and surrogate markers will further enhance this field of study[7]. In this short period of time, PK/PD modeling helps scientists from all disciplines that are involved in drug development to combine their efforts in an efficient and coordinated manner in order to understand the underlying principles of drug action and behavior. This diverse data can be captured in a knowledge base that can be expressed mathematically and can be applied to answer various hypothetical scenarios. It is aiding the evolution of drug development process from an empirical, check-box or cookbook approach to a question-driven sophisticated process. The acceptance of PK/PD modeling and its growing importance can be highlighted by the issuance of draft guidance on concentration-effect relationship by the US FDA. This will hopefully spur more research in the PK/PD arena, leading to the conduct of fewer, but more informative trials.

Explanation of PK/PD terms used:

Abbreviation	***Explanations***
A	Hybrid constant
AUC	Area under the concentration.time-time curve
AUC	Area under the concentration-time curve
AUC_E	Area under the effect-time curve
C_e	Concentration at the effect site
C_{max}	Maximal concentration
C_{maxE}	Maximal effect
C_m	Free methylprednisolone concentration
C_{min}	Minimum concentration
Cl/CL	Clearance
Css	Steady-state plasma concentration

D	Dose
D_g	Drug
D_{50}	Dose necessary to produce and maintain 50% of the maximum effect
DR	Drug-receptor complex
DR_{50}	Dosing rate that produces and maintain 50% of the maximum effect
EEG	Electroencephalogram
EC_{50} /E_{50}	Concentration that Produces 50% of the Effect
E_{max}	Maximum effect
E_o	Baseline effect
F	Bioavailability
F_b	Fraction bound to plasma proteins=1-Fu
fn	Function
FDA	United States Food and Drug Administration
F_u	Fraction unbound to plasma proteins
GI	Gastrointestinal
HIV	Human Immunodeficiency Virus
I(t)	Inhibition function
IM	Intramuscular administration
IND	Investigation New Drug
iv/IV	Intravenous administration
K, Ke	Elimination rate constant from the central compartment
K_a	Absorption rate constant
K_d	Equilibrium constant between drug, receptor and drug-receptor complex
K_{eo}	Elimination rate from the effect compartment
k_{in}	Input constant
k_{out}	Output constant
L	Liter
MAT	Mean absorption time
MEC	Mean effect concentration at the time re-medication is required
mL	Milliliter
MRT	Mean residence time
n	Sigmoidity factor that determines the shape of the curve or shape factor
NDA	New Drug Application
PD	Pharmacodynamics
PK	Pharmacokinetics
p.o	Per os; oral administration
Prn	Pro re nata; as needed
R	Response in relation to time
R_c	Free receptors
R,S	R and S enantiomers
R_{tot}	Total number of receptors
S	Slope
Sc	subcutanoues
S(t)	Stimulation function
T, t	Time
T_{max}	Time when maximal concentration is observed
T_{maxE}	Time when maximal effect is observed
WBC	White blood cells
W_c	Warfarin concentration

References

1. Meibohm B, Derendorf H. Basic concepts of pharmacokinetic/pharmacodynamic (PK/PD) modelling. Int J Clin Pharmacol Ther 1997 Oct;35(10):401-13.
2. Derendorf H, Meibohm B. Modeling of pharmacokinetic/pharmacodynamic (PK/PD) relationships: concepts and perspectives. Pharm Res. 1999 Feb;16(2):176-85.
3. Gabrielsson J and Weiner D. Pharmacokinetic and pharmacodynamic data analysis: Concepts and applications. 3rd Edition., Apotekarsocieteten, Swedish pharmaceutical society, 2000.

4. Kwon Y. Handbook of essential pharmacokinetics, pharmacodynamics and drug metabolism for industrial scientists. Kluwer Academic/Plenum Publishers, New York, 2001.
5. Physician's Desk Reference, (56), Medical Economics Company, Inc., Montvale, NJ 2002.
6. Lesko LJ, Rowland M, Peck CC, Blaschke TF. Optimizing the science of drug development: opportunities for better candidate selection and accelerated evaluation in humans. Pharm Res. 2000 Nov;17(11):1335-44. Review.
7. Hochhaus G, Barrett JS, Derendorf H. Evolution of pharmacokinetics and pharmacokinetic/dynamic correlations during the 20th century. J Clin Pharmacol. 2000 Sep;40(9):908-17.
8. Teorell T. Kinetics of distribution of substances administered to the body I. The extravascular modes of administration. Arch Int Pharmacodyn Ther 1937: 57: 202-225.
9. De Jongh SE, Wijnans M. The influence of divided doses of drugs on the duration of effect and integral of effect. Acta Physiol Pharmacol Neerl 1950: 1: 237-255.
10. Garrett ER, Thomas R, Wallach D, Always C. Psicofuranine: kinetics and mechanisms in vivo with the application of the analog computer. J Pharmac Exp Ther 1960: 140: 106-118.
11. Levy G. Kinetics of pharmacological effects. Clin Pharm Ther 1966: 7: 362-372.
12. Wagner JG. Kinetics of pharmacologic response. I. Proposed relationships between response and drug concentration in the intact animal and man J Theor Biol 1968: 20: 173-201.
13. Sheiner LB. Computer aided long-term anticoagulation therapy. Comput Biomed Res 1969: 10: 22-35.
14. Gibaldi M, Perrier D. Pharmacokinetics. New York: Marcel and Dekker, 1975.
15. Holford NHG, Sheiner LB. Understanding the dose-effect relationship: clinical application of pharmacokinetic-pharmacodynamic models. Clinical Pharmacokinetics 1981: 6: 429-453.
16. Boudinot FD, D'Ambrosio R, Jusko WJ. Receptor-mediated pharmacodynamics of prednisolone in rat. Journal of Pharmacokinetics and Biopharmaceutics 1986: 14: 469-493.
17. Peck CC, Barr WH, Benet LZ, Collins J, Desjardins RE, Frust DE, et al. Opportunities for integration of pharmacokinetics, pharmacodynamics, and toxicokinetics in rational drug development. Pharm Sci 1992: 81: 605-610.
18. Sheiner LB, Ludden TM. Population pharmacokinetics and pharmacodynamics. Annu Rev Pharmacol Toxicol 1992: 32: 185-209.
19. Derendorf H and Hochhaus G. Handbook of pharmacokinetic/pharmacodynamic correlation. CRC Press, Boca Raton, 1995.
20. Bruno R, Vivier N, Verginol J: A popuation pharmacokinetic model for Docetaxel (Taxotere): model building and validation. J Pharmacokinetic Biopharm 1996: 24; 153.
21. US FDA Guidance for Industry. Population Pharmacokinetics (Issued 2/1999, Posted 2/10/1999).
22. US FDA Guidance for Industry. Exposure-Response Relationships: Study Design, Data Analysis, and Regulatory Applications. DRAFT GUIDANCE. U.S. Department of Health and Human Services, Food and Drug Administration, Center for Drug Evaluation and Research (CDER), Center for Biologics Evaluation and Research (CBER), March 2002.
23. Gibaldi M, Perrier D. Pharmacokinetics. 2nd edition, New York: Marcel Dekker, 1982.
24. Rowland M, Tozer TN. Clinical Pharmacokinetics, Concepts and Application (third edition), Williams and Wilkins, Baltimore, 1995.
25. Barrett JS. Population pharmacokinetics in Pharmacokinetics in drug discovery, Schoenwald RD eds., CRC Press, 2002.
26. Colburn WA. Optimizing the use of biomarkers, surrogate endpoints, and clinical endpoints for more efficient drug development. Journal of Clinical Pharmacology 2000: 40, 1419-1427.
27. Meihbohm B., Derendorf H. Pharmacokinetic/pharmacodynamic studies in drug product development. Minireview. Journal of Pharmaceutical Sciences 2002; 18-31.
28. Chaikin P, Rhodes GR, Bruno R, Rohatagi S, Natarajan C. Pharmacokinetics/pharmacodynamics in drug development: an industrial perspective. J Clin Pharmacol. 2000 Dec;40(12 Pt 2):1428-38. Review.
29. CAST investigators (Cardiac Arrhythmia Suppression Trial Investigators): Preliminary report: effect of encainide and flecainide on mortality in a rondamized trial of arrhythmia suppression after myocardial infarction. N Engl J Med 1989: 321: 406-412.
30. The international Chronic Granulomatus Disease Cooperative Study Group: A controlled trial of interferon gamma to prevent infection in chronic granulomatous disease. N Engl J Med 1991: 324: 509-526.
31. Reidenberg MM, Goodman H, Erle H, Gray G, Lorenzo B, Leipzig RM, Meyer BR, Drayer DE. Hydromorphone levels and pain control in patients with severe chronic pain. Clin Pharmacol Ther 1988: 44: 376.
32. Mawer GE, Ahmad R, Dobbs SM, Tooth JA.. Experience with a gentamicin nomogram. Postgrad Med J. 1974 Nov;50 Suppl 7:31-2.

33. Weaver ML, Tanzer JM, Kramer PA.. Pilocarpine disposition and salivary flow responses following intravenous administration to dogs. Pharm Res. 1992 Aug;9(8):1064-9.
34. McDevitt DG, Shand DG. Plasma concentrations and the time-course of beta blockade due to propranolol. Clin Pharmacol Ther. 1975 Dec;18(06):708-13.
35. Lalonde RL, Straka RJ, Pieper JA, Bottorff MB, Mirvis DM. Propranolol pharmacodynamic modeling using unbound and total concentrations in healthy volunteers. J Pharmacokinet Biopharm. 1987 Dec;15(6):569-82.
36. Dingemanse J, Van Bree JBMM and Danhof M. Pharmacokinetic modeling of anticonvulsant response of Phenobarbital in rats. J Pharmacol Exp Ther 1989: 249: 601.
37. Schaefer HG, Heining R, Ahr G, Adelmann H, Tetzioff W, Kubimann J 1997. Pharmacokinetic-pharmacodynamic modelling as a tool to evaluate the clinical relevance of a drug-food interaction for nisoldipine controlled-release dosage form. Eur J. Clin Pharmacol 51: 473-480.
38. Suri A, Grundy BL, Derendorf H. Pharmacokinetics and pharmacodynamics of enantiomers of ibuprofen and flurbiprofen after oral administration. Int J Clin Pharmacol Ther. 1997 Jan;35(1):1-8.
39. Holford NHG, Sheiner LB. Understanding the dose-effect relationship: clinical application of pharmacokinetic-pharmacodynamic models. Clin. Pharmacokinet. 1981; 6: 429-453.
40. Rohatagi S. PhD Thesis. Pharmacokinetic and Pharmacodynamic modelling of methylprednisolone and prednisolone after single and multiple administration, 1995.
41. Dalla Costa T, Nolting A, Rand K, Derendorf H. Pharmacokinetic-pharmacodynamic modelling of the in vitro antiinfective effect of piperacillin-tazobactam combinations. Int J Clin Pharmacol Ther. 1997 Oct;35(10):426-33.
42. Jonkers RE, Braat MC, Koopmans RP, van Boxtel CJ. Pharmacodynamic modelling of the drug-induced downregulation of a beta 2-adrenoceptor mediated response and lack of restoration of receptor function after a single high dose of prednisone. Eur J Clin Pharmacol. 1995;49(1-2):37-44.
43. van den Meiracker AH, Manint Veld, AJ, Boomsma F, Fishberg DJ, Molinoff PB, Schalekamp MADH. Hemodynamic and beta-adrenergic receptor adaptation during long-term beta-adrenoreceptor blockade. Circulation 1989: 80: 903.
44. Antal EJ, Pyne DA, Starz KE, Smith RB. Probability models in pharmacodynamic analysis of clinical trials, in Pharmacokinetics and pharmacodynamics. Vol 2, Currents problems, potential solutions, Kroboth PD, Smith RB, Juhl RP eds., Harvey Whitney Books, Cincinnati, 1988, 219.
45. Holford, N.H.G. and Sheiner, L.B., Kinetics of pharmacologic response., in Pharmacokinetics: Theory and Methodology, Rowland, M. and Tucker, G. Editors.,Pergamon Press, New York, 1985, 189-212.
46. Matthews JC. Fundamentals of receptor, enzyme and transport kinetics. CRC Press, Boca Raton, 1993.
47. Derendorf H, Hochhaus G, Mollmann H, Barth J, Krieg M, Tunn S, Mollmann C. Receptor-based pharmacokinetic-pharmacodynamic analysis of corticosteroids. J Clin Pharmacol. 1993 Feb;33(2):115-23.
48. Shahin M, Iyengar SS, Rao RM. Computers in simulation and modeling of complex biological systems. CRC Press, Boca Raton, 1985.
49. Kac M. Some mathematical models in science. Science. 1969 Nov 7;166(906):695-9.
50. Holford NHG. Clinical pharmacokinetics and pharmacodynamics of warfarin: Understanding the dose-effect relationship. Clinical Pharmacokinetics 1986: 11: 483-504.
51. Mollmann H, Hochhaus G, Rohatagi S, Barth J, Derendorf H. Pharmacokinetic/pharmacodynamic evaluation of deflazacort in comparison to methylprednisolone and prednisolone. Pharm Res. 1995 Jul;12(7):1096-100.
52. Gisleskog PO, Hermann D, Hammarlund-Udenaes M, Karlsson MO. Validation of a population pharmacokinetic/pharmacodynamic model for 5a-reductase inhibitors. Eur J Pharm Sci 1999: 8: 291-299.
53. Gobburu JVS, Lawrence J. Application of resampling techniques to estimate exact significance levels for covariate selection during nonlinear mixed effects model building: Some inferences. Pharm Res 2002: 19(1): 92-98.
54. Rohatagi S, Bye A, Mackie AE and Derendorf H. Mathematical Modeling of Cortisol Circadian Rhythm and Cortisol Suppression. Eur. J. Pharm. Sci. 4: 341-350, 1996.
55. Lalonde RL, O'Rear TL, Wainer IW, Drda KD, Herring VL, Bottorff MB. Labetalol pharmacokinetics and pharmacodynamics: evidence of stereoselective disposition. Clin Pharmacol Ther. 1990 Nov;48(5):509-19.
56. Zannikos PN, Rohatagi S, Jensen BK, DePhillips S, Massignon D, Calic F, Sibille M, Kirkesseli S. Pharmacokinetics, Pharmacodynamics and Safety of a Platelet GPIIb/IIIa Antagonist, RGD891, Following Intravenous Administration in Healthy Male Volunteers. J. Clin Pharmacol., 40: 1245-1256, 2000.
57. Grabowski H, Vernon J, DiMasi JA. Returns on research and development for 1990s new drug introductions. Pharmacoeconomics 2002;20 Suppl 3:11-29.
58. DiMasi JA. The value of improving the productivity of the drug development process: faster times and better decisions. Pharmacoeconomics 2002;20 Suppl 3:1-10.

59. Peck CC, Barr WH, Benet, LZ, Collins D, R.E., Furst DE, Harter JG, Levy G, Ludden T, Rodman JH, Sanathanan L, Schentag JJ, Shah VP, Sheiner LB, Skelly JP, Stanski DR, Temple RJ, Vishwanathan CT, Weissinger J and Yacobi A. Opportunities for the integration of pharmacokinetics, pharmacodynamics, and toxicokinetics in rational drug development, Drug Development 1994: 34, 111-119.
60. Rolan PE, Mercer AJ, Wootton R, Posner J. Pharmacokinetics and pharmacodynamics of tucresol, an antisickling agent, in healthy volunteers. Br J Clin Pharmacol 1995: 39: 375-380.
61. Narjes H, Muller TH, Weisenberger H, Guth B, Brickl R. Inhibition of platelet aggregation as a surrogate marker. J Clin Pharmacol 1997: 37: 59S-64S.
62. Barrett JS, Yu J, Kapil R, Padovani P, Brown F, Ebling WF, Corjay MH, Reilly TM, Bozarth JM, Mousa SA, Pieniaszek HJ Jr. Disposition and exposure of fibrogen receptor antagonist XV459 on alphaIIbeta3 binding sites in the guinea pig. Biopharm Drug Dispos 1999: 20: 309-318.
63. Benincosa LJ, Chow FS, Tobia LP, Kwok DC, Davis CB, Jusko WJ. Pharmacokinetics and pharmacodynamics of a humanized monoclonal antibody to faction IX in cynomolgus monkeys. J Pharmacol Exp Ther 2000: 292: 810-816.
64. Gobburu JV, Tenhoor C, Rogge MC, Frazier DE Jr, Thomas D, Benjamin C, Hess DM, Jusko WJ. Pharmacokinetics/dynamics of 5c8, a monoclonal antibody to CD154 (CD40 ligand) suppression of an immune response in monkeys. J Pharmacol Exp Ther 1998: 286: 925-930.
65. Van Der Graaf PH, Van Schaick EA, Math-ot RA, Ijzerman AP, Danhof M. Mechanism-based pharmacokinetic-pharmacodynamic modeling of the effects of N6-cyclopentyladenosine analogs on heart rate in rat: Estimation of in vivo operational affinity and efficacy at adenosine A1 receptors. J Pharmacol Exp Ther 1997: 283: 809-816.
66. Cox EH, Kerbusch T, Van der Graaf PH, Danhof M. Pharmacokinetic-pharmacodynamic modeling of electroencephalogram effect of synthetic opioids in rats: Correlation with the interaction at the mu-opioid receptor. J Pharmacol Exp Ther 1998: 284: 1095-1103.
67. Mandema JW, Tukker E, Danhof M. In vivo characterization of the pharmacodynamic interaction of a benzodiazepine agonist and antagonist: midazolam and flumazeni. J Pharmacol Exp Ther 1992: 260: 36-44.
68. Barrett JS, Hochadel TJ, Morales RJ, Rohatagi S, DeWitt KE, Watson SK, DiSanto AR. Pharmacokinetics and Safety of a Selegiline Transdermal System Relative to Single-Dose Oral Administration in the Elderly. Am J Ther. 1996 Oct;3(10):688-698.
69. Bauer JA, Balthasar JP, Fung HL. Application of pharmacodynamic modeling for designing time-variant dosing regimens to overcome nitroglycerine tolerance in experimental heart failure. Pharm Res 1997: 14: 1140-1145.
70. Khor SP, McCarthy K, Dupont M, Murray K, Timony G. Pharmacokinetics, pharmacodynamics, allometry and dose selection of rPSGL-Ig for Phase I trial. J. Pharmacol Exp Ther 293: 618-624.
71. Pechstein B, Nagaraja NV, Hermann R, Romeis P, Locher M, Derendorf H. Pharmacokinetic-pharmacodynamic modeling of testosterone and luteinizing hormone suppression by cetrorelix in healthy volunteers. J Clin Pharmacol. 2000 Mar;40(3):266-74.
72. Radwanski E, Chakraborty A, Van Wart S, Huhn RD, Cutler DL, Affrime MB, Jusko WJ. Pharmacokinetics and leukocyte responses of recombinant human interleukin-10. Pharm Res. 1998 Dec;15(12):1895-1901.
73. Schaefer HG, Heinig R, Ahr G, Adelmann H, Tetzloff W, Kuhlmann J. Pharmacokinetic-pharmacodynamic modeling as a tool to evaluate the clinical relevance of a drug-food interaction for a nisoldipine controlled-release dosage form. Eur J Clin Pharmacol 1997: 51: 473-480.
74. Argenti D, Vaccaro SK, Shah B, Gillen, Rohatagi S, Jensen BK. Effect of food and gender on the pharmacokinetics of RP 73401, a phosphodiesterase IV inhibitor. Int J Clin Pharmacol Ther. 2000 Dec;38(12):588-94.
75. Rohatagi S, Calic F, Harding N, Ozoux ML, Bouriot JP, Kirkesseli S, DeLeij L, Jensen BK. Pharmacokinetics, pharmacodynamics, and safety of inhaled cyclosporin A (ADI628) after single and repeated administration in healthy male and female subjects and asthmatic patients. J Clin Pharmacol. 2000 Nov;40(11):1211-26.
76. Marino MT, Schuster BG, Brueckner RP, Lin E, Kaminskis A, Lasseter KC. Population pharmacokinetics and pharmacodynamics of pyridostigmine bromide for prophylaxis against nerve agents in humans. J Clin Pharmacol 1998: 38: 227-235.
77. Salazar D, Much D, Nichola P, Seibold J, Shindler D, Slugg P. A pharmacokinetic-pharmacodynamic model model of d-sotalol Q-Tc prolongation during intravenous administration to healthy subjects. J Clin Pharmacol 1997: 37: 799-809.
78. Anderson BJ, Holford NH, Woolard GA, Kanagasundaram S, Mahadevan M. Perioperative pharmacodynamics of acetaminophen analgesia in children. Anesthesiology 1999: 90: 411-421.

79. Derendorf H, Hochhaus G, Krishnaswami S, Meibohm B, Mollmann H. Optimized therapeutic ratio of inhaled corticosteroids using retrometabolism. Pharmazie. 2000 Mar;55(3):223-7.
80. Lipworth BJ. The problem of dose-response and therapeutic ratio of inhaled steroids. Am J Respir Crit Care Med. 2001 Jun;163(7):1758.
81. Raueneau I, Laveille C, Jochemsen R, Resplandy G, Funck-Brentano C, Jaillon P. Pharmacokinetic-pharmacodynamic modeling of the effects of ivabradine, a direct sinus node inhibitor, on heart rate in healthy volunteers. Clin Pharmacol Ther 1998: 64: 192-203.
82. van den Berg BT, Derks MG, Koolen MG, Braat MC, Butter JJ, van Boxtel CJ. Pharmacokinetic/pharmacdynamic modeling of the eosinopenic and hypokalemic effects of formoterol and theophylline combination combination in healthy men. Pulm Pharmacol Ther 12: 185-192.
83. Belani Cp, Kearns CM, Zuhowski EG, Erkmen K, Hiponia D, Zacharski D, Engstrom C, Ramanathan RK, Caozzoli MJ, Aisner J, Egorin MJ. Phase I trial, including pharmacokinetic and pharmacodynamic correlations, of combination paclitaxel and carboplatin in patients with metastatic non-small-cell lung cancer. J Clin Oncol 1999: 17: 676-684.
84. Lehne G, Nordal KP, Midtvedt K, Goggin T, Brosstad F. Increased potency and decreased elimination of lamifiban, a GPIIb-IIIa antagonist, in patients with severe renal dysfunction. Thromb Haemost 1998: 79: 1119-1125.
85. Mathot RA, Geus WP. Pharmacodynamic modeling of the acid inhibitory effect of ranitidine in patients in an intensive care unit during prolonged dosing: characterization of tolerance. Clin Pharmacol Ther 1999: 66: 140-151.
86. Lemmens HJ, Dyck JB, Shafer SL, Stanski DR. Pharmacokinetic-pharmacodynamic modeling in drug development: application to the investigational opioid trefentanil. Clin Pharmacol Ther 1994: 56: 261-271.
87. Jonsson A, Chan JC, Rydberg T, Vaaler S, Hallengren B, Cockram CS, Critchley JA, Melander A. Effects and pharmacokinetics of oral glibenclamide and glipizide in Caucasian and Chinese patients with type-2 diabetes. Eur J Clin Pharmacol. 2000 Dec;56(9-10):711-4.
88. US FDA Guidance for Industry. Bioavailability and Bioequivalence Studies for Orally Administered Drug Products — General Considerations (Issued 10/2000, Posted 8/22/2002).
89. US FDA Guidance for Industry. Statistical Approaches to Establishing Bioequivalence (Issued 2/2001, Posted 2/1/2001).
90. US FDA Guidance for Industry. Waiver of In Vivo Bioavailability and Bioequivalence Studies for Immediate-Release Solid Oral Dosage Forms Based on a Biopharmaceutics Classification System. (Issued 8/2000, Posted 8/31/2000).
91. US FDA Guidance for Industry. Bioavailability and Bioequivalence Studies for Orally Administered Drug Products — General Considerations (Issued 7/2002, Posted 7/2002).
92. US FDA Guidance for Industry. Clozapine Tablets in Vivo Bioequivalence and in Vitro Dissolution Testing (Issued 11/15/1996, Reposted 10/15/1998).
93. US FDA Guidance for Industry. Phenytoin/Phenytion Sodium (capsules, tablets, suspension) In Vivo Bioequivalence and In Vitro Dissolution Testing (Issued 3/4/1994, Posted 3/2/1998).
94. US FDA Guidance for Industry. Potassium Chloride (slow-release tablets and capsules) In Vivo Bioequivalence and In Vitro Dissolution Testing (Revised 6/6/1994, Posted 6/22/1998).
95. Rohatagi S, Arya V, Zech K, Jensen BK, Barrett S. Population Pharmacokinetic/Pharmacodynamics of Ciclesonide. J.Clin. Pharmacol. 2003 (in press).
96. Zannikos PN, Rohatagi S, Jensen BK, DePhillips S, and Rhodes GR. Pharmacokinetics and Concentration-Effect Analysis of Intravenous RGD891, A Platelet GPIIb/IIIa Antagonist, Using Mixed Effects Modeling (NONMEM). J. Clin Pharmacol., 2000: 40: 1129-1140.
97. Gieschke R, Reigner BG, Steimer JL. Exploring clinical study design by computer simulation based on pharmacokinetic/pharmacodynamic modelling. Int J Clin Pharmacol Ther. 1997 Oct;35(10):469-74. Review.
98. Jackson JC. A pharmacokinetic-pharmacodynamic model of chemotherapy of human immunodeficiency virus infection that relates development of drug resistance to treatment intensity. J Pharmacokinet Biopharm 1997: 25: 713-730.
99. Gieschke R, Reigner BG, Steimer JL. Exploring clinical study design by computer simulation based on pharmacokinetic/pharmacodynamic modelling. Int J Clin Pharmacol Ther. 1997 Oct;35(10):469-74. Review.

POPULATION PHARMACOKINETIC AND PHARMACODYNAMIC MODELING

Roger Jelliffe[1*], Alan Schumitzky[1], Aida Bustad[1], Michael Van Guilder[1], Xin Wang[1], and Robert Leary[2]

1. INTRODUCTION

As we acquire experience with the clinical and pharmacokinetic behavior of a drug, it is very desirable to capture this experience and its related information in the form of a population pharmacokinetic model, and then to relate the behavior of the model to the clinical effects of the drug or to a linked pharmacodynamic model. The purpose of population modeling is thus to describe and capture our experience with the behavior of a drug in a certain group or population of patients or subjects in a manner that will be useful for the treatment of future patients. This is true not only for clinical patient care, but also to optimize each step of drug development, to develop the optimal understanding of drug behavior, so that the next step can be taken most intelligently.

The traditional method of Naive Pooling has been used for population modeling when experiments are performed on animals, for example, which must be sacrificed to obtain a single data point per subject. Data from all subjects is then pooled as if it came from one single subject. One then can estimate pharmacokinetic parameter values, but cannot estimate any of the variability that exists between the various subjects making up the population. This method has generally been supplanted by the more informative methods described below.

2. PARAMETRIC POPULATION MODELING METHODS

A variety of parametric population modeling methods exist and have been very well described in [1]. They obtain means and standard deviations (SD's) for the

* [1]Laboratory of Applied Pharmacokinetics, University of Southern California School of Medicine, Los Angeles, CA, USA; [2]San Diego Supercomputer Center, University of California, San Diego, CA, USA.

pharmacokinetic parameters and correlations (and covariances) between them. Only a few of these will be described in this chapter, and quite briefly.

2.1. The Standard Two-Stage (S2S) approach

This approach involves first using a method such as weighted nonlinear least squares to obtain pharmacokinetic model parameter estimates for each individual patient. Correlations between the parameters may also be obtained, based on the individual parameter values in the various subjects. In the second and final step, the population mean, SD, and correlation coefficients in the sample of people studied are then computed for each of the pharmacokinetic parameters. This method usually requires at least one serum concentration data point for each parameter to be estimated.

One can also examine the frequency distributions of the individual parameter values to see if they are Gaussian or not. In this latter setting, the S2S method can also be regarded as being, in a sense, nonparametric as well, since no assumptions need to be made concerning the shape of the frequency distribution of the various individual parameter values. The method is basically parametric, however, as it gathers together the individual results into Gaussian summary parameter values of means, SD's, and correlations for each of the model parameters. This is what is meant by parametric population modeling - the frequency distributions of parameters in the model are described in terms of the parameters of an assumed function (with its specific distribution parameters such as means and SDs) that describes the assumed shape/class of the model parameter distributions. In this case, since the parameter distributions are usually assumed to be Gaussian or lognormal, these other distribution parameters are the means, SD's, variances and covariances, of the various pharmacokinetic – pharmacodynamic (PK / PD) parameters of the structural PK / PD model employed.

2.2. The Iterative Two-stage Bayesian (IT2B) method

This method can start by using the S2S mean parameter values as obtained above, and their SD's. On the other hand, one can set up <u>any</u> reasonable initial estimate of the population mean parameter values and their SD's. In the IT2B method, one uses these initial selected parameter means and SD's as the Bayesian priors, and then examines the individual patient data to obtain each patient's maximum a posteriori probability (MAP) Bayesian posterior parameter values, using the MAP Bayesian procedure in current wide use [2]. It uses the First Order Conditional Expectation (FOCE) approximation to calculate the log–likelihood of the population parameter values given the population raw data and the weighting scheme employed in analyzing the data.

With this method, one can iteratively recompute the population means and SD's of the parameter values found. For example, one can use these S2S summary population parameter values (see above) once again, now as initial Bayesian population priors, for another MAP Bayesian analysis of the data. One can once again obtain each patient's new MAP Bayesian values. This process can then continue iteratively indefinitely. The procedure ends when a convergence criterion is reached. The IT2B method is less subject to the problems of local minima often found when fitting data by least squares. In addition, it does not require as many serum concentration data points per patient (as few as only one per patient), and so is much more efficient in this respect. The Global

Two Stage (G2S) method is a further refinement of the S2S and the IT2B in which the covariance and correlations between the parameters are also estimated.

2.3. The Parametric EM method

This method is also an iterative method. The letters EM stand for the two steps in each iteration of 1) computing a conditional expectation (E) and 2) the maximization (M) of a conditional likelihood, resulting in a set of parameter values which are more likely than those in the previous iteration. The process continues until a convergence criterion is met. The results with the parametric EM method for the various pharmacokinetic parameter distributions are again given in terms of the model parameter means, SD's, and correlations, or means, variances, and covariances [3, 4]. As is the case with the IT2B method, an approximation such as FOCE is used to compute the conditional likelihoods to avoid computationally intensive numerical integrations.

2.4. NONMEM

True population modeling began with the Nonlinear Mixed Effects Model with first-order approximation (NONMEM) of Beal and Sheiner [5-7]. The overall population, even if it has only 1 data point per subject, almost always supplies enough data for this approach, if the various data points are spread throughout the dosage interval so that dynamic information about the behavior of the drug can be obtained. The NONMEM method estimates means, SD's, and covariances of population parameter values. However, it has sometimes given different answers from other methods. It is also a parametric method, and gives its results in terms of parameter means and variances.

This method was the first true population modeling program, as it eliminated the need for having at least one data point for each patient for each parameter to be estimated. It estimates both fixed effects (those containing only a single point value for a parameter, such as a parameter mean), and those containing random distributions, such as the random variability of a model parameter about its mean. This random variability is characterized by SD's, and covariances or correlations. The name "mixed" is used because the method estimates both types of model parameters, fixed and random. The method can function with as few samples as one per patient. While this method is in wide use, it lacks the desirable property of mathematical consistency [8-10]. Earlier (FO) versions of this method have at times given results which differed considerably from those of other methods [11,12]. Subsequent First Order Conditional Expectation (FOCE) versions of this method have shown improved behavior. This will be discussed further toward the end of this chapter. Other variations on this approach are those of Lindsdtrom and Bates [13], and Vonesh and Carter [14].

3. ANALYZING ASSAY AND ENVIRONMENTAL SOURCES OF ERROR

3.1. Determining the Assay Error Polynomial

In analyzing any data set, it is useful to assign a measure of credibility to each data point to be fitted or analyzed. In the IT2B program of the USC*PACK collection [16,42], for example, one is encouraged first to determine the error pattern of the assay quite specifically, by determining several representative assay measurements in at least quadruplicate, and to find the standard deviation (SD) of each of these points. One can measure, in at least quadruplicate, a blank sample, a low one, an intermediate one, a high one, and a very high one. One can then fit the relationship between the serum concentration (or other response) and the SD with which it has been measured, with a polynomial of up to third order, so that one can then compute the Fisher information, for example, as a useful measure of the credibility of each serum concentration data point [15,16]. One can then express the relationship as

$$SD = A_0 + A_1C + A_2C^2 + A_3C^3$$

where SD is the assay SD, A_0 through A_3 are the coefficients of the polynomial, C is the measured concentration, C^2 is the concentration squared, and C^3 is the concentration cubed. A representative plot of such a relationship, using a second order polynomial to describe the error pattern of an EMIT assay of gentamicin, is shown in Figure 1.

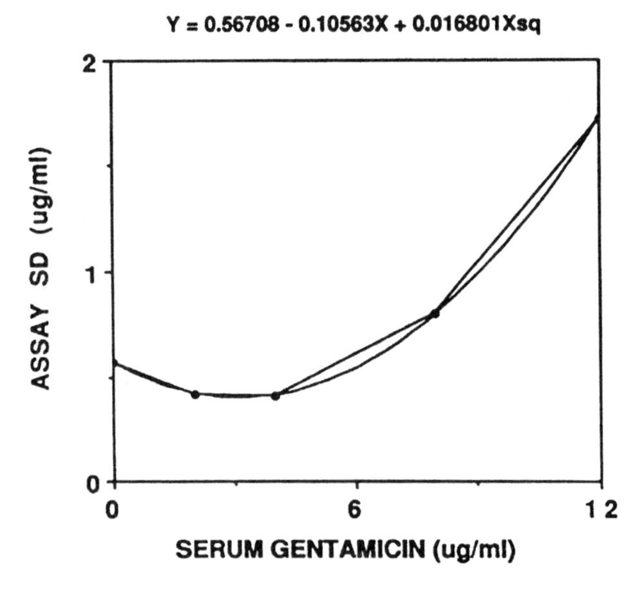

Figure 1. Graph of the relationship between serum Gentamicin concentrations, measured by Emit assay in at least quadruplicate (the dots) and the standard deviations (SD's) of the measurements. The relationship is captured by the polynomial equation shown at the top. Y = assay SD, X = measured serum concentration, Xsq = square of serum concentration.

3.2 Determining the Environmental Error

In addition, a parameter which we have called gamma, a further measure of all the other environmental sources of intra-individual variability, can also be computed. It is used in the USC*PACK IT2B program as a multiplier of each of the coefficients of the assay error polynomial as described above. The nominal value of gamma is 1.0, indicating that there is no other source of variability that the assay error pattern itself. Gamma is therefore usually greater than 1.0. It includes not only the various environmental errors such as those in preparing and administering the doses, recording the times at which the doses were given, and recording the times at which the serum samples were obtained, but also the errors in which the structural model used fails to describe the true events completely (model misspecification), and also any possible changes in the model parameter values over time, due to the changing status of the patient during the period of data analysis. Gamma is thus an overall measure of all the other sources of intraindividual variability besides the assay error. In this way, one can calculate how much of the total SD is due to the assay SD, and how much is due to the remaining overall environmental SD.

Determining gamma will help to explain the environmental variability found in any fit. If gamma is small, it suggests that the sum of the environmental sources of noise is small. If it is large, it suggests that the overall environmental noise, the total effect of all the other factors mentioned above, is large.

However, most of these other sources are not really sources of measurement noise, but are rather due to noise in the differential equations describing the behavior of the drug. The environmental sources are most correctly described as sources of process noise rather than measurement noise. The problem is that it is difficult to estimate process noise, as it requires stochastic differential equations, which contain these other noise terms. However, no software for estimating process noise in pharmacokinetic models exists at present, to our knowledge.

The IT2B program can also be used to compute estimates of the various combined assay error and environmental polynomial error coefficients, if one has no knowledge of what the assay error pattern is, or if the measurement is one which is impossible to replicate to determine its error. In this case, gamma is not determined separately, but is included in the various other polynomial coefficients.

4. MAKING A PARAMETRIC (IT2B) POPULATION MODEL

The following results are taken from a representative run of the IT2B program. The original patient data files were made using the USC*PACK clinical software. The following illustrative results are taken from data obtained by Dr. Dmiter Terziivanov in Sofia, Bulgaria [17], on 17 patients who received intramuscular Amikacin, 1000 mg, every 24 hours for 5 or 6 days. For each patient, two clusters of serum concentrations were measured, one on the first day and the other on the 5th or 6th day, approximately 5 samples in each cluster. Creatinine clearance (CCr) was estimated from data of age, gender, serum creatinine, height and weight [18]. Serum concentrations were measured by a bioassay method. The assay error pattern was described by a polynomial in which the assay SD = 0.12834 + 0.045645C, where C is the serum concentration. The assay SD of a blank was therefore 0.12834 ug/ml, and the subsequent coefficient of variation was 4.5645%. In this particular analysis, gamma was found to be 3.2158, showing that the SD of the environmental noise was about 3.2 times that of the assay SD, or conversely, that the assay SD was about 1/3 of the total noise SD.

The initial (very first) parameter estimates, and their SD's, were set at: Ka (the absorption rate constant) = 3.0 ± 3.0 hr^{-1}, Ks (the increment of elimination rate constant per unit of creatinine clearance in ml/min / $1.73M^2$) = 0.004 ± 0.004 hr^{-1}, and Vs1 (the apparent central volume of distribution) = 0.3 ± 0.3 l/kg. The nonrenal intercept of the elimination rate constant (Ki) was held fixed at 0.0069315 hr^{-1}, so that the elimination rate constant = Ki + Ks1 x creatinine clearance, and the serum half-time, when CCr = 0, is fixed at 100 hours.

The following results were obtained with the USC*PACK IT2B program. The IT2B program converged, on this data set, on the 1053th iteration. The population mean values for the parameters Ka, Ks1, and Vs1 found were 1.349 hr^{-1}, 0.00326 hr^{-1}, and 0.2579 L/kg respectively. The medians were 1.352 hr^{-1}, 0.00327hr^{-1}, and 0.2591 L/kg respectively. The population parameter standard deviations were 0.062 hr^{-1}, 0.000485 hr^{-1}, and 0.0350 L/kg respectively, yielding coefficients of variation of 4.55, 14.83, and 13.86 percent respectively.

The individual MAP Bayesian distributions of Ka, Ks1, and Vs1 are shown in Figures 2 through 4. While the distributions of Ka and Vs1 are fairly Gaussian, that of Ks1 is skewed to the left. The joint distribution of Ks and Vs in shown in Figure 5, which shows an extremely high positive correlation between the two parameters, consistent with their population parameter correlation coefficient of +0.991. That between Ka and Ks1 was similar, +0.924, and that between Ka and Vs1 was also very high at +0.950. These probably spuriously high correlations are similar to those found by Leary [27], which are discussed in section 11 below.

Figures 6 and 7 are scattergrams of predicted versus measured serum concentrations. Figure 6 shows the predictions based on the population parameter medians and the doses each subject received. In contrast, Figure 7 shows the predictions made using each subject's individual MAP Bayesian posterior parameter values to predict only his/her own measured serum concentrations. The improved predictions in Figure 7 are due to the removal of the population inter-individual variability, as perceived by the IT2B program. The remaining smaller scatter is due to the intraindividual variability resulting not only from the assay error, but also to the other sources of noise in the system, such as the errors in preparation and administration of the various doses, errors in recording the times the doses were given and the serum samples drawn, and the mis-specification of the pharmacokinetic model used. The results shown in Figure 7 show that the study was done with reasonable precision.

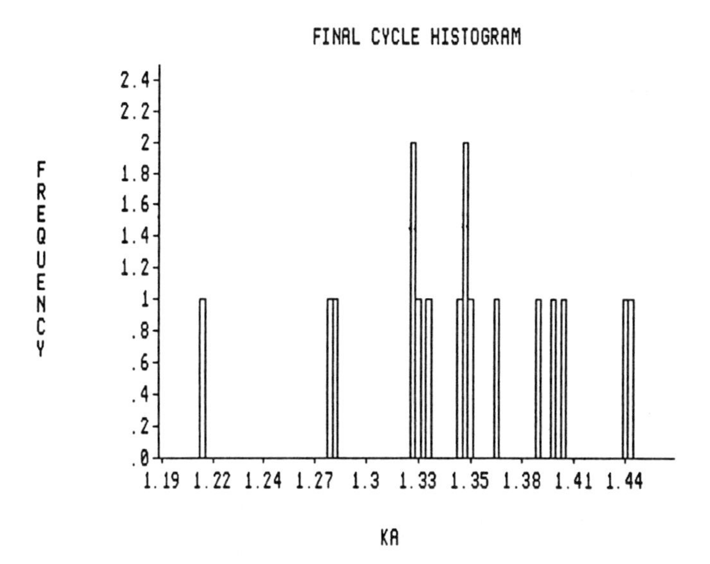

Figure 2. Graph of the marginal frequency of population parameter Ka. The plot is divided into 100 cells over the range from 1.19 to 1.47 (horizontal axis). The frequency of the patient parameter values in each cell is shown on the vertical. See text for discussion.

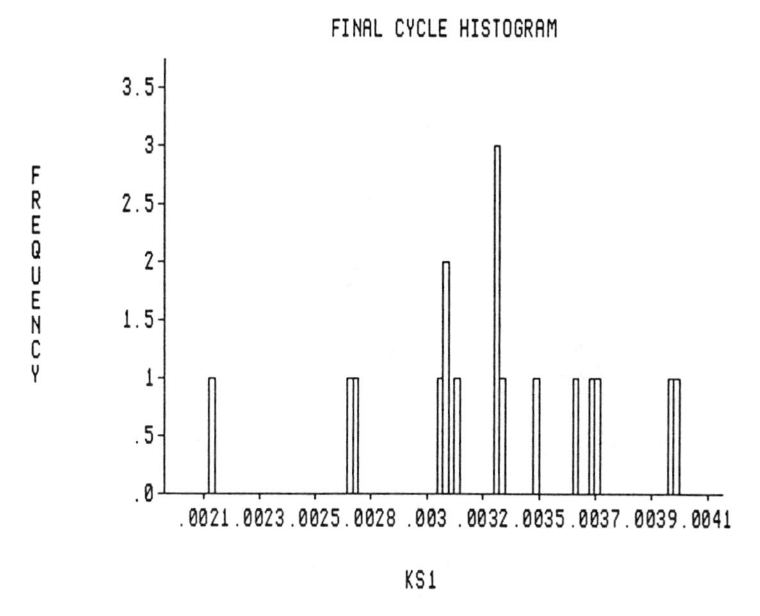

Figure 3. Graph of the marginal frequency of population parameter Ks1. The plot is divided into 100 cells over the range from 0.0019 to 0.0041 (horizontal axis). The frequency of the patient parameter values in each cell is shown on the vertical. See text for discussion.

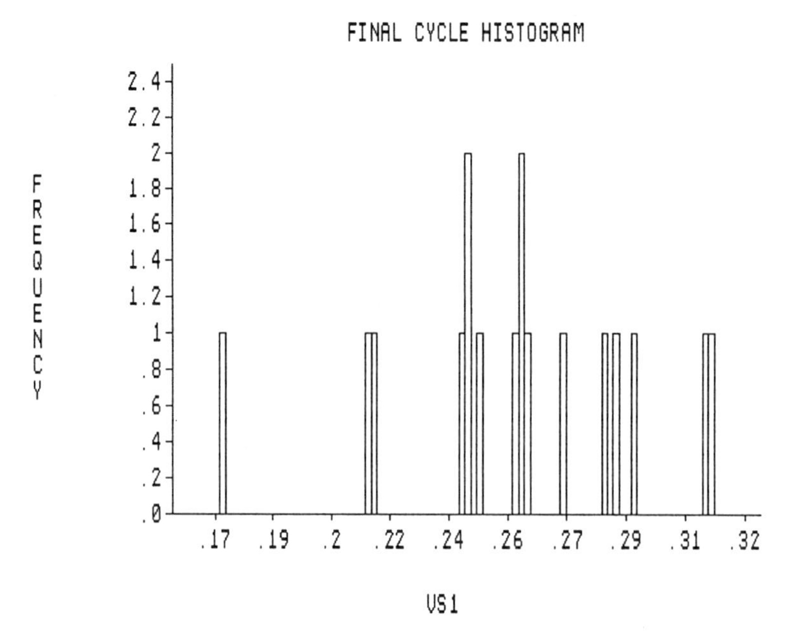

Figure 4. Graph of the marginal frequency of population parameter VS1. The plot is divided into 100 cells over the range from 0.15 to 0.32 (horizontal axis). The frequency of the patient parameter values in each cell is shown on the vertical. See text for discussion.

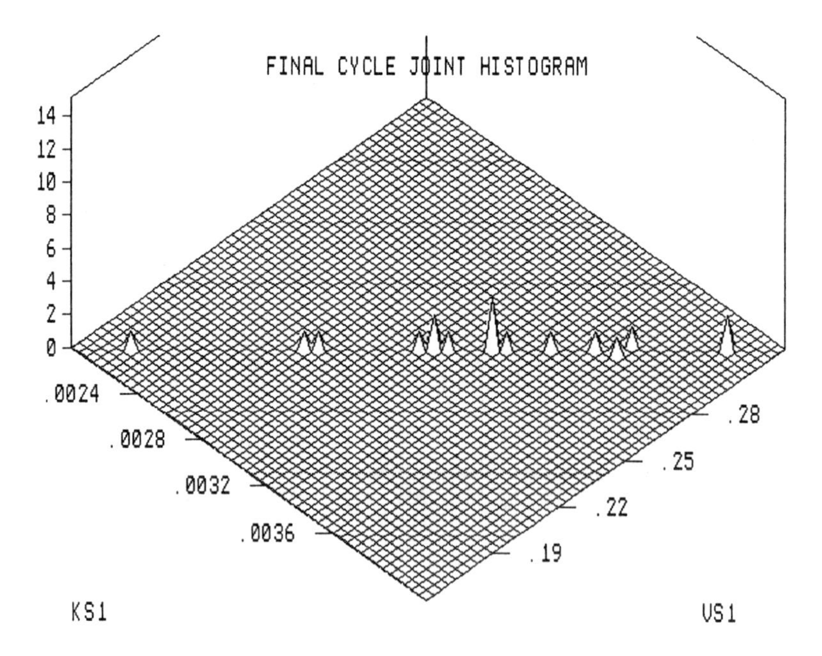

Figure 5. Graph of the joint marginal frequency of population parameter Ks1 and Vs1. The plot is divided into 50 by 50 cells over the ranges stated in Figures 2 - 4. The frequency of the patient parameter values in each cell is shown on the vertical. Note the extremely high (probably spurious) correlation between the parameters. The correlation coefficient was 0.991. See text for discussion.

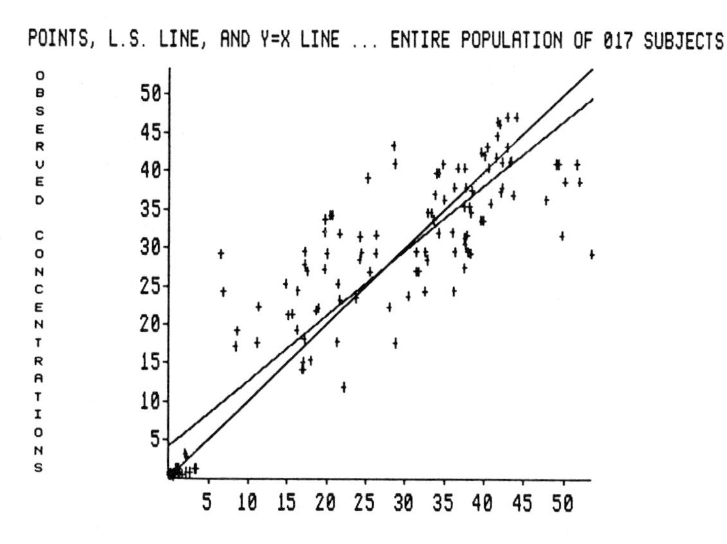

Figure 6. Scattergram of relationship between predicted serum concentrations (horizontal) and measured ones (vertical), based on median population parameter values.

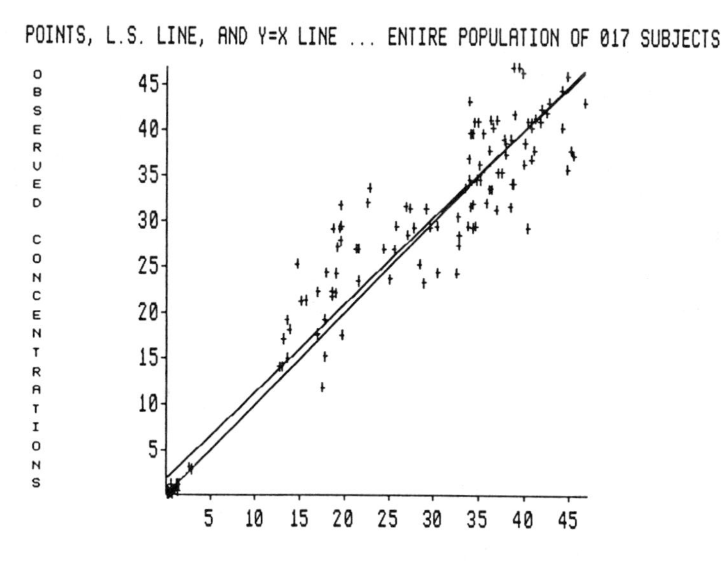

Figure 7. Plot of relationship between predicted serum concentrations (horizontal) and measured ones (vertical), based on each subject's own maximum aposteriori probability (MAP) Bayesian posterior parameter values, where each subject predicts only his/her own measured concentrations.

The IT2B method of population modeling is a useful one, and is based on the widely used and robust strategy of MAP Bayesian individualization of pharmacokinetic models. Its weaknesses, like those of any parametric method, are that it only perceives population parameter values in terms of their means, medians, variances, and correlations. The actual parameter distributions are usually not of this type. Lognormal assumptions have often been made, but the actual parameter distributions are frequently not of that form either. In addition, the true log-likelihood (not the FOCE approximation) of the distribution of the individual subjects' parameter values in the population was also computed exactly, just as if the discrete collection of individual parameter values had come from an NPEM or NPAG analysis (see below) for direct comparison with their results. It was found, for this IT2B analysis, to be -389.548 (see Table 1, below).

5. LARGER AND NONLINEAR IT2B POPULATION MODELS

Similar software for IT2B population modeling of large and nonlinear PK/PD models has been implemented on the IBM "Blue Horizon" parallel computer at the San Diego Supercomputer Center (SDSC), as a research resource for such work, on a Linux cluster of Dell machines in our laboratory, and on a larger Linux cluster at USC. The user uses a PC program on his machine in the USC*PACK collection to specify the data files to be analyzed and the instructions for the analysis. This becomes an input file for the parallel machine or cluster. One also either writes the linear or nonlinear ordinary differential equations for the specific structural PK/PD model to be used, or employs the BOXES program in the USC*PACK collection [42], placing boxes on the screen for the compartments and connecting them with arrows to represent the various types of pathways involved. The differential equations of the PK/PD model are then generated automatically and stored in a model file.

These two files are then sent to the parallel machine or cluster via a secure protocol, over the Web. The model file, which is in Fortran source code, is compiled and linked. The analysis is performed using the desired number of processors. A differential equation solver (VODE) is employed. The results are then sent back by email to the user's PC where they are examined just as in Figures 2 through 7 above. Thus one can now make large and nonlinear models of a drug, with multiple responses such as serum concentrations and various drug effects [20].

6. STRENGTHS AND WEAKNESSES OF PARAMETRIC POPULATION MODELING

The major strength of the parametric population modeling approaches has been their ability to separate inter-individual variability in the population from intra-individual variability in the individual subjects (gamma, for example), and also from variability due to the assay error itself. Because of this, it still seems best, for the present, to begin making a population model of a drug by using a parametric method such as IT2B. First, though, one should estimate the assay error pattern explicitly, obtaining the assay error polynomial as described above [15,16]. Then, having that assay error polynomial, one can use a parametric method such as IT2B to find gamma, to determine the overall intraindividual variability, and to know what fraction of that is actually due to the assay itself, and what is due to the environmental uncertainty. Then, one is in a position to overcome the weaknesses of these approaches (see below) by using this information in making a nonparametric population (NP) model.

One weakness of the parametric methods is that they generally have lacked the desirable property of mathematical consistency, which is a real strength of the nonparametric methods discussed later on [21-23]. In addition, the parametric methods make parametric assumptions about the shape of the parameter distributions, and do not take into account the entire actual shape of the distributions, as the nonparametric methods do [24,25]. Further, they give only single point summary parameter estimates such as the mean, median, or mode of each overall parameter distribution. Much discussion has taken place about which of these is the better estimate.

The major clinical weaknesses of the most widely used FOCE parametric approaches are that they only obtain the single point estimates of the parameter distributions, and that they are not statistically consistent, have poor statistical convergence, and poor statistical efficiency [27]. This will be discussed further on in more detail later on.

Further, when one uses such a model to develop a dosage regimen for a patient to achieve a desired target goal at a desired target time, the regimen is simply the one which should hit the desired target exactly. There is no method to estimate in advance the degree to which the regimen will fail to hit the target, as there is only a single model, with each parameter consisting of only a single point estimate. Because of this, with parametric population models, the only course of action that can be taken (the next dosage regimen) based on that model, to hit a desired target goal, is based only on the central tendencies of the various parameter distributions, and not on the entire distribution itself. Such action (the dosage regimen) is therefore not designed to achieve the target goals optimally.

7. THE SEPARATION PRINCIPLE

The separation or heuristic certainty equivalence principle states that whenever the behavior of a system is controlled by separating the control process into:

1. Getting the best single point parameter estimates, and then,
2. Using those single point estimates to control the system,

the task of control is usually done suboptimally [26], as there is no specific performance criterion that is optimized. This is the major weakness of using parametric population models for designing drug dosage regimens. The parameter distributions are often neither Gaussian or symmetrical, and measures of central tendency are not optimal in computing the dosage regimens, and may occasionally be dangerous, as when the volume of distribution, for example, may be at the 70th percentile of the distribution. Other thing being equal, that would result in 70% of the concentrations being above a desired target goal, and only 30% being below. Because of this, when single point parameter estimates are used to compute dosage regimens, one cannot estimate the expected precision with which a given dosage regimen will hit the target, and this is usually results in a suboptimal (less than maximally precise) dosage regimen. The resulting regimen is based only on the central tendencies of the various parameter distributions, and not on the shapes of the entire distributions themselves, which are often non Gaussian and multimodal. This problem will be discussed more fully in the chapter on clinical applications.

One may ask why we make models – to simply capture such single point estimates, or to take some useful and practical action based on the information obtained from the modeling process? It is useful and practical to supplement the knowledge of the assay error, empirically determined before starting the modeling, with information about the intra-individual variability (gamma, above) obtained from a parametric IT2B population model. Having this information, one can then proceed to make a nonparametric population model which can overcome the difficulties presented by the separation principle stated above. This will enhance the process of drug development, as it will help to optimize the design of dosage regimens for various populations of patients, to achieve the desired goals most precisely. It will also optimize the process of designing dosage regimens for various clinical trials, or designing the next step in drug development optimally. Again, this will be discussed in the chapter on clinical applications.

8. NONPARAMETRIC POPULATION MODELING

There is great variability among patients with regard to their pharmacokinetic parameter values. Nevertheless, we have become accustomed to using selected single numbers to summarize such diverse behavior. For example, we have usually used the population mean or median parameter values as the best single number to describe the central tendencies of their distributions, and the standard deviation (SD) to describe the dispersion of values about the central tendency. It has been customary to focus on such single numbers as summaries of experience with subjects or patients, rather than to consider the entire collection of our varied experiences with each individual patient. We will now examine newer nonparametric methods which can give us richer and more likely information from the raw population data.

What is meant here by the word nonparametric? Most of the time, when we gather data and summarize it statistically, we have been accustomed to obtaining a single parameter value to summarize the central tendency of a distribution such as the mean, median, or mode, and another single parameter value to describe the dispersion about this central tendency, such as the standard deviation (SD). The usual reason for this is that many events in statistics have a normal or Gaussian distribution, and that the mean and the SD are the two parameters in the equation which describe the shape of a Gaussian distribution explicitly. Because of this, describing a distribution parametrically, in terms of its mean and SD, is very common. Indeed, the entire concept of analysis of variance is based on the assumption that the shape of the parameter distributions in the system are Gaussian, and are therefore best described by means, SD's, and covariances. A great body of experience has been brought to bear to describe pharmacokinetic models parametrically, in this way. Much of this is described in an excellent review given in [1].

On the other hand, if the structural PK/PD model could be exactly known, if each individual subject's pharmacokinetic parameter values in a given population could also somehow be truly and exactly known, and if, for example, we were examining two such typical parameters such as volume of distribution (V), and elimination rate constant (K), then the truly optimal joint population distribution of these parameter values would be the entire collection of each individual patient's exactly known parameter values. All subpopulations, and those in between, would be truly known as well (perhaps not yet explicitly recognized or classified), but nevertheless located and quantified.

Obviously though, the individual subject parameter values can never be known exactly. They must be estimated from the data of doses given and serum concentrations measured, and in the setting of environmental uncertainty, as described above.

In the nonparametric approach, the maximum likelihood parameter distributions one obtains are discrete spikes, up to one for each subject studied in the population [24,25,27]. The location of each spike (support point) reflects its set of estimated parameter values. The height of the spike represents the estimated probability of that individual set of estimated parameter values. The likelihood of the entire collection of support points can be computed and compared under similar conditions (see Table 1 further on for an example). No summary parameters such as mean or SD will be any more likely, given the data of dosage and serum concentrations, than the actual collection of all the estimated individual discrete points, each one having certain parameter values such as V and K, for example, and the estimated probability associated with each combined point of V and K [24,25,27].

This is what is meant by the word nonparametric in this sense. It is not to be confused with the noncompartmental modeling approach based on statistical moments, which is often also called nonparametric. The NP approach always has a specific structural model. In addition, as will be seen further below, the NP estimates of the parameter means SD's, and correlations are at least as reliable as those obtained by parametric methods, and not infrequently are significantly better than those obtained using the FOCE approximation [27]. The actual shape of the discrete NP parameter distribution is totally determined by the raw data of the subjects studied in the population and the error pattern used, and not by any assumed equation describing the assumed shape (Gaussian, lognormal, etc.) of the parameter distributions. The NP methods estimate the collection of support points and their probabilities which is most likely (maximum likelihood), that which specifically maximizes the likelihood or log-likelihood function. It is interesting that in a great number of papers which have described parametric population models, the likelihood of the results obtained (as opposed to common indices of "goodness of fit") is usually not reported.

Many patient populations actually are made up of genetically determined clusters or subpopulations. For example, there may be fast and slow metabolizers of a drug. The relative proportions of fast, in between, and slow subjects may vary from one population (Caucasian people, for example) to another (Asian people, for example) [28] Describing such a distribution of clusters optimally is not possible with a normal or lognormal distribution.

Since it is not possible to know each patient's values exactly in real life, we study a sample of patients requiring therapy with a drug (the most relevant population sample) by giving the drug and measuring serum concentrations and/or other responses. Lindsay [29] and Mallet [24] were the first to show that the optimal solution to the population modeling problem is actually a discrete (not continuous), spiky (not smooth) probability distribution in which no preconceived parametric assumptions (such as Gaussian, lognormal, multimodal, or other) have to be made about its shape. The nonparametric maximum likelihood (NPML) estimate of the population joint parameter density or distribution (analogous to the entire collection of each patient's exactly known parameter values described above), whatever its shape or distribution turns out to be, is supported by up to N discrete points for the N patients in the population studied. Each such support point is a collection of estimated single numbered parameter values, one for each parameter such as V, K, etc., along with an estimate of the probability associated with each such combination. The probabilities of all the various points add up to 1.0. The NPML methods of Mallet [24], like the parametric NONMEM method, can also function with only one sample per patient. However, as mentioned, the nonparametric parameter distributions may have any shape, and this depends only on the actual subject data.

The means, SD's, and other common statistical summary parameters can easily be obtained as well, from the entire discrete distribution. The only assumption made about the shape of the discrete parameter distributions is that, for each model parameter, the shape, whatever it is, is the same for all subjects in the population. Because of this, the method is capable of discovering unsuspected subpopulations of subjects such as fast and slow metabolizers, without recourse to other descriptors or covariates, and without recourse to individual Bayesian posterior parameter estimates [25].

A similar nonparametric EM (NPEM) method was developed by Schumitzky [25,30]. It is an iterative EM method like the parametric EM method, but is nonparametric. Like the NPML method, it also can function with only one sample per patient. Like the NPML method, it also does not have to make any parametric assumptions about the shape of the joint probability distribution. It also computes the entire discrete joint density or distribution of points. In contrast to the NPML method, though, the NPEM method obtains a continuous (although very spiky) distribution. This distribution becomes discrete in the limit, after an infinite number of iterations. Within each iteration, the NPEM method examines the patient data and develops a more and more spiky (and more likely) joint distribution. In the limit, the spikes become up to one discrete support point for each subject studied, just as with the NPML method.

As with the NPML method, the NPEM population joint distribution also becomes a collection of discrete support points, each of which contains a set of parameter values, and each of which has a certain probability. Both the NPML and the NPEM methods have been shown to converge to essentially the same results [31]. Both the NPML and the NPEM methods are proven under suitable hypotheses to have the desirable property of mathematical consistency [8,23,27,32].

Figures 8 through 10 illustrate the ability of the nonparametric approach, as shown by the NPEM algorithm, to discover unsuspected subpopulations [25]. The NPML method of Mallet has similar capability.

Figure 8 shows a carefully constructed simulated population of patients, which is actually a bimodal distribution consisting of two subpopulations. Half were "fast" and half were "slow" metabolizers of a drug. While they all had the same volume of distribution, they had two different elimination rate constants. There was no correlation between the two parameters.

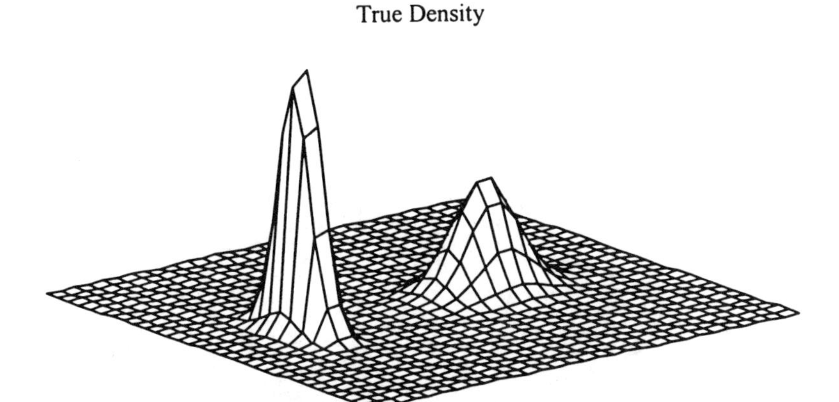

Figure 8. The true pharmacokinetic population joint density from which the 20 samples were taken. If the bottom (or 0.0,0.0) corner is "Home plate", then the axis toward third base is that of the volume of distribution V, while that toward first base is the elimination rate constant K. The vertical axis is the relative probability of each parameter pair. Note that there are actually two subpopulations, with two clusters of distributions for K. V and K are uncorrelated.

From this simulated population, twenty hypothetical patients were sampled at random. Their parameter values were therefore known exactly. Figure 9 shows these sampled patients' exactly known parameter values as they appear when smoothed and graphed in the same manner as shown in Figure 8. Figure 9 therefore shows the true population parameter distribution that any population modeling method should now discover.

These twenty hypothetical patients then each were "given" a simulated single dose of a drug having one compartment behavior, and five simulated serum samples were drawn at uniformly spaced times after the very short intravenous infusion. The simulated assay SD was ± 0.4 concentration units. The simulated data were then presented to the NPEM algorithm and computer program [25,30], as well as to a parametric population modeling method such as NONMEM or IT2B.

Figure 10 shows the results of the NPEM analysis, again smoothed and graphed as in Figure 8. The NPEM program clearly detected and located the two subpopulations of patients. Figure 10 is similar in shape to the known original population joint distribution shown in Figure 9.

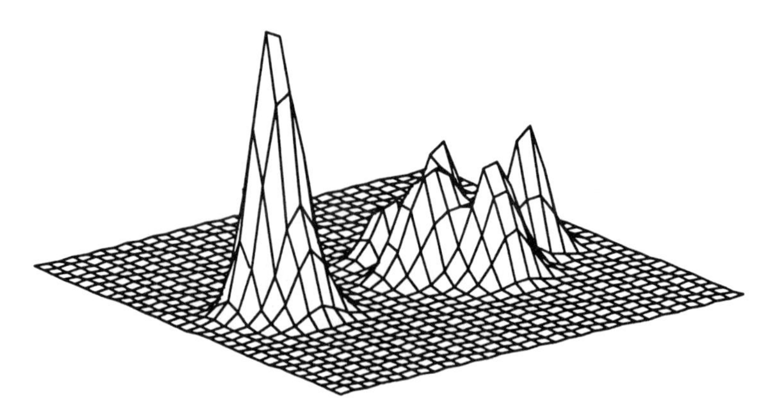

Figure 9. Graph, smoothed as in Figure 8, of the actual parameter values in the twenty sampled patients. The axes are as in Figure 8. This is the true population distribution that NPEM, or any other method, should now discover.

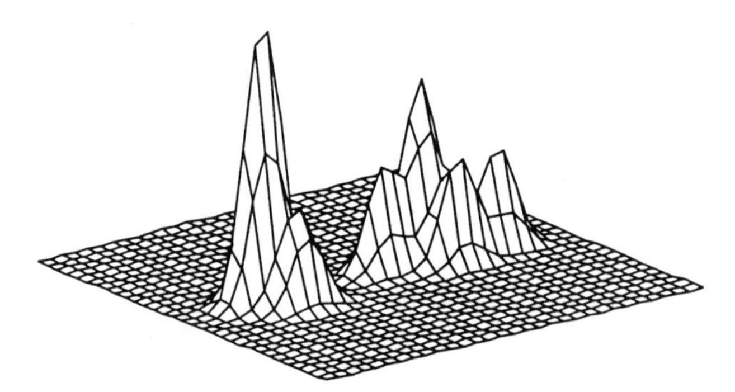

Figure 10. Smoothed estimated population joint density obtained with NPEM, using all five serum concentrations. Axes as in Figure 8. Compare this figure with Figure 9.

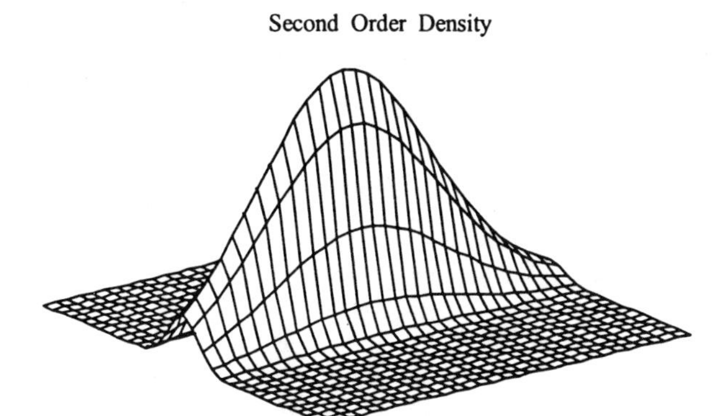

Figure 11. Plot of the population density (the second order density) as perceived by a theoretically optimal parametric method. Axes as in Figure 8. Compare this figure with Figure 9. The two subpopulations are not discovered at all. The true parameter distributions are perceived with great error.

In contrast, Figure 11 shows how a parametric method such as IT2B, the parametric EM, or NONMEM perceives the population shown in Figure 9. The second order density in the figure is the one obtained by a parametric method, which obtains means, variances, and covariances or correlations. Note that the mean is actually where there are no subjects at all. Parametric methods thus cannot discover subpopulations without additional aid. They give an entirely different impression of the population behavior of the drug. One also gets an impression in Figure 11 of much greater variability between patients than actually exists among the two fairly tightly grouped subpopulations, which is shown in Figures 9 and 10.

9. MAKING A NONPARAMETRIC POPULATION MODEL

The same data of the 17 patients receiving intramuscular Amikacin, described earlier in "Making a Parametric Population Model" above, was also analyzed using both the NPEM and the nonparametric adaptive grid (NPAG, a later version of NPEM with improved speed, precision, and convergence rate) software [27]. As before, the parameters were Ka, the absorptive rate constant from the intramuscular injection site, Vs, the volume of distribution in L/kg, and Ks, the increment of elimination rate constant per unit of creatinine clearance. Initial ranges for these parameters were set at 0 to 6 hr-1 for Ka, 0 to 0.6 l/kg for Vs, and 0 to 0.008 for Ks. Gamma was set at the value previously obtained with the IT2B analysis.

The results are summarized in Table 1, below, where they are also compared with the previous results from the IT2B program.

Table 1. Parameter values (mean, median, percent coefficient of variation -CV%, and log likelihood obtained with the IT2B, the NPEM, and the NPAG programs. Ka = absorptive rate constant (hr^{-1}), VS1, apparent central volume of distribution (L/kg), KS1, increment of elimination rate constant (hr^{-1} per unit of creatinine clearance). CV% is less with IT2B, but so it the Log likelihood, which is better with NPEM and NPAG.

		IT2B	**NPEM**	**NPAG**
Mean	Ka	1.349	1.408	1.380
	VS1	0.258	0.259	0.258
	KS1	0.003258	0.003271	0.003275
Median/CV%	Ka	1.352/4.55	1.363/20.42	1.333/21.24
	VS1	0.2591/13.86	0.2488/17.44	0.2537/17.38
	KS1	0.003273/14.83	0.003371/15.53	0.003183/15.76
Log – Likelihood		-389.548	-374.790	-374.326

When comparing the results of these IT2B, NPEM, and NPAG analyses, at first glance there seems to be little to choose between them. The parameter values are all quite similar. In fact, the population percent coefficients of variation (CV%) are clearly least with the IT2B program. This might suggest that the population parameters are estimated with greater precision with that program. However, the likelihood of the results is clearly least with the IT2B program, as shown in Table 1. It is greater with NPEM, and just a bit better still with NPAG. In addition, as shown in Figures 12 through 14, very high correlations were found between all parameters with IT2B. This was not the case with the two nonparametric methods. Because of this, the very high correlations between all pairs of parameter seen with IT2B is probably spurious. The general finding is that the likelihood is significantly greater with the nonparametric methods, and the smaller population CV% found with IT2B is therefore probably due to its constraining assumption that the parameters must have Gaussian distributions. The price paid for this Gaussian assumption is seen in the lower likelihood of the results obtained. In contrast, NPEM and NPAG, because they are not restricted to the assumption of Gaussian parameter distributions, were better able to detect the full diversity in the population parameter distributions, and were able to obtain more likely results. Notice also that the parameter distributions in Figures 12-14 are not Gaussian, but skewed. The specific marginal distributions of Ka, Ks1 and Vs1 obtained with NPEM and NPAG are not shown because of space considerations.

Many papers using parametric population modeling have not reported the likelihood of their results, but have restricted themselves to reporting only the common indices of "goodness of fit". The reason for this is that IT2B and other methods such as NONMEM, which use the FOCE (first order, conditional expectation) approximation, compute only an approximate value of the likelihood. This is usually not reported. In contrast, the nonparametric methods compute the likelihood value exactly.

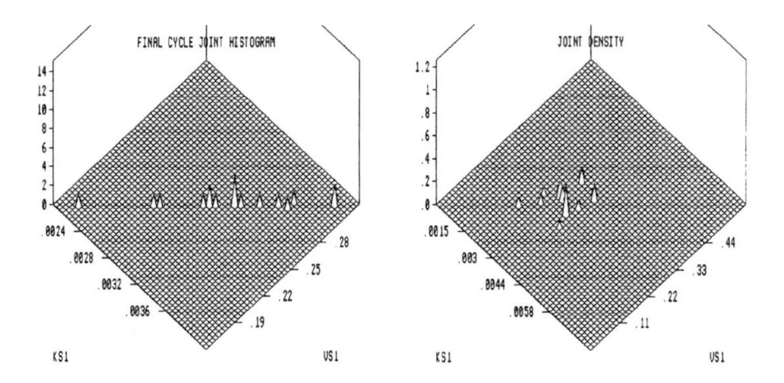

Figure 12. Joint marginal density obtained for KS1 and VS1 with IT2B, left, and NPAG, right. Note the very high, and probably incorrect, correlation between the parameters seen with the IT2B program.

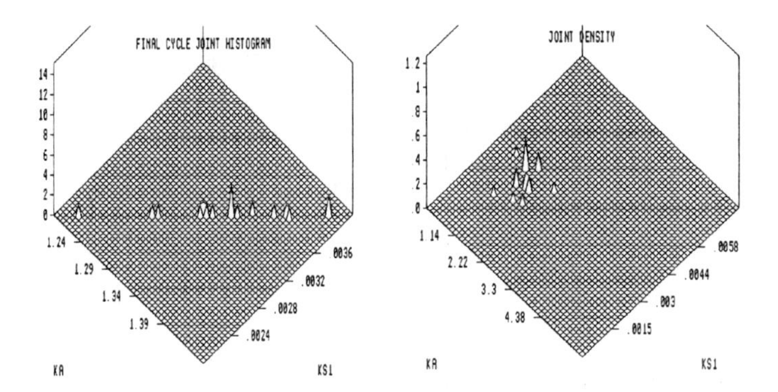

Figure 13. Joint marginal density obtained for KA and KS1 with IT2B, left, and NPAG, right. Note the very high, and probably incorrect, correlation between the parameters seen with the IT2B program.

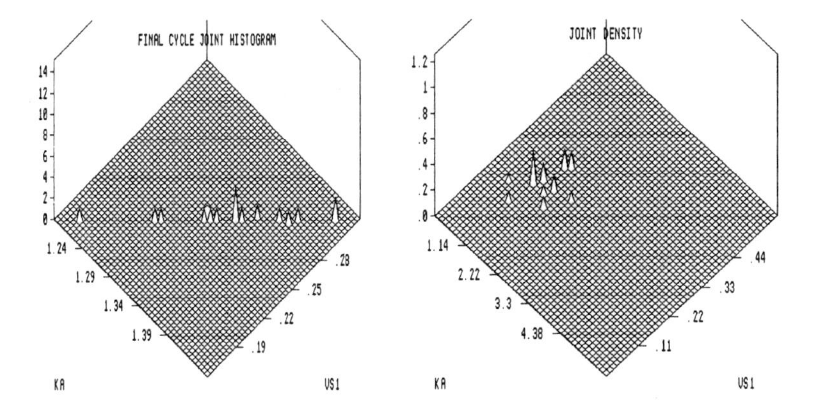

Figure 14. Joint marginal density obtained for KA and VS1 with IT2B, left, and NPAG, right. Note the very high, and probably incorrect, correlation between the parameters seen with the IT2B program.

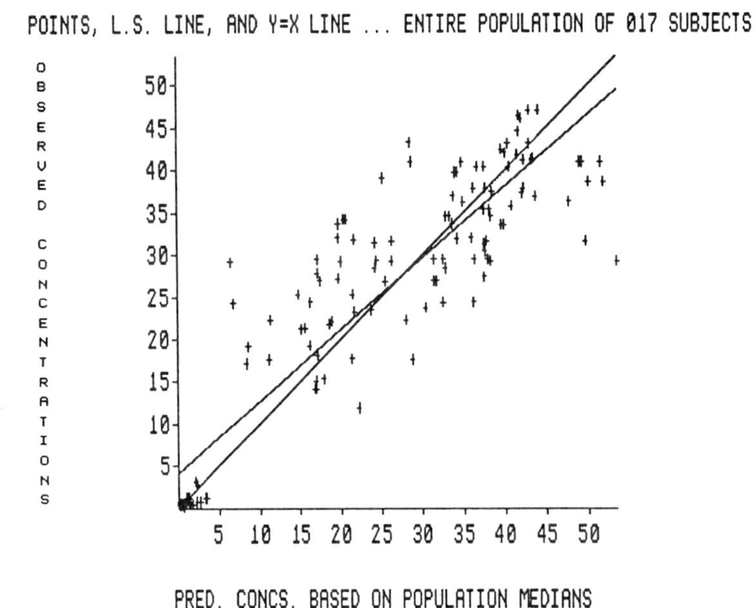

Figure 15. Comparison of predicted and measured serum concentrations based on the population median parameter values, using the IT2B program.

Figures 15 through 18 show the scatterplots of the above analysis of such fits using the IT2B and the NPAG programs.

An analysis of the scatterplots of estimated versus measured serum concentrations, and also those of the NPEM program (not shown), reveals that R^2, the coefficient of the determination (the fraction of the variance explained by the regression relationship between the estimated and the measured data), was greatest with the nonparametric programs. The mean error was the least, and the mean squared error was also the least, showing that the scatterplots actually were more correlated, less biased, and more precise with the two nonparametric modeling methods (Table 2).

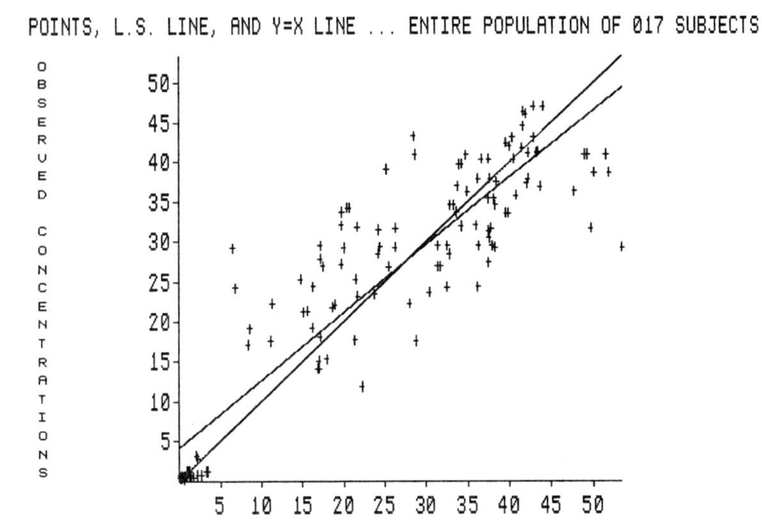

Figure 16. Comparison of predicted and measured serum concentrations based on the population median parameter values, using the NPAG program.

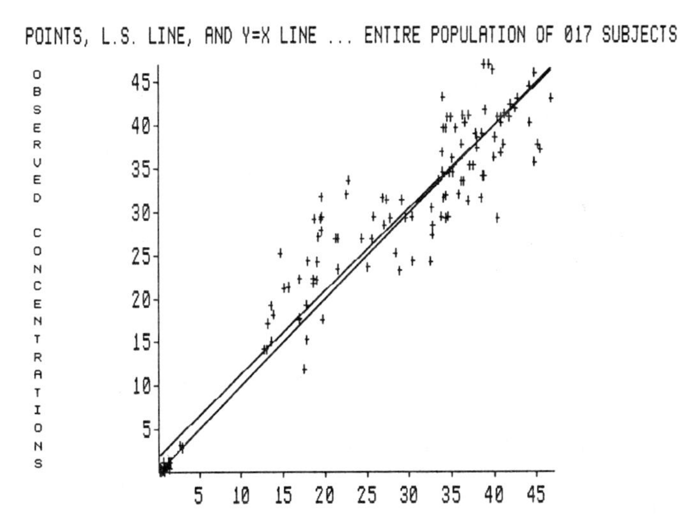

Figure 17. Comparison of predicted and measured serum concentrations based on each subject's individual maximum aposteriori probability (MAP) Bayesian posterior parameter values, predicting only his/her own serum data, using the IT2B program.

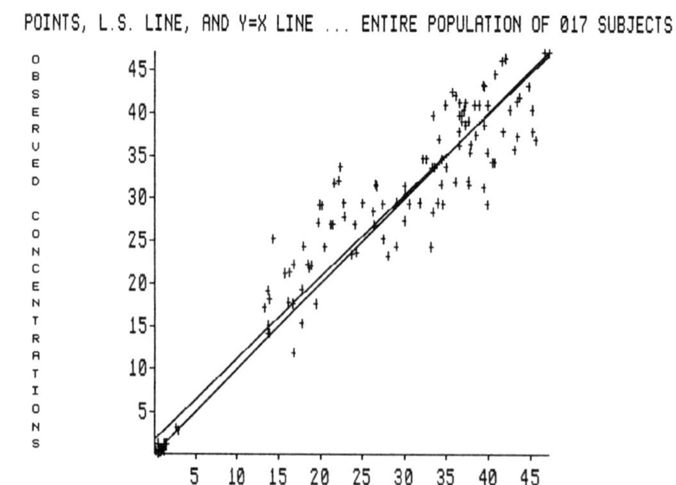

Figure 18. Comparison of predicted and measured serum concentrations based on the median value of each subject's individual nonparametric Bayesian posterior parameter distributions, predicting only his/her own serum data, using the NPAG program.

Table 2. Analysis of estimated versus measured serum concentrations based on population median parameter values, using the IT2B, NPEM, and NPAG programs. R^2: square of the correlation coefficient, ME: mean error, MSE: mean squared error.

	IT2B	NPEM	NPAG
R^2 =	0.814	0.879	0.880
ME =	-0.575	-0.751	0.169
MSE =	48.69	29.01	29.70

9.1 Large and Nonlinear Nonparametric PK/PD Population Models

Large and nonlinear nonparametric population modeling software for combined pharmacokinetic and pharmacodynamic modeling has now been implemented on the large IBM "Blue Horizon" parallel machine at the San Diego Supercomputer Center (SDSC), as a research resource for such work. The user describes the specific structural model and pathways using PC software now in the USC*PACK collection [42]. The model file, the patient data files, and the instructions are sent to the SDSC machine, to a 3-cpu Linux cluster of Dell PC's in our laboratory, or to a larger cluster at USC, over the Web. The analysis is done, and the results are sent back by email to the user's PC, where they are examined just as with the NPEM program described above [39]. More recently, many of these analyses can be done on a single cpu machine such as any PC, though this may require more computer time.

10. NEW DEVELOPMENTS IN NONPARAMETRIC POPULATION MODELING

A significant improvement in nonparametric modeling has recently been made by Leary [27]. He pointed out that the NPEM strategy, based on a large grid covering the parameter space, is computationally quite intensive. He showed that for an analysis of a 5 parameter model, for example, the number of grid points required for a given percent resolution grows greatly with the desired degree of precision, as shown in the table below.

Table 3. Relationship between percent resolution for each parameter of a 5 parameter model, and the number of fixed grid points needed to achieve that resolution.

RESOLUTION	GRID POINTS
10%	100,000
5%	3.2 million
2%	310 million
1%	10 billion

Leary also showed that for a 5 parameter model with only 8 subjects, the likelihood of the results correlated strongly with the number of grid points used in the computations. Using the IBM "Blue Horizon" parallel computer with 1152 processors at the San Diego Supercomputer Center the following relations in Table 4 were obtained between the grid size, computer time, memory required, and the log-likelihood of the results [27].

Table 4. Relationship between Grid Size, Computer time (CPU hours), Memory required, and Log-likelihood of the results obtained with the NPEM program.

Grid Size	CPU hours	Memory (Mbytes)	Log-likelihood
10,000 points	0.1	0.7	-7221.2
40,000	0.5	2.5	-572.8
160,000	2.0	10	-543.7
640,000	7.9	40	-506.2
2,560,000	30.8	160	-462.1
10,280,000	121.8	640	-454.5
40.960,000	501.2	2500	-437.3
164,000,000	2037.4	10000	-433.1

The quality of the result (the log-likelihood) thus depends in large part on the number of grid points used in the analysis. A powerful machine and much computer time are needed to obtain good results with NPEM.

In contrast, Leary has now developed a new nonparametric "adaptive grid" (NPAG) procedure, which combined with an interior point algorithm, has made significant advances in the quality, speed and memory requirements for the analysis

[27]. The method begins with a much smaller and coarser grid, often as low as 5000 grid points for a problem similar to that above. After this is solved, the grid is refined by adding perturbations (extra grid points, about 10 for each previous solution support point), near them. Then the problem is solved again. Once again, new grid points are placed near the previous solution points. This process then continues iteratively, using decreasing perturbations, adaptively obtaining finer and finer resolution of the grid, until a convergence criterion is met.

The outcome has been much improvement in quality of the results, with far less overall computational time and effort. For example, the above problem using NPEM, using 256 processors, took 2037 processor-hours on the IBM Blue Horizon (at the time the fastest non-classified computer in the world), used 10000 Mbytes of memory, and achieved a likelihood of -433.1. In contrast, NPAG, running on a single processor 833 MHz Dell PC, used only 6 Mbytes of memory, took only 1.7 processor hours, and obtained a result with a likelihood of -433.0. NPAG thus greatly reduces the computational time and memory requirements compared to NPEM, and now permits many tasks to be done on a notebook PC that used to require the large parallel machine using NPEM. It also permits still more complex analyses to be done on the larger parallel machine, tasks that previously would have been impossible using NPEM, such as a large 15 parameter model of an antibiotic dosage regimen and its ability to prevent the emergence of resistant organisms. Furthermore, gamma, the measure of intra-individual variability other than the assay error polynomial, can also now be computed directly in the NPAG software. A new version of the NPAG program with this feature is currently being implemented.

11. STRENGTHS AND WEAKNESSES OF PARAMETRIC AND NONPARAMETRIC POPULATION MODELING METHODS

The main strength of the parametric approaches has been that their ability currently to separate inter-individual from intra-individual variability in the population. In the IT2B program, this has been extended to the determination on the environmental noise by first using an explicit polynomial that reflects the relationship between the assay SD and the measured concentration. Then the remaining intraindividual variability can be ascribed to the other sources of environmental noise as discussed earlier.

There has been a general consensus that nonparametric methods are better when it is known that the model parameter distributions are non-Gaussian. However, there has also been a general impression that the nonparametric methods may be less efficient, especially when the parameter distributions are truly Gaussian, and that parameter means and variances may be more reliable for such Gaussian distributions, when parametric modeling methods are used.

This question has recently been examined by Leary [27]. He did a careful study on a simulated population in which the parameter distributions were carefully determined to be Gaussian. A one-compartment, two-parameter model was used, with parameters apparent volume of distribution V and elimination rate constant K. The mean V was 1.1, with standard deviation 0.25. Mean K was 1.0, also with SD = 0.25. The correlation coefficient was set at three different values in three different simulated scenarios: -0.6, 0.0, and +0.6. A single unit intravenous bolus dose was given at time zero, and two simulated serum concentrations were used, each with a 10 %

observational error. Populations ranging in size from 25 to 800 simulated subjects were studied.

The study objectives were to evaluate the consistency, efficiency, and asymptotic convergence rate of the NPAG program, and to compare it with an approximate parametric method using the FOCE approximation (the IT2B program) and with another parametric EM (PEM) method developed by Schumitzky [4], but using the more recent Faure low discrepancy sequence integration method [27]. Over 1000 replications were done to evaluate bias and efficiency.

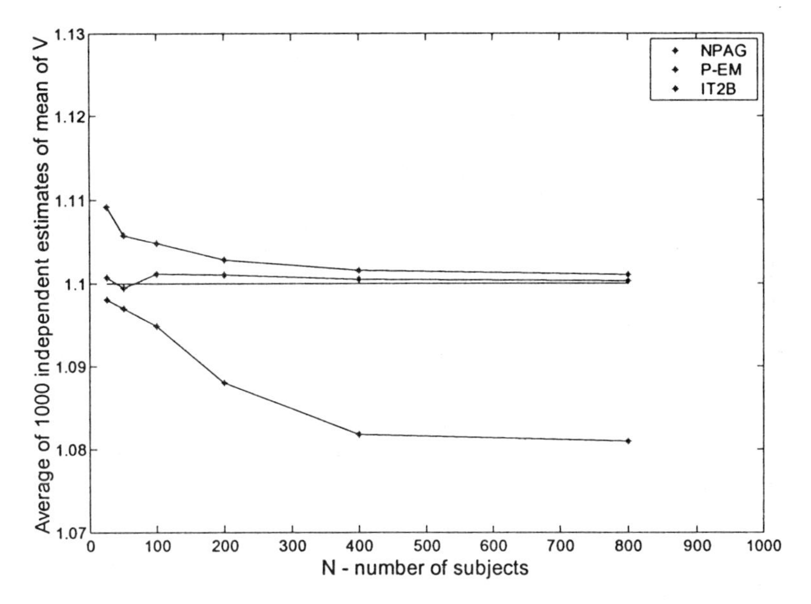

Figure 19. Consistency of estimators of mean of V. NPAG and PEM are consistent. The estimates approach the true value as the number of subjects increases. FOCE IT2B is not consistent.

Figure 19 shows that the NPAG and the PEM programs have consistent behavior. As the number of subjects in the simulated population increased from 25 to 800, the estimated value of the mean of V became closer and closer to the true value, while the IT2B program with the FOCE approximation did not.

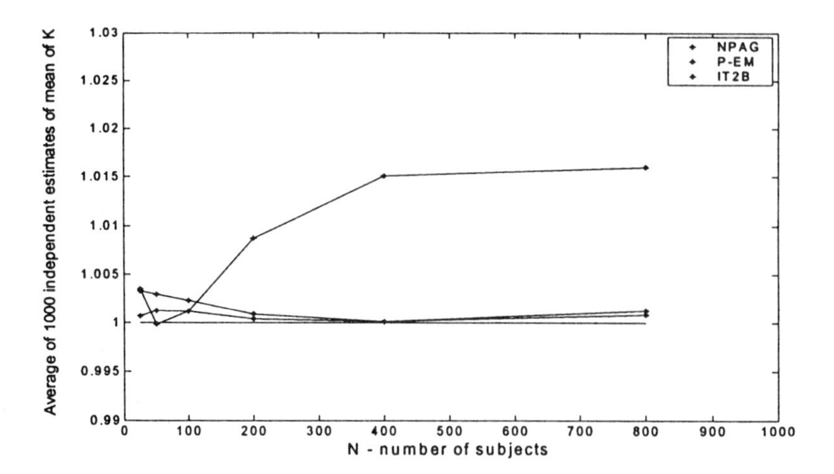

Figure 20. Consistency of estimators of mean of K. NPAG and PEM are consistent. True value of mean K = 1.0. Again, estimates with NPAG and PEM approach the true value as the number of subjects increases. The FOCE IT2B is not consistent.

Figure 20 shows that the same was true for the estimation of K. The estimates of K with NPAG and PEM approached the true value as the number of subjects in the population increased, while the IT2B FOCE estimates again drifted away from the true value.

Figure 21 shows the same behavior for the estimates of the SD of K. NPAG and PEM had consistent behavior whlie the FOCE IT2B again drifted away. Similar behavior was present in the estimation of the SD of V (not shown).

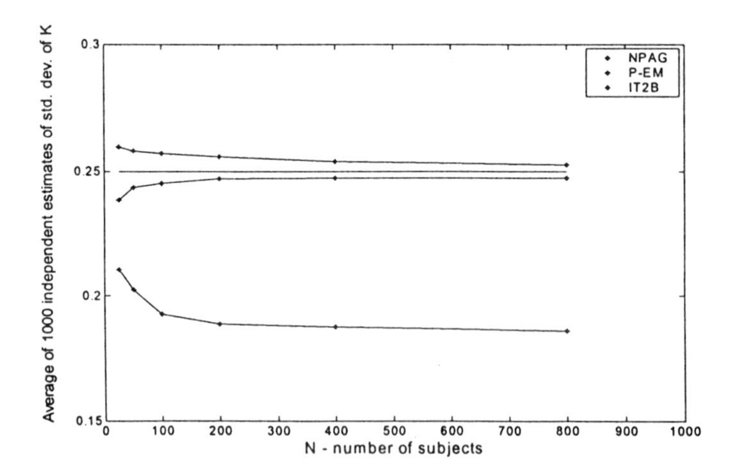

Figure 21. Consistency of estimators of SD of K. NPAG and PEM are consistent. Results approach the true value of SD of K = 0.25 as the number of subjects increases. FOCE IT2B is not consistent. Results actually drift away from the true value with increasing subjects.

Figure 22 shows that the estimation of the correlation coefficient between V and K with NPEM and NPAG was consistent, approaching the true value more and more closely as the number of subjects increased, while the FOCE IT2B started at about zero instead of the true –0.6, and then increased up to +0.2. Similar behavior was seen when the correlation coefficient was 0.0, and also when it was +0.6.

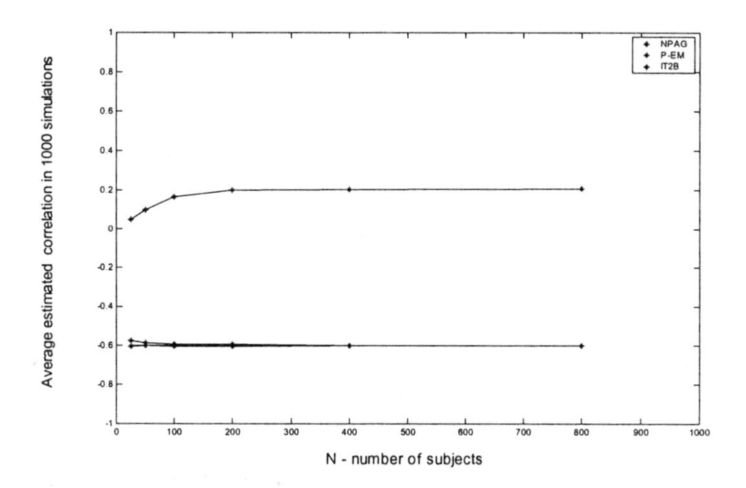

Figure 22. Consistency of estimators of correlation coefficient between V and K. NPAG and PEM are consistent. FOCE IT2B is not, and is severely biased.

In the further examination of statistical efficiency, Figure 23 shows that the efficiency with NPAG and PEM both were about 0.8 throughout, while that of the FOCE IT2B was much less, starting at about 0.4 with 25 subjects, decreasing to less than 0.1 for 800 subjects.

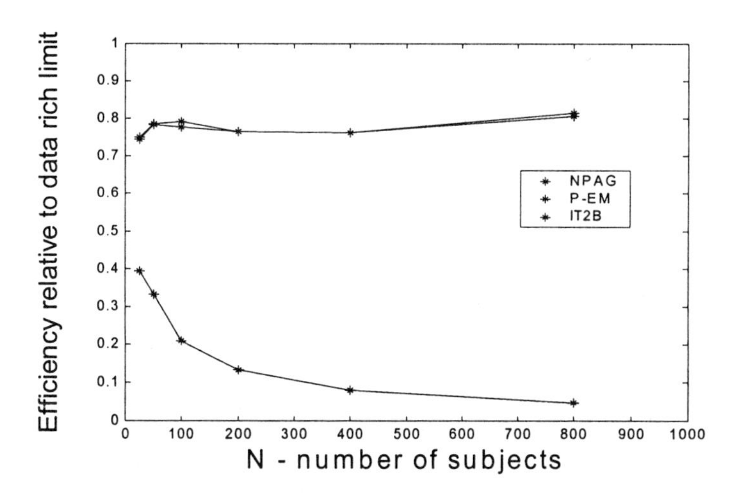

Figure 23. Statistical efficiencies of NPAG and PEM are nearly identical, and are much greater than that of the FOCE IT2B.

Furthermore, as shown in Figure 24, the asymptotic convergence rate was close to theoretical with NPEM and NPAG, while that of the FOCE IT2B was very much less. In order to decrease the SD of a parameter estimate by half, 4 times as many subjects were required by NPAG and PEM, as is consistent with asymptotic theory, while 16 times as many were required by the FOCE IT2B program.

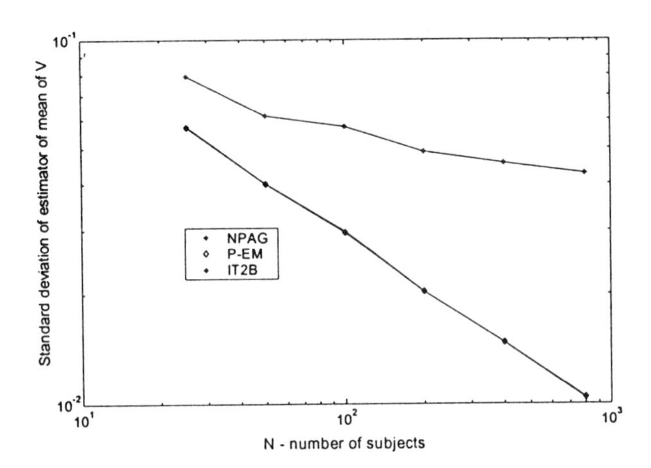

Figure 24. Asymptotic convergence rate of IT2B is much less than that of NPAG and PEM. For NPAG and PEM, the SD decreases by half with 4 times the number of subjects. For the FOCE IT2B, fully 16 times the number of subjects are required for the same SD decrease by half.

In summary, NPAG and PEM, which have exact or very low discrepancy computations of the likelihood, had quite consistent behavior, good efficiency, and good asymptotic convergence. In contrast, the IT2B program, which used the FOCE approximation, suffered a loss of consistency, with a small (1 – 2%) bias for the mean parameter values, moderate (20 – 30%) bias for the SD's, and severe bias for the correlation coefficients, as described above. Further, the NPAG and PEM programs were quite efficient, much more so than the FOCE IT2B, and had much better statistical convergence, close to theoretical.

One weakness of the nonparametric population modeling approach is that until recently there has been no feature to separate the various sources of variability into their respective components – the inter-individual variability due to the diversity among the subjects in the ways they handle the drug, and the intra-individual variability due to the assay and the environmental errors. Current nonparametric methods have not resolved these things. That, however, is what the parametric methods do very well, as described in other papers in this collection. Nevertheless, this deficiency has recently been overcome by Leary [27]. An improved version incorporating this feature is now being implemented.

Another weakness of current nonparametric approaches is that there is no method to obtain confidence limits for these nonparametric distributions. There is nothing analogous to the "standard error of the mean" in parametric distributions. However, since the NP parameter means and variances are at least as good as those seen with P methods [27], confidence intervals based on these means and variances can be used with at least the same justification as those based on parametric modeling methods.

However, to get confidence intervals for the entire nonparametric parameter distributions, one must use the much more computationally intensive bootstrap methods to obtain them.

The strengths of nonparametric approaches are many. First, they have the desirable properties of mathematical consistency, good statistical efficiency, and good asymptotic convergence [8,10,27,32]. Second, no assumptions about the shape of the parameter distributions need to be made. Because of this, nonparametric methods can detect, without additional aid from covariates or descriptors, previously unsuspected subpopulations of patients, as shown in Figures 8-11. Third, instead of obtaining only single-point parameter estimates, one gets multiple estimates, up to one for each subject studied. This is why the nonparametric approach is more informative. It comes the closest to the ideal of the best that could ever be done, namely to obtain the collection of each subject's exactly known parameter values. Fourth, the multiple sets of parameter values provide a tool to circumvent the separation principle [40] and to calculate and optimize the predicted precision with which any candidate dosage regimen is predicted to hit a desired target goal at a desired time. The nonparametric population models thus permit "multiple model" design of dosage regimens to optimize a specific performance criterion [37,40, 41]. The development of such optimally precise dosage regimens, using nonparametric population models, will be discussed in a separate chapter. Table 5 below summarizes the various current strengths and weaknesses of the parametric and nonparametric modeling approaches.

Table 5. Strengths and weaknesses of parametric and nonparametric population modeling methods.

Method	
Parametric	**Nonparametric**
Strengths	
• Get intraindividual variability. • Get parameter confidence intervals.	• Get results with max likelihood. • Don't need Gaussian assumptions. • Consistent, efficient, convergent. • Best quality means, SD's, etc. • Best suited for maximally precise dosage regimens. • Can use conf intervals from Gaussian theory just as well as parametric methods can.
Weaknesses	
• Not consistent, efficient, or convergent. • Constrained by Gaussian assumptions. • Not suited for optimally precise dosage regimens.	• No confidence intervals for the full nonparametric distributions.

11.1 Current Optimal Strategies in Population Modeling

The optimal strategy for making clinically useful population PK/PD models, until the estimation of Gamma is implemented in NPAG, currently appears to be the following sequence of steps:

1. determine the assay error pattern explicitly and obtain the assay error polynomial.
2. use a parametric population modeling program such as IT2B, to obtain gamma.
3. having both of the above, use a nonparametric population modeling program such as NPAG to obtain the most likely entire discrete joint parameter distribution.

This sequence of steps currently appears to make optimal use of the information about the assay SD, often 1/3 to 1/2 of the overall environmental SD, and the raw data present in the population studied, to obtain the most probable parameter distributions. The approach also appears to provide optimal tools to develop dosage regimens to achieve desired target goals with maximum precision. This is useful in drug development for designing maximally precise dosage regimens for specific populations of patients, and clinically, for developing the most precise initial regimen for individual patients where the margin of safety for the drug in question is small and the dosage must be carefully individualized. The nonparametric population models, with their multiple support points, are extremely well suited for the new "multiple model" method of dosage design [35,41], which can specifically evaluate the weighted squared error of the failure of a dosage regimen to achieve a desired target goal, and optimize the regimen to minimize this error. This will be described in more detail in another chapter.

Acknowledgements

Supported by US Government grants LM 05401, RR 01629, RR11526, and GM65619.

References

1. Variability in Drug Therapy: Description, Estimation, and Control. Ed by Rowland M, Sheiner L, and Steimer JL. Raven Press, New York, 1985.
2. Sheiner L, Beal S, Rosenberg B, and Marathe V: Forecasting Individual Pharmacokinetics. Clin. Pharmacol. Therap. 26: 294-305, 1979.
3. Aarons L: The Estimation of Population Pharmacokinetic Parameters using an EM Algorithm. Comput. Methods and Programs in Biomed. 41: 9-16, 1993.
4. Schumitzky A: EM Algorithms and Two Stage Methods in Pharmacokinetic Population Analysis. In: Advanced Methods of Pharmacokinetic and Pharmacodynamic Systems Analysis II. D. Z.D'Argenio, ed., Plenum Press, New York, 1995, pp. 145-160.
5. Beal S, and Sheiner L: NONMEM User's Guide I. Users Basic Guide. Division of Clinical Pharmacology, University of California, San Francisco, 1979.
6. Sheiner L: The population Approach to Pharmacokinetic Data Analysis: Rationale and Standard Data Analysis Methods. Drug Metab. Rev. 15: 153-171, 1984.
7. Beal S: Population Pharmacokinetic Data and Parameter Estimation Based on their First Two Statistical Moments. Drug Metab. Rev. 15: 173-193, 1984.

8. De Groot M: Probability and Statistics, 2nd edition, 1986, reprinted 1989, Addison-Wesley, Reading MA, pp. 334-336.
9. Spieler G and Schumitzky A: Asymptotic Properties of Extended Least Squares Estimators with Approximate Models. Technical Report 92-4, Laboratory of Applied Pharmacokinetics, University of Southern California School of Medicine, 1992.
10. Spieler G and Schumitzky A: Asymptotic Properties of Extended Least Squares Estimates with Application to Population Pharmacokinetics. Proceedings of the American Statistical Society, Biopharmaceutical Section, 1993, pp. 177-182.
11. Rodman J and Silverstein K: Comparison of Two Stage (TS) and First Order (FO) Methods for Estimation of Population Parameters in an Intensive Pharmacokinetic (PK) Study. Clin. Pharmacol. Therap. 47: 151, 1990.
12. Maire P, Barbaut X, Girard P, Mallet A, Jelliffe R, and Berod T: Preliminary results of three methods for population pharmacokinetic analysis (NONMEM, NPML, NPEM) of amikacin in geriatric and general medicine patients. Int. J. Biomed. Comput., 36: 139-141, 1994.
13. Lindstrom M and Bates D: Nonlinear Mixed-Effects Models for Repeated Measures Data. Biometrics, 46: 673-687, 1990.
14: Vonesh E and Carter R: Mixed Effects Nonlinear Regressions for Unbalanced Repeated Measures. Biometrics, 48: 1-17, 1992.
15. Jelliffe R: Explicit Determination of laboratory assay error patterns: a useful aid in therapeutic drug monitoring. No. DM 89-4 (DM56). Drug. Monit. Toxicol. 10: (4) 1-6, 1989.
16. Jelliffe R, Schumitzky A, Van Guilder M, Liu M, Hu L, Maire P, Gomis P, Barbaut X, and Tahani B: Individualizing Drug Dosage Regimens: Roles of Population Pharmacokinetic and Dynamic Models, Bayesian Fitting, and Adaptive Control. Therap. Drug Monit. 15: 380-393, 1993.
17. Bustad A, Jelliffe R, and Terziivanov D: A comparison of Parametric and Nonparametric Methods of Population Pharmacokinetic Modeling. A poster presentation at the Annual Meetings of the American Society for Clinical Pharmacology and Therapeutics, Atlanta, GA, March 26, 2002.
18. Jelliffe R: Estimation of Creatinine Clearance in Patients with Unstable Renal Function, without a Urine Specimen. Am. J. Nephrol. 22: 320-324, 2002.
19. Jazwinski A: Stochastic Processes and Filtering Theory. Academic Press, New York, 1970.
20. Van Guilder M, Leary R, Schumitzky A, Wang X, Vinks S, and Jelliffe R: Nonlinear Nonparametric Population Modeling on a Supercomputer. Presented at the 1997 ACM/IEEE SC97 Conference, San Jose CA, November 15-21, 1997.
21. De Groot M: Probability and Statistics, 2nd edition, 1986, reprinted 1989, Addison-Wesley, Reading MA, pp. 334-336.
22. Spieler G and Schumitzky A: Asymptotic Properties of Extended Least Squares Estimators with Approximate Models. Technical Report 92-4, Laboratory of Applied Pharmacokinetics, University of Southern California School of Medicine, 1992.
23. Spieler G and Schumitzky A: Asymptotic Properties of Extended Least Squares Estimates with Application to Population Pharmacokinetics. Proceedings of the American Statistical Society, Biopharmaceutical Section, 1993, pp. 177-182.
24. Mallet A: A Maximum Likelihood Estimation Method for Random Coefficient Regression Models. Biometrika. 73: 645-656, 1986.
25. Schumitzky A: The Nonparametric Maximum Likelihood Approach to Pharmacokinetic Population Analysis. Proceedings of the 1993 Western Simulation Multiconference - Simulation for Health Care. Society for Computer Simulation, 1993, pp 95-100.
26. Bertsekas D: Dynamic Programming: deterministic and stochastic models. Englewood Cliffs (NJ): Prentice-Hall, pp.144-146, 1987.
27. Leary R, Jelliffe R, Schumitzky A, and Van Guilder M: A Unified Parametric/Nonparametric Approach to Population PK/PD Modeling. Presented at the Annual Meeting of the Population Approach Group in Europe, Paris, France, June 6-7, 2002.
28. Bertilsson L: Geographic/Interracial Differences in Polymorphic Drug Oxidation. Clin. Pharmacokinet. 29: 192-209, 1995.
29. Lindsay B: The Geometry of Mixture Likelihoods: A General Theory. Ann. Statist. 11: 86-94, 1983.
30. Schumitzky A: Nonparametric EM Algorithms for Estimating Prior Distributions. App. Math. and Computation. 45: 143-157, 1991.
31. Maire P, Barbaut X, Girard P, Mallet A, Jelliffe R, and Berod T: Preliminary results of three methods for population pharmacokinetic analysis (NONMEM, NPML, NPEM) of amikacin in geriatric and general medicine patients. Int. J. Biomed. Comput., 36: 139-141, 1994.

32. Spieler G and Schumitzky A: Asymptotic Properties of Extended Least Squares Estimates with Application to Population Pharmacokinetics. Proceedings of the American Statistical Society, Biopharmaceutical Section, 1993, pp. 177-182.
33. Hurst A, Yoshinaga M, Mitani G, Foo K, Jelliffe R, and Harrison E: Application of a Bayesian Method to Monitor and Adjust Vancomycin Dosage Regimens. Antimicrob. Agents and Chemotherap. 34: 1165-1171, 1990.
34. Bayard D, Milman M, and Schumitzky A: Design of Dosage Regimens: A Multiple Model Stochastic Approach. Int. J. Biomed. Comput. 36: 103-115, 1994.
35. Bayard D, Jelliffe R, Schumitzky A, Milman M, and Van Guilder M: Precision Drug Dosage Regimens using Multiple Model Adaptive Control: Theory and Application to Simulated Vancomycin Therapy. in Selected Topics in Mathematical Physics, Professor R. Vasudevan Memorial Volume, ed. by Sridhar R, Srinavasa Rao K, and Vasudevan Lakshminarayanan, Allied Publishers Inc., Madras, 1995, pp. 407-426.
36. Mallet A, Mentre F, Giles J, Kelman A, Thompson A, Bryson S, and Whiting B: Handling Covariates in Population Pharmacokinetics with an Application to Gentamicin. Biomed. Meas. Infor. Contr. 2: 138-146, 1988.
37 Taright N, Mentre F, Mallet A, and Jouvent R: Nonparametric Estimation of Population Characteristics of the Kinetics of Lithium from Observational and Experimental Data: Individualization of Chronic Dosing Regimen Using a New Bayesian Approach. Therap. Drug Monit. 16: 258-269, 1994.
38. Jerling M: Population Kinetics of Antidepressant and Neuroleptic Drugs. Studies of Therapeutic Drug Monitoring data to Evaluate Kinetic Variability, Drug Interactions, Nonlinear Kinetics, and the Influence of Genetic Factors. Ph. D. Thesis, Division of Clinical Pharmacology, Department of Medical Laboratory Sciences and Technology, Karolinska Institute at Huddinge University Hospital, Stockholm, Sweden, 1995, pp 28-29.
39. Van Guilder M, Leary R, Schumitzky A, Wang X, Vinks S, and Jelliffe R: Nonlinear Nonparametric Population Modeling on a Supercomputer. Presented at the 1997 ACM/IEEE SC97 Conference, San Jose CA, November 15-21, 1997.
40. Bertsekas D: Dynamic Programming: deterministic and stochastic models. Englewood Cliffs (NJ): Prentice-Hall, pp.144-146, 1987.
41. Jelliffe R, Schumitzky A, Bayard D, Milman M, Van Guilder M, Wang X, Jiang F, Barbaut X, and Maire P: Model-Based, Goal-Oriented, Individualised Drug Therapy: Linkage of Population Modelling, New "Multiple Model" Dosage Design, Bayesian Feedback and Individualised Target Goals. Clin. Pharmacokinet. 34: 57-77, 1998.
42. Jelliffe R, Schumitzky A, Van Guilder M, and Jiang F: User Manual for Version 10.7 of the USC*PACK Collection of PC Programs, USC Laboratory of Applied Pharmacokinetics, USC School of Medicine, Los Angeles, CA, December 1, 1995.

ROLE OF PRECLINICAL METABOLISM AND PHARMACOKINETICS IN THE DEVELOPMENT OF CELECOXIB

Susan K. Paulson[1] and Timothy J. Maziasz[2]

1. INTRODUCTION

Nonsteroidal anti-inflammatory drugs (NSAIDs) are the principle therapy for the treatment of pain, fever, and inflammation (e.g. the signs and symptoms of arthritic disease). The use of these drugs are associated with significant toxicities in the gastrointestinal (GI) tract, along with renal and platelet side effects (bleeding) (Borda and Koff, 1992). Smith and Willis (1971) and Vane (1971) proposed over 30 years ago that the mechanism of action of NSAIDs was due to the blockade of the production of prostaglandins via inhibition of the enzyme cylooxygenase (COX). The clinical significance of the discovery of two cyclooxygenases (COX-1 and COX-2) in the early 1990s along with their significant role in physiological function and disease has been amply reviewed (Needleman and Isakson, 1997; Hawkey et al., 1999; Silverstein et al., 2000). The subsequent development and approval of inhibitors that target COX-2 relative to COX-1 *in vivo* brought into clinical use new drugs that obviated the toxicological limitations of older NSAIDs. Celecoxib (Celebrex®), the first drug approved based on this COX-2 selectivity, entered clinical use in 1999. In the context of this chapter, COX-2-selective refers to the extent to which a drug inhibits COX-2 *in vivo* relative to COX-1.

The structures of the new COX-2-selective drugs (e.g. celecoxib, rofecoxib, and valdecoxib) are markedly different from those of the earlier NSAIDs. Celecoxib is a diaryl pyrazole that belongs to the general class of diaryl heterocyclic inhibitors (Figure 1). Its unique structure differs from salicylates, pyrazolones, fenamic acid, acetic and propionic acid, and oxicam-based NSAIDs. Given the lack of precedence with this structure, it is now not surprising that some new experiences were encountered during the initial ADME characterizations of this molecule.

[1] Susan K. Paulson, NeoPharm Inc., Lake Forest, Illinois 60045.

[2] Timothy J. Maziasz, Pharmacia, Skokie, Illinois 60077.

Applications of Pharmacokinetic Principles in Drug Development
Edited by Krishna, Kluwer Academic/Plenum Publishers, 2004

This chapter presents a case study of the role played by metabolism and pharmacokinetics in the development of celecoxib. The emphasis will be the application of pharmacokinetic principles throughout the preclinical development of celecoxib along with the strategies utilized to streamline the development process. This discussion will include the approaches used to determine toxicokinetics, estimate clinical dose, predict drug-drug interactions and evaluate potential for food effects on drug absorption. In addition, the discovery of polymorphisms in the metabolism of celecoxib in four different species (mouse, rat, dog and human) and their significance for the preclinical program for this drug is addressed here.

2. METABOLISM

The integration of *in vitro* ADME studies into drug discovery and development programs rapidly increased over the last decade. *In vitro* metabolism studies are now routinely used to provide rationale for selection of animal species for toxicology studies, to determine the metabolic pathway for human, identify the oxidative enzymes involved in drug metabolism, predict drug-drug interactions and evaluate the impact of a polymorphism on the pharmacokinetic variability. At the time that celecoxib entered development the usefulness of these tools was discussed in depth in the literature (Wrighton et al., 1993; Wrighton et al., 1995). Below is a discussion of how these technologies were incorporated into the celecoxib drug development program.

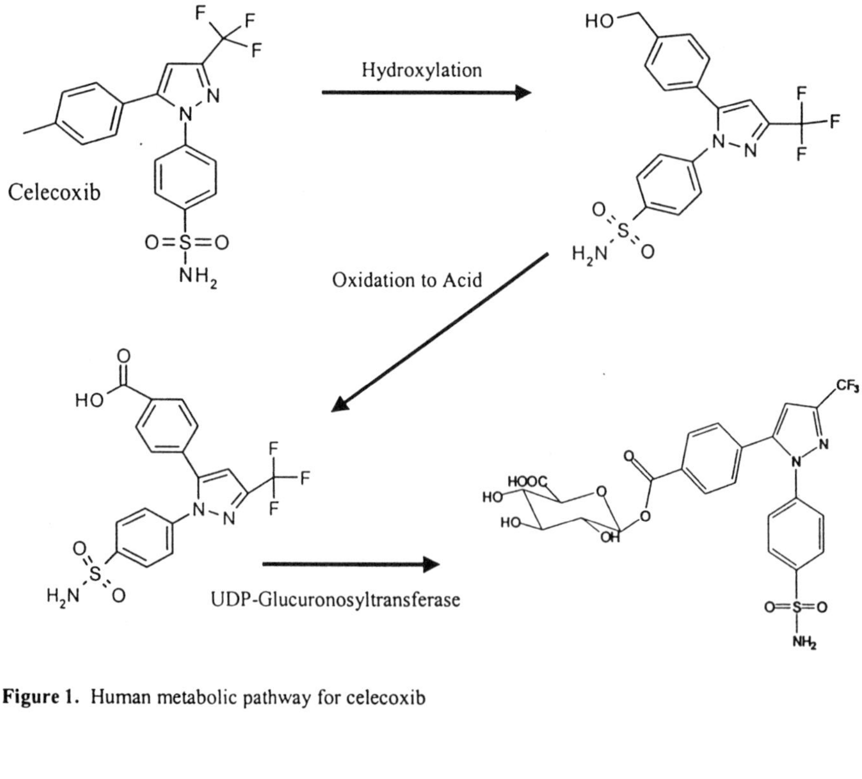

Celecoxib

Hydroxylation

Oxidation to Acid

UDP-Glucuronosyltransferase

Figure 1. Human metabolic pathway for celecoxib

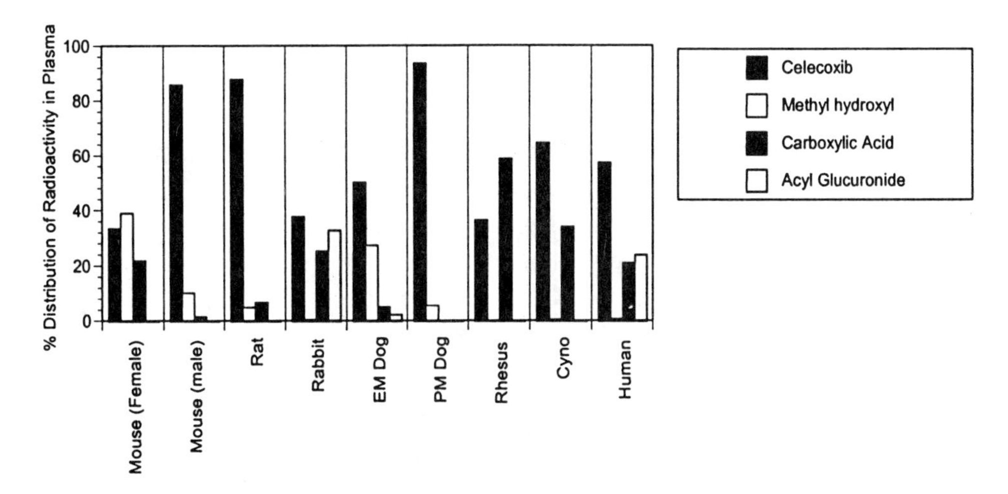

Figure 2. Metabolite profiles in plasma after a single iv dose of 1, 1 and 5 mg/kg [^{14}C]celecoxib to cynomolgus monkey, rhesus monkey and dog, respectively and at the approximate Cmax after oral doses of 10 mg/kg, 35 mg/kg and 300 mg [^{14}C]celecoxib to rabbit, mouse and human, respectively.

2.1. Celecoxib Metabolic Pathway and Comparative Metabolism

Comparative metabolism studies allow for the rational selection of species for toxicology studies and for use in allometric scaling. The metabolic profile of celecoxib was first determined *in vitro* using human liver microsomes prior to selection of celecoxib for development. The major route of metabolism was shown to be methyl hydroxylation followed by further oxidation of the alcohol to the carboxylic acid (Figure 1, Paulson et al 2000a). Once celecoxib entered development, both *in vitro* studies and radiolabeled ADME studies confirmed that the metabolites produced by human liver microsomes were the same as produced by all species being considered for use in toxicology, reproductive toxicology and carcinogenicity studies (Figure 2, Paulson et al., 2000bc). An additional metabolite found in rabbit *in vivo* but not human liver microsomes was a phenylhydroxy metabolite (Zhang et al., 2000).

Subsequently, the metabolic profile of celecoxib was confirmed in human in a radiolabeled ADME study conducted early in clinical development. The major metabolites of celecoxib seen in healthy male volunteers confirmed that the methyl hydroxylation was the primary route of elimination of celecoxib; phenyl ring hydroxylation was found not to occur in humans (Paulson et al., 2000a; Karim et al., 1997). However, a glucuronide conjugate of the carboxylic acid metabolite was a metabolite identified in the human ADME study that had not been previously observed in animals. Subsequently, a biliary excretion study in rats and an ADME study in EM dogs were conducted demonstrating that species produced the glucuronide conjugate of the carboxylic acid metabolite (Paulson et al., 2000bc).

The major route of excretion of celecoxib was biliary (Figure 3). The major excretion product was the carboxylic acid metabolite. Very little celecoxib was excreted unchanged after iv administration to rat (~1%), dog (<1%) or monkey (<1%) or after oral administration to mouse (<1% for female; 8% for male), rabbit (<1%) or healthy male volunteers (~2%). In general, metabolic pathway of celecoxib is similar across species and celecoxib appears as the major circulating component. Collectively, the species used for toxicology assessments and allometric scaling generated all the metabolites identified in healthy male volunteers.

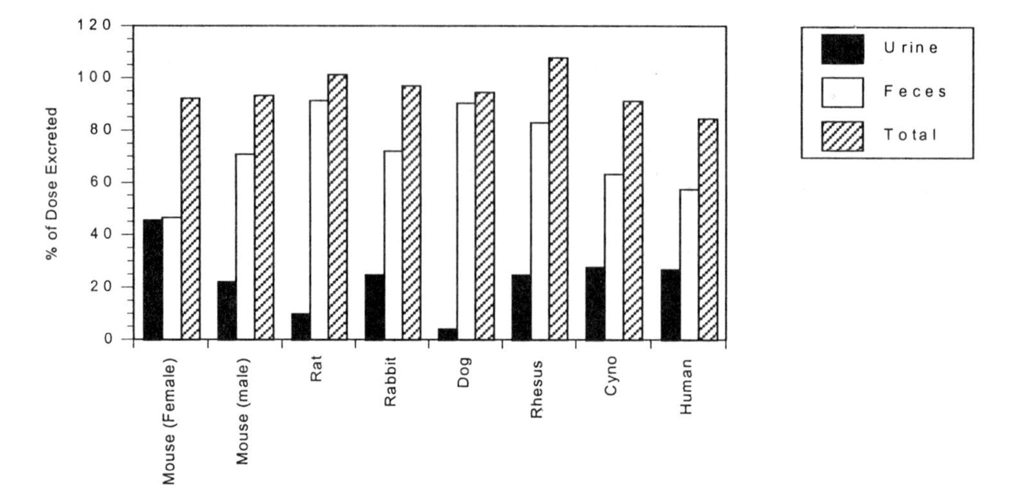

Figure 3. Excretion of radioactivity in urine and feces after a single iv dose of 1, 1 and 5 mg/kg [^{14}C]celecoxib to cynomolgus monkey, rhesus monkey and dog, respectively and oral doses of 10 mg/kg, 35 mg/kg and 300 mg [^{14}C]celecoxib to rabbit, mouse and human, respectively.

2.2. Prediction of Drug-Drug Interactions

Celecoxib entered development phase about the time *in vitro* metabolism tools were being developed and utilized to identify the cytochrome P450 isozyme involved in metabolism of new drugs and to predict the ability of a drug to inhibit metabolism (Wrighton et al., 1993; Tucker, 1992). The identity of the CYP2C9 as the cytochrome P450 isoform involved in the hydroxylation celecoxib was determined using (1) regression analysis with selected CYP450 marker substrate probes; (2) inhibition studies with chemical inhibitors and (3) metabolism with human cDNA-espressed CYP450s (Karim et al., 1997). The role of CYP2C9 in the metabolism of celecoxib in humans was supported by a drug-drug interaction study with the CYP2C9-specific inhibitor, fluconazole. Co-administration of celecoxib with the CYP2C9 inhibitor fluconazole to healthy volunteers increased the AUC and Cmax of 134% and 68%, respectively; while co-administration with ketoconazole (a CYP3A4 inhibitor) produced only minimal effect on celecoxib plasma concentration profiles (Karim et al., 1998). These *in vitro* studies provided some rationale for the conduct of subsequent drug-drug interactions studies of celecoxib with phenytoin, warfarin and tolbutamide (FDA NDA #20998, 1998). The roles of major role of CYP2C9 in the metabolism of celecoxib have also been recently reported by Tang et al., 2000.

There is a genetic polymorphism in the coding region of the CYP2C9 gene. Single substitutions have been described specifically CYP2C9*2 ($Arg_{144}Cys$) and CYP2C9*3 ($Ile_{359}Leu$) (Romkes et al., 1991; Kimura et al., 1987). The CYP2C9*3 allele has been shown associated with slower metabolism of CYP2C9 substrates, tolbutamide, phenytoin and warfarin (Furuya et al., 1995, Sullivan-Klose et al., 1996). Among the approximately 500-1000 healthy subjects treated with 200 mg celecoxib in the clinical development program, 5 subjects exhibited outlier plasma concentrations that were 5-9-fold greater than the mean. Two of the five subjects were genotyped and found with the CYP2C9*3 allele (FDA NDA #20998, 1998, Davies et al., 2000). This allele has been shown to be poor metabolizing for celecoxib by additional *in vitro* studies (Tang et al., 2001). These data further underscore the importance of CYP2C9 in the metabolism of celecoxib and the impact of genetic polymorphism on circulating celecoxib concentrations.

Celecoxib was examined for its ability to inhibit cytochrome P450 isoform-specific catalytic activity associated with CYP2C9, CYP2C19, CYP2D6 and CYP3A4 (FDA NDA #20998, 1998). Celecoxib was not a potent *in vitro* inhibitor of CYP2C9, CYP2C19 or CYP3A4 (Table 1). Celecoxib appears to be a moderately potent *in vitro* inhibitor of CYP2D6 with an apparent Ki that was approximately 3-fold higher than plasma celecoxib concentrations observed in human patients given 200 mg. However, a subsequent drug-drug interaction study conducted of celecoxib with dextromethorphan, a substrate for CYP2D6, demonstrated that celecoxib is not a clinically relevant inhibitor of CYP2D6 (Karim et al., 2000).

Table 1. Apparent Ki of celecoxib and chemical inhibitors on CYP Isoform activity

CYP Isoform	Marker Activity	Inhibitor	Apparent Ki (μM)
CYP2C9	Tolbutamide 4-hydroxylation	Celecoxib Sulphaphenazole	44.4 0.585
CYP2C19	S-Mephenytoin 4'-hydroxylation	Celecoxib Omeprazole	17.8 5.64
CYP2D6	Bufuralol 1'-hydroxylation	Celecoxib Quinidine	4.19 0.466
CYP3A4	Testosterone 6β-hydroxylation	Celecoxib Ketoconazole	106 0.0483

3. POLYMORPHISMS AND CELECOXIB TOXICOKINETICS

Toxicology studies are conducted early in the drug development process to assess the potential hazards of the drug candidate before clinical testing. Safety assessment studies with celecoxib began with the compound's entry into development in mid-1994. The general toxicology studies utilized the rat and dog as the rodent and non-rodent species, respectively. Dose ranging for the mouse cancer bioassay was also initiated early in development.

In order to extrapolate to human any findings in animals given a new chemical entity (NCE), it is important to understand the inter-species relationship between systemic exposure to NCE and toxicity (i.e., toxicokinetics). One of the major challenges of conducting toxicology studies is to identify doses and dose regimens that provide adequate systemic exposure to the drug candidate for the duration of the study. Ideally, animals in toxicology and carcinogenicity studies should be treated with dose regimens of a compound that produce steady state concentrations that are multiples of the anticipated therapeutic exposure in patients. In addition, steady state plasma concentrations should be targeted for the duration of evaluations that can be up to 6 months for rodents and 1 year for non-rodents in chronic toxicology studies, and up to 2 years in rodent cancer bioassays. Therefore, it is necessary to understand the behavior of the drug candidate at doses higher than those typically evaluated in pharmacology models. This is typically done in separate studies for each species prior to the initiation of pivotal toxicity studies. The unique toxicokinetics of celecoxib presented an interesting challenge to establishing ideal dosing regimens, primarily due to (1) the discovery of polymorphism in the canine clearance of celecoxib, and (2) the existence of gender dimorphisms in celecoxib clearance in the rat and mouse.

3.1 Dog Studies

The metabolic profile of celecoxib observed in dogs in bioavailability studies was similar to human metabolism *in vitro*, which supported use of this species in non-rodent toxicology evaluations (Karim et al., 1997; Paulson et al., 2000a). Although the primary objective of the preclinical safety characterizations was to support early clinical testing, it was understood that these studies would also provide early proof-of-concept in animals of the improved gastrointestinal (GI) safety profile predicted for the mechanism of COX-2-selective inhibition. Since the canine GI tract is very sensitive to prostaglandin inhibition, it is unable to tolerate even sub-therapeutic doses of NSAIDs (Taylor and Crawford, 1968; Spyridakis et al., 1986; Kore, 1990); therefore, dogs provide a sentinel species for detecting GI mucosal injury, and for testing the hypothesis of improved GI safety profile of COX-2-selective inhibitors, like celecoxib. It was predicted that doses of celecoxib that produce analgesic/anti-inflammatory exposures would not produce the GI mucosal injury associated with classical non-selective NSAIDs.

The initial toxicology study in beagle dogs was a 4-week oral study. Celecoxib was administered in single daily doses of 0, 25, 50, 100 or 250 mg/kg. The low dose (25 mg/kg) was selected to target systemic exposures to celecoxib that were just above plasma concentrations of celecoxib that were anticipated to produce maximal anti-inflammatory benefit in the clinic. The rat adjuvant arthritis model, among the various laboratory models used in inflammation research, has been used to predict clinical efficacy of anti-inflammatory agents based on a mechanism of action of prostaglandin inhibition (Dubinsky etal., 1987). Therefore, the celecoxib AUC at the ED_{80} for the rat adjuvant arthritis was selected as the targeted systemic exposure for the low dose in the toxicity study. Preliminary pharmacokinetic data in dog, indicated that a dose of 25 mg/kg/day should produce an AUC about 2-3 times the ED_{80} in the rat adjuvant arthritis model. Preliminary pharmacokinetic data in the dog also indicated that maximal absorption of drug administered in capsule form was at a dose of $\geq$ 100 mg/kg/day, so a dose of 250 mg/kg/day was expected to provide maximal possible exposure following capsule administration. It was projected that the highest dose (250 mg/kg) would produce plasma concentrations 10 to 20-fold greater than estimated therapeutic levels. This range of projected exposures was deemed appropriate to meet the goals of the study.

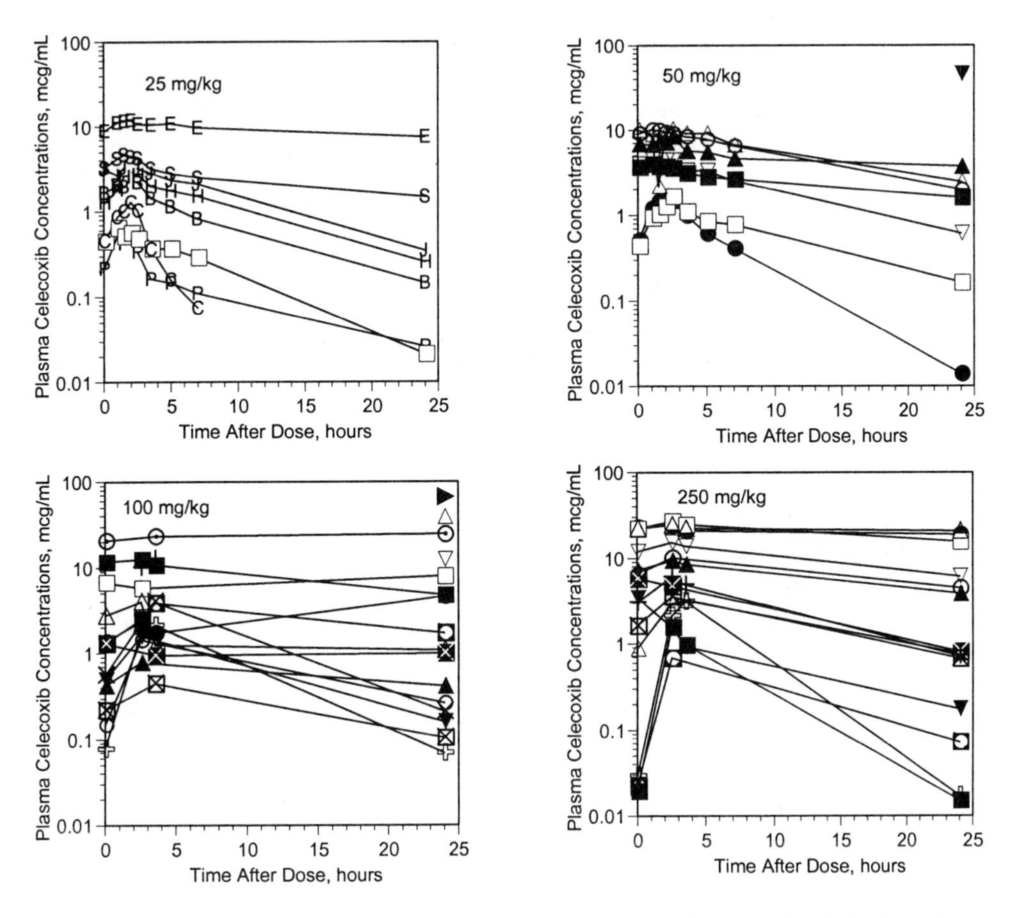

Figure 4. Plasma concentration of celecoxib on Day 27 in dogs given 25 and 50 mg celecoxib/kg/day and on Day 15 in dogs given 100 and 250 mg celecoxib/kg/day in a 4 week toxicity study. Plasma celecoxib concentrations are also plotted from blood samples from 5 dogs subject to unscheduled sacrificed on Days 11-14 of the study.

The lowest dosage of celecoxib (25 mg/kg/day) was tolerated and defined the No-Observed Effect Level (NOEL) for the study; however, the higher dosages produced mucosal injury (erosions, ulceration, and perforation) of the small intestine, which is pathognomonic of the toxicity of NSAIDs. Five animals given dosages of 50-250 mg/kg/day were found in a moribund state thus necessitating unscheduled necropsies, which subsequently confirmed severe GI mucosal injury, including perforation of the small intestine with exsanguinations.

GI toxicity is consistent with inhibition of COX-1, however, its incidence in a study with a COX-2-selective inhibitor was initially surprising. Perhaps more surprising was the extremely variable systemic exposure in every group and that plasma concentrations varied up to 50-fold even at the NOEL (Figure 4). The greatest variability was observed in the 100 mg/kg dose group where Cmax values for celecoxib range from 1.1 - 66.4 μg/mL. A toxicology study conducted in parallel in rats showed plasma concentrations to reach a steady state and be well defined with dosage, whereas steady state plasma concentrations of celecoxib were less apparent in dogs. Marked accumulation of celecoxib in plasma, beyond that expected from the half-life of celecoxib in dog (~9 hours) occurred in several animals. Figure 5 shows the marked accumulation of celecoxib in Dog No. 2 given 25 mg celecoxib/kg and Dogs No. 1-4 given 100 mg celecoxib/kg. There was also evidence that accumulation of celecoxib may have precipitated the GI mucosal injury in the five dogs subjected to unscheduled necropsy early in the study as evidenced by their plasma concentrations (ranging between 12.4 to 66.4 μg/mL), which were sufficient to produce non-selective inhibition of COX-1. Even at the NOEL (25 mg/kg/day), plasma concentrations of celecoxib in one dog had reached ~10 μg/mL by the last day of the study (Figure 4).

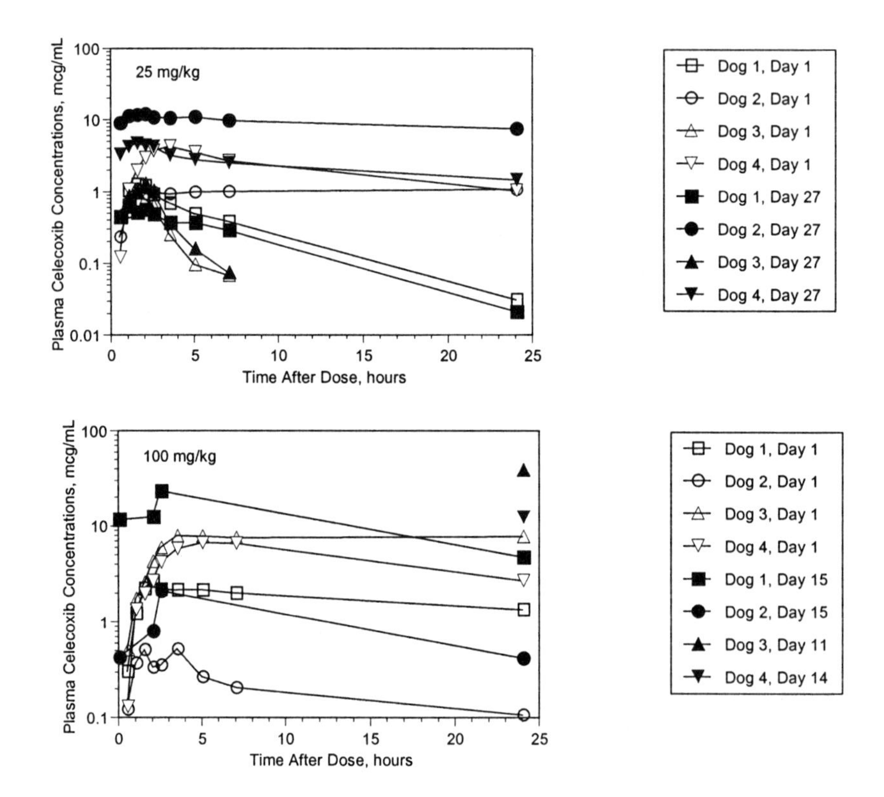

Figure 5. Plasma concentrations of celecoxib on Days 1 and 27 of treatment in four dogs given 25 mg/kg (upper panel) and in four dogs given 100 mg/kg (lower panel) celecoxib once daily for 11, 14 or 15 days. Plasma concentration data from Day 1 of dosing are represented by open symbols; closed symbols show plasma concentration data after repeated dosing.

The results from the 4-week study posed a problem for the successful conduct of longer term toxicology studies; specifically, a dosing regimen was needed for dogs that would adequate provide steady state exposures for toxicological evaluation without causing accumulation and GI mucosal injury. Clearly, it was necessary to understand the reason for the variability of celecoxib plasma concentrations in dogs. Moreover, because of the potential value of the canine as a sentinel species for validating the improved GI safety profile of COX-2-selective inhibitors, there was strong motivation to continue studies with dogs and not pursue another non-rodent species (ie. monkeys).

Initial studies initiated to investigate the mechanism(s) of the high variability in celecoxib plasma concentrations, focused on characterizing celecoxib clearance in larger numbers of dogs. After celecoxib was given to 38 dogs (19 male & 19 female) a polymorphism was revealed that was characterized by phenotypes of poor (PM) or extensive metabolizers (EM) of celecoxib (Figure 6, left panel). This polymorphism was subsequently found to occur with equal incidence across gender and independently of age (Paulson et al., 1999a). The distribution of the two pheotypes in the populations was 45% of the EM phenotype and 53.5% of the PM phenotype; 1.65% of the population studied could not be adequately characterized (Paulson et al., 1999a). Further characterization of the two phenotypes with a study designed to determine *in vivo* K_m of celecoxib revealed that saturation of hepatic clearance of celecoxib, and the potential for accumulation, occurred at relatively low plasma concentrations in the PM phenotype (K_m equal to approximately 2.1 μg/mL). Based on these findings, it was conceivable that the administration of high dosages of celecoxib to PM dogs saturated celecoxib metabolism resulting in plasma accumulation observed in the 4-week study. Moreover, an unequal distribution of the two phenotypes across dose groups may have further obscured the relationship of plasma levels with both dose and toxicity.

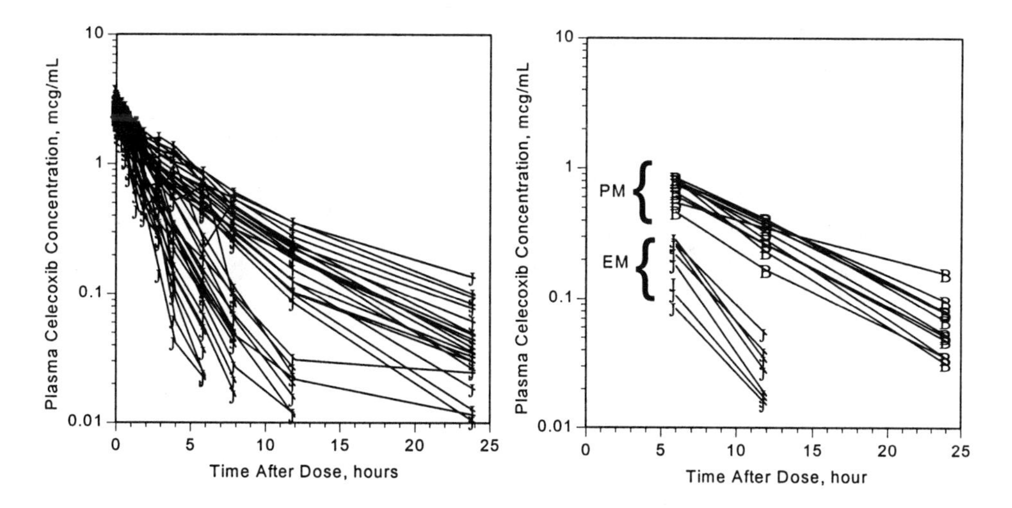

Figure 6. (left panel) Plasma concentrations of celecoxib to 38 dogs after i.v. administration of 5 mg celecoxib/kg body weight. **(right panel)** Plasma concentrations of celecoxib in dogs 6, 12 and 24 hours after i.v. administration of 5 mg celecoxib/kg body weight.

A rapid pre-screening procedure was developed to characterize individual dogs with respect to their celecoxib clearance. This procedure involved iv administration of a 5 mg/kg screening dose followed by pharmacokinetic determination of the rate of disappearance of celecoxib from plasma over a period of 24 hours. The results from a representative prescreen is shown in Figure 6 (right panel). All dogs that were used in subsequent sub-chronic and chronic toxicology studies with celecoxib were screened for phenotype. By equally distributing the two phenotypes within each dosage group, the ability to discern relationships between toxicity, plasma concentrations, and clearance was sharpened. Because the phenotypes were defined only with respect to celecoxib clearance, it was possible that other differences existed between the two sub-populations of dogs that were not known at the time. To ensure that the human risk assessment remained robust, the decision was made to include both phenotypes in toxicology studies, although clearly the PM phenotype, by virtue of greater exposure, was expected to be pivotal for defining any toxicity as well as the safety margins.

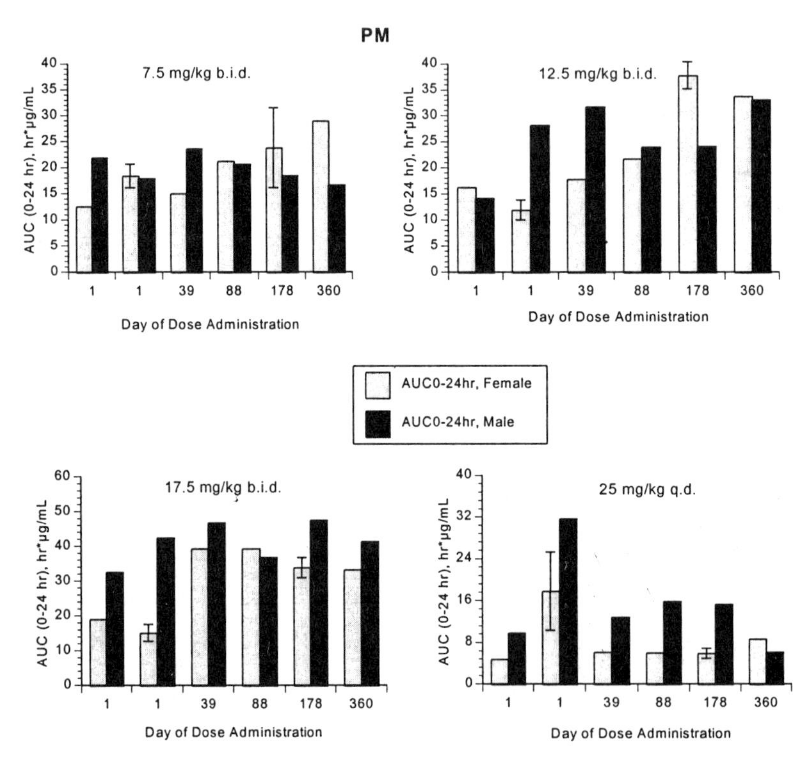

Figure 7. $AUC_{(0\text{-}24hr)}$ values obtained from PM dogs given celecoxib bid or qd for up to 1-year in toxicity studies.

A split-dosing regimen was also used in longer-term toxicity studies in dogs to maximize celecoxib exposures while avoiding plasma concentrations of drug that might exceed the K_m and cause accumulation in PM animals. The combination of the pre-screen for phenotype identification and split-dosing regimen helped achieve steady-state concentrations of celecoxib in the PM dogs throughout the entire duration of the 1-year chronic toxicity study (Figure 7). In addition, using the twice daily dosing regimen a robust systemic exposure to test article was still achieved in these longer-term toxicity studies that at the highest dose was ~15-fold greater than the systemic exposure required to produced maximal anti-inflammatory efficacy in animal models. Although given the same doses as PM animal, EM dogs had lower exposure to celecoxib (Figure 8).

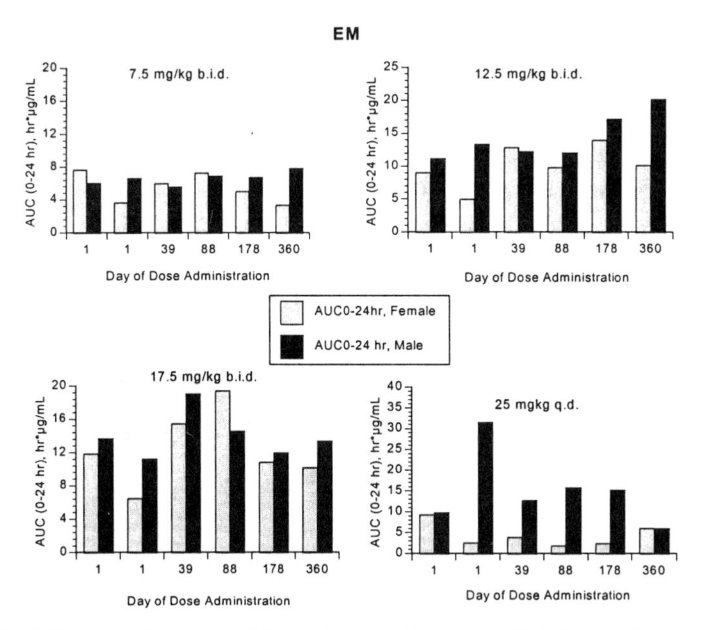

Figure 8. $AUC_{(0\text{-}24hr)}$ values obtained from EM dogs given celecoxib bid or qd for up to 1-year in toxicity studies.

Retrospective analysis of the prescreening data from 240 dogs used in the toxicological evaluation of celecoxib has revealed possibly four potential phenotypes in celecoxib clearance (Figure 9). Differences in the *in vitro* rate of metabolism of celecoxib were seen with molecular characterizations of liver cytochrome P-450 isolated from the two original phenotypes (Paulson et al., 1999a). Although it became clear that the CYP2D15 subfamily was important in the canine metabolism of celecoxib, the contribution of this isoform to the celecoxib polymorphisms has yet to be proven (Paulson et al., 1999a). Other canine polymorphisms have been described in the literature such as the polymorphism in the expression of the CYP2C41 (Blaisdell et al., 1998). The possibility that this isoform or other isoform(s) may contribute to the celecoxib polymorphism, remains untested.

The discovery of a canine polymorphism early in the nonclinical safety assessment of celecoxib had important consequences for the continued use of dogs in establishing the GI safety profile of the novel COX-2-selective inhibitors. The ability to recognize and understand this polymorphism, as well as develop dosing strategies in response to it, were significant factors in the successful completion of the nonclinical safety program for celecoxib.

The canine continues to be an important species in new drug development. Dogs are used as models for testing the pharmacological efficacy of potential drug candidates, but more often, as the non-rodent species in nonclinical safety and toxicology evaluations.

Because of this, it is likely that canine polymorphisms will continue to be encountered in drug development. Clearly, the impact of such polymorphisms must be considered prior to conducting safety assessment studies in dogs. Additional molecular investigations of canine biotransformation pathways are warranted to help identify potential polymorphisms as well as address their relevance for human drug metabolism and clearance.

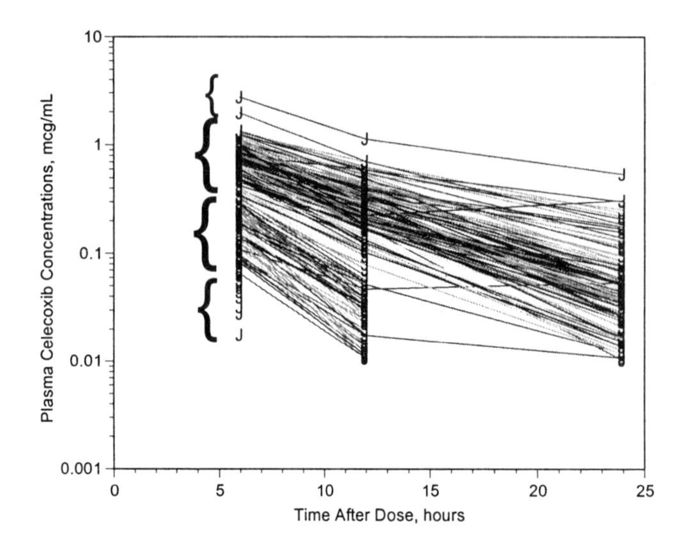

Figure 9. Plasma concentrations of celecoxib in 240 dogs prescreened for EM and PM phenotype. The dogs were given a single iv dose of celecoxib at 5 mg/kg. Plasma concentrations of celecoxib were measured at 6, 12 and 24 hours after dosing.

3.2. Mouse and Rat Studies

Other, less complex, polymorphisms were encountered with celecoxib in mouse and rat. The initial pharmacokinetic studies in rats revealed a gender dimorphism in the clearance of celecoxib, which is faster in males than females. The half-life in male rats is ~ 4 hours compared to 14 hours in females (Paulson et al., 2000b). Gender differences of this kind are relatively common in rats, and can be attributed to gender-specific expression of rat CYP2C and CYP3A genes (Waxman et al., 1985).

The reverse of the rat polymorphism is seen in the mouse, female mice clear celecoxib at a higher rate than males (Paulson et al., 2000c). Like the rat, this is also attributed to gender-specific expression of cytochrome P450 enzymes. Gender differences in the expression of hepatic CYP2A4 has been reported (Burkhart et al., 1985).

Although gender dimorphism was observed in both mice and rats, different strategies were employed for dosing in the respective cancer bioassays in these species. Celecoxib plasma concentrations in mice remain proportional in both genders over a wide dosage range (150 - 1000 mg/kg/day) (Figure 10, left panel). Therefore, it was relatively simple to establish different doses for male (75, 150 and 300 mg/kg) and female (150, 300 and 1000 mg/kg) mice that produced comparable systemic exposures (Figure 10, right panel).

Plasma concentrations in rats are not proportional with dosage in either gender due to saturation of absorption with increasing dosage (Figure 11). The higher exposures achieved with dosages above 400 mg/kg did produce death secondary to GI mucosal injury in female rats in other toxicology studies. Given the lack of dose proportionality, attempts to match systemic exposures in rats as done in mice would have been difficult. Therefore, male and female rats were administered identical doses with the highest dosage (400 mg/kg) producing saturation of systemic exposure in males and maximally tolerated exposures in females.

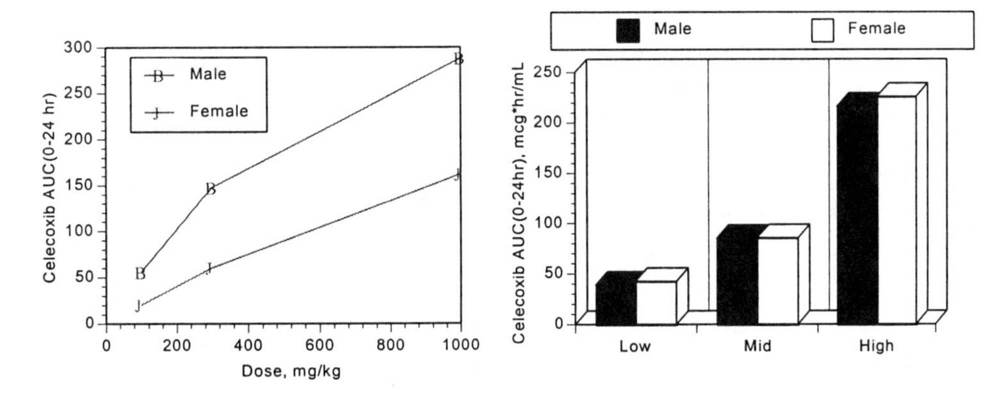

Figure 10. **(Left Panel)** The relationship between dose and celecoxib AUC(0-24h) in male and female mice given celecoxib at 150, 300 and 1000 mg/kg by diet admix in a 2-week dose-ranging study. **(Right panel)** The relationship between dose and celecoxib AUC(0-24h) in male and female mice given celecoxib by diet admix at low (75 mg/kg for male; 150 mg/kg for female), mid (150 mg/kg for male and 300 mg/kg for female) and high (300 mg/kg for male and 1000 mg/kg for female) doses in a 13 week dose-ranging study.

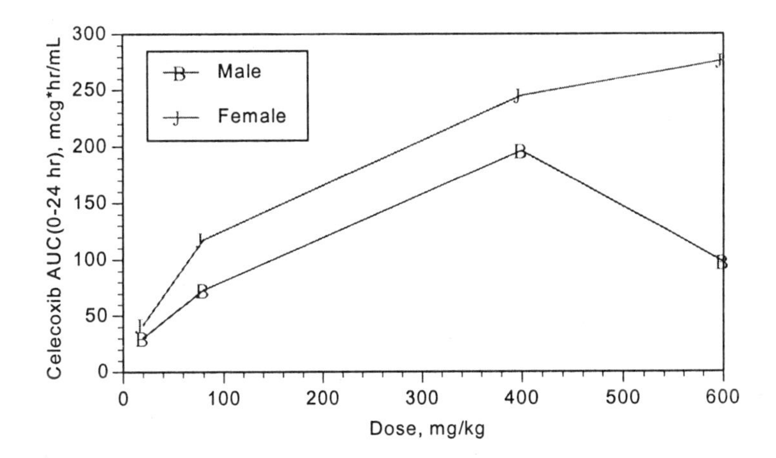

Figure 11. The relationship between dose and celecoxib AUC(0-24h) in male and female rats given celecoxib at 20, 80, 400 and 600 mg/kg by diet admix in a 4-week toxicity study.

4. POLYMORPHISM, ALLOMETRIC SCALING AND CLINICAL DOSE PREDICTION

The clinical dose for celecoxib was projected using the equation below; where AUC_{ED80} is area under the celecoxib plasma concentration-time curve at the ED_{80} in the rat adjuvant arthritis model, Cl_{human} is the human clearance of celecoxib predicted by allometric scaling, and F is the estimated fraction of drug absorbed.

$$\text{Dose} = (AUC_{ED80})(Cl_{human})/F$$

Before estimating clinical doses using the equation above, a target exposure in the form of an area under the concentration-time curve (AUC) for the desired clinical benefit needed to be defined. The rat adjuvant arthritis model which has been successfully used to predict clinical efficacy of anti-inflammatory agents based on a mechanism of action of prostaglandin inhibition (Dubinsky et al., 1987) was selected. The ED_{80} for reduction of limb swelling by celecoxib in this model was found to be ~1.4 mg/kg (Figure 12, left panel), and the corresponding systemic exposure (Cmax and $AUC_{0\text{-}24hr}$) for this dosage were 200 ng/mL and 2.29 μg/mL*hr, respectively (Figure 12, right panel). The ED_{80} was selected as the target because it allow for near maximal anti-inflammatory efficacy in animals being just at the asymptote on the dose-response curve. These systemic exposures were somewhat lower than the plasma concentrations that have been shown to be fully therapeutic in arthritis patients at the 200 mg dose with a Cmax of 705 ng/mL (CV = 38%) (Goldenberg et al., 1999). These exposures are closer to those achieved with the approved 100 mg dose of celecoxib. Since free plasma concentrations are relevant for biologic activity, it was also necessary to show that human and rat exhibit similar plasma protein binding prior to use of the adjuvant arthritis model in the human dose prediction (Paulson et al., 1999b).

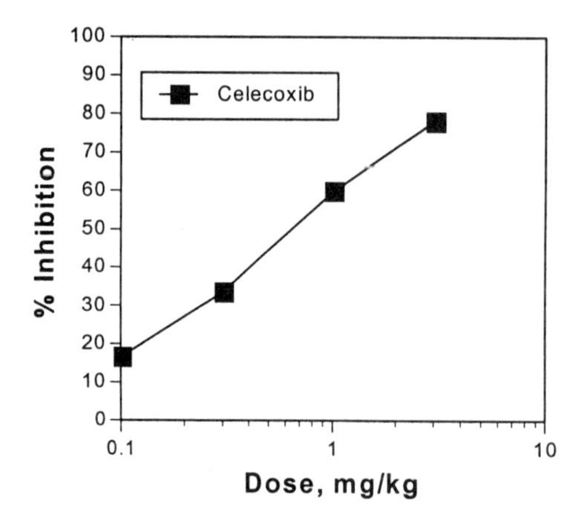

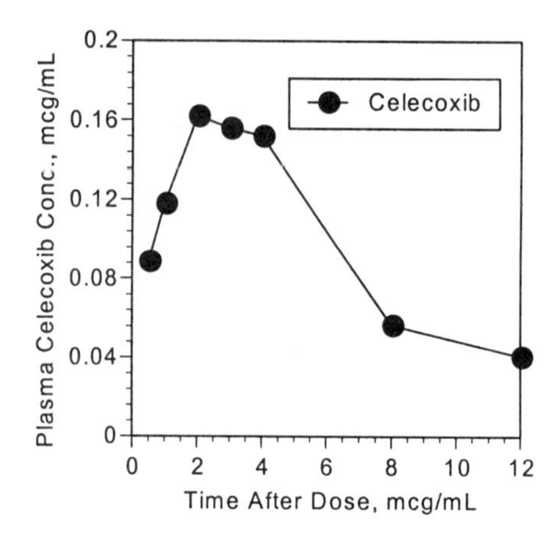

Figure 12. (Left Panel) Dose-response curve for celecoxib in the rat adjuvant arthritis model. **(Right Panel)** Celecoxib plasma concentrations at the ED_{80} (0.7 mg/kg bid) in the rat adjuvant arthritis model.

Second, an additional component of the equation used to estimate clinical dose is F, or the fractional oral bioavailability. This value was estimated from dog bioavailability studies that showed near complete absorption of celecoxib when given as solution to about 25-30% absorption when given as neat chemical in a capsule. To allow for uncertainty in the estimated human bioavailability of the celecoxib dose form, a range of 0.25-1 was used for F.

Table 2. The iv pharmacokinetic parameters of celecoxib in animals

Species	Sex	Dose mg/kg	$T_{1/2}$ h	Cl mL/min/kg	Vd L/kg	Vd_{ss} L/kg	$AUC_{0-\infty}$ μg*h/mL
Rat	Male	1	3.73	7.76	2.51	ND	2.15
Rat	Female	1	14.0	1.99	2.42	ND	8.38
G. Pig	Male	6	1.16	20.5	1.98	ND	5.49
C. Monkey	Female	1	1.66	22.7	3.58	3.22	0.736
R. Monkey	Female	1	1.50	17.8	2.73	2.34	0.957
Dog[a]	Female	5	8.84	3.08	2.42	ND	31.2
Dog[b]	EM	5	1.72	18.2	2.50	2.10	5.05
Dog[b]	PM	5	5.18	7.15	3.13	2.37	12.1

ND – Not determined
EM – Extensive metabolizer of celecoxib (males and females)
PM – Poor metabolizer of celecoxib (males and females)
[a] Data from these female dogs were used in the initial allometric prediction of human clearance.

Celecoxib clearance was determined by allometric scaling. Allometric scaling is commonly used to estimate human clearance of potential drug candidates as a first step in defining doses for initial clinical trials (Lave et al., 1999). The accuracy of allometric scaling for predicting human clearance has been the subject of several reviews (Boxenbaum et al., 1984; Mahmood et al., 1999). In the case of celecoxib, the existence of the animal polymorphisms in clearance provided another challenge in the use of allometric scaling to accurately predict human clearance. Allometric scaling requires commitment of considerable resources for bioanalytical assay development and conduct of animal iv pharmacokinetic studies, so the selection of species to be use requires careful consideration. In the case of celecoxib, human clearance was scaled using clearance data from five species, including rat, guinea pig, cynomolgus monkey, rhesus monkey and dog (Table 2). The species selected for allometric scaling were those that were to be used in future safety assessment studies or that were currently being used as animal models of COX-2 in the discovery program. Clearance data from rat, guinea pig and dog were used in the initial prediction of human clearance. Subsequent refinement of the clearance prediction utilized clearance data from cynomolgus monkeys and rhesus monkeys. The doses selected for the iv pharmacokinetic studies targeted plasma concentrations that produced anti-inflammatory efficacy in the rat adjuvant arthritis model and were linear with respect to celecoxib pharmacokinetics. In addition, metabolism studies were conducted in each species to demonstrate similarity to the human metabolism of celecoxib before it was included in the allometric scaling.

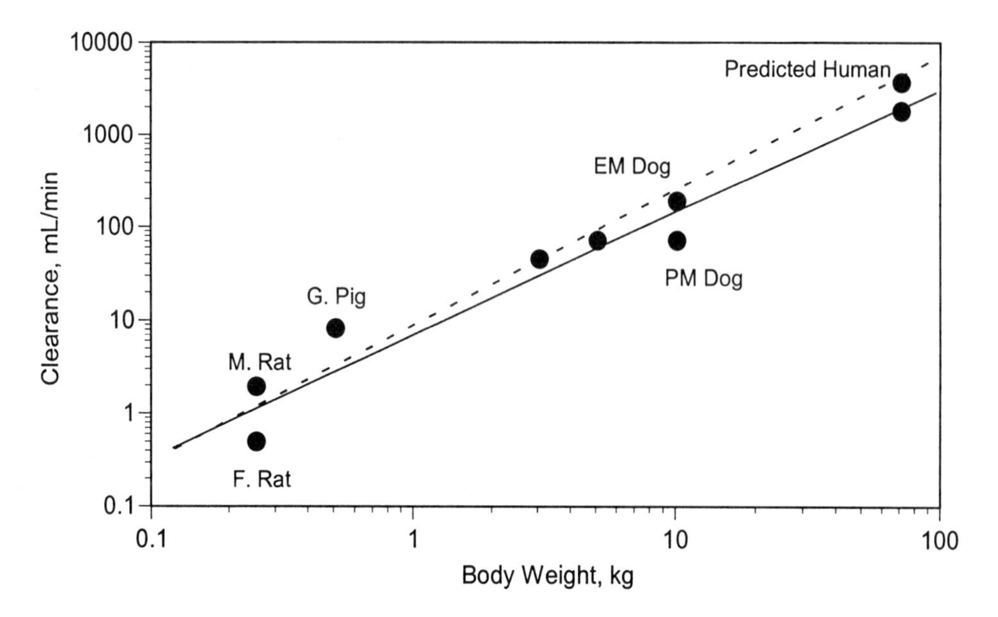

Figure 13. Allometric relationship between clearance and species body weight for celecoxib. The solid lines represent the fit of data using the PM dog and the dashed line represents the fit of the data using the EM dog.

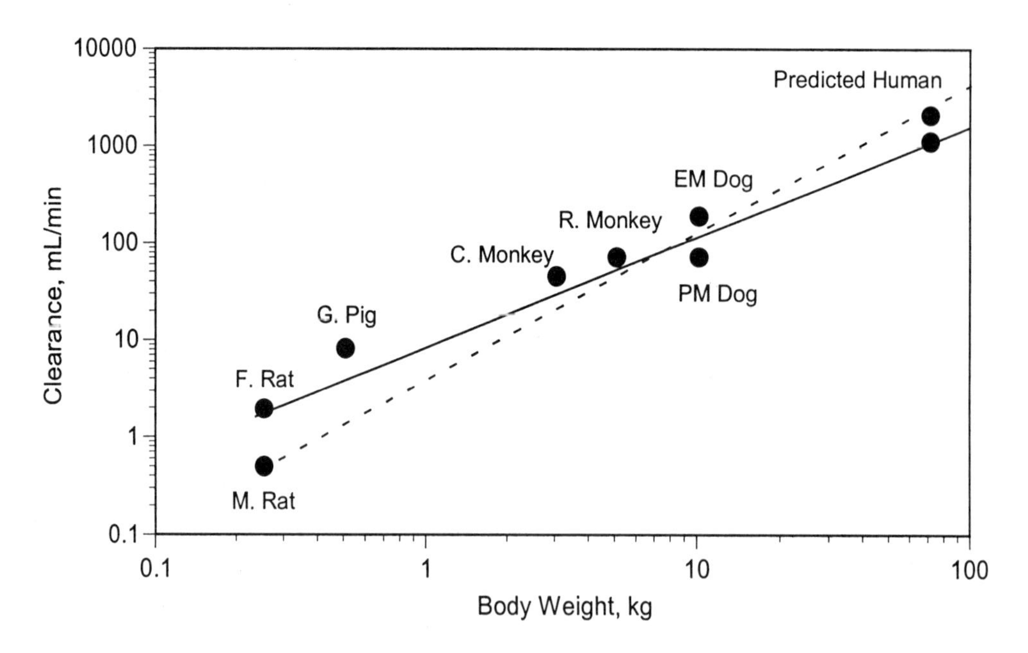

Figure 14. Allometric relationship between clearance and species body weight for celecoxib. The solid lines represent the fit of data using the male rat and the dashed line represents the fit of the data using the female rat.

Table 3. Predicted human clearance values for celecoxib using allometric scaling

Dog Phenotype & Rat Gender Used in Allometric Scaling[a]	Predicted Human Celecoxib Clearance ml/min	Predicted Celecoxib Clinical Dose mg/day
EM Dog, M & F Rat	3850	500 - 2000
PM Dog, M & F Rat	1832	250 - 1000
EM & PM Dog, M Rat	1126	150 - 600
EM & PM Dog, F Rat	2110	300 - 1200

[a]Guinea pig, Rhesus monkey and Cynomolgus monkey were used in allometric scaling in addition to the gender of rat and phenotype of dog listed.

The initial prediction of human celecoxib clearance by the allometric scaling was ~550 mL/min. This value very closely matched the apparent plasma clearance (Cl/F) of celecoxib in humans which was reported to be about 500 mL/min (see Product Label for Celecoxib). An oral dose of celecoxib is extensively metabolized with only ~2% of the excreted as unchanged drug. Therefore, the Cl/F may be close to the true clearance if it is assumed that celecoxib is completely absorbed and metabolized systemically.

At the time the first prediction was made the dog polymorphism had yet to be discovered (Paulson et al., 1999b), and therefore, was not factored into the initial allometric estimate of human celecoxib clearance. Also, clearance data were only available from male rat. Retrospectively, it appears that the dogs used in the initial estimates were of the PM phenotype. If clearance data from EM dogs and both genders of rat had been used, the estimated human clearance would have been overestimated by about 7-fold at 3850 mL/min as depicted by the dashed line in Figure 13 and in Table 3. The gender polymorphism in rats also can influence the allometrically-scaled prediction of human clearance as shown in Figure 14 and Table 3. Although clearance values from male rats were used in the initial predictions, different values for human celecoxib clearance are obtained depending if male rats or female rats are used in the scaling (Figure 14, Table 3). The existence of polymorphism both known and unknown can clearly effect the clearance values predicted by allometric scaling. This case study of celecoxib clearly illustrates the limitations of allometric scaling of clearance data for compounds that are extensively metabolized.

Using this approach, the clinical dose range was originally predicted to be about 100-400 mg/day, and the recommended clinical doses of celecoxib for the treatment of OA and RA were subsequently shown in clinical trials to be 100-200 mg/day. However, these dose predictions can vastly change depending on the species polymorphisms that are encountered during development (Table 3).

5. BIOAVAILABILITY AND FOOD EFFECTS

Early studies that characterized the absorption of celecoxib were conducted using dogs as the primary model. The advantages of this species as a model for oral absorption include the ability to use almost all dosage human forms and pharmacokinetic profiling, including crossover, can be easily obtained from one animal. Although there are similarities in the anatomy of the gastrointestinal tracts of dog and human i.e., stomach volume and small intestine length, differences in the gastric emptying time and small intestine transit time between the two species can affect the rate and extent of drug absorption (Ritschel, 1987; Dressman, 1986).

One of the first bioavailability (BA) studies conducted with celecoxib in dogs compared to a solution dose given intravenously or orally with capsule containing neat chemical in dog. The absolute bioavailability of celecoxib was low when given as neat chemical in a gelatin capsule (~22% in the EM and 40% in the PM phenotypes, respectively). Absorption was greater or near complete when given as a solution to both phenotypes (EM ~64%, PM ~90%) (Paulson et al., 2001). The higher BA in PM versus EM dogs suggests that some first pass metabolism occurs in EM animals.

Celecoxib is classified as Biopharmaceutical Class II. Its octanol/water partition coefficient is on the order of 10^3 and its aqueous solubility is as low as 3-7 μg/mL. The limited absorption of celecoxib in dogs when given as a solid is consistent with a poorly soluble drug for which bioavailability is dissolution rate limited; whereas the near complete absorption of celecoxib from solution reflects high permeability. These characteristics increase the potential for a positive effect of food on the absorption of celecoxib (Karim 1996). Food tends to increase the extent of absorption of poorly soluble drugs by increasing GI residence time, and by increasing both fluid volume for dissolution and secretions, such as bile, that enhance drug dissolution (Karim 1996; Fleisher et al., 1999).

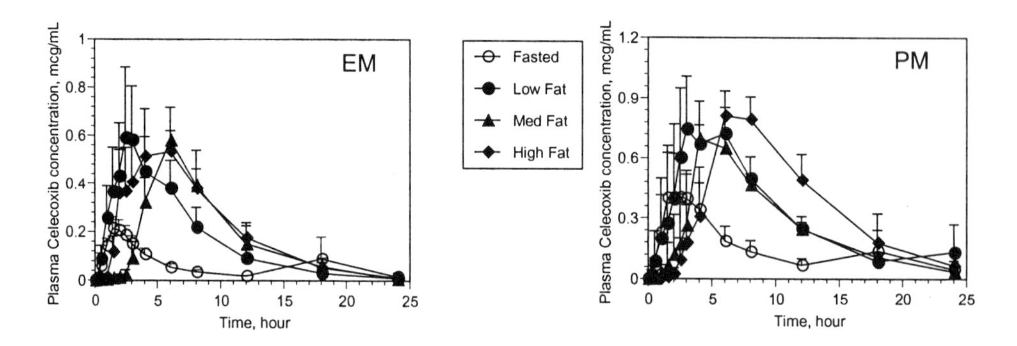

Figure 15. Mean (± SEM) celecoxib plasma concentrations in male (left panel) and female (right panel) beagle dogs given orally 5 mg/kg celecoxib in a gelatin capsule under fasting conditions or in the presence of a low, medium or high fat meal.

Early identification of food effect in nonclinical or in first in man studies can facilitate subsequent clinical development of a drug. The potential food effect was characterized prior to the first in man study with celecoxib using both EM and PM dogs. Although food was found to delay the absorption of celecoxib, the extent of absorption was increased by 3- to 5-fold in dogs (Figure 15). Subsequent clinical data showed the dog to be a poor model for predicting the effect of food on the human absorption of celecoxib. In humans, a high-fat meal only minimally affected the extent of celecoxib absorption. The relative BA was 110.7% when given with high-fat breakfast compared to under fasting conditions (Paulson et al., 2001). Although food effect studies in animals can provide an early indication of the potential in humans, the limits of animal models for absorption can result in over or under prediction of effects in humans.

6. SUMMARY

New technologies and approaches to help improve characterizations of drug metabolism became available and widely used around the time Celecoxib entered development in mid-1994. The application of these new technologies was pivotal in understanding polymorphisms observed in animals during the development of Celecoxib and their human relevance.

The probability remains high that animal and human metabolic polymorphisms will continue to be discovered as drug discovery and design explores new mechanisms of action and molecules with novel structures. Successful drug development will required earlier recognition and characterization of polymorphisms that can potentially have significant impact on laboratory and clinical evaluations of drug candidates..

7. REFERENCES

Bazan NG, Marcheselli VL, Mukherjee PK, Lukiw WJ, Gordon WC, Zhang D. COX-2 in brain and retina: Role in neuronal survival. In: Selective COX-2 Inhibitors: Pharmacology, Clinical Effects and Therapeutic Potential, J. Vane and J Botting (editors). Lancaster UK and London UK: Kluwer Academic Publishers and William Harvey Press; 1998. p. 47-54.

Blaisdell J, Goldstein JA, Bai SA. Isolation of a new canine cytochrome P450 cDNA from the cytochrome P450 2C subfamily (CYP2C41) and evidence for polymorphic differences in its expression. Drug Metab Dispos. 1998;26(Suppl 3):278-283.

Borda IT and Koff RS. NSAIDs: A Profile of Adverse Effects. Philadelphia, PA: Hanley and Belfus Inc.; 1992. p. 1-240.

Boxenbaum H. Interspecies pharmacokinetic scaling and the evolutionary-comparative paradigm. Drug Metab Rev. 1984;15:1071-1121.

Burkhart BA, Harada N, Negishi M. Sexual dimorphism of testosterone 15 alpha-hydroxylase mRNA levels in mouse liver. cDNA cloning and regulation. J Biol Chem. 1985;260:15357-15361.

Campbell DB. Are we doing too many animal biodisposition investigations before phase I studies in man: A re-evaluation of the timing and extent of ADME studies. Eur J Drug Metab Pharmacokinet. 1994;19(3):283-293.

Davies NM, McLachlan AM, Day RO, Williams KM. Clinical pharmacokinetics and pharmacodynamics of Celecoxib: A selective cyclooxygenase inhibitor. Clin Pharmacokinet. 2000;38(3):225-242.

Dressman JB. Comparison of canine and human gastrointestinal physiology. Pharm Res 3(3):123-131, 1986.

Dubinsky B, Gebre-Mariam S, Capetola RJ, Rosenthale ME. The antialgesic drugs: Human therapeutic correlates of their potency in laboratory animal models of hyperalgesia. Agents Actions. 1987;20(1-2):50-60.

Fleisher D, Li C, Zhou Y, Pao LH, Karim A. Drug, meal and formulation interactions influencing drug absorption after oral administration: Clinical implications. Clin Pharmacokinet. 1999;36(3):233-254.

Food and Drug Administration. New drug application #20998: clinical pharmacology/biopharmaceutics review section celecoxib. Bethesda, MD: FDA; 1998.

Furuya H., Fernandez-Salguero P. Gregory W, Taber H, Steward A, Gonzalez FJ et al. Genetic polymorphism of CYP2C9 and its effect on warfarin maintenance dose requirement in patients undergoing anticoagulation therapy. Pharmacogenetics. 1995;5(6):389-392.

Goldenberg MM. Celecoxib, a selective cyclooxygenase-2 inhibitor for the treatment of rheumatoid arthritis and osteoarthritis. Clin Ther. 1999;21(9):1497-1513.

Hawkey CJ. COX-2 inhibitors. Lancet. 1999;353(9149):307-314.

Karim A. Importance of food effect studies early in drug development. In: Midha KK, Nagai T, editors. Bioavailability, Bioequivalence, and Pharmacokinetic Studies. Tokyo: business Center for Academic Societies Japan; 1996. p. 221-29.

Karim A, Tolbert D, Burton E, Piergies A, Harper K, Paulson S, et al. SC-58635 (Celecoxib): A highly selective inhibitor of cyclooxygenase-2, disposition kinetics in man and identification of its major CYP450 isozyme in its biotransformation. Pharm Res. 1997;14(Suppl 11):S617.

Karim A, Tolbert D, Piergies A, Bradford D, Slater M, Paulson S. Celecoxib biotransformation: Fluconazole but not ketoconazole substantially increases celecoxib exposure in man. Pharm Sci. 1998;(Suppl 1):S626.

Karim A, Bradford D, Slater M, Wallemark C, Laurent A. Effect of selective cyclooxygenase-2inhibitor celecoxib on the systemic exposure of cytochrome P450 2D6 metabolized probe drug dextromethorphan. AAPS Pharm Sci. 2000; 2(4)Suppl 1:

Kimura S, Pastewka J, Gelboin HV, Gonzalez FJ. cDNA and amino acid sequences of two members of the human P450IIC gene subfamily. Nucleic Acids Res. 1987;15(23):10053-10054.

Kore AM. Toxicology of nonsteroidal anti-inflammatory drugs. Vet Clin North Am Small Anim Pract. 1990;20(2):419-430.

Lave T, Coassolo P and Reigner B. Prediction of hepatic metabolic clearance based on interspecies allometric scaling techniques and *in vitro-in vivo* correlations. Clin Pharmacokinetic. 1999;36(3):211-231.

Mahmood I, Balian JD. The pharmacokinetic principles behind scaling from preclinical results to phase I protocols. Clin. Pharmacokinet. 1999;36(1):1-11.

Needleman P, Isakson PC. The discovery and function of COX-2. J Rheumatol. 1997;24(Suppl 49):6-8.

Paulson SK, Engel L, Reitz B, Bolten S, Burton EG, Maziasz TJ, et al. Evidence for polymorphism in the canine metabolism of the cyclooxygenase 2 inhibitor, celecoxib. Drug Metab Dispos. 1999a;27(10):1133-1142.

Paulson SK, Kaprak TA, Gresk CJ, Fast DM, Baratta MT, Burton EG, et al. Plasma protein binding of celecoxib in mice, rat rabbit, dog and human. Biopharm Drug Dispos. 1999b;20:293-299.

Paulson SK, Hribar JD, Liu NWK, Hajdu E, Bible RH, Piergies A, etal. Metabolism and excretion of [^{14}C]celecoxib in healthy male volunteers. Drug Metab Dispos. 2000a;28(3):308-314.

Paulson SK, Zhang JY, Breau AP, Hribar JD, Liu NWK, Jessen SM, etal. Pharmacokinetics, tissue distribution, metabolism and excretion of celecoxib in rats. Drug Metab Dispos. 2000b;28:514-521.

Paulson SK, Zhang JY, Jessen SM, Lawal Y, Liu NWK, Dudkowshi CM, etal. Comparison of celecoxib metabolism and excretion in mouse, rabbit, dog cynomolgus monkey and rhesus monkey. Xenobiotica. 2000c;30(7):731-744.

Paulson SK, Vaughn MB, Jessen SM, Lawal Y, Gresk CJ, Yan B, etal. Pharmacokinetics of celecoxib after oral administration in dogs and humans: Effect of food and site of absorption. J Pharm Exp Ther. 2001;297(2):638-645.

Ritschel WA. *In vivo* animal models for bioavailability assessment. STP Pharma. 1987;3(2):125-141.

Romkes M, Faletto MB, Blaisdell JA, Raucy JL, Goldstein JA. Cloning and expression of complementary DNAs for multiple members of the P450 IIC subfamily. Biochemistry. 1991;30(13):3247-3255.

Silverstein FE, Faich G, Golstein JL, Simon LS, Pincus T, Wehelton A. Gastrointestinal toxicity with Celecoxib versus nonsteroidal anti-inflammatory drugs for osteoarthritis and rheumatoid arthritis: the CLASS study, a randomized controlled trial. J Am Med Assoc. 2000;284:1247-1255.

Smith JH, Willis AL. Aspirin selectively inhibits prostaglandin production in human platelets. Nat New Biol. 1971;231:235-237.

Spyridakis LK, Bacia JJ, Barsanti JA, Brown SA. Ibuprofen toxicosis in a dog. J. Am Vet Med Assoc. 1986;189(8):918-919.

Sullivan-Klose TH, Ghanayem BI, Bell DA, Zhang Z-Y, Kaminsky LS, Shenfield GM, etal. The role of the CYP2C9-Leu359 allelic variant in the tolbutamide polymorphism. Pharmacogenetics. 1996;6(4):341-349.

Tang C, Shou M, Mei Q, Rushmore TH, Rodrigues AD. Major role of human liver microsomal cytochrome P450 2C9 (CYP2C9) in the oxidative metabolism of celecoxib, a novel cyclooxygenase-II inhibitor. J Pharm Exp Ther. 2000;293(2):453-459.

Tang C, Shou M, Rushmore TH, Mei Q, Sandhu P, Woolf EJ, etal. In-vitro metabolism of celecoxib, a cyclooxygenase-2 inhibitor, by allelic variant forms of human liver microsomal cytochrome P450 2C9: Correlation with CYP2C9 genotype and in-vivo pharmacokinetics. Pharmacogenetics. 2001;11(3):223-235.

Taylor LA, Crawford LM. Aspirin-induced gastrointestinal lesions in dogs. J Am Vet Med Assoc. 1968;152(6):617-619.

Tucker GT. The rational selection of drug interaction studies: Implications of recent advances in drug metabolism. Int J Clin Pharmacol Ther Toxicol. 1992;30:550-553.

Vane JR. Inhibition of prostaglandin synthesis as a mechanism of action for aspirin-like drugs. Nat New Biol. 1971;231(25):232-235.

Waxman DJ, Dannan GA, Guengerich FP. Regulation of rat hepatic cytochrome P450: age-dependent expression, hormonal imprinting, and xenobiotic inducibility of sex-specific isoenzymes. Biochemistry. 1985;24:4409-4417.

Wrighton SA, Vandenbranden M, Stevens JC, Shipley IA. *In vitro* methods for assessing human hepatic drug metabolism: their use in drug development. Drug Metab Rev. 1993;25:453-484.

Wrighton SA, Ring BJ, VandenBranden M. The use of *in vitro* metabolism techniques in the planning and interpretation of drug safety studies. Toxicol Pathol. 1995;23(2):199-208.

Zhang YJ, Wang Y, Dudkowski C, Yang D, Chang M, Yuan J, et al. Characterization of metabolites of celecoxib in rabbits by liquid chromatographic-tandem mass spectrometry. J Mass Spectrom. 2000;35(11):1259-1270.

THE ROLE OF CLINICAL PHARMACOLOGY AND OF PHARMACOKINETICS IN THE DEVELOPMENT OF ALENDRONATE – A BONE RESORPTION INHIBITOR

Arturo G. Porras and Barry J. Gertz*

1. INTRODUCTION

The skeleton is the central architectural feature of vertebrates: It distinguishes them from other creatures, it provides structural support and protection to important soft tissues, and it serves as mechanical pivot and provides the mechanical advantage to muscles that makes possible the versatility of movement characteristic of these organisms. In addition to its mechanical functions, bone tissue, which constitutes the bulk of the skeletal mass, carries out various physiological functions. Amongst these, hematopoiesis and mineral metabolism stand out.

The solid phase of bone is composed of an organic matrix and of crystalline salts deposited on the matrix. The organic matrix is composeded primarily of collagen fibers (90%) in a three-dimensional web and gives bone its tensile and torsional strengths. The crystalline salts are primarily calcium phosphates in the form of hydroxyapatite and they give bone its compressive strength[1].

Two common diseases of adult human bone are osteoporosis and Paget's disease, both of which affect the mechanical capabilities of the skeleton.

Osteoporosis is a generalized, progressive loss of bone tissue accompanied by deterioration of this tissue's architecture, leading to an increased risk of fracture. It is common enough to constitute a major health hazard in industrialized countries. Women, the elderly, smokers and people of European and Asian ancestry all face a heightened risk for the disease, though men, nonsmokers and people of other ancestries are also among the sufferers and it can also be found, occasionally, in the young. Osteoporosis is often characterized symptomatically by vertebral collapse (with associated increase in spinal curvature, loss of height and backache), and by wrist and hip fractures arising from minimal trauma. Particularly at risk are postmenopausal women, in whom yearly bone

* Arturo G. Porras, Merck Research Laboratories, West Point, PA, USA. Barry J. Gertz, Merck Research Laboratories, Rahway, NJ, USA

loss averages about 2 to 4% in the years immediately following menopause and continues at about 1% in their later years[1]. Loss of bone mass is greater in trabecular than in cortical bone, giving rise to the characteristic pattern of fractures observed in this disease.

Paget's disease of bone (Osteitis Deformans) is a chronic disorder of the adult skeleton in which localized areas of hyperactive bone are replaced by a softened and enlarged osseous structure. It is relatively common, with an incidence of about 3% in people older than 40 and about 50% in women age 70 or older[2]. It shows a 3:2 pattern of male prevalence[3]. It displays excessive bone turnover at the bone sites involved, showing increased osteoclastic activity and corresponding accelerated osteoblastic repair which produces coarsely woven thickened lamellae and trabeculae. Histologically, the disorder is characterized by extensive vascularity, increased marrow fibrosis and intense cellular activity with irregular areas of lamellar bone interspaced by 'cement' lines typically arranged in a mosaic pattern. This mosaic layering of collagen results in structurally enlarged and weakened bone, even though it is heavily calcified.

With the exception of healing bones following a fracture, the rates of deposition and resorption of bone are balanced in adults, so that the total mass of bone remains constant. In osteoporosis, however, the rate of resorption is greater than the rate of deposition such that a steady rate of bone loss ensues. Whether there is also a primary reduction in bone formation due to a loss of 'osteoblast vigor', is a matter of debate. The net effect, though, is a progressive loss of bone tissue which substantially reduces its mechanical strength, leading to fractures caused by minimal trauma, which are frequently disabling and may give rise to fatal complications. In Paget's disease, the osteoclastic and osteoblastic activities are not only accelerated but unbalanced in alternating directions, the net result of which is a layering of osteoblastic bone with interspersed areas of fibrosis with a resultant (possible) increase in bone volume and loss of structural integrity. The deformations thus induced may produce a variety of symptoms including bone pain, compression neuropathy leading to pain, hearing loss, spinal stenosis, paresis, or paraplegia, deformities from bowing of the long bones or secondary osteoarthritis of the adjacent joints, and pathological fractures.

Previous to development of alendronate as a treatment for osteoporosis, the disease was recognized, in western countries, to present a substantial health challenge given the increasing life expectancies, particularly for women, which resulted in ever increasing percentages of the population being at risk for osteoporosis.[4,5] Therapies available at that time included dietary-calcium and Vitamin-D supplementation, estrogen-replacement therapy for osteoporosis; plicamycin for severe cases of Paget's disease; and calcitonin and etidronate for both conditions, none considered to give fully satisfactory therapeutic results without an associated increase in the risk for other conditions: Efficacy of Vitamin D and of calcium supplementation lacked documentation.[4–7] Calcitonin, a peptide, is expensive to manufacture, must be given subcutaneously or intranasally and, at that time, its effectiveness had not been clearly established[8]. Increased risks of breast and uterine cancer are associated with estrogen replacement therapy.[9,10,11] The introduction of more effective and safer treatments was, therefore, readily appreciated as a priority that could provide a substantial advance in public health. One of the bisphosphonates, etidronate, had been proposed as treatment for various bone diseases but its use in osteoporosis was considered controversial or, at least, not yet established[12].

2. MEDICAL NEED

2.1 Therapeutic Need

Osteoporosis can develop silently over many years and then come to the forefront with devastating effects on the life of the sufferer. The primary clinical outcome of this disease, and often the first signal of its existence, is fracture, with significant resultant morbidity (pain, height loss, kyphoscoliosis, disability) and mortality[13]. The attendant societal costs are high in all countries where it has been studied[14]. The Osteoporosis Society of the United Kingdom has reported that more women die from osteoporotic fracture than from cancers of the cervix, uterus and breast combined[15], and in the United States, osteoporotic fracture-related costs are over 10 billion dollar per year[16].

BMD is the most powerful predictor of fracture risk, with the risk of osteoporotic fractures inversely related to BMD.[17,18,19] In fact, BMD accounts for 75 to 85% on the variance in bone strength[20].

Of the therapies available to restore lost bone mass before alendronate became available, none was both clearly efficacious and well tolerated. For example, while estrogen therapy is effective at preventing further bone loss and is associated with modest increases in bone mass, it is not effective at restoring lost bone mass to normal levels[21,22]. Estrogen is associated with an increased risk of major adverse experiences such as endometrial cancer (from unopposed use) and possibly breast cancer[21] and its use is compromised by poor compliance due to additional side effects such as mastodynia, fluid retention, nausea, gallbladder disease, hypertension, uterine fibroids, vaginal bleeding, cholestatic jaundice and edema.[23,24] Likewise, calcitonin therapy has generally been shown to result in transient increases in bone mass with efficacy not sustained beyond 12 to 24 months[25]. Therapy with calcitonins has been associated with several adverse effects including nausea, vomiting, flushing, urinary frequency and dysgeusia[26]. Thus, it was agreed that a safe, well-tolerated and clearly efficacious pharmacological therapy for the treatment of osteoporosis in postmenopausal women would constitute a significant therapeutic advance in the management of this disease.

The pathophysiologic mechanism underlying postmenopausal osteoporosis involve a relative increase of osteoclastic bone formation in the setting of increased bone turnover.[16,27] Therefore, inhibition of bone resorption represents a rational target for its treatment[16]. Alendronate is a specific inhibitor of osteoclast-mediated bone resorption without direct effects on bone formation even at many multiples of the recommended clinical dose. Unlike estrogen or calcitonin, therapy with alendronate not only prevents further bone loss but also produces significant increases in bone mass at all of the clinically important sites (spine, hip and total skeleton), increases that are progressive through 24 months of treatment. Importantly, the new bone formed under treatment with alendronate has been demonstrated to have normal strength in animals[28-35] and to be qualitatively normal in man,[33-36] thus predicting a marked reduction in fracture risk. Because its actions are specific to the skeleton and are nonhormonal, therapy with alendronate has been shown largely to avert adverse effects associated with other forms of anti-resorptive therapy such as estrogen or calcitonin. All of these aspects of the scientific rationale for the use of alendronate in the treatment of osteoporosis in postmenopausal women will be discussed in greater detail.

2.2 Prevalence of Postmenopausal Osteoporosis

The prevalence of osteoporosis depends upon its definition. Applying the accepted definition of a decrease in bone mineral density (BMD) associated with a substantially increase risk of fracture,[37,38] the prevalence of osteoporosis has been estimated to be greater than 50% in postmenopausal women.[39] Furthermore, it is estimated that approximately 40% of 50-year-old women of European descent will experience at least one osteoporosis related fracture of the hip, spine or wrist during their lifetimes.

The Consensus Conference on Osteoporosis most current at the time of development of alendronate (1991) had defined osteoporosis as a "systemic skeletal disease characterized by low bone mass and microarchitectural deterioration of bone tissue, with a consequent increase in bone fragility and susceptibility to fracture."[38] The appropriateness of this definition is supported by the numerous longitudinal studies of women in various countries using different measures of BMD that have shown a consistently strong, positive association between decreasing BMD and increasing fracture risk for both vertebral and non-vertebral fractures. In most studies assessing the predictive value of BMD on fracture risk, a decrease of 1 SD in bone mass was associated with at least a 100% increase in the incidence of fractures, with some studies showing as much as a 320% increase in fracture risk.[40] The association between bone mass and fracture risk is much stronger than the well recognized association between cholesterol and heart disease, which shows only a 25% increase in heart attack risk for every 1 SD increase in serum cholesterol.[19] Defining osteoporosis as a BMD which is at least 1 SD below the normal peak value would encompass more than 50% of postmenopausal women in the industrialized countries of Europe,[41,42] Asia[43] and the Americas.[44] Using a more stringent definition of a bone density more than 2 SD below the mean for normal young women still includes 50% of women by age 65.[45] Thus the prevalence of osteoporosis is quite high. Based on the WHO definition of osteoporosis (χ 2.5SD below normal BMD) up to 45% of postmenopausal women are thought to be at risk for ostoeporosis at one or more skeletal sites.[46]

2.3 Clinical and Socioeconomic Importance of Osteoporosis

At the time of development of alendronate the demographic trend of aging populations was already firmly established. In the absence of additional therapeutic choices or of aggressive treatment it was expected that increases in osteoporosis-related fractures would increase markedly around the world. For example, the rapidly aging populations in Europe were expected to yield a doubling or tripling of hip fractures in various countries in that continent by the end of the twentieth century.[14] Age-specific rates of osteoporotic fracture had doubled during the 25 years leading to 1980 in the United Kingdom, and almost one in four British women living to the age of 90 was expected to suffer a hip fracture.[47] Another expression of this age-related increase in risk of hip fracture is that, from age 50 to 90, the risk in women increases 50-fold (and the corresponding risk of vertebral fracture increases 15- to 30-fold).[48] The number of new femoral neck fractures in Finland doubled between 1970 and 1985.[49] In Norway, an exponential rise in the age-specific incidence of hip fractures had been reported for both sexes in the decade between 1978-79 and 1988-89.[50] Similar trends had been reported for Belgium,[51] Sweden,[52,53] Spain,[54] Italy[55] and Canada.[56] In Italy, the incidences of hip fracture is 251 per 100,000 in women 50 years or older. The fracture rate rises sharply with age such that in patients of both sexes aged 85 years and older the incidences is 2,311 per 100,000.[57] The increasing incidence of hip fractures is not restricted to Europe and Canada, however. In Malaysia, the hip fracture incidence doubled for 1981 to 1989.[58] The world-wide incidence of hip fractures, estimated at 1.66 million in 1990, is projected to increase to 3.94 million in 2025 and 6.26 million in 2050.[58] In the United States alone, osteoporosis is responsible for more than 1.5 million fractures annually,[16] of which hip and wrist fractures each account for approximately 20%, vertebral fractures for 40% and fractures of the humerus, ribs, pelvis, ankle and digits accounting for the bulk of the remaining 20%.[14,13]

Clinically diagnosed vertebral fractures cause pain, height loss, kyphoscoliosis and disability and may be associated with excess mortality.[13,60] Hip fractures are particularly catastrophic, with 12-month excess mortality rates reported to be as high as 20 to 33%. Hip fractures result in long-term nursing home care for up to 33% of affected patients, and only 25 to 50% of hip fracture survivors recover their pre-fracture levels of function.[15]

This morbidity and mortality translate into high costs to society for medical treatments with hip fractures estimated to cost more than 200 billion lire annually in Italy (not including the cost of surgery, drugs, rehabilitation and loss of income)[55] and over 10 billion dollars per year in the United State, with estimates as high as 62 billion dollars by 2020.[14] Similar trends for increasing incidence and cost are predicted for the European community[14] and even steeper increases are predicted for Latin America and Asia.[59] These projections, then, served to highlight the urgent need for efficacious and well-tolerated treatment such as alendronate might be able to provide.

3. MECHANISM OF ACTION

The solidity of bone belies the constant biochemical activity being carried out on the very foundational structure of this tissue. Bone is continuously being absorbed and deposited by osteoclasts and osteoblasts, respectively. In healthy bone, osteoclasts are active in about 1% of all bone surfaces at any given time and are usually found in highly concentrated masses which usually resorb that surface of the bone which is in contact with them, developing tunnels which may be as wide as one millimeter and several millimeters in length.[61] After about three weeks of resorptive activity, the osteoclasts abandon the bone surface and are replaced by osteoblasts, which proceed to deposit new bone for a number of months until the tunnel developed by the osteoclasts is filled, except for a small tunnel left behind through which the blood vessels that sustained the osteoblasts flowed. This tunnel is known as the Haversian canal. A schematic of the steps involved in normal bone turnover is shown in Figure 1.

In healthy young adults, the rates of bone resorption and deposition are balanced with no net gain or loss of bone through this process except for bone growth in the form of bone repair following fractures or in response to stress.[61] In osteoporosis, though, this balance becomes altered in favor of bone resorption leading to slow but steady loss of bone mass and consequent loss of bone strength which favors fractures even under moderate force loads.

Osteoclasts accomplish their resorptive tasks by attaching themselves to the bone surface and developing numerous *villi*, which form a 'ruffled border' where the cells come into contact with the bone matrix. This ruffled border secretes organic acids, including citric and lactic acids, lowering the pH of the extracellular medium in contact with the inorganic phase of bone. The hydroxyapatite (calcium phosphate), highly insoluble at prevailing physiological pH is mobilized into solution and taken up by the cell. Various proteolytic enzymes released from the osteoclast lysosomes then digest the collagen exposed by the removal of the mineral phase. Disruption of this process could tilt the balance between resorption and deposition of bone in osteoporotics in favor of formation and thereby retard or prevent the inexorable loss of bone mass to which these patients are subject.

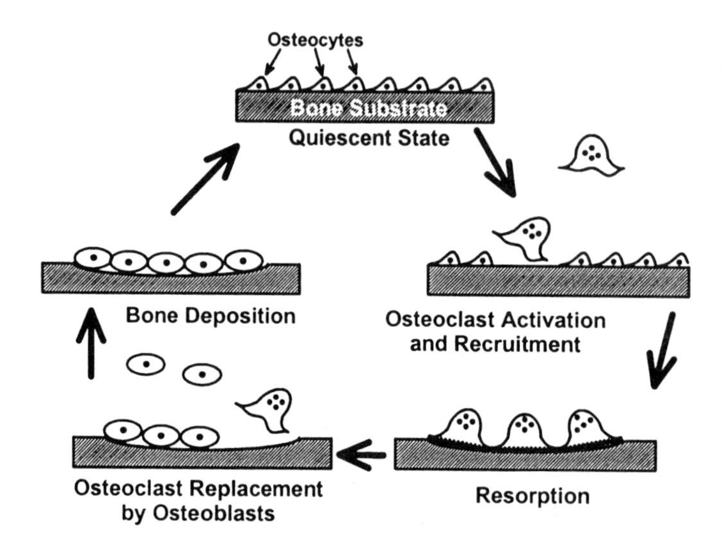

Figure 1. Schematic of Bone Turnover at the Cellular Level

Early work by Fleisch and others[62,63] had shown that pyrophosphate binds very strongly to calcium phosphate inhibiting both crystallization and dissolution of this material, *in vitro*, though they could not detect any effect on *in-vivo* bone resorption.[63] This difference is likely to arise from pyrophosphate itself being a naturally occurring metabolite, its quantities homeostatically controlled. Administered pyrophosphate is likely promptly hydrolyzed *in vivo*. Bisphosphonates, though, are analogs of pyrophosphate which, similarly to pyrophosphate, act as inhibitors of dissolution and crystallization of calcium phosphates *in-vitro*[64,65] but, in contrast, are stable *in vivo*. It was, therefore, reasoned that by stabilizing the surface of hydroxyapatite in bone, they would be capable of inhibiting the accelerated bone resorption characteristic of osteoporosis without affecting other organs.

Though the mechanistic reasoning for the potential therapeutic properties of pyrophosphate turned out to be wrong, bisphosphonates are, nonetheless, targeted to bone by the high affinity for hydroxyapatite provided by their structural similarity to pyrophosphate. Their individual effects, though, differ due to the uniqueness of the side chain of each individual bisphosphonate.[66] At the bone surface, these compounds inhibit osteoclast-mediated bone resorption by mechanisms which at the time had not been defined but which have since been demonstrated to consist of inhibition of two different pathways depending on the structure of the bisphosphonate. Aminobisphosphonates such as alendronate and risedronate act as specific inhibitors of farnesyl diphosphate synthase[67,68] inhibiting the isoprenoid-biosynthesis pathway and thereby inhibiting protein prenylation by depletion of the intracellular pool of geranylgeranyl diphosphate levels, which in turn leads to osteoclast inactivation and inhibition of resorptive activities.[67,68] Other bisphosphonates, such as clodronate and etidronate, induce osteoclast apoptosis after being metabolized to ATP analogs.[69] In general, in spite of the apoptotic mechanism by which non-amino bisphosphonates operate, bisphosphonates do not appear to be grossly cytotoxic *in vivo*. At the histological level, their mechanism of action appears to involve binding to the hydroxyapatite of bone followed by local release as a result of osteoclast acidification in the region of the cell's ruffled border, uptake by the osteoclast, and subsequent inhibition of cellular activity through suppression of protein prenylation which then induces loss of the ruffled border and inhibition of bone resorption.[66,69] Other data suggest a possible role for bisphosphonate-mediated inhibition of osteoclast recruitment or differentiation in decreasing bone resorption.[66] These mechanisms may operate alongside each other and help explain the short and long term effects of the bisphosphonates *in vivo*.

4. RATIONALE FOR DEVELOPMENT

At the time of development of alendronate, other bisphosphonates were already in use for treatment of bone disease but presented assorted therapeutic difficulties. Individual bisphosphonates may be distinguished through their selectivity for inhibition of bone resorption relative to direct inhibition of bone mineralization and bone formation, potency, gastrointestinal tolerability and the cytotoxicity which contributes to osteoclast inhibitory activity. Even today, many of the available bisphosphonates show limitations in several of these respects. Alendronate, however, appeared at the time to possess an excellent balance of characteristics for the management of osteoporosis. Alendronate (in rats) deposits preferentially at sites of bone resorption[71] and has a selectivity for inhibition of bone resorption relative to inhibition of bone mineralization of more than 100-fold.[72] Thus, it was understood that when given at therapeutic doses (10 mg for osteoporosis, 40 mg for Paget's disease) alendronate carries no risk of inducing osteomalacia in patients. In the clinical development studies, the therapeutic doses showed no increased GI risk though postmarketing experience hs produced reports of gastric and duodenal ulcers. Preclinical data showed that alendronate does not exhibit the cytotoxicity associated with etidronate and clodronate.[71,73] In summary, alendronate appeared to have all the characteristics of a promising new agent for the treatment of bone diseases associated with disturbances of turnover.

5. PRECLINICAL PHARMACOKINETICS AND PHARMACODYNAMICS

5.1 Preclinical Pharmacology

Before development of alendronate for treatment of osteoporosis, mainly based on work with etidronate, the effectiveness of bisphosphonates was suspected but had not been clearly established. Etidronate, however, was known to inhibit bone mineralization at high doses as does a promoter of bone formation, fluoride, which had been tried as anti-osteoporotic therapy but had been found to give mixed effects, enhancing bone deposition at lower doses by increasing osteoblastic activity[74,75] but producing detrimental effects on bone structure at higher doses.[76] Since the efficacy and safety of bisphosphonates in general were in doubt, it was necessary to establish them for alendronate which was believed likely to be more efficacious than etidronate and was expected to lack the deleterious effects on bone quality and turnover rates. To document the efficacy of alendronate in rats, the bone remodeling of male Sprague-Dawley which had undergone immobilization by unilateral sciatic neurectomy and were given doses up to 1.0 mg/kg alendronate was examined. Alendronate proved effective in inhibiting bone loss due to immobilization in a dose-dependent manner and effectively prevented the bone loss caused by the immobilization.[77] Having established that alendronate is capable of preventing loss of bone mass, it was important to show that the preservation of bone density resulted in preservation of bone strength. This was particularly important in lieu of the lack of correlation which had been observed for fluoride, which was understood to arise from a loss of structural integrity in bone. Ovariectomized Sprague-Dawley rats received vehicle or 0.28, 2.8, or 28 μg/kg of alendronate twice weekly through subcutaneous injection for up to 13 months. An additional group of control animals received neither surgery nor drug. The stiffness, yield, and ultimate loads of the femoral midshaft, the sixth lumbar (L6) vertebra, and the femoral neck were determined and the geometric properties of the cortical bone measured from digitized images of the tibial diaphysis at the level of the synostosis. In general, alendronate treated animals were shown to preserve their bone strength in a dose-related manner and to maintain the geometric integrity of their cortical bone. Microscopic examination of bone structure showed that alendronate helps preserve the spicular structure of bone, consistent with its ability to maintain bone strength.[78]

To examine the mode of action of alendronate, newborn rats were administered [^{3}H]alendronate by injection of (0.4 mg/kg). One day after administration, 72% of the osteoclastic surface, 2% of the bone forming, and 13% of all other surfaces were found to be densely labeled. Silver grains were seen above the osteoclasts and no other cells demonstrating an intimate association between osteoclast activity and alendronate deposition. Six days after administration, the label was found to be buried inside the bone, 600-1,000 microns away from the epiphyseal plate, indicating that normal growth and matrix deposition occur on the bone in which alendronate has been sequestered.[79] The mode of action at the cellular level was examined by infusing osteoclasts from adult animals, treated with 0.4 mg/kg alendronate subcutaneously for 2 days, with parathyroid hormone-related peptide. Cells were found to lack ruffled border but not clear zone. Additionally, alendronate was found to bind to bone particles, *in vitro*, with a $K_d \approx 1$ mM with a capacity of 100 nmol/mg at pH 7. At pH 3.5 binding was reduced by ~50%[79].

Altogether, these findings suggest that newly dosed alendronate binds to actively resorbing surfaces and is incorporated into the mineral phase of bone. During remodeling, alendronate is locally released through acidification by the osteoclast and its increased concentration stops resorption and membrane ruffling, without destroying the osteoclast.[79]

Similar to the results observed in rats, alendronate prevented the increase in bone turnover and increased both bone volume and bone strength in vertebrae of ovariectomized baboons, which show bone changes similar to those seen in ovariectomized women. In a comparative study between alendronate and etidronate, alendronate was found to be 1000-fold more potent in inhibiting bone resorption and had at least a 1000-fold higher safety margin with respect to inhibition of mineralization and osteomalacia.[80]

In general, it was considered important for the development of alendronate to establish, in as many models as possible, that the drug shows activity beneficial to bone density and is conserving of bone structure.

Models based on hyperthyroidism which leads to increased bone turnover and osteopenia showed that excess thyroid hormone induces cancellous bone loss associated with high bone turnover in the rat, though not so in vertebra, presumably because of lower turnover rates in this particular bone, and that this bone loss is prevented by alendronate through the inhibition of osteoclastic activity.[81-83]

Models based on the ability of prostaglandin E_2 (PGE_2) to stimulate both bone resorption and formation in experimental animals, leading to augmentation of trabecular and cortical bone yielded results similar to those found with other systems. Groups of female Sprague-Dawley rats were ovariectomized at the age of 6 months and treated with either vehicle, PGE_2 (3 mg/kg/day), alendronate (0.8 μg/kg/day), or PGE_2 and alendronate. Another group served as non-ovariectomized untreated controls. Histomorphometric analysis of 6-10 microns-thick tibial sections after *in-vivo* fluorochrome double labeling showed that treatment with PGE_2 alone increases the rates of endocortical mineral deposition and of bone formation, stimulates production of bone trabeculae in the marrow cavity, and increases cortical porosity. Combined alendronate and PGE_2 treatment prevented the resorption induced by PGE_2 but not the stimulation of bone formation on endocortical and periosteal surfaces and resulted in a significant increase in cortical thickness.[29]

The efficacy of alendronate, in reducing alveolar bone loss caused by experimental periodontitis in cynomolgus monkeys was examined in adult monkeys which had periodontitis induced by ligation of the mandibular molar teeth and inoculation with *Porphyromonas gingivalis*. The contralateral, homologous teeth, which were not ligated, served as controls. The monkeys received, either saline as placebo treatment or alendronate doses of 0.05 or 0.25 mg/kg, intravenously, every 2 weeks for 16 weeks. The placebo-treated animals experienced significant bone loss while those treated with alendronate at 0.05 mg/kg presented significantly-reduced bone loss associated with the experimental periodontitis at both sites. In contrast, the dose of 0.25 mg/kg presented mixed results as it turned out to be ineffective or only slightly effective in attenuating alveolar bone loss. Histomorphometric results correlated closely with those of the radiographic analysis of the same experiment. These results clearly supported the efficacy of alendronate in palliating bone loss as well as indicating that alendronate could be used to reduce the loss of alveolar support associated with periodontitis.[30]

5.2 Preclinical Pharmacokinetics of Bisphosphonates other than Alendronate

The pharmacokinetic and pharmacodynamic properties of bisphosphonates are difficult to study because these agents are highly charged in solution, form insoluble salts with divalent cations at physiological pH (>6.5) and incorporate into the mineral phase of bone where they take up long term residence until mobilized by acts of bone remodeling. Alternatively, acquiring insights into the pharmacology of these compounds is facilitated because the kinetic and dynamic behaviors of the various bisphosphonates present numerous common features. It is, therefore, instructive to examine the behavior of bisphosphonates side by side.

One significant common feature of bisphosphonates is poor oral absorption in all animal species studied. The oral absorption of pamidronate in rats is approximately 0.5%[84]. The oral absorption of etidronate in rats, rabbits, and monkeys is less than 10% and is quite variable, both between and within species[85]. Absorption of oral etidronate has been characterized in substantial detail. The primary sites of absorption of this compound in the rat appear to be the jejunum[86] and the stomach.[85] Oral absorption is greater in young rats and dogs than in the corresponding adults.[85] Oral absorption and disposition of [^{14}C]etidronate were shown to be linear with dose in fasted rats at doses of 0.1 to 10 mg given by intragastric administration.[85] Ingestion of bisphosphonates with food decreases oral absorption.[87]

The distribution and retention of bisphosphonates within the body has been studied in animals using radiolabeled compounds. In rats and mice, [^{14}C]clodronate disappeared promptly from plasma following an IV dose[88] and was measurable in bone, spleen, and, to a lesser extent, in liver up to 90 days postdose.[89] Some radioactivity was detected in the spleen, thymus, and small intestine of mice and in the spleen of rats 12 months after dosing, suggesting uptake of clodronate by the reticuloendothelial system of rodents. Sustained high levels of radioactivity were detected in bones through 12 months.[90] Using [^{14}C]pamidronate in rats, radioactivity was retained in the bones (25 to 30% of dose) and in the liver (3% of the dose at lower doses, up to 30% at higher doses).[84] Some of the deposition of clodronate and pamidronate within the stomach, liver, and spleen[89-92] was later found to be due to the formation of insoluble complexes with calcium[91] or with iron released by hemolysis.[93] Bisphosphonates are resistant to biochemical hydrolysis. Indeed, no evidence of metabolism has been detected following administration of etidronate to rats and dogs[85] or pamidronate to rats.[84] However, the various side chains of this class of compounds may represent an opportunity for metabolism,[66] though it is likely that the highly charged nature of the bisphosphonate moiety prevents access of these compounds to the sites of metabolism. Urinary excretion is the major route of elimination of bisphosphonates, with the balance of the dose retained in the body, largely in the skeleton. Approximately half the dose of etidronate administered intravenously in rats and dogs is excreted in the urine and the other half retained in the skeleton.[85] Renal excretion occurs both by glomerular filtration and by tubular secretion, the latter by means of a pathway distinct from the classic acid and base pathways.[94] Enterohepatic circulation of etidronate[85] and pamidronate[84] in rats is negligible. The biological half life of pamidronate in the bones of rats is approximately one year.[84] The terminal half life of clodronate and etidronate in rodents was estimated to be at least three and at least four months, respectively.[87,89]

5.3 Preclinical Pharmacokinetics of Alendronate

The absorption, distribution, metabolism, and excretion of alendronate were investigated extensively in rat, dog, and monkey. Plasma protein binding was investigated *in-vitro* for these species and for man.

5.3.1. Metabolism

The potential metabolism of alendronate was studied in rats and dogs administered radiolabeled [^{14}C]alendronate.[95,96] The radioactivity excreted in urine from both species and that which immobilized in the bones of rats was extracted and compared chromatographically with authentic [^{3}H]alendronate. No evidence of metabolism in either species could be found.

5.3.2. Binding to Plasma Proteins

Binding to plasma proteins varies by species and is influenced by pH and by the concentrations of drug (>1 μg/mL) and of calcium.[95,96,97] The fraction unbound at concentrations of 1 μg/mL of alendronate is approximately 4, 80, 38, and 22%, respectively, in rat, dog, monkey,[97] and man[98] at physiological pH (7.4) and calcium concentration (~2.5 nM). The unbound fraction is independent of drug concentrations up to the highes value tested (1 μg/mL). The higher plasma concentrations are not relevant clinically. Bound alendronate in rat plasma increases by about two-fold (50 to 98%) in the pH range 6.6 to 8.6 ([Ca] = 2.5 mM).[97,99] Increases in calcium concentration result in greater binding of the drug to rat albumin (at physiologic pH), from negligible at zero calcium concentration to greater than 90% bound at 2.5 mM calcium (physiological calcium concentration is 2.3 to 2.75 mM). Addition of EDTA decreases this effect, showing that calcium plays an important role in the binding of alendronate to plasma proteins in rats. Binding of alendronate to human plasma proteins parallels the results obtained with rat plasma with respect to drug concentration, pH, and calcium concentration. In addition, binding to purified serum albumin is stronger than to human plasma, suggesting the presence of one or more displacers of alendronate in human plasma.[98]

5.3.3. Renal Excretion

Renal clearance (Cl_r) of alendronate at low IV doses (1 mg/kg) is comparable to glomerular filtration rate (GFR) in rats (Cl_r = 7.4 mL/min/kg).[100] Since more than 90% of drug in plasma is bound to plasma proteins, this high clearance suggests that the drug is actively secreted by the renal tubules. Renal clearance decreases about twofold when the dose is raised to 20 mg/kg, with no change in GFR.[100] Renal clearance of alendronate in rats decreases with increasing drug concentrations in plasma suggesting the involvement of active transport in the renal excretion of alendronate. Cimetidine, quinine, probenecid, and p-aminohippuric acid, however, have no effect on renal excretion, showing that neither of the classic anionic or cationic transport systems is involved.[100] Etidronate does inhibit excretion in a dose-dependent manner, however, suggesting that bisphosphonates are excreted through a separate renal transport system likely associated with phosphate metabolism.[100]

Loss of renal function leads to proportional decreases in renal clearance of alendronate. The renal clearance of alendronate in uremic rats (GFR = 0.71 mL/min/kg was approximately 1 mL/min/kg, compared with 12 mL/min/kg in healthy rats (GFR = 7.4 mL/min/kg). The main consequence of this decrease was a 55% increase in the amount of alendronate deposited in bone of uremic rats compared with controls.[100]

5.3.4. *Tissue Distribution*

Following IV administration, alendronate is widely distributed to the calcified and non-calcified tissues of rats.[97] A large fraction of the dose distributes promptly to soft tissues (63% by five minutes after administration). Kidney, liver, lung, spleen, and stomach show the most uptake among the soft tissues. Residence in soft tissues, however, is short lived. One hour after administration of a 1 mg/kg dose, only ~ 5% of total dose can still be found in soft tissues. At an extremely high dose (30 mg/kg), clearance of drug from soft tissues is slower (23% of dose remains 48 hours postdose) quite possibly as a result of precipitation in the form of Ca or Mg salts (or other multivalent cations).

About 60 to 70% of an IV dose is found in bone one hour after administration (1 mg/kg IV). This amount remains essentially unchanged 71 hours later. Concentrations in bone decline in a biphasic manner over an 80-day period. Half lives of 30 and 200 days, respectively, were estimated for the two phases observed in rats. The terminal half life in dogs was estimated to be greater than 1000 days.[97]

As demonstrated in rats, alendronate does not distribute homogeneously in bones.[100] Actively remodeling surfaces (i.e., epiphyses of long bones) accumulate more drug than do other surfaces.

The clearance of alendronate to bone (uptake) is comparable to plasma flow through this tissue, indicating bone's strong ability to extract this drug (extraction ratio in bones ≈ 0.8). Acute IV administration of very high doses (50 mg/kg) to rats appears to saturate the instantaneous uptake capability of bone (~60% of amount projected from lower doses).[100] Repeated administration of smaller doses, up to a cumulative total dose of 35 mg/kg, however, did not yield any evidence of saturation, indicating that the overall carrying capacity of bone for alendronate had not been reached. Uptake of alendronate in bone is higher in younger rats (two months old) but is not different between 12- and 20-month old rats. At lower doses, male rats accumulate more drug in bone than females (~30%), but the difference disappears at higher doses.

Hyper- and hypocalcemia in rats result in a three to fourfold decrease and a 60 to 70% increase, respectively, in bone uptake of alendronate.[101] The effects of plasma calcium on uptake are postulated to arise from its influence on bone turnover and on the plasma binding of alendronate. Renal failure (GFR <2.7 mL/min/kg) increases bone uptake of alendronate by about 50%.

5.3.5. *Absorption*

Alendronate is poorly absorbed in rats, dogs, and monkeys.[96,102] Drug concentrations in plasma following oral administration are undetectably low, precluding the use of plasma area under the curve (AUC) in evaluating oral absorption. To estimate drug absorption and skeletal retention, drug deposition was examined in bone following oral (^{14}C-labeled) and IV (^{3}H-labeled) administration. This method is valid given that alendronate is not metabolized, is largely deposited in bone (60 to 70% of systemic dose), and shows a long residence in this tissue. Bioavailabilities were estimated at 0.7, 1.8, and 1.7% in rat, dog, and monkey, respectively. More detailed examination in rats shows absorption of alendronate occurs more readily in the upper regions of the small intestine (jejunum>duodenum>ileum). Food decreases bioavailability substantially in rats. EDTA was found to enhance absorption of an alendronate dose of 8 mg/kg in rats in a dose-related manner as the EDTA dose increased from 1.2 to 18.6 mg/kg. This suggests that the formation of insoluble divalent cation (Ca^{++}, Mg^{++}) salts may prevent the absorption of alendronate from the GI tract. More than proportional increases in absorption were observed as the dose of alendronate increased from 2 to 40 mg/kg (supporting this conjecture), presumably due to saturation of this binding.

6. CLINICAL PHARMACOLOGY STUDIES

6.1 Pharmacodynamic Measurements

Clinical pharmacology studies for preliminary assessment of drug candidates are generally carried out over short periods. In the case of bisphosphonates, however, efficacy was expected to consist of altering the balance of bone turnover and positive measurements clear enough to stand out from background were not expected to be attainable until after months, or maybe years, of dosing for the indications for which these drugs were intended. Bone resorption, however, since it involves the dismantling of the crystalline and organic phase of bone, produces various biochemical markers which can be detected in the urine and, therefore, be monitored for the possible effects of bisphosphonates in bone turnover. Changes in these markers were expected to be much faster than changes in bone density or structure and, therefore, it was deemed that use of the proper models should allow biochemical assessment of alendronate efficacy in a time frame consistent with the needs of a development program. It is important to bear in mind that the pharmacodynamic results in man were to be seen in light of the preclinical pharmacology in which it had been shown that alendronate diminishes bone turnover with a consequent increase in BMD and a resultant increase in bone strength which, it was presumed, would lead to a smaller likelihood of fracture in osteoporosis patients. Alternatively in patients with Paget's disease of bone, slower bone turnover (with slower bone deposition indirectly produced by slower bone resorption) would reduce the rate at which abnormal bone tissue accumulates. Direct evidence of alendronate's ability to decrease the morbidity associated with these two diseases was not to be obtained until much later in the program.

6.2 Summary of the Clinical Pharmacology Program

Two early studies, a single dose one and a multiple dose one were carried out in patients with Paget's disease, which is characterized by focal areas of pathologically accelerated bone turnover. This acceleration provided a better opportunity to evaluate biochemical efficacy in a reasonable time. A six-week study was carried out in women in early menopause to examine the time course of onset and reversal of the biochemical effects of alendronate at various doses. Additionally, the duration of biological effects was examined following administration of a 30-mg intravenous dose. Given that the half life of alendronate was estimated from preclinical results to be fairly long (months or years), it was thought that the effects of the drug might be much longer lived than it is typical for drugs with shorter half lives. The studies in postmenopausal women were carried out in subjects of European and Asian origin since the prevalence of osteoporosis in women of sub-Saharan African descent is known to be much lower than in other populations. Measurements of mineral homeostasis for all these studies included serum concentrations of calcium, phosphate, parathyroid hormone, 1,25-dihydroxyvitamin D and 25-hydroxyvitamin D. The rate of bone resorption was monitored by following changes in urinary excretion of calcium, hydroxyproline, lysylpyridinoline (also referred to as deoxypyridinoline), and hydroxylysylpyridinoline (also referred to as pyridinoline), all corrected for creatinine excretion to allow for differences in renal function. Subsequently, urinary excretion of N-telopeptide (arising from breakdown of collagen crosslinks) was used as a marker for bone resorption. Serum alkaline phosphatase and osteocalcin were used as measures of bone formation.

6.3 Studies in Patients with Paget's Disease of Bone

In the first of the two studies carried out in patients with Paget's disease of bone, the biochemical activity and safety of alendronate were examined in five patients. During a 25-day period each patient received five doses of alendronate (5-, 10-, 25-, 50-, and 75-mg capsules) in a sequentially increasing manner with two doses of placebo randomly interspersed in each subject's regimen for a total of seven administrations. The patients and the investigators were both blinded to active treatment and placebo. Patients fasted from midnight prior to administration of each treatment to one hour afterwards. Serum calcium, phosphate and alkaline phosphatase were measure pre-study, pretreatment on each study day and post-study. Urinary excretions of hydroxyproline and calcium were measured in fasting two-hour urine collections pre-study, pretreatment on study Days 1 and 4 and post-study. No evidence of safety concerns that would have precluded advancement to higher doses was detected either biochemically or by clinical examination.

Following the escalating dose study, the safety and the biochemical activity of alendronate when administered as multiple oral doses were examined in a double-blind placebo-controlled study in which 22 patients with Paget's disease of bone received either alendronate capsules or placebo for three consecutive weeks, five days/week. Ten of the patients were men and twelve were women, all falling in the age range of 48 to 76 years old. Twelve of the patients were randomized to receive 25 mg/day for Week 1, 50 mg/day for Week 2 and 100 mg/day for Week 3 for a total dose of 875 mg. The remaining ten patients constituted a parallel control group and were administered placebo on all treatment days. Serum alkaline phosphatase activity and fasting urinary excretion of hydroxyproline and of calcium were monitored as biochemical markers of bone turnover. Baseline values for urinary hydroxyproline excretion were higher than in normal populations as a consequence of the higher bone turnover characteristic of the disease but were not significantly different between the treated and placebo groups. A consistent and significant within-treatment reduction form baseline was observed by the second week of therapy in the alendronate group (median % change of –33%, $p<0.01$)

with significant differences from placebos seen as early as Day 5 of Week 2 as shown in Figure 2.

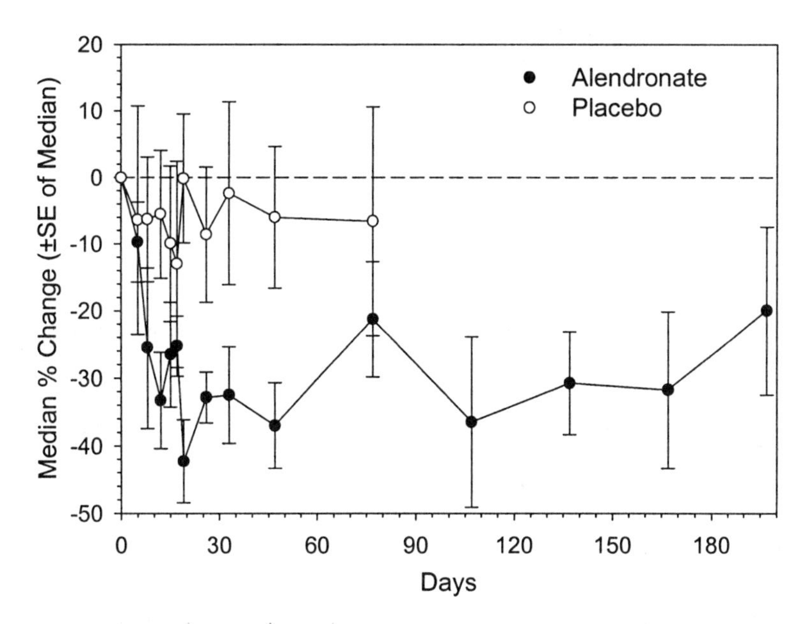

Figure 2. Median (SE of Median) % Change in Urinary Hydroxyproline Corrected for Urinary Creatinine from Baseline in Patients with Paget's Disease of Bone.

While a trend toward reduction in calcium excretion was suggested by the data, the substantial variability observed obscured any differences between the alendronate- and placebo-treated groups and the difference was, therefore, not found to be significant. Elevated alkaline phosphatase activity is characteristic of the accelerated bone turnover of patients with Paget's disease and effective therapy would be expected to correct or ameliorate this abnormality. As shown in Figure 3, treatment with alendronate significantly reduced serum alkaline phosphatase activity. The maximum reduction from baseline was observed at two months post-treatment (43% vs. placebo, $p<0.001$). Within-treatment comparisons suggested that a significant reduction may have occurred as early as one week post-treatment (median reduction from baseline = 11%, $p<0.01$). This parameter had not returned to baseline at five months post-treatment. The apparent delay in the change in alkaline phosphatase activity as compared with that of urinary hydroxyproline is characteristic of the therapeutic response of pagetic patients to anti-resorptive therapy. The primary drug effect is reflected in the index of bone resorption, urinary hydroxyproline, which is then succeeded by a reduction in the index of bone formation, alkaline phosphatase. This follows from the observation that while bone affected by Paget's disease has a marked increase in bone turnover, the processes of bone resorption and formation generally remain coupled. Thus, inhibition of bone resorption is followed by a reduction of bone formation.

Significant reduction in median serum calcium concentration was observed from Day 5 on the second week of treatment through one week post-treatment. The maximum median reduction from baseline was 0.58 mg/dL on Day 3 of Week 3. No patient showed symptoms of hypocalcemia and median concentrations of calcium stabilized by the end of Week 2 in spite of continued therapy with 100 mg alendronate per day.

6.4 Examination of Long-Term Effect on Biochemical Markers in Women with Postmenopausal Osteoporosis

Some measures of biochemical efficacy were included in an open-label, uncontrolled study carried out primarily to examine the terminal elimination half life of alendronate which was expected to be very long. In this study, twenty one women (aged 56 to 75) received intravenous doses of 30 mg administered as daily doses of 7.5 mg for four consecutive days. The dose was administered in a divided manner for safety reasons, to avoid the possibility of an acute-phase reaction. Individual intravenous doses of 10 mg had been previously administered safely. While the results suggested that the duration of antiresorptive effects of bone, which may result from a relatively large parenteral dose given over a brief interval, the absence of placebo-treated controls and the attrition of patients over time prevented the drawing of firm pharmacodynamic conclusions from this study. The median (SD) baseline value for urinary excretion of hydroxyproline was 32.76 (11.53) mg/g of creatinine, a value generally within the normal range (<35 mg/g). Significant median percent decreases from baseline were observed across Days 4 through 8 in the range of 32 to 46% (within-group $p<0.05$). Median percent decreases from baseline ranged from 24 to 53% across the 2- to 24-month time points with the reductions being less consistently significant after Month 8.

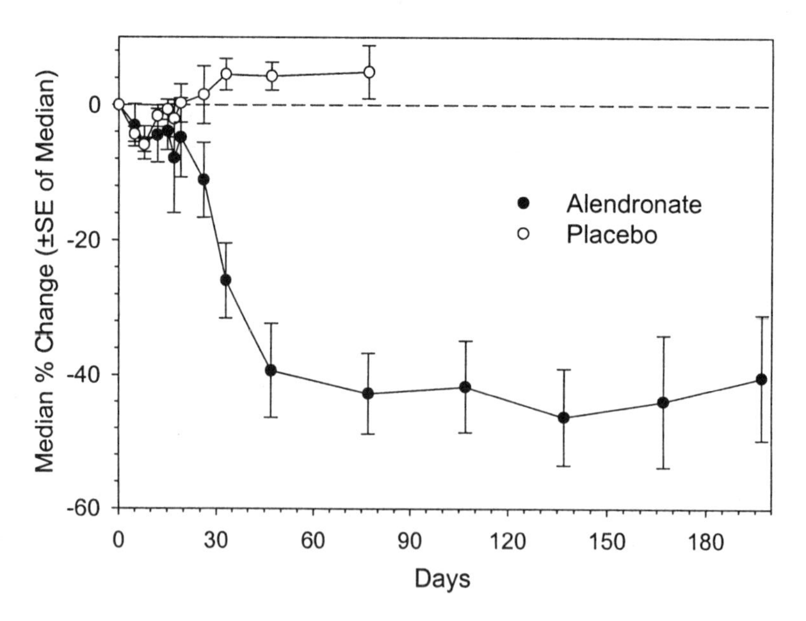

Figure 3. Median (SE of Median) % Change in Alkaline Phosphatase from Baseline in Patients with Paget's Disease of Bone

For urinary excretion of calcium, the median baseline value was 0.09 (0.05) mg/dL of glomerular filtrate. Median percent decreases were found to be significant ($p<0.05$) at nearly all times. The maximum median percent change from baseline was a decrease of 72% at Day 7. This reduction was attenuated to the range of 28% to 40% between Months 6 and 15. By Month 24, a less than significant median decrease-from-baseline of 15% was observed ($p>0.2$).

For serum alkaline phosphatase, the baseline median value was 80.0 (31.6) U/L. Significant median decreases from baseline were observed from Month 2 through Month 15 ($p<0.05$). The maximum median change from baseline was a 21 U/L decrease at Month 4. By Month 15, the median reduction had contracted to 10 U/L and by Month 24 to 1.0 U/L.

In summary, the biochemical indices of bone turnover suggest that after a total intravenous dose of 30 mg of alendronate, there is a slow resolution of the effect of alendronate on bone resorption over the 24 months that the subjects were followed up. Unfortunately, since no parallel control group was included in this study, these results have to be treated with caution. In spite of the long residence of alendronate in bone (terminal $t_{½}$ $\geq$10.9 yrs.) the biochemical effects begin to wane over a period of months following the end of therapy. This loss of biochemical efficacy may very well reflect the incorporation of alendronate in the mineral phase of bone which then becomes inaccessible to the inhibition until the specific spot in which the drug is located undergoes another round of bone resorption.

6.5 Studies in Women in Early Menopause

The time course of the pharmacodynamic effects of alendronate was better characterized in a large double-blind, placebo-controlled, two-center study in which oral doses of alendronate were administered for six weeks to women early in menopause. Sixty five women between the ages of 41 and 58 years whose last menstrual period had occurred sometime between six months and three years before the beginning of the study were studied. Seventeen women received placebo and three groups of 16 women each received alendronate 5, 20, or 40 mg once daily over the six-week treatment period. Treatment was preceded by a two-week run-in period and followed by a 30-week run-out period. The effectiveness of alendronate as an inhibitor of bone resorption was assessed using urinary excretions of hydroxyproline, calcium lysylpyridinoline, hydroxylysyl-pyridinoline and crosslinked N-telopeptides, all corrected for urinary creatinine. The crosslinked N-telopeptides and lysylpyridinoline may be the most specific markers for turnover of bone collagen and, in particular, are specific for the Type 1 collagen of bone. The urinary excretion of crosslinked N-telopeptides and pyridinoline are sensitive and specific markers of bone resorption. Serum alkaline phosphatase activity and osteocalcin concentrations were monitored as indices of bone formation. The effects of alendronate on mineral homeostasis were monitored using albumin-corrected serum calcium concentration and serum concentrations of phosphorus, parathyroid hormone, 1,25-dihydroxyvitamin D and 25-hydroxyvitamin D. At pre-study and at Week 36, bone mineral density (BMD) for the lumbar vertebrae (mean of L2, L3 and L4) was determined using the Hologic-1000 densitometer. The median albumin-corrected serum calcium concentration was found to be significantly lower after treatment with 40-mg alendronate compared with placebo at the three- and six-week time points. The median

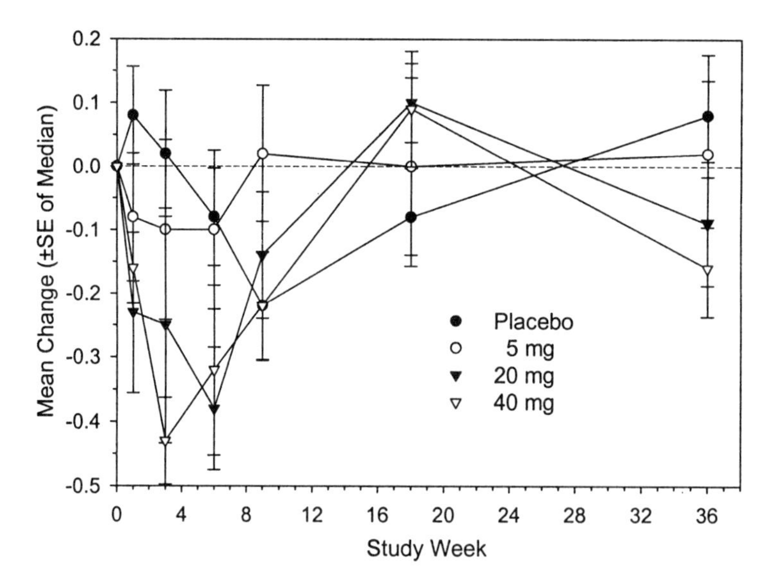

Figure 4. Median (SE of Median) Change in Serum Calcium from Baseline in Women in Early Menopause (mg/dL, Corrected for Albumin) Following Dosing for 6 Weeks with Alendronate of Placebo

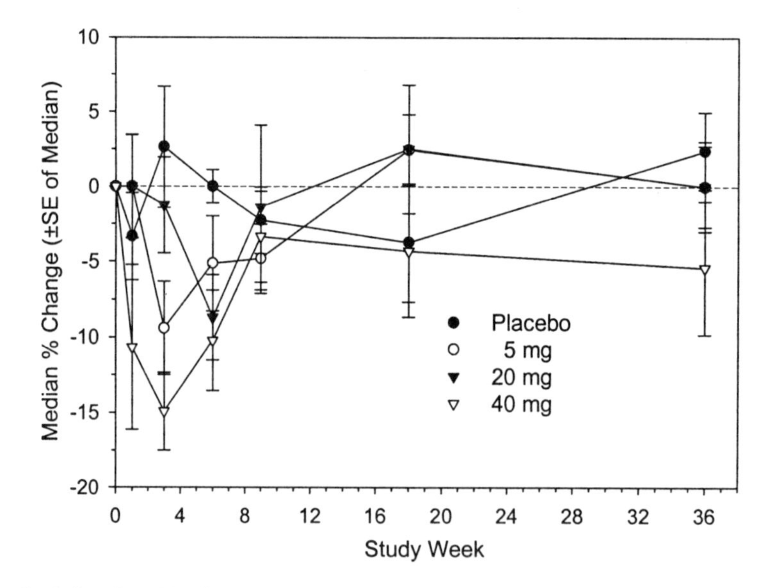

Figure 5. Median (SE of Median) % Change in Serum Phosphate from Baseline in Women in Early Menopause Following Dosing for 6 Weeks with Alendronate of Placebo

change in albumin-corrected serum calcium concentration over time is depicted in Figure 4. The median percent decrease in serum phosphorus after treatment with 40-mg alendronate was consistently greater than those observed after treatment with placebo at each week (Figure 5). At Week 6, significant differences were detected between 20 mg and placebo and also between 40 mg and placebo. The median serum PTH concentrations increased over baseline at Week 6 after each alendronate treatment regimen, while a small decrease was noted in the placebo group (Figure 6). At Weeks 6 and 9, each pairwise comparison with placebo was significant ($p<0.05$). By Week 36 there were no persistent differences in PTH concentration between treatment groups. Serum 1,25-dihydroxyvitamin D and 25-hydroxyvitamin D concentrations showed little or no difference between treatment groups.

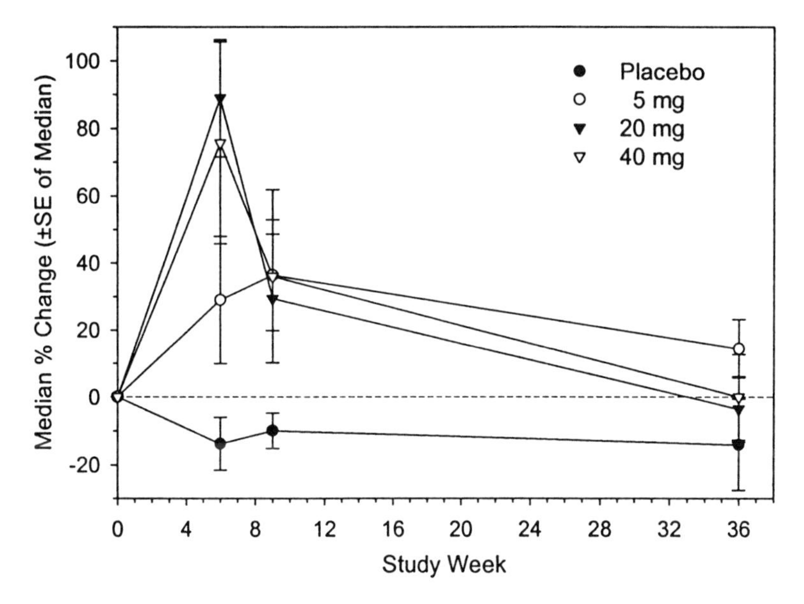

Figure 6. Median (SE of Median) % Change in Serum Parathyroid Hormone from Baseline in Women in Early Menopause Following Dosing for 6 Weeks with Alendronate of Placebo

The effectiveness of alendronate as an inhibitor of bone resorption was assessed using urinary excretion of calcium, hydroxyproline, and pyridinoline, each corrected for creatinine. The median two-hour fasting urinary calcium excretion corrected for creatinine was reduced by Week 3 of alendronate treatment at all dose levels as shown in Figure 7. The median decrease in the two-hour fasting urinary calcium excretion was significant for each dose of alendronate when compared with placebo at Week 3 ($p<0.01$). At Week 6, significant differences were detected for 20 and 40 mg as compared with placebo. The median percent reduction in two-hour fasting urinary hydroxyproline excretion corrected for creatinine was consistently greater after treatment with 20 and 40 mg as compared to placebo and 5 mg over Weeks 3 through 36 (Figure 8). Significant differences were detected between 20 mg and placebo and also between 40 mg and placebo at Week 3, 9 and 18.

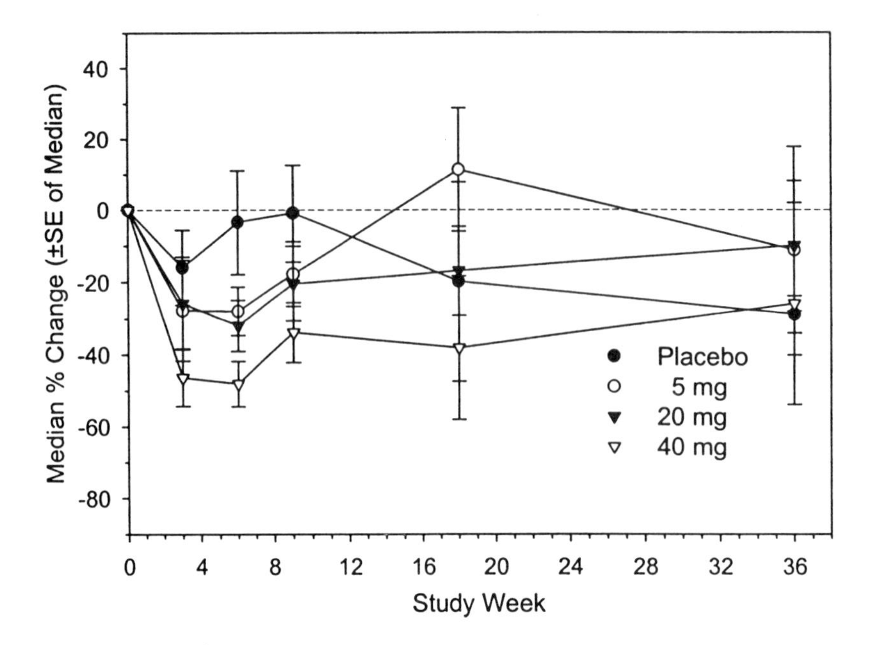

Figure 7. Median (SE of Median) % Change in Urinary Calcium from Baseline in Women in Early Menopause Following Dosing for 6 Weeks with Alendronate or Placebo

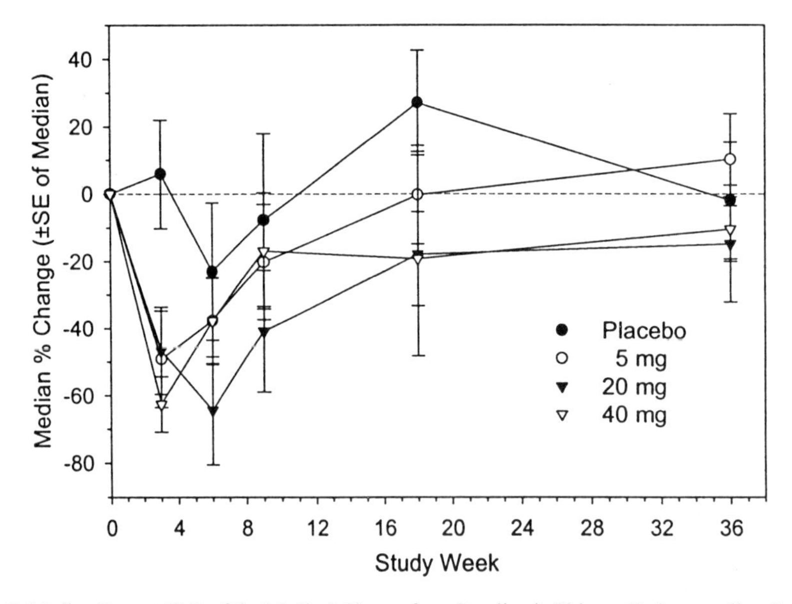

Figure 8. Median Percent (S.E. of the Median) Change from Baseline in Urinary Hydroxyproline Corrected for Urinary Creatinine (mg/g) with Time in Women Early in Menopause Following Alendronate (5, 20 or 40 mg) or Placebo for Six Weeks

Lysylpyridinoline may be the most specific marker for the turnover of bone collagen. Lysylpyridinoline excretion, corrected for creatinine, decreased in a dose-related manner with 40 mg consistently yielding the largest median percent decrease from baseline through Week 18 (Figure 9). Significant or nearly significant differences were detected between 20 mg and placebo and also between 40 mg and placebo at Weeks 3, 6 and 9. Similar changes were seen in the percent change from baseline for total pyridinolines (lysylpyridinoline plus hydroxylysylpyridinoline) corrected for creatinine. Following therapy, pyridinolines returned toward baseline values during the run-out period. Similar changes were observed for the urinary excretion of the crosslinked N-telopeptides.

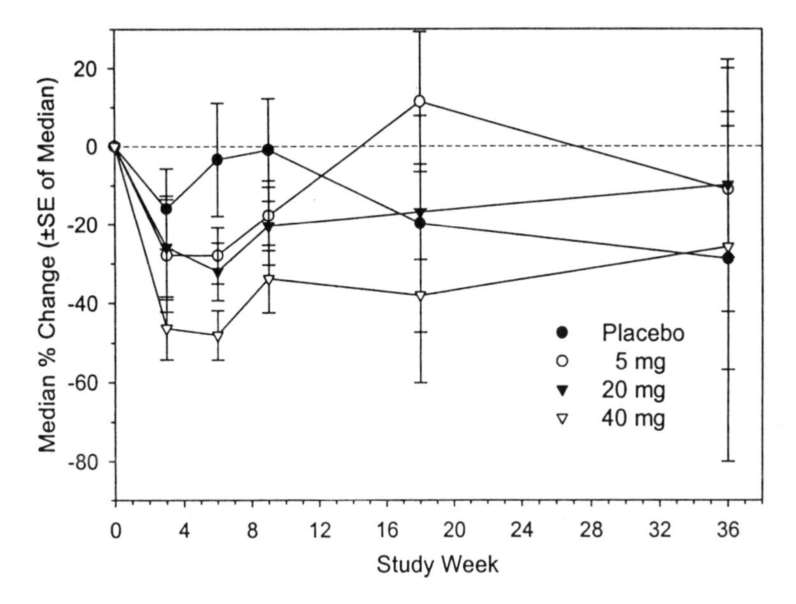

Figure 9. Median Percent (S.E. of Median) Change form Baseline in Urinary Lysyslpyridinoline Corrected for Urinary Creatinine (pmoles/mg) in Women Early in Menopause Following Alendronate (5, 20 or 40 mg) or Placebo for Six Weeks

Serum alkaline phosphatase activity and osteocalcin concentrations were monitored as indices of bone formation. As would be anticipated from the effect of a primary antiresorptive agent and from the coupling of bone resorption and bone formation, there was a delay in the changes in alkaline phosphatase relative to the suppression of the indices of bone resorption. There appeared to be little or no difference between the treatment groups in alkaline phosphatase % change from baseline before Week 9 when a dose-related decrease from baseline was observed. By Week 36, these values had returned to or above baseline values as seen in Figure 10.

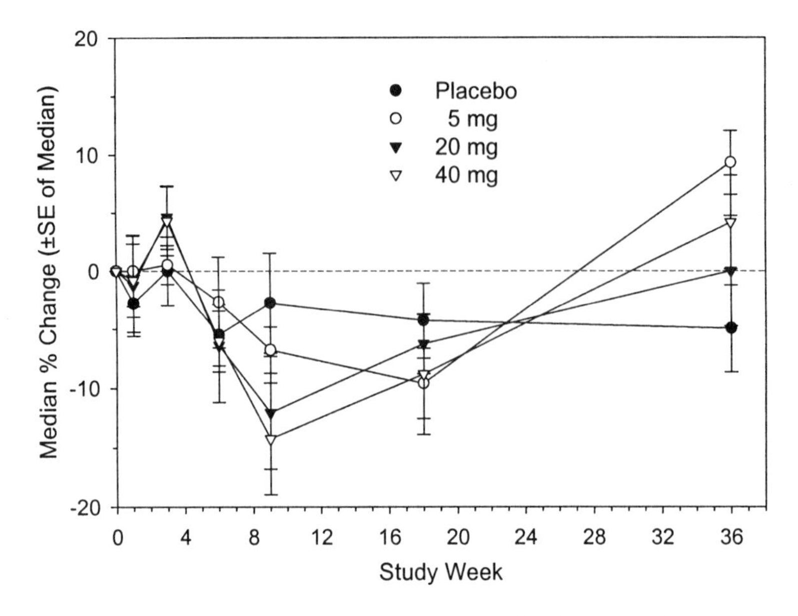

Figure 10. Median Percent (S.E. of Median) Change form Baseline Serum Alkaline Phosphatase (U/L) in Women Early in Menopause Following Alendronate (5, 20 or 40 mg) or Placebo for Six Weeks

Dose-related decreases in osteocalcin were observed at Weeks 6 and 9, when the largest % decreases from baseline were noted after treatment with 40 mg. In addition, the 20-mg treatment was found to be significantly different from placebo and from 5 mg at Week 9. Only the 40-mg group was significantly different from placebo by Week 36. These results are shown in Figure 11

The median percent change from baseline in lumbar spine BMD (average of L2 to L4) at Week 36 showed a significant dose-related trend toward higher values (% increase) at Week 36 ($p = 0.0002$) where the median percent changes were -2.3, -1.2, +0.7 and +1.2% following placebo, 5-, 20-, and 40-mg doses, respectively.

In summary, treatment with alendronate produced biochemical changes consistent with dose-dependent inhibition of bone resorption, including small reductions in serum calcium and phosphate concentrations and suppression of urinary calcium hydroxyproline and pyridinoline excretion as early as Week 3 of treatment. These changes are exactly what would be expected from a compound that inhibits osteoclast-mediated bone resorption. As would also be expected, this suppression of bone resorption eventually resulted in a reduction of bone formation as indicated by the subsequent fall in serum alkaline phosphatase activity and osteocalcin concentrations.

These data indicate that alendronate therapy produces a dose dependent suppression of bone turnover in women early in menopause similar to other antiresorptive agents such as estrogen. The magnitude of the changes seen with alendronate treatment were similar to those observed following short-term (one to six months) estrogen administration to a similar population. Following discontinuation of alendronate therapy, these biochemical changes returned toward baseline values.

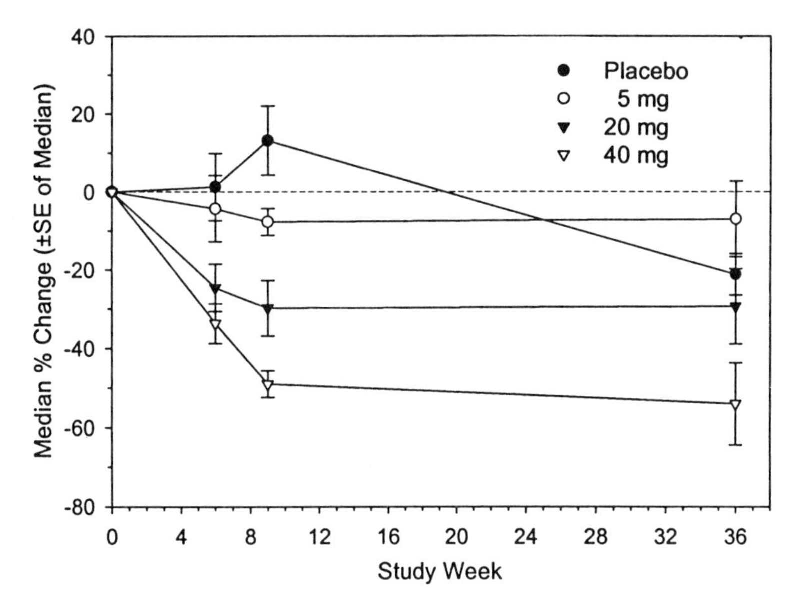

Figure 11. Median Percent (S.E. of Median) Change from Baseline Serum Osteocalcin (ng/mL) in Women Early in Menopause Following Alendronate (5, 20 or 40 mg) or Placebo for Six Weeks

A reduction of the alendronate-induced changes was evident as early as three weeks post-treatment for most parameters. This occurred despite the long retention time of the drug in the skeleton consistent with the hypothesis that following discontinuation of therapy, the non-resorbed alendronate may be subsumed under newly formed bone and become biochemically inaccessible.

On the strength of of the safety and efficacy results from the Clinical Pharmacology program for alendronate, the long term safety and efficacy were studied in a series of Phase III studies.

In a three-year, randomized, double-blind, multicenter study, 478 postmenopausal women with osteoporosis received one of the following four treatments: placebo, 5 or 10 mg/day alendronate for 3 years or 20 mg/day alendronate for 2 years followed by 5 mg/day for 1 year (20/5 mg). In addition, all patients received 500 mg/day of calcium. Bone mineral density (BMD) was measured by dual energy X-ray absorptiometry. Increases in BMD of the lumbar spine, femoral neck and trochanter were observed following three years of dosing with alendronate for all doses. Increases with the 10-mg dose were significantly greater than observed with either the placebo or the 5-mg/day dose while the mixed treatment (20 mg for 2 years/5 mg for 1 year) was found to be similarly effective to the 10-mg/day dose. Bone turnover was seen to decrease ($p<0.01$) as estimated by urinary excretion of deoxypyridinoline (resorption marker) and serum alkaline phosphatase and osteocalcin (formation markers) and loss of stature was reduced by a mean 41% in the treated subjects (vs. placebo). The safety profile for all doses of alendronate was similar to that of placebo. These results reinforced the standing of alendronate as a potential advance in the treatment of osteoporosis.[103]

In a second study to ascertain the positive effects of long-term daily oral alendronate therapy on bone mass in postmenopausal women with osteoporosis, 516 postmenopausal women aged 45-80 years (spine BMD 2.5 SD or more below the mean for young premenopausal women) received the same treatments as in the previous study (placebo, 5 or 10 mg/day alendronate for 3 years or 20 mg/day alendronate for two years followed by 5 mg/day for one year +500 mg/day calcium for all). The results from this study were consistent with those of the previous with 7.4%, 5.5% and 7.2% increases in BMD for the 10-mg/day group vs. 0.6, 0.7, and 0.4% for the placebo group at the spine, femoral neck and trochanter, respectively. Efficacy of the 20/5 mg alendronate treatment was similar while the 5-mg dose was less effective. BMD continued to increase over the entire 3-year study duration in the alendronate-treated groups and, compared with the other doses, 10-mg alendronate presented the largest gains in BMD during the 3rd year of the study. Changes in biochemical markers of bone turnover and mineral homeostasis confirmed the effect of alendronate to decrease bone turnover to a new steady-state level. The safety and tolerability of alendronate were comparable with those of placebo. In short, the results of the second study endorsed the conclusion of efficacy and safety of alendronated derived from the first[104].

To confirm that alendronate reduces fracture risk not just in postmenopausal women with vertebral fractures, but also in those without, 4432 women aged 54 to 81 years with a femoral neck BMD of 0.68 g/cm2 or less but no vertebral fractures were randomized to either placebo or 5-mg/day alendronate for two years followed by 10-mg/day for antoehr two years[105]. : Clinical fractures confirmed by X-ray, new vertebral deformities detected by morphometric measurements on radiographs, and BMD measured by dual x-ray absorptiometry were followed as the main outcomes. Over the course of the four years of the study, alendronete safely increased BMD and significantly increase the risk of clinical fracture among women with osteoporosis[105]. In postmenopausal women with preexisting vertebral fracture, alendronate therapy for 3 years reduced the number of days of bed disability and days of limited activity caused by back pain[106]. Treatment with alendronate was also shown to reduce the risk of multiple symptomatic fractures during the treatment period (4.3 years averag). The reductions were consistent across prespecified sub-groups. This effect is evident early in treatment and is sustained[107].

In a followup study,[108] it was hypothesized that since growth hormone increases bone turnover and stimulates osteoblast activity, the combination of a growth-hormone secretagogue with a potent inhibitor of bone resorption would maintain a higher rate of bone formation than would be possible with either agent alone, thereby generating greater enhancement of bone mineral density in patients with osteoporosis. To examine this hypothesis, the individual and combined effects of a growth hormone secretagogue (MK-677, Merck & Co., Whitehouse Station, NJ) and alendronate on insulin-like growth factor I levels and on biochemical markers of bone formation (osteocalcin and bone-specific alkaline phosphatase) and resorption [urinary N-telopeptide cross-links (NT_x)] were examined for 12 months and BMD for 18 months in a multicenter, randomized, double blind, placebo-controlled study in which 292 women (64-85 yr. old) with low femoral neck BMD were randomly assigned in a 3:3:1:1 ratio to one of four daily treatment groups for 12 months: 1)- 25-mg MK-677 plus 10-mg alendronate, 2)- 10-mg alendronate, 3)-; 25-mg MK-677 or 4)- placebo (double dummy). Patients assigned to groups 3 and 4 were switched at the end of 12 months to treatment 1 for the remaining six months of the study. Patients in groups 1 and 2 maintained their originally assigned therapy for the duration of the study. All patients also received 500 mg/day of calcium.

At the end of 12 months, patients treated with MK-677, alone or with alendronate showed increases from baseline in levels of the insulin-like growth factor I (39% and 45%, respectively, $p<0.05$ vs. placebo.

The primary results, except for BMD, are provided for month 12. MK-677, with or without alendronate, increased insulin-like growth factor I levels from baseline (39% and 45%; $P < 0.05$ vs. placebo). MK-677 increased osteocalcin and urinary NT_x by 22% and 41%, on the average, respectively ($P < 0.05$ vs. placebo). MK-677 and alendronate mitigated the reduction in bone formation compared with alendronate alone based on mean relative changes in serum osteocalcin (-40% vs. -54%; $P < 0.05$, combination vs. alendronate) and reduced the effect of alendronate on resorption (NTx) as well (-52% vs. -61%; $P < 0.05$, combination vs. alendronate). MK-677 plus alendronate increased BMD at the femoral neck (4.2% vs. 2.5% for alendronate; $P < 0.05$). However, similar enhancement was not seen with MK-677 plus alendronate in BMD of the lumbar spine, total hip, or total body compared with alendronate alone. Growth hormone-mediated side effects were noted in the groups receiving MK-677, although adverse events resulting in discontinuation from the study were relatively infrequent. In conclusion, the anabolic effect of growth hormone, as produced through the growth hormone secretagogue MK-677, attenuated the indirect suppressive effect of alendronate on bone formation, but did not translate into significant increases in BMD at sites other than the femoral neck. Although the femoral neck is an important site for fracture prevention, the lack of enhancement in bone mass at other sites compared with that seen with alendronate alone is a concern when weighed against the potential side effects of enhanced growth hormone secretion.

7. PHARMACOKINETICS IN MAN

As evidenced from the preclinical results for the various bisphosphonates including alendronate, pharmacokinetics of these compounds are very similar to each other's and also amongst the different mammalian species. This similarity is presumed to arise from the presence of the bisphosphonic acid group which confers on these molecules their therapeutic capabilities and which also dominates their chemistry giving them a high solubility at high pH or in the absence of multivalent cations, poor absorption (because of their high polarity), general lack of metabolism (through an inability to travel across cell membranes), high renal clearance and their ability to bind to the mineral phase of bone and remain in the body for very long times. Due to the sparsity of information which can be generated for any single one of these drugs, a discussion of the kinetics of other bisphosphonates at the time of development of alendronate should help put into perspective the findings for this compound.

7.1 Clinical Pharmacokinetics of Bisphosphonates other than Alendronate

The pharmacokinetics of bisphosphonates in man have been characterized to a very limited extent. Bisphosphonates are difficult to measure analytically in biological fluids (especially plasma), and their disposition properties make it difficult to fully characterize their pharmacokinetic behavior. Due to the analytical limitations, little is known about plasma kinetics at therapeutic doses. Most of the pharmacokinetic information has been derived from urinary excretion data. The plasma pharmacokinetics of risedronate have been examined to a reasonable extent through the use of an immunoassay sensitive enough for this task but they have not yet been discussed extensively in the literature and are, therefore, not generally publicly available.

Following IV administration, etidronate plasma concentrations decline rapidly[109]. Following 60 mg pamidronate infused intravenously over one to four hours to breast-cancer patients, plasma concentrations also declined rapidly over the course of a few hours.[110,111] Bisphosphonates appear to be moderately protein bound in man. The free fraction of clodronate was estimated at 60 to 70%[112,113] and that of etidronate at about 80%[113]. No evidence of metabolism of bisphosphonates in man has been documented. Clodronate appears intact in the urine[112] and similar results have been seen with etidronate[109]. Neither compound is excreted in the bile.[109,112] Elimination of bisphosphonates in man appears to occur exclusively through renal excretion, similar to what is seen in preclinical studies. Following IV administration to healthy subjects or cancer patients, urinary excretion averaged 73 to 80% for clodronate,[112,114] 35 to 52% for etidronate,[109,115-117] and 30 to 50% for pamidronate.[111,118] Renal excretion of circulating bisphosphonates is rapid and appears to be independent of dose. More than 70% of an IV dose of clodronate given to healthy men was excreted in urine within 48 hours. Fractional excretion did not vary over the dose range of 3 to 10 mg/kg[114].

The mean renal clearance of etidronate, clodronate, and pamidronate in man has been reported as 1.52 mL/min/kg, 1.15 to 1.4 mL/min/kg, and ~1.0 mL/min/kg, respectively.[117-120] Renal clearance has been found to be 1.7-fold greater than, and approximately equal to, GFR for clodronate and pamidronate, respectively.[110,112] Together with the extent of protein binding of these drugs, this suggests that renal secretion may be present in man, as well as in animals.

The fraction of a systemic dose of bisphosphonate that is not eliminated by the kidney over the first 48 hours after administration, by analogy to animal results, is thought to be sequestered in bone, where it resides for a long time.

The whole body retention of a dose of pamidronate (i.e., that which was not excreted in the urine immediately following treatment) was similar whether the drug was infused over 4 or 24 hours.[111] Four infusions of pamidronate (60 mg repeated at four- to five-week intervals) resulted in a constant whole body retention of the dose for each of the two to four doses, suggesting that its pharmacokinetics were not affected by previous exposure[111].

The oral absorptions of etidronate, clodronate, and pamidronate have been estimated in fasting man, respectively as 1.5 to 3.5%,[115-117,121] 1 to 2.2%[112,117,119] and about 0.3%,[110] respectively. Concurrent food effectively eliminates oral absorption of etidronate.[121]

7.2 Rationale for Biopharmaceutics Development Program for Alendronate

Biopharmaceutics studies are typically conducted in healthy young males. Alendronate was known from preclinical studies to adhere strongly to bone and to reside there for a long time. In addition, alendronate was intended to reduce bone resorption. Therefore, pharmacokinetic studies were conducted in subjects representative of the ultimate target populations. In at least some of these studies, there was also the potential for the subjects to benefit from dosing with alendronate. In addition, studies in these populations enabled an assessment of biochemical efficacy parameters reflective of bone turnover and mineral metabolism. Study populations were chosen according to the design and objectives of the study.[124]

An oral single rising dose, safety, and tolerability study, an oral, incremental, multiple-dose, safety, and tolerability study and an intravenous multiple-dose safety and tolerability study were carried out in patients with Paget's disease. This disease is characterized by focal areas of pathologically accelerated bone turnover and permitted study of the effect of alendronate on biochemical markers of bone turnover. A bioavailability study was carried out in healthy men and a study of the disposition of [^{14}C]alendronate was carried out in women with breast cancer metastatic to the skeleton. The 10-mg IV dose used was judged to be of potential therapeutic value to these subjects. All other pharmacokinetic studies examined postmenopausal women with or without osteoporosis.

Urinary recovery of alendronate was used to investigate the pharmacokinetic behavior of alendronate in all but the metabolic disposition study. Alendronate is a potent bisphosphonate and very low plasma concentrations are expected following oral doses to be used for the treatment of osteoporosis. These concentrations are not measurable with the assay currently available for alendronate in plasma.[122,123] The dose of 10-mg IV employed in the [^{14}C]alendronate disposition study was, however, high enough for analytical detection of alendronate in plasma. Preclinical studies have shown that about 50% of an IV dose in rats and dogs binds to bone and resides there for a long time (terminal $t_{½}$ = 200 days for rats, >1000 days for dogs).[96] The remainder of the drug is excreted unchanged in urine in a short time (<48 hours). No evidence of metabolism was found in either dogs or rats.[95,96] Alendronate was assumed to behave similarly in man. Appropriate studies to verify these assumptions were performed and will be discussed below.

Because of the very long residence of alendronate in bone, the possibility was anticipated that crossover study designs might be inappropriate given that previous exposure to alendronate might alter the pharmacokinetic behavior of subsequent doses. To rule out this possibility, a multiple-dose study was carried out in postmenopausal women to validate the use of crossover designs.

7.3. Pharmacokinetics of Alendronate in Man

7.3.1. Disposition of [^{14}C]Alendronate

Investigation of the disposition of radiolabeled alendronate presented an unusual challenge. The very long retention (years) of this drug in bone (as predicted from preclinical results) and the amount of radioactivity required for detection made it inadvisable to perform such a study in healthy subjects. Patients with metastatic bone disease, in contrast, might derive therapeutic benefit from administration of large amounts of alendronate and the long residence of radiolabeled alendronate in their bones probably poses very little risk because of their underlying disease state and should not interfere with other treatment. The open-label study of the disposition of single doses of [^{14}C] labeled alendronate, therefore, was conducted in 12 patients with breast cancer that had metastasized to the skeleton.[125]

The subjects were divided into two panels of six and eight subjects, respectively (two subjects from the first panel were re-evaluated in the second).[125] The second panel (including the repeat participation of two subjects from the first panel) was enrolled because there were apparent discrepancies in the urinary recovery of radiolabeled material and unlabeled alendronate in some subjects from the first panel. Each subject received a single IV infusion (over two hours) of [^{14}C]alendronate (10 mg, approximately 26 μCi) labeled in the C(4) position. In the first panel, plasma and urine samples were collected predose and for 36 and 72 hours, respectively, following drug administration. All feces were collected for five days following the dose.[125] Urine for only 48 hours and plasma for 12 hours were collected from the second panel of subjects. Concentrations of alendronate were measured in plasma (first panel only) and urine samples by the HPLC methods described earlier. Radioactivity was measured in plasma (first panel only), urine, and feces samples by standard scintillation counting.

Mean profiles for concentrations of alendronate and radioactivity in plasma are shown in Figure 12. Individual plasma concentration *vs.* time curves for alendronate and radioactivity were essentially superimposable for all subjects. This is strong evidence against metabolites of alendronate circulating in plasma.[125] Radioactivity and concentrations of alendronate were measurable in plasma for approximately 12 hours. Urinary excretion was measurable for at least 72 hours. Therefore, disposition phases exist after the first 12 hours that are not detectable in plasma, and the AUC and area under the first moment curve (AUMC) cannot be extrapolated to infinity. Renal clearance (Cl_r) was estimated from partial (0 to 12 hours) AUC and urinary excretion measurements. Plasma clearance (Cl_p) and volume of distribution (V_{dss}) were estimated by noncompartmental analysis from the partial (0 to 12 hours) AUC and AUMC values[126]:

$$Cl_p = Dose/AUC$$
$$V_{dss} = Dose \cdot AUMC/AUC^2 - Dose \cdot \tau \cdot /(2 \cdot AUC)$$

Values estimated from partial areas constitute bounded estimates of Cl_p (upper bound) and V_{dss} (lower bound). Specifically, values of V_{dss} do not include distribution of the drug to a deep compartment (assumed to be bone). Since the alendronate distributed to this compartment was not appreciably released back into circulation over the course of the study, it was assumed that it does not contribute to the estimates of AUC and AUMC leading to bounded values for Cl_p and V_{dss}.

Geometric mean Cl_r of alendronate was 71 mL/min. The geometric mean V_{dss} and Cl_p were estimated to be at least 28 L and no larger than 199 mL/min, respectively.

The geometric mean plasma concentration at the end of the two-hour infusion was 265 ng/mL, which decreased rapidly to less than 5% of this value by six hours postdose.[125] The V_{dss} is relatively small, suggesting that distribution of alendronate to soft tissues is not extensive.

Renal clearance accounts for at least one-third of apparent plasma clearance. Considering that plasma clearance is likely to be overestimated, this finding is consistent both with the observation that about 40% of an intravenously administered dose is recovered in urine initially and with the assumption that drug is removed to bone at nearly the same rate.

No radioactivity was recovered in the feces of three patients, and the amounts recovered from the other three (0.2, 0.4, 2.0%) were judged to be the result of contamination by urine. Mean urinary recovery of alendronate was 45% of the dose.[125] The mean urinary recovery of radioactivity was 47% of the dose.[125] Most of the radioactivity recovered in urine was recovered in the first 12 hours after administration. The remaining drug is presumably sequestered in the skeleton, where it exerts its therapeutic effect. Preclinical studies using alendronate, and clinical studies with bone scanning agents such as methylene bisphosphonate, support the contention that whole body retention is virtually synonymous with skeletal retention.[87]

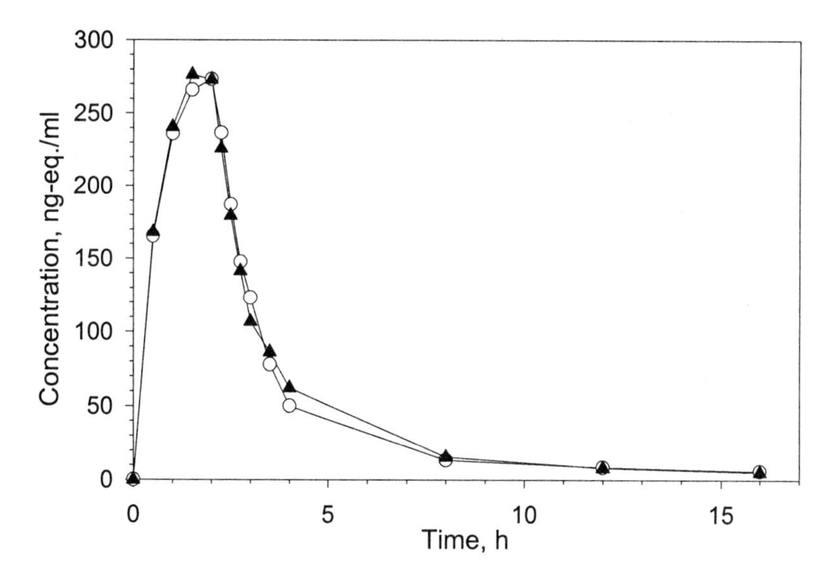

Figure 12. Mean Concentrations of Alendronate and Radioactivity in Plasma Following Administration of 10 mg of [^{14}C]Alendronate (26 μCi)

Since no radioactivity is excreted in feces, elimination of the drug through biliary secretion is negligible. Therefore, excretion of alendronate or any potential metabolites occurs exclusively through the kidney. The excretion of alendronate in this study[125] accounts for essentially all (96%) of the radioactivity excreted in urine. This finding is in agreement with a failure to fine evidence of alendronate metabolism in animal species.[96] It appears then, that alendronate in man is either not metabolized or at most metabolized to a negligible extent.

7.3.2. Pharmacokinetics Following Intravenous Administration

Intravenous doses of 50 to 400 μg, assuming an oral bioavailability of approximately 1% based on clinical studies with other bisphosphonates (0.3 to 3.5%),[110,112,115,116,117,119,121] would produce systemic exposure similar to that following the 5 to 40 mg oral doses, which were anticipated to be therapeutic for the various indications of alendronate. Doses administered as IV reference in the pharmacokinetic studies were within this range (either 125 or 250 μg). Because plasma levels would be below the reliable limit of quantitation at these doses, all pharmacokinetic parameters were characterized based on urinary excretion of alendronate in these studies. Urinary excretion was also documented after higher IV doses of alendronate (7.5 and 10 mg).

7.3.2.1 Single Dose. In several studies, single IV doses were administered to healthy and osteoporotic postmenopausal subjects and to breast-cancer patients with bone metastases.[125] IV doses administered in these studies were 20 μg, 125 μg, 250 μg, 7.5 mg, or 10 mg. Total urinary output was collected for 24 to 72 hours following administration to determine excretion of alendronate. The mean urinary excretion rates observed in split urine collections for the 36 hours following a 250 μg IV dose in one of these studies are shown in Figure 13. The shape of this profile is representative of those observed with other doses.

Following administration, a substantial fraction of the dose is promptly excreted in urine (~40% in the first eight hours). The rate of excretion decreases considerably thereafter, but alendronate continues to be quantifiable for at least 72 hours. Approximately 5% of the administered dose is excreted between 8 and 72 hours. Generally, recoveries through the first 72 hours accounted for 31 to 56% of administered dose. Although 40 to 60% of a dose still remained in the body at 72 hours postdose, the urinary recoveries became very small (Figure 13). Thus, excretion of alendronate after this period must occur at a much slower rate.

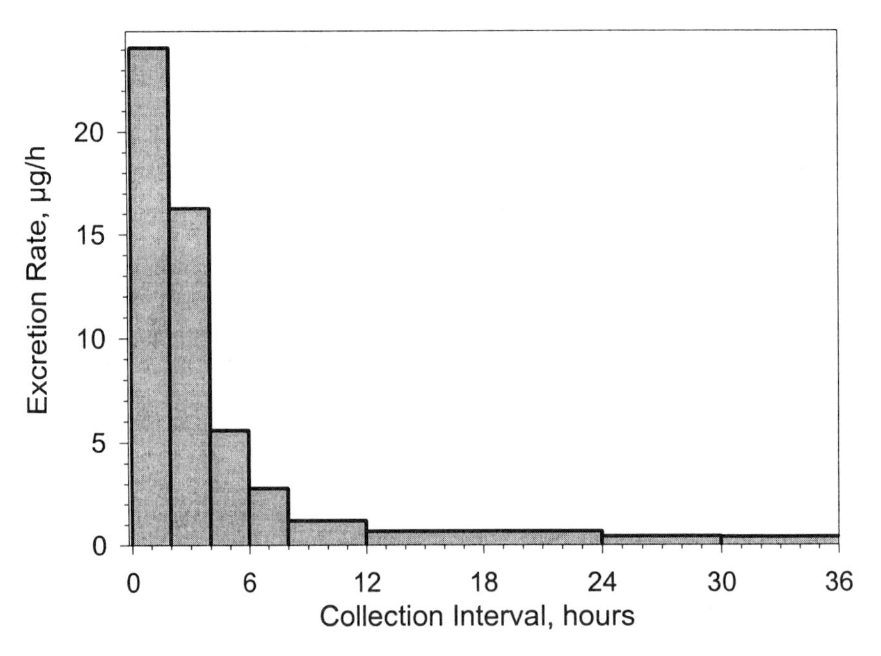

Figure 13: Mean Urinary Excretion Rate for Each Collection Interval Following Administration of a Single 150 μg Intravenous Doses.

The fraction of administered dose excreted in the urine during the initial 24 to 36 hours following IV infusion was found to be independent of dose and of the duration of infusion. The disposition of alendronate thus appears to be consistent over a wide range of single IV doses, from 20 µg to 10 mg, and uninfluenced by the duration of infusion at higher doses.

The disposition of alendronate in healthy postmenopausal women, as reflected by urinary excretion (31 to 56% following IV administration), appeared similar to that in women with breast cancer metastatic to the skeleton who were treated only with hormonal therapy.

7.3.2.2. Multiple Dose. Pharmacokinetic studies (e.g., bioavailability, dose-proportionality, dosage-form comparisons) are usually conducted with crossover designs. This is generally possible because most drugs do not have residual effects on the absorption or disposition of subsequent doses and are cleared from the body in a matter of days. As previously discussed, the latter is not true for alendronate. Conceivably, administration of alendronate might alter the bone uptake of subsequent doses, thereby altering the urinary excretion patterns from those observed in subjects who had never previously received this compound. To establish whether previous exposure to alendronate altered the short-term (72 hours or less) renal elimination of an intravenously administered dose, and thus determine whether it would be valid to conduct crossover studies with this drug, 10 postmenopausal subjects were given repeated IV infusions of alendronate in an open, multiple-dose study,[125] and total urinary output was collected following the first and last dose to compare excretions of alendronate. Excretion following administration of 7.5 mg IV doses for four consecutive days[125] was also examined to investigate the potential effect of alendronate on its own systemic disposition at a much higher dose (The total amount administered is comparable to dosing 10 mg/day orally for about 430 days).

To avoid the variability that characterizes the poor oral absorption of alendronate, IV administration was chosen to study the effects of repeated doses. The IV dose used, 125 µg, was expected to provide systemic loads similar to those of doses which were thought to be clinically beneficial and at the time were under investigation in Phase III osteoporosis studies, assuming an oral bioavailability on the order of 0.5 to 1%. Thus, the results of this study were expected to be applicable to multiple oral doses of alendronate. On Day 1 of the study, subjects were given a single I.V .infusion of 125 µg alendronate (10 µg/mL x 6.25 mL/h x 2 hours). Urine was collected for two hours predose and for 36 hours following the beginning of IV administration. On Days 4 to 8, subjects were given daily IV infusions of 125 µg alendronate (same regimen as for first dose). Drug was not administered again for 10 days. On Day 18, the subjects were administered the last IV dose and urine was collected as before. Alendronate was not detected in urine specimens collected preceding the first dose. Urinary excretion of alendronate over the two hours immediately preceding the last dose was small relative to the quantities excreted postdose and averaged 0.3 µg (equivalent to <3.6 µg over 24 hours). Results were similar whether the predose contribution to alendronate excretion for the seventh dose was taken into account or not. The geometric mean urinary excretions for the 36 hours following administration of the first and last dose were 68.3

and 61.3 μg, respectively. The geometric mean fraction of dose excreted (0 to 36 hours) was 56.3% (95% CI 53.5, 59.2) of the first dose and 50.2% (95% CI 39.7,63.5) of the last dose. The geometric mean ratio of last to first dose was 0.9, which was not significantly different from 1 ($p > 0.200$). These results justified the use of crossover designs in pharmacokinetic investigations.

A study to investigate the elimination half-life of alendronate[125] also allowed an assessment of the consistency of urinary excretion following multiple doses. In this second study, where four daily IV doses of alendronate (7.5 mg daily) were administered to postmenopausal women, the geometric mean (95% CI) urinary excretion of alendronate over the 24 hours following administration of the first dose was 44.3% (36.7, 51.3) of the dose.

Cumulative fractional excretion of the total dose (30 mg) by 48 hours after the fourth dose (Day 6) averaged 53.6% (48.5, 58.8).[127] Since the latter value measures excretion during a longer collection period it was anticipated to be greater, nonetheless the disposition of the first 7.5 mg dose was not significantly different from the total dose of 30 mg.

7.4 Long-Term Elimination

No adequate study to estimate the terminal elimination half life of those bisphosphonates that are currently in clinical use, including etidronate, clodronate, zolendronate and pamidronate, had been reported at the time of development of alendronate.

This statement is, in our estimation, generally true to date, since the elimination half life reported in the risedronate label does not reflect the kinetics of bone remodeling which we have concluded dominate the elimination of that fraction of the dose still in the body 24 hours post-administration.

Published data in animals suggest that pamidronate and clodronate have terminal half lives in animals greater than one year[84] or 90 days[89] respectively; the limited sampling intervals indicate these are likely to be gross underestimates.

In an attempt to estimate the terminal half life of alendronate, IV doses of alendronate sodium were given to patients with postmenopausal osteoporosis in an open-label study.[127] Subjects received a 7.5-mg IV dose of alendronate sodium in 500 mL of saline, infused at a constant rate over 12 hours, on each of four consecutive days for a total dose of 30 mg. Urine was collected for two hours predose and in 24-hour intervals postdose. All urine was collected for the seven days following administration of the first dose. Twenty-four-hour urine specimens were then collected at predetermined intervals for the subsequent 18 to 24 months. Quantities excreted over the first week were measured directly. Cumulative excretion at later times was estimated from 24-hour collections by integration of the measured daily excretion. Amounts remaining in the body were then estimated by subtraction of the amount excreted in the urine from the administered dose. Analysis of the urinary excretion of alendronate, and the estimation of terminal half life, was performed only in the 11 (of a total of 21) patients in the trial who had sufficiently complete urine collections.

Mean cumulative urinary excretions are plotted in Figure 14. As expected from previous results, about 54% of the total dose is excreted in the first week after the beginning of administration. This is followed by much slower excretion phases that, after 540 days, accounted for 70% of the dose (mean total urinary excretion = 21.0 mg). Mean quantities of alendronate remaining in the body (dose minus excretion to that point) are shown in Figure 15.

Elimination from a linear pharmacokinetic system can be reasonably approximated by a sum of n exponentials: $y = \Sigma A_i e^{-k_i t}$ (i=1 to n), with empirical pre-exponential values (A_i) and exponential factors (k_i), where Dose = ΣA_i. As time passes, the faster-decaying phases contribute less and less to the profile ($A_i e^{-k_i t} \rightarrow 0$ for all $i < n$) and the curve comes to be dominated by the longest-lived exponential (smallest k_i, henceforth designated as k_ω).

At this point it is possible to estimate the value of the terminal elimination rate from a point measurement, provided that a method exists for estimating the value of the corresponding pre-exponential (A_ω).

At the terminal stage of elimination, the various compartments in the body have come to a quasi-equilibrium such that elimination (at this point) is adequately described as a one-compartment process.

$A_\omega e^{-k_\omega t_o}$, is then equivalent to the amount of drug remaining in the body (A_o) at an arbitrary time (t_o) after the onset of mono-exponential decay. Excretion over a given time interval Δt ($= t_1 - t_o$) is then given by:

$$\Delta y = A_\omega(e^{-k_\omega t_o} - e^{-k_\omega t_1}) \qquad \text{[eq. I]}$$

which can be rearranged as:

$$\Delta y / (A_\omega e^{-k_\omega t_o}) = (1 - e^{-k_\omega \Delta t}) \qquad \text{[eq. II]}$$

$\Delta y / Ao$ is then the fractional excretion (f) observed during the interval Δt, and k_ω can be estimated from [eq. II][127].

Equation II was used to obtain point estimates of the terminal half life over the interval of 300 to 540 days following dosing. These results, however, should be regarded with caution because, unless data are collected for a time greater than four times the terminal $t_{½}$, it is difficult to establish when contributions from faster - decaying phases have ceased, and, therefore, values for k_ω may be overestimated and should be considered as upper bounds for the actual value[127].

Terminal half lives of alendronate were estimated during the 300 to 540 days after administration. The mean $t_{½}$ estimated at this point was 10.9 years, but, as explained earlier, this value represents a lowest bound estimate.

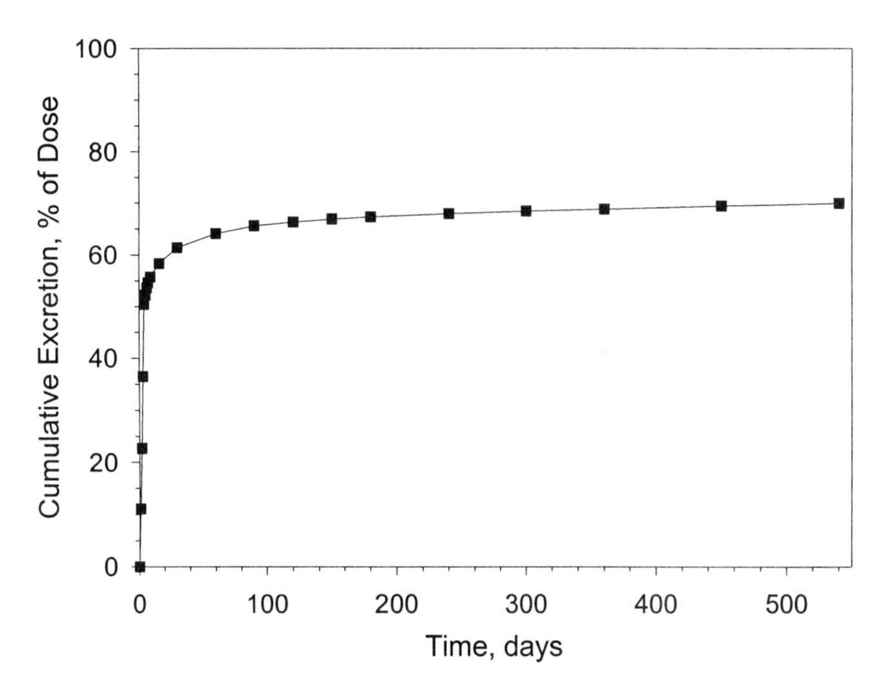

Figure 14: Mean Cumulative Urinary Excretion of Alendronate Following Intravenous Administration of 7.5 mg/Day for Four Days (as % of Total Dose)

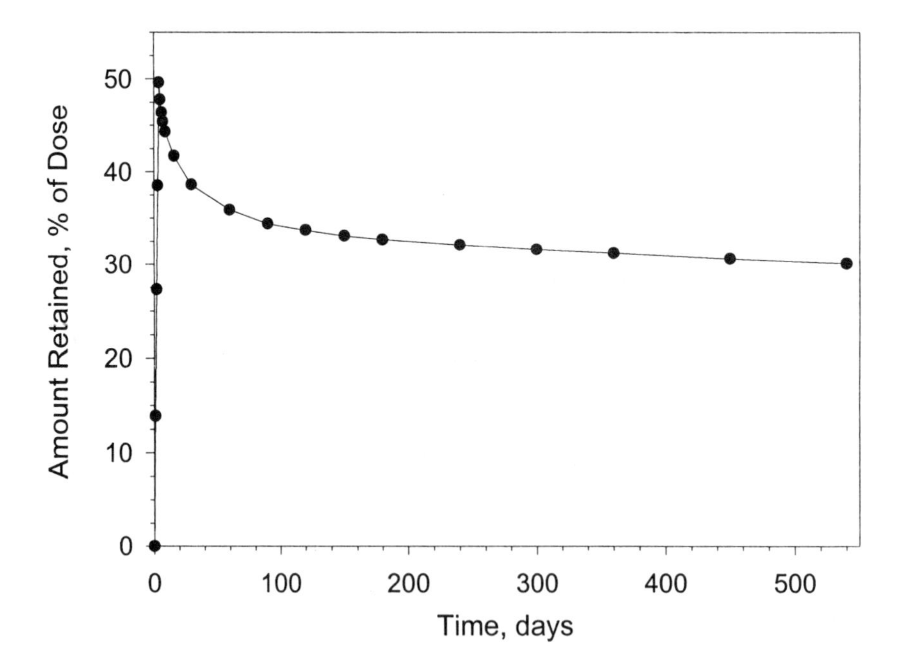

Figure 15. Mean Amounts of Alendronate Remaining in the Body Following Intravenous Administration of 7.5 mg/Day for Four Days (as % of Total Dose)

7.5 Oral Administration — Dose Proportionality and Bioavailability

Bioavailability of alendronate was evaluated in three studies in postmenopausal women. The first study was a pilot bioavailability study in which seven subjects with postmenopausal osteoporosis between the ages of 48 and 73 were given 0.25 mg IV and 50 mg as an oral capsule of alendronate in a double-blind, placebo-controlled, crossover fashion[128]. Subjects were dosed fasting, two hours before breakfast, and urine fractions were collected over the succeeding 72 hours for analysis of alendronate. Mean urinary recoveries were 41 and 0.19% of dose following the IV and oral treatments, respectively. Mean oral bioavailability was 0.41%, 90% C.I. = (0.23, 0.72)[128].

In the second study, the oral bioavailability of alendronate and the linearity of its disposition were evaluated in 15 postmenopausal subjects who were given single-dose treatments of 250 μg IV and 5, 10, 40, and 80 mg oral tablets in an open-label, five-way, crossover study[128]. Subjects were dosed in the early morning after an overnight fast and not fed for two hours after oral dosing or one hour after the IV dose. Total urines were collected in increments for 36 hours after dosing and analyzed for alendronate contents. The mean cumulative excretion curves following each of the oral doses are shown in Figure 16. Urinary excretion profiles following each oral dose display essentially identical shapes to those curves seen following IV administration. A comparison of dose-adjusted urinary excretion curves is shown in Figure 17. Most of the urinary excretion of alendronate occurred in the first 8 to 12 hours after dosing and was followed by a more steady, but greatly diminished, rate of excretion. Concentrations of alendronate in urine were still detectable at 36 hours following administration of all doses. Total amounts excreted increased with dose in a linear fashion as depicted in Figure 18. Geometric mean (S.D.) urinary excretion following IV administration was 42%[128] of dose. Excretions following oral doses averaged 0.30, 0.36, 0.35, and 0.31% of dose for the 5, 10, 40, and 80 mg doses[128]. Oral bioavailability was constant with dose and averaged 0.76% overall[128].

In the third study, the oral bioavailability and linearity of disposition of alendronate following administration of the various final production process tablets (2.5, 5, 10, and 40 mg) were evaluated in 20 healthy postmenopausal women who were given single doses of IV (125 ıg) and oral (2 x 2.5 mg, 1 x 5 mg, 1 x 10 mg, and 1 x 40 mg tablets) alendronate[128]. These tablets were used in Phase III studies of efficacy and safety. Treatments were administered early in the morning following an overnight fast and two hours before breakfast in a five-way crossover fashion with a seven-day washout between periods. Urine was collected predose and for 36 hours postdose and analyzed for alendronate content. The results were similar to the study described above, mean total amounts of alendronate excreted in urine increased linearly with dose. Geometric mean (95% C.I.) urinary excretion following IV administration was 38.4% (30.5, 48.2) of dose[128].

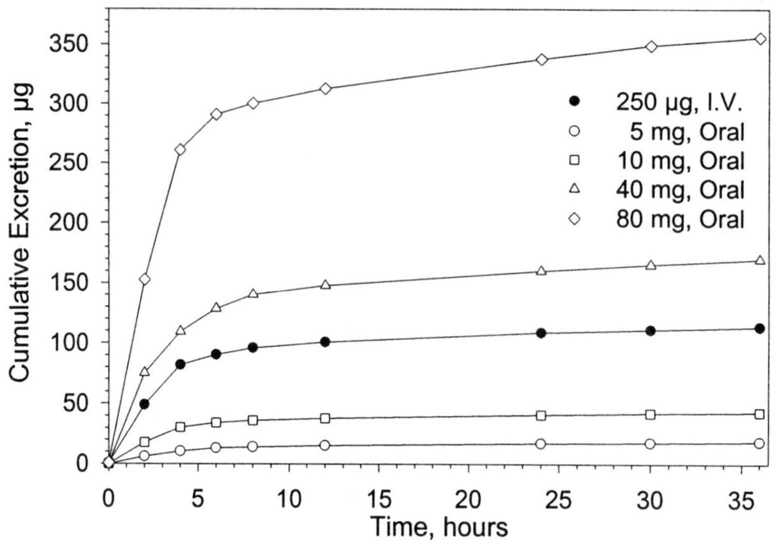

Figure 16. Mean Cumulative Excretions of Alendronate in Urine Following Oral and Intravenous Administrations.

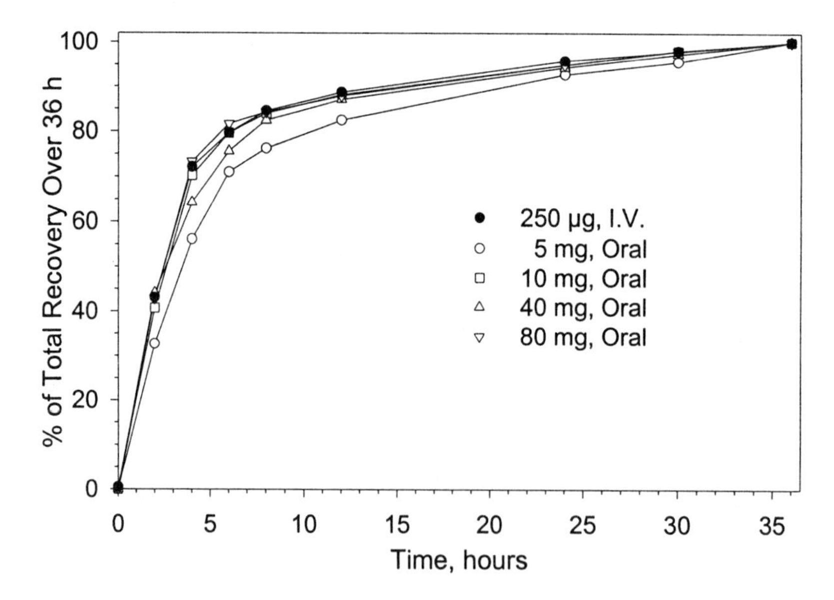

Figure 17. Mean Cumulative Excretions of Alendronate as % of Total Excreted in the 36 Hours Following Administration.

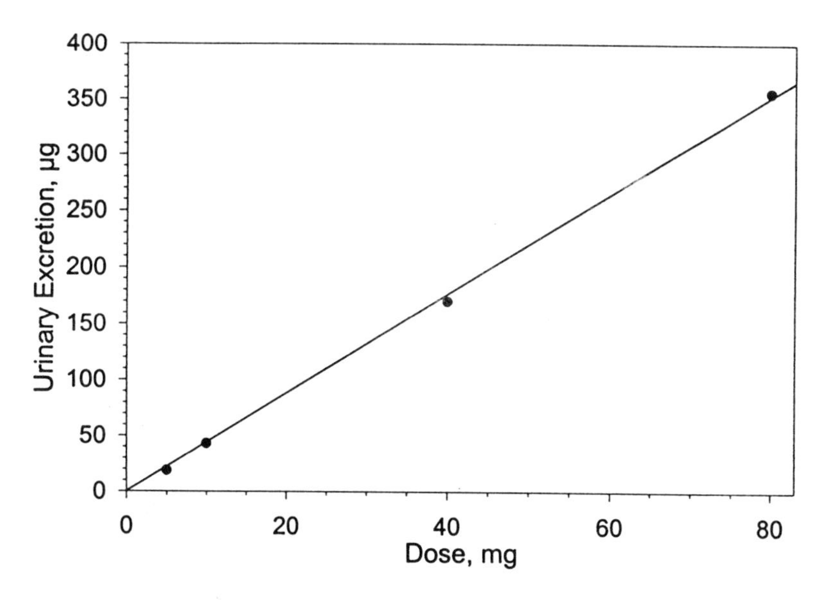

Figure 18. Mean Excretions of Alendronate as a Function of Oral Dose

Excretion following oral doses averaged 0.305, 0.243, 0.298, and 0.229% of dose for the 2 x 2.5, 5, 10, and 40 mg doses, respectively[128]. Oral bioavailability averaged 0.69% overall and was independent of dose[128].

Evaluation of the variability of alendronate absorption across the dose range of 5 to 80 mg provided estimates of within-subject and between-subject variability that were relatively similar (within-subject and pooled between-subject C.V. = 55 and 75%, respectively)[128]. In the other study, across the dose range of 5 to 40 mg, the within-subject and between-subject coefficients of variation were similar, i.e., 63 and 77%, respectively[128]. Thus, postmenopausal women absorb, on average, 0.69 to 0.76% of a dose of alendronate.

Additionally, the oral bioavailability of alendronate was studied in sixteen healthy male volunteers who were given single doses of oral (10 mg) and IV (125 µg) alendronate in an open-label, randomized, two-period, crossover study[128]. Eight of the men were between the ages of 21 and 60 years and eight were between 60 and 76 years old. Subjects were dosed in the early morning after an overnight fast and not fed or given anything to drink (except water) for two hours after dosing. Total urine was collected for the two hours before and for the 24 hours following dosing and analyzed for alendronate. Urinary excretions of alendronate observed in the male subjects are similar to the pattern observed in women. The bulk of the excretion occurred in the 8 hours following either oral of IV administration and urinary recovery of alendronate over the 24 hours monitored[128]. Total excretions averaged 42% of dose [90% C.I. = (33.6, 53.1)] for the 125 µg IV and ~0.25% of dose for the 10-mg tablet [90% C.I. = (0.20, 0.31)]. The bioavailability of the oral dose in men[128] averaged 0.59% [90% C.I. = (0.43,0.81)].

The oral bioavailability of alendronate is very low (<1%) under fasting conditions. Since alendronate is a charged molecule that forms insoluble complexes with multivalent cations such as calcium, magnesium, iron, and aluminum, the potential exists for interactions in the G.I. tract to decrease its oral absorption. In addition, pH has important effects on the aqueous solubility of alendronate, as well as its potential to form insoluble complexes with multivalent cations. The potential effects of the timing of meals with and without calcium supplementation, and of beverages other than water taken with the dose, were studied in healthy postmenopausal women. Since the incidence of achlorhydria and hypochlorhydria increases with age[129], the effect of changes in gastric acidity on oral absorption was investigated.

Bisphosphonates, in general, form highly insoluble complexes with multivalent cations commonly found in food, drinks, antacids, and multivitamins, raising concerns that eating before, or shortly after, administration of alendronate would diminish its bioavailability. Additionally, calcium supplements are often administered to osteoporotic patients to help treat the disease. As described previously, food diminishes the bioavailability of alendronate in animals. Food has also been reported to decrease the oral bioavailability of other bisphosphonates in humans as well as in animals.[87,121] Initially, patients in the clinical program were required to wait two hours after dosing alendronate before having breakfast. Since a two-hour wait before breakfast was thought to be burdensome, studies were carried out to investigate the effect of allowing meals closer to the dose.

Two studies were conducted to investigate the effect of the timing of meals, with or without calcium supplementation, on the oral absorption of alendronate. Fifteen postmenopausal women were given alendronate in an open-label, balanced, randomized, crossover study[128]. Subjects fasted overnight and then were given single 20-mg doses of alendronate, followed by meals, with or without calcium supplements, at various times. An additional 49 postmenopausal women were given single 10-mg oral doses of alendronate in a second open-label, balanced, randomized, crossover study[128]. Subjects fasted overnight and were dosed at various times relative to meal intake. The composition of the breakfast was somewhat different in the two studies. Total urine was collected incrementally for 36 hours following each treatment in both studies.

Administration of alendronate two hours before the meal (without calcium supplement) was chosen as the reference treatment in both studies, since two hours was recommended for other bisphosphonates and was judged to be the maximum acceptable wait before breakfast[128]. Alendronate absorption was found to be lower when drug administration is less than two hours before breakfast. Waiting to eat breakfast for only one or one-half hour after dosing diminished absorption, on average, by about 40% (30 and 46%, respectively). The difference between waiting one or one-half hour was not significant in either study, although in one of them[128], alendronate was not detectable in a greater proportion of urine specimens from treatment one-half hour before breakfast than from treatment one hour before breakfast.

Supplementation of the meal (one or two hours postdose) with 1 g of elemental calcium had no significant effect on absorption of alendronate beyond the effect of the meal by itself. Dosing immediately following, or up to two hours after, breakfast dramatically decreases alendronate absorption. It was estimated that absorption was only 10 to 15% that of the reference treatment, and many of the urine samples from these treatments had concentrations too low to measure. Thus, alendronate must not be given concurrently with food or within at least two hours following a meal.

With regard to alendronate dosing, ingestion of calcium supplements should follow the same restrictions as for food. While advising patients to take alendronate 30 minutes prior to any food or liquid other than water results in a 30 to 46% reduction in absorption (relative to a two-hour wait before breakfast), this provides a reasonable recommendation for chronic administration.

The clinical efficacy trials for this program were conducted with alendronate dosed after an overnight fast and from one-half to two hours before breakfast, depending on the study.

To investigate the effect of beverages other than water on the absorption of alendronate, 42 healthy postmenopausal women were given single, 10 mg oral doses of alendronate with water, orange juice, or black coffee in an open-label, balanced, randomized, three-period crossover study[128]. Subjects were dosed after an overnight fast with a single 10 mg tablet accompanied by 240 mL of orange juice (with no calcium supplement), black coffee (without cream or sweetener), or tap water in each study period. A standardized meal was provided two hours after dosing. Urine was collected for two hours before and 24 hours following administration of each treatment and analyzed for alendronate content. Geometric mean quantities recovered were 19.2, 7.43, and 6.77 μg following administration of alendronate with water, coffee, and orange juice, respectively. Geometric mean (90% C.I.) excretion ratios following administration with coffee or orange juice with reference to administration with water averaged 0.39 (0.32, 0.47) and 0.35 (0.29, 0.43), respectively[128].

Administration of alendronate with coffee or orange juice instead of tap water appears to decrease bioavailability of the drug by about 60%. In conjunction with previous results describing the effect of food on absorption of alendronate, it would seem that administration of the drug with any food or beverage, other than tap water, would risk a substantial decrease in drug absorption.

In summary, alendronate should be optimally taken with water, after an overnight fast, and ingestion of all other food and liquid (except water) delayed by at least one-half hour.

7.6. Effect of Gastric pH on the Absorption of Alendronate

To investigate the effect of altered gastric pH on the oral absorption of alendronate, 10 postmenopausal subjects were given simultaneous single oral and IV doses of alendronate in an open-label, randomized, two-period, crossover study[128]. Gastric pH was elevated in one of the two treatment periods by an infusion of ranitidine. Intravenous ([^{14}C] labeled) alendronate was administered to simultaneously monitor its systemic disposition. The urinary excretions of total and radiolabeled alendronate were measured to permit assessment of systemic disposition and oral absorption of alendronate under conditions of basal (pH$<$2) and elevated (pH$>$6) gastric pH.

Doses were administered as follows: After an overnight fast, a nasogastric probe was inserted for continuous monitoring of gastric pH. Gastric pH was monitored for two hours prior to administration of alendronate and for four hours afterward. An IV infusion of 0.9% sodium chloride or ranitidine (90 mg in 97.5 mL 0.9% sodium chloride at 15 mL/h) was begun 1.5 hours prior to alendronate dosing and continued for a total of 5.5 hours (four hours after alendronate dosing). During the period in which ranitidine was administered, alendronate was administered only if the gastric pH was greater than or equal to 4, and during the period in which ranitidine was not administered, alendronate was given only if gastric pH was less than 2. Alendronate (40 mg) was dosed orally 1.5 hours after the start of the saline infusion, followed immediately by [^{14}C]alendronate (0.5 μCi, 20 μg in 3 mL 0.9% saline) administered intravenously over 15 minutes. Urine was collected predose and for 30 hours following alendronate dosing for the determination of alendronate concentration and radioactivity. Mass-equivalents of excreted alendronate arising from IV administration were obtained directly from urinary excretion of radioactivity, since alendronate appears not to be metabolized[125].

Ranitidine maintained the mean gastric pH at approximately six for the four-hour period following alendronate dose. When saline alone was infused, gastric pH remained below 2. Quantities of radiolabeled alendronate excreted in the urine were 6.8 and 6.0 ıg, respectively[128]. There was no significant difference between these two geometric means. Their ratio of 1.14 was not significantly different from 1.0. The similarity in excretion of [^{14}C]alendronate from IV administration observed when the subjects were given ranitidine compared with when they were not indicates that there are no effects of ranitidine on the systemic disposition of alendronate[128]. The geometric mean urinary excretion of alendronate derived from the oral dose was 146.4 and 59.1 µg for the treatment arms with and without ranitidine, respectively. The ratio of these geometric means was 2.5, which was significantly different from 1.0 Geometric mean oral bioavailability (S.D.) of alendronate was 1.07% (1.11) for the arm in which ranitidine was given and 0.49% (0.51) for the arm in which it was not, showing higher absorption when gastric pH was approximately 6. The ratio of these geometric means was 2.18[128].

In the ranitidine arm of the study, subjects showed a distinctly higher gastric pH for the four hours following dosing, compared with those in the control arm. The estimated oral bioavailability of alendronate was approximately two fold higher when mean gastric pH was maintained at approximately 6 with ranitidine infusion, relative to the bioavailability observed during the saline infusion. The difference could not be attributed to a difference in systemic disposition induced by ranitidine. It is unlikely that this increase in apparent oral bioavailability due to an elevation of gastric pH will have an important effect on the efficacy or safety profile of alendronate. Clinical experience to date indicates that alendronate has a relatively flat dose-response curve with regard to bone mineral density and systemic adverse effects at doses of 10 and 20 mg/day[124]. Alendronate has shown good tolerability on chronic dosing with up to 40 mg/day in patients with osteoporosis or Paget's disease of bone.

8. DRUG-DRUG INTERACTIONS

The low absorption, rapid renal elimination, poor protein binding and lack of metabolism of alendronate led us to expect that the possibility of interactions between alendronate and other drugs was remote. Therefore, no *in-vivo* drug-drug interaction studies were carried out with this compound. However, the potential for such interactions was examined based on kinetic considerations were carefully consider as follows:

8.1 Potential Interactions Involving Oral Absorption

Most potential interactions between alendronate and other compounds that could influence oral absorption are those that result from the formation of insoluble complexes. While drugs or conditions that influence G.I. motility may affect the oral absorption of alendronate, most of these potential interactions can be avoided by appropriate timing of drug administration. Alendronate, taken in the morning on an empty stomach, is likely to be timed near the trough plasma concentrations of other drugs, including those which influence gastrointestinal motility. Alendronate is known not to influence G.I. motility in mice[102] and thus is not anticipated to affect the absorption of other drugs through such a mechanism.

Provided that alendronate is administered in the fasting state and other drugs, food, and mineral supplements are delayed by at least one-half hour, adequate oral absorption of alendronate should occur. This provides a practical recommendation for dosing, which is facilitated by alendronate's new once-a-week regimen.

8.2 Potential Interactions Involving Systemic Disposition, Metabolism, or Elimination

Alendronate is present in plasma at very low concentrations over the therapeutic range of doses, distributes over extracellular fluid volume, and is removed rapidly from the systemic circulation. Retention within the body is the result of skeletal sequestration and not of accumulation in soft tissues. Elimination is by renal excretion; biliary excretion is negligible and no metabolites have been identified.

Alendronate is bound to proteins in human plasma; approximately 78% is bound at alendronate concentrations lower than 1 μg/mL. Interactions involving displacement from protein binding are considered clinically important for drugs that have a narrow therapeutic index, such as warfarin. However, at therapeutic doses of alendronate, plasma concentrations will be very low (in the nM range) and insufficient to occupy more than a minor fraction of available binding sites on plasma proteins at their physiologic concentrations (α_1-acid-glycoprotein, 12.5 to 37.5 μM, γ-globulins, 33 to 103 μM, or albumin, 530 to 758 μM). Given the low concentration of alendronate and the moderate degree of protein binding, it is not expected to displace other drugs from their binding sites to a clinically meaningful extent, regardless of relative affinities.

The features of the systemic disposition of alendronate enumerated above limit the potential mechanisms by which alendronate could interact with other drugs. Systemic interactions fall broadly into the categories of inhibition or induction of metabolism, of competition for transport mechanisms or active sites, or of binding to plasma proteins. Clinical data are consistent with preclinical studies that show alendronate distributes minimally to non-skeletal tissues, where it might interfere with the metabolism of other drugs. There is no evidence for metabolites of alendronate in humans[125]. Alendronate does not induce Cytochrome P-450 in rats. The very low plasma concentrations, limited distribution, lack of metabolism, and rapid removal from the systemic circulation by sequestration in bone and urinary excretion make it very unlikely that alendronate induces or inhibits the metabolism of other drugs. Drug interactions with anticonvulsants as a class most frequently involve enhancement or inhibition of metabolism; alendronate is, therefore, unlikely to interact with members of this class.

Available data indicate that elimination of alendronate is exclusively through renal excretion.[66,125] Thus, interactions involving biliary elimination are unlikely. Given the low circulating concentrations of the drug, and the fact that 50 to 60% of systemically available drug is presumably taken up by the skeleton, even a substantial reduction in renal elimination of alendronate would lead to only a relatively minor fractional increase in the amount of drug taken up by bone, which over a large dose range appears itself to be non-saturable. Given the relatively flat dose-response curve above 10 mg, this appears to be of relatively minor importance. Patients with chronic renal insufficiency (creatinine clearance less than 35 mL/min) have been generally excluded from clinical trials to avoid the potential for any nephrotoxicity suggested by earlier trials with other bisphosphonates, and thus should not be exposed to the drug. There is no experience in patients with end-stage renal disease.

Interactions with diuretics are unlikely. Most diuretics are acidic and therefore excreted renally through glomerular filtration or through the acid secretory channel. Alendronate appears to be excreted by a distinctly separate renal transport system. Furthermore, alendronate has not been noted to affect disposition of sodium or potassium, another relatively common basis for interactions with diuretics.

When the use of a drug with substantial potential for renal toxicity, such as cisplatin or aminoglycosides, is necessary, it is probable that renal function will be monitored closely. Provided renal function is not seriously impaired, it is likely that alendronate can continue to be safely administered. If renal function should seriously deteriorate, there is no clinical experience to specifically guide the use of alendronate. However, if it were considered to be in the best interests of the patient to continue alendronate, it may be appropriate to reduce the dose and administer drug with careful monitoring of renal function.

Alendronate is unlikely to interact with other drugs through changes in protein binding, metabolism, or biliary or renal excretion. These categories include the basis for many of the interactions with heparin, nonsteroidal anti-inflammatory drugs (NSAIDs), and glucocorticoids. The dose of alendronate for treatment of osteoporosis, 10 mg, has shown a low incidence of G.I. adverse effects, which, with the exception of abdominal pain and dysphagia, are not significantly different than with placebo. Furthermore, there was no greater incidence of G.I. adverse experiences in patients taking NSAIDs and 10-mg alendronate concomitantly.

Alendronate administered to reproductively mature female dogs in doses up to 1 mg/kg for three years showed no adverse effects on bone. No clinical studies of concurrent estrogen and alendronate therapy have been done.

9. Summary of Pharmacokinetic and Pharmacodynamic Findings on Alendronate

- Following IV administration, alendronate distributes rapidly from plasma. About half of the dose is excreted in urine within the first few days postdose, and the rest is retained within the body, presumably in the skeleton, from which it is eliminated slowly.
- Alendronate is eliminated exclusively through renal excretion, i.e., biliary elimination is negligible.
- No metabolites of alendronate have been identified. Metabolism is either negligible or nonexistent.
- The terminal half life of alendronate is greater than 10 years.
- Despite the long half life of elimination, the pharmacokinetic behavior of alendronate is not affected significantly by previous administration; thus, it is appropriate to conduct crossover studies.
- Disposition of alendronate is independent of dose, with oral bioavailability averaging about 0.76% over the range of 5 to 80 mg in postmenopausal women. The oral bioavailability in men is similar.

- Elevated gastric pH, as may be seen with hypo- or achlorhydria, is associated with an approximately two-fold increase in oral absorption of alendronate. This change is not anticipated to substantially affect the safety or efficacy of the drug.
- Food decreases oral absorption of alendronate. As compared with consuming a standard breakfast two hours after a dose, absorption is moderately reduced by consuming food one hour postdose, with no additional significant reduction when the meal is provided one-half hour after drug. Absorption is greatly reduced by consuming the meal with, and up to at least two hours prior to the dose.
- Administration of a 1 g elemental calcium supplement with the standard breakfast at one or two hours postdose does not alter the apparent oral absorption of alendronate beyond that produced by the meal itself.
- Beverages other than water substantially decrease oral absorption of alendronate when taken with the dose.
- Drug interactions are not expected to be observed based on the extremely low plasma concentration, the absence of metabolism and the means of systemic disposition.

10. CONCLUSION

Development of alendronate for use in postmenopausal osteoporosis and Paget's disease of bone was a challenging process because of the physicochemical characteristics of this compound, which it shares with other bisphosphonates in general. Through a combination of the examination of pharmacodynamic and of pharmacokinetic parameters it was possible to take advantage of the great similarities in kinetic and dynamic behavior of this drug in different species. Development of other bisphosphonates for therapeutic use can be accomplished in a similar manner.

REFERENCES

1 Cooper, C.: Bone Mass Throughout Life: Bone Growth and Involution, In: Osteoporosis, Pathogenesis and Management, Chapter 1, Francis, R. M., ed., Kluwer Academic Publishers, Boston, pp. 1-26, 1990.

2 Nordin, B. E. C.: Osteoporosis with Particular Reference to the Menopause in the Osteoporotic Syndrome Detection, Prevention, and Treatment (Avioli, L.V., Ed.) Grune and Stratton, N. Y., 1983, Ch. 2.

3 The Merck Manual, 15th Ed. (Berkow, R., Fletcher, A. V., Eds.) Merck, Sharp & Dohme Res. Lab., Rahway, NJ, 1987, ch. 112.

4 Christiansen, C., The different routes of administration and the effect of hormone replacement therapy on osteoporosis. Fertil. Steril. 1994 Dec; 62(6 Suppl 2): 152S-156S

5 Alden, J. C., Osteoporosis--a Review. Clin. Ther. 1989; 11(1): 3-14

6 Orwoll, E. S.: Oviatt, S. K.; McClung, M. R.; Deftos, L. J., and Sexton, G.: The Rate of Bone Mineral Loss in Normal Men and the Effects of Calcium and Cholecalciferol Supplementation. Annals of Internal Medicine 112(1): 29-34, 1990.

7 Riis, B.; Thomsen, K., and Christiansen, C.: Does Calcium Supplementation Prevent Postmenopausal Bone Loss? A Double-Blind, Controlled Clinical Study. New Eng. J. Med. 316(4): 173-177, 1987.

8 Overgaard, K.; Riis, B. J.; Christiansen, C., and Hansen, M. A.: Effect of Salcatonin Given Intranasally on Early Postmenopausal Bone Loss. Brit. Med. J. 299: 477-479, 1989.

9 Grady, D.; Rubin, S. M.; Petitti, D. B.; Fox, C. S.; Black, D.; Ettinger, B.; Ernster, V. L., and Cummings, S. R.: Hormone Therapy to Prevent Disease and Prolong Life in Postmenopausal Women. Ann. of Int. Med. 117(12): 1016-1037, 1992.

10 Jacobs, H. S. and Loeffler, F. E.: Postmenopausal Hormone Replacement Therapy. Brit. Med. J. 305: 1403-1408, 1992.

11 Barrett-Connor, E.: Risks and Benefits of Replacement Estrogen. Ann. Rev. Med. 43: 239-251, 1992.
12 Prestwood,KM; Pilbeam,CC, and Raisz,LG, Treatment of Osteoporosis., Ann. Rev. Med. 1995; 46: 249-56
13 Cooper, C. and Melton, L.J., Epidemiology of Osteoporosis., Trends Endocrinol. Metab. 3(6): 224 - 229, 1992
14 Cummings, S.R., Rubin, S.M., Black, D., The Future of Hip Fractures in the United States. Numbers, Costs, and Potential Effects of Postmenopausal Estrogen., Clin. Orthop. 1990 Mar., (252): 163 - 6
15 Avioli, L.V., Significance of osteoporosis: a growing international health care problem., Calcif. Tissue. Int. 1991., 49 Suppl: S5 - 7
16 Riggs, B.L., Melton, L.J 3rd., The prevention and treatment of osteoporosis., N. Engl. J. Med. 1992 Aug 27., 327(9): 620 - 7
17 Cooper, C., Wickham, C., Walsh, K., Appendicular skeletal status and hip fracture in the elderly: 14 - year prospective data., Bone. 1991; 12(5): 361 - 4
18 Cummings, S.R., Black, D.M., Nevitt, M.C., Browner, W.S., Cauley, J.A., Genant, H.K., Mascioli, S.R., Scott, J.C., Seeley, D.G., Steiger, P., et.al., Appendicular bone density and age predict hip fracture in women. The Study of Osteoporotic Fractures Research Group., JAMA. 1990 Feb 2; 263(5): 665 - 8
19 Hui, S.L., Slemenda, C.W., Johnston, C.C Jr., Age and bone mass as predictors of fracture in a prospective study., J. Clin. Invest. 1988 Jun; 81(6): 1804 - 9
20 Melton, L.J., Chao, E.Y.S. and Lane, J., Biomechanical Aspects of Fracture. Chapter 4, in Osteoporosis, Ethiology, Diagnosis and Management, Riggs, G.L. and Melton, L.J., eds., Raven Press, NY, 1988, pp. 111 - 131
21 Barrett-Connor, E., Risks and Benefits of Replacement Estrogen., Annu. Rev. Med. 1992; 43: 239 - 51
22 Lindsay, R., Hart, D.M., Forrest, C. and Baird, C., Prevention of Spinal Osteoporosis in Oophorectomised Women., Lancet. 1980 Nov 29; 2(8205): 1151-4
23 Ryan, P.J., Harrison, R., Blake, G.M., Fogelman, I., Compliance With Hormone Replacement Therapy (HRT) After Screening For Post Menopausal Osteoporosis., Br. J. Obstet. Gynaecol. 1992 Apr; 99(4): 325 - 8
24 U.S. Package Circular for PREMARIN, Physician's Desk Reference, 2624 - 2626, 1993
25 Szucs, J., Horvath, C., Kollin, E., Szathmari, M., Hollo, I., Three-Year Calcitonin Combination Therapy For Postmenopausal Osteoporosis With Crush Fractures of the Spine., Calcif. Tissue. Int. 1992 Jan; 50(1): 7 - 10
26 U.S. Package Circular for MIACALCIN, Physician's Desk Reference, 2017 - 2018, 1992
27 Parfitt, A.M., The Coupling of Bone Formation to Bone Resorption: A Critical Analysis of the Concept and of Its Relevance to the Pathogenesis of Osteoporosis., Metab. Bone. Dis. Relat. Res. 1982; 4(1): 1 - 6
28 Balena, R., Toolan, B.C., Shea, M., Markatos, A., Myers, E.R., Lee, S.C., Opas, E.E., Seedor, J.G., Klein, H., Frankenfield, D., et. al., The Effects of 2-Year Treatment With the Aminobisphosphonate Alendronate on Bone Metabolism, Bone Histomorphometry, and Bone Strength in Ovariectomized Nonhuman Primates., J. Clin. Invest. 1993 Dec; 92(6): 2577-86
29 Lauritzen, D B; Balena, R; Shea, M; Seedor, J G; Markatos, A; Le, H M; Toolan, B C; Myers, E R; Rodan, G A; Hayes, W C ; Effects of Combined Prostaglandin and Alendronate Treatment on the Histomorphometry and Biomechanical Properties of Bone in Ovariectomized Rats., J Bone Miner Res. 1993 Jul; 8(7): 871 9
30 Weinreb, M; Quartuccio, H; Seedor, J G; Aufdemorte, T B; Brunsvold, M; Chaves, E; Kornman, K S; Rodan, G A ; Histomorphometrical Analysis of the Effects of the Bisphosphonate Alendronate on Bone Loss Caused By Experimental Periodontitis in Monkeys., J Periodontal Res. 1994 Jan; 29(1): 35 40
31 Balena, R; Markatos, A; Seedor, J G; Gentile, M; Stark, C; Peter, C P; Rodan, G A ; Long-Term Safety of the Aminobisphosphonate Alendronate in Adult Dogs. II. Histomorphometric Analysis of the L5 Vertebrae., J Pharmacol Exp Ther. 1996 Jan; 276(1): 277 - 83
32 Peter, C P; Cook, W O; Nunamaker, D M; Provost, M T; Seedor, J G; Rodan, G A ; Effect of Alendronate on Fracture Healing and Bone Remodeling in Dogs., J Orthop Res. 1996 Jan; 14(1): 74 - 9
33 Reid, I. R; Nicholson, G. C; Weinstein, R. S; Hosking, D. J; Cundy, T; Kotowicz, M. A; Murphy, W. A Jr; Yeap, S; Dufresne, S; Lombardi, A; Musliner, T. A; Thompson, D. E; Yates, A. J Biochemical and Radiologic Improvement in Paget's Disease of Bone Treated With Alendronate: A Randomized, Placebo-Controlled Trial.
35 Chavassieux, P. M; Arlot, M. E; Reda, C; Wei, L; Yates, A. J; Meunier, P. J., Histomorphometric Assessment of the Long-Term Effects of Alendronate on Bone Quality and Remodeling in Patients With Osteoporosis.,

34 Bone, H. G; Downs, R. W Jr; Tucci, J. R; Harris, S. T; Weinstein, R. S; Licata, A. A; McClung, M. R; Kimmel, D. B; Gertz, B. J; Hale, E; Polvino, W. J., Dose-Response Relationships For Alendronate Treatment in Osteoporotic Elderly Women. Alendronate Elderly Osteoporosis Study Centers.,

36 McClung, M; Clemmesen, B; Daifotis, A; Gilchrist, N. L; Eisman, J; Weinstein, R. S; Fuleihan. G. el. H; Reda, C; Yates, A. J; Ravn, P., Alendronate Prevents Postmenopausal Bone Loss in Women Without Osteoporosis. A Double-Blind, Randomized, Controlled Trial. Alendronate Osteoporosis Prevention Study Group.,

37 Kanis, J.A., Geusens, P., Christiansen, C., Guidelines for Clinical Trials in Osteoporosis. A Position Paper of the European Foundation for Osteoporosis and Bone Disease., Osteoporos. Int. 1991 Jun., 1(3): 182 - 8

38 Consensus Development Conference: Diagnosis, Prophylaxis and Treatment of Osteoporosis. Amer. J. Med., 94: 646-650, 1993

39 Chrischilles, E.A., Butler, C.D., Davis, C.S., Wallace, R.B., A Model of Lifetime Osteoporosis Impact., Arch. Intern. Med. 1991 Oct., 151(10): 2026. 32

40 Gardsell, P., Johnell, O., Nilsson, B.E., Gullberg, B., Predicting Various Fragility Fractures in Women By Forearm Bone Densitometry: A Follow-Up Study., Calcif. Tissue. Int. 1993 May; 52(5): 348 - 53

41 Haddaway, M.J., Davie, M.W., McCall, I.W., Bone Mineral Density in Healthy Normal Women and Reproducibility of Measurements in Spine and Hip Using Dual-Energy X-Ray Absorptiometry., Br. J. Radiol. 1992 Mar; 65(771): 213 - 7

42 Wuster, C., Duckeck, G., Ugurel, A., Lojen, M., Minne, H.W., Ziegler, R., Bone Mass of Spine and Forearm in Osteoporosis and in German Normals: Influences of Sex, Age and Anthropometric Parameters., Eur. J. Clin. Invest. 1992 May; 22(5): 366 - 70

43 Kin, K., Kushida, K., Yamazaki, K., Okamoto, S., Inoue, T., Bone Mineral Density of the Spine in Normal Japanese Subjects Using Dual-Energy X-Ray Absorptiometry: Effect of Obesity and Menopausal Status., Calcif. Tissue. Int. 1991 Aug; 49(2): 101 - 6

44 Steiger, P., Cummings, S.R., Black, D.M., Spencer, N.E., Genant, H.K., Age-Related Decrements in Bone Mineral Density in Women Over 65., J. Bone. Miner. Res. 1992 Jun; 7(6): 625 - 32

45 Nordin, B.E., Morris, H.A., Osteoporosis and Vitamin D., J. Cell. Biochem. 1992 May; 49(1): 19 - 25

46 Siris, E.S, Miller, P.D, Barrett Connor, E., Faulkner, K.G., Wehren, L.E., Abbott, T.A., Berger, M.L., Santora, A., C., and Sherwood, L.M., Identification and Fracture Outcomes of Undiagnosed Low Bone Mineral Density in Postmenopausal Women: Results from the National Osteoporosis Risk Assessment, .J. Am. Med. Assoc. 2001 Dec 12; 286(22): 2815-22

47 Law, M.R., Wald, N.J., Meade, T.W., Strategies For Prevention of Osteoporosis and Hip Fracture., BMJ. 1991 Aug 24; 303(6800): 453 - 9

48 Kanis, J.A. and McCloskey, E. V., Epidemiology of Vertebral Osteoporosis., Bone. 1992; 13 Suppl 2: S1-10

49 Simonen, O., Incidence of Femoral Neck Fractures: Senile Osteoporosis in Finland in the Years 1970 - 1985., Calcif. Tissue. Int. 1991; 49 Suppl: S8 - 10

50 Falch, J.A., Kaastad, T.S., Bohler, G., Espeland, J., Sundsvold, O.J., Secular Increase and Geographical Differences in Hip Fracture Incidence in Norway., Bone. 1993 Jul-Aug; 14(4): 643 - 5

51 Nagant-de-Deuxchaisnes, C., Devogelaer, J. P., Increase in the Incidence of Hip Fractures and of the Ratio of Trochanteric to Cervical Hip Fractures in Belgium., Calcif. Tissue. Int. 1988 Mar; 42(3): 201 - 3

52 Hedlund, R., Lindgren, U., Ahlbom, A., Age- and Sex-Specific Incidence of Femoral Neck and Trochanteric Fractures. An Analysis Based on 20,538 Fractures in Stockholm County, Sweden, 1972 - 1981., Clin. Orthop. 1987 Sep; (222): 132 - 9

53 Obrant, K.J., Bengner, U., Johnell, O., Nilsson, B.E., Sernbo, I., Increasing Age-Adjusted Risk of Fragility Fractures: A Sign of Increasing Osteoporosis in Successive Generations?, Calcif. Tissue. Int. 1989 Mar; 44(3): 157 - 67

54 Lizaur-Utrilla, A., Puchades-Orts, A., Sanchez-del-Campo, F., Anta-Barrio, J., Gutierrez-Carbonell, P., Epidemiology of Trochanteric Fractures of the Femur in Alicante, Spain, 1974 - 1982., Clin. Orthop. 1987 May; (218): 24 - 31

55 Mazzuoli, G.F., Gennari, C., Passeri, M., Celi, F.S., Acca, M., Camporeale, A., Pioli, G., Pedrazzoni, M., Incidence of Hip Fracture: An Italian Survey., Osteoporos. Int. 1993; 3 Suppl 1: 8 - 9

56 Martin, A.D., Silverthorn, K.G., and Houston, C.S., Age-Specific Increase in Hip Fractures in Canada, Osteoporosis, Crhistiansen, C, Johanse, J.S. AND Riis, B.J. ed, Osteopress ApS, Copenhagen, 1987, PP 111 - 112

57 Gennari, C., Epidemiology and Financial Aspects of Osteoporosis., Calcium Reg. and Bone Metabol., 1987, 9:897 - 899

58 Lee, C.M., Sidhu, J.S., Pan, K.L., Hip Fracture Incidence in Malaysia 1981 - 1989., Acta. Orthop. Scand. 1993 Apr; 64(2): 178 - 80

59 Cooper, C., Campion, G., Melton, L.J 3rd., Hip Fractures in the Elderly: A World-Wide Projection., Osteoporos. Int. 1992 Nov; 2(6): 285 - 9

60 Cooper, C., Atkinson, E.J., Jacobsen, S.J., O'Fallon, W.M., Melton, L.J 3rd., Population-Based Study of Survival After Osteoporotic Fractures., Am. J. Epidemiol. 1993 May 1; 137(9): 1001 - 5

61 Gutyon, A.C. and Hall, J.E. Textbook of medical physiology, 10th Ed., W. B. Saunders Company, Philadelphia, 2000.

62 Fleisch, H; Russell, R. G; Straumann, F, Effect of Pyrophosphate on Hydroxyapatite and Its Implications in Calcium Homeostasis., Nature. 1966 Nov 26; 212(65): 901-3

63 Fleisch, H; Maerki, J; Russell, R. G, Effect of Pyrophosphate on Dissolution of Hydroxyapatite and Its Possible Importance in Calcium Homeostasis., Proc-Soc-Exp-Biol-Med. 1966 Jun; 122(2): 317-20

64 Fleisch, H; Russell, R. G; Bisaz, S; Casey, P. A; Muhlbauer, R. C, The Influence of Pyrophosphate Analogues (Diphosphonates) on the Precipitation and Dissolution., Calcif. Tissue-Res. 1968; Suppl:10-10a

65 Fleisch, H; Russell, R. G; Francis, M. D, Diphosphonates Inhibit Hydroxyapatite Dissolution In-Vitro and Bone Resorption in Tissue Culture and In-Vivo., Science. 1969 Sep 19; 165(899): 1262-4

66 Fleisch, H.: Bisphosphonates: Pharmacology and Use in the Treatment of Tumour-Induced Hypercalcaemic and Metastatic Bone Disease. Drug 42(6): 919-944, 1991.

67 Bergstrom, J D, Bostedor, R G, Masarachia, P J, Reszka, A A, Rodan, G, Alendronate Is a Specific, Nanomolar Inhibitor of Farnesyl Diphosphate Synthase., Arch Biochem Biophys. 2000 Jan 1, 373(1): 231 41

68 Fisher, J E, Rodan, G A, Reszka, A A, In-Vivo Effects of Bisphosphonates on the Osteoclast Mevalonate Pathway., Endocrinology. 2000 Dec, 141(12): 4793 6

69 Reszka, A. A; Halasy-Nagy, J; Rodan, G. A, Nitrogen-Bisphosphonates Block Retinoblastoma Phosphorylation and Cell Growth By Inhibiting the Cholesterol Biosynthetic Pathway in a Keratinocyte Model For Esophageal Irritation., Mol. Pharmacol. 2001 Feb; 59(2): 193-202

70 Rodan GA, Mechanisms of Action of Bisphosphonates, Annual Review of Pharmacology and Toxicology. 1998; 38 : 375 388

71 Sato, M.; Grasser, W.; Endo, N.; Akins, R., and Simmons, H.: Bisphosphonate Action: Alendronate Localization in Rat Bone and Effects on Osteoclast Ultrastructure. J. Clin. Invest. 88: 2095-2105, 1991.

72 Comparison of MK-217 to Etidronate in Preclinical Studies, MRL, February 20, 1992 [Nonclinical Pharmacology and Toxicology Documentation, Nonclinical Pharmacodynamics, Reference 30].

73 Flanagan, A. M. and Chambers, T. J.: Dichloromethylenebisphosphonate (CL2MBP) Inhibits Bone Reabsorption Through Injury to Osteoclasts That Reabsorb CL2MBP-Coated Bone. Bone and Mineral 6: 33-43; 1989.

74 Lundy, M. W; Stauffer, M; Wergedal, J. E; Baylink, D. J; Featherstone, J. D; Hodgson, S. F; Riggs, B. L, Histomorphometric Analysis of Iliac Crest Bone Biopsies in Placebo-Treated Versus Fluoride-Treated Subjects., Osteoporos-Int. 1995 Mar; 5(2): 115-29

75 Cranney, A, Guyatt, G, Krolicki, N, Welch, V, Griffith, L, Adachi, J D, Shea, B, Tugwell, P, and Wells, G: A Meta Analysis of Etidronate For the Treatment of Postmenopausal Osteoporosis., Osteoporos Int. 2001; 12(2): 140 51

76 Riggs, BL, Hodgson, SF, O'Fallon, WM, Chao, EYS, Wahner, HW, Muhs, JM, Cedel, SL and Melton LJ III. Effect of Fluoride Treatment on the Fracture Rate in Postmenopausal Women with Osteoporosi, N. Eng. J. Med., 1990, 322-802-809

77 Thompson, D D; Seedor, J G; Weinreb, M; Rosini, S; Rodan, G A ; Aminohydroxybutane Bisphosphonate Inhibits Bone Loss Due to Immobilization in Rats., J Bone Miner Res. 1990 Mar; 5(3): 279 86

78 Toolan, B C; Shea, M; Myers, E R; Borchers, R E; Seedor, J G; Quartuccio, H; Rodan, G; Hayes, W C ; Effects of 4-Amino-1-Hydroxybutylidene Bisphosphonate on Bone Biomechanics in Rats., J Bone Miner Res. 1992 Dec; 7(12): 1399 406

79 Sato, M; Grasser, W; Endo, N; Akins, R; Simmons, H; Thompson, D D; Golub, E; Rodan, G A ; Bisphosphonate Action. Alendronate Localization in Rat Bone and Effects on Osteoclast Ultrastructure., J Clin Invest. 1991 Dec; 88(6): 2095 105

80 Rodan, G A; Seedor, J G; Balena, R ; Preclinical Pharmacology of Alendronate., Osteoporos Int. 1993; 3 Suppl 3: S7 12

81 Yamamoto, M; Markatos, A; Seedor, J G; Masarachia, P; Gentile, M; Rodan, G A; Balena, R ; The Effects of the Aminobisphosphonate Alendronate on Thyroid Hormone-Induced Osteopenia in Rats., Calcif Tissue Int. 1993 Oct; 53(4): 278 82

82 Rodan, G A; Balena, R ; Bisphosphonates in the Treatment of Metabolic Bone Diseases., Ann Med. 1993 Aug; 25(4): 373 8

83 Balena, R; Markatos, A; Gentile, M; Masarachia, P; Seedor, J G; Rodan, G A; Yamamoto, M ; The Aminobisphosphonate Alendronate Inhibits Bone Loss Induced By Thyroid Hormone in the Rat. Comparison Between Effects on Tibiae and Vertebrae., Bone. 1993 May Jun; 14(3): 499 504
84 Wingen, F. and Schmähl, D.: Pharmacokinetics of the Osteotropic Diphosphonate 3-Amino-1-Hydroxypropane-1,1-Diphosphonic Acid in Mammals. Arzneim. Forsch./Drug Res. 37(9): 1037-1042, 1987.
85 Michael, W. R.; King, W. R., and Wakim, J. M.: Metabolism of Disodium Ethane-1-Hydroxy-1,1-Diphosphonate (Disodium Etidronate) in the Rat, Rabbit, Dog and Monkey. Toxicology and Applied Pharmacology 21: 503-515, 1972.
86 Gural, R. P.: Pharmacokinetics and Gastrointestinal Absorption Behavior of Etidronate. Abstract of Dissertation, University of Kentucky, 1975.
87 Francis, M. D. and Martodam, R. R.: Chemical, Biochemical, and Medicinal Properties of the Diphosphonates, In: The Role of Phosphonates in Living Systems, Chap. 4, pp. 55-96.
88 Lauren, L.; Österman, T., and Karhi, T.: Pharmacokinetics of Clodronate After Single Intravenous, Intramuscular and Subcutaneous Injections in Rats. Pharmacology and Toxicology 69: 365-368, 1991.
89 Mönkkönen, J.; Ylitalo, P.; Elo, H. A., and Airaksinen, M. M.: Distribution of [14C]Clodronate (Dichloromethylene Bisphosphonate) Disodium in Mice. Toxicology and Applied Pharmacology 89: 287-292, 1987.
90 Mönkkönen, J.: A One Year Follow-Up Study of the Distribution of [14C]Clodronate in Mice and Rats. Pharmacol-ogy and Toxicology 62: 51-53, 1988.
91 Wingen, F. and Schmähl, D.: Distribution of 3-Amino-1-Hydroxypropane-1,1-Diphosphonic Acid in Rats and Effects on Rat Osteosarcoma. Arzneim. Forsch./Drug Res. 35(10): 1565-1571, 1985.
92 Larsson, A. and Rohlin, M.: In-Vivo Distribution of 14C-Labeled Ethylene-1-Hydroxy-1,1-Diphosphonate in Normal and Treated Young Rats. An Autoradiographic and Ultra-structural Study. Toxicology and Applied Pharmacology 52: 391-399, 1980.
93 Mönkkönen, J.; Urtti, A.; Paronen, P.; Elo, H. A., and Ylitalo, P.: The Uptake of Clodronate (Dichloromethylene Bisphosphonate) by Macrophages In-Vivo and In-Vitro. Drug Metabolism and Disposition 17(6): 690-693, 1989.
94 Troehler, U.; Bonjour, J. P., and Fleisch, H.: Renal Secretion of Diphosphonates in Rats. Kidney International 8: 6-13, 1975.
95 Lin, J.H., Bisphosphonates: A Review of their Pharmacokinetic Properties, Bone-. Feb. 1996; 18 (2) : 75-85
96 Lin, J.H; Duggan, D.E; Chen, I.W; Ellsworth, R.L, Physiological Disposition of Alendronate, A Potent Anti-Osteolytic Bisphosphonate, in Laboratory Animals., Drug Metab. Dispos. 1991 Sep-Oct; 19(5): 926-32
97 Lin, J.H; Chen, I.W; deLuna, F.A, Nonlinear Kinetics of Alendronate. Plasma Protein Binding and Bone Uptake., Drug Metab. Dispos. 1994 May-Jun; 22(3): 400-5
98 Porras, A.G; Holland, S.D; Gertz, B.J, Pharmacokinetics of Alendronate., Clin. Pharmacokinet. 1999 May; 36(5): 315-28
99 Lin, J.H; Chen, I .W; deLuna, F.A; Hichens, M, Role of Calcium in Plasma Protein Binding and Renal Handling of Alendronate in Hypo- and Hypercalcemic Rats., J. Pharmacol. Exp. Ther. 1993 Nov; 267(2): 670-5
100 Kino, I; Kato, Y; Lin, J.H; Sugiyama, Y, Renal Handling of Biphosphonate Alendronate in Rats., Biopharm. Drug. Dispos. 1999 May; 20(4): 193-8
101 Lin, J.H; Chen, I.W; Deluna, F.A; Hichens, M, Renal Handling of Alendronate in Rats. An Uncharacterized Renal Transport System., Drug Metab. Dispos. 1992 Jul-Aug; 20(4): 608-13
102 Lin, J.H., Chen, I.W., and deLuna, F.A., On the Absorption of Alendronate in Rats, J. Pham. Sci. 1994, 83(12), 1741-6
103 Tucci, J.R., Tonino, R.P., Emkey, R.D., Peverly, C.A., Kher, U., Santora, A.C, 2nd , Effect of Three Years of Oral Alendronate Treatment In Postmenopausal Women with Osteoporosis, Am. J. Med. 1996 Nov., 101(5): 488-501.
104 Devogelaer, J.P., Broll, H., Correa-Rotter, R., Cumming, D.C., De-Deuxchaisnes, C.N., Geusens, P., Hosking, D., Jaeger, P., Kaufman, J.M., Leite, M., Leon, J., Liberman, U., Menkes, C.J., Meunier, P.J., Reid, I., Rodriguez, J., Romanowicz, A., Seeman, E., Vermeulen, A., Hirsch, L.J., Lombardi, A., Plezia, K., Santora, A.C., Yates, A.J. and Yuan, W., Oral Alendronate Induces Progressive Increases in Bone Mass of the Spine, Hip, and Total Body over 3 Years in Postmenopausal Women with Osteoporosis, Bone,1996 Feb., 18(2): 141-50.
105 Nevitt, M.C., Thompson, D.E., Black, D.M., Rubin, S.R., Ensrud, K., Yates, A.J., and Cummings, S.R , Effect of Alendronate on Limited-Activity Days and Bed-Disability Days Caused by Back Pain In

Postmenopausal Women with Existing Vertebral Fractures. Fracture Intervention Trial Research Group. Arch. Intern. Med. 2000 Jan 10., 160(1): 77-85.

106 Cummings, S.R., Black, D.M., Thompson, D.E., Applegate, W.B., Barrett-Connor, E., Musliner, T.A., Palermo, L., Prineas, R., Rubin, S.M., Scott, J.C., Vogt, T., Wallace, R., Yates, A.J., and LaCroix, A.Z, Effect of Alendronate on Risk of Fracture in Women with Low Bone Density but Without Vertebral Fractures: Results from the Fracture Intervention Trial. JAMA. 1998 Dec 23-30., 280(24): 2077-82.

107 Levis, S., Quandt, S.A., Thompson, D., Scott, J., Schneider, D.L., Ross, P.D., Black, D., Suryawanshi, S., Hochberg, M., and Yates, J, Alendronate Reduces the Risk of Multiple Symptomatic Fractures: Results from the Fracture Intervention Trial. J. Am. Geriatr. Soc. 2002 Mar., 50(3): 409-15.

108 Greenspan, S.L., Holland, S., Maitland-Ramsey, L., Poku, M., Freeman, A., Yuan, W., Kher, U., and Gertz, B., Alendronate stimulation of nocturnal parathyroid hormone secretion: a mechanism to explain the continued improvement in bone mineral density accompanying alendronate therapy. Proc. Assoc. Am. Physicians. 1996 May., 108(3): 230-8.

109 Caniggia, A. and Gennari, C.: Kinetics and Intestinal Absorption of [^{32}P]EHDP in Man. Calcif. Tissue Res. 22(Suppl): 428-429, 1977.

110 Daley-Yates, P. T.; Dodwell, D. J.; Pongchaidecha, M.; Coleman, R. E., and Howell, A.: The Clearance and Bioavailability of Pamidronate in Patients With Breast Cancer and Bone Metastases. Calcif. Tissue Int. 49: 433-435, 1991.

111 Leyvraz, S.; Hess, U.; Flesch, G.; Bauer, J.; Hauffe, S.; Ford, J. M., and Burckhardt, P.: Pharmacokinetics of Pamidronate in Patients With Bone Metastases. JNCI 84(10): 788-792, 1992.

112 Pentikäinen, P. J.; Elomaa, I.; Nurmi, A.-K., and Kärkkäinen, S.: Pharmacokinetics of Clodronate in Patients With Metastatic Breast Cancer. International J. of Clinical Pharmacology, Therapy and Toxicology 27(5): 222-228, 1989.

113 Wiedmer, W. H.; Zbinden, A. M.; Trechsel, U., and Fleisch, H.: Ultrafiltrability and Chromatographic Properties of Pyrophosphate, 1-Hydroxyethylidene-1,l-Bisphosphonate, and Dichloromethylenebisphosphonate in Aqueous Buffers and in Human Plasma. Calcif. Tissue Int. 35: 397-400, 1983.

114 Conrad, K. A. and Lee, S. M.: Clodronate Kinetics and Dynamics. Clin. Pharmacol. Ther. 30(1): 114-120, 1981.

115 Gural, R. P.: Pharmacokinetics and Gastrointestinal Absorption Behavior of Etidronate. Abstract of Disserta-tion, Universi-ty of Kentucky, 1975.

116 Recker, R. R. and Saville, P. D.: Intestinal Absorption of Disodium Ethane-1-Hydroxy-1,1-Diphosphonate (Disodium Etidronate) Using a Deconvolution Technique. Toxicology and Applied Pharmacology 24: 580-589, 1973.

117 Powell, J. H. and DeMark, B. R.: Clinical Pharmacokinetics of Diphosphonates. in: Bone Resorption, Metastasis, and Diphosphonates, Garattini, S. ed., Raven Press, New York, 1985, pp. 41-49.

118 Fitton, A. and McTavish, D.: Pamidronate: A Review of its Pharmacological Properties and Therapeutic Efficacy in Resorptive Bone Disease. Drugs 41(2): 289-318, 1991.

119 Yakatan, G. J.; Poynor, W. J.; Talbert, R. L.; Floyd, B. F.; Slough, C. L.; Ampulski, R. S., and Benedict, J. J.: Clodronate Kinetics and Bioavailability. Clin. Pharmacol. Ther. 31(3): 402-410, 1982.

120 Hanhijärvi, H.; Elomaa, I.; Karlsson, M., and Lauren, L.: Pharmacokinetics of Disodium Clodronate After Daily Intravenous Infusions During Five Consecutive Days. International J. of Clinical Pharmacology, Therapy and Toxicology 27(12): 602-606, 1989.

121 Fogelman, I.; Smith, L.; Mazess, R.; Wilson, M. A., and Bevan, J. A.: Absorption of Oral Diphosphonate in Normal Subjects. Clinical Endocrinology 24: 57-62, 1986.

122 Kline, W. F.; Matuszewski, B. K., and Bayne, W. F.: Determina-tion of 4-Amino-1-Hydroxybutane-1,1-Bisphosphonic Acid in Urine by Automated Pre-Column Derivatization With 2,3-Naphthalene Dicarboxyaldehyde and High-Performance Liquid Chromatography With Fluorescence Detection. J. Chromatogr. (Biomed. Appl.) 534: 139-149, 1990.

123 Kline, W. F. and Matuszewski, B. K.: Improved Determination of the Bisphosphonate Alendronate in Human Plasma and Urine by Automated Precolumn Derivatization and High Performance Liquid Chromatography With Fluorescence and Electrochemical Detection. J. Chromatogr. (Biomed. Appl.) 583: 183-193, 1992.

124 Porras, A.G., Holland, S.D., Gertz, B.J., Pharmacokinetics of Alendronate, Clin. Pharmacokinet. May 1999, 36 (5) : 315-32

125 Cocquyt, V., Kline, W.F., Gertz, B.J., Van Belle, S.J.P., Holland, S.D., DeSmet, M., Quan, H., Vyas, K.P., Zhang, K.Y.E., De Greve, J., Porras, A.G., Pharmacokinetics of Intravenous Alendronate, J. Clin. Pharm., Apr 1999; 39 (4) : 385-393

126 Gibaldi, M. and Perrier, D.: Chapter 11: Noncompartmental Analysis Based on Statistical Moment Theory, in: Pharmacokinetics. 2nd ed., Marcel Dekker, Inc., New York and Basel, 1982, pp. 409-417.

127 Khan, S.A., Kanis, J.A., Vasikaran, S., Kline, W.F., Matuszewski, B.K., McCloskey, E.V., Beneton, M.N.C., Gertz, B.J., Sciberras, D.G., Holland, S.D., Orgee, J., Coombes, G.M., Rogers, S.R., Porras, A.G, Elimination and Biochemical Responses to Intravenous Alendronate in Postmenopausal Osteoporosis, J. Bone Mineral Res., Oct 1997., 12 (10) : 1700-1707

128 Gertz, B.J., Holland, S.H., Kline, W.F., Matuszewski, B.K., Freeman, A., Quan, H., Lasseter, K.C., Mucklow, J.C., Porras, A.G., Studies of the Oral Bioavailability of Alendronate, Clin. Pharm. Ther. Sep 1995., 58 (3) : 288-298

129 Altman, D. F.: The Effect of Age on Gastrointestinal Function. Sleisenger, M. H., Fordtran, J. S., eds., In: Gastrointestinal Disease, Volume 1, Pathophysiology, Diagnosis, Management, 4th ed., W. B. Saunders Company, Philadelphia, pp. 162-169.

OPTIMIZING INDIVIDUALIZED DOSAGE REGIMENS OF POTENTIALLY TOXIC DRUGS

Roger W. Jelliffe[1*], Alan Schumitzky[1], Robert Leary[2], Andreas Botnen[3], Ashutosh Gandhi[1], Pascal Maire[4], Xavier Barbaut[5], Nathalie Bleyzac[4], and Irina Bondareva[6]

1. INTRODUCTION

The end product of drug development is the use of the drug in clinical therapy. When a drug has a narrow margin of therapeutic safety, we must steer its dosage between one that is too low, and likely to be ineffective on the one hand, or too high, and likely to be toxic, on the other. We must carefully plan and individualize the dosage for each patient, to achieve some desired target goal such as a serum concentration, or its profile over time. We must then observe the patient, and if needed, monitor serum concentrations at appropriate intervals. These intervals should be frequent enough so we can evaluate the patient when there are relatively small changes in the total amount of drug in the body between observations, so that if toxicity develops, we detect it in an early stage of its development so we can make the appropriate adjustment in dosage early, rather than later, after toxicity has become more severe and dangerous.

It is not useful to talk about dosage individualization without saying with respect to what. It is commonly said that one should individualize dosage to body weight and renal function, for example. But again, to what specific end, toward what specific goal? This is usually not explicitly stated. We usually do this, however, to control either the total amount of drug in the patient's body, or the serum concentration, for example, at a desired specific target value, usually within some general target "therapeutic range" of serum concentrations where most patients (but not all) do well, and where the incidence of toxicity is acceptably low.

* [1]Laboratory of Applied Pharmacokinetics, University of Southern California School of Medicine, Los Angeles CA, USA; [2]San Diego Supercomputer Center, University of California, San Diego CA, USA; [3]Center for Bioinformatics, University of Oslo, Norway; [4]Hospices Civils de Lyon, France ; [5]Hospice de Beaune, France ; [6]Institute of Physical and Chemical Medicine, Moscow, Russia.

Applications of Pharmacokinetic Principles in Drug Development
Edited by Krishna, Kluwer Academic/Plenum Publishers, 2004

However, this is approach is appropriate only for the initial regimen, and it still ignores the opportunity to be gentle, moderate, or aggressive in the approach to the patient, according to each individual patient's need for the drug.

The expected incidence of toxicity should be no greater that that which is appropriate for the patient's need for the drug. In many cases, if the need for the drug is not great, or is not acute, the target goal should first be one that is associated with a low incidence (risk) of toxicity, leading to a gentle dosage regimen. Based on the patient's response, the target goal can then be revised upward and a higher dosage given to achieve it. This, for example, is behind the "start low, go up slow" dosage policy so well advocated by Cohen [1].

On the other hand, a firmer approach may be clinically indicated. If the patient has an acute and significant need for the drug, the "start low, go slow" approach is not warranted, and is not safe. Here a higher target goal must be selected, one which is more likely to be effective, and a greater risk of toxicity will have to be accepted in order to achieve such a higher goal.

The target goal must therefore be selected individually for each patient, according to that patient's individual need for the drug at that time. In this way, one can then develop a gentle approach to one patient, but a more firm or aggressive approach to another, as each patient's need dictates. Clearly, things are not the same at the bottom of the therapeutic range as at the top, or even further, if it is necessary to go to a still higher target if the patient's need dictates. Examples of this approach are the acceptance of a certain risk of toxicity with cancer chemotherapy, therapy for AIDS, and with the risk of toxicity with digoxin, aminoglycosides, vancomycin, and transplant chemotherapy, for example.

2. SET INDIVIDUALIZED TARGET GOALS FOR EACH PATIENT

The concept of a general "therapeutic range" of serum drug concentrations is therefore only a generalization. It is an overall range in which most patients, but certainly not all, do well. One must always check each individual patient to see if he or she is doing not only well, but optimally, on clinical grounds, regardless of whatever the serum concentration is actually found to be. This approach is quite different from much clinical teaching, but still is similar to what many clinicians give lip service to – "look at the patient, not just the serum concentration".

2.1. Problems with "Therapeutic Ranges"

Figure 1 shows the usual means by which therapeutic ranges appear to have been obtained. It is interesting that these ranges have never, to the authors' knowledge, been defined in a specific, quantitative, and explicitly described manner, but instead have simply been described as regions below which therapy is generally "ineffective", and above which "significant" toxicity has been observed. It is usually stated that first, there is a "significant" incidence of therapeutic effects with increasing serum drug concentrations. This defines the beginning of the therapeutic range. Later on the incidence of toxic effects also becomes "significant", and the "toxic range" has been entered.. For example, Evans [2] presents a definition of therapeutic range as "a range of drug concentrations within which the probability of the desired clinical response is relatively high and the probability of unacceptable toxicity is relatively low" However, the therapeutic range has never, to our knowledge, been defined quantitatively. The eye is drawn to the bends in each line in Figure 1, and the classification of the apparent "therapeutic range" has been developed, published, and accepted without apparent criticism or further thought. However, this procedure does not consider the need to develop a gentle dosage regimen for a patient who needs only a gentle touch, or a more aggressive one for a patient who really needs the dosage "pushed". Another problem has been that special populations of patients have not had their special needs recognized. For example, it is well known that patients with atrial fibrillation need higher serum concentrations, usually averaging 2.0 ng/ml, for full control of their ventricular rate, and yet this has not been followed up by setting a special therapeutic range for them. Special types of patients, the elderly, for example, may need special target serum concentrations selected for them.

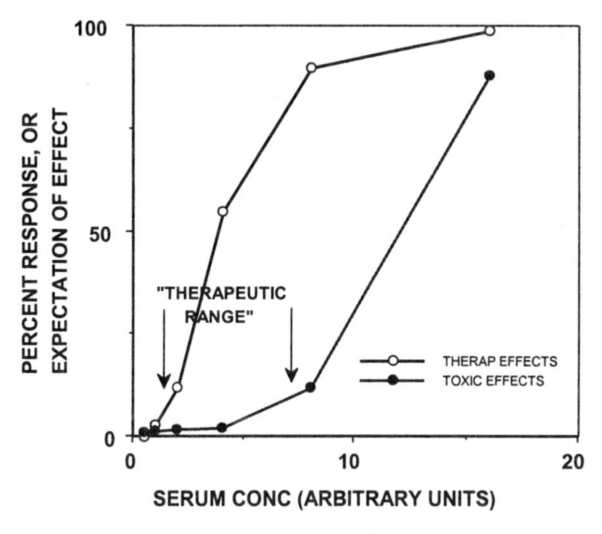

Figure 1. General relationships usually found between serum drug concentrations and the incidence of therapeutic and toxic effects. The eye is drawn to the bends in the curves, and the therapeutic range is classified in relation to these bends. This qualitative procedure of classification discards the important quantitative relationship of the incidence of toxic effects versus serum concentration.

In addition, the risks and benefits of using a therapeutic range or window are quite complex. One could, for example, develop a dosage regimen to maximize the probability of having the patient's serum concentration be within some desirable therapeutic window. This sounds good at first, but the decision-making process rapidly becomes quite complex. One must also weigh the benefit and the probability of a desirable response against the risks, and their probabilities, of being outside that window, either below or above it.

Those risks associated with being below the window are usually associated with lack of therapeutic effect. Those associated with being above the window are usually associated with toxicity. Each of these outcomes - subtherapeutic, therapeutic, or toxic- thus has not only its associated probability but also its own positive or negative quantitative utility function of goodness or badness. Optimizing such a complex set of probabilities and utilities becomes a most complex process, and is poorly amenable to rigorous clinical decision analysis, especially at the bedside.

2.2 Setting specific target goals based on need.

A more intuitive and individualized approach is one in which the clinician evaluates the magnitude of each patient's individual clinical need for the drug in question, and selects an estimated risk of toxicity which is felt on clinical grounds to be justified by the patient's need. In this approach, there is no window of neutrality about a target, as in a "therapeutic range. Based on the relationship between serum concentration and incidence of toxicity shown in Figure 1, for example, one selects a specific target serum concentration goal to be achieved for a specific patient. One does not want the patient to run any greater risk of toxicity than is justified by the patient's clinical need for the drug. Within that constraint, however, one wants to give the patient as much drug as possible, to get the maximum benefit. This approach provides the rationale for selecting a specific target serum concentration goal, rather than some wider window, and then to attempt to achieve that target goal with the greatest possible precision, just as if one were shooting at any other target.

In this approach, the risks of being just slightly above the desired target goal are only minimally different from those associated with being just slightly below it, in the sense of an infinitesimal difference in calculus. Because of this, it appears easier, and also more intuitive for a clinician, to choose a desired target goal rather than some wider window, and then to attempt to achieve or hit that selected target goal with the greatest precision (least error) possible.

Without selecting such a specific target goal, there can be no truly individualized precise drug therapy. Individualized drug therapy therefore begins by setting a specific individualized target goal for each patient. The task of the clinician is to select, and then to hit, the desired target goal as precisely as possible. As soon as the initial regimen is given, the clinician's task is then to observe the patient's clinical response at appropriately frequent intervals, and to reevaluate whether the target goal was hit precisely enough or not, whether the target goal was correctly chosen or not, or if it should be changed and a new dosage regimen developed to hit the new target goal. This is the basis of the "target - oriented, model - based, approach to individualized drug dosage for each individual patient [3].

3. THE NEED FOR MODELS

Pharmacokinetic models, like other mathematical models, condense huge collections of experience into a form that can be easily grasped and understood. Newton's equations condensed wide ranges of experience and extensive data into his superb models of celestial mechanics. He did not make hypotheses. He simply discovered and then described with models these widely scattered relationships and events. These models are very scientific, in the deepest sense of the word.

In exactly the same way, pharmacokinetic models, like all models, can be extremely useful. No pharmacokinetic model is an exact description of reality, which is always more complex. Nevertheless, these models describe in useful, quantitative terms, the behavior of drugs when they are given to patients – their absorption, serum concentrations, pathways and exchange rates to and from other compartments, and their various therapeutic and toxic effects. Often these processes and pathways may have important relationships to other clinical descriptors or covariates such as body weight, age, renal function, smoking status, gender, and genetic CYP450 makeup, for example.

Pharmacokinetic and pharmacodynamic (PK/PD) models also provide the tool to apply that recorded past experience to the care of new patients. Past experience with drug behavior is now usually stored in the form of a population PK/PD model which is then used to design the initial dosage regimen for the next patient who appears to belong to that particular population.

The dosage regimen to achieve the therapeutic target goal is computed and given. The patient is then monitored both clinically and by measuring serum concentrations. The serum concentrations are used not only to note if they are within some general "therapeutic range", but most importantly to make a specific model of the behavior of the drug in that individual patient, based on information from the population PK/PD model, and using Bayes' theorem to develop an individualized model that best describes the behavior of the drug in that individual patient, using both the patient's individual data of serum concentrations, and balancing the patient's individual data against the more general information from the population model.

In striking this balance, the relative credibility of both kinds of information must be weighed. Since individual data is often sparse, especially at the beginning of an experience with a patient, it never explains the entire picture, and it is usefully supplemented with the general population information. The more individual data is obtained, the more it dominates the picture. In this manner, using Bayes' theorem, an individual model can be made, which has the property of being able to predict future serum concentrations at least somewhat better that models made without considering population information, using weighted nonlinear least squares, for example, which consider only the individual patient data.

One can then see what the patient's probable serum concentrations were at all other times when they where not measured, even when he or she was not at all in a steady state, even during highly unstable clinical situations in very acutely ill patients with great changes in their clinical status and in their renal function over time. One can also reconstruct and see graphically see the computed concentrations of drugs in a peripheral nonserum compartment or in various effect compartments. The patient's individualized model permits one also to make dosage adjustments without having to wait for a steady state before sampling serum concentrations, to take into consideration practical clinical situations, such as handling data of different dosage regimens (with unequal doses and unequal dosing intervals), and totally arbitrary time intervals between drug doses and blood sampling.

These important physiological, pharmacological, pharmacokinetic and pharmacodynamic relationships cannot be seen or inferred at all without such models. In addition, the use of models gets around the need to wait for a steady state before monitoring serum concentrations.. By comparing the clinical behavior of the patient with the behavior of the patient's model, one can evaluate the patient's clinical sensitivity to the drug, and can adjust the target goal appropriately. For digoxin, for example, the inotropic effect of the drug correlates best with the computed concentrations of the drug in the peripheral compartment (μg/kg of body weight, for example) rather than with the serum concentrations. The excellent model made by Reuning and colleagues for digoxin [4] has been highly useful clinically [5].

4. CURRENT BAYESIAN INDIVIDUALIZATION OF DRUG DOSAGE REGIMENS

The Reverend Thomas Bayes, who died in 1761, was a mathematically inclined minister, and it is said that he was interested in "seeing through" the operations of chance, to better understand God's design for the world. He described how we learn from life, by revising our expectations based upon our experience. His theorem describes in quantitative terms the important sequential relationship between:

1. the estimated probabilities of certain events (for example, in PK/PD terms, a patient's apparent volume of distribution, and the clearance or rate constant for elimination of a drug), that are present in a patient before we have had any chance to know anything about the patient's response to the drug (or the serum concentrations), because we have not given him/her the drug yet, and then

2. the measured serum concentrations that are found in that patient, and their precision, and then

3. the revised (Bayesian conditional posterior) probabilities of these PK/PD parameter values after the new information is obtained and evaluated (posterior to the new information).

In the beginning, before the drug is given, all we have is our past experience about the behavior of the drug in similar patients. This is why one of our most important tasks as clinicians is to store our experiences with patients in a form that can be used in the future to apply that experience optimally, usually using Bayes' theorem above, to the care of the next similar patient. This is why it is important to make population PK/PD models of the behavior of the drug in the actual patients we treat, not just in research clinical trials, to obtain and store that important clinical past experience optimally.

All this decision making must be done before (prior to) giving the drug and before (prior to) being able to observe anything about the patient's own individual clinical response and serum concentrations. Furthermore, since fitting a pharmacokinetic model only to the patient's data is not optimal without supplementing it with information of general past experience as well, the Bayesian approach describes quantitatively the sequential relationship between prior probabilities (those present before obtaining any new information), the new information (serum concentrations, for example, and their precision), and the revised (posterior) probabilities after that information is taken into account.

Bayes' theorem can be used to describe this sequence of pharmacokinetic and pharmacodynamic events. It is widely used throughout the scientific and the military communities. The great majority of flight control, fire control, and missile guidance systems are Bayesian adaptive controllers. We use the same approach here to control the behavior of the patient as we treat him or her. We do this by selecting desired target therapeutic goals, by using the population model to compute the initial regimen to best hit the desired target, by monitoring the patient with serum concentrations, using that data to make an individual Bayesian posterior model, then by re-evaluating on clinical grounds whether or not the correct target goal was chosen (and changing it if needed), and finally by computing the new and adjusted dosage regimen to best hit the selected target once again. This cycle can then repeat as often as clinically indicated.. This therapeutic process is called Bayesian adaptive individualization and control of drug dosage regimens.

The Maximum Aposteriori Probability (MAP) Bayesian approach to individualization of drug dosage regimens was introduced to the pharmacokinetic community by Sheiner et al. [6]. In this approach, parametric population models are used as the Bayesian priors. In these models, the parameters in the structural model (the apparent volume of distribution, clearances, rate constants, and their variances, are described by other parameters of their means, standard deviations (SD's), and the correlations between them. The credibility of these population models (their parameter SD's) is then evaluated in relationship to the SD's of the measured serum concentrations as they are obtained. The contribution of these two types of data and their SD's to the MAP Bayesian posterior individualized patient model is shown in the MAP Bayesian objective function below,

$$\sum \frac{(C_{obs} - C_{mod})^2}{SD^2_{(Cobs)}} + \sum \frac{(P_{pop} - P_{mod})^2}{SD^2_{(Ppop)}} \qquad (1)$$

where C_{obs} is the collection of observed serum concentrations, $SD_{(Cobs)}$ is the collection of their respective SD's, and C_{mod} is the model estimate of each serum concentration at the time it was obtained. Similarly, P_{pop} is the collection of the various population model parameter values, $SD^2_{(Ppop)}$ is the collection of their respective SD's, and P_{mod} is the collection of the Bayesian posterior model parameter values. Each data point ccan be given a weight according to its Fisher information, the reciprocal of its variance (the square of the SD), and so is each population parameter value, according to its variance. Population models in which there is greater diversity, and therefore greater variance, contribute less to the individualized model than do population models having smaller variances. Similarly, a precise assay will draw the fitting procedure more closely to the observed concentrations, and a less precise assay will do the opposite. The more serum data are obtained, the more that information dominates the determination of the MAP Bayesian posterior parameter values (P_{mod}) in the patient's individualized pharmacokinetic model.

In contrast with older methods of fitting of PK/PD data, the MAP Bayesian method can fit using only a single serum concentration data point if needed. This is because the MAP Bayesian procedure already has one data point for each parameter. Those data points are the collection of the population parameter values themselves. Because of this, the MAP Bayesian procedure can start to fit with only a single serum concentration. This feature of the MAP Bayesian method allows one to handle the often very poor and sparse data usually present in clinical strategies of therapeutic drug monitoring and dosage adjustment.

Having made the patient's individualized model, one then uses it to reconstruct the past behavior of the drug in the patient during his therapy to date. One can examine a graphical plot of the behavior of this model over the duration of the past therapy. One can thus evaluate the clinical sensitivity of the patient to the drug, by looking at the patient clinically and comparing the patient's clinical behavior with that of the patient's individualized pharmacokinetic model. In that way, one can evaluate whether the initial target goal was well chosen or not. One can choose a different goal if needed, and once again one can compute the dosage regimen to achieve it. In this way, the model can be individualized and dosage can continue to be adjusted to the patient's body weight, renal function, and available serum concentrations, for example, to achieve the desired target goal, usually with increasing precision during the course of the patient's therapy.

5. COMPARISON WITH OTHER METHODS OF FITTING DATA

The MAP Bayesian fitting procedure has been shown to be generally better in predicting future serum concentrations than the method of weighted nonlinear least squares. MAP Bayesian fitting is also significantly better than the earlier traditional but now obsolete method of linear regression on the logarithms of the concentrations (see below).

5.1 Weighted Nonlinear least Squares Regression

The conventional weighted least squares regression procedure is not quite so smart as the MAP Bayesian one, because its objective function is less complete, and has only the left hand side of the MAP Bayesian objective function, as shown below.

$$\sum \frac{(\mathrm{Cobs} - \mathrm{C\ mod})^2}{\mathrm{SD}^2\,(\mathrm{Cobs})} \qquad (2)$$

Because of this, only the patient's serum data are considered in the fitting procedure, and this information is not supplemented by the additional information from the population parameter values. Because of this, fitted models made using weighted nonlinear least squares have been shown to predict future serum concentrations slightly less well than those made using MAP Bayesian fitting [7].

Like the MAP Bayesian procedure, this method can fit the model to data of doses and serum concentrations acquired over many dose intervals, usually the patient's entire dosage history. There is no longer any reason to do the traditional "single dose" pharmacokinetic study. Further, there is no need for the patient to be in a steady state or for the serum data to be only post-distributional. Studies and population pharmacokinetic / pharmacodynamic modeling can be done on the actual patients being treated, as they are receiving their therapy. This is a second, and very important, function of therapeutic drug monitoring. The algorithm of Nelder and Mead [8] is a good one for fitting the data in both the least squares and the MAP Bayesian fitting procedures. A very useful nonmathematical description of this method has been given in BYTE magazine [9].

Secondly, like the MAP Bayesian method, weighted nonlinear least squares can provide correct weighting of serum concentration data according to its credibility or Fisher information [10]. It thus has the potential for obtaining good estimates of the pharmacokinetic parameter values.

However, this method cannot take into account population information that is generally known about how that drug usually behaves in patients like the individual patient under consideration. As the procedure moves from the starting population parameter values to others which fit the data better, it discards all the general information used to begin the fitting procedure, instead of supplementing it with the individual patient's data. Since no fitting procedure ever explains the entire relationship between doses given and concentrations found, discarding the general population information is a suboptimal feature. It may well be because of this feature that the nonlinear least-squares method, while "fitting" serum concentration data "best", has been shown to be a slightly poorer predictor of subsequent serum concentrations than the MAP Bayesian method [7]. In contrast to the MAP Bayesian procedure, this method, like linear least squares regression (see below), requires at least one serum concentration for each parameter to be fitted, or at least two serum concentrations in the models considered here, as will be discussed further below. The MAP Bayesian method, in contrast, can begin to fit using only a single serum concentration data point if needed. This is because the MAP Bayesian procedure already has one data point for each parameter. Those data points are the collection of the population parameter values themselves. Because of this, the MAP Bayesian procedure can start to fit with only a single additional data point, the very first serum concentration. This is an important feature that is extremely helpful in the “practical” aspects of therapeutic drug monitoring for patient care.

5.2 Linear Least Squares Regression

Another method used to fit serum concentrations to make individual patient models has been the old traditional but now obsolete method of linear regression on the logarithms of the serum concentrations (see below). This method was the traditional one in which a pharmacokinetic model (restricted to only a single compartment) was fitted to data obtained during only a single dose interval, specifically to the logarithms of the serum concentrations. No weighting of the serum data was used. The method was simple, and it has been widely implemented on hand calculators. It was generally the community standard for monitoring serum gentamicin concentrations ever since Sawchuk and Zaske showed its utility to individualize aminoglycoside dosage regimens [11].

The method requires at least 2 serum concentrations. It cannot handle anything more than a 1-compartment pharmacokinetic model. Distribution of the drug after a dose must be complete before a “meaningful” serum sample can be obtained, where the ratio between the serum concentration and that in any other compartment is constant. The method takes advantage of the fact that one can linearize the solution of a first-order linear differential equation for such a model if one transforms the serum concentration values to their logarithms. However, the method has three important weaknesses.

First, the method can only fit serum concentration data acquired during a single dose interval. It discards all previous serum data (and all previous information about the patient) whenever a new set of serum concentrations is obtained. There is therefore a loss of continuity each time new serum data are analyzed. This method is the most wasteful of any in its use of serum concentration data, as the useful life span of a serum sample is shorter here than with any of the other methods which do not have to discard old data, but can integrate it with more recent data from other dose intervals, as the nonlinear least squares and MAP Bayesian procedures can do.

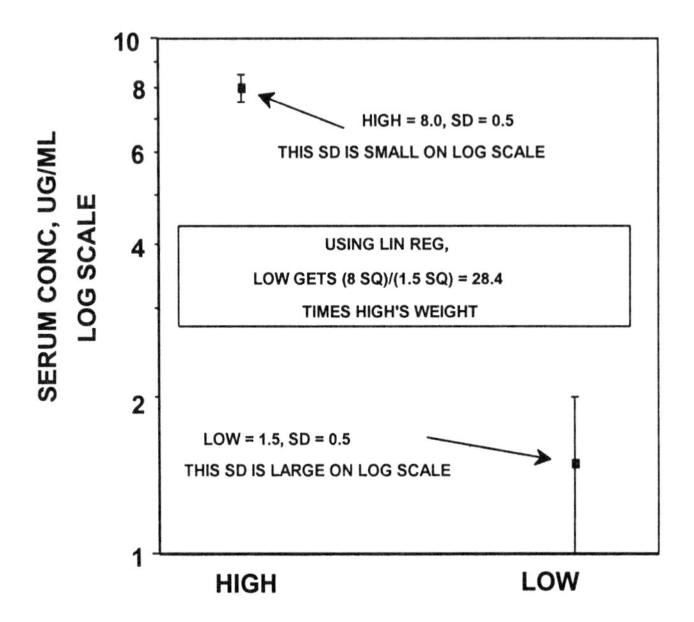

Figure 2. Error pattern assumed using fitting by linear regression on logarithms of serum concentrations. Note the much greater weighting given to the lower concentrations.

Second, linear regression contains the assumption that the assay error is a constant percent of the measured concentrations. The lower the concentration, the more accurately it is assumed to be known. Because of this, if the assay has any other error pattern over its working range (and it almost always does), this method greatly overestimates the credibility of low serum concentrations over high ones. This can be seen if one considers two serum samples, one of 8.0 μg/ml for example, and one of 1.5 μg/ml, as shown in Figure 2. One usually wishes to attach approximately equal credibility (weight) to these data points. One might thus assume that their laboratory error is approximately equal. Since the Fisher information (an index of credibility) of a data point having a normally distributed error is proportional to the reciprocal of the variance of that data point, the relative weights given by linear regression to serum concentrations of 8.0 and 1.5 μg/ml would be proportional to the reciprocal of their squares [7,10]. Because of this, the method of linear least squares, which assumes that the error bars are <u>equal</u> <u>on the logarithmic scale</u>, arbitrarily gives the value of 1.5 μg/ml a weight of $8^2/1.5^2 = 64/2.25 =$ 28.4 times the weight of the concentration of 8.0 μg/ml. A concentration of 0.1 has 100 times the weight of a concentration of 1.0, and 10,000 times the weight of a concentration of 10.0 units! Because of this assumption, this method often obtains model parameter values that are significantly different from those obtained by other methods [7]. Third, this method ignores all population data, and therefore all past general experience, concerning the behavior of the drug.

5.3 Conclusions

The MAP Bayesian method [6] appears to be the best of these three [7]. As with nonlinear least squares, it can provide correct weighting of serum concentration data according to the known laboratory assay error, and it can analyze such data over many dose intervals. In addition, it supplements population data (general knowledge) with specific information about each patient, instead of discarding it. Because of this, the method has been a slightly better predictor of future serum concentrations [7]. Lastly, the method requires only a single serum concentration to begin the analysis, no matter how many parameters are present in the population pharmacokinetic model. As more serum concentrations are obtained, the fitted model gradually becomes less of a population model and more of a patient-specific model. Both general and patient-specific data are combined intelligently in the MAP Bayesian procedure to provide the most probable single-point estimates of the parameter values given both types of data and their respective standard deviations.

Finally, one other fitting procedure, now coming on the scene, holds promise of doing still better than the MAP Bayesian method. This is the "Multiple Model" method of fitting data and designing drug dosage regimens [12]. It is a stochastic rather than a deterministic method, and is based on nonparametric population models [13,14] and their individualized Bayesian posterior pharmacokinetic models. This method of dosage design will be discussed more fully later on in this chapter.

6. EXAMPLES OF MAP BAYESIAN TARGET-ORIENTED, MODEL – BASED, APPROACHES TO PATIENT CARE

6.1 Gentamicin Therapy

With a 1-compartment pharmacokinetic model in which the elimination rate constant (Kel) was composed of a nonrenal component (Knr) and a renal component having a slope (Kslope) relationship to creatinine clearance (CCr) so that Kel = Knr + Kslope x CCr, the MAP Bayesian procedure resulted in significantly better prediction of future serum concentrations (see Figure 3) than predictions made using linear regression (see Figure 4). In contrast to most patients in the literature, who may have either normal or reduced renal function but whose renal function is stable, many patients in the above study were highly unstable and had changing renal function, to a quite significant degree, during their therapy [7].

Because the software used in that study [7,15] was specifically designed to operate in the presence of significant changes in renal function from dose to dose, it has also been useful in the analysis and management of aminoglycoside therapy for patients who must undergo periodic hemodialysis.

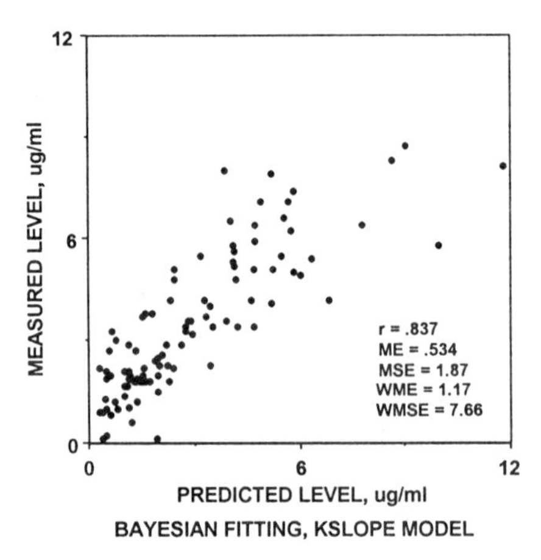

Figure 3. - *Predicted versus measured serum Gentamicin concentrations found with M.A.P. Bayesian fitting and the Kslope model. r = correlation coefficient, ME = mean error, MSE = mean squared error. WME = mean weighted error. WMSE = weighted mean squared error. See text for discussion.*

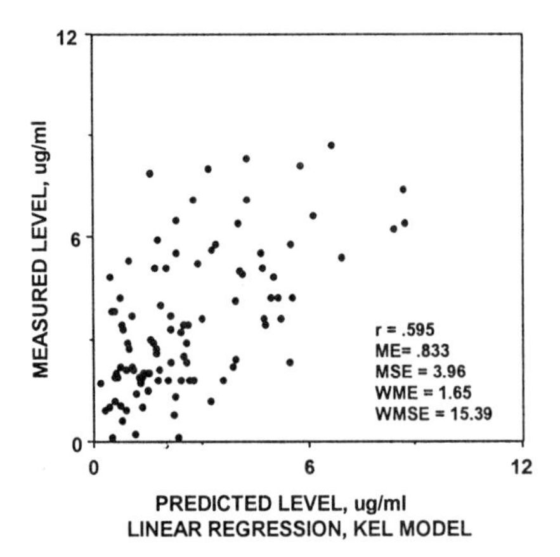

Figure 4. - *Predicted versus measured serum concentrations found with linear regression on the logarithms of the serum concentrations. Other symbols as in Figure 3.*

Managing aminoglycoside therapy can also be done quite well in patients who must be on periodic hemodialysis. The key is to get serum samples of both the drug and of serum creatinine concentrations both before and after dialysis. For current dialysis equipment, and for the aminoglycoside antibiotics, the apparent increase in creatinine clearance during dialysis is about 50 ml/min above the patient's baseline value.

The baseline value of CCr can be estimated by examining the most recent rising pair of serum creatinine values, after one dialysis for example, and just before the next one. This is why the pair of serum creatinine samples pre- and post- dialysis is useful. In addition, when a patient goes on dialysis, one can record this as giving a dose of the drug, where the amount of the dose is 0.0 mg. With this dose of zero, the infusion time can be stated as very short, 0.1 hr, for example, and the creatinine clearance can be directly entered as being 50 ml/min above that of the patient's baseline. Finally, when the patient goes off the dialysis, another dose of zero is entered at that time, and the patient's creatinine clearance is set back to the baseline value [18].

6.2 Timing the Aminoglycoside Dose and the Dialysis

Another corollary for dialysis patients is that while most of them are given their dose of drug soon after the dialysis, this can cause a problem. The serum aminoglycoside concentrations in such patients have extremely long half-times, and these patients are often the ones who have the greatest incidence of renal toxicity and otoxoxicity, because their serum concentrations stay so high for so long after each dose, even though the doses themselves are adjusted to keep the total area under the serum concentration curve at an appropriate value constrained by the desirable target peak and trough goals.

Instead of this, it may be more prudent and useful to give the dose before dialysis, about 2 or 3 hours before dialysis. In this case, one gets the desired peak value. Then the dialysis helps to mimic the renal function of a patient with more normal (or less abnormal) renal function, reducing the serum concentration more rapidly, and helps to achieve a serum concentration profile somewhat more like that of a patent with better renal function.

7. CLINICAL STUDIES OF OUTCOME AND COST

7.1. Gentamicin Therapy

Probably the best examination to date of the utility of the MAP Bayesian approach to individualize drug dosage regimens for patients has been the work of van Lent-Evers et al [3]. They compared the model-based, target goal approach to aminoglycoside therapy with a more conventional therapeutic drug monitoring strategy. The mean peak and trough concentrations in the study group were 10.6 ± 2.9 μg/ml and 0.7 ± 0.6 μg/ml respectively versus 7.6 ± 2.2 and 1.4 ± 1.3 μg/ml respectively, both significant differences. The peaks were significantly higher and the troughs significantly lower in the study group. Overall mortality was 9 of 105 (9%) in the study group versus 18 of 127 (14%) in the control group, not a significant difference (P = 0.26). However, in those patients who had obvious infections present on admission, mortality was only 1 of 48 in the study group versus 9 of 62 in the control group, a significant difference (p = 0.023). In addition, nephrotoxicity was only 2.9% in the study group versus 13.4% in the control group.

While the clinical outcome was significantly improved (more effective, less toxic) with the use of this model-based, target-oriented approach to monitoring and dosage individualization, it was interesting to see that the hospital stay was also significantly reduced, from 26.3 ± 2.9 days overall in the control group to 20.0 ± 1.4 days in the study group (p = 0.045). For patients with infections present on admission, the stay was similarly reduced, from 18.0 ± 1.4 days in the control group to 12.6 ± 0.8 days in the study group. Thus in both patient groups, those with and also without clearcut infections on admission, hospital stay was reduced by about 6 days with the use of this approach to serum concentration monitoring and model–based dosage individualization.

Further, despite the added effort and cost to implement this therapeutic approach, the overall cost per patient was reduced from 16,882 ± 17,721 Dutch florins in the control group to 13,125 ± 9,267, a significant difference ($p < 0.05$). In the patients with infections on admission, the cost was reduced from 11,743 ± 7,437 Dutch florins to 8,883 ± 3,778 florins, an even more significant difference ($p < 0.001$). Thus in a sizeable group of patents, the model based, target oriented method of monitoring and individualizing aminoglycoside dosage regimens not only resulted in better outcomes, but also in shorter hospital stays, at a net cost savings of about $1000 per patient [3].

7.2. Amikacin Therapy

MAP Bayesian target-oriented, model-based adaptive control has been used to manage amikacin therapy in geriatric patients, often for extended periods, by Maire et al [16]. In their patients, whose renal function was often quite reduced but who were generally clinically stable, visibly better prediction (and therefore control) of serum concentrations was seen with MAP Bayesian analysis than with their unfitted population model].

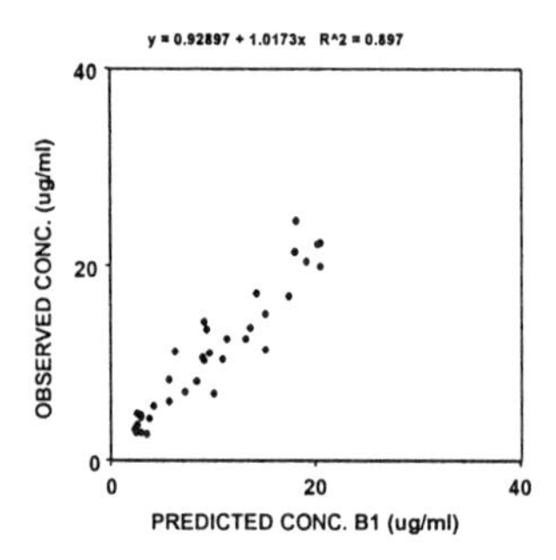

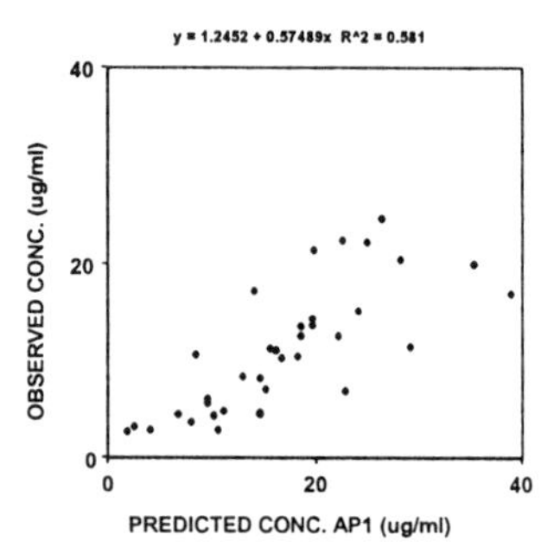

Figure 5. Left: Predicted versus measured serum Amikacin concentrations found with M.A.P. Bayesian fitting, 1 compartment Kslope model (B1). Right: Predicted versus measured serum Amikacin concentrations found with A Priori population 1 compartment Kslope model (AP1).

The results of Maire et al. [16] in these clinically more stable patients are shown in Figure 5, left. They are better than those found in the gentamicin patients with unstable renal function [7] shown in Figure 3 above. Further, Figure 5, right, shows the much poorer predictions based simply on the population model for Amikacin, without any fitting to the serum data.

7.3. Vancomycin Therapy

Vancomycin therapy was evaluated by Hurst et al [17] using a two - compartment (central plus peripheral compartment) model. Using traditional linear regression, extremely poor prediction was found, as shown in Figure 6, left. In contrast, the 2 compartment model, coupled with MAP Bayesian fitting, led to significantly better prediction of future serum concentrations than did the linear regression method, as shown in Figure 6, right.

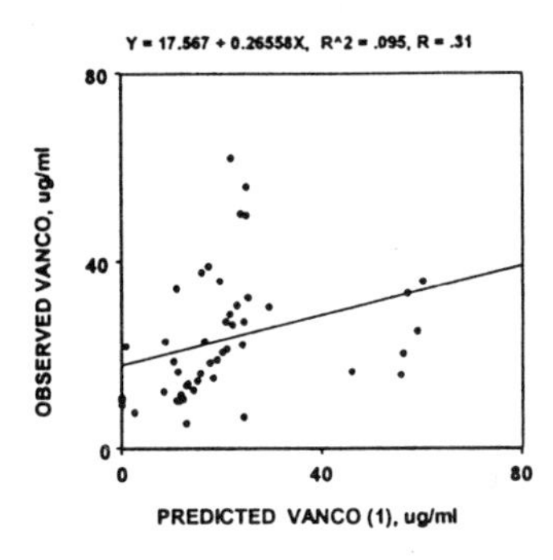

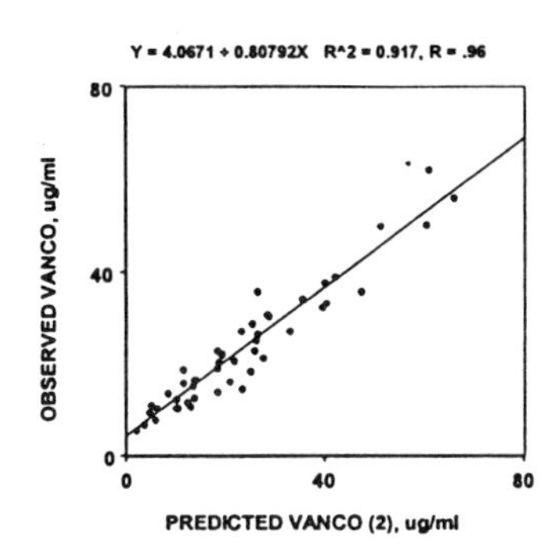

Figure 6. Left: Predicted versus measured serum Vancomycin concentrations found with Linear regression. Right: - Predicted versus measured serum Vancomycin concentrations found with a 2 compartment Kslope model and MAP Bayesian fitting).

7.4. Digoxin Therapy

The digoxin population model used in the USC*PACK MAP Bayesian software [18] is based on that described by Reuning, Sams, and Notari [4]. That two - compartment model uses both a central (serum) and a peripheral (nonserum) compartment. Computed concentrations of drug in the peripheral compartment correlate much better with inotropic effect than do serum concentrations [4,19]. The USC*PACK digoxin software [18] not only uses this model, but also develops dosage regimens to achieve desired target goals in either the central (serum concentration) compartment or in the peripheral (tissue or effect) compartment.

The following example is illustrative. A 58 year old man developed rapid atrial fibrillation at another center, after missing his usual daily dose of 0.25 mg. He was clinically titrated with several intravenous doses of digoxin, and converted to sinus rhythm. The problem then was to select a successful dosage regimen for the patient. He was placed back on his original oral maintenance dosage. After a day, atrial fibrillation recurred, showing that his digoxin requirements had changed. He again was titrated with several doses of intravenous digoxin and again converted to sinus rhythm. Again, the problem was to select a successful dosage regimen for the patient. Once again, he was placed on his original oral maintenance dosage, and once again, after about two days, atrial fibrillation recurred. For a third time he was titrated with several intravenous doses of digoxin, and for a third time he converted to sinus rhythm. A week of hospital time had been consumed during this phase of his care. The same question remained – now that sinus rhythm had been restored, what digoxin dosage regimen should this patient receive?

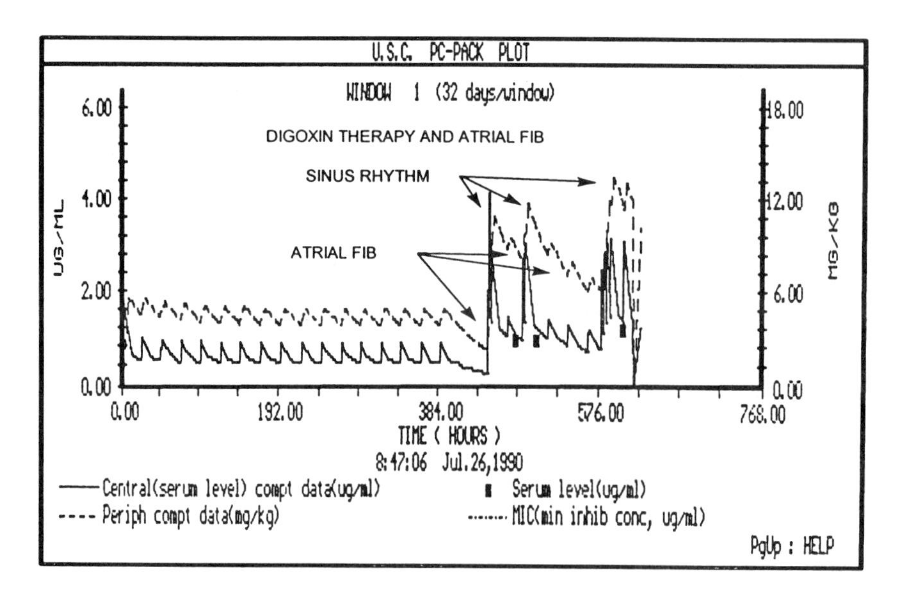

Figure 7. Screen plot of patient with atrial fibrillation who was successfully converted to sinus rhythm with IV digoxin three separate times, but who relapsed into atrial fibrillation twice when put back on his previous maintenance dose. Sinus rhythm was consistently present when peripheral body glycoside concentrations were 10-13 ug/kg (right hand scale, and not mg/kg as labeled). Selection of a therapeutic goal of 11.5 ug/kg in the peripheral compartment led to a dosage regimen of 0.5 and 0.625 mg/day. On that regimen, the patient could be discharged home in sinus rhythm and was still in sinus rhythm when seen in clinic 2 weeks later.

At this point the USC*PACK MAP Bayesian digoxin software [18] was used, in a telephone consultation, to analyze the patient's situation. The raw data of three serum concentrations, all taken during the post-distributional phase after a dose, showed almost no correlation with the patient's clinical behavior. As shown in Figure 7, he was back in atrial fibrillation when the first serum concentration of 1.0 ng/ml was obtained (the first black rectangle) and had converted again to sinus rhythm when the second and third serum concentrations of 1.0 and 1.2 ng/ml were obtained (the 2nd and 3rd black rectangles) with a lapse back into atrial fibrillation in between them. An important question is: how can it be that the patient was in atrial fibrillation at one time with a serum concentration of 1.0 ng/ml, and in sinus rhythm at another time, with exactly the same serum concentration? It is experiences of this type that have made many cardiologists feel that serum monitoring of digoxin is not useful.

Two questions need to be asked here. 1) Was the patient in a steady state at the time the serum samples were drawn? The answer clearly in no – the patient was not in a steady state at all when either of the two samples was drawn. 2) Were the serum samples obtained at the same time after the dose? Again, the answer is no. Because of this, there is no justification for using the conventional interpretation of serum concentrations which is based on these premises. Without the use of a model, the raw data of the serum concentrations is of little use.

However, when the 2 - compartment digoxin population model was fitted to the data of his various doses and these serum concentrations, the resulting fitted model was very informative, and quite good correlation was seen between the computed peripheral tissue concentrations and the patient's clinical status.

As shown in Figure 7, sinus rhythm was present in this patient whenever his peripheral compartment tissue concentrations were between 10.0 to 13.0 μg/kg. Based on this correlation, a therapeutic goal of 11.5 μg/kg was chosen for the desired peripheral compartment peak body concentration. The resulting regimen to achieve the target goal was 0.25 mg for the first day, and then averaged 0.57 mg/day.

Following this consultation, he was given 0.25 mg the first day, and then placed on a maintenance regimen of 0.5 and 0.625 mg on alternating days. On this regimen he remained in sinus rhythm. He was able to leave the hospital in sinus rhythm, and was still in sinus rhythm without evidence of toxicity when seen in the clinic 2 weeks later.

When one considers that an entire week had elapsed before the above consultation, with three successful conversions to sinus rhythm but two failures to arrive at a useful regimen to maintain that conversion,, all involving the patients, the physicians', and the ward personnel's time, effort, and money, one can see that the above pharmacokinetic consultation and recommendation was extremely cost effective.

7.5. Lidocaine Therapy

A target-oriented, model-based approach was used to manage lidocaine therapy [20]. Patients with myocardial infarcts who had arrhythmias requiring Lidocaine were retrospectively evaluated. Seventy eight patients received conventional nonpharmacokinetically oriented therapy, and an equal number of patients received pharmacokinetically designed, target-oriented, model-based infusion regimens. Of the conventional therapy patients, eight developed ventricular fibrillation, one became toxic, and 33 required additional lidocaine to control their arrhythmias. In contrast, only two patients receiving the target and model based regimens developed ventricular fibrillation, a suggestive though not significant difference ($P=0.12$), one patient became toxic, and the two who developed ventricular fibrillation were the only patients who required more lidocaine to control their arrhythmias ($p<0.001$). The pharmacokinetically designed regimens not only achieved more effective serum concentrations [19] but also suggestively reduced the incidence of ventricular fibrillation, and significantly reduced the incidence of breakthrough arrhythmias [20].

7.6. Busulfan Therapy

In a comparison with an equal number of historical control subjects by Bleyzac et el. [21], children who underwent bone marrow transplantation for various hematologic problems and malignancies had their Busulfan therapy given as a standard regimen during the process of transplantation versus having it given using MAP Bayesian adaptive control. Twenty nine patients composed each group. The patients having Bufulfan given by MAP Bayesian adaptive control had an incidence of veno-occlusive disease reduced from 24.1 to 3.4 percent ($p<0.05$). In addition, graft failure was reduced from 12.0 percent to zero percent. Furthermore, survival was increased from 65.5 to 82.8 percent. Because of this, MAP Bayesian adaptive control of the Busulfan dosage regimens in this study made visible and significant improvements in the care of these patients [21].

8. WHY WE REALLY MONITOR SERUM CONCENTRATIONS: FOR MODEL-BASED, TARGET-ORIENTED INDIVIDUALIZED DRUG THERAPY

Traditional approaches to therapeutic drug monitoring were originally designed for use only in steady state situations, and usually employed only 1 - compartment models. They developed dosage regimens only for such steady state situations, and were oriented to keeping serum concentrations within a general therapeutic range rather than to achieving a specific target goal for a specific individual patient. Such approaches made it impossible to deal with patients in some of their most important clinical moments, as, for example, during changing renal function or dialysis, or when certain "golden clinical moments" must be captured, and a dosage regimen developed to achieve and maintain a desired target goal immediately, without waiting for a steady state, as in the case of the above patient receiving digoxin.

The above patient on digoxin shows how truly individualized drug therapy begins with clinical selection of a specific therapeutic goal for each patient, based on that individual patient's need for the drug. One then should achieve that goal with the greatest possible precision, without any zone of indifference about it. The approach to that patient was highly cost-effective, when compared to the fact that an entire week of hospital time was spent in the previous attempts at dosage adjustment without the aid of a model - based, target - oriented method.

That patient's case also emphasizes the fact that one does not use serum concentrations simply to see whether or not they are in some general "therapeutic range", nor even to correlate them with the patient's clinical behavior, although that is often possible, but significantly not so in this patient. This patient clearly shows that the real reason for monitoring serum concentrations is rather to find out how each patient actually handles the drug, how the drug (and its model) really behaves in each individual patient, especially in non-steady-state situations, and to correlate the behavior of each patient's fitted model with his/her own clinical behavior. Only then can one optimally evaluate each patient's clinical sensitivity to, and specific need for, a drug. MAP Bayesian adaptive control, in the context of model based, target-oriented individualized drug therapy, brings a precision and capability to drug dosage which is not possible with older obsolete approaches based on linear regression or simply on raw data of the serum concentrations alone.

8.1 Optimal monitoring strategies

The issue of what are the best times to obtain serum concentrations is also important. Often samples are obtained at the trough, just before the next dose, after distribution is complete, and the errors in recording the time at which the previous dose was given and the time at which the sample was drawn make the least difference in the value of the measured serum concentration. It is not generally realized that because of this, one has deliberately separated the time containing the least information concerning the actual behavior of the drug. There is minimal information about the processes of absorption, distribution, elimination, and their relationship to the actual time course and profile of the serum concentrations, and to the time course of the drug effects.

It is often much better to obtain serum samples when they contain the most information about the various processes described above. One can use a model, and can make small variations in the model parameter values, and note their effect upon the profile of the serum concentrations. At what time do changes in the model parameter values cause the greatest changes in the serum concentration profile? These are the times when the serum concentrations are maximally sensitive to changes in the various parameter values. These are the times when getting the serum samples lets one best :see through" the many clinical uncertainties, and best understand the behavior of the drug, by permitting the most precise parameter estimates to be made for the chosen model. These times can be calculated using the well-known D-optimal sampling strategies, based on the work of D'Argenio, for example [37].

These strategies can easily be employed in routine clinical care. For the aminoglycosides, for example, one can start by getting a peak sample, out of the opposite arm at the end of the intravenous infusion. It is then useful to wait, whatever the dose interval is, until about 21 hours into the regimen, when the patient's creatinine clearance at least 40 ml.min/1.73 M^2. Because of this, it is easy to center the patient's aminoglycoside doses about three hours after routine morning blood drawing time. In that way it is east to make the routine blood sample at that time be quite close to a D-optimal sample. In general, considerations of D-optimal sampling strategies also suggest that it is useful to obtain at least one sample for each parameter to be fitted in the patient' model when doing therapeutic drug monitoring.

8.2 More general comments

Further, we need to monitor drug therapy better in general. It is distressing to see patients with multidrug resistant TB, for example, dosed without such monitoring. Since many patients with multidrug resistant TB absorb the drugs poorly, it wastes the patient's lives by treating them with an unmonitored regimen and waiting to see if their sputum smears and cultures eventually become negative. It is much more useful to know early in the course of therapy whether the serum concentrations achieved on a given regimen are likely to be effective.

We spend a great deal of money of expensive treatments for patients with cancer and AIDS, and we follow the viral load, and the measures of hematological toxicity. But we are not yet optimizing this process, and we should. We treat cancer patients with methotrexate to a desired area under the serum curve (AUC), but we usually do not ask if that AUC is really optimal for each individual patient. Monitoring serum concentrations and determining their relationship to the hematocrit, leukocyte count, and platelet count, for example, would permit therapy to be optimized within the constraints of tolerable measures of toxicity. There is a great deal to be done in this area!

9. SPECIAL CASES: ENTERING INITIAL CONDITIONS - CHANGING POPULATION MODELS DURING THE FITTING PROCEDURE.

Most pharmacokinetic analyses deal with patients (and their pharmacokinetic models), who have had stable values for their various parameters such as volume of distribution, rate constants, clearances, etc.. However, this is not always so, even though one can express a rate constant as an intercept plus a slope times a descriptor of elimination such as creatinine clearance or cardiac index [34], so that renal function or cardiac index can change from dose to dose during therapy, and the patient's drug model can keep up with these changes as they take place.

Probably the most serious problem in analyzing pharmacokinetic data in patients is caused by sudden significant changes in a patient's volume of distribution (Vd) of the central (serum concentration) compartment, without any change in any currently known clinical descriptor.. It is generally known, for example, that patients in an ICU setting have larger values for the Vd of gentamicin and other aminoglycosides than do general medical patients. Indeed, young very healthy people who suddenly require an aminoglycoside for a perforated or gangrenous appendix often have even smaller values for Vd [18].

9.1. An Aminoglycoside Patient with a sudden Change in Clinical Status and Volume of Distribution

An interesting 54 year old woman in Christchurch, New Zealand, was seen through the courtesy of Dr. Evan Begg in the fall of 1991. She was 69 in tall, weighed 80 kg, and her serum creatinine on admission was 0.7 mg/dL. She had a pyelonephritis, and was receiving tobramycin 80 mg approximately every 8 hours. She had a measured peak serum concentration of 4.6 and a trough of 0.4 ug/ml respectively, and had been felt by all to be having a satisfactory clinical response. During this time, her Vd was 0.18 l/kg, based on those two serum samples. However, on about the 6th day, she suddenly and most unexpectedly relapsed and went into clear-cut septic shock.

Following her surprising relapse on therapy, she was aggressively treated with much larger doses, 300 mg every 12 hours during this time. Her serum concentrations rose to peaks of 10.1 μg/ml. During this period of sudden septic shock, her serum creatinine also rose, from 0.7 to 3.7, and her estimated CCr fell to 18 ml/min/1.73m^2. After about another 10 days she improved. At that time, her serum tobramycin concentrations rose to a peak of 16, and it was necessary to sharply reduce the dose to 140 mg about every 12 to 24 hours. Her serum creatinine fell to 1.1 to 1.3 mg/dL, and her CCr rose to 57 ml/min/1.73m^2.

It was simply not possible to get a good MAP Bayesian fit to all the serum data over the entire time period. Most samples were obtained during her second, sickest phase, and they dominated the fit. The ones at the beginning, prior to the sepsis, and at the end, after her improvement, were not at all well fitted.

Because of this, the data was divided into three parts - an initial one before her relapse into sepsis, a second one when she was septic, and a third one following improvement, but before it was felt safe to discontinue therapy. Each data set was fitted separately, using the USC*PACK programs [18].

During the first data set, the first 6 days, when her clinical behavior was that of a general medical patient, not gravely ill, her Vd was 0.18 L/kg as described above. The problem then was to pass on the ending values of the serum and peripheral compartment concentrations as initial conditions for the fitting process for the second data set. This was done, using that feature of the USC*PACK clinical software [18], which was developed specifically for this purpose.

A major change in her Vd was then seen when fitting the data obtained during the second, septic, phase. The Vd rose from 0.18L/kg in the previous phase to 0.51 L/kg, and the Kslope, the increment of elimination rate constant per unit of CCr, fell to zero. However, the Kcp, the rate constant from serum to peripheral compartment, rose to 0.255 hr^{-1}, suggesting that she was "third-spacing" the tobramycin somewhere. The ending concentrations in the central (serum) compartment for this data set were 2.09 μg/ml, and for the peripheral compartment were a very high 44.1 μg/kg.

These ending values were then passed on to the third part of her data set, that of recovery. During this time the serum peaks were 16 and 12 μg/ml, and the dose was reduced to 140 mg every 12-24 hours. Her Vd during this third phase, that of recovery, when she was no longer seriously ill, had fallen greatly to 0.15 L/kg, close to her previous initial value as a general medical patient.

The ability to enter stated initial conditions permitted changing population models during the overall fitting procedure, and allowed intelligent analysis of this patient's data, especially as quite significant concentrations were present not only in the central (serum) compartment, but also in the peripheral compartment, during the transition from the patient's second to the third, recovery, phase.

At the Cleveland Clinic, Drs Marcus Haug and Peter Slugg [22] have spoken of "Vd collapse", when the Vd would drop from a larger to a smaller value. They showed that this change was a clinical indicator of incipient recovery of the patient. The present patient not only demonstrated such Vd collapse later on, as she got better, but also its opposite, Vd expansion, as she made the earlier transition from being a general medical patient with a pyelonephritis to a seriously ill ICU patient with life-threatening septic shock.

We see, therefore, that not only do different populations of aminoglycoside patients have different values of Vd, but that each individual patient goes through these transitions, as demonstrated by this patient. The analysis of this patient's data was greatly facilitated, and indeed was only possible, using the MAP Bayesian approach, by breaking the dosage history up into several parts. Each part was then analyzed, and the ending concentrations from one part were passed on to the next data set as initial conditions or concentrations of drug present prior to the first dose given in the next data set, with the appropriate population model, if needed, as well.

9.2. A Patient on Digoxin when Quinidine was Added.

Another example of the utility of using initial conditions, provided through the courtesy of Dr. Marcus Haug, is that of a 72 year old woman, 4 ft 10 in tall, who weighed only 75 pounds. She was admitted to another hospital with congestive heart failure and atrial fibrillation. Her estimated creatinine clearance on admission was 38 $ml/min/1.73m^2$, falling to 23 after admission. She had been receiving 0.25 mg of digoxin daily. This was continued after admission to the hospital.

A serum digoxin concentration was 1.8 ng/ml on admission. Following this, her serum creatinine rose to 1.8 mg/dL, and her digoxin concentration after 5 days rose to 2.5 ng/ml. Her digoxin was stopped, though she had no clinical manifestations of toxicity. The next day her serum concentration had fallen to 2.0, and the next day it was down to 1.4 ng/ml, as shown in Figure 8.

At this point her ventricular rate with her atrial fibrillation had become rapid again, and she was restarted on her digoxin, again at 0.25 mg/day, to control the rate. However, quinidine was also added to her regimen at the same time. Her creatinine clearance at that time was 22 ml/min/1.73m^2.

Five days later her serum digoxin concentration was measured and found to be 7.6 ng/ml for a trough, and 10.0 ng/ml two hours after the next dose was given.

What was going on here? She again had no clinical evidence of toxicity. Was all of this due to the digoxin - quinidine reaction? Was it a problem of digoxin - like material appearing in the assay as a result of her poor renal function? Was there something else in addition?

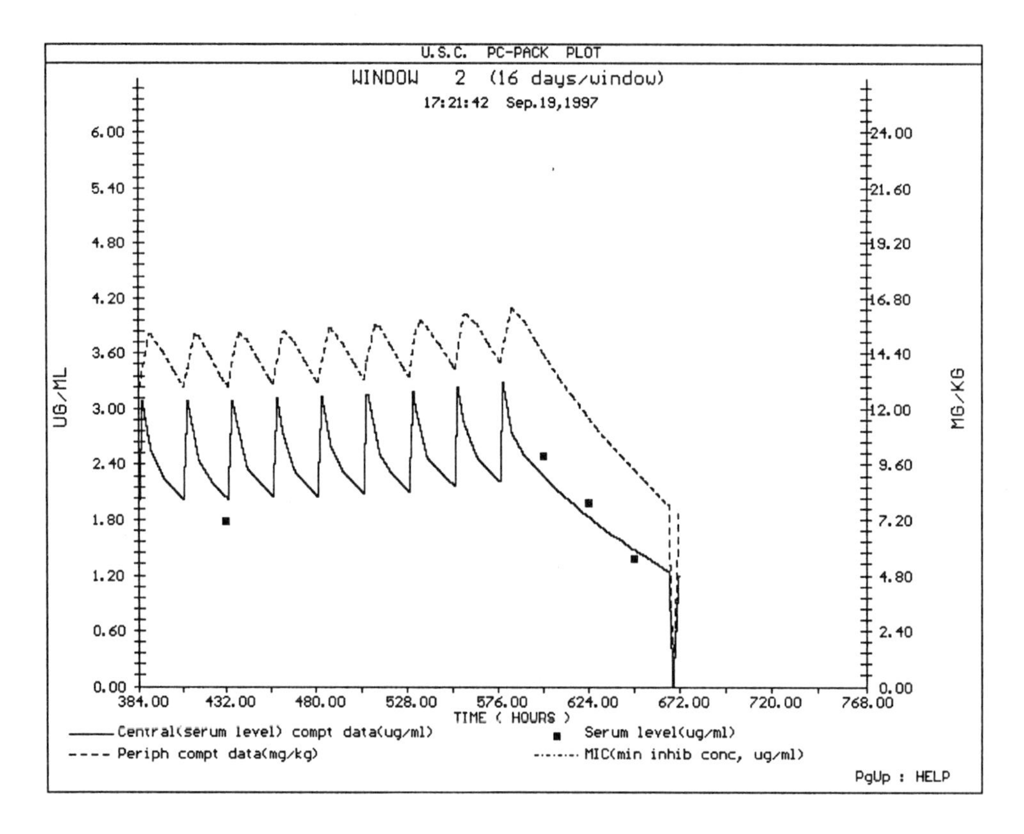

Figure 8. Plot of serum and peripheral compartment digoxin concentrations of patient admitted receiving digoxin. She was receiving 0.25 mg of digoxin daily, ands weighed only 75 lb. Solid rectangles - measured serum concentrations. Solid line and left hand scale - digoxin serum concentrations. Dashed line and right hand scale - digoxin peripheral (nonserum) compartment concentrations. Using the MAP Bayesian approach, the population model for digoxin was fitted to the patient's data of doses and serum concentrations. Serum concentrations rose as her renal function worsened. Digoxin was stopped after the serum concentration of 2.5 ng/ml was obtained, after which her serum concentrations fell to 1.4, and, in the fitted model, finally to 1.19ng/ml at the end of this plot, when digoxin was begun again, but along with quinidine.

The clinical problem was analyzed as follows. First, her original dosage history on digoxin alone was fitted to her serum concentrations, using the 2-compartment population model for digoxin made from the work of Reuning, Sams, and Notari [4,18]. This included the three measured serum concentrations. At the end of that part of her history, just before her first dose of quinidine was added, her fitted and predicted central compartment (serum) concentration was 1.19 ng/ml, and her peripheral (nonserum) compartment concentration was 7.58 μg/kg.

These two ending values from this first phase of her analysis were passed on as initial concentrations of drug already present in those compartments of her pharmacokinetic model at the time her digoxin was restarted, but now with quinidine as well.

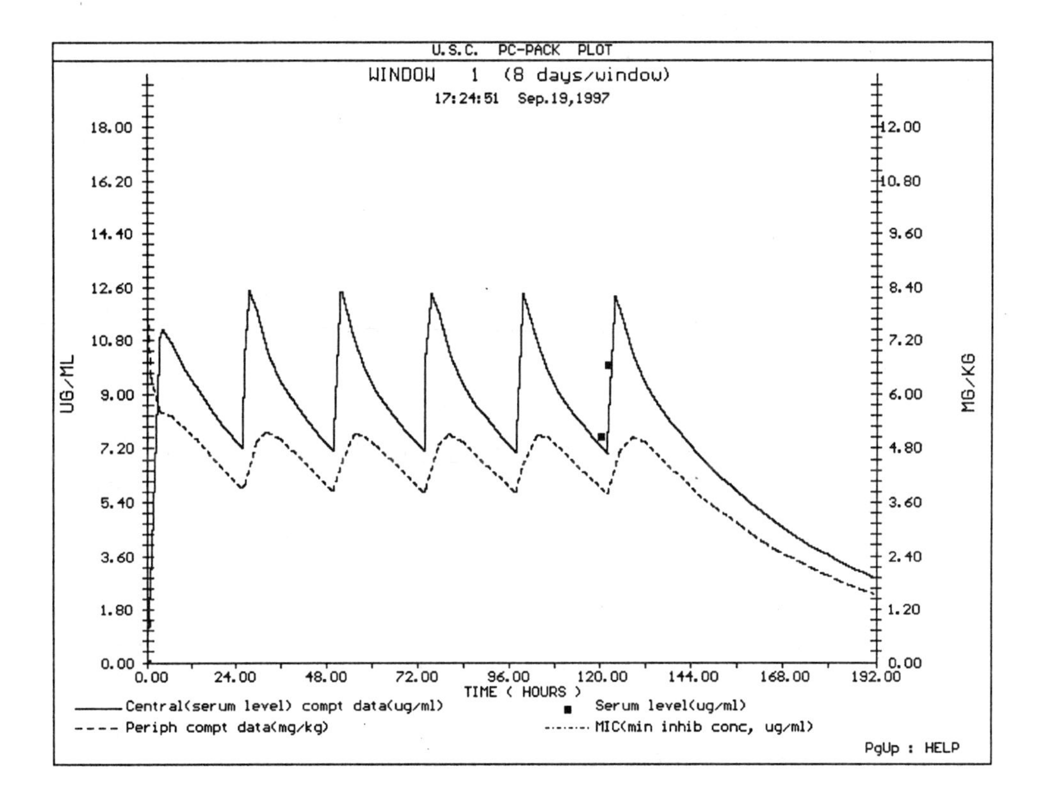

Figure 9. Plot of serum and peripheral compartment digoxin concentrations of patient admitted receiving digoxin. In this plot, digoxin was restarted at 0.25 mg/day, but along with quinidine. Solid rectangles - measured serum concentrations. Solid line and left hand scale - digoxin serum concentrations predicted using the population model for digoxin with quinidine [18]. Dashed line and right hand scale - predicted digoxin peripheral (nonserum) compartment concentrations. This plot begins with initial conditions equal to the final concentrations found at the end of the plot in Figure 10.

A population model for digoxin with quinidine [18] was now used. This model was not fitted to her subsequent serum concentrations, but merely used to supply predictions of those high measured concentrations. If the prediction was good, the interpretation would be that the interaction would quantitatively account for the measured concentrations found. If not, then another explanation would have to be considered.

As shown in Figure 9, the predicted concentration of 7.2 ng/ml closely corresponded to the measured one of 7.6 ng/ml. In addition, the measured concentration of 10.0 ng/ml was predicted as 9.9 ng/ml. Because of these good predictions, it was felt that the digoxin-quinidine interaction adequately explained the measured concentrations found, and that no other alternative explanation was needed.

This is a good example of how pharmacokinetic analyses can be used to evaluate clinical experiences with drugs, and can provide strong evidence for or against a particular clinical question or issue, much more than a clinical opinion made only judgmentally, unsupported by the quantitative evidence which can be obtained from such a model.

The use of initial conditions was the key to being able to change from one population model to another in the middle of this patient's clinical history. In the same way, one can make the transition from regular theophylline to a long-acting preparation, or from conventional carbamazepine and valproate to their sustained-release formulations, for example.

With the use of initial conditions, one can thus follow the patient as he or she goes from one clinical situation to another, passing on the information from one data set to another.

10. LINKED PHARMACODYNAMIC MODELS: DIFFUSION OF DRUGS INTO ENDOCARDIAL VEGETATIONS, AND MODELS OF POSTANTIBIOTIC EFFECT AND BACTERIAL GROWTH AND KILL

In this section we will describe the linkage of nonlinear pharmacodynamic models to the basic linear pharmacokinetic model, and show some applications in clinical software describing drug diffusion into endocardial vegetations, the simulation of a postantibiotic effect, and the modeling of bacterial growth in the absence of a drug and its kill by an antibiotic.

10.1 MODELING DRUG DIFFUSION

A problem in the treatment of patients with infectious endocarditis is that it is difficult to estimate whether or not the drug is able to kill the organisms all the way into the center of a vegetation. Because of this, a diffusion model was made of this process.

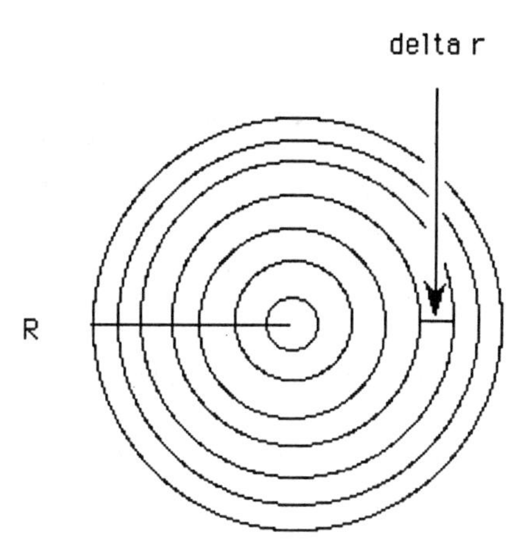

Figure 10. Diagram of the concentric layers of the spherical model of the endocardial vegetation.

A spherical shape was assumed for the vegetation, as shown in Figure 10, and it was modeled having several concentric layers, with diffusion taking place from layer to layer, delta r, as shown in Figure 10. The sphere was assumed to be homogeneous, with equal diffusion in all directions, and as having a constant coefficient of diffusion throughout. The diffusion was assumed to be dependent on the concentration of drug in the surrounding medium, such as the serum concentration, and its time course. The diameter of endocardial vegetations can be measured by transesophageal echocardiography.

The following equation was used to model the diffusion:

$$\frac{\partial C}{\partial t} = \frac{1}{r^2} \times \frac{\partial}{\partial r} \left[D \times r^2 \times \frac{\partial C}{\partial r} \right] \tag{3}$$

where C represents the concentration in the sphere at time t, at a distance r from the center of the sphere, and D represents the coefficient of diffusion in the sphere, and x indicates multiplication.

When D is assumed constant, the equation becomes

$$\frac{\partial C}{\partial t} = D \times \left[\frac{\partial^2 C}{\partial r^2} + \frac{2}{r} \times \frac{\partial C}{\partial r} \right] \tag{4}$$

The vegetation is assumed to be continuously immersed in the surrounding medium, and the drug concentration in that medium is assumed to attain a value which results in equilibrium with the very outer layer of the sphere. The medium then undergoes the changes in concentration with time that constitute the serum level time course. This time course is thus presented as the input to the spherical model [23].

The diffusion coefficient found by Bayer, Crowell, et al. for aminoglycosides in experimental endocarditis [24,25] was used. The model has become part of the USC*PACK clinical programs for individualizing drug dosage regimens [18].

The model can also be used to simulate behavior inside an abscess, and, by appropriate choice of sphere diameter and diffusion coefficient, to simulate the post-antibiotic effect of a certain desired duration.

10.1.1 Examples: Simulated Endocardial Vegetations of Various Diameters

Suppose one were to develop an amikacin dosage regimen for a hypothetical 65 year old man, 70 in tall, weighing 70 kg, with a serum creatinine of 1.0 mg/dL.

Let us assume that he has a vegetation seen by echocardiography on his aortic valve that might be either 0.5, 1.0, or 2.0 cm in diameter. We wish to examine the ability of an amikacin regimen designed to achieve serum peaks of 45 μg/ml and troughs of approximately 5.0 μg/ml to reach effective concentrations within the vegetation in these three cases.

Let us apply the findings of Bayer et el [24,25] to compute the time course of probable amikacin concentrations in the center of these three vegetations of different diameters, to examine their possible ability to kill an organism having an estimated minimum inhibitory concentration (MIC) of 8.0 μg/ml, for example.

Using the Amikacin program in the USC*PACK collection [18], let us estimate, from the patient's age, gender, height, weight, and serum creatinine concentration, that his creatinine clearance (CCr) is about 69 ml/min/1.73M^2. This method of estimating CCr is described elsewhere [15].

We enter the target goal for the peak serum concentration of 45 μg/ml and an initial trough concentration of about 5.0 μg/ml. The ideal dose interval to achieve that peak and trough exactly, adjusted for the patient's renal function, employing a planned duration of the IV infusion of 0.5 hr, turns out to be 10.231 hrs.

Let us approximate this in a practical manner by choosing a dose interval of 12 hrs. The dosage regimen to achieve the peak goal with such a dose interval is, when revised to practical amounts, 850 mg for the first dose, followed by 750 mg every 12 hrs thereafter.

On this regimen, predicted serum concentrations are 43 μg/ml for the peak and 3.2 μg/ml for the trough. The peak is 542 % of the stated MIC, and serum concentrations are predicted to be at least the MIC for 66 % of each dose interval. The AUC/MIC ratio for the first 24 hours is 48.8.

The plot of these predicted serum concentrations is shown in Figure 11.

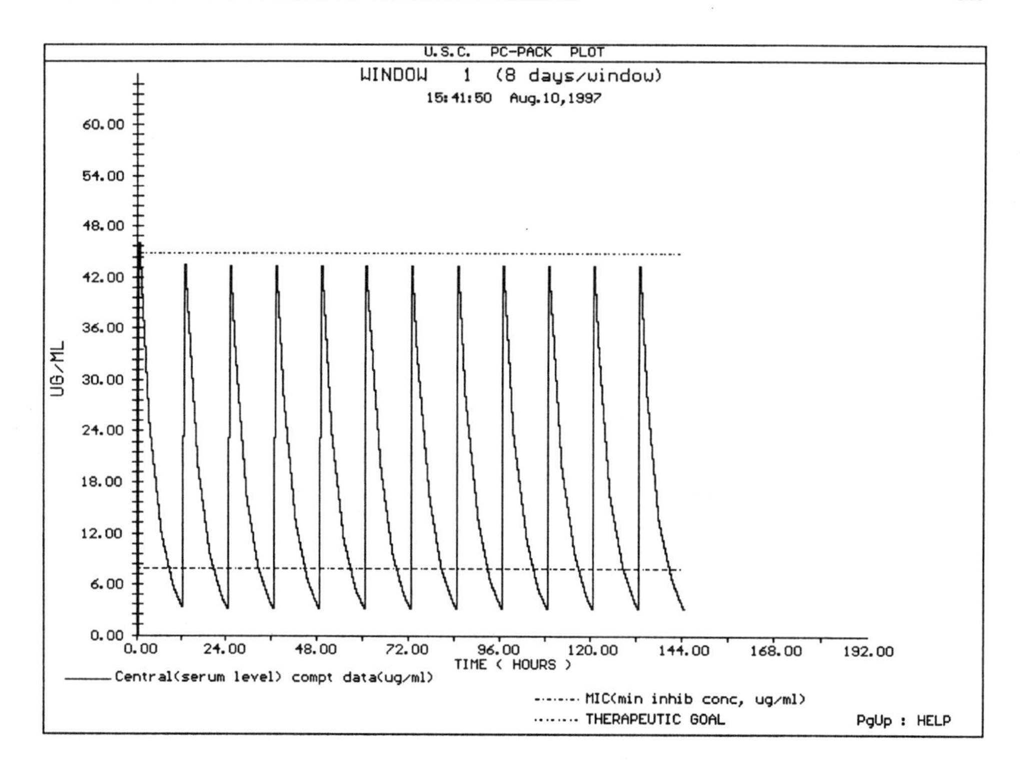

Figure 11. Predicted time course (the first 6 days) of serum Amikacin concentrations for the patient described. Upper horizontal dotted line - initial stated target peak serum concentration of 45 ug/ml. Lower horizontal dashed line - the estimated organism MIC of 8.0 ug/ml.

The important question now is whether or not this predicted serum concentration profile will result in adequate penetration of the vegetation in each of the three cases, and whether or not the regimen will kill effectively there, as well as in the central (serum level) compartment.

Figure 12 now shows the predicted amikacin concentrations in the center of the simulated vegetation having a diameter of 0.5 cm.

As shown, concentrations rise rapidly above the MIC and stay there, suggesting that the above regimen should probably be able to kill organisms having an MIC of about 8.0 μg/ml fairly promptly in the center of the vegetation. The time lag of concentrations in the center of the sphere is modest, about 3-4 hrs, behind the serum concentrations.

On the other hand, if the vegetation were 1.0 cm in diameter instead, the drug would take about 12 hours to diffuse to the center and reach the MIC, and the rise and fall of drug concentrations would be much more damped, as shown in Figure 13.

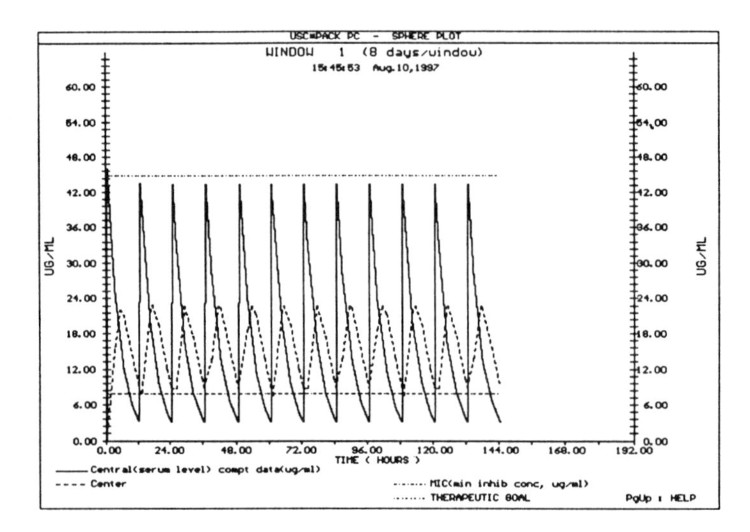

Figure 12. Predicted time course (the first 6 days) of Amikacin concentrations (dashed line) in the center of a simulated endocardial vegetation of 0.5 cm. Solid line - Predicted serum concentrations, and other lines and symbols as in Figure 11. The predicted endocardial concentrations rise promptly, and are consistently above the estimated MIC of 8.0 ug/ml.

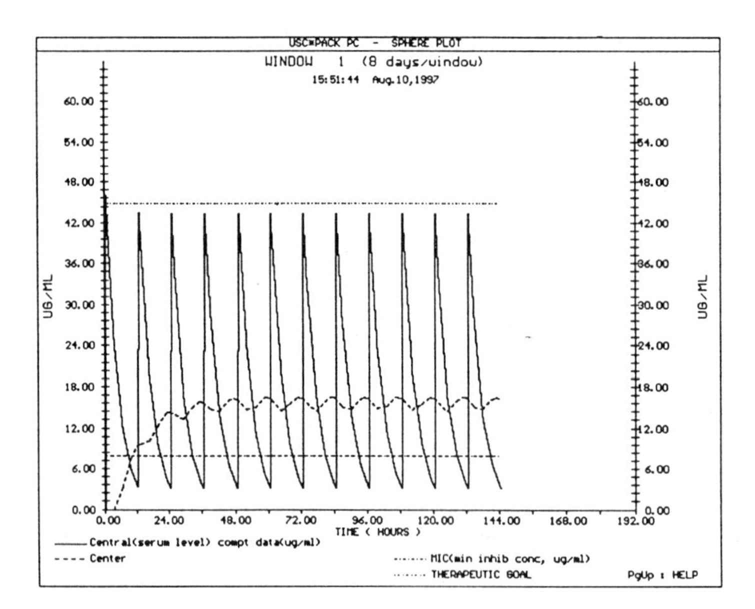

Figure 13. Predicted time course (the first 6 days) of Amikacin concentrations (dashed line) in the center of a simulated endocardial vegetation of 1.0 cm. Solid line - Predicted serum concentrations, and other lines and symbols as in Figure 11. Predicted endocardial concentrations rise more slowly, are more damped, with smaller oscillations from peak to trough, but once the estimated MIC is reached, are consistently above 8.0 ug/ml.

Further, if the diameter of the vegetation were 2.0 cm, all this would take still longer, and the time course of the computed concentrations in the center would be as shown in Figure 14. The drug would take considerably longer, about 48 hours, to reach the MIC in the center of the vegetation, and significant growth of organisms might well take place before that. For every doubling of the diameter of the sphere, the equations show that it will take 4 times as long (the square of the ratio of the diameters) to reach an equal concentration in the center of the sphere.

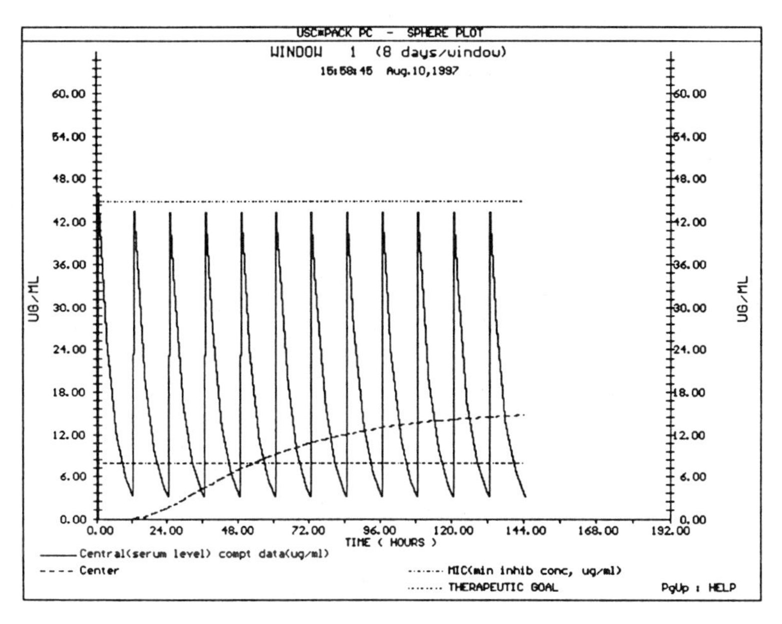

Figure 14. Predicted time course (the first 6 days) of Amikacin concentrations (dashed line) in the center of a simulated endocardial vegetation of 2.0 cm. Solid line - Predicted serum concentrations, and other lines and symbols as in Figure 11. The predicted endocardial concentrations rise much more slowly and are much more damped, with essentially no oscillations from peak to trough. Once the estimated MIC is reached, the concentrations are consistently above 8.0 ug/ml, but two full days are required before the MIC is reached.

10.1.2 Another Example: Simulating an Abscess or a Post-Antibiotic Effect.

Another use for such a spherical model might, of course, be an abscess. If we could know the diffusion coefficient into abscesses of different sizes, we could similarly begin to model and compute the concentrations of drug diffusing into the abscess. There might well be different diffusion coefficients through the wall, into the bulk of the abscess, and into the center. All this is theoretically capable of being modeled. Effects of oxygen tension and pH upon bacterial growth and response to drugs can also be determined by careful needle aspiration done at carefully documented times just prior to incision and drainage of them, with careful cultures and determination of pH, pO2, numbers of viable organisms, and rates of growth and kill from different parts of the abscesses. In this way, useful models of events taking place within an abscess can be made.One can see visually, in Figure 14, for example, why abscesses much over 1 cm in diameter usually need to be incised and drained, as the diffusion into abscesses is very likely to be poorer that that seen here for endocardial vegetations.

Figure 15 shows computed drug concentrations in the center of a small sphere simulating a microorganism having a diameter of 0.1 micron, 3 simulated layers of diffusion, and a diffusion coefficient of 1.5 x 10^{-14}. This particular sphere has the property that in its center, the concentrations of drug fall below the MIC about 6 hrs after the serum levels do, thus simulating (without making any suggestions or conclusions about mechanism of action) a post-antibiotic effect of about 6.0 hrs, as the organisms will not begin to grow again for about 6 hrs after the serum concentrations fall below the MIC.

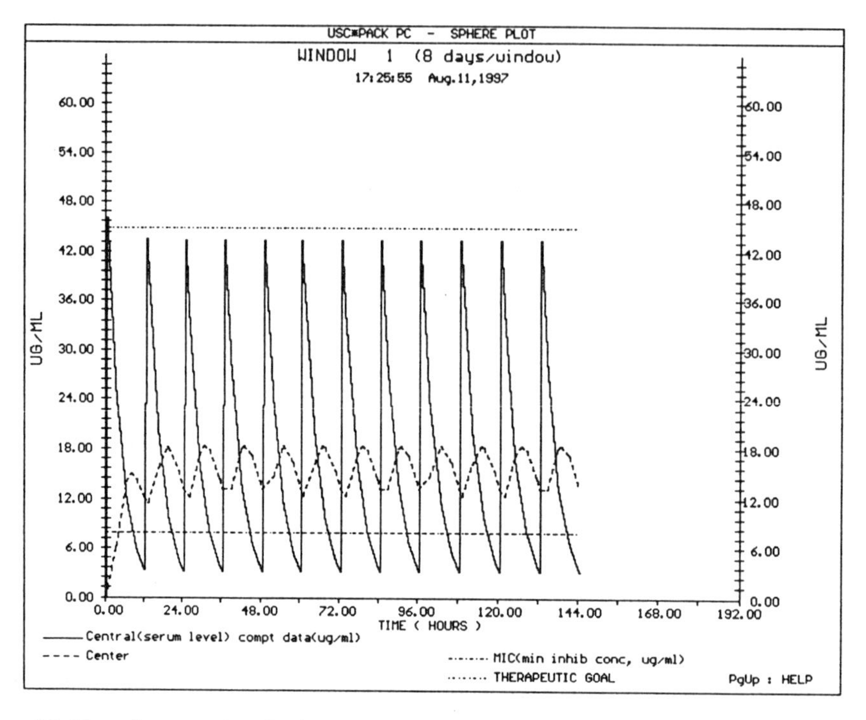

Figure 15. Plot of computed amikacin concentrations (the first 6 days) in the center of a simulated microorganism. Diffusion coefficients in the very small sphere diffusion model are adjusted so that concentrations in the center of the organism lag behind the serum concentrations and, if they fall below the MIC, would do so approximately 6 hrs after the serum concentrations do, thus simulating a post-antibiotic effect of about 6 hrs.

The effects of these computed concentrations in the center of these spheres will be discussed below in the section on modeling bacterial growth and kill. We see here that the process of diffusion into and out of spherical porous objects such as endocardial vegetations and small microorganisms can be described with reasonably simple models. The equations describing this process are the same as those for release of drug from a sustained-release preparation formulation.

11. MODELING BACTERIAL GROWTH AND KILL: CLINICAL APPLICATIONS

11.1 General Considerations

Let us assume that an organism is in its logarithmic phase of growth in the absence of any antibiotic. It will have a rate constant for this growth, and a doubling time. The killing effect of the antibiotic can be modeled as a Michaelis-Menten or Hill model. The model generates a rate constant for this effect. The rate of growth or kill of an organism depends upon the difference between these two rate constants. The killing effect will be determined by the Emax, representing the maximum possible rate constant for kill, the EC50, the concentration at which the effect is half maximal, and the time course of the serum concentrations achieved with the dosage regimen the patient is given. Both the growth rate constant and the Emax can be found from available data in the literature for various organisms. The general growth versus kill equation is

$$\frac{dB}{dt} = (K_g - K_k) \times B \qquad (5)$$

and

$$Kk = \frac{(E_{max} \times C_t^n)}{(EC_{50}^n + C_t^n)} \qquad (6)$$

where B is the number of organisms (set to 1 relative unit at the start of therapy), K_g is the rate constant for growth, K_k is the rate constant for killing, Emax is the maximum possible effect (rate of killing), EC_{50} is the concentration at which the killing rate is half maximal, n is the Hill or sigmoidicity coefficient, and C_t is the concentration at the site of the effect (serum, peripheral compartment, effect compartment, or in the center of a spherical model of diffusion), at any time t, and x indicates multiplication.

The EC_{50} can be found from the measured (or clinically estimated) minimum inhibitory concentration (MIC) of the organism. This relationship was developed by Zhi et al. [26], and also independently by Schumitzky [27]. The MIC is modeled as a rate of kill that is equal to but opposite in direction to the rate constant for growth. The MIC thus offsets growth, and at the MIC there is neither net growth nor decrease in the number of organisms. At the MIC,

$$\frac{dB}{dt} = 0, \quad \text{and } K_k = - K_g \qquad (7)$$

and

$$MIC = \left(\frac{K_g \times EC_{50}^n}{E_{max} - K_g}\right)^{1/n} \qquad (8)$$

In this way, the EC_{50} can be found from the MIC, and vice versa.

The input to this effect model can be from either the central or the peripheral compartment concentrations of a pharmacokinetic model, or from the center (or any other layer) of one of the spherical models of diffusion. The sphere may represent an endocardial vegetation, an abscess, or even a small microorganism. In the latter case, one can adjust the sphere diameter and the diffusion coefficient so that the concentrations in the center of the small sphere lag behind the serum concentrations and cross below the MIC about 6 hours after the serum concentrations do, to simulate a post-antibiotic effect of about 6 hours, for example. The effect relationship was modeled by Bouvier D'Ivoire and Maire [28], from data obtained from Craig and Ebert [29], for pseudomonas and an aminoglycoside.

Let us first examine these effect models with relationship to the dosage regimen of amikacin developed in the previous section on analyzing concentrations in spherical diffusion models. In that section we had considered a hypothetical 65 year old man, 70 in tall, weighing 70 kg, having a serum creatinine of 1.0 mg/dL. We also assumed that he had a vegetation on his aortic valve, seen by echocardiography, that might be either 0.5, 1.0, or 2.0 cm in diameter, and we wanted to examine the ability of an amikacin regimen designed to achieve serum peaks of 45 and troughs of about 5 μg/ml to reach effective concentrations within the vegetation in these three cases. We applied the findings of Bayer et al [24,25] to predict the time course of amikacin concentrations in the center of the above three different vegetations. Let us now examine the results of these analyses.

The patient's dosage regimen consisted of an initial dose of 850 mg of amikacin followed by 750 mg every 12 hours thereafter. On that regimen, predicted serum concentrations were 43 μg/ml for the peak and 3.2 for the trough, possibly a bit low, as the MIC of the organism was stated to be 8.0 μg/ml. The computed concentration of amikacin in the various vegetations was shown in the previous paper. Figure 16 is a plot not only of the predicted time course (the first six days) of serum amikacin concentrations for the patient described here, but also of its ability to kill microorganisms using the model made by Bouvier D'Ivoire and Maire [28], based on the data of Craig and Ebert [29]. In Figure 16 there is no assumption of any post-antibiotic effect. The serum concentration profile alone is presented as the input to the bactericidal effect model.

The model always assumes an initial inoculum of one relative unit of organisms. The scale of the relative number of organisms is shown on the right side of Figure 16, while the scale of the serum concentrations is on the left. As shown in the figure, the serum concentration profile resulting from that regimen appears to be able to kill such an organism well in this particular patient. As the serum concentrations fall below the MIC with the first dose, however, the organisms begin to grow again, but the second dose kills them again, with slight regrowth once again toward the end of that dose interval. The third dose reduces the number of organisms essentially to zero. Use of this effect model suggests that such a serum concentration profile should be effective in killing an organism having an MIC of 8.0 μg/ml, even though the serum concentrations are below the MIC about one third of the time, as the high peaks are effective in the killing.

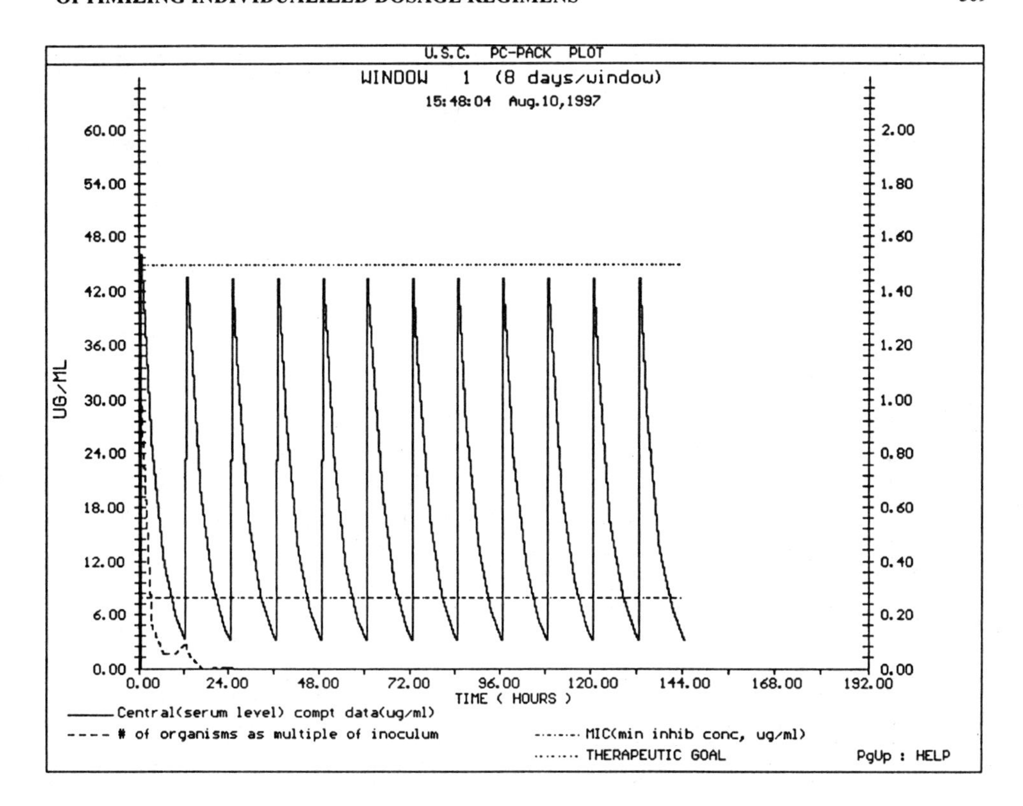

Figure 16. Predicted Killing effect of the regimen. Input from the central (serum) compartment profile of serum concentrations. The regimen is likely to kill well for a bloodstream infection (sepsis). Solid line and left hand scale - serum concentrations. Dashed line and right hand scale - relative numbers of organisms, with 1.0 relative unit present at the start of therapy. Upper horizontal dotted and dashed line - original peak serum goal of therapy. Lower horizontal dashed line: the patient's MIC of 8.0 ug/ml.

Figure 17 now shows the computed results in the center of the simulated endocardial vegetation having a diameter of 0.5 cm. The solid line shows the computed time course of amikacin which has diffused into the center of the vegetation. The dotted line again represents the computed relative number of organisms. Here the organisms grow almost 4-fold, to almost four relative units, before the concentrations in the center of the vegetation reach the MIC, after which killing begins. The organisms are reduced essentially to zero by 24 hours, suggesting that such a regimen would probably kill well in the center of an endocardial vegetation of approximately 0.5 cm diameter.

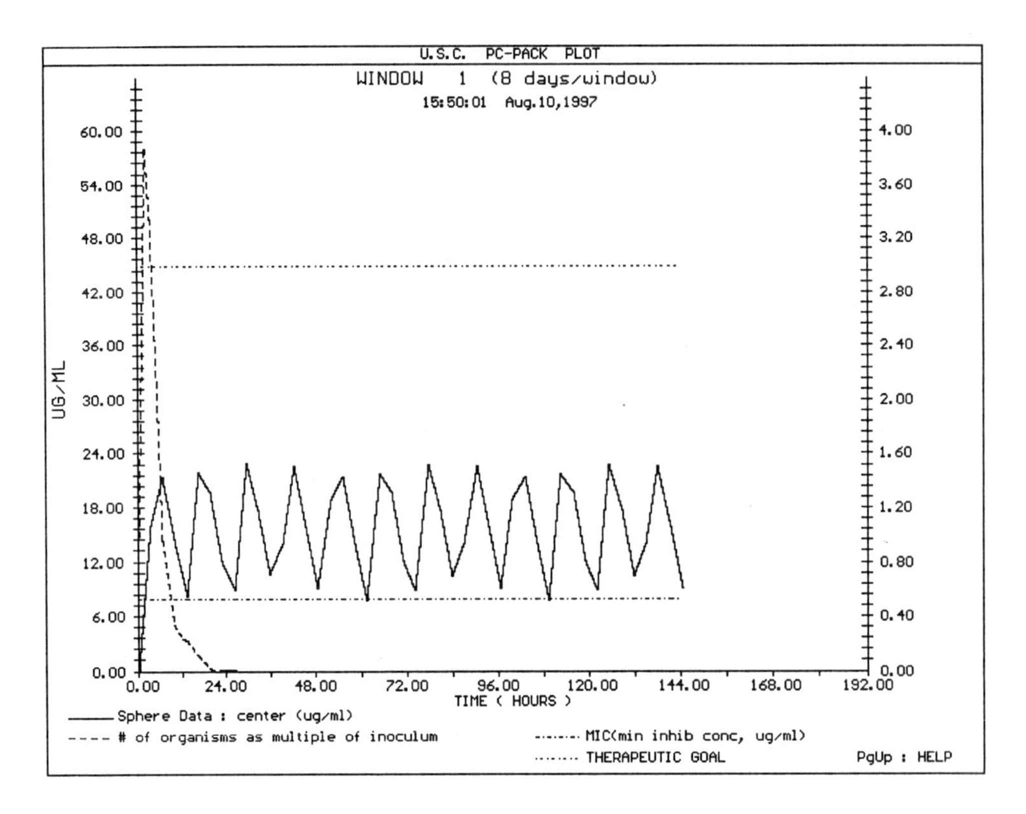

Figure 17. Killing effect as predicted in the center of the 0.5 cm diameter vegetation. Good and fairly prompt killing is seen. Solid line and left hand scale - drug concentrations in the center of the vegetation. Dashed line and right hand scale - relative numbers of organisms, with 1.0 relative unit present at the start of therapy. Upper horizontal dotted and dashed line - original peak serum goal of therapy. Lower horizontal dashed line: the patient's MIC of 8.0 ug/ml.

Figure 18 is a similar plot, but for the simulated vegetation having a diameter of 1.0 cm. Things here are not quite so good. There is a significant lag time of about 3 to 4 hours before any visible concentrations are reached in the center of the vegetation. The MIC is not reached until approximately 10 hours. During that time, the number of organisms has increased from one to approximately 150 relative units. However, after the MIC is reached, killing begins, although not quite so rapidly as with the smaller vegetation, due to the slower rate of rise of drug concentration in the center of the larger vegetation. However, killing appears to be essentially complete after approximately 40 hours. This suggests that the above regimen may be adequate to kill in the center of a 1.0 cm vegetation, but probably does so with less confidence of success than with the vegetation of 0.5 cm diameter. The doubt about this is suggested by the slower rate of killing and by the longer time required to reduce the number of organisms essentially to zero.

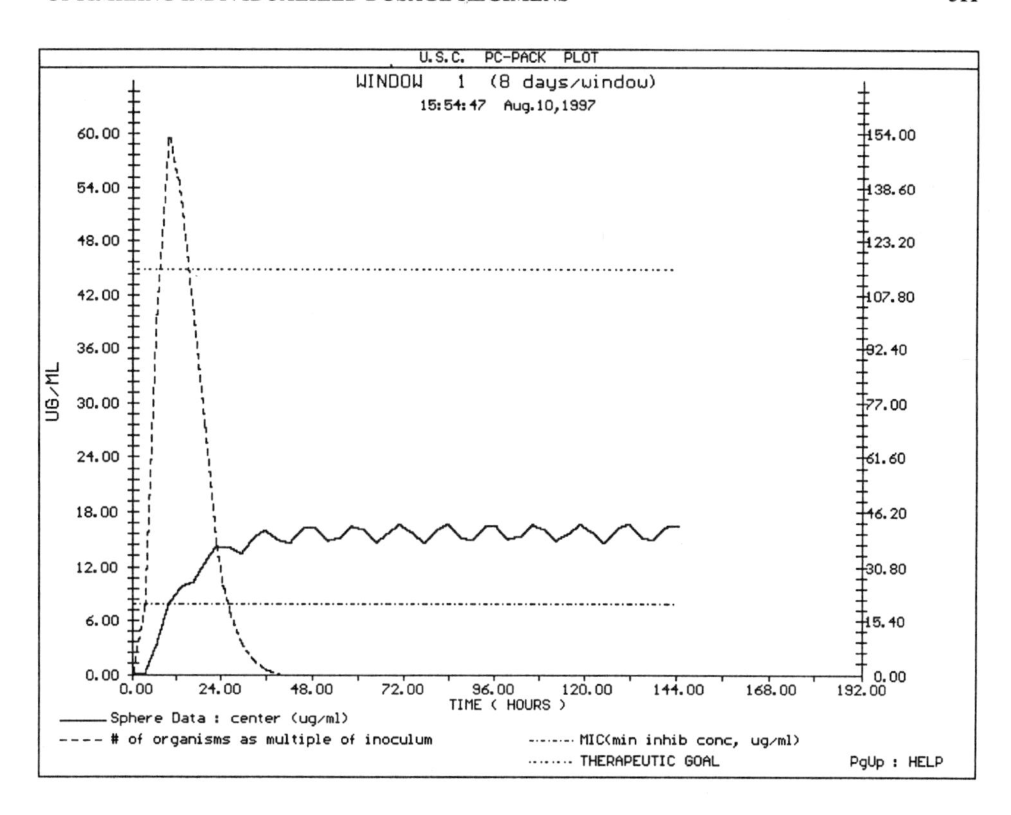

Figure 18. Killing effect computed for the center of a simulated vegetation with 1.0 cm diameter. The effect is delayed due to the slower diffusion into the center, but finally became adequate. Solid line and left hand scale - drug concentrations in the center of the vegetation. Dashed line and right hand scale - relative numbers of organisms, with 1.0 relative unit present at the start of therapy. Upper horizontal dotted and dashed line - original peak serum goal of therapy. Lower horizontal dashed line: the patient's MIC of 8.0 ug/ml.

As shown in Figure 19, things are much worse for the simulated endocardial vegetation of 2.0 cm. Diffusion into the center is a great deal (4 times) slower. Visible concentrations are not achieved until after approximately 12 hours, and about 48 hours are required before they reach the MIC. During this time, the number of organisms has increased astronomically, from 1 relative unit to over 1 million such units. However, after about five days, due to the continued presence of drug concentrations in the center of the vegetation approaching 12 to 15 μg/ml, killing in fact does seem to take place, and after about six days the number of organisms appears to be close to zero. However the behavior of this model strongly suggests that such a dosage regimen might very likely be inadequate in the center of a 2.0 cm simulated vegetation and might, at a minimum, require much more aggressive therapy with higher doses and serum concentrations, surgery, much more prolonged therapy, or all of these.

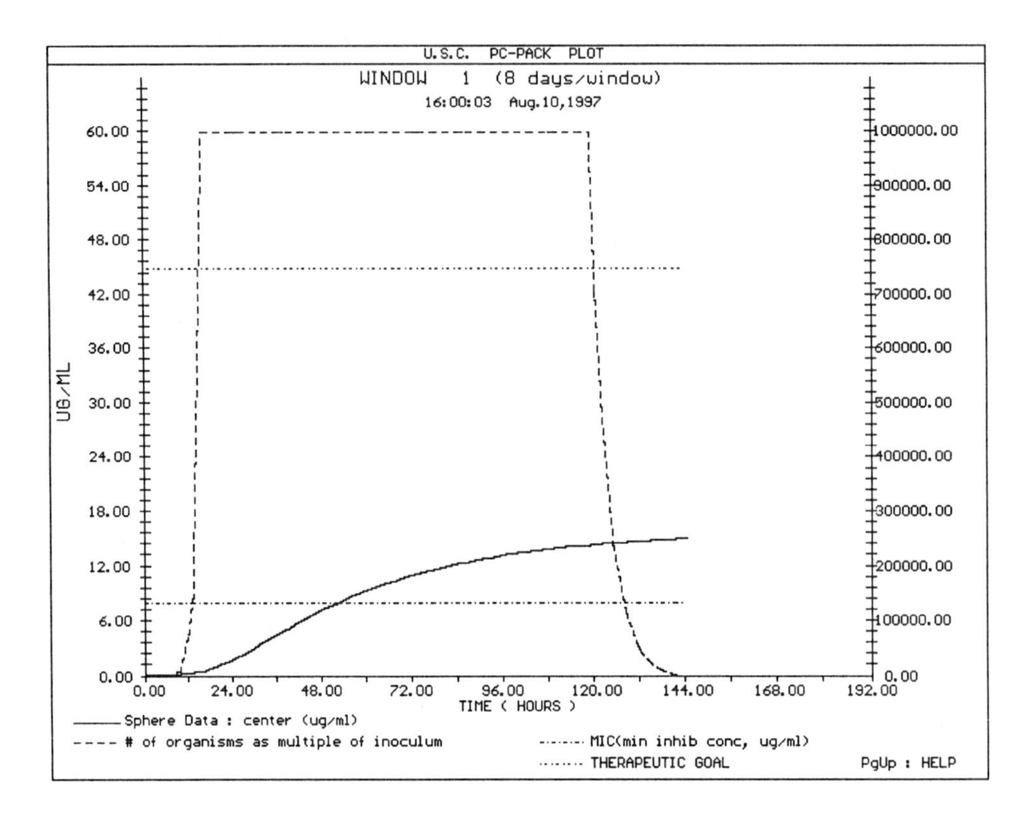

Figure 19. Killing effect computed for the center of a 2.0 cm simulated vegetation. Diffusion to the center is much prolonged while bacterial growth continues. Killing is delayed very significantly. Solid line and left hand scale - drug concentrations in the center of the vegetation. Dashed line and right hand scale - relative numbers of organisms, with 1.0 relative unit present at the start of therapy. Upper horizontal dotted and dashed line - original peak serum goal of therapy. Lower horizontal dashed line: the patient's MIC of 8.0 ug/ml.

Figure 20 shows the computed concentrations in the small hypothetical microorganism used in the previous paper to simulate the time course of the post-antibiotic effect (PAE). There is a lag of about 6 hours between the fall of the serum concentrations and that of the concentrations in the center of this hypothetical microorganism. Because of this, if the dosage interval were to be greater so that the concentrations in the hypothetical microorganism would fall below the MIC, one would see that they would do so approximately six hours after the serum concentrations fall below the MIC, thus simulating a post-antibiotic effect of approximately six hours.

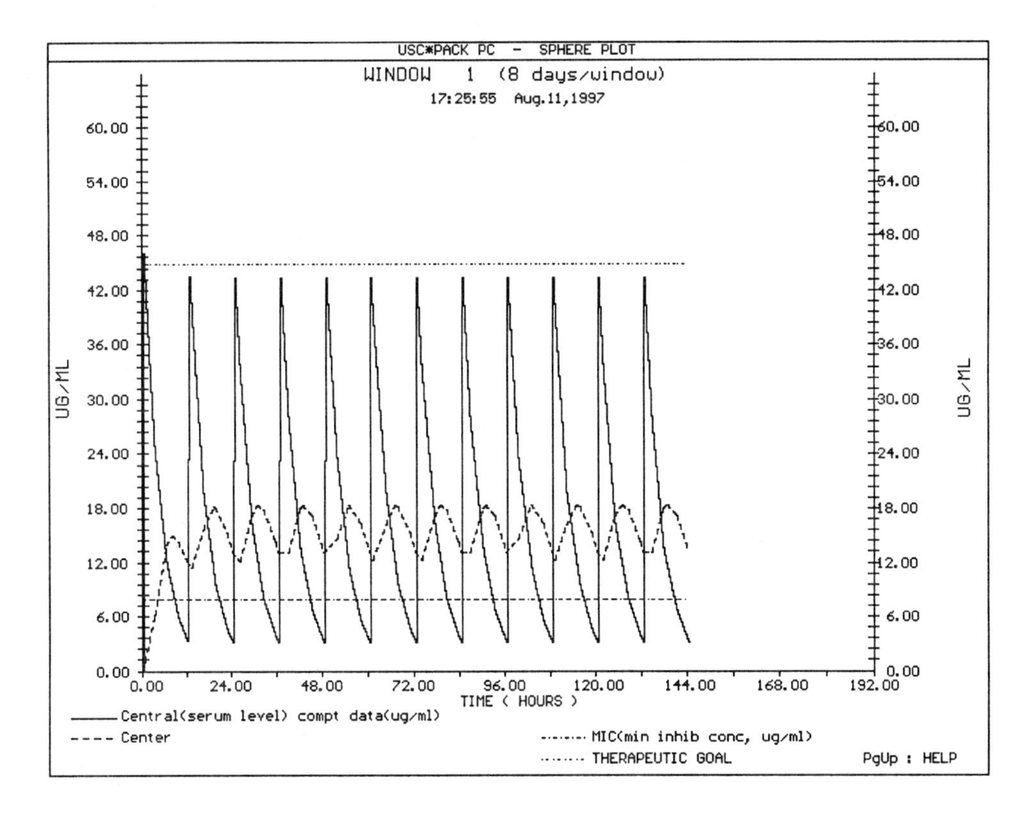

Figure 20. Computed amikacin concentrations in the center of a hypothetical microorganism in which concentrations fall below the MIC about 6 hr after the serum concentrations do, thus simulating (regardless of mechanism) a post-antibiotic effect of about 6 hrs. Solid line and left hand scale - serum drug concentrations. Dashed line and right hand scale - computed concentrations in the center of the microorganism simulating the post-antibiotic effect. Upper horizontal dotted and dashed line - original peak serum goal of therapy. Lower horizontal dashed line: the patient's MIC of 8.0 ug/ml..

What would be the contribution (if any) of such a PAE to overall therapy? As shown in Figure 21, the outcome is not very different from that shown in Figure 16. In both cases, killing is rapid and prompt. One can see that due to the diffusion model, there may be a delay of approximately six hours before the concentrations in the hypothetical microorganism reach the MIC. During that time, the number of organisms has grown from 1 to about 5.4 relative units. However, after that time, the concentrations are always above the MIC, and killing at a significant rate begins and continues, with the organisms being reduced essentially to zero by about 36 hours.

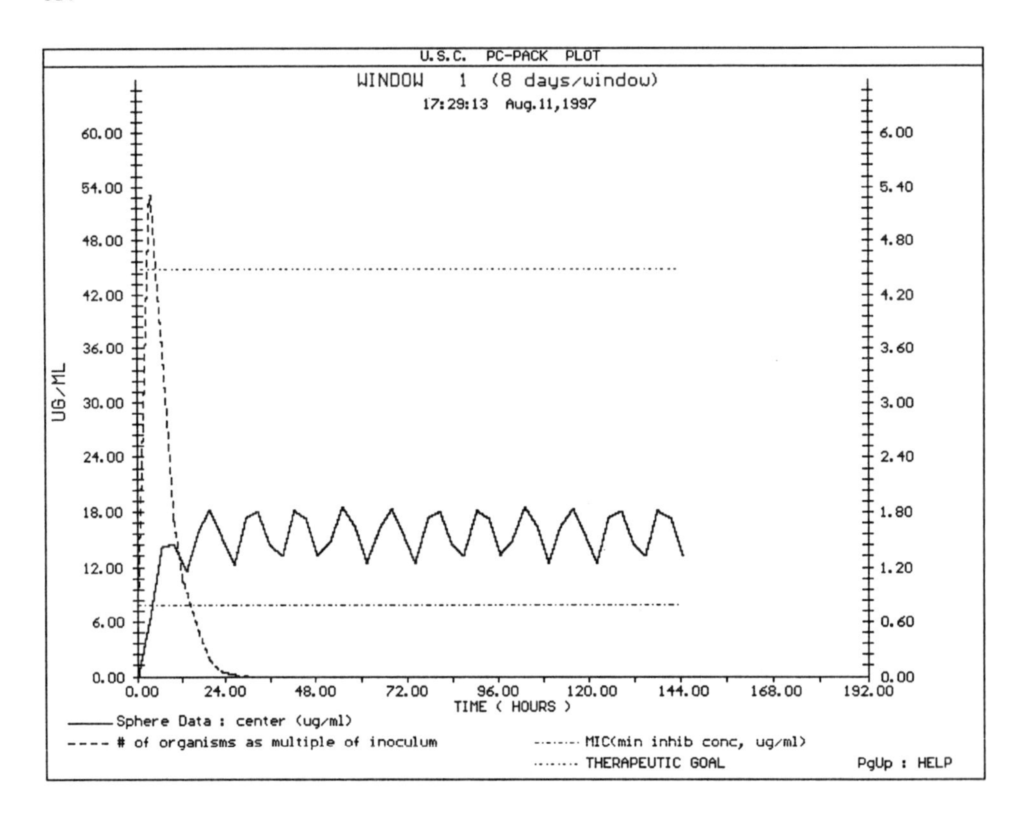

Figure 21. Killing effect predicted for the simulated post-antibiotic effect of 6 hrs, using the computed concentrations in the center of the simulated microorganism as input to the effect model. Solid line and left hand scale - drug concentrations in the center of the microorganism simulating the post-antibiotic effect. Dashed line and right hand scale - relative numbers of organisms, with 1.0 relative unit present at the start of therapy. Upper horizontal dotted and dashed line - original peak serum goal of therapy. Lower horizontal dashed line: the patient's MIC of 8.0 ug/ml.

11.2 An Interesting Case

These diffusion and effect models were of interest when they were used to analyze retrospectively the data obtained much earlier, back in 1991, from the patient described earlier, in section 9.1, from Christchurch, New Zealand, seen through the courtesy of Dr. Evan Begg. She had a pyelonephritis and received tobramycin, 80 mg approximately every 8-12 hours. She was having a satisfactory clinical response to therapy when, on about the 6th day of therapy, she suddenly and unexpectedly relapsed and went into septic shock. She then received much more aggressive tobramycin, and eventually recovered. The MIC of her organism was 2.0 μg/ml. The analysis of bacterial growth and kill described below was not done until several years later, after the models of diffusion and growth and kill had been developed.

Figure 22 shows the computed concentrations of drug in the center of the 0.1 micron sphere representing a hypothetical organism having a PAE of 6 hours, in the patient's first phase, when she appeared to be a general medical patient (not an ICU patient) having a satisfactory clinical response to her tobramycin therapy. However, at the end of this time

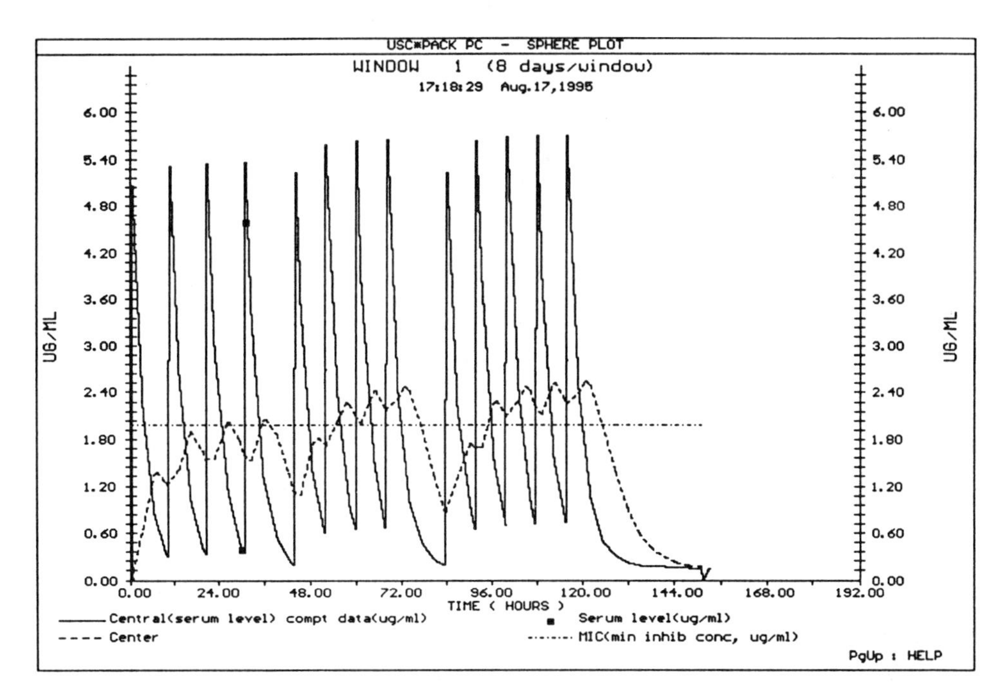

Figure 22. Patient receiving Tobramycin. This figure shows the measured serum concentrations (small rectangles), and the Bayesian fitted model. Small solid rectangles - measured serum concentrations. Solid line and left hand scale - fitted serum drug concentrations. Dashed line and right hand scale - concentrations in the small organism simulating the post-antibiotic effect. Horizontal dashed line: the patient's MIC of 2.0 ug/ml.

(from 120-148 hours into therapy) she unexpectedly relapsed on about day 6 of therapy, went into septic shock, and clearly became an ICU patient.

Note the damped response in the center of the small sphere to the sharp peaks and troughs of the serum concentrations. While one may have many different views as to what the mechanism of the PAE is, this diffusion model appears to do a reasonable job of describing the effect itself. As data accumulate about diffusion into endocardial vegetations and abscesses, this diffusion model will permit modeling of these events during a patient's clinical care in a way that is now becoming possible, as illustrated in the present case.

Now, consider the computed rate of growth and kill of her organisms during her treatment. We have a hint in that her peak serum concentrations were low, only about 5.0 μg/ml. Her measured serum peak was 4.5 μg/ml, and her trough was 0.4 μg/ml. Figure 23 describes the growth and kill of the organisms in response to events in her serum concentration compartment, while Figure 24 shows the same events as viewed with the sphere model simulating the PAE.

Note in both figures that there appear to be few organisms present at the outset of therapy. Growth becomes visible in Figure 23 after the first hiatus between doses, and then becomes exponential during day 6 of therapy, after the last dose on that plot, which

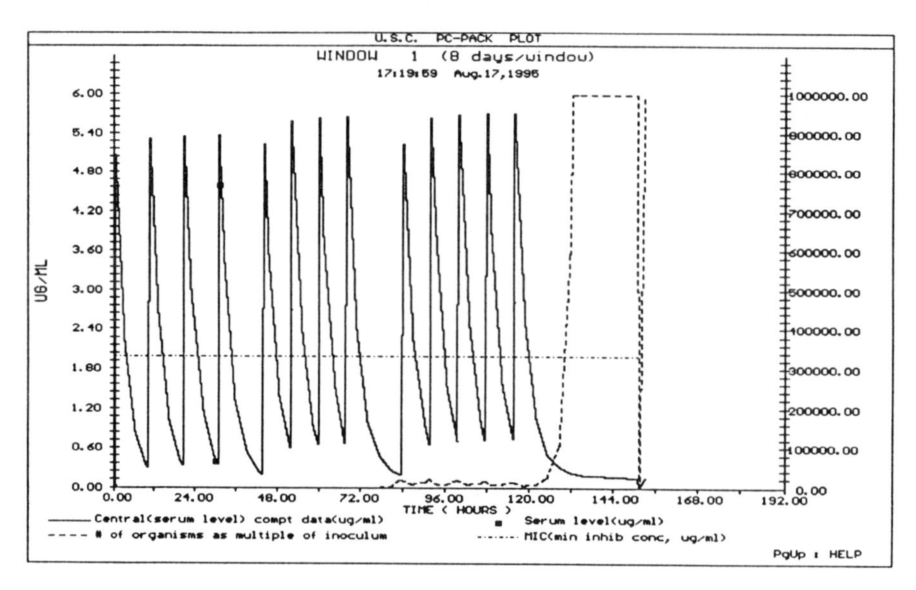

Figure 23. Patient receiving Tobramycin. This figure shows the measured serum concentrations (small solid rectangles), and her individualized Bayesian fitted model. Solid line and left hand scale: fitted serum drug concentrations. Dashed line and right hand scale: relative numbers of organisms. The plot always begins with 1.0 relative units of organism. Horizontal dashed line: the patient's MIC of 2.0 ug/ml.

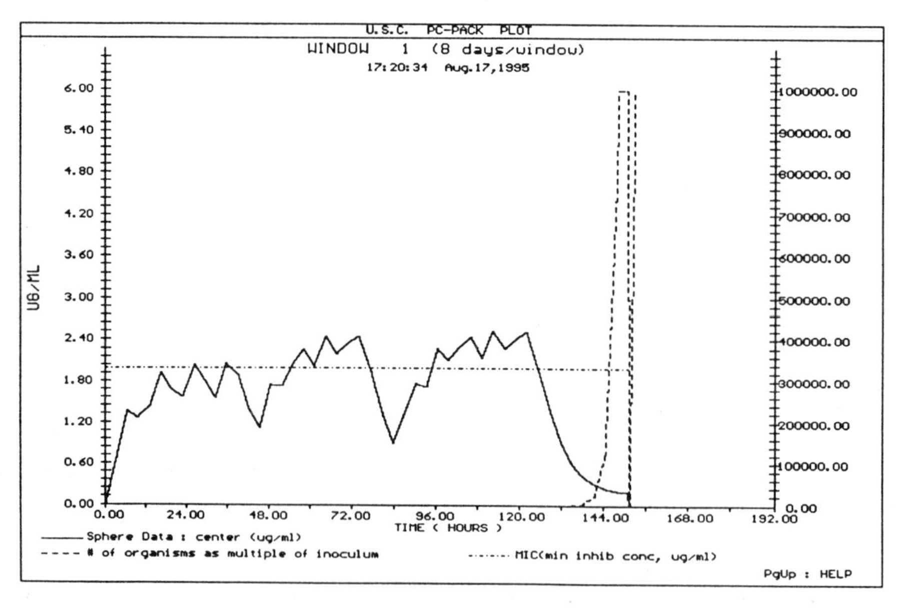

Figure 24. Graph of effects found with the model simulating the PAE of about 6 hours. Solid line and left hand scale: drug concentrations in the microorganism simulating the post-antibiotic effect. Dashed line and right hand scale: relative numbers of organisms. The plot always begins with 1.0 relative units of organism, as shown on the right hand scale. Horizontal dashed line: the patient's MIC of 2.0 ug/ml.

ends just before the next dose which was given during his next period, that of septic shock. Figure 24 extends this examination to the possible contribution of the PAE, where the concentrations in the center of the small sphere are evaluated with respect to their ability to kill the organisms. Note that the regimen also appeared to be effective at first, but that the organisms grew out exponentially when the concentrations fell below the MIC for a significant time, but slightly later than in Figure 23, showing events without any PAE. The exponential growth of organisms escaping from control as shown in Figures 23 and 24 correlated well with the patient's clinical relapse at just that time, with development of septic shock.

Figure 25 shows the subsequent course of this patient in her next phase of acute urosepsis. There was essentially no carry-over of drug from the last dose shown in Figures 22 to 24 to the patient's next dose, which was given at time zero in Figure 25. Figure 25 shows the many serum concentrations measured during this second phase of her hospital course. It also shows the results of Bayesian fitting based on the population model of tobramycin for ICU patients [18], with its much larger central volume of distribution (now that she had become a seriously ill ICU patient) in both her central (serum) and peripheral (nonserum) compartments. During this time, the patient's serum creatinine also rose from 0.7 to 3.7 mg/dL. One can see that it took about two days, as new serum concentrations were obtained, for ward personnel to react to her suddenly much increased volume of distribution (from 0.18 to 0.51 L/kg), and to her much decreased renal function, and to give her the much larger doses required to achieve effective peak serum concentrations, despite her rising serum creatinine. Note also that her trough concentrations rose from about 0.3 up to 2.0 μg/ml during this time, so that the time that serum concentrations were below the MIC was greatly reduced.

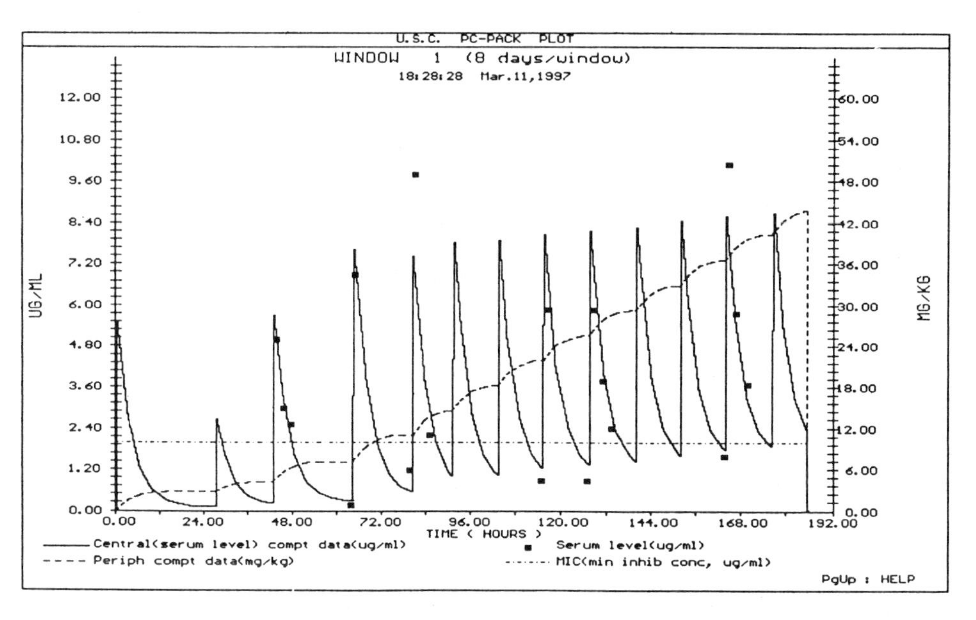

Figure 25. Plot of serum and peripheral compartment concentrations during the time of the patient's sepsis and his recovery. Small solid rectangles - measured serum concentrations. Solid line and left hand scale - fitted serum concentrations. Dashed line and right hand scale - peripheral compartment concentrations, also fitted from the serum data. Horizontal dashed line: the patient's MIC of 2.0 ug/ml.

Figure 26 shows the plot of the computed bacterial growth and kill based on the input from the serum concentration profile during this time, without any aid from the simulated PAE. The organisms grow out of control in the first two days (again correlating with the patient's relapse into sepsis the day before, and the time required for ward personnel to perceive the problem and to adjust her dosage sharply upward to achieve serum peaks in the range of 7 to 9 μg/ml). As these higher and more effective concentrations were achieved, however, bacterial killing could finally be seen at about the sixth day in this figure, and appeared to be effective after that, thus correlating with the patient's subsequent clinical recovery.

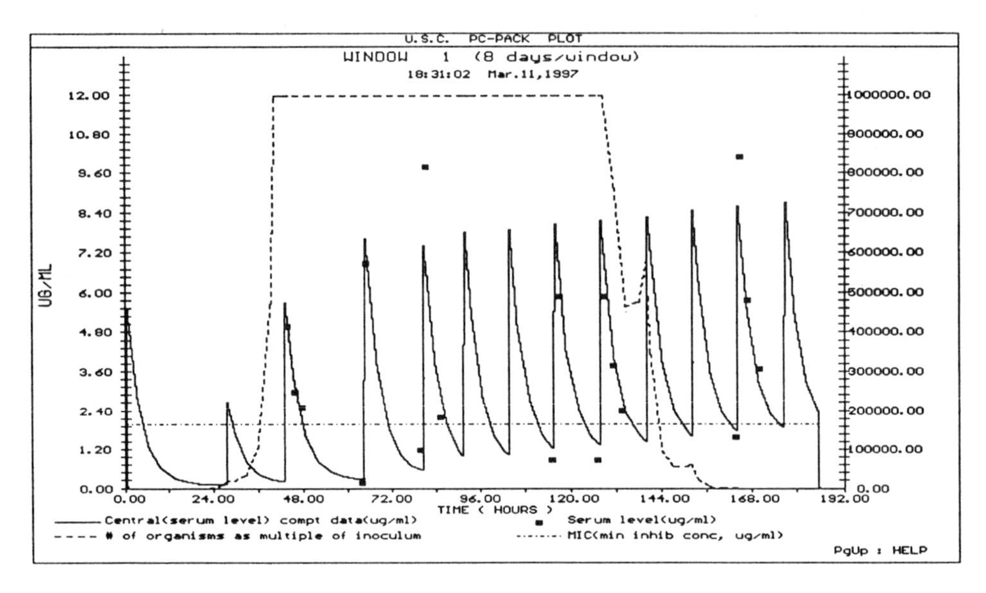

Figure 26. Plot of the effect on growth and kill using input from the serum concentration profile. The organisms grow out of control when serum concentrations are lower, but kill again when they are higher. These events correlated well with the patient's relapse at the beginning of the plot, and his recovery about one week later. Small solid rectangles - measured serum concentrations. Solid line and left hand scale - serum concentrations. Dashed line and right hand scale - relative numbers of organisms, with 1.0 relative unit present at the start of therapy. Horizontal dashed line: the patient's MIC of 2.0 ug/ml.

Figure 27 below shows the same events, but now using the diffusion model of the small microorganism and its simulated PAE. Using this model, the concentrations in the center of the small simulated microorganism do not exceed the MIC until almost 72 hours. Significant killing can be seen to begin slightly earlier in this figure, at about 110 hours in the figure, compared to about 130 hours (about 1 day later) in Figure 26, and appeared to be effective after that.

In general, models of bacterial growth and kill permit one to incorporate known in vitro data of the logarithmic growth rate of the organism and the maximum kill rate achieved with the antibiotic, to integrate it with data of the MIC of each individual patient's organism, and to model the growth and decline of the relative numbers of organisms. These Zhi models have correlated well, in this patient, with her unexpected relapse from having an apparently satisfactory response to therapy to becoming a

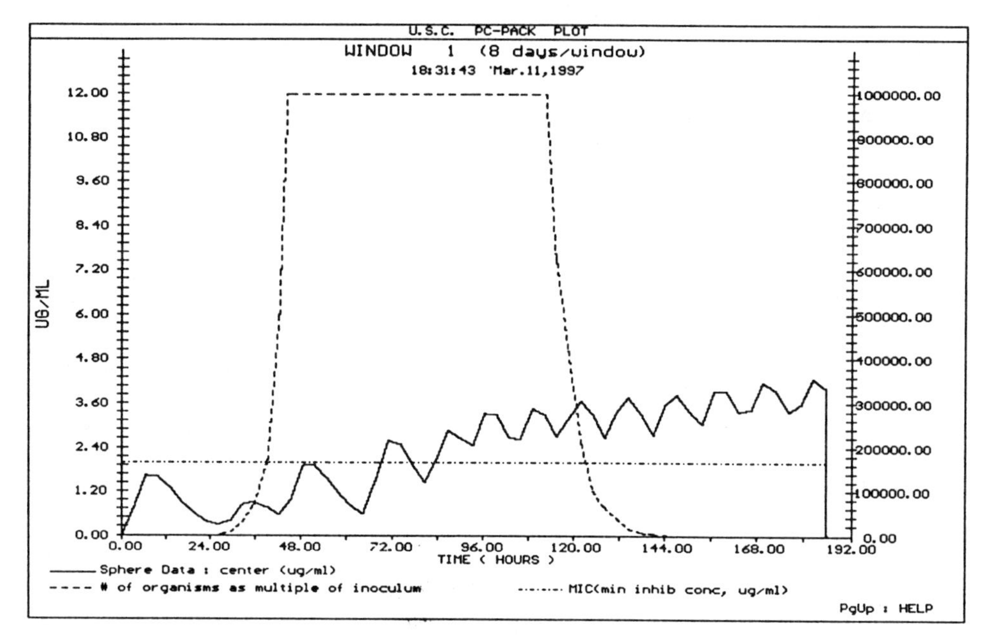

Figure 27. Plot of the effect on growth and kill using input from the center of the organism, using its computed concentrations as input to the effect model. The post-antibiotic effect helps somewhat to delay the relapse and to augment the kill. Solid line and left hand scale - computed concentrations in center of microorganism simulating the post-antibiotic effect. Dashed line and right hand scale - relative numbers of organisms, with 1.0 relative unit present at the start of therapy. Horizontal dashed line: the patient's MIC of 2.0 ug/ml.

seriously ill patient with septic shock, and with her subsequent recovery later on, as effective serum concentrations were achieved and maintained.

The Zhi model does not describe the decline of bacterial growth rate seen over time, reaching a maximum number of organisms, as found by Mouton, Vinks, and Punt [30]. The organisms are always assumed to be in their logarithmic growth phase, the maximum possible. In addition, the model does not account for the increase in bacterial resistance found over time, and the emergence of resistant organisms. However, one can use the maximum possible MIC which the emerging resistant organism is expected to reach, and examine the behavior of the model. In this case, the Zhi model becomes a useful example of the "worst case" scenario, with the resistant organisms being so from the very beginning of therapy, and with the logarithmic growth rate always being in effect, never slackening. If a given dosage regimen, generating a certain serum concentration profile, can kill well using the Zhi model, one might expect it probably to do at least as well in clinical circumstances, where the growth rate may (or may not) slacken with time and may (but may not) reach a maximum number of organisms, and the resistant organisms emerge more slowly with time.

Clearly, further work in this area is needed, but models of this type are beginning to provide a useful new way to perceive, analyze, and evaluate the efficacy of antibiotic therapy. Similar approaches may also be useful in analyzing therapy of patients with AIDS, using the PCR assays, and with cancer.

12. LIMITATIONS OF CURRENT MAP BAYESIAN ADAPTIVE CONTROL

The maximum aposteriori probability (MAP) Bayesian approach to adaptive control and dosage individualization is straightforward and robust. However, it does not represent an optimal approach to dosage individualization. It has two significant limitations.

The first one is that the pharmacokinetic model parameter values used to describe the behavior of the drug are assumed to be either normally or log-normally distributed. This is often not so. Many drugs, for example, have clusters of both rapid and slow metabolizers within the population, and therefore may well have multimodal population parameter distributions for the elimination rate constant. Furthermore, the volume of distribution for drugs such as the aminoglycosides is affected by the patient's clinical state as a general medical patient or a patient in an intensive care unit, for example. Because of this, parameter distributions are often asymmetrical, neither normally or lognormally distributed, and are therefore not optimally described by mean, median, or mode values. This point reflects the significant problems associated with making parametric population models, and with using mean or median parameter values to develop dosage regimens. The problem is largely overcome by making nonparametric population models which describe the entire joint parameter distribution within the population, with up to one support point (set of parameter values, and its estimated probability) for each subject studied in the population [13,14].

The second limitation is that there is no tool in the MAP Bayesian strategy to estimate and predict the precision with which a desired dosage regimen developed to hit a desired target goal actually is likely to do so. The method lacks a vital performance criterion.

The separation, or heuristic certainty equivalence, principle is well known among the stochastic control community, but less so among the pharmacokinetic community. It states [31] that when the task of controlling the behavior of a system is separated into the steps of:

1. Obtaining the best single point parameter values in the model describing the behavior of the system, and then,

2. Using these single point values to design the inputs to control the system,

that, the task of hitting the target goal is usually performed suboptimally. Yet this is exactly what the MAP Bayesian, and all methods which estimate single values for each model parameter, do.

There is no performance criterion to optimize in the MAP Bayesian dosage strategy (such as the estimated precision with which the desired target will be hit, for example) as there is only one set of parameter values, and the target is simply assumed to be hit exactly.

13. OVERCOMING THE LIMITATIONS: "MULTIPLE MODEL" DESIGN OF MAXIMALLY PRECISE DRUG DOSAGE REGIMENS

The two limitations above are overcome by the combination of nonparametric population models [13,14] and the "multiple model" design of dosage regimens [32]. Nonparametric population models have been discussed in another chapter. Their strength is that they are consistent, statistically efficient, and have good properties of statistical convergence [35]. They are not limited by the assumption that the parameter distributions must be Gaussian or lognormal, as in parametric methods. Instead of simply estimating parameter means, variances, and correlations between them, as point estimates of a distribution, the nonparametric methods estimate the entire parameter distributions themselves. These distributions are discrete, not continuous. They consist of discrete sets of parameter estimates, along with an estimate of the probability of each set [13,14]. Up to one set (support point of the distribution) is obtained for each subject studied in the population. This closely approaches the ideal population model (which can never be attained), which would consist of the correct structural model of the drug system, along with the exact value of each parameter in each subject if it would somehow be possible to know those values.

When a parametric population model is used as the Bayesian prior to design an initial dosage regimen for the next patient one encounters, one usually has only a single estimated value for each parameter. Because of this, only one prediction of future concentrations can be made. The action taken is therefore based only on the estimates of

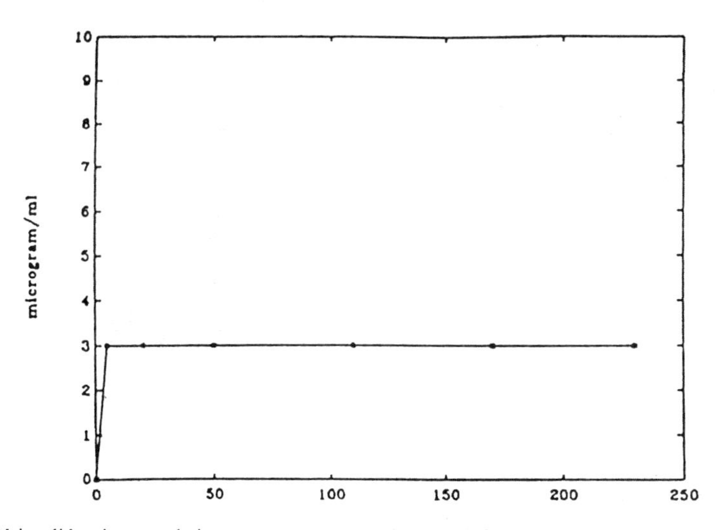

Figure 28. Using lidocaine population mean parameter values, an infusion regimen designed to achieve and maintain a target goal of 3 ug/ml does so exactly when the patient, as here, has exactly the mean population parameter values.

the central tendencies of the parameter distributions, and not on the entire parameter distributions, which may well be multimodal, due to genetic polymorphism in the distribution and metabolism of drugs. The dosage regimen is simply assumed to achieve the target goal exactly, as shown in Figure 28. Figure 28 shows the results of an infusion regimen of lidocaine, based on the mean population parameter values for that drug, which was designed to achieve and maintain a target serum concentration of 3 μg/ml. As shown, this regimen, based on the single mean population parameter values, hits the target exactly, but only when the patient has parameter values which are exactly the population mean values. However, as shown in Figure 29, when the regimen used in Figure 28 was given to the combination of the actual 81 diverse nonparametric population support points from which these mean parameter values were obtained, an extremely wide distribution of predicted serum concentrations was seen, due to the diversity in the nonparametric population support points from which the mean parameter values were obtained, representing the diversity in the parameter values in the patient population. The predicted serum concentrations actually covered much more than the usual therapeutic range of 2 to 6 μg/ml.

In contrast, if one has a nonparametric population model [13,14], with its multiple sets of model parameter values (81 in this case), one can make multiple predictions, instead of only one, forward into the future from any candidate dosage regimen which is "given" to all the models in the population discrete joint density. The richer and more likely population parameter joint density reflects much better the actual diversity among the subjects studied in the past population.

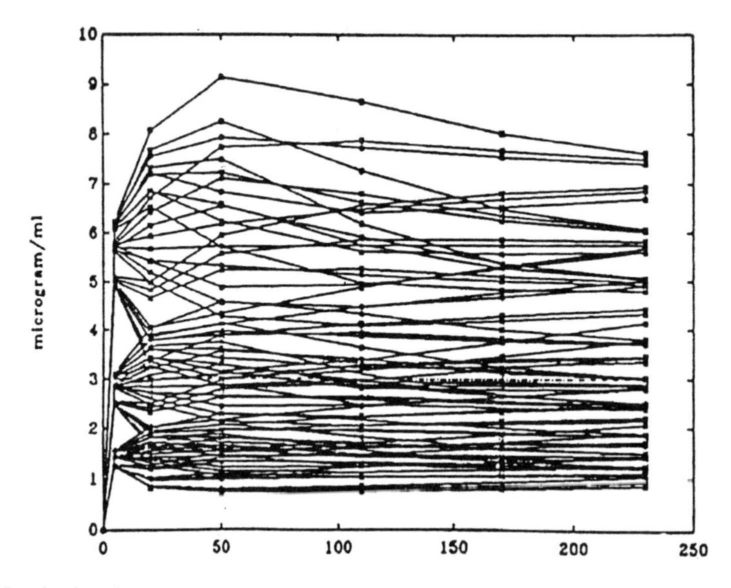

Figure 29. Result when the above lidocaine infusion based on population mean parameter values is given to the 81 diverse support points from which the population mean values were obtained. Great diversity in the predicted responses is seen.

Based on these multiple models in the population (the discrete joint density), one can compute the weighted squared error with which any candidate regimen is predicted to fail to achieve the desired target goal at a target time. Other regimens can then be considered, and the optimal regimen can be found which is specifically designed to achieve the desired target goal with the least weighted squared error [32].

This approach, using the multiple models of the patient provided by the nonparametric population model, avoids much of the limitations of the separation principle. This is the real strength of the combination of nonparametric population models coupled with "multiple model" dosage design [32].

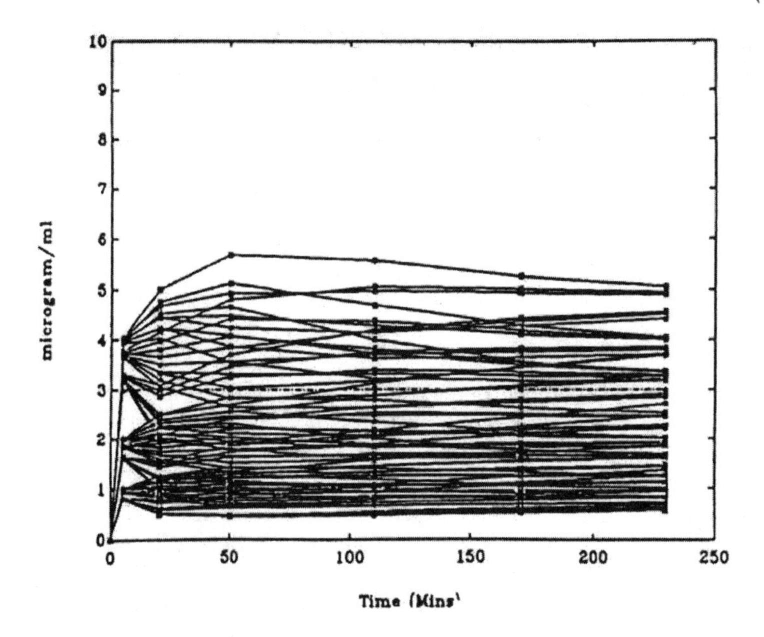

Figure 30. Predicted response of the 81 support points (models) when the regimen obtained by multiple model dosage design is given. The target is achieved with visibly greater, and optimal, precision.

As shown in Figure 30, the multiple model (MM) dosage regimen, based on the same nonparametric population model with its 81 support points, obtained a much more precise achievement of the target goal, because it was specially designed to do so. The error in the achievement of the therapeutic target goal is much less, and the dispersion of predicted serum concentrations about the target goal is much less.

13.1 Obtaining "Multiple Model" Bayesian Posterior Joint Parameter Distributions

With the MAP Bayesian approach to posterior parameter values, the single most likely value for each parameter is obtained when they altogether minimize the objective function shown in equation (1). In contrast, the MM Bayesian approach, using the nonparametric joint densities, preserves the multiple sets of population parameter values, but specifically recomputes their Bayesian posterior probability, based upon the serum concentrations obtained. Those combinations of parameter values that predict the measured concentrations well become more probable. Those that predict them less well become less so. In this way, the probabilities of all the nonparametric population model support points become revised, using Bayes' theorem [33]. A smaller number of significant points, or perhaps even only one, is usually obtained. When the regimen for the next cycle is developed, these revised models, containing their revised MM Bayesian posterior probabilities, are used to develop it. The regimen is again specifically designed to achieve the desired target goal with maximum precision (minimum weighted squared error).

13.2 Other Bayesian Approaches

Three sequential Bayesian approaches have also been used by us to incorporate feedback from measured serum concentration data. The first is the sequential MAP Bayesian approach, in which the MAP posterior parameter values are sequentially updated after each serum concentration data point is obtained. This procedure improves the tracking of the behavior of the drug through each data set. However, at the end of each full feedback cycle, (after each new full cluster of data points), at the time the next regimen is to be developed, this method has learned no more with respect to developing the next new dosage regimen, than if it had fitted all the data together at once, even though it estimates changing MAP Bayesian parameter values sequentially. This is because the method only estimates a single set of parameter values that fit the objective function best, over all the data points.

The second approach is the sequential MM Bayesian one [33]. Here the MM Bayesian posterior joint density is also sequentially updated after each data point. Still, at the end of each feedback cycle, this procedure similarly has learned no more with respect to developing the next dosage regimen than if all the data in that cluster were fitted simultaneously. The procedure is still looking for a hypothetical single model (support point, set of parameter values) which best describes all the data. When this fails to be the case, combinations of support points are found which fit best. Still, the procedure estimates a fixed and unchanging single model, or combination of models (support points), which best fit the data, even though the posteriors are fitted sequentially.

A third approach is the interacting multiple model (IMM) approach [36]. This method permits the true patient being sought for actually to jump from one model or support point to another during the sequential Bayesian analysis. Because of this, the IMM method, originally designed to track missiles and aircraft taking evasive action, permits detection of changing pharmacokinetic parameter densities during the sequential analysis procedure. It thus provides an improved method to track the changing parameter densities and the behavior of a patient during the evolution of his/her clinical therapy. For example, it permits an improved ability to detect and to quantify changes in the volume of distribution of aminoglycoside drugs during changes in a patient's clinical status which are not captured by the use of conventional clinical descriptors.

Using carefully simulated models in which the true parameter values changed during the data collection, the integrated total error in tracking a simulated patient was very similar with the sequential MAP and sequential MM Bayesian procedures. However, the integrated total error of the sequential IMM procedure was only about one half that of the other two [36].

13.3 Clinical Applications

Nonparametric population parameter joint densities, MM dosage design and IMM Bayesian posterior joint densities appear to offer significant improvements in the ability to track the behavior of drugs in patients throughout their care, especially when the patients are unstable and have changing parameter values. These approaches also develop dosage regimens which are specifically designed to achieve target goals with maximum precision. These methods make optimal use of all information contained in the past population data, coupled with whatever current data of feedback may be available up to that point, to develop that patient's most precise dosage regimen.

13.4 Analyzing the Tobramycin Patient with MM and IMM Sequential Bayesian Methods. Implementation into Clinical Software

The above tools have now been implemented in clinical software, the MM-USCPACK package [33]. Figure 31 shows the plot of the fit to the data from the patient described in sections 9.1 and 11.2, and in Figure 22, when analyzed using the sequential MM Bayesian approach. The data was very poorly fitted, due to the changing parameter values in this highly unstable patient, as she changed from being a general medical patient with a pyelonephritis before 150 hours to an acutely ill and highly unstable patient with severe septic shock after that time.

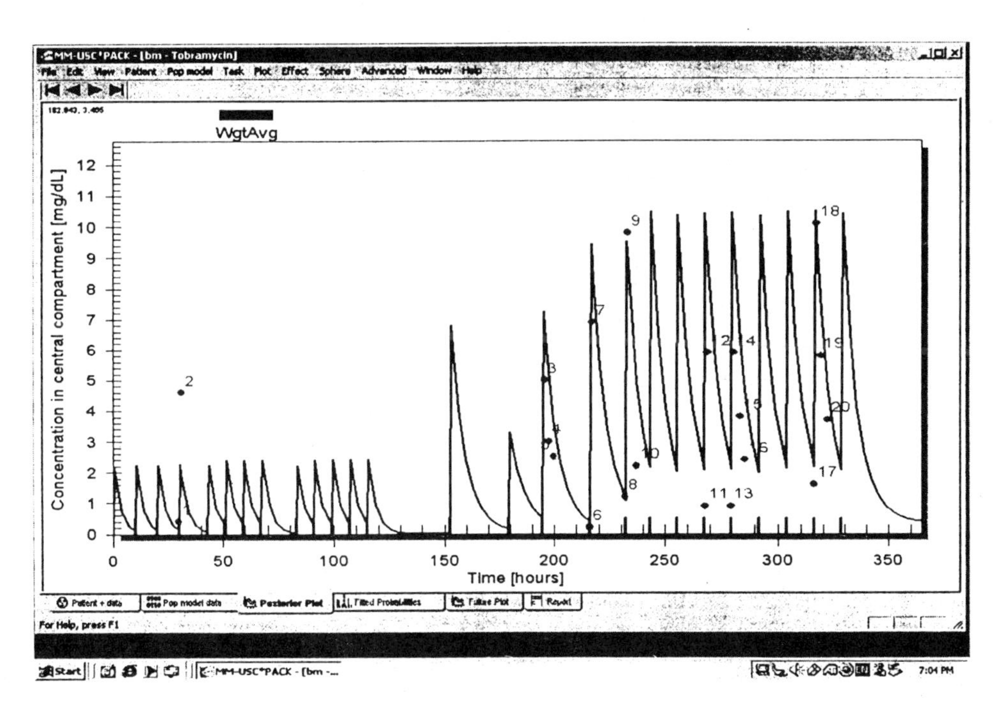

Figure 31. Fit to data of patient described in section 9.1, 11.2, and Figure 22, analyzed with the MM Bayesian approach. Note the very poor fit to the data, due to the patent's changing parameter values as her clinical status changed significantly, going from someone with a pyelonephritis before 150 hours, to someone with clearcut septic shock afterward, becoming an acutely and severely ill intensive care patient.

On the other hand, Figure 32 shows the result when the IMM algorithm was used to analyze the patient's data. The fit was greatly improved, and the IMM algorithm was able to track the changing behavior of tobramycin in this acutely and severely ill, highly unstable patient.

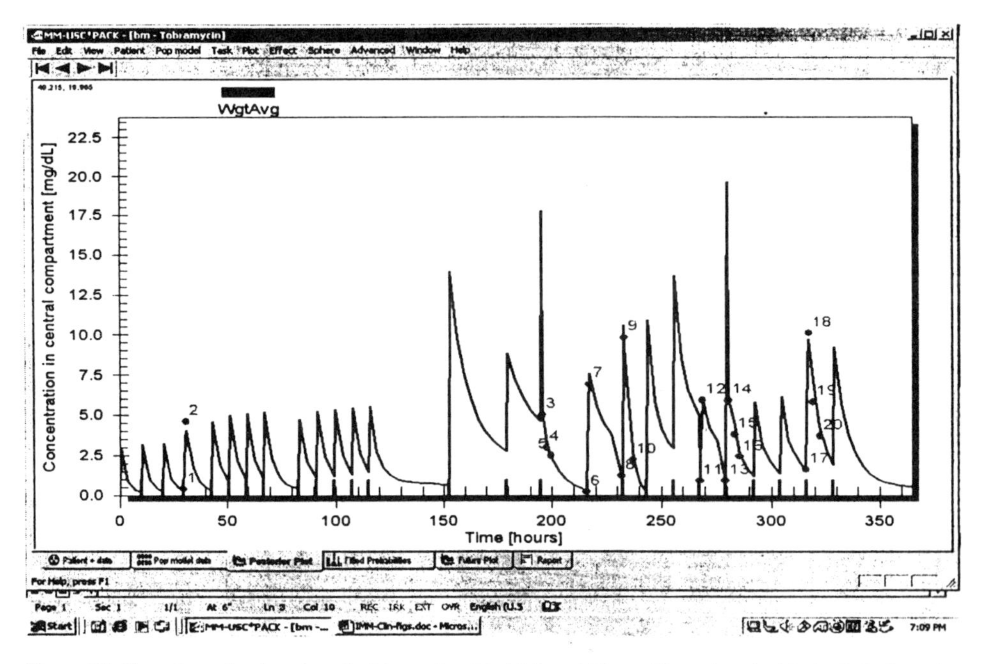

Figure 32. Fit to data of patient described in section 9.1, 11.2, and Figure 22, analyzed with the IMM Bayesian approach. Note the very much improved fit to the data, as the IMM approach tracks the changing parameter values taking place in this acutely ill and highly unstable patient.

14. THE FUTURE OF INDIVIDUALIZED DRUG THERAPY

The clinical software above incorporates all the strengths of the nonparametric population models and the multiple model dosage design, and the MM and IMM Bayesian analysis of individual patient data. Work is now under way to develop similar MM dosage designers for the large and often nonlinear models of drugs having that behavior, such as Phenytoin, Carbamezapine, and many drugs used in the treatment of patients with AIDS, transplants, and cancer, for example. Nonparametric population models can be made now of large, nonlinear, interacting and multiple drug systems such as those found in the combination chemotherapy of many of the above patients, where the concentrations of one drug may either compete with a metabolite, or may increase or decrease the rate of metabolism of another drug. As MM dosage design is developed for such large multidrug systems, it will be possible, for the first time, to develop truly coordinated, maximally precise combination chemotherapy for patients with such problems, maximizing effectiveness while constraining toxicity within specifically selected quantifiable limits.

Acknowledgments

Supported by US Government grants LM 05401, RR 01629, RR 11526, and GM65619.

References

1. Cohen J: Make Your Medicine Safe. ISBN: 0-380-79075-0. Avon Books Inc., New York, 1998.
2. Evans W: General Principles of Applied Pharmacokinetics, Chapter 1, in Applied Pharmacokinetics: Principles of Therapeutic Drug Monitoring, ed. By Evans W, Schentag J, and Jusko W. Applied Therapeutics Inc., Vancouver B.C., 1992.
3. van Lent-Evers N, Mathot R, Geus W, van Hout B, and Vinks A: Impact of Goal-Oriented and Model-Based Clinical Pharmacokinetic Dosing of Aminoglycosides on Clinical Outcome: A Cost-Effectiveness Analysis. Therap. Drug Monit. 221: 63-73, 1999.
4. Reuning R, Sams R, and Notari R: Role of Pharmacokinetics in Drug Dosage Adjustment. 1. Pharmacologic Effects, Kinetics, and Apparent Volume of Distribution of Digoxin. J. Clin. Pharmacol. 13: 127-141, 1973.
5. Jelliffe R, Schumitzky A, Van Guilder M, Liu M, Hu L, Maire P, Gomis P, Barbaut X, and Tahani B: Individualizing Drug Dosage Regimens: Roles of Population Pharmacokinetic and Dynamic Models, Bayesian Fitting, and Adaptive Control. Therap. Drug Monit., 15: 380-393, 1993.
6. Sheiner LB, Beal S, Rosenberg B, and Marathe V: et al.: Forecasting Individual Pharmacokinetics. Clin. Pharmacol. Ther., 31: 294-305, 1979.
7. Jelliffe R, Iglesias T, Hurst A, Foo K, and Rodriguez J: Individualizing Gentamicin Dosage Regimens: A Comparative Review of Selected Models, Data Fitting Methods, and Monitoring Strategies. Clin. Pharmacokinet. 21: 461- 478, 1991.
8. Nelder JA and Mead R: A Simplex Method for Function Minimization. Computer Journal 7: 308-313, 1965.
9. Caceci MS and Cacheris WP: Fitting Curves to Data: The Simplex Algorithm is the Answer. BYTE Magazine, May 1984, pp. 340-362.
10. De Groot MH: Probability and Statistics, Second Edition, Addison - Wesley Publishing Co., Reading, MA, 1989, pp. 422-423.
11. Sawchuk R and Zaske D: Pharmacokinetics of Dosing Regimens which Utilize Multiple Intravenous Infusions: Gentamicin in Burn Patients. J. Pharmacokin. Biopharm 4: 183-195, 1976.
12. Jelliffe R, Schumitzky A, Bayard D, Milman M, Van Guilder M, Wang X, Jiang F, Barbaut X, and Maire P: Model-Based, Goal-Oriented, Individualised Drug Therapy: Linkage of Population Modelling, New "Multiple Model" Dosage Design, Bayesian Feedback, and Individualized Target Goals. Clin. Pharmacokinet. 34: 57-77, 1998.
13. Mallet A: A Maximum Likelihood Estimation Method for Random Coefficient Regression Models. Biometrika. 73: 645-656, 1986.
14. Schumitzky A: Nonparametric EM Algorithms for Estimating Prior Distributions. App. Math. and Computation. 45: 143-157, 1991.
15. Jelliffe R : Estimation of Creatinine Clearance in Patients with Unstable Renal Function, without a Urine Specimen. Am. J. Nephrol. 22: 320-324, 2002.
16. Maire P, Jelliffe R, Dumarest C, Roux D, Breant V, Charpiat B, Vermeulen E, Brazier J, and Courpron P: Controle Adaptatif Optimal des Posologies: Experience des Aminosides en Geriatrie. in Information et Medicaments. Comptes Rendus du Colloque AIM-IF et IRT, Paris, December 1989, ed. by Venot A and Degoulet P, Volume 2 of Informatique et Sante, directed by Degoulet P, Springer Verlag, Paris, 154-169, 1989.
17. Hurst A, Yoshinaga M, Mitani G, Foo K, Jelliffe R, and Harrison E.: Application of a Bayesian Method to Monitor and Adjust Vancomycin Dosage Regimens. Antimicrob. Agents Chemother., 34; 1165-1171, 1990.
18. Jelliffe R, Schumitzky A, Van Guilder M, and Jiang F: User Manual for Version 10.7 of the USC*PACK Collection of PC Programs. December 1, 1995. Laboratory of Applied Pharmacokinetics, University of Southern California School of Medicine, Los Angeles, CA.
19. Jelliffe R: Clinical Applications of Pharmacokinetics and Control Theory: Planning, Monitoring, and Adjusting Dosage Regimens of Aminoglycosides, Lidocaine, Digitoxin, and Digoxin. In Maronde R, ed: Topics in Clinical Pharmacology and Therapeutics, Sprnger-Verlag, New York, 1986, pp. 26-82.
20. Rodman J, Jelliffe R, Kolb E, Tuey D, de Guzman M, Wagers P, and Haywood L: Clinical Studies with Computer-Assisted Initial Lidocaine Therapy. Arch. Int. Med. 144: 703-709, 1984.
21. Bleyzac N, Souillet G, Magron P, Janoly A, Martin P, Bertrand Y, Galambrun C, Dai Q, Maire P, Jelliffe R, and Aulagner G: Improved Clinical Outcome of Paediatric Bone Marrow Recipients

using a Test Dose and Bayesian Pharmacokinetic Individualization of Busulfan Dosage Regimens. Bone Marrow
Transplantation 28: 743-751, 2001.

22. Marcus Haug, Pharm.D., and Peter Slugg, M.D., Personal communication.
23. Maire P, Barbaut X, Vergnaud JM, El Brouzi M, Confesson M, Pivot C, Chuzeville M, Ivanoff N, Brazier J, and Jelliffe R: Computation of Drug Concentrations in Endocardial Vegetations in Patients during Antibiotic Therapy. Int. J. Bio-Med. Comput., 36: 77-85, 1994.
24. Bayer A, Crowell D, Yih J, Bradley D, and Norman D: Comparative Pharmacokinetics and Pharmacodynamics of Amikacin and Ceftazidime in Tricuspid and Aortic Vegetations in Experimental *Pseudomonas* Endocarditis. J. Infect. Dis., 158: 355-359, 1988.
25. Bayer A, Crowell D, Nast C, Norman D, and Borelli R: Intravegetation Antimicrobial Distribution in Aortic Endocarditis Analyzed by Computer-Generated Model: Implications for Treatment. Chest, 97: 611-617, 1990.
26. Zhi J, Nightingale CH, and Quintiliani R: Microbial Pharmacodynamics of Pipericillin in Neutropenic Mice of Systemic Infection due to Pseudomonas Aeruginosa. J Pharmacokin. Biopharm. 4: 355-375, 1988.
27. Schumitzky A: personal communication.
28. Bouvier D'Ivoire M, and Maire P: Dosage Regimens of Antibacterials: Implications of a Pharmacokinetic/Pharmacodynamic Model. Drug Invest. 11: 229-239, 1996.
29. Craig W, and Ebert S: Killing and Regrowth of Bacteria in Vitro: a Review. Scand J Infect Dis Suppl 74: 63-70, 1991.
30. Mouton J, Vinks AATMM, and Punt N: Pharmacokinetic-Pharmacodynamic Modeling of Ceftazidime during Continuous and Intermittent Infusion. Chapter 6, pp 95-110, in the Ph.D. Thesis of Vinks AATMM: Strategies for Pharmacokinetic Optimization of Continuous Infusion Therapy of Ceftazidime and Aztreonam in Patients with Cystic Fibrosis, November, 1996.
31. Bertsekas D: Dynamic Programming: deterministic and stochastic models. Englewood Cliffs (NJ): Prentice-Hall, pp. 144-146, 1987.
32. Jelliffe R, Bayard D, Milman M, Van Guilder M, and Schumitzky A: Achieving Target Goals most Precisely using Nonparametric Compartmental Models and "Multiple Model" Design of Dosage Regimens. Therap. Drug Monit. 22: 346-353, 2000.
33. Jelliffe R, Bayard D, Schumitzky A, Milman M, Jiang F, Leonov S, Gandhi A, and Botnen A: A New Clinical Software Package for Multiple Model (MM) Design of Drug Dosage Regimens for Planning, Monitoring, and Adjusting Optimally Individualized Drug Therapy for Patients. Presented at the 4th International Meeting on Mathematical Modeling, Technical University of Vienna, Vienna, Austria, February 6, 2003.
34. Jelliffe R: A Mathematical Analysis of Digitalis Kinetics in Patients with Normal and Reduced Renal Function. Math. Biosci. 1: 305-325, 1967.
35. Leary R, Jelliffe R, Schumitzky A, and Van Guilder M: A Unified Parametric/Nonparametric Approach to Population PK/PD Modeling. Presented at the Annual Meeting of the Population Approach Group in Europe, Paris, France, June 6-7, 2002.
36. Bayard D and Jelliffe R: Bayesian Estimation of Posterior Densities for Pharmacokinetic Models having Changing Parameter Values. Presented at the Annual Meetings of the Society for Computer Simulation, San Diego CA, January 23-27, 2000. Published in the Proceedings, Health Sciences Simulation, pp. 75-83.
37. D'Argenio D: Optimal Sampling Times for Pharmacokinetic Experiments. J. Pharmacokin. Biopharm., 9: 739-756, 1981.

APPENDIX

Note. For abbreviations and explanations in Table A.1., the reader is referred to the chapter on Pharmacokinetic/Pharmacodynamic modeling in drug development authored by Shashank Rohatagi, Nancy E. Martin, Jeffrey S. Barrett.

Table A.1. PK/PD Models and Derived Parameters Used in Various Therapeutic Areas/Drug Class: A Partial List.

Drug/Class/Indication[1]	Dosing Regimen[1]	Pharmacokinetic Parameters [1,2,3]	PD-Model[3]	PD-Marker[3]	Pharmacodynamic Parameters	References
Anesthetics:Neuromuscular Blocking Agents						
d-tubocrarine (Skeletal muscle relaxant)*	0.5-0.6 mg/kg IM, IV Q 40-60 min	V_d: 0.39±0.14 L/kg CL: 1.9± 0.6 mL/min/kg $T_{1/2}$: 2.0± 1.1h	Indirect Link Sigmoid E_{max} Model	Neromuscular block % Muscular Paralysis	E_{max}: 100% E_{50}: 0.60±0.22 mcg/mL N: - $T_{1/2}$Keo=4.7±1.2 min	Stanski DR, Ham J, Miller RD, Sheiner LB. Pharmacokinetics and pharmacodynamics of d-tubocurarinne in man. Anesthesiology 1972: 36: 213.
Pancuronium (Skeletal muscle relaxant)	0.06-0.1 mg/kg IV maintain 0.01 mg/kg Q 60-100 min.	V_d: 0.26±0.07 L/kg CL: 1.8±0.4 mL/min/kg	Indirect Link Sigmoid E_{max} Model	Neromuscular block % Muscular Paralysis	E_{max}: 100% E_{50}: 0.21±0.08 mcg/mL N: 5.48±1.64 $T_{1/2}$Keo: 1.43 min	Evans MA, Shanks CA, Brown KF, Triggs EJ. Pharmacokinetics and pharmacodynamic modeling with pancuronium. Eur J Clin Pharmacol. 1984: 26: 243.
Vecuronium (Skeletal muscle relaxant)	0.08-0.1 mg/kg IV then 0.01-0.015 mg/kg Q 25-40 min.	V_d: 0.21±0.08 L/kg CL: 3.0±0.1 mL/min/kg	Indirect Link Sigmoid E_{max} Model	Neromuscular block % Muscular Paralysis	E_{max}: 100% E_{50}: 0.11±0.02mcg/mL N: - $T_{1/2}$Keo: 6.1 min	Sohn YJ, Bencini AF, Scaf AHJ, Kersten UW, Agoston S. Comparative pharmacokinetics and dynamics of vecuronium and pancuronium in anaesthetized patients. Anesth Analg 1986: 65: 233.
Atracurium (Skeletal muscle relaxant in renal/ hepatic impaired patients)	0.4-0.5 mg/kg IV then 0.08-0.1 mg/kg Q 20-45 min to relax skeletal muscles	V_d: 0.16±0.06L/kg CL: 6.2±2.0 mL/min/kg	Indirect Link Sigmoid E_{max} Model	Neromuscular block % Muscular Paralysis	E_{max}: 100% E_{50}: 0.24±0.04mcg/mL N: 3.4±0.37 $T_{1/2}$Keo: 5.1 min	Donati F, Varin F, Ducharme J, Gill SS, Theoret Y, Bevan DR. Pharmacokinetics and pharmacodynamics of atracurium obtained with arterial and venous blood samples. Clin Pharmacol Ther 1991: 49: 515.
Doxacurium (Skeletal muscle relaxant, maintain hemodynamic stability	0.05-0.08 mg/kg IV then 0.0005-0.01 mg/kg to		Indirect Link Sigmoid E_{max} Model	Neromuscular block % Muscular Paralysis	E_{max}: 100% E_{50}: 34.9±12.0 ng/mL N: 3.3±0.7 $T_{1/2}$Keo=10.7±4.4 min	Schmidt VD, Lai A, Dressner DL, Basta SJ, James CD, Wargin WA, Savarese JJ. Pharmacodynamic modeling of doxacurium in young adult and elderly patients. Phar Res 1991: 8: S-276.
Pipecuronium	0.07-0.085 mg/kg then 0.01-0.015 Q 50 –120 min	V_d: 0.35±0.08 CL: L/kg 2.1±0.5mL/min/kg	Indirect Link Sigmoid E_{max} Model	Neromuscular block % Muscular Paralysis	E_{max}: 100% E_{50}: 63 ng/mL N: - $T_{1/2}$Keo: -	Ornstein E, Matteo RS, Schwartz AE, Jamdar SC, Diaz J. Pharmacokinetics and pharmacodynamics of pipecurium bromide (Arduan) in elderly surgical patient
Mivacurium	0.15-0.25 mg/kg IV bolus then 0.1 mg/kg Q 15 min.		Indirect Link Sigmoid Emax Model	Neromuscular block % Muscular Paralysis	E_{max}: 100% E_{50}: 67.3±30.4 ng/mL N: 4.4±1.3 $T_{1/2}$Keo: 4.9±2.3 min	Phillips L, Schmidt VD, Lien CA, Embree PB, James CD, Lai A, Savarese JJ. Mivacurium sterioisomer pharmacodynamics. Pharm Res 1992: 9: S-356.

1. Lacy CF, Armstrong LL, Goldman MP, Lance LL. Drug information handbook. 1999-2000.
2. Hardman JG, Limbard LE, Gilman AG. The pharmacological basis of therapeutics. 10th edition, McGraw Hill, New York. 2001.
3. Derendorf H and Hochhaus G. Handbook of pharmacokinetic/pharmacodynamic correlation. CRC Press, Boca Raton, 1995.

Drug/Class/Indication[1]	Dosing Regimen[1]	Pharmacokinetic Parameters[1,2,3]	PD-Model[3]	PD-Marker[3]	Pharmacodynamic Parameters	References
Cholinesterase Inhibitors						
Rivastigmine	1-6mg po BID	F: 22-119% V_d: 236 L CL: 126 L/hr	Sigmoid E_{max} Model	Cognitive change (measured by the Computerised Neuropsychological Test Battery i.e CNTB score	E_{max}: 100% E_{50}: 5.42 mcg/L N=0.5 score/year	Gobburu JVS, Tammara V, et al. Pharmacokinetic-Pharmacodynamic Modeling of Rivastigmine, a Cholinesterase Inhibitor, in Patients with Alzheimer's Disease. J Clin Pharmacol 2001;41:1082-1090.
Antipyretics						
Acetaminophen*	12.5 mg/kg po	F: 88% V_d: 0.15 ±0.02 L/kg CL: 0.75±0.2 mL/min/kg $T_{1/2}$: 2.0±0.5 h	Sigmoid E_{max} Model	Neromuscular block % Muscular Paralysis	E_{50}: 4.63 ± 0.39 mcg/mL	Brown RD, et al. Integrated pharmacokinetic-pharmacodynamic model for acetaminophen, ibuprofen, and placebo antipyresis in children. J Pharmacokinet Biopharm. 1998 Oct;26(5):559-79.
Ibuprofen*	5-10 ng/kg po	F: 80% V_d: 0.95 ±0.12 L/kg CL: 5.0±1.4 mL/min/kg $T_{1/2}$: 2.0±0.4 h	Sigmoid E_{max} Model	Neromuscular block % Muscular Paralysis	E_{50}: 11.33± 1.35 mcg/mL	Brown RD, et al. Integrated pharmacokinetic-pharmacodynamic model for acetaminophen, ibuprofen, and placebo antipyresis in children. J Pharmacokinet Biopharm. 1998 Oct;26(5):559-79.
Opioid Anesthetics						
Fentanyl (anesthetic)	0.5-1.0 mcg/kg/dose IM, IV; 25mcg/h transdermal	V_d: 4.0±0.4 L/kg CL: 13±2 mL/min/kg	Indirect Link Sigmoid E_{max} Model	Electroencephalogram (EEG)	E_{max}: 14.1±1.8 Hz E_{50}: 6.9±1.5 ng/mL N: 4.9±1.0 $T_{1/2}Keo$=6.4±1.3 min E_0: 19.2±1.6 Hz	Scott JC, Ponganis KV, Stanski DR. EEG Quantification of narcotic effect: the comparative pharmacodynamic effect of fentanyl and alfentanil. Anesthesiology 1991: 74: 34.
Alfentanil (anesthetic)	8-40 mcg/kg IV	V_d: 0.8±0.3L/kg CL: 6.7±2.4 mL/min/kg	Indirect Link Sigmoid E_{max} Model	Electroencephalogram (EEG)	E_{max}: 14.7±3.1 Hz E_{50}: 520±163 ng/mL N: 4.8±1.5 $T_{1/2}Keo$=1.1±0.3min E_0: 20.1±3.4 Hz	Scott JC, Ponganis KV, Stanski DR. EEG Quantification of narcotic effect: the comparative pharmacodynamic effect of fentanyl and alfentanil. Anesthesiology 1985: 62:.
Sufentanil (for analgisia /anesthetic)	1-2mcg/kg IV and 10-25 mcg prn	V_d: 1.7±0.6 L/kg CL: 12.7±2.5 mL/min/kg	Indirect Link Sigmoid E_{max} Model	Electroencephalogram (EEG)	E_{max}: 14.7±3.1 Hz E_{50}: 0.68±0.31 ng/mL N: 3.1±.09 $T_{1/2}Keo$=6.2±2.8min E_0: 24.3±2.7 Hz	Scott JC, Cooke JC, Stanski DR. Electroencephalographic quantification of opioid effect: comparative pharmacodynamic effect of fentanyl and sufentanil. Anesthesiology 1991: 74: 34.

Drug/Class/Indication[1]	Dosing Regimen[1]	Pharmacokinetic Parameters[1,2,3]	PD-Model[3]	PD-Marker[3]	Pharmacodynamic Parameters	References
Benzodiazepines-sedatives, anticonvulsants						
Diazepam (anxiety, sedation, or muscle relaxation)	2-10 mg p.o. 2-4 times/day, 2-10 mg IM, IV Q 3-4h or 5-10 mg IV Q 10-20 min, up to 30 mg /8h	F: 100% V_d: 1.1±0.3 L/kg CL: 0.38±0.06 mL/min/kg	Sigmoid E_{max} Model	SA-TV: aperiodic signal	E_{max}: 134±57 E_{50}: 958±200 ng/mL N: 1.6±.0.6 $T_{1/2}$Keo=1.6±0.5 min E_0: 39± 17	Buhrer M, Maitre PO, Crevoisier C, Stanski DR. Electroencephalographic effects of benzodiadepines II. Pharmacodynamic modeling of the electroencephalographic effects of midazolam and diazepam. Clin Pharmacol Ther 1990: 48: 555.
				PSAb: Power spectrum analysis of alpha waves	E_{max}: 20.2% E_{50}: 270 ng/mL N: 2.1	Greenblatt DJ, Ehrenberg BL, Gunderman J, Locniskar A, Scavone JM, Harmatz JS, Shader RI. Pharmacokinetic and encephalographic study of intravenous diazepam, midazolam and placebo. Clin Pharmacol Ther 1989: 45: 356.
Midazolam (sedation)	Preoperative sedation: 0.07-0.08 mg/kg IM or 0.02-0.04 mg/kg IV Q 5 min. up to 0.1—0.2mg/kg	F: 44% V_d: 1.1±0.06 L/kg CL: 6.6±1.8 mL/min/kg	Indirect Link Sigmoid E_{max} Model	SA-TV: aperiodic signal.	E_{max}: 125±13 E_{50}: 152±48 ng/mL N: 1.9±.0.2 $T_{1/2}$Keo=4.8±1.5 min E_0: 40± 18	Pharmacokinetic and encephalographic study of intravenous diazepam, midazolam and placebo. Clin Pharmacol Ther 1989: 45: 356.
			Indirect Link Sigmoid E_{max} Model	SA-TNW: aperiodic signal analysis of total number of waves.	E_{max}: 9.7±1.5 E_{50}: 290±98 ng/mL N: 3.1±.1.0 $T_{1/2}$Keo=1.7±0.7 min E_0: 0.7± 0.3	Briemer LTM, Hennis PJ, Burm AGL, Danhof M, Bovill JG, Spierdijk J, Vletter AA. Quantification of the EEG effect of midazolam by aperiodic analysis in volunteers: pharmacokinetic/pharmacodynamic modeling. Clin Pharmacokinet 1990: 18: 245.
			Sigmoid E_{max} Model	PSA-b: power spectrum analysis of beta-waves	E_{max}: 22.4% EC_{50}: 35 ng/mL N: 2.2	Greenblatt DJ, Ehrenberg BL, Gunderman J, Locniskar A, Scavone JM, Harmatz JS, Shader RI. Pharmacokinetic and encephalographic study of intravenous diazepam, midazolam and placebo. Clin Pharmacol Ther 1989: 45: 356.
Oxazepam	10-30 mg po q6h	F: 100% Fb: 86-99% T1/2: 2.8-5.7 h	Sigmoid E_{max} Model	Anticonvulsant in the PTZ model	EC_{50}: 99 ± 19 ng/mL Receptor binding to GABA: 86 ± 15 ng/mL	Mandema JW, Sansom LN, Dios-Vieitez, MC, Hollander-Jansen M, and Danhof M. Pharmacokinetic-pharmacodynamic modeling of the EEG effects of benzodiazepines. Correlation with receptor binding and with anticonvulsant activity. J Pharmacol Exp Ther 1991: 257: 472.
Lorazepam	2 to 4 mg po daily	Fb: 88-92% V_d: 1.3 L/kg T1/2: 10-20 h	Sigmoid E_{max} Model	CNS impairment measured by computerized tracking system (TRKN)	E_{max}: 418 EC_{50}: 35.8 ng/mL N: 6.29 $T_{1/2}$Keo=0.43 h	Gupta SK, Ellenwood EH, Nikaido AM, Heatherly DG. Simultaneous modeling of the pharmacokinetic and pharmacodynamic properties of benzodiazepines I. Lorazepam. J Pharmacokinet Biopharm 1990: 18: 89.

Drug/Class/Indication[1]	Dosing Regimen[1]	Pharmacokinetic Parameters[1,2,3]	PD-Model[3]	PD-Marker[3]	Pharmacodynamic Parameters	References
Triazolam	0.125 –0.25 mg po daily	Fb: 89-94% V_d: 0.8-1.3 L/kg T1/2: 2.3 h	Sigmoid E_{max} Model	CNS impairment measured by DSS (psychomotor performance) test score	EC_{50} 4.65 ng/mL	Smith RB, Kroboth PD, Varner PD. Pharmacodynamics of triazolam after intravenous administration. J Clin Pharmacol 1987: 27: 971.
Alprazolam	0.25-1mg po TID	Fb: 80% V_d: 1.1 L/kg T1/2: 12-15 h	Sigmoid E_{max} Model	CNS impairment measured by DSS (psychomotor performance) test score	EC_{50}: 10 ng/mL	Schmith VD, Piraino B, Smith RB, Kroboth PD. Alprazolam in end stage renal disease II. Pharmacodynamics, Clin Pharmacol Ther 1992: 51: 533.
Flumazenil (benzodiazepine antagonist)	0.2 mg/kg over 15 sec. with maximum cumulative dose of 1 mg	F: 80% V_d: 0.95 L/kg CL: high based on liver blood flow $T_{1/2}$: 241-79 min	Competitive antagonist of Midazolam using E_{max} model	EEG	E_{50} Midazolam : 30 ng/mL. E_{50} flumazenil: 26 ng/ml but has no effect on EEG alone	Mandema JW, Tukker E, Danhof M. In vivo characterization of pharmacodynamic interaction of a benzodiazepine agonist and antagonist: midazolam and flumazenil. J Pharmacol Exp Ther 1992: 260: 36.
Anticonvulsant agents						
Phenobarbital	30-120 mg po daily in 2-3 divided doses	F: 100 ± 11% Fb: 51 ± 3% V_d: 0.54 ± 0.03 L/kg CL: 0.062 ± 0.013 ml/min/kg $T_{1/2}$: 99 ± 18 h	Sigmoid E_{max} Model	Anticonvulsant PTZ model (pentylenetetrazol threshold concentration)	E_{max}: 120 mg/L of PTZ EC_{50} (free): 44 ± 5 mg/L	Dingemanse J, Van Bree JBMM and Danhof M. Pharmacokinetic modeling of anticonvulsant response of Phenobarbital in rats. J Pharmacol Exp Ther 1989: 249: 601
Anesthetic agents						
Propofol (sedative, general anesthetic)	2-2.5 mg/kg iv until onset of induction	V_d: 1.7±0.7 L/kg CL: 276±5 mL/min/kg $T_{1/2}$: 3.5±1.2 h	E_{max} Model Fixed Effect	% of patients that were awake after surgery adequate anesthesia for surgery	EC_{50}: 0.95-1.07 mcg/mL Mean Concentrations > 4.01 mcg/mL for major surgery and > 2.97 mcg/mL for non-major surgery	Vuyk J, Engbers FHM, Lemmons HJM, Burm HGL, Vletter AA, Gladines MPRR, Bovil JG. Pharmacodynamics of propofol in femal patients. Anesthesiology 1992: 77: 3. Shafer A, Doze VA, Shafer SL, White PF. Pharmacokinetics and pharmacodynamics of propofol infusions during general anesthesia. Anesthilogy 1988: 69: 348.
Etomidate (general anesthesia, hynotic)	0.2-0.6 mg/kg over 30-60 sec.	F_b: 76% V_d: 3.6-4.5 L/kg $T_{1/2}$: 2.6 h	Inhibitory Sigmoid E_{max}-Model	Median frequency calculated from spectral analysis of EEG	E_{50}: thiopental/70	Arden JR, Holley FO, Stanski DR, Ebling WF. Dose potency comparison of thiopental and stomidate. Anesthesiology 1985: 63: A286.
Ketamine (R,S) (general anesthesia, hynotic)	1-4.5 mg/kg iv or 3-8 mg/kg im in adults or 6-10 mg/kg oral in children,		Inhibitory Sigmoid E_{max}-Model	Median frequency calculated from spectral analysis of EEG	E_{50}: (R-): 1.8 mcg/mL E_{50}: (S+): 0.8 mcg/mL E_{50}: (R,S): 2.0 mcg/mL E_{max} was different for the three analytes	Schuttler J, Stanski DR, White PF, Trevor AJ, Horai Y, Verotta D, Sheiner LB. Pharmacodynamic modeling of the EEG effects of ketamine and its enantiomers in man. J Pharmacokinet Biopharm 1987: 71: 334.

Drug/Class/Indication[1]	Dosing Regimen[1]	Pharmacokinetic Parameters[1,2,3]	PD-Model[3]	PD-Marker[3]	Pharmacodynamic Parameters	References
Opioid Analgesics						
Morphine (analgesia)	15-30mg PO every 8-12h; 10 mg/dose IM, IV, SC every 4h; 0.8-10mg/h IV cont. infusion; 5mg epidural up to 10mg/24h; 10-20mg rectal every 4h	F: 24±12% Fb: 35±2% V_d: 3.3±0.9 L/kg CL: 24±10 mL/min/kg $T_{1/2}$: 1.9±0.5 h	Logistic/Probit analysis	Cumulative frequency of patients attaining comfort score of 2.	E_{50}: 12.5-13.1 ng/mL N: 3.64-6.89	Gourlay GK, Willis RJ, Lamberty J. A double-blind comparison of the efficacy of methadone and morphine in postoperative pain control. Anesthesiology 1986: 64: 322.
			Fixed Effect	Mean effect concentration (MEC) which is the concentration at the time remedication is required.	MEC: 15±5 ng/mL	Dahlstrom B, Tamsen A, Paalzow L, Hartvig P. Patient controlled analgesic therapy III. Pharmacokinetics and analgesic plasma concentrations of morphine. Clin Pharmacokinet 1982: 266: 7.
Methadone (analgesia)	Analgesia: 2.5-10 mg PO, IM, SC every 3-8h; Detoxification: 15-40mg/day PO; Opiate dependence: 20-120mg/day	F: 92±21% Fb: 89±2.9% V_d: 3.6±1.2 L/kg CL: 2.3±1.2 mL/min/kg $T_{1/2}$: 27±12 h	Logistic/Probit analysis	Cumulative frequency of patients attaining comfort score of 2.	E_{50}: 50.4 ng/mL N: 4.28	
			Fixed Effect	Mean effect concentration (MEC) which is the concentration at the time remedication is required.	MEC: 59±24 ng/mL	Dahlstrom B, Tamsen A, Paalzow L, Hartvig P. Patient controlled analgesic therapy III. Pharmacokinetics and analgesic plasma concentrations of morphine. Clin Pharmacokinet 1982: 266: 7.
Ketobemidone (analgesia)			Logistic/Probit analysis	Cumulative frequency of patients attaining comfort score of 2.	E_{50}: 22.5 ng/mL N: 4.64	
			Fixed Effect	Mean effect concentration (MEC) which is the concentration at the time remedication is required.	MEC: 25±11 ng/mL	Tamsen A, Bondesson U, Dahlstrom B, Hartvig P. Patient-controlled analgesic therapy III. Pharmacokientics and analgesic plasma concentrations of ketomidone. Clin Pharmacokinetic 1982: 266: 7.

Drug/Class/Indication[1]	Dosing Regimen[1]	Pharmacokinetic Parameters[1,2,3]	PD-Model[3]	PD-Marker[3]	Pharmacodynamic Parameters	References
Meperidine (analgesia)	50-150 mg/dose PO, IM, IV, SC every 3-4 hours	F: 52±3% F_b: 58±9% V_d: 4.4±0.9 L/kg CL: 17±5 mL/min/kg $T_{1/2}$: 3.2±0.8 h	Logistic/Probit analysis	Cumulative frequency of patients attaining comfort score of 2.	E_{50}: 253 ng/mL N: 4.38	Glynn CJ, Mather LE, Cousins MJ, Graham JR, Wilson PR. Peridural meperidine in humans: analgetic response, pharmacokinetics and transmission into CSF. Anesthesiology 1981: 55: 520.
			Fixed Effect	Mean effect concentration (MEC) which is the concentration at the time remedication is required.	MEC: 311±188 ng/mL	
Hydromorphone (analgesic, antitussive)	1-4 mg Q4-6 h po, im, iv sc, prn	F: 42±23% F_b: 7.1% V_d: 2.90±1.31 L/kg CL: 14.6±7.6 mL/min/kg $T_{1/2}$: 2.4±0.6 h	Fixed Effect	Pain Control	Hydromorphone concentrations> 4 ng/mL provided good pain control in patients with bone or soft tissue injury but not effective in neuropathic pain.	Reidenberg MM, Goodman H, Erle H, Gray G, Lorenzo B, Leipzig RM, Meyer BR, Drayer DE. Hydromorphone levels and pain control in patients with severe chronic pain. Clin Pharmacol Ther 1988: 44: 376.
Nonopioid Analgesics						
Acetaminophen	325-500 mg po q 6 h; up to 4 g/day	F: 88±15% F_b: 20% V_d: 0.95± 0.12L/kg CL: 5.0±1.4 mL/min/kg $T_{1/2}$: 2.0±0.4 h	E_{max}-Link Model	Pain control/ Analgesia	EC_{50}: 134-267 mcg/mL $T_{1/2}$Keo=7-13 min	Colburn WA. Simultaneous pharmacokinetic and pharmacodynamic modeling. J Pharmacokinet Biopharm. 1981: 9: 367.
Nonsteroidal Anti-Inflammatory Drugs-Analgesic effects						
Aspirin	325-650 mg po q 4-6 h up to 4 g/daily	F: 68 ± 3% F_b: 49% V_d: 0.15±0.03 L/kg CL: 9.3±1.1 mL/min/kg $T_{1/2}$: 0.25±0.03 h	Sigmoid E_{max}-Link Model	Pain model with third molar surgery	E_{max}: 100% EC_{50}: 5.3 mcg/mL N: 1.35 $T_{1/2}$Keo=20 min	Velagapudi R, Harter JC, Bruecker R, Peck CC. Pharmacokinetic pharmacodynamic models in analgesic study design, in advances in pain research and therapy. Vol 18. Max M, Portenoy R, and Laska E, Eds. Raven Press, New York 1991, 559.
Indomethacin	0.2 mg/kg in neonates for patent ductus arteriosus	F: 98±21 % F_b: 90% V_d: 0.29±0.04 L/kg CL:1.4 ± 0.2mL/min/kg $T_{1/2}$: 2.4±0.2 h	Fixed Effect	Patency of ductus arteriosus	Effect seen in neonate at concentrations above 200 mcg/L; No correlation for analgesia	Berash A, Hickey D, Graham T, Stahlam M, Oates J, and Cotton RB. Pharmacokinetics of indomethacin in the neonate: relation of plasma indomethacin levels to response of ductus arteriosus. N Eng J Med 1981:305:67.

Drug/Class/Indication[1]	Dosing Regimen[1]	Pharmacokinetic Parameters[1,2,3]	PD-Model[3]	PD-Marker[3]	Pharmacodynamic Parameters	References
Ibuprofen	400-800 mg po tid or qid with maximum dose of 3.2 g/d	F: 80% F_b: 99% V_d: 0.15±0.02 L/kg CL: 0.75±0.20 mL/min/kg $T_{1/2}$: 2 ± 0.5 h	E_{max}-Model	Pain relief after dental surgery	EC_{50}: 30 mcg/mL	Laska EM, Sunshine A, Marrero I, Olsen N, Siegel C, and Mc Cormick N. The correlation between blood levels of ibuprofen and clinical analgesic response. Clin Pharmacol Ther 1986: 40: 1.
			E_{max}-Model with time delay (T_{lag})	Antiinflammatory effect linked with platelet aggregation	EC_{50}: 3 mcg/mL T_{lag}: 2-6 min	Cox SR, VenbderLugt JT, Gumbleton TJ, Smith RB. Relationships between thromboxane production, platelet aggregability, and serum concentrations of ibuprofen and flurbiprofen. Clin Pharmacol Ther 1987: 41: 510.
Flurbiprofen	200-300 mg/day po in 2, 3, or 4 divided doses	F: 92% F_b: 99.5% V_d: 0.15± 0.02L/kg CL: 0.35± 0.09mL/min/kg $T_{1/2}$: 5.5± 1.4 h	E_{max}-Model with time delay (T_{lag})	Antiinflammatory effect linked with platelet aggregation	EC_{50}: 0.2 mcg/mL T_{lag}: 2-6 min	Cox SR, VenbderLugt JT, Gumbleton TJ, Smith RB. Relationships between thromboxane production, platelet aggregability, and serum concentrations of ibuprofen and flurbiprofen. Clin Pharmacol Ther 1987: 41: 510.
Tolmentin	600 – 1800 mg po daily in divided doses	F_b: 99% $T_{1/2}$: 5 h	No model identified	Synovial fluid prostaglandin E	No correlation observed as recommended doses too high	Dromgoole SH, First DE, Desiraju RK, Nayak RK, Jirshenbaum MA, and Paulus HE. Tolmentin kinetics and synovial prostaglandin E levels in rheumatoid arthritis. Clin Pharmacol Ther 1982: 32: 371.
MAO Inhibitor						
Selegiline	5 mg po twice daily at breakfast and lunch	F: negligible F_b: 94% V_d: 1.9 L/kg CL: 1500 mL/min/kg $T_{1/2}$: 1.9 ± 1.0 h	E_{max} Model	Platelet MAO_B inhibiton	E_{max}: 100% E_{50}: 23 ng/mL MAO_A activity as defined by plasma MHPG concentration was 1000-fold lower	Pressor Response to Tyramine After 24-Hour Application Of A Selegiline Transdermal System In Healthy Males. J.S. Barrett, T. Hochadel, S. Rohatagi, K.E. DeWitt, S.K. Watson, J. Darnow, A.J. Azzaro and A.R. DiSanto. J. Clin Pharmacol., 37: 238-247, 1997.

Drug/Class/Indication[1]	Dosing Regimen[1]	Pharmacokinetic Parameters[1,2,3]	PD-Model[3]	PD-Marker[3]	Pharmacodynamic Parameters	References
Antibiotics						
Ampicillin	500 mg po Q 6 h	F: 50% F_b: 15-25% $T_{1/2}$: 1-1.8 h	Modified E_{max} Model N_{max}=Max number of bacteria without dilution N_t=Measured number of bacteria Δ=Change K=elimination rate constant BAUC=Ratio of area under the bacterial concentration time curve and the initial inoculum size $N_t' = \frac{N_{max} \bullet N_t}{(N_{max} - N_t) \bullet e^{-k \bullet t} + N_t}$	*E.Coli* bacterial survival dilution factor	ΔN_{max}: -3.6log CFU/mL ΔN_t: 1.9 BAUC: 106 log CFU*h/mL CFU: Colony forming unit	White CA, Toothaker RD, Smith AL, Slattery JT. In vitro evaluation of the determinants of the bactericidal activity of ampicillin dosing regimens against Escherichia Coli. Antimicrob Agents Chemother 1989: 33: 1046.
Rifampicin	600 mg per day	F_b: 60-90% V_d: 0.97 ± 0.36 L/kg CL: 3.5 ± 1.6 mL/min/kg $T_{1/2}$: 3.5 ± 0.8h	Sigmoidal E_{max} Model	*S. aureus* bacterial survival dilution factor	E_{50}: 66-75 mg/mL	Hoogerterp JJ, Mattie H, Krul AM, van Furth R. The efficacy of rifampicin against *Staphylococcus aureus* in vitro and in an experimental infection in normal and granluocytopenic mice. Scand J Infect Dis 1998: 20: 649.
Gentamicin	1-2.5 mg /kg per dose iv/im q 8h	F_{IM}: 100% F_b: <10% V_d: 0.31 ± 0.10 L/kg CL: 0.82*Creatinine clearance + 0.11 mL/min/kg $T_{1/2}$: 2-3 h	Sigmoidal E_{max} Model	*K. pneumoniae* bacterial survival dilution factor	E_{max}: 4.87 ± 0.15 EC_{50}: 2.34 ±0.26 mg/kg	Leggett JE, Fntin B, Ebert S, Totsuka K, Vogelman B, Calame W, Mattie H, Craig WA. Comparative antibiotics dose-effect relations at several dosing intervals in murine pneumonitis and thigh infection models. J Infect Dis 1989: 15: 281.
Ceftazidime	500-2000 mg iv/im q 8	F_{IM}: 91% F_b: 21 ± 6% V_d: 0.23 ± 0.02 L/kg CL: ~Creatinine clearance mL/min/kg $T_{1/2}$: 1.6 ± 0.1h	Sigmoidal E_{max} Model	*K. pneumoniae* bacterial survival dilution factor	E_{max}: 4.69 ± 0.30 EC_{50}: 61.4 ± 8.8 mg/kg	Leggett JE, Fntin B, Ebert S, Totsuka K, Vogelman B, Calame W, Mattie H, Craig WA. Comparative antibiotics dose-effect relations at several dosing intervals in murine pneumonitis and thigh infection models. J Infect Dis 1989: 15: 281.
Netilmicin	1.5-2 mg/kg per dose mg iv/im q 8	$T_{1/2}$: 2-3 h	Sigmoidal E_{max} Model	*K. pneumoniae* bacterial survival dilution factor	E_{max}: 4.87 ± 0.15 EC_{50}: 2.34 ±0.26 mg/kg	Leggett JE, Fntin B, Ebert S, Totsuka K, Vogelman B, Calame W, Mattie H, Craig WA. Comparative antibiotics dose-effect relations at several dosing intervals in murine pneumonitis and thigh infection models. J Infect Dis 1989: 15: 281.

Drug/Class/Indication[1]	Dosing Regimen[1]	Pharmacokinetic Parameters[1,2,3]	PD-Model[3]	PD-Marker[3]	Pharmacodynamic Parameters	References
Bisphoaphonates						
Pamidronate (osteoporosis)	15 mg/day IV infusion	V_d: 29.5 L CL: 181 ± 78 mL/min	E_{max} Model	Urinary hydroxyproline		Cremers S, et al. A PK and PD model for intravenous bisphosphonate (pamidronate) in osteoporosis. Eur J Clin Pharmacol 2002;57:883-890.
ACE Inhibitors						
Cilazapril (Cilazeprilat)	2.5 mg po QD	F: 77.5% V_d: 25.4 L CL: 14.28 L/h	Sigmoid E_{max} Model	Inhibition of plasma angiotensin-converting enzyme (ACE) activity	E_{50}= $E_{50cpatopril}$/8	Belz GG, Kirch W, Kleinbloesem CH. Angiotensin-converting enzyme inhibitors: relationship between pharmacodynamics and pharmacokinetics. Clin Pharmacokinetics 1988: 15: 295.
Captopril	12.5-150 mg po TID	F: 60-75% F_b: 25-30% V_d: 0.7 L/kg $T_{1/2}$: 1.9 h	Sigmoid E_{max} Model	Inhibition of plasma angiotensin-converting enzyme (ACE) activity	E_{50}= $E_{50cpatopril}$	Belz GG, Kirch W, Kleinbloesem CH. Angiotensin-converting enzyme inhibitors: relationship between pharmacodynamics and pharmacokinetics. Clin Pharmacokinetics 1988: 15: 295.
Lisinopril	10-40 mg po QD	F: 25±10% F_b: 0 % V_d: 2.4±1.1.4 L/kg CL: 4.2±2.2 mL/min/kg $T_{1/2}$: 12 h	Sigmoid E_{max} Model	Inhibition of plasma angiotensin-converting enzyme (ACE) activity	E_{50}= $E_{50cpatopril}$/8	Belz GG, Kirch W, Kleinbloesem CH. Angiotensin-converting enzyme inhibitors: relationship between pharmacodynamics and pharmacokinetics. Clin Pharmacokinetics 1988: 15: 295.
Enalapril	10-40 mg or divided dose	F: 41±15% F_b: 50-60% V_d: 1.7±0.7 L/kg CL: 4.9±1.5 mL/min/kg $T_{1/2}$: 11 h	Sigmoid E_{max} Model	Inhibition of plasma angiotensin-converting enzyme (ACE) activity	E_{50}= $E_{50cpatopril}$/2	Belz GG, Kirch W, Kleinbloesem CH. Angiotensin-converting enzyme inhibitors: relationship between pharmacodynamics and pharmacokinetics. Clin Pharmacokinetics 1988: 15: 295.
Perindopril (Perindoprilat)	4-16 mg po QD	F: 65-95% F_b: 60% CL: 9.36 L/h $T_{1/2}$: 25-30 h	Sigmoid E_{max} Model	Inhibition of plasma angiotensin-converting enzyme (ACE) activity		Belz GG, Kirch W, Kleinbloesem CH. Angiotensin-converting enzyme inhibitors: relationship between pharmacodynamics and pharmacokinetics. Clin Pharmacokinetics 1988: 15: 295.
Ramipril (Ramiprilat)	2.5-20 mg po QD	F: 50-60% F_b: 56% CL: 6 L/h $T_{1/2}$: 2.4±0.6 h	Sigmoid E_{max} Model	Inhibition of plasma angiotensin-converting enzyme (ACE) activity	E_{50}= $E_{50cpatopril}$/8	Belz GG, Kirch W, Kleinbloesem CH. Angiotensin-converting enzyme inhibitors: relationship between pharmacodynamics and pharmacokinetics. Clin Pharmacokinetics 1988: 15: 295.
Angiotensin II Receptor Antagonists						
Losartan	25-100 mg po QD	F: 35.8 ± 15.5% Fb: 98.7% Vd: 0.45 ± 0.24 L/kg CL: 8.1 ± 1.8 ml/min/kg T1/2: 2.5 ± 1 h	E_{max} Link Model	Systolic blood pressure (SBP) Diastolic blood pressure (DBP)	E_{max}: 27.2 mmHg E_{50}: 228 mcg/L K_{eo}=5.9 h^{-1} E_{max}: 23.3 mmHg E_{50}: 159 mcg/L K_{eo}=6.1 h^{-1}	Csajka C, Buclin, T, Fattinger K, Brunner HR, Biollaz J. Population Pharmacokinetic-Pharmacodynamic Modelling of Angiotensin Receptor Blockade in Healthy Volunteers. Clin Pharmacokin 2002: 41 (2): 137-152.

Drug/Class/Indication[1]	Dosing Regimen[1]	Pharmacokinetic Parameters[1,2,3]	PD-Model[3]	PD-Marker[3]	Pharmacodynamic Parameters	References
Candesartan	4-32 mg po QD	F: 42% Fb: 99.8% Vd: 0.13 L/kg CL: 0.37 ml/min/kg T1/2: 9.7 h	E_{max} Link Model	Systolic blood pressure (SBP) Diastolic blood pressure (DBP)	E_{max}: 30.2 mmHg E_{50}: 12.2 mcg/L K_{eo} =0.4 h^{-1} E_{max}: 25.1 mmHg E_{50}: 8 mcg/L K_{eo} =0.3 h^{-1}	Csajka C, Buclin, T, Fattinger K, Brunner HR, Biollaz J. Population Pharmacokinetic-Pharmacodynamic Modelling of Angiotensin Receptor Blockade in Healthy Volunteers. Clin Pharmacokin 2002: 41 (2): 137-152.
Irbesartan	150-300 mg po QD	F: 60-80% F_b: 90% Vd: 0.72 ± 0.20 L/kg CL: 2.12 ± 0.54 ml/min/kg $T_{1/2}$: 13 ± 6.2 h	E_{max} Link Model	Systolic blood pressure (SBP) Diastolic blood pressure (DBP)	E_{max}: 28.7mmHg E_{50}: 124 mcg/L K_{eo} =0.8 h^{-1} E_{max}: 26.8 mmHg E_{50}: 89 mcg/L K_{eo} =0.8 h^{-1}	Csajka C, Buclin, T, Fattinger K, Brunner HR, Biollaz J. Population Pharmacokinetic-Pharmacodynamic Modelling of Angiotensin Receptor Blockade in Healthy Volunteers. Clin Pharmacokin 2002: 41 (2): 137-152.
GPIIb/IIIa antagonist						
Klerval	Oral and intravenous administration 5-120 mcg/kg	V_d: 23.0-25.9 L CL: 11.2-15.5 L/h $T_{1/2}$: 1.2-2.1 h	E_{max}-Model	Inhibition of platelet aggregation	E_{max}: 100% E_{50}: 58 ng/mL	Pharmacokinetics, Pharmacodynamics and Safety of a Platelet GPIIb/IIIa Antagonist, RGD891, Following Intravenous Administration in Healthy Male Volunteers. P.N. Zannikos, S. Rohatagi, B.K. Jensen, S. DePhillips, D. Massignon, F. Calic, Michel Sibille and Stephane Kirkesseli. J. Clin Pharmacol., 40: 1245-1256, 2000. Pharmacokinetics and Concentration-Effect Analysis of Intravenous RGD891, A Platelet GPIIb/IIIa Antagonist, Using Mixed Effects Modeling (NONMEM). P.N. Zannikos, S. Rohatagi, B.K. Jensen, S. DePhillips, and G. R. Rhodes. J. Clin Pharmacol., 40: 1129-1140, 2000.
Abciximab (angioplasty)	0.25 mg/kg iv bolus: 0.125 mcg/kg/min for 12 h	CL: 183 +/- 72 ml/min $T_{1/2}$: 0.5 h	E_{max}-Model	Receptor occupancy % inhibition of platelet aggregation 5-micromol/L adenosine diphosphate (ADP) % inhibition of platelet aggregation 20-micromol/L ADP	E_{50}: 72.8 ± 6.4 ng/mL E_{50}: 68.9 ± 9.2 ng/mL E_{50}: 141 ± 16.8 ng/mL	Abernethy DR, Pezzullo J, Mascelli MA, Frederick B, Kleiman NS, Freedman J. Pharmacodynamics of abciximab during angioplasty: comparison to healthy subjects. Clin Pharmacol Ther. 2002 Mar;71(3):186-95.

Drug/Class/Indication[1]	Dosing Regimen[1]	Pharmacokinetic Parameters[1,2,3]	PD-Model[3]	PD-Marker[3]	Pharmacodynamic Parameters	References
Calcium Channel Blocker						
Verapamil (d+,l-) (antiangina, antiarrhythmia, antihypertensive)	5-10 mg iv 80-120 mg Q8h	F: 22±8% F_b: 90±2% (d+ 2.5-fold higher F than l-) V_d: 5.0±2.1 L/kg CL: 15±6 mL/min/kg (d+ is 0.5-fold lower than l-) $T_{1/2}$: 4.0±1.5 h	E_{max}-Model	PR-Interval prolongation	E_{50}(l)=E_{50} (d)/11 E_{50},po (d,l)=120±20 ng/mL E_{50},iv (d,l)=40±25 ng/mL	Echizen H, Vogelgesang B, Eichelbaum M. Effects of d,l-verapamil on artrioventricular conduction in relation to its stereoselective first-pass metabolism. Clin Pharmacol Ther 1985: 38: 71.
Nisoldipine	20-60 mg po qd	F: 5% $T_{1/2}$: 7-12 h	E_{max}-Model	Supine Systolic Blood Pressure (SBP) change from baseline	E_{max}: 27.9% E_{50}: 0.99-2.62 mcg/mL	Schaefer HG, Heinig R, Ahr G, Adelmann H, Tetzloff W, Kuhlmann J. Pharmacokinetic-pharmacodynamic modelling as a tool to evaluate the clinical relevance of a drug-food interaction for a nisoldipine controlled-release dosage form. Eur J Clin Pharmacol. 1997;51(6):473-80
				Supine Diastolic Blood Pressure (DBP) change from baseline	E_{max}: 36.4% E_{50}: 0.99-2.62 ng/mL	
Proton Pump Inhibitors						
Lansoprazole	15-30 mg po qd	F: 81% F_b: 97% V_d: 0.35±0.05 L/kg CL: 6.23± 1.6 mL/min/kg $T_{1/2}$: 0.9± 0.44 h	Modified E_{max} Model K: apparent reaction rate constant k: apparent turnover rate constant k/K x fp: apparent dissociation constant corrected for plasma free fraction (fp) t1/2: half-life of effect	H/K ATPase	K : 0.339 ± 0.002 mM/h k : 0.0537 ± 0.0006 h^{-1} k/K x fp: 1 nM t1/2 effect: 12.9 h	Katashima M, Yamamoto K, Tokuma Y, Hata T, Sawada Y, Iga T. Comparative pharmacokinetic/ pharmacodynamic analysis of proton pump inhibitors omeprazole, lansoprazole and pantoprazole, in humans. Eur J Drug Metab Pharmacokinet. 1998 Jan-Mar;23(1):19-26.
Omeprazole	20-40 mg/day qd	F: 53±29 % F_b: 95% V_d: 0.34±0.09 L/kg CL: 7.5± 2.7 mL/min/kg $T_{1/2}$: 0.7± 0.5 h	Modified E_{max} Model K: apparent reaction rate constant k: apparent turnover rate constant k/K x fp: apparent dissociation constant corrected for plasma free fraction (fp) t1/2: half-life of effect	H/K ATPase	K : 1.34 ± 0.17 mM/h k : 0.0252 ± 0.0019 h^{-1} k/K x fp: 1 nM t1/2 effect: 27.5 h	Katashima M, Yamamoto K, Tokuma Y, Hata T, Sawada Y, Iga T. Comparative pharmacokinetic/ pharmacodynamic analysis of proton pump inhibitors omeprazole, lansoprazole and pantoprazole, in humans. Eur J Drug Metab Pharmacokinet. 1998 Jan-Mar;23(1):19-26.

Drug/Class/Indication[1]	Dosing Regimen[1]	Pharmacokinetic Parameters[1,2,3]	PD-Model[3]	PD-Marker[3]	Pharmacodynamic Parameters	References
Pantoprazole	40 mg po qd	F: 77% F_b: 98% V_d: 11-24 L CL: 7.5± 2.7 mL/min/kg $T_{1/2}$: 1 h	Modified E_{max} Model K: apparent reaction rate constant k: apparent turnover rate constant k/K x fp: apparent dissociation constant corrected for plasma free fraction (fp) t1/2: half-life of effect	H/K ATPase	K : 0.134 ± 0.006 mM/h k : 0.0151 ± 0.0002 h^{-1} k/K x fp: 2.3 nM t1/2 effect: 45.9 h	Katashima M, Yamamoto K, Tokuma Y, Hata T, Sawada Y, Iga T. Comparative pharmacokinetic/ pharmacodynamic analysis of proton pump inhibitors omeprazole, lansoprazole and pantoprazole, in humans. Eur J Drug Metab Pharmacokinet. 1998 Jan-Mar;23(1):19-26.
Antihistamines: H_1Blockers						
Noberastine	10 mg po	V_d/F: 1928± 32 L $T_{1/2}$: 15 h	Sigmoid E_{max} Link Model	Placebo corrected histamine induced wheal inhibition	E_{ma}: 90% E_{50}: 1.2±0.2 ng/mL N: 1.0±.0.2 Keo=0.39±0.13 h^{-1}	Heykants et al. The in vivo study of drug action. Van Boxtel CJ, Holford NHG, Danhof, eds. Elsevier: 1992: 335.
				Placebo corrected histamine induced flare inhibition	E_{max}: 90% E_{50}: 0.3±0.1 ng/mL N: 0.8±.0.2 Keo=0.09±0.03 h^{-1}	
$Beta_2$-Adrenergic Agonists: Bronchiodilators						
Fenoterol	IV bolus, infusion, nasal	V_d: 140 ± 70 L CL: 63 ± 24 L/h $T_{1/2}$: 3.3 ± 1.2 h	Sigmoid E_{max} Link Model	Airway resistance (RAW)	E_{50}: 0.38 ng/mL N: 1 Keo=12 h^{-1}	Hochhaus G, Schmidt EW, Rominger KL, Mollmann H. Pharmacokinetic/dynamic correlation of pulmonary and cardiac effects of fenoterol in asthmatic patients after different routes of administration. Pharm Res 1992: 9: 291.
				Heart rate (HR)	E_{50}: 1.1 ng/mL N: 1 K_{eo}:12 h^{-1}	
Albuterol	2-4 mg/dose Oral 3-4 times/day up to 32 mg/day; 1-2 inhalations every 4-6 h up to 12 inhalations/day or oral sublingual	V_d: 156 ± 38L CL: 29 ± 7 L/h $T_{1/2}$: 3.9 ± 0.8 h	Sigmoid E_{max} Model	Forced expiratory volume in 1 sec (FEV1)	E_{50}: 5 ng/mL N: 2	Hochhaus G, Mollman H. PK/PD analysis of albuterol action: application to a comparative assessment of β_2-adrenergic drugs. Eur J Pharm Sci 1993: 1.
				Heart rate (HR)	E_{50}: 18 ng/mL N: 1	

Drug/Class/Indication[1]	Dosing Regimen[1]	Pharmacokinetic Parameters[1,2,3]	PD-Model[3]	PD-Marker[3]	Pharmacodynamic Parameters	References
Terbutaline	0.25 mg/dose Subcutaneous repeated once at 15-30 minutes or subcutaneous or 2.5-10 mcg/min Infusion effective maximum dose 17.5-30 mcg/min	Vd: 79 ± 27L CL: 22 ± 6.7 L/h T1/2: 2.5 ± 0.5 h	Sigmoid E_{max} Link Model	Airway resistance (RAW)	E_{50}: 3.7 ng/mL N: 1 Keo=7.8 h^{-1}	Oosterhuis B, Braat P, Roos VM, Werner J and van Boxtel CJ. Pharmacokinetic-pharmacodynamic modeling of terbutaline bronchodilation is asthma. Clin Pharmacol Ther 1986: 40: 469.
				Forced expiratory volume in 1 sec (FEV1)	E_{50}: 2.3 ng/mL N: 1 Keo=5.5 h^{-1}	
			E_{max} Model	Heart rate (HR)	E_{50}: 11 ng/mL	
Corticosteroids (anti-inflammatory/asthma) *RRA= relative glucocorticoid receptor affinity [dexamethasone=100]						
Prednisone	5-60 mg po daily in 1 to 4 divided doses	F: 80% F(inhaled): N/A F_b: 75% Vd: 70L CL: 15L/h $T_{1/2}$: 3.2 h	Indirect Response Model	Lymphocyte Suppression (LS) Granulocyte Induction (GI) Cortisol Suppression (CS)	E_{50LS}:18.6 ± 8.0 ng/mL E_{50GI}: 16.7 ± 6.8 ng/mL $E_{max\ LS,CS}$: 100% $E_{max\ GI}$: 0.68 ± 0.12% K_{out}: 0.35 ± 0.07 h^{-1} RRA*: 16	Mollmann H, Hochhaus G, Rohatagi S, Barth J, Derendorf H. Pharmacokinetic/pharmacodynamic evaluation of deflazacort in comparison to methylprednisolone and prednisolone. Pharm Res. 1995 Jul;12(7):1096-100.
Methylprednisolone	2-60 mg po daily in 1 to 4 divided doses	F: 99% F(inhaled): F_b: 77% Vd: 80L CL: 21L/h $T_{1/2}$: 2.6 h	Indirect Response Model	Lymphocyte Suppression (LS) Granulocyte Induction (GI) Cortisol Suppression (CS)	E_{50LS}:4.8 ± 2.7 ng/mL E_{50GI}: 3.6 ± 2.0 ng/mL $E_{max\ LS,CS}$: 100% $E_{max\ GI}$: 0.68 ± 0.12% K_{out}: 0.35 ± 0.07 h^{-1} RRA*: 42	Mollmann H, Hochhaus G, Rohatagi S, Barth J, Derendorf H. Pharmacokinetic/pharmacodynamic evaluation of deflazacort in comparison to methylprednisolone and prednisolone. Pharm Res. 1995 Jul;12(7):1096-100.
17-Beclomethasone monopropionate (administered as Beclomethasone diproprionate)	168 – 840 mcg inhaled in one or two divided doses	F: 26% F(inhaled): 36% F_b: Vd: 424L CL: 120 L/h $T_{1/2}$: 6.5 h	Indirect Response Model	Lymphocyte Suppression (LS) Granulocyte Induction (GI) Cortisol Suppression (CS)	RRA*: 1345	Rohatagi S. PhD Thesis. Pharmacokinetic and Pharmacodynamic modelling of methylprednisolone and prednisolone after single and multiple administration, 1995.
Budesonide	200-600 mcg inhaled in one or two divided doses	F: 11% F(inhaled): 28% F_b: 88% Vd: 183-217.5L CL: 80-84 L/h $T_{1/2}$: 2.8 h	Indirect Response Model	Lymphocyte Suppression (LS) Granulocyte Induction (GI) Cortisol Suppression (CS)	RRA*: 935	Rohatagi S. PhD Thesis. Pharmacokinetic and Pharmacodynamic modelling of methylprednisolone and prednisolone after single and multiple administration, 1995.

Drug/Class/Indication[1]	Dosing Regimen[1]	Pharmacokinetic Parameters[1,2,3]	PD-Model[3]	PD-Marker[3]	Pharmacodynamic Parameters	References
Flunisolide	500-2000 mcg inhaled in one or two divided doses	F: 7% F(inhaled): 39% F_b: 80% Vd: 61-96L CL: 56-65 L/h $T_{1/2}$: 1.2-1.9h	Indirect Response Model	Lymphocyte Suppression (LS) Granulocyte Induction (GI) Cortisol Suppression (CS)	RRA*: 180	Rohatagi S. PhD Thesis. Pharmacokinetic and Pharmacodynamic modelling of methylprednisolone and prednisolone after single and multiple administration, 1995.
Fluticasone Propionate	88-660 mcg inhaled in two divided doses	F: 1% F(inhaled): 26% F_b: 90% Vd: 318L CL: 65-69L/h $T_{1/2}$: 11.5	Indirect Response Model	Lymphocyte Suppression (LS) Granulocyte Induction (GI) Cortisol Suppression (CS)	$E_{50totalCS}$:0.34 ± 0.16 ng/mL $E_{max\,CS}$: 100% RRA*: 1800	Dynamic Modeling of Cortisol Reduction After Inhaled Administration of Fluticasone Propionate. S. Rohatagi, A. Bye, C. Falcoz, A.E. Mackie, B. Meibohm, H. Möllmann and H. Derendorf. J. Clin. Pharmacol. 36: 938-941, 1996.
Triamcinolone Acetonide	400 – 2000 mcg inhaled in one or two divided doses	F: 23% F(inhaled): 22% F_b: 71% V_d: 103L CL: 37 L/h $T_{1/2}$: 2.0 h	Indirect Response Model	Lymphocyte Suppression (LS) Granulocyte Induction (GI) Cortisol Suppression (CS)	E_{50LS}: 0.24 ± 0.7 ng/mL E_{50GI}: 0.22 ± 0.06 ng/mL E_{50CS}: 0.14 ± 0.04ng/mL $E_{max\,LS}$: 55% $E_{max\,CS}$: 100% $E_{max\,GI}$: 60 ± 7% K_{out}: 0.64 ± 0.13 h^{-1} RRA: 233-346	Pharmacokinetic/Phamcodynamic Evaluation of Triamcinolone Acetonide After Intravenous, Oral and Inhaled Administration. S. Rohatagi, G. Hochhaus, H. Möllmann, J. Barth, H. Sourgens, M. Erdmann and H. Derendorf. J. Clin. Pharmacol. 35: 1187-1193, 1995.
Beta-Blockers						
Propranolol (antiarrhythmia)	10-30 mg po Q 6-8h	F: 26 ± 10% F_b: 87 ± 6% V_d: 4.3 ± 0.6 L/kg CL: 16 ± 5 mL/min/kg $T_{1/2}$: 3.9 ± 0.4 h	E_{max} Model	Exercise Induced Tachycardia	E_{max}: 83 ± 6 beats/min E_{50}: 18.1 ± 4.3 ng/mL	Brynne L, Paalzow LK, Karlsson MO. Mechanism-based modeling of rebound tachycardia after chronic l-propranolol infusion in spontaneous hypertensive rats. J Pharmacol Exp Ther. 1999 Aug;290(2):664-71.
			Linear Model	Exercise Induced Tachycardia	Slope=28.9 ± 2.8 beats*min/(min* mcg)	
Metoprolol (antiarrhythmia)	100-450 mg/day in 2-3 divided doses	F: 38 ± 14% F_b: 11 ± 1% V_d: 4.2 ± 0.7 L/kg CL: 15 ± 3 mL/min/kg $T_{1/2}$: 3.2 ± 0.2 h	E_{max} Model	Exercise Induced Tachycardia	E_{max}: 103 ± 6 beats/min E_{50}: 50.6 ± 15.2 ng/mL	Brynne L, Paalzow LK, Karlsson MO. Mechanism-based modeling of rebound tachycardia after chronic l-propranolol infusion in spontaneous hypertensive rats. J Pharmacol Exp Ther. 1999 Aug;290(2):664-71.
			Linear Model	Exercise Induced Tachycardia	Slope=4.48 ± 0.39 beats*min/(min* mcg)	
Immunosuppressants						
Cyclosporin (Transplant)	15mg/kg/day po	F: 28 ± 18% F_b: 93 ± 2% V_d: 4.5 L/kg CL: 5.7 mL/min/kg $T_{1/2}$: 10.7 h	Fixed Effect	Nephrotoxicity	Nephrotoxicity>200 ng/mL for steady-state trough concentrations	Rohatagi S, Calic F, Harding N, Ozoux M-L, Kirkesseli S, DeLeij L, Jensen. BK Pharmacokinetics, Pharmacodynamics and Safety of Inhaled Cyclosporin A (ADI 628) After Single and Repeated Administration in Healthy Male and Female Subjects And Asthmatic Patients. J Clin Pharmacol., 2000:40: 1211-1226.
			Fixed Effect	Immunosuppression	Immunosuppression>20 0 ng/mL for steady-state trough concentrations	

Drug/Class/Indication[1]	Dosing Regimen[1]	Pharmacokinetic Parameters[1,2,3]	PD-Model[3]	PD-Marker[3]	Pharmacodynamic Parameters	References
Oncology						
Daunorubicin (acute nonlymphocytic leukemia, lymphoma)	30-60 mg/m^2 iv for 3-5 days, repeat dose in 3-4 weeks (protocol based)	V_d: 40 L/kg $T_{1/2}$: 24-48 h	Fixed Effect Model	Myeloblast: erythrocyte activity ratio	Responders to complete hematoloic response have greater myleblost: erythrocyte activity ratio	Greene W, Huffman D, Wiernik PH, Schmiff S, Benjamin R, Bachur N. High-dose daunorubicin therapy for acute nonlymphocytic leukemia: correlation of response and toxicity with pharmacokinetics and intracellular daunorubicin reductase activity. Cancer 1972: 30: 1419.
Doxorubicin (breast cancer, antineoplastic agent)	60-75 mg/m^2 iv single, repeat dose every 21 days (protocol based)	F: 5% F_b: 76% V_d: 682±433 L/m^2 CL: 666±339 mL/min/ m^2 $T_{1/2}$: 26±17 h	Linear Model	Short-term tumor response	Plasma AUCα short-term tumor response (r^2=0.67)	Robert J, Illiadis A, Hoerni B, Cano JP, Durand M and Langrade C. Pharmacokinetics of adriamycin in patients with patients with breast cancer: correlation between pharamcokinetics parameters and short-term clinical response. Eur j Cancer Clin Oncol 1982: 18-739.
			Linear Model	Nadir White blood cells (WBC)	Css α Nadir WBC (r^2=-0.28)	Ackland SP, Ratain MJ, Vogelzang NJ, Choi KE, Ruane M, Sinkule JA. Pharmacokinetics and pharmacodynamics of long-term continuous-infusion doxorubicin. Clin Pharamcol Ther 1989: 40: 45.
5-Fluorouracil (treatment of solid tumors such as breast cancer)	400-500 mg/m^2/day iv for 4-5 days	F_b: 94% V_d: 35±21 L/kg CL: 9.6±6.9 mL/min/kg $T_{1/2}$: 53±41 h	E_{max} and exponential	Frequency of Stomatitis	$C_{ss}50$ (r^2=-0.85)	Trump DL, Egorin MJ, Forrest A, Wilson JKV, Remick S, Tutsch KD. Pharmacokinetic and pharmacodynamic analysis of fluorocil during 72 hour continuous infusion with and without dipyridamole. J Clin Oncol 1991: 9: 2027.
Cytarabine (lymphomas and leukemias)	200 mg/m^2/day iv for 5 days at 2 week intervals	F: <20% F_b: 13% V_d: 3.0±1.9 L/kg CL: 13±4 mL/min/kg $T_{1/2}$: 2.6±0.6 h	Fixed-Effect /logistic regression	Remission rate	Higher complete remission rate > 75 mM concentration	Plunkett W, Iacoboni S, Estey E, Danhouser L, Lilliemark JO, Keating MJ. Pharmacologically directed ara-C therapy for refractory leukemia. Semin Oncol 1985: 12 (suppl 3): 20
Cisplatin	50-70 mg/m^2 iv every 3-4 weeks	V_d: 0.28±0.07 L/kg CL: 6.3±1.2 mL/min/kg $T_{1/2}$: 0.53±0.10 h	Fixed Effect	Response to Therapy Gastrointestinal toxicity	AUC > 507 ± 187 responded to therapy AUC > 451 had greater gastrointestinal toxicity	Desoize B, Marechal F, Millart H, Cattan A. Correlationof clinical pharmacokinetic parameters of cisplatin with efficacy and toxicity. Biomed Pharmacother 1991: 45: 203.

INDEX

AAPS, 23
Absorbable dose, D_{abs}, 61
Absorption, 144, 177–178, 251
ACAT, 149, 237
Accuracy/precision, 28
Acetaminophen, 80–81, 211, 531
Active metabolite, 13, 90
Acyl glucuronide, 39, 40, 42
ADME, 1, 3, 54, 75
Ah, 119
AIC, 355
Albuterol, 541
Alendronate, 427
  biopharmaceutics, 452
  clinical pharmacology, 439
  dose proportionality, 461
  drug-drug interactions, 466
  human pharmacokinetics, 451
  mechanism of action, 431
  preclinical pharmacokinetics, 437
  preclinical pharmacology, 434
Alfentanil, 531
Allometric scaling (*see* interspecies scaling), 99, 138
Alprazolam, 533
Alzheimer, 1, 264–265
Amikacin therapy, 490
Analytical qualifiers, 29
ANDA, 209
Angiotensin, 13
Animal exposure, 7
Animal models, 9
APCI, 22
API, 22
Apolysichrons, 142
Aptamers, 10
Artificial organ systems, 10
Assay error, 375
Atorvastatin, 32, 77–78
Atracurium, 530
Autoinduction, 316

BA, 226
BA/BE, 3, 225
Basiliximab, 256
Bayes' theorem or Bayesian, 374, 481, 483
BCS, 196,198–199, 228, 230–231
BE, 226
Benzodiazepines, 261
Beta-blockers, 543
Bile-duct cannulated, 87
Biliary clearance, 163
Bioanalytical methods, 21
Bioavailability, 197
Biological license application, 6
Biological matrix, 28
Biomarker, 3, 12, 14, 338, 351
Biotechnology, 8, 10
Biowaivers, 225, 234
Bisphosphonates, 436
BMD, 429
Bootstrap, 401
Boxenbaum approach, 160
Brain weight or BRW, 99, 143, 159
Bridging data package, 280
Bridging studies, 364
Bupropion, 36
Busulfan therapy, 493

^{14}C mass effects, 36
Caco-2, 237, 313
Caloric load, 203
Captopril, 538
CAR, 119
Carbohydrate meals, 200
Carcinogenicity, 3
Carrier-mediated, 55, 200
Cassette dosing, 29, 87
CAT, 57–58, 60, 63
CDER, 248
Cefprozil, 274
Celecoxib, 405
  allometric scaling, 418

Celecoxib (*continued*)
  clinical dose prediction, 418
  drug-drug interactions, 408
  food effects, 421
  metabolism, 406
  PM, EM, 413
  toxicokinetics, 409
  2C9, 408–409
21CFR, 225
Chemical derivatization, 21
Child-Pugh, 293
Chromatographic, 22, 28
Chronotherapeutic, 189
CI-1007, 104
CID, 25
Ciprofloxacin, 289
Clearance, 75
Clinical development, 2, 7
Clinical trial simulation, 365
CLND, 43, 44
Clockwise hysteresis, 344
Clozapine, 45
CMC, 226, 228
cMOAT, 206
Cocktail, 320
Co-elution, 36
Common technical document, 17
Compartment modeling, 53–54, 335
Corticosteroids, 349, 542
Cortisol, 359, 366
Covalent binding, 88, 91
COX, 405
Creatinine clearance, 378
CRF, 8
CRIMS, 45
Cryopreserved, 85
Cyclosporine, 252, 543
CYP, 77, 80, 307
  induction, 311
  inhibition, 308
  2C19, 277
  2C9, 277
  2D6, 9, 82, 277
  3A4, 180–181

D_{50}, 349
Data transformation, 353
Data warehousing, 8, 10
Dedrick plot, 142–143
Demographics, 246
Detection limit, 29
Dialysis, 286
Diazepam, 37, 532
Diffusion coefficient, 59–60
Diffusion rate limited, 134
Digoxin therapy, 491
Direct injection, 24
Direct link, 341
Direct response, 341
Disopyramide, 215
Dispersion model, 62
Dissolution, 54–55, 60
Dissolution time, 61
Distribution, 151, 252
DMP-728, 32
DMPK, 133
Dofetilide, 326
Dosage form, 55, 195, 201
Dose, 201
Dose adjustment, 246, 288, 298
Dose individualization, 477
Dose proportionality, 3
Dose selection, 167
Doxacurium, 530
Doxorubicin, 544
DR_{50}, 349
Drug development, 1
Drug-disease interaction, 364
Drug-drug interaction, 9, 75, 105, 307, 364
Drug
  flux, 196
  interaction, 3
  interactions, elderly, 264
  metabolism, 75
  metabolizing enzyme, 75–76
  particle dissolution, 59
  property, 196
  solubility, 196, 198
Drug-food effects, 196
Drug-nutrient interaction, 195
Duodenum, 55–56, 181

e-ADME, 71
EC_{50}, 336
Effective permeability, 61
Efficacy bridging, 280
Efflux transporters, 57
Electrospray ionization (ESI), 22
Elimination, 252
Emax, 340, 346–347
End of phase I meeting, 2
End of phase II meeting, 2
Endogenous ligand, 13
Enterion, 178, 182–184
Enzyme induction, 115
Enzyme inhibition, 107, 111
Epimers, 32
Ethnicity, 246, 275
Everolimus, 295
Exposure, 337
Exposure-response, elderly, 263
Extended release, 188
Extensive metabolizers, 102
Extraction ratio, 155

F_2, 239
FaSSIF, 237
Fat meals, 199
FDA, 23, 54
FDA modernization Act or FDAMA, 5, 248
Fentanyl, 531
FeSSIF, 237
FEV_1, 360
Fexofenadine, 320
Fick's law, 59

First in human, 3, 363
First order kinetics, 55
First-pass metabolism, 55, 196, 205
Fisher information, 376, 483
Fixed effect, 339
Fluid volume, 202
Flumazenil, 533
Fluvoxamine, 270
FO, 375
FOCE, 374
Food effect, 363
Food effect BA/BE, 209
Food-drug interaction, 195
Fosinopril, 25–27
Fosinoprilat, 25–27
Fraction absorbed, 57, 146–148
Functional proteomics, 8

Gabapentin, 216
Ganciclovir, 61, 62
Gastric emptying, 55, 196–197, 203
Gastric pH, 55, 214
Gastroplus™, 149–150
Gauss-Newton, 354
GC, 21
GC/MS, 22
Gender effect, 363
Gentamicin therapy, 487, 489, 537
Geriatric labeling guidance, 266
Geriatric population, 246, 260
GI residence time, 203
GI tract, 145
GLP, 12, 30, 315
Gonadal steroids, 271
Goodness of fit, 354
Grapefruit juice, 327

Hard link, 343
Hepatic clearance, 86, 96, 98–99
Hepatic extraction ratio, 86, 98
Hepatic function, 293
Hepatic impairment, 291
Hepatic metabolism, 62
Hepatocytes, 35, 84–85
HepG2, 85, 119
HER2, 9
Herceptin®, 9
High frequency capsule, 182
HPLC, 21
HPMC, 215
HTS, 8
Human drug absorption (HDA), 177
Human metabolic clearance prediction, 96
Human oral bioavailability, 53–54
Hysteresis loop, 357

Ibuprofen, 531, 536
IC_{50} and K_i, 109
Ileum, 55–56, 181
IMMC, 203
Immediate release (IR), 53, 225
Impurities, 31
In silico, 8, 53

IND, 2, 76
Indinavir, 213
Indirect link, 341
Indirect response, 342–343
Indomethacin, 41, 535
Induction, 75, 77
Influx transporters, 57
Inhibition, 75, 77
Inhibition kinetics, 108
Initial estimates, 354
Innovation, 11
In-source hydrolysis, 28
Intelisite capsule, 182
International Conference on Harmonization (ICH), 4
Interspecies scaling, 133, 137
Intestinal absorption, 56, 196
Intestinal efflux, 55, 179, 186
Intestinal metabolism, 180
Intestinal transit, 55–56
Intestine, 55
Intrinsic clearance, 97, 154–155
Investigator's brochure, 1–2
IR, 191
Irbesartan, 539
Isolated perfused liver, 86
Isomers, 32
IVIVC, 64, 157, 227, 229–231
IX-SPE, 32

Jejunum, 56, 181

K_a, 336, 379
Kaletra, 329
Kallynochrons, 142
K_e, 336
K_{eo}, 346
Ketamine, 533
Ketoconazole, 318
K_i, 309,311
K_m, 309
K_p, 153

Labeling, 195, 323
Labile, 38
Lactation, 267, 274
LC/MS/MS, 23
LCM, 187
Lead identification, 8
Lead optimization, 8
Lidocaine therapy, 493
Line extension, 225, 365
Linear model, 339, 355
Link model, 357
Linked pharmacodynamic models, 500
Lipinski's Rule of Five, 66
Liposomal, 10
Liquid scintillation counting, 45
Liver
  microsomes, 35
  necrosis, 94
  slices, 85
Local minima, 374

Δ log D, 68
Log P, 66
Logistic regression, 345
Log-likelihood, 374
Log-linear, 339, 355
Lopinavir, 328
Lorazepam, 532
Losartan, 538
Low clearance drugs, 96
LY303366, 216
Lysylpyridinoline, 447

MALDI, 38
MAP, 374, 483
Matrix, 96
Matrix effects, 36
Maturation, 251
Maximum life span, 99
Maximum likelihood, 385
MDR, 164
Meal, 195
Meal type, 204
Meal viscosity, 204
Mechanism-based inactivation, 113
Metabolic
  clearance, 154
  profiling, 77
  stability, 77
Metabolism dependent inhibition, 113
Metabolite, 21
  identification, 89
  quantitation, 29
  standards, 30
Methadone, 534
Methylprednisolone, 356
Mibefradil, 113
MIC, 336
Microsomal protein, 98
Microsome, 83
Midazolam, 259, 318, 532
Mivacurium, 530
MLP, 143, 159
MLR, 65
Model, 54
Model misspecification, 377
Model validation, 358
Modeling and simulation, 11, 14
Modeling drug diffusion, 500
Modified release, 233
Morphine, 534
MRI, 8, 10
MRP, 206
Multiple model, 523

Nanotechnology, 8, 10
NDA, 2, 76, 209
Nefazodone, 325
Nelder-Mead, 354
New molecular entity (NME), 1, 185
N-in-1, 29, 87
Nitrazepam, 281
NMR, 30–31
NOAEL, 167
Non-clinical, 2
Noncompartmental, 335
Non-linear pharmacokinetics, 86
NONMEM, 354, 375
Nonparametric model, 383–384
NPAG, 390
NPEM, 386
NPML, 386
NSAIDs, 405
NTI, 229, 232
NTR, 232

OATP, 164, 308
OCT, 164
Omapatrilat, 39
Omeprazole, 279, 540
Oral drug product, 196
Organ impairment, 246, 263
Osteocalcin, 447
Osteoporosis, 427, 430
Oxazepam, 532

Paget's disease, 428
Pancuronium, 530
Paracellular, 56, 179, 200
Paracetamol, 36
Parallel tube model, 62
Parameter space, 395
Parametric population modeling, 374
Parkinson's disease, 1
PBPK, 133,134
Pediatric drug formulation, 254
Pediatric labeling, 259
Pediatrics, 246, 247
P_{eff}, 57, 61
Perfusion rate limited, 134
Peristaltic waves, 55
Permeability, 55, 60, 196, 198, 226, 236
Permeability coefficient, 136
Permeation, 179
PET, 8, 10
p-glycoprotein (or Pgp; *see* Transporter), 179, 180, 206, 308, 314
Pharmacodynamic, 3, 333
Pharmacoeconomics, 4
Pharmacoepidemiology, 4
Pharmacogenetics, 8
Pharmacogenomics, 8
Pharmacokinetic, 3, 333
Pharmacology/toxicology, 6
Phase I enzymes, 77, 82
Phase II enzymes, 77, 79, 82
Phenobarbital, 533
Phenytoin, 80–81, 212
Physiological time scales, 140
Physiologically based direct scaling, 156
Pioglitazone, 324
Pipecuronium, 530
PK/PD, 10–12, 333, 351, 374, 481
pKa, 32
Plasma drug concentration, 12
Plasma protein binding, 13
Polymorphism, 102

Poor metabolizers, 102
Population, 373
Population PK/PD, 7, 481
Post-column photolysis, 36
Post-marketing studies, 4, 361
Potency, 63
Power, 355
PPAR, 119
Pravastatin, 40
Pre-clinical development, 2, 76, 361
Pregnancy, 267, 273
Pre-IND, 2
Pre-NDA meeting, 2
Probe inhibitors, 310
Probe substrates, 310, 315
Probenecid, 41
Product extension, 4
Product labeling, 245
Proof of concept, 360
Protein binding, 137
Protein meals, 200
Protein precipitation, 36
PXR, 119

QSAR, 54, 66
QSBR, 54, 65, 68, 70
QT prolongation, 307

R, S-enantiomer, 41
Reaction phenotyping, 77, 85, 101, 312
Reactive metabolites, 77
Reactive thiol, 39
RBC/plasma partitioning, 39
Receptor binding, 13
Region-dependent absorption, 207
Regioselectivity, 83
Remifentanil, 40
Renal clearance, 165
Renal function, 283
Renal impairment, 282
Residence time, 56
Retention time, 28
Rezulin (troglitazone), 120
Rifampicin, 537
Ritonavir, 213
Rivastigmine, 264, 531
Roxifiban, 31–33
Roxifiban metabolites, 31–35
Run times, 23

S9, 35
St. John's wort, 320, 326–327
Saquinavir, 213
Saturable, 57
SAX, 32
Scaling factors, 98
Schwartz criteria, 355
Selectivity, 22
Selegiline, 536
SELEX, 10
Sensitivity, 22
Separation principle, 383
Sex, 246
Sigmoidal Emax, 340, 357
Sigmoidicity, 507
Sildenafil, 325
Soft link, 343
Solid dosage form, 226
Sparse sampling, 13
Special populations, 245
Specificity, 22
SPECT, 8, 10
SRM, 25
Stable isotope internal standards, 32
Standard two-stage, 374
Starting dose, 167
Stationary phase, 24
Structural model, 353
Subcellular fractions, 83
Sufentanil, 531
Suicide inhibition, 113
SUPAC, 227–228, 233
Surrogate end points, 9, 338, 351

Tamoxifen, 36
Target-oriented, model-based approach, 487
Temporal, 12
TGA, 31
Therapeutic drug monitoring, 495
Therapeutic index, 364
Therapeutic range, 478
Tight junction, 56
Time invariant, 344
Time variant, 344
Tissue binding, 151
Tissue-to-plasma, 137
TOF, 44
Tolerance model, 344
Toxic metabolite, 90
Toxicokinetics, 3
Toxicology, 2, 3
Transcellular, 56–57, 179
Transporter, 54, 313
Transporter/absorption, 9
Triazolam, 533
Triple-quadrupole technology, 22
Troglitazone, 77
Trough, 13
Tubocurarine, 530
Two-peak, ranitidine, 58

Ultrarapid metatabolizers, 102
Unbound drug, 137
Unstable metabolites, 38
Urinary hydroxyproline, 441
US FDA (see FDA), 5
USC*PACK, 375, 377

Valdecoxib, 324
Vancomycin therapy, 490
Vecuronium, 530
Verapamil, 540
Vmax, 97
VODE, 382
Vss, 151
Vssu, 153

Warfarin, 355
Well stirred model, 62
Women in bioequivalence studies, 268

Xenobiotic, 76

Zwitterion, 31–32